Innovations in Computational Chemistry

THEORETICAL AND QUANTUM CHEMISTRY AT THE DAWN OF THE 21ST CENTURY

Edited by
Ramon Carbó-Dorca, PhD
Tanmoy Chakraborty, PhD

APPLE ACADEMIC PRESS

Apple Academic Press Inc.
3333 Mistwell Crescent
Oakville, ON L6L 0A2 Canada

Apple Academic Press Inc.
9 Spinnaker Way
Waretown, NJ 08758 USA

Library and Archives Canada Cataloguing in Publication

Theoretical and quantum chemistry at the dawn of the 21st century / edited by Ramon Carbó-Dorca, PhD, Tanmoy Chakraborty, PhD.
(Innovations in computational chemistry book series)
Includes bibliographical references and index.
Issued in print and electronic formats.
ISBN 978-1-77188-682-6 (hardcover).--ISBN 978-1-351-17096-3 (PDF)
1. Quantum chemistry. 2. Chemistry, Physical and theoretical. I. Carbó, Ramon, editor II. Chakraborty, Tanmoy, editor III. Series: Innovations in computational chemistry book series
QD462.T54 2018   541'.28   C2017-907975-1   C2017-907976-X

Library of Congress Cataloging-in-Publication Data

Names: Carbó, Ramon, editor. | Chakraborty, Tanmoy, editor.
Title: Theoretical and quantum chemistry at the dawn of the 21st century / editors, Ramon Carbó-Dorca, PhD, Tanmoy Chakraborty, PhD.
Description: Toronto : Apple Academic Press, 2018. | Series: Innovations in computational chem-istry | Includes bibliographical references and index.
Identifiers: LCCN 2017058983 (print) | LCCN 2018000708 (ebook) | ISBN 9781351170963 (ebook) | ISBN 9781771886826 (hardcover : alk. paper)
Subjects: LCSH: Chemistry, Physical and theoretical. | Quantum chemistry.
Classification: LCC QD453.3 (ebook) | LCC QD453.3 .T44 2018 (print) | DDC 541--dc23
LC record available at https://lccn.loc.gov/2017058983

Apple Academic Press also publishes its books in a variety of electronic formats. Some content that appears in print may not be available in electronic format. For information about Apple Academic Press products, visit our website at **www.appleacademicpress.com** and the CRC Press website at **www.crcpress.com**

Cover graphic: Pep Camps, Girona Artist

THEORETICAL AND QUANTUM CHEMISTRY AT THE DAWN OF THE 21ST CENTURY

CONTENTS

ABOUT THE EDITORS

Ramon Carbó-Dorca, PhD
Emeritus Professor, Institute of Computational Chemistry and Catalysis, University of Girona, Catalonia, Spain

Ramon Carbó-Dorca, PhD, is Emeritus Professor of the Institute of Computational Chemistry and Catalysis at the University of Girona, Catalonia, Spain. He was formerly working as a director of the Institute, which he also founded. He was previously affiliated with Barcelona's Institut Químic de Sarrià and the Universitat Autònoma as well as with Ghent University in Belgium. He was a visiting professor at the University of Alberta, Canada; and at the universities in Tromsoe (Norway); Pisa (Italy); Saskatoon, Saskatchewan (Canada); Hyderabad (India), and Tokyo (Japan). He has organized several international meetings on theoretical chemistry. As a prolific author, he has published 334 papers and 17 books, including research monographs, teaching classroom volumes, and scientific divulgation studies on quantum chemistry, linear algebra, and quantum similarity.

In addition, Dr. Carbó-Dorca is a member of the editorial boards of several journals. In 1980, he introduced the concepts structuring the theoretical body associated with quantum similarity and quantum quantitative structure-property relationships. Several of his contributions are aimed at integrating quantum chemistry with appropriate algorithms that can be automatically programmed for computations in parallel. He has been honored with the Narcís Monturiol medal of the Generalitat de Catalunya.

Tanmoy Chakraborty, PhD
Associate Professor, Department of Chemistry, Manipal University, Jaipur, India

Tanmoy Chakraborty, PhD, is now working as Associate Professor and HoD in the Department of Chemistry at Manipal University Jaipur, India. He has been working in the challenging field of computational and theoretical chemistry for the last six years. He has completed his PhD from the

University of Kalyani, West-Bengal, India, in the field of application of QSAR/QSPR methodology in the bioactive molecules. He has published many research papers in peer-reviewed international journals with high impact factors. Dr. Chakraborty is serving as an international editorial board member of the *International Journal of Chemoinformatics and Chemical Engineering*. He is also a reviewer of the *World Journal of Condensed Matter Physics* (WJCMP). Dr. Tanmoy Chakraborty is the recipient of a prestigious Paromeswar Mallik Smawarak Padak, from Hooghly Mohsin College, Chinsurah (University of Burdwan), in 2002.

INNOVATIONS IN COMPUTATIONAL CHEMISTRY BOOK SERIES

Editor-in-Chief: Dr. Tanmoy Chakraborty
Associate Professor, Department of Chemistry
Manipal University Jaipur |http://jaipur.manipal.edu/muj.html
Dehmi kalan | Jaipur| Rajasthan | 303007
Contacts: (91) 9166721541
Email: tanmoychem@gmail.com; tanmoy.chakraborty@jaipur.manipal.edu

Computational chemistry is one of the important domains of modern-day research, which has a wide spectrum of applications in the practice. Theoretical and computational approaches of chemical sciences have opened up a new dimension of research through which we have not only been able to correlate any physico-chemical experimental properties in terms of computational descriptors but we can also predict new models. Computational chemistry can be defined simply as "the use of computers to aid chemical inquiry." Under the heading of computational chemistry, a wide range of research domains exist, but it includes mainly:

- molecular mechanics (classical Newtonian physics)
- molecular dynamics
- semi-empirical molecular orbital theory
- ab initio molecular orbital theory
- density functional theory
- quantitative structure-activity relationships
- graphical representation of structures/properties

From the quantum physics, the theory of chemistry emerges. To turn the theory into any prediction, a lot of computation is required. Computation's ability to make accurate predictions of experimental measurements is a good test of the validity of a theory.

Computational chemistry has a spectrum of applications. It can provide a way of obtaining information that would be very difficult, expensive, or time-consuming to obtain experimentally otherwise. Through computational chemistry, we can study a system's behavior at high temperature and pressure. Different structural aspects at atomic scale can also be predicted in terms of computational descriptors.

In this new book series, Innovations in Computational Chemistry, we aim to cover the important applications of computational chemistry in diverse fields. The basic theory and methodology will also be covered in volumes. Different computational approaches including molecular mechanics and molecular dynamics will definitely be the major portions of this venture.

The probable coverage of topics in the series would be as follows:

1. Computational Chemistry in QSAR
2. Analysis of Nano-Materials Invoking Computational Methodologies
3. Conceptual Density Functional Theory and Its Real Field Applications
4. Theoretical Study of Chemical Reaction
5. Molecular Simulation and Its Biological Applications
6. Approach of Computational Technique in the Domain of Renewable Energy
7. Theoretical Study of Corrosion Science
8. Study of Material Science in Terms of Computational Technique
9. Impact of Electron Density Distribution on Chemical Bonding and Properties
10. Study of Different of Periodic Descriptors Invoking Theoretical Approach

We invite you to participate in this emerging topic by contributing to this book series either with individual chapters or as the editor/author of a complete book. There is no fee for inclusion of a chapter. Author(s) interested may submit a short proposal [Title, authors' names and contact information, summary, keywords] of the chapter by email.

Manuscript preparation: AAP requires an e-copy of the full chapter submitted in Word on a letter-sized paper with no line numbering, line spacing of 1.5, in Times New Roman 11-point font size with margins top/bottom of 1", margins left/right of 1.25", and editable equations with an equation editor. A typical chapter will include the title, author(s) with affiliation/email/mailing address, introduction, literature review if applicable, theoretical approach if applicable, materials and methods, results and discussion, conclusions, summary, detailed list of keywords, references using numbering system and alphabetical order, any appendix, etc. Also, we prefer a numbering system for citations in the body of a chapter. Tables and figures should be included in the text at right place just after their citation. Specialized chapters need not follow this format.

Our team expects to work with you closely. We can assure you the best quality production. AAP handles all processes from copyediting to marketing at no charge to authors and editors. All contributing authors will receive pdf copy of the volume that includes their chapter.

For more information and to submit your proposal for a chapter or a book, please contact the book series editor, Dr. TanmoyChakraborty, at tanmoychem@gmail.com; tanmoy.chakraborty@jaipur.manipal.edu

BOOKS IN THE SERIES

• Theoretical and Quantum Chemistry at the Dawn of the 21st Century
  Editors: Ramon Carbó-Dorca, PhD, and Tanmoy Chakraborty, PhD

LIST OF CONTRIBUTORS

Guillaume Acke
Department of Inorganic and Physical Chemistry, Ghent University, Krijgslaan 281 (S3), 9000 Gent, Belgium

Neil L. Allan
School of Chemistry, University of Bristol, Cantock's Close, Bristol BS81TS, UK

M. Alonso
Eenheid Algemene Chemie (ALGC), Vrije Universiteit Brussel (VUB), Pleinlaan 2, 1050 Brussels, Belgium

Adam Archer
School of Chemistry, University of Bristol, Cantock's Close, Bristol BS81TS, UK

Paul W. Ayers
Department of Chemistry and Chemical Biology, McMaster University, Hamilton, Ontario, Canada

Emili Besalú
Department of Chemistry and Institute of Computational Chemistry and Catalysis, University of Girona, C/Maria Aurèlia Capmany 69, 17003 Girona, Spain

Josep Maria Bofill
Universitat de Barcelona, Departament de Química Inorgànica i Orgànica, Secció de Química Orgànica, Universitat de Barcelona, and Institut de Química Teòrica i Computacional, Universitat de Barcelona, (IQTCUB), Martí i Franquès, 1, 08028 Barcelona, Spain, E-mail: jmbofill@ub.edu

Patrick Bultinck
Department of Inorganic and Physical Chemistry, Ghent University, Krijgslaan 281 (S3), 9000 Gent, Belgium

Debajit Chakraborty
Department of Chemistry and Chemical Biology, McMaster University, Hamilton, Ontario, L8P4Z2, Canada

Debdutta Chakraborty
Department of Chemistry and Center for Theoretical Studies, Indian Institute of Technology, Kharagpur – 721302, West Bengal, India, Tel.: +91-3222-283304, Fax: +91-3222-255303

Tanmoy Chakraborty
Department of Chemistry, Manipal University Jaipur, DehmiKalan, Jaipur, India, E-mail: Tanmoy.chakraborty@jaipur.manipal.edu; tanmoychem@gmail.com

Pratim Kumar Chattaraj
Department of Chemistry and Center for Theoretical Studies, Indian Institute of Technology, Kharagpur – 721302, West Bengal, India, Tel.: +91-3222-283304, Fax: +91-3222-255303, E-mail: pkc@chem.iitkgp.ernet.in

Jerzy Cioslowski
Institute of Physics, University of Szczecin, Wielkopolska 15, 70-451 Szczecin, Poland

J. Contreras-García
Sorbonne Universités, UPMC Univ Paris 06,CNRS, Laboratoire de Chimie Théorique (LCT), 4 Place Jussieu, 75252 Paris Cedex05, France

David L. Cooper
Department of Chemistry, University of Liverpool, Liverpool L69 7ZD, United Kingdom

Ernesto Estrada
Department of Mathematics and Statistics, University of Strathclyde, 26 Richmond Street, Glasgow, G11XH, UK

Stijn Fias
Department of Chemistry and Chemical Biology, McMaster University, Hamilton, Ontario, L8P4Z2, Canada

P. Geerlings
Eenheid Algemene Chemie (ALGC), Vrije Universiteit Brussel (VUB), Pleinlaan 2, 1050 Brussels, Belgium

Enric Gibert
Pharmacelera S. L., Pl. Pau Vila 1, Edifici Palau de Mar, Sector 1, Nucli C, 08039 Barcelona, Spain

Tiziana Ginex
Department of Nutrition, Food Sciences and Gastronomy, Campus Torribera, School of Pharmacy and Institute of Biomedicine, University of Barcelona, Av. Prat de la Riba, 171, 08921, Santa Coloma de Gramenet, Spain, E-mail: tiziana.ginex@gmail.com

Ramon Goñi
Life Sciences Department. Barcelona Supercomputing Center, Barcelona, Spain, Joint BSC-IRB Program in Computational Biology, Barcelona, Spain

Farnaz Heidar-Zadeh
Department of Chemistry and Chemical Biology, McMaster University, Hamilton, Ontario, Canada / Department of Inorganic and Physical Chemistry, Ghent University, Krijgslaan 281 (S3), 9000 Gent, Belgium / Center for Molecular Modeling, Ghent University, Technologiepark 903, 9052 Zwijnaarde, Belgium

Enric Herrero
Pharmacelera S. L., Pl. Pau Vila 1, Edifici Palau de Mar, Sector 1, Nucli C, 08039 Barcelona, Spain

Kimihiko Hirao
Director, Computational Chemistry Unit, RIKEN Advanced Institute for Computational Science, Kobe, Hyogo 650-0047, Japan, Tel./Fax: +81-78-940-5555, E-mail: hirao@riken.jp

J. Vicente de Julián-Ortiz
Drug Design and Molecular Connectivity Investigation Unit, Department of Physical Chemistry, Pharmacy Faculty, University of Valencia, Av V. Andrés Estellés 0, 46100 Burjassot, Valencia, Spain, Proto QSAR SL, Scientific Park of the University of Valencia, C/Catedrático Agustín Escardino 9, 46980 Paterna, Valencia, Spain

Ajay Kumar
Department of Mechatronics, Manipal University Jaipur, DehmiKalan, Jaipur, India

Laurence Leherte
Department of Chemistry, Namur Medicine and Drug Innovation Center (NAMEDIC), University of Namur, Rue de Bruxelles 61, B-5000 Namur, Belgium, Tel.: +32-81-72-45-60, E-mail: laurence.leherte@unamur.be

F. Javier Luque
Department of Nutrition, Food Sciences and Gastronomy, Campus Torribera, School of Pharmacy and Institute of Biomedicine, University of Barcelona, Av. Prat de la Riba, 171, 08921, Santa Coloma de Gramenet, Spain, E-mail: fjluque@ub.edu

Paul G. Mezey
Canada Research Chair in Scientific Modeling and Simulation, Department of Chemistry, and Department of Physics and Physical Oceanography, Memorial University of Newfoundland 283 Prince Philip Drive, St. John's, NL, A1B3X7, Canada, Tel.: 1-709-749 8768, Fax: 1-709-864-3702, E-mail: paul.mezey@gmail.com

Chris E. Mohn
Centre for Earth Evolution and Dynamics, University of Oslo, Postbox 1048, Blindern, N-0315 Oslo, Norway

Saradamoni Mondal
School of Chemistry, University of Hyderabad, Hyderabad – 500046, India

Camelia Muñoz-Caro
SciCom research group.Escuela Superior de Informática. Universidad de Castilla-La Mancha, Paseo de la Universidad 4, 13004 Ciudad Real. Spain

Alfonso Niño
SciCom research group. Escuela Superior de Informática. Universidad de Castilla-La Mancha, Paseo de la Universidad 4, 13004 Ciudad Real. Spain

Modesto Orozco
Joint BSC-IRB Program in Computational Biology, Barcelona, Spain, Institute for Research in Biomedicine (IRB Barcelona), The Barcelona Institute of Science and Technology, Barcelona, Spain, Department of Biochemistry and Biomedicine, Faculty of Biology, University of Barcelona, Barcelona, Spain, E-mail: modesto.orozco@irbbarcelona.org

Juan Jesus Perez
Department of Chemical Engineering, Universitat Politecnica de Catalunya, Barcelona Tech. Av. Diagonal, 647, 08028, Barcelona, Spain

Lionello Pogliani
Drug Design and Molecular Connectivity Investigation Unit, Department of Physical Chemistry, Pharmacy Faculty, University of Valencia, Av V. Andrés Estellés 0, 46100 Burjassot, Valencia, Spain, MOLware SL, C/Burriana 36-3, 46005 Valencia, Spain

Robert Ponec
Institute of Chemical Process Fundamentals, Czech Academy of Sciences, Prague 6, Suchdol 2, 165 02 Czech Republic

Frank De Proft
Eenheid Algemene Chemie (ALGC), Vrije Universiteit Brussel (VUB), Pleinlaan 2, 1050 Brussels, Belgium

Wolfgang Quapp
Mathematisches Institut, Universität Leipzig, PF 100920, D-04009 Leipzig, Germany, E-mail: quapp@uni-leipzig.de

Prabhat Ranjan
Department of Mechatronics, Manipal University Jaipur, Dehmi Kalan, Jaipur, India

Sebastián Reyes
SciCom research group. Escuela Superior de Informática, Universidad de Castilla-La Mancha, Paseo de la Universidad 4, 13004 Ciudad Real. Spain

Jaime Rubio-Martinez
Department of Physical Chemistry, Faculty of Chemistry, Universitat de Barcelona and the
Institut de Recerca en Quimica Teoricai Computacional (IQTCUB). Matii Franques 1–3, 08028
Barcelona, Spain

K. D. Sen
School of Chemistry, University of Hyderabad, Hyderabad – 500046, India,
E-mail: kds77@uohyd.ac.in

Miquel Solà
The Institute of Computational Chemistry and Catalysis (IQCC) and Department of Chemistry,
University of Girona, C/Maria Aurèlia Campmany, 69, E-17003-Girona, Catalonia, Spain

Anton J. Stasyuk
The Institute of Computational Chemistry and Catalysis (IQCC) and Department of Chemistry,
University of Girona, C/Maria Aurèlia Campmany, 69, E-17003-Girona, Catalonia, Spain

Takao Tsuneda
Professor, Fuel Cell Nanomaterials Center, University of Yamanashi, Kofu – 4000021, Japan,
Tel./Fax: +81-55-254-7139, E-mail: ttsuneda@yamanashi.ac.jp

T. Woller
Eenheid Algemene Chemie (ALGC), Vrije Universiteit Brussel (VUB), Pleinlaan 2, 1050 Brussels,
Belgium

LIST OF ABBREVIATIONS

ABC	Ascona B-DNA consortium
ADMP	atom-centered density matrix propagation
AIEE	aggregation induced emission enhancement
AIM	atoms-in a-molecule
ANNs	artificial neural networks
ANO-S	atomic natural orbital small basis set
APSG	antisymmetrized product of strongly orthogonal geminals
ASA	atomic shell approximation
BCP	bond critical point
BFS	breath-first-search
BL2O	balanced leave-2-out
CAS	complete-active-space
CASPT2	CAS second-order perturbation theory
CASSCF	complete active space self-consistent field
CBO	covalent bond order
CD	charge density
CDFT	conceptual density functional theory
CÉCI	Consortium des Équipements de CalculIntensif
CG	coarse grain
CI	configuration interaction
CLS	classical least-squares
COSMO	conductor-like screening model
CP	conjugation pathway
CP	critical points
CPKS	coupled-perturbed Kohn-Sham
CSD	crystallographic structural database
CSGT	continuous set of gauge transformations
CSNs	chemical space networks
CT	charge transfer
CV	cross-validation
CWT	continuous wavelet transform
DAFHs	domain-averaged Fermi holes

DCM	dichloromethane
DEM	diffusion equation method
DFT	density functional theory
DMABN	dimethylaminobenzonitrile
DMSO	dimethylsulfoxide
DWT	discrete wavelet transform
ECFP	extended-connectivity fingerprints
ED	electron density
EDA	energy decomposition analysis
ELF	electron localization function
EM	electron microscopy
EOM-CC	equation of motion coupled-cluster
EOM-CCSD	equation-of-motion coupled cluster model with singles and doubles
ER	Erdös-Rényi
ESIPT	excited state intramolecular proton transfer
FCI	full configuration interaction
FF	force field
FNF	false negatives fraction
FPF	false positives fraction
FWT	fast wavelet transform
GAMESS	general atomic and molecular electronic structure system
GB	generalized born
GGA	generalized-gradient-approximation
GPUs	graphical processing units
HEDT	holographic electron density theorem
HF	Hartree-Fock
HINT	hydropathic interactions
HLW	high-level waste
HMLP	heuristic molecular lipophilic potential
HMO	Hückel molecular orbital
HOMA	harmonic oscillator model of aromaticity
HOMO	highest occupied molecular orbital
HPC	high-performance computer
HSE	Heyd-Scuseria-Ernzerhof
HTS	high-throughput screening
ISE	isomerization stabilization energies
ITS	internal test sets

LC	long-range corrected
LD	linear discriminant
LMNN	large margin nearest neighbor
LMO	leave-many-out
LNOs	localized natural orbitals
LOAELs	lowest observed adverse effect levels
LOO	leave-one-out
LS	least squares
LUMO	lowest unoccupied molecular orbital
MC	Monte Carlo
MC-FEP	Monte-Carlo free energy perturbation
MCQDPT	MC quasi-degenerate perturbation theory
MCS	maximum common substructure
MCSCF	multiconfigurational self-consistent-field
MD	molecular dynamics
MHP	maximum hardness principle
MLP	molecular lipophilicity potential
MLR	multilinear regression
MMC	mahalanobis metric for clustering
MOX	mixed oxide
MPP	minimum polarizability principle
MRA	multi-resolution analysis
MRMP	multireference Møller-Plesset
MS-CASPT2	multistate CASPT2
MSM	Markov state models
MST	Miertus-Scrocco-Tomasi
MUE	mean unsigned errors
NBO	natural bond orbitals
NCI	noncovalent interaction
NICS	nucleus-independent chemical shift
NLO	nonlinear optical
NOF	natural orbital functional
NPA	natural population analysis
NSERC	Natural Sciences and Engineering Research Council of Canada
OLS	orthogonal least-squares
PAHs	polycyclic aromatic hydrocarbons
PASA	promolecular atomic shell approximation

PAW	projector-augmented wave
PCA	principal component analysis
PCM	polarizable continuum model
PDB	protein data bank
PES	potential energy hyper-surfaces
PKA	primary knock on atom
PLS	partial least squares
PRSIC	pseudospectral regional self-interaction correction
PTCI	PlateformeTechnologique de CalculIntensif
QED	quantum electrodynamics
QI	quantum interference
QM	quantum mechanical
QM/MM	quantum mechanical/molecular mechanical
QMSM	quantum molecular similarity measure
QSAR	quantitative structure activity relationships
QTAIM	quantum theory of atoms in molecules
RDM	reduced density matrix
RESP	restrained electrostatic potential
REXMD	replica exchange molecular dynamics
RHF	restricted Hartree-Fock
RMS	root-mean-square deviations
ROC	receiver operating characteristic
RPA	random phase approximation
RPCM	reduced point charge models
RRT-MD	real real-time MD
RSIC	regional self-interaction correction
SAC-CI	symmetry-adapted-cluster CI
SARI	structure-activity relationship index
SAS	solvent accessible surface
SC	spin-coupled
SCF	self-consistent field
SCP	screened Coulomb potentials
SCRF	self-consistent reaction field
SDP	semidefinite program
SEDI	shared-electron distribution index
SF	spin-flip
SHED	Shannon entropy descriptors
SMD	smoothed molecular dynamics

SOMOs	singly occupied molecular orbitals
SSIR	superposing significant interaction rules
SVM	support vector machines
TDDFT	time-dependent density functional theory
TDHF	time-dependent HF
TDKS	time-dependent Kohn-Sham
THF	tetrahydrofuran
TNF	true negatives fraction
TPA	two-photon absorption
TPF	true positives fraction
TS	transition states
TSED	Taylor series expansion of electronic density
TTF-CA	tetrathiafulvalene-p-chloranil
VB	valence bond
VQ	vector quantization
VRML	virtual reality markup language
WBI	Wiberg bond index
WMRA	wavelet multi-resolution analysis
WT	wavelet transform
XAFS	X-ray absorption hyper-fine spectrum
ZORA	zero order regular approximation

PREFACE

A few months ago, when I was approaching one of what I call my sequence of versions, my 75.0 version nearby at hand, it occurred to me that it might be of interest to gather several papers of my old scientific colleagues and acquaintances. Practically, for this purpose, I thought, I could approach all of those who have attended the Girona Workshops or Seminars, held under the shelter of the University of Girona. Such scientific gatherings started around one third of a century ago and still are running, although now with other scientific interests than the original ones of several years ago.

While thanking the Gods of Old for the good news about my renewed version status, I was thinking of the fact that, arriving to a 75.0 one, it is not a trivial occurrence. It occurred to me that perhaps one could construct a substitute bearing with the same spirit of the Girona Workshops, but constructed in some other way. Therefore, I suggested to Professor Tanmoy Chakraborty from Jaipur University, the possibility of planning a book with him as co-editor, where each of my colleagues of old, in case that he or she could or will, might contribute a chapter. When Tanmoy managed to convince a publisher to edit such a book, half of the job was accomplished.

On the day of my 75.0 version, I sent several mails to my known fellow scientists, asking for a contribution to the book. Many agreed, and a few of them, sorrowfully due to varied difficulties, were not able to contribute. Some have enlarged the number of book chapters, encouraging young scientists to send significant work. Many have gathered with other colleagues to write their contributions.

The result is both a Spanish local assembly of recent research in many domains, but also a wide international showing of the state of affairs concerning theoretical chemistry in general, within the circles of mathematical, computational, and quantum chemistry, a book published at this very dawn's end of 21st century, presenting an assembly of some current research in the field.

At the very moment of writing this introduction, I am dealing with my version 76.5. The resultant collection of book chapters, at closing time for contributions, has convinced me that chemistry out of classical laboratory works still might follow new paths and try to crack old problems.

I hope everyone can read this book as a written substitute of and as well as a continuation by another way of the Girona Workshops. Plenty of the original and wide free spirit of research in theoretical chemistry.

—Ramon Carbó-Dorca, PhD
SQQM-CERT and IQQC
University in Girona, Spain

CHAPTER 1

THEORETICAL ANALYSIS: ELECTRONIC, RAMAN, VIBRATIONAL, AND MAGNETIC PROPERTIES OF Cu$_n$Ag (n = 1–12) NANOALLOY CLUSTERS

PRABHAT RANJAN,[1] TANMOY CHAKRABORTY,[2] and AJAY KUMAR[1]

[1]*Department of Mechatronics, Manipal University Jaipur, Dehmi Kalan, Jaipur, India*

[2]*Department of Chemistry, Manipal University Jaipur, Dehmi Kalan, Jaipur, India, E-mail: Tanmoy.chakraborty@jaipur.manipal.edu; tanmoychem@gmail.com*

CONTENTS

ABSTRACT

The studies of bi-metallic nanoalloy clusters have received a lot of attention because of its wide applications in the field of biomedicine, optics, catalysis, electronics, and optoelectronics. A deep theoretical insight is required to explain the physico-chemical properties of such compounds. Among such nanoalloy clusters, the compound formed between copper and silver is of immense importance because of their marked electronic, optical, and catalytic properties. Density functional theory (DFT) is one of the most successful approaches of quantum mechanics to study the electronic properties of materials. In this chapter, we have investigated Cu_nAg (n = 1–12) nanoalloy clusters in terms of conceptual DFT-based descriptors. We have also analyzed relative stability, vibrational, and Raman spectra along with the ferroelectricity behavior and magnetic susceptibility of Cu_nAg clusters; and compared the results with pure Cu_{n+1} (n = 1–12) clusters. The computed HOMO-LUMO energy gap, fragmentation energy, and second difference in energies of Cu_nAg clusters show interesting odd–even oscillation behaviors. The Raman and vibrational spectra displays many significant peaks, showing that wavelengths of the maximum Raman peaks of Cu_nAg and Cu_n clusters decreases with the cluster size. The result reveals that pure Cu_n clusters have larger wavelengths and wavenumbers of maximum intensity of Raman and vibrational spectra, respectively, as compare to Cu_nAg clusters. The wavenumbers and intensities of the maximum vibrational spectra of Cu-Ag and pure Cu clusters also vary with the cluster size. The computed Raman and vibrational spectra of Cu_n and Cu_nAg clusters may be useful for investigating size, composition, and structures of clusters. Ferroelectricity and magnetic susceptibility of Cu-Ag and pure Cu clusters are also studied showing their new potential applications in nonlinear optical devices.

1.1   INTRODUCTION

In recent years, the studies based on nanoalloy clusters has become important part in science and engineering due to its large scale applications in technological domain. Nanoalloy clusters consist of two or more molecules from different groups or same atoms in the periodic table. It has been already reported that nanoalloy clusters show interesting physico-chemical properties, which are different from bulk materials and initiate from their large surface-to-volume ratio and quantum-size effects [1–4]. The variation in size,

$_s$structure, and composition of clusters results in change the optical, cataly-sis, magnetic, and electronic properties [5–7]. Therefore, it may be suitable to design building block of nanodevices using these systems. Clusters with well-defined structure and composition may lead to some better alternatives. The study of noble metal clusters particularly Cu, Ag, and Au has been focus of experimental as well as theoretical research due to its numerous techno-logical applications [8–14].

Bimetallic nanoalloy clusters have emerged as important research topic due to its large applications in solid-state chemistry, material sci-ence, catalysis, electronics, biomedicine, and plasmonic technologies [15–20]. Group 11 metal (Cu, Ag and Au) clusters possess filled d orbitals and one unpaired electron in s shell [21]. This electronic arrangement is responsible to reproduce the exactly similar shell effects [22–26], which are experimentally observed in the alkali metal clusters [27–29]. Bishea et al. [30] have investigated diatomic CuAg clusters using spectroscopy technique. They have analyzed four band systems in the range of 20,000–27,000 cm^{-1}. James et al. [31] have studied adiabatic ionization potential of neutral bimetallic CuAg, and dissociation energy and frequency of cat-ionic CuAg$^+$ clusters using photoionization spectroscopy technique. The electronic properties and structures of mixed Cu-Ag clusters have been well-documented in literature using density functional theory (DFT) and it is mentioned that Ag atoms prefer to be at edge positions in tetramers and pentamers clusters [1, 2, 4, 15, 32–42]. Li et al. [15, 39–41] have investi-gated UV-visible, Raman and vibrational spectra of Ag_n, $Ag_{n-1}Cu$ ($n = 2$–8, 13, 20, 32), and Cu_mAg_n ($m + n = 7$ and 13) clusters using DFT. The com-puted wavelength, intensity, and range of Raman and vibrational spectra of pure Ag and Ag-Cu clusters vary with the size of clusters. The Raman and vibrational spectroscopy can be helpful in defining size, composition, and geometries of clusters. Investigation of structure, charge distribution, sta-bility, and electronic properties of bimetallic neutral and cationic $Cu_{n-1}Ag$ ($n = 2$–8) clusters has been done by Jiang et al. [42]. Recently, we have studied electronic and optical properties of mixed and impurity-doped bimetallic CuAg, AgAu, and CuAu clusters using DFT methodology [43–47]. Although, most of the studies have been done on the electronic, struc-tural, and optical properties of pure Cu, Ag, and mixed CuAg clusters, the reports on relative stabilities, electronic properties, Raman and vibrational spectra of Ag-doped Cu_nAg clusters are very limited till date.

In the present venture, we have investigated DFT-based global descriptors viz. HOMO-LUMO energy gap, hardness, softness, electronegativity,
electrophilicity index, quadrupole moment, and dipole moment of bi-metallic Cu_nAg ($n = 1$–12) clusters. The binding energy, fragmentation energy,
second-order difference in energy, Raman and vibrational spectra, bond
order and magnetic susceptibility of Cu_nAg clusters have been analyzed
and compared with the pure Cu_{n+1} ($n = 1$–12) clusters. The HOMO-LUMO
energy gaps, fragmentation energy, and second-order difference in energy
show interesting odd–even alternation behavior with cluster size. Our computed bond length and vibrational frequency of Cu_2, Ag_2, and CuAg are in
very well agreement with the experimental results. An attempt has been
made to correlate the properties of the compounds with their computational
counterparts.

1.2 COMPUTATIONAL DETAILS

In the last couple of years, DFT has been dominant and effective computational technique for bimetallic and multi-metallic clusters. DFT methods
are open to many innovative fields in material science, physics, chemistry,
surface science, nanotechnology, biology, and earth sciences [48]. Among
all the DFT approximations, the hybrid functional Becke's three parameter
Lee-Yang-Parr (B3LYP) exchange correlation functional has been proven
very efficient and used successfully for mixed and impurity-doped metal
clusters [8, 49–51]. The basis set LanL2dz has a high accuracy for metallic
clusters, which has been recently analyzed by researchers [8, 42, 52]. In this
study, all the modeling and structural optimization of compounds have been
performed using Gaussian 03 software package [53] within DFT framework. For optimization purpose, B3LYP exchange correlation with basis
set LanL2dz has been adopted. The used computation methodology in this
chapter is based on the molecular orbital approach, using linear combination
of atomic orbitals. Z-axis has been chosen for the spin polarization axis. The
quadrupole moment of molecule is calculated in terms of analytical integration methodology.

Invoking Koopmans' approximation [54], ionization energy (I),
and electron affinity (A) of all the nanoalloys have been computed as
following:

$$I = -\varepsilon_{\text{HOMO}} \tag{1}$$

$$A = -\varepsilon_{\text{LUMO}} \tag{2}$$

Thereafter, using I and A, the conceptual DFT-based descriptors viz. electronegativity (χ), global hardness (η), molecular softness (S), and electrophilicity index (ω) have been computed. The equations used for such calculations are as given under:

$$\chi = -\mu = \frac{I + A}{2} \tag{3}$$

where, μ represents the chemical potential of the system.

$$\eta = \frac{I - A}{2} \tag{4}$$

$$S = \frac{1}{2\eta} \tag{5}$$

$$\omega = \frac{\mu^2}{2\eta} \tag{6}$$

1.3 RESULTS AND DISCUSSIONS

1.3.1 HOMO-LUMO ENERGY GAP AND DFT-BASED DESCRIPTORS

Computational analysis of bi-metallic Cu_nAg ($n = 1$–12) nanoalloy clusters has been performed invoking electronic structure theory. The orbital energies in the form of highest occupied molecular orbital (HOMO)-lowest unoccupied molecular orbital (LUMO) gap along with computed DFT-based descriptors viz. electronegativity, hardness, softness, and electrophilicity index have been analyzed and reported in Table 1.1. The quadrupole moment along with different axes are also reported in the Table 1.2. Table 1.1 reveals that HOMO-LUMO energy gap of clusters run hand in hand with their computed hardness values. As the frontier orbital energy gap

TABLE 1.1 Computed DFT-Based Descriptors of Cu_nAg ($n = 1–12$) Nanoalloy Clusters in eV

Species	HOMO-LUMO gap	Electronegativity	Hardness	Softness	Electrophilicity index
CuAg	3.055	4.050	1.527	0.327	5.368
Cu_2Ag	1.143	3.374	0.571	0.875	9.961
Cu_3Ag	1.877	3.823	0.918	0.532	7.784
Cu_4Ag	1.523	3.183	0.771	0.656	6.651
Cu_5Ag	3.156	3.891	1.578	0.316	4.796
Cu_6Ag	2.095	3.333	1.047	0.477	5.302
Cu_7Ag	2.122	3.891	1.061	0.471	7.134
Cu_8Ag	1.170	3.414	0.685	0.854	9.967
Cu_9Ag	1.714	3.550	0.857	0.583	7.355
$Cu_{10}Ag$	1.333	4.013	0.667	0.750	12.081
$Cu_{11}Ag$	1.414	3.918	0.707	0.706	10.850
$Cu_{12}Ag$	1.033	3.701	0.516	0.967	13.245

TABLE 1.2 Quadrupole Moment of Computed DFT-Based Descriptors Cu_nAg ($n = 1–12$) Nanoalloy Clusters in Debye-Ang

Species	Quad-XX	Quad-XY	Quad-XZ	Quad-YY	Quad-YZ	Quad-ZZ
CuAg	−34.089	0.000	0.000	−34.089	0.000	−26.934
Cu_2Ag	−47.487	0.000	0.000	−40.371	0.000	−50.693
Cu_3Ag	−60.671	0.000	0.000	−53.958	0.000	−61.644
Cu_4Ag	−74.143	0.000	0.000	−70.441	0.000	−64.137
Cu_5Ag	−87.190	0.000	0.000	−79.331	0.000	−78.578
Cu_6Ag	−100.561	0.000	0.000	−89.317	0.000	−0.439
Cu_7Ag	−111.747	−0.011	−0.051	8.348	−0.000	−5.201
Cu_8Ag	−125.272	4.771	0.000	−118.369	0.000	−128.398
Cu_9Ag	−141.298	−0.033	0.188	−134.514	−0.019	0.511
$Cu_{10}Ag$	−134.564	−2.216	0.000	−121.903	0.000	−140.445
$Cu_{11}Ag$	−135.592	−0.024	0.195	−159.377	−0.035	−167.062
$Cu_{12}Ag$	−179.925	0.000	0.000	−165.805	0.000	−166.724

increases, their hardness value increases. From the experimental point of view, it is an expected trend that the clusters having highest HOMO-LUMO energy gap will be the least prone to response against any external perturbation. From the Table 1.1, it is concluded that the Cu_5Ag cluster has maximum HOMO-LUMO energy gap (3.156 eV), whereas $Cu_{12}Ag$ cluster possess the lowest energy gap (1.033 eV). Though there is no such available quantitative data for optical properties of aforesaid clusters, we can tacitly assume that there must be a direct relationship between optical properties of Cu-Ag clusters with their computed HOMO-LUMO gap. The assumption is based on the fact that optical properties of materials are interrelated with flow of electrons within the systems, which in turn depend on the difference between the distance of valence and conduction band [55]. On that basis, we may conclude that optical properties of bimetallic nano clusters exhibit a direct relationship with hardness values. Similarly, the softness data exhibits an inverse relationship towards the experimental optical properties. The linear correlation between HOMO-LUMO gaps along with their computed softness is lucidly plotted in the Figure 1.1. The high value of regression coefficient ($R^2 = 0.895$) observed in the Figure 1.1, validates our predicted model.

We have investigated the chemical stabilities of Cu_nAg ($n = 1–12$); in terms of HOMO-LUMO energy gap. In a molecular system chemical stability is enhanced with large HOMO-LUMO gap [62]. The HOMO-LUMO energy gap as of any compound shows an odd–even oscillation behavior, which is depicted in the Figure 1.2. The clusters with an even number of total atoms possess larger energy gap as compared to the neighbor clusters having odd number of atoms, which is an expected trend for

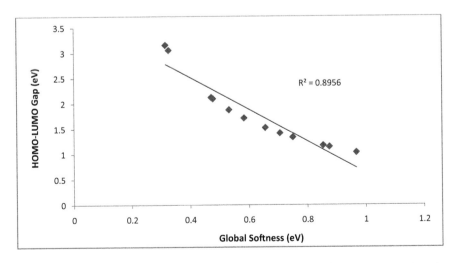

FIGURE 1.1   A linear correlation between global softness *vs.* HOMO-LUMO gap of Cu_nAg ($n = 1–12$) nanoalloy clusters.

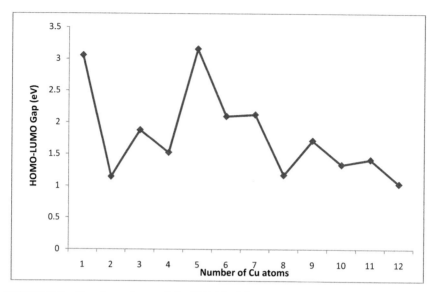

FIGURE 1.2 Size dependence of the HOMO-LUMO gaps of lowest energy structure of Cu_nAg ($n = 1–12$) nanoalloy clusters.

closed or open shell systems [56–59]. The stability of even numbered cluster is actually outcome of their closed electronic configuration, which always produces extra stability. It shows that these nanoalloy clusters possess enhanced chemical stability and may be suitable for building block of nanodevices [56].

The quadrupole charge separation is represented in form of quadrupole moment according to Buckingham convention in the Table 1.2. The quadrupole moment values in different axes are represented in Debye-Ang. The quadrupole moment can be used as an index to define probable electrostatic interactions of the molecules [60].

In order to apply our computed data in the real field, a bond length of some of the species have been computed. A comparative analysis has been reported between experimental bond length and our data in the Table 1.3. A close agreement in magnitude between experimental data and our computed bond length is reflected form Table 1.3. The computed bond length of pure Cu_n ($n = 3–9$) clusters is also in agreement with previous data reported by Kabir et al. [62]. It supports and validates our computational analysis. Table 1.3 reveals that the bond lengths between metal–metal follow the order Ag-Ag > Ag-Cu > Cu-Cu, which is consistent with the previous reports [15].

TABLE 1.3 The Calculated Average Bond Length of the Species Cu_n ($n = 1-10$) and CuAg

Species	Computed average bond length (Å)	Kabir [62] Scale (Å)	Experimental bond length (Å)
Ag_2	2.51	–	2.53 [61]
Cu_2	2.18	–	2.22 [56, 63]
Cu_3	2.25	2.24	–
Cu_4	2.23	2.23	–
Cu_5	2.37	2.38	–
Cu_6	2.41	2.39	–
Cu_7	2.42	2.45	–
Cu_8	2.40	2.41	–
Cu_9	2.42	2.41	–
CuAg	2.33	2.33	2.37 [30]

1.3.2 *RELATIVE STABILITIES AND BOND ORDER PARAMETER*

As per the cluster physics, the binding energy $E_b(n)$, fragmentation energies $D(n)$, and the second-order difference in energy $\Delta_2 E(n)$ have marked influence on the relative stability of a particular system [56]. In order to compare the degree of relative stability of Ag doped Cu clusters $Cu_n Ag$ ($n = 1-12$) we have also computed $E_b(n)$, $D(n)$, and $\Delta_2 E(n)$ of pure copper clusters Cu_{n+1} ($n = 1-12$).

For calculating $E_b(n)$, $D(n)$, and $\Delta_2 E(n)$ of Cu_{n+1} ($n = 1-12$), we have used the following formula:

$$E_b(n+1) = \frac{[(n+1)E(Cu) - E(Cu_{n+1})]}{n+1} \tag{7}$$

$$D(n+1, n) = E(Cu_n) + E(Cu) - E(Cu_{n+1}) \tag{8}$$

$$\Delta_2 E(n+1) = E(Cu_n) + E(Cu_{n+2}) - 2E(Cu_{n+1}) \tag{9}$$

For $Cu_n Ag$ ($n = 1-12$) clusters, $E_b(n)$, $D(n)$, and $\Delta_2 E(n)$ are calculated as follows:

$$E_b(n) = \frac{E(Ag) + nE(Cu) - E(Cu_nAg)}{n+1} \tag{10}$$

$$D(n, n-1) = E(Cu_{n-1}Ag) + E(Cu) - E(Cu_nAg) \tag{11}$$

$$\Delta_2 E(n) = E(Cu_{n-1}Ag) + E(Cu_{n+1}Ag) - 2E(Cu_nAg) \tag{12}$$

where, $E(Cu)$, $E(Ag)$, $E(Cu_nAg)$, $E(Cu_{n-1}Ag)$, $E(Cu_{n+1}Ag)$, and $E(Cu_n)$ are the total energies of the most stable Cu, Ag, Cu_nAg, $Cu_{n-1}Ag$, $Cu_{n+1}Ag$, and Cu_n clusters, respectively.

The variation of calculated binding energies (E_b) for the most stable isomers of Cu_{n+1} and Cu_nAg (n = 1–12) nanoalloy clusters are shown in Figure 1.3. The binding energy of pure and doped clusters is increasing with the number of atoms in the cluster. It indicates that clusters continue to gain energy during the growth process. It is observed that binding energies of the lowest energy structure of Cu_nAg are larger than the corresponding pure Cu_{n+1} clusters, which shows that the impurity atom, i.e., Silver may enhance the stability of pure copper nanoalloy clusters. The computed results of pure and doped clusters follow the similar trend with previous reports [56, 64–67].

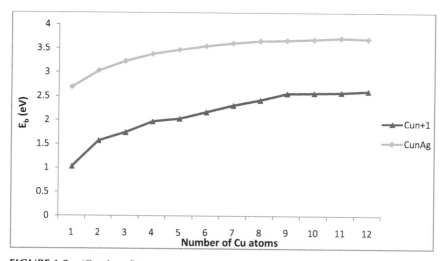

FIGURE 1.3   (Continued).

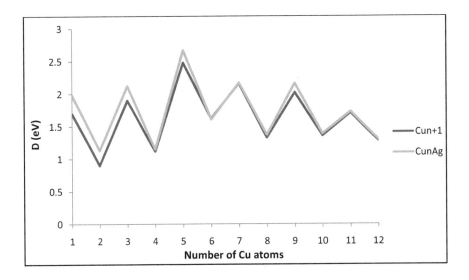

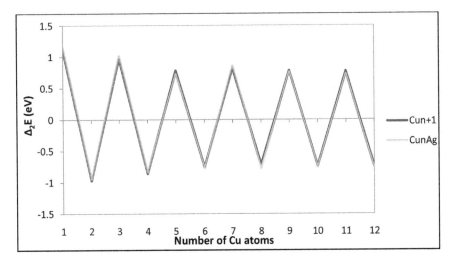

FIGURE 1.3 Size dependence of binding energy (E_b), fragmentation energy (D), and second-order in energy ($\Delta_2 E$) for the most stable cluster Cu_{n+1} and $Cu_n Ag$ ($n = 1–12$) nanoalloy clusters.

The size dependence of the fragmentation energy (D) and the second order difference in energies ($\Delta_2 E$) for Cu_{n+1} and $Cu_n Ag$ ($n = 1–12$) clusters have been studied and reported in Figure 1.3. The fragmentation energy and

second order difference in energies shows an interesting odd–even oscillation behavior as a function of cluster size, these energies are highly sensitive quantities towards the structure and geometry of the nanoalloy clusters
and they exhibit odd–even oscillatory phenomena. Figure 1.3 reveals that
Fragmentation energy and Second order difference in energies of total
number of even number clusters is larger than total number of odd-number
clusters. This confirms that nanoalloy clusters with even number of atoms
are more stable than their neighbor odd number of atoms. For pure clusters
Cu_{n+1} and doped cluster Cu_nAg, maximum value of fragmentation energy are
observed at $n = 5$, indicating that clusters Cu_6 with D = 2.48 eV and Cu_5Ag
with D = 2.67 eV possess maximum stability, which follow the similar trend
for pure and doped clusters calculated by Li-Ping et al. [56] and Tafoughalt
and Samah [67].

In the view of binding energy, fragmentation energy, and second order
difference in energy, we can say that most stable structure has been observed
for Cu_5Ag cluster in this molecular system.

In this report, we have also analyzed bond order parameter (σ) of Cu_nAg
($n = 1$–12) clusters. It is a factor, which provides quantitative measure information of segregation in nanoalloy clusters. Positive value of σ suggests the
segregation of atoms in the cluster, though it shows disorderly mixed clusters when the value of σ is zero. The negative value of σ indicates the mixed
and onion like phases of clusters [1].

The bond order parameter of Cu_nAg ($n = 1$–12) cluster is defined as
[15]:

$$\sigma = \frac{N_{Ag-Ag} + N_{Cu-Cu} - N_{Ag-Cu}}{N_{Ag-Ag} + N_{Cu-Cu} + N_{Ag-Cu}} \tag{13}$$

where, N_{Ag-Ag}, N_{Cu-Cu}, and N_{Ag-Cu} are the number of nearest-neighbor bonds
between Ag-Ag, Cu-Cu, and Ag-Cu, respectively.

Figure 1.4 displays the bond order parameter as function of the lowest
energy structure of Cu_nAg ($n = 1$–12) clusters. The value of σ is positive for
all the clusters except Cu_2Ag and Cu_4Ag. The negative value of σ indicates
that these nanoalloy clusters are disorderly mixed or alloyed clusters. Our
computed data for cluster Cu_6Ag is in accordance with the previous result
calculated by Li et al. [40].

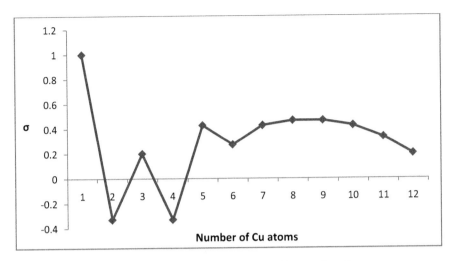

FIGURE 1.4   Bond order parameter of Cu_nAg ($n = 1-12$) nanoalloy clusters.

1.3.3 VIBRATIONAL AND RAMAN SPECTRA OF CU_NAG AND CU_{N+1} (n = 1–12) NANOALLOY CLUSTERS

The vibrational spectra of Cu_nAg clusters are analyzed and compared with pure Cu_{n+1} ($n = 1-12$) clusters, the same is reported in Figure 1.5. There is only one peak observed for vibrational spectra of CuAg (0.0016 Km/mol at a frequency, $f = 216.79$ cm^{-1}), however no vibrational spectra for Cu_2 cluster has been observed. The calculated spectra of CuAg cluster are in well-agreement with the previous result calculated by Li et al. [41]. The vibrational spectra of Cu_2Ag and Cu_3 also have one intense peak. The vibrational spectrum of Cu_2Ag has maximum peak 0.735 Km/mol at $f = 200.05$ cm^{-1}, whereas Cu_3 has maximum peak 0.7947 Km/mol with $f = 241.944$ cm^{-1}. The wavenumber of the peak of Cu_2Ag is smaller than that of Cu_3 cluster. The intensity of main peak of Cu_3Ag cluster is three times larger than that of Cu_4 cluster. At larger wavenumber, the Cu_4 cluster has two strong peaks but Cu_3Ag has only one peak. The maximum peak of the Cu_3Ag cluster is stronger than that of Cu_4 cluster. The Cu_4Ag cluster has two peaks in the range of 140–235 cm^{-1}, whereas pure cluster Cu_5 has no peak in this range. The intensity of main peak of the Cu_4Ag cluster is stronger than that of Cu_5 cluster but the wavenumber of pure Cu_5 cluster is larger than that of Cu_4Ag cluster. The vibration spectrum of Cu_5Ag and Cu_6 clusters lie at large wavenumber, but the intensity of main peak of the Cu_5Ag

Spectra of CuAg

Spectra of Cu$_2$

Spectra of Cu$_2$Ag

Spectra of Cu$_3$

Spectra of Cu$_3$Ag

Spectra of Cu$_4$:

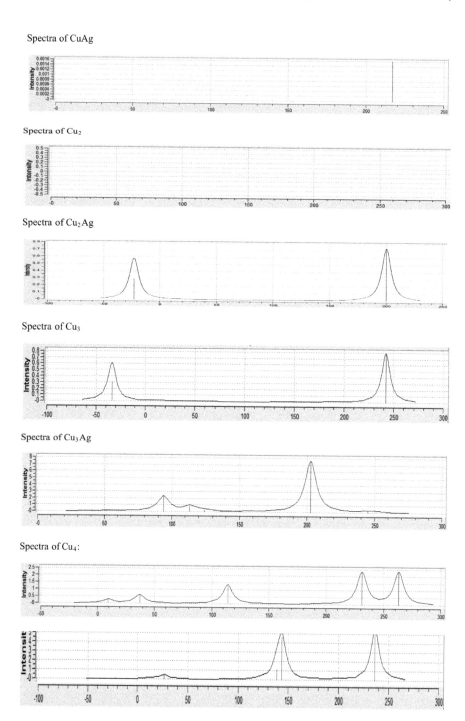

FIGURE 1.5   (Continued).

Spectra of Cu$_5$:

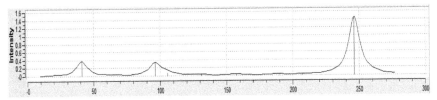

Spectra of Cu$_5$Ag:

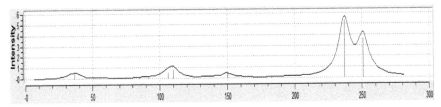

Spectra of Cu$_6$:

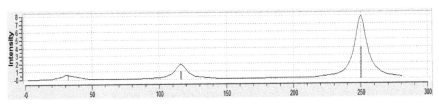

Spectra of Cu$_6$Ag:

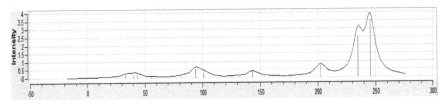

Spectra of Cu$_7$:

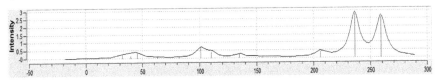

FIGURE 1.5   (Continued).

Spectra of Cu_7Ag:

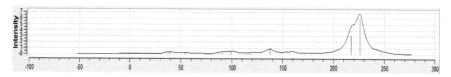

Spectra of Cu_8:

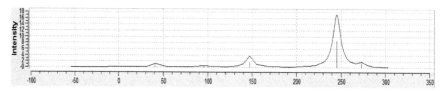

Spectra of Cu_8Ag:

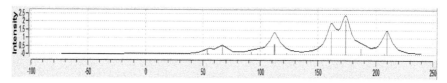

Spectra of Cu_9:

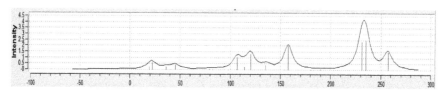

Spectra of Cu_9Ag:

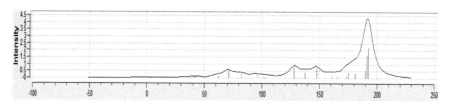

Spectra of Cu_{10}:

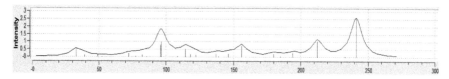

FIGURE 1.5 (Continued).

Spectra of $Cu_{10}Ag$:

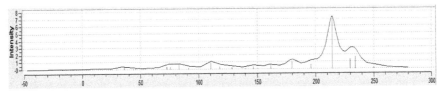

Spectra of Cu_{11}:

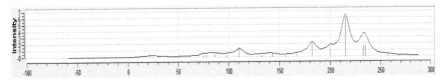

Spectra of $Cu_{11}Ag$:

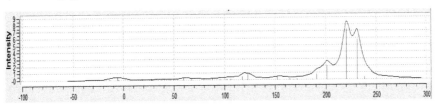

Spectra of Cu_{12}:

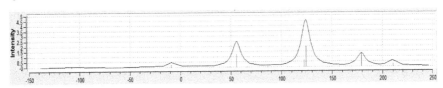

Spectra of $Cu_{12}Ag$

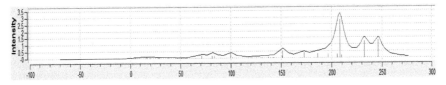

Spectra of Cu_{13}

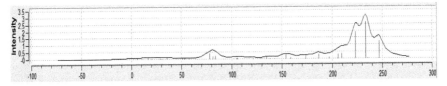

FIGURE 1.5   Vibrational spectra of Cu_nAg and Cu_{n+1} (n = 1–12) nanoalloy clusters.

cluster is stronger than that of the Cu_6 cluster. The vibrational spectra of Cu_6Ag cluster are very much likely to Cu_7 cluster till wavenumber 200 cm⁻¹. Both clusters have two strong peaks at large wavenumbers. The peak of vibrational spectra is strong in the case of Cu_6Ag cluster as compare to Cu_7. However, wavenumber of Cu_7 cluster is larger than that of Cu_6Ag cluster. The intensity and wavenumber of Cu_8 is larger than that of the Cu_7Ag cluster. The cluster Cu_8 has maximum vibrational spectrum 8.60 Km/mol at $f = 245.27$ cm⁻¹, whereas the cluster Cu_7Ag has peak (5.99 Km/mol) at 225.91 cm⁻¹. Pure Cu_9 cluster has three strong peaks at larger wavenumbers (at 230.88, 234.86, and 257.19 cm⁻¹), but the Cu_8Ag cluster has no peak at this range. The intensity of main peak and wavenumber is larger in pure cluster Cu_9 than that of Cu_8Ag. The pure cluster Cu_{10} has two strong spectra at larger wavenumber 210–245 cm⁻¹ but doped cluster Cu_9Ag has no peak at this range. The intensity and wavenumber of main peak of pure cluster Cu_{10} is larger than that of Cu_9Ag cluster. The vibrational spectra of $Cu_{10}Ag$ cluster is similar to Cu_{11} cluster. The wavenumber of $Cu_{10}Ag$ is smaller than that of pure Cu_{11} cluster but its intensity is larger than that of Cu_{11} cluster. The vibrational spectra of $Cu_{11}Ag$ have two main peaks at larger wavenumber whereas there is no peak present in this range for pure Cu_{12} cluster. The wavenumber and intensity of the main peak of the $Cu_{11}Ag$ cluster is larger than that of Cu_{12} cluster. However, at lower wavenumber the intensity of pure cluster Cu_{12} is more as compare to $Cu_{11}Ag$ cluster. The cluster $Cu_{12}Ag$ and Cu_{13} have strong peaks at larger wavenumber. The wavenumber of the main peak of $Cu_{12}Ag$ is smaller than that of Cu_{13} cluster but its intensity is larger than that of pure Cu_{13} cluster.

Figure 1.6 displays the Raman activity of Ag doped Cu_n Cu_nAg and pure copper Cu_{n+1} ($n = 1–12$) nanoalloy clusters. The Raman spectra have been simulated by the wavelength of incident light at 514.5 nm and 10 K. All Raman spectra of nanoalloy clusters show many significant vibration modes. There is only one intense peak observed for CuAg (57.15 A⁴/amu at $f = 216.79$ cm⁻¹) and Cu_2 (60.76 A⁴/amu at $f = 256.02$ cm⁻¹) clusters, which is in agreement with the previous result calculated by Li et al. [41]. The wavelength and intensity of Raman spectra of Cu_2 cluster is larger than that of CuAg cluster. The Raman spectra of Cu_3 and Cu_2Ag have also only one intense peak. The cluster Cu_3 has spectrum 83.05 A⁴/amu at $f = 149.57$ cm⁻¹ and the cluster Cu_2Ag has 94.88 A⁴/amu at $f = 148.30$ cm⁻¹. The intensity of Raman spectra of Cu_2Ag is stronger than Cu_3 cluster. The cluster Cu_3Ag, with symmetry point group C_{2v} appears in the range of

Raman Spectra of CuAg:

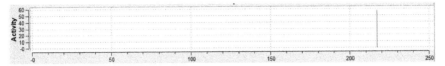

Raman Spectra of Cu_2

Raman Spectra of Cu_2Ag

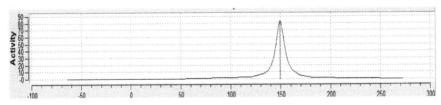

Raman Spectra of Cu_3:

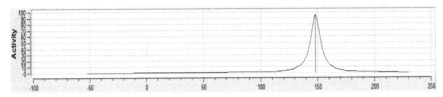

Raman Spectra of Cu_3:

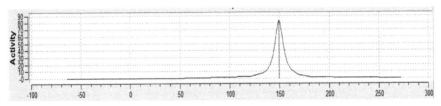

Raman Spectra of Cu_4

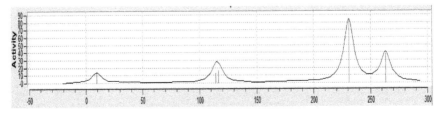

FIGURE 1.6   (Continued).

Raman Spectra of Cu_4

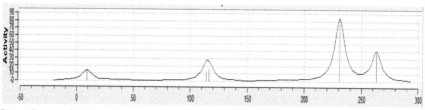

Raman Spectra of Cu_5

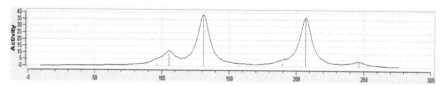

Raman Spectra of Cu_5Ag

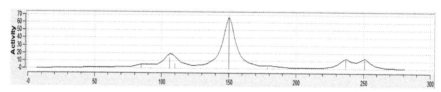

Raman Spectra of Cu_6

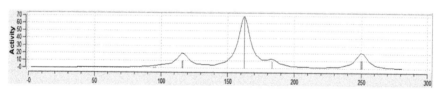

Raman Spectra of Cu_6

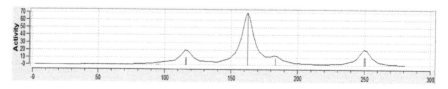

Raman Spectra of Cu_7

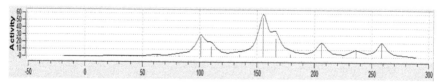

FIGURE 1.6   (Continued).

Raman Spectra of Cu₇Ag

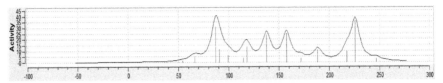

Raman Spectra of Cu₈

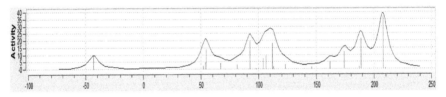

Raman Spectra of Cu₈Ag

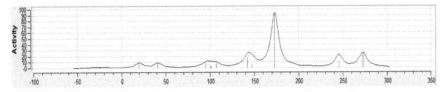

Raman Spectra of Cu₉

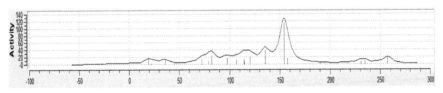

Raman Spectra of Cu₉Ag

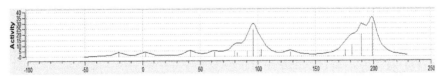

Raman Spectra of Cu₁₀

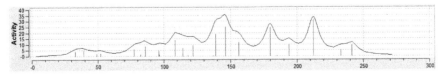

FIGURE 1.6   (Continued).

Raman Spectra of $Cu_{10}Ag$:

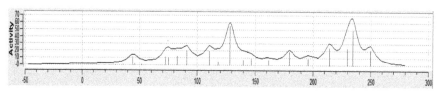

Raman Spectra of Cu_{11}:

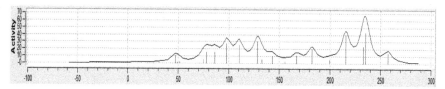

Raman Spectra of $Cu_{11}Ag$:

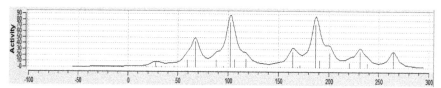

Raman Spectra of Cu_{12}:

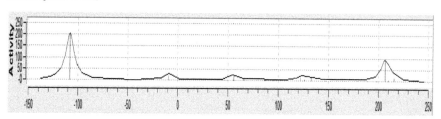

Raman Spectra of $Cu_{12}Ag$:

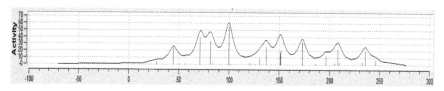

Raman Spectra of Cu_{13}:

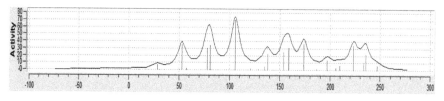

FIGURE 1.6   Raman spectra of Cu_nAg and Cu_{n+1} ($n = 1–12$) nanoalloy clusters.

0–275 cm-1, which has two strong peaks. The strongest peak of intensity 38.96 A^4/amu is observed at f = 124.059 cm^{-1} for Cu$_3$Ag cluster. The cluster Cu$_4$ with symmetry group C$_s$ appears in the range of 0–293 cm^{-1}, which has maximum intensity of 83.113 A^4/amu at longer wavelength f = 230.97 cm^{-1}. The intensity of spectra is high in the case of Cu$_4$ cluster; however, Raman spectra of Cu$_3$Ag is strong at smaller wavelength. The cluster Cu$_4$Ag with symmetry group D$_{2h}$ has three strong peaks (at f = 104, 132 and 202 cm^{-1}). The strongest peak of Raman spectra 163.65 A^4/amu lie at f = 104.235 cm^{-1} for Cu$_4$Ag cluster. Raman spectra of Cu$_5$ cluster with C$_1$ symmetry group has maximum intensity 38.05 A^4/amu at f = 131.07 cm^{-1}. The cluster Cu$_5$ has total nine peaks, including strong intensity at longer wavelength at 207 cm^{-1} and of weak intensity at 246.3 cm^{-1}. The intensity of Cu$_4$Ag cluster is very strong as compare to Cu$_5$ cluster. The Raman spectra of cluster Cu$_5$Ag is having similar pattern of pure Cu$_6$ cluster. The maximum intensity of pure cluster Cu$_6$ is little higher than Cu$_5$Ag but at longer wavelength (250 cm^{-1}). The Raman spectra of Cu$_6$Ag and Cu$_7$ clusters have similar peaks at short wavelengths. The maximum Raman strengths observed at 143.17 and 155.29 cm^{-1} for Cu$_6$Ag and Cu$_7$ cluster, respectively. The Raman strength of main peak of Cu$_6$Ag cluster is stronger than pure Cu$_7$ cluster but its wavelength is shorter than that of pure Cu$_7$ cluster. The cluster Cu$_7$Ag has several peaks at lower as well as higher wavelengths. The maximum intensity of Raman spectrum of Cu$_7$Ag cluster is weaker than that of Cu$_8$ cluster. In the case of Cu$_8$ cluster, maximum peak of spectra 93.24 A^4/amu is found at 172.94 cm^{-1}, but in the case of Ag doped Cu$_7$ cluster Cu$_7$Ag, maximum peak 36.003 A^4/amu is observed at f = 225.94 cm^{-1}. The wavelength of Raman spectrum of Cu$_8$ cluster is larger than that of Cu$_7$Ag cluster. However, at shorter wavelength cluster Cu$_7$Ag has strong intensity as compare to Cu$_8$ cluster. Our computed spectra of Cu$_7$Ag and Cu$_8$ clusters are in agreement with the previous result reported by Li et al. [35]. The cluster Cu$_8$Ag has three peaks at lower wavelength ranging between 50–120 cm^{-1} and three peaks at larger wavelength ranging between 170–210 cm^{-1}. The maximum Raman intensity of cluster Cu$_8$Ag 36.85 A^4/amu is observed at 207.55 cm^{-1}. The cluster Cu$_9$ has many Raman spectra peaks in the range of 19–260 cm^{-1}, maximum peak 114.65 A^4/amu is at 154.09 cm^{-1}. The intensity of Raman spectra and wavelength of pure cluster Cu$_9$ is larger than that of cluster Cu$_8$Ag. The wavelengths of Raman spectra are in the range of 2–200 cm^{-1} and 32–241 cm^{-1} for clusters Cu$_9$Ag and Cu$_{10}$, respectively. The pure cluster Cu$_{10}$ has one strong peak and two weak peaks present at longer wavelength whereas there is no peak present in this range for cluster Cu_9Ag. The cluster Cu_{10} has a high Raman intensity and larger wavelength than that of Cu_9Ag cluster. However, at lower wavelength cluster Cu_9Ag has a high intensity (23.84 A^4/amu at 95.94 cm^{-1}) than Cu_{10} cluster. The clusters $Cu_{10}Ag$ and Cu_{11} have various peaks at lower and higher wavelength. The main peak of $Cu_{10}Ag$ cluster has strong intensity than that of Cu_{11} but the wavelength of Cu_{11} cluster is larger than $Cu_{10}Ag$ cluster. The Raman spectra of $Cu_{11}Ag$ and Cu_{12} clusters have many peaks in the range of 25–270 cm^{-1} and 50–215 cm^{-1}, respectively. There are three additional peaks present in the $Cu_{11}Ag$ cluster at the larger wavelength compare with the Cu_{12} cluster. The wavelength and intensity of the main Raman peak of the $Cu_{11}Ag$ cluster is smaller than that of pure cluster Cu_{12}. The Raman spectra of $Cu_{12}Ag$ is having similar pattern of pure Cu_{13} cluster. Both the clusters spectra are in the range of 4–246 cm^{-1}. The Raman strength and wavelength of main peak of Cu_{13} cluster is stronger than that of doped $Cu_{12}Ag$ cluster.

The intensities and wavelengths of maximum Raman peaks of pure Cu_{n+1} and Cu_nAg ($n = 1–12$) nanoalloy clusters are reported in Table 1.4. The result reveals that the wavelengths of the maximum Raman peaks of Cu_nAg clusters decreases with increasing the number of Cu atoms, except, cluster from Cu_4Ag to Cu_5Ag and from Cu_6Ag to Cu_7Ag. The wavelengths of maximum

TABLE 1.4   Intensities and Wavelengths of Maximum Raman Peak of Cu_nAg and Pure Cu_{n+1} Clusters ($n = 1–12$)

Species	Intensities (A⁴/amu)	Wavelength (cm⁻¹)	Species	Intensities (A⁴/amu)	Wavelength (cm⁻¹)
$CuAg$	57.15	216.79	Cu_2	60.76	256.02
Cu_2Ag	94.88	148.30	Cu_3	83.05	149.57
Cu_3Ag	38.96	124.05	Cu_4	83.10	230.95
Cu_4Ag	163.65	104.23	Cu_5	38.05	131.07
Cu_5Ag	66.27	150.21	Cu_6	68.46	162.30
Cu_6Ag	63.17	143.17	Cu_7	53.92	155.29
Cu_7Ag	36.00	225.91	Cu_8	93.24	172.94
Cu_8Ag	36.85	207.55	Cu_9	114.65	154.09
Cu_9Ag	29.72	199.21	Cu_{10}	31.48	212.41
$Cu_{10}Ag$	54.90	128.36	Cu_{11}	43.23	235.24
$Cu_{11}Ag$	75.81	102.22	Cu_{12}	86.78	206.67
$Cu_{12}Ag$	54.85	99.67	Cu_{13}	69.68	105.76

Raman peaks of pure Cu_n clusters are large as compare to Ag-doped Cu_nAg clusters, except Cu_7Ag and Cu_8Ag clusters. Overall, the intensities of the maximum Raman peaks of pure Cu clusters are stronger than that of CuAg clusters, except Cu_2Ag, Cu_4Ag, and $Cu_{10}Ag$ clusters.

Table 1.5 displays the wavenumber and intensities of maximum vibrational spectra of pure Cu_{n+1} and Cu_nAg ($n = 1$–12) nanoalloy clusters. The wavenumbers of the maximum spectra of Cu-Ag clusters swing with increasing cluster size, when clusters change from Cu_6Ag to Cu_8Ag, from $Cu_{11}Ag$ to $Cu_{12}Ag$, their wavenumber decreases; however, when clusters changes from Cu_2Ag to Cu_6Ag, and from Cu_8Ag to $Cu_{11}Ag$, their wavenumber increases. The intensities of the maximum peaks of Cu-Ag clusters fluctuate with increasing cluster size, when clusters change from Cu_3Ag to Cu_6Ag, from Cu_7Ag to Cu_8Ag, from $Cu_{11}Ag$ to $Cu_{12}Ag$, their strength decreases; when clusters change from Cu_2Ag to Cu_3Ag, from Cu_6Ag to Cu_7Ag, from Cu_8Ag to Cu_9Ag and from Cu_9Ag to $Cu_{11}Ag$, their strength increases. The wavenumbers and intensities of maximum vibrational spectra of pure Cu_n clusters exhibits interesting odd–even oscillation behaviors, indicating even number of clusters show high intensity and larger wavenumber as compare to their neighbor odd number of clusters.

TABLE 1.5 Intensities and Wavenumbers of the Maximum Vibrational Peaks of Cu_nAg and Pure Cu_{n+1} Clusters ($n = 1$–12)

Species	Intensities (km/mol)	Wavenumbers (cm^{-1})	Species	Intensities (Km/mol)	Wavenumbers (cm^{-1})
CuAg	0.00	216.79	Cu_2	0.00	256.02
Cu_2Ag	0.73	200.05	Cu_3	0.79	241.94
Cu_3Ag	7.66	202.65	Cu_4	2.33	263.08
Cu_4Ag	5.55	235.99	Cu_5	1.45	246.30
Cu_5Ag	5.29	236.90	Cu_6	3.95	250.65
Cu_6Ag	3.47	245.53	Cu_7	2.76	236.11
Cu_7Ag	5.99	225.91	Cu_8	8.60	245.27
Cu_8Ag	2.11	174.03	Cu_9	2.42	234.86
Cu_9Ag	2.19	192.42	Cu_{10}	2.55	241.16
$Cu_{10}Ag$	6.95	214.14	Cu_{11}	6.10	215.28
$Cu_{11}Ag$	7.16	220.53	Cu_{12}	1.96	124.25
$Cu_{12}Ag$	2.76	208.06	Cu_{13}	2.62	233.55

It has been observed that different clusters show different Raman and vibrational spectrum. Raman and vibrational spectrum of nanoalloy clusters show some characteristics, which relate with size, geometry, and composition of clusters. The computed result is in agreement with the previous results reported by Li et al. [40]. Therefore, we can possibly determine the size, structure and composition of the clusters and compare the theoretical and experimental spectra of these clusters [40, 41]. We have also compared the computed vibrational frequency of the species namely, Cu_2, CuAg, and Ag_2 with the experimental results [30], the same is reported in the Table 1.6. A close agreement between experimental and computed data is reflected from Table 1.6, which supports and validates our computational analysis.

1.3.4 FERROELECTRIC PROPERTIES

The dipole moment of pure clusters, Cu_{n+1} and doped clusters, Cu_nAg ($n = 1-12$) is reported in Table 1.7. The result indicates that nanoalloy clusters with low HOMO-LUMO gap in the range of 1.00–1.90 eV have large dipole moment (above 1 Debye). The Ag doped Cu_n clusters, namely CuAg, Cu_5Ag, Cu_6Ag, and Cu_7Ag, which have large HOMO-LUMO gap (more than 2.0 eV) have low dipole moment (less than 1.0 Debye). In the case of pure copper clusters, Cu_{n+1} ($n = 1-12$) similar trend has been observed. The result reveals that $Cu_{12}Ag$ cluster have maximum dipole moment with lowest HOMO-LUMO gap in this molecular system. As this is an expected trend from the experimental point of view also that the ferroelectric materials are soft, flexible, light weight, and can be easily processed at relative low temperature and printed onto soft substrates. This form of metallic ferroelectricity is a cluster property. Moreover, not only are the dipole moments large, they are ubiquitous and in all cases show several distinct properties [69]. As expected, the ferroelectric properties of Cu_nAg and Cu_{n+1} ($n = 1-12$)

TABLE 1.6   The Computed Vibrational Frequency of Cu_2, CuAg, and Ag_2 (in cm^{-1})

Species	Computed vibrational frequency	Experimental vibrational frequency [30]
Cu_2	256	265
CuAg	216	232
Ag_2	181	192

TABLE 1.7   Dipole Moment of Cu_nAg and Cu_{n+1} ($n = 1$–12) Nanoalloy Clusters

Species	Dipole Moment (Debye)	Species	Dipole Moment (Debye)
CuAg	0.455	Cu_2	0.516
Cu_2Ag	1.235	Cu_3	0.587
Cu_3Ag	3.182	Cu_4	1.709
Cu_4Ag	2.007	Cu_5	1.515
Cu_5Ag	0.663	Cu_6	0.557
Cu_6Ag	0.678	Cu_7	0.698
Cu_7Ag	0.275	Cu_8	0.933
Cu_8Ag	1.258	Cu_9	1.418
Cu_9Ag	1.352	Cu_{10}	1.042
$Cu_{10}Ag$	1.081	Cu_{11}	1.023
$Cu_{11}Ag$	1.116	Cu_{12}	1.405
$Cu_{12}Ag$	3.316	Cu_{13}	3.538

nanoalloy clusters are very much sensitive to structural symmetry or asymmetry [15]. These ferroelectric materials may be suitable for the applications of sensing, actuation, energy harvesting, and data storage. [68].

1.3.5   MAGNETIC SUSCEPTIBILITY

In this study, magnetic susceptibility tensors of nanoalloy clusters Cu_nAg and Cu_{n+1} ($n = 1$–12) have been performed using continuous set of gauge transformations (CSGT) method and the same is reported in Table 1.8. The magnetic susceptibility is isotropic in nature for most of the clusters, reflecting the global symmetry of these systems. For most of the nanoalloy clusters, the absolute value of isotropic properties increasing with the number of atoms for this molecular system, except for some clusters [70]. With the exception of Cu_2Ag, all nanoalloy clusters studied are diamagnetic in nature. The result indicates that the cluster Cu_2Ag with C_{2v} symmetry group is paramagnetic in nature with isotropic value 423.741 cgs-ppm. This is due to the cluster geometry, which is triangular with silver atom occupying the center of the structure and leading to a larger value of paramagnetic current relatively to the diamagnetic current.

These nanoalloy clusters also show interesting relation between HOMO-LUMO gap and anisotropic value of magnetic susceptibility. The result

TABLE 1.8   Magnetic Susceptibility of Cu_nAg and Cu_{n+1} (n = 1–12) Nanoalloy Clusters

Species	Symmetry	Magnetic susceptibility (cgs-ppm)		Species	Symmetry	Magnetic susceptibility (cgs-ppm)	
		Isotropic	Anisotropy			Isotropic	Anisotropy
CuAg	C_{2v}	−63.150	4.206	Cu_2	D_{2h}	−50.415	3.186
Cu_2Ag	C_{2v}	423.741	1495.070	Cu_3	D_{3h}	−70.662	0.557
Cu_3Ag	C_{2v}	−76.375	66.841	Cu_4	C_s	−78.639	35.467
Cu_4Ag	D_{2h}	−134.557	37.611	Cu_5	C_1	−113.758	13.239
Cu_5Ag	C_{2v}	−157.332	21.218	Cu_6	C_1	−143.385	18.358
Cu_6Ag	C_1	−171.624	8.846	Cu_7	C_2	−158.237	6.613
Cu_7Ag	C_1	−12.682	509.23	Cu_8	C_1	−140.447	89.698
Cu_8Ag	C_2	−109.117	151.25	Cu_9	C_2	−161.016	80.093
Cu_9Ag	C_1	−170.121	122.725	Cu_{10}	C_1	−227.672	146.462
$Cu_{10}Ag$	C_s	−200.594	66.910	Cu_{11}	C_s	−192.680	57.904
$Cu_{11}Ag$	C_1	−310.390	84.009	Cu_{12}	D_{2d}	−273.856	623.955
$Cu_{12}Ag$	C_{2v}	−356.199	112.113	Cu_{13}	C_{2v}	−342.098	115.995

reveals that in the case of Cu_nAg (n = 1–12), clusters having low HOMO-LUMO gap (1.00–1.90 eV) exhibits large anisotropic value ranging between 60–1500 cgs-ppm and large HOMO-LUMO gap (above 2 eV) have low anisotropic value between 0–59 cgs-ppm, except Cu_7Ag. Similar trend has been observed in the case of pure Cu_{n+1} clusters, except Cu_8. The Cu_2Ag nanoalloy cluster, which is paramagnetic in nature, has a high value of anisotropic and isotropic susceptibility with low HOMO-LUMO gap. These results indicate that magnetic susceptibility and HOMO-LUMO gap are related to the nanoalloy cluster geometry and structural symmetry.

1.4   CONCLUSION

Due to the diverse nature of applications, bi-metallic nanoalloy clusters have gained enormous importance. Nanoalloy clusters containing group 11 elements, namely copper and silver, exhibits a noticeable electronic, optical, and magnetic property. In this report, we have analyzed Cu-Ag bimetallic nanoalloy Cu_nAg (n = 1–12) clusters in terms of conceptual DFT-based descriptors. We have also investigated the relative stability, vibrational, and Raman spectra, ferroelectricity behavior and magnetic susceptibility of Cu_nAg clusters and compared the result with pure Cu_{n+1} (n = 1–12) clusters. It is distinctly observed that HOMO-LUMO gaps of Cu-Ag nanoalloy clusters maintain direct relationship with their computed hardness values. From experimental point of view this is an expected trend, as the molecule possess the highest band gap, it will be least prone to response against any external perturbation. It means it will be least reactive. The high value of linear regression analysis (R^2 = 0.89) has been observed between HOMO-LUMO gap along with their computed softness, which validates our predicted model. The binding energies of the lowest energy structure of doped cluster Cu_nAg are larger than the corresponding pure Cu_{n+1} clusters, which indicates that the impurity atom may enhance the stability of pure copper nanoalloy clusters. The HOMO-LUMO energy gap, fragmentation energy and second order difference in energies of pure Cu and Cu-Ag clusters show interesting odd–even oscillatory phenomenon. This confirms that nanoalloy clusters with even number of atoms are more stable than their neighbor odd number of atoms. Raman and vibrational spectra of pure Cu and Cu-Ag nanoalloy clusters have been also studied using DFT methodology. The wavelengths of the maximum Raman peaks of Cu_nAg clusters decreases with increasing cluster size, however exceptions are observed in the case of clusters from Cu_4Ag to Cu_5Ag and from Cu_6Ag to Cu_7Ag where, wavelength increases with the cluster size. The wavelengths of maximum Raman peaks of pure Cu_n clusters are larger as compare to the doped Cu_nAg clusters, except Cu_7Ag and Cu_8Ag clusters. Overall, the intensities of the maximum Raman peaks of pure Cu clusters are stronger than that of Cu-Ag clusters, except Cu_2Ag, Cu_4Ag and $Cu_{10}Ag$ clusters. The wavenumbers of the maximum vibrational spectra of Cu-Ag clusters swing with increasing cluster size, when clusters change from Cu_6Ag to Cu_8Ag, from $Cu_{11}Ag$ to $Cu_{12}Ag$, their wavenumber decreases; when clusters changes from Cu_2Ag to Cu_6Ag, from Cu_8Ag to $Cu_{11}Ag$, their wavenumber increases. The intensities of the maximum vibrational spectra of Cu-Ag clusters fluctuate with increasing cluster size, when clusters change from Cu_3Ag to Cu_6Ag, from Cu_7Ag to Cu_8Ag, from $Cu_{11}Ag$ to $Cu_{12}Ag$, their strength decreases; when clusters change from Cu_2Ag to Cu_3Ag, from Cu_6Ag to Cu_7Ag, from Cu_8Ag to Cu_9Ag, from Cu_9Ag to $Cu_{11}Ag$, their strength increases. The wavenumbers and intensities of maximum vibrational spectra of pure Cu_n clusters exhibits interesting odd–even oscillation behaviors, indicating even number of clusters show high intensity and larger wavenumber as compare to their neighbor odd number of clusters. The CuAg and pure Cu cluster also have interesting relation between HOMO-LUMO gap and dipole moment, which reveal that nanoalloy clusters with low HOMO-LUMO energy gap have large dipole moment (above 1 Debye) and vice versa. The result reveals that $Cu_{12}Ag$ cluster have maximum dipole moment with lowest HOMO-LUMO gap in this molecular system. The magnetic susceptibility of Cu_nAg and pure Cu_{n+1} (n = 1–12) nanoalloy clusters have been computed using CSGT technique. With the exception of Cu_2Ag, all nanoalloy clusters studied are diamagnetic in nature. The result indicates that the cluster Cu_2Ag with C_{2v} symmetry group is paramagnetic in nature. This is due to the cluster geometry, which is triangular with silver atom occupying the center of the structure and leading to a larger value of paramagnetic current relatively to the diamagnetic current. Our computed bond lengths and vibrational frequencies for the species namely, Cu_2, Ag_2 and CuAg are numerically adjacent to the experimental values.

KEYWORDS

- **bi-metallic nanoalloy**
- **density functional theory**
- **HOMO-LUMO gap**
- **magnetic susceptibility**
- **Raman and vibrational spectra**

REFERENCES

1. Molayem, M., Grigoryan, G. V., & Springborg, M., (2011). Global minimum structures and magic clusters of Cu_mAg_n nanoalloys, *J. Phys. Chem. C, 115,* 22148.
2. Ferrando, R., Jellinek, J., & Johnston, R. L., (2008). Nanoalloys: from theory to applications of alloy clusters and nanoparticles, *Chem. Rev, 108,* 845.
3. Baletto, F., & Ferrando, R., (2005). Structural properties of nanoclusters: energetic, thermodynamics, and kinetic, *Rev. Mod. Phys, 77,* 371.
4. Rossi, G., Rapallo, A., Mottet, C., Fortunelli, A., Baletto, F., & Ferrando, R., (2004). Magic polyicosahedral core-shell clusters, *Phys. Rev. Lett, 93,* 105503.
5. Zabet-Khosousi, A., & Dhirani, A.-A., (2008). Charge transport in nanoparticle assemblies, *Chem. Rev, 108,* 4072.
6. Ghosh, S. K., & Pal, T., (2007). Interparticle coupling effect on the surface plasmon resonance of gold nanoparticles: from theory to applications, *Chem. Rev., 107,* 4797.

7. Chaudhuri, R. G., & Paria, S., (2012). Core/shell nanoparticle: classes, properties, synthesis mechanisms, characterization, and applications, *Chem. Rev., 112*, 2373.

8. Wang, H. Q., Kuang, X. Y., & Li, H. F., (2010). Density functional study of structural and electronic properties of bimetallic copper-gold clusters: comparison with pure and doped gold clusters, *Phys. Chem. Chem. Phys, 12*, 5156.

9. Taylor, K. J., Pettiette-Hall, C. L., Cheshnovsky, O, & Smalley, R. E., (1992). *J. Chem. Phys*, Ultraviolet photoelectron spectra of coinage metal clusters, *96*, 3319.

10. Cheeseman, M. A., & Eyler, J. R., (1992). Ionization potentials and reactivity of coinage metal clusters, *J. Phys. Chem, 96*, 1082.

11. Fernandez, E. M., Soler, J. M., Garzon, I. L., & Balbas, L. C., (2004). Trends in the structure and bonding of noble metal clusters, *Phys. Rev. B: Condens. Mater. Phys, 70*, 165403.

12. Quinn, B. M., Dekker, C., & Lemay, S. G., (2005). Electrodeposition of noble metal nanoparticles on carbon nanotubes, *J. Am. Chem. Soc., 127*, 6146.

13. Eustis, S., & El-Sayed, M. A., (2006). Why gold nanoparticles are more precious than pretty gold: Noble metal surface plasmon resonance and its enhancement of the radiative and nonradiative properties of nanocrystals of different shapes, *Chem. Soc. Rev, 35*, 209.

14. Cottancin E., Celep, G., Lerme, J., Pellarin, M., Huntzinger, J. R., Vialle, J. L., et al., (2006). Optical properties of noble metal clusters as a function of the size: comparison between experiments and a semi-quantal theory, *Theor. Chem. Acc, 116*, 514.

15. Li, W., & Chen, F., (2013). A density functional theory study of structural, electronic, optical and magnetic properties of A-Cu nanoalloys, *J. Nanopart. Res, 15*, 1809.

16. Darbha, G. K., Ray, A., & Ray, P. C., (2007). Gold nanoparticle-based miniaturized nanomaterial surface energy transfer probe for rapid and ultrasensitive detection of mercury in soil, water and fish, *ACS Nano, 1*, 208.

17. Dreaden, E. C., Mackey, M. A., Huang, X. H., Kang, B., & El-Sayed, M. A., (2011). Beating cancer in multiple ways using nanogold, *Chem. Soc. Rev, 40*, 3391.

18. Brongersma, M. L., Hartman, J. W., & Atwater, H. A., (2000). Electromagnetic energy transfer and switching in nanoparticle chain arrays below diffraction limit, *Phys. Rev. B, 62*, 16356.

19. Mirkin, C. A., Letsinger, R. L., Mucic, R. C., & Storhoff, J. J., (1996). A DNA-based method for rationally assembling nanoparticle into macroscopic materials, *Nature, 382*, 607.

20. Quinten, M., Leitner, A., Krenn, J. R., & Aussenegg F. R., (1998). Electromagnetic energy transport via linear chains of silver nanoparticles, *Opt. Lett., 23*, 1331.

21. Alonso, J. A., (2000). Electronic and atomic structure, and magnetism of transition-metal clusters, *Chem. Rev., 100*, 637.

22. Katakuse, I., Ichihara, T., Fujita, Y., Matsuo, T., Sakurai, T., & Matsuda, H., (1985). Mass distributions of copper, silver and gold clusters and electronic shell structure, *Int. J. Mass Spectrom. Ion Processes, 67*, 229.

23. Katakuse, I., Ichihara, T., Fujita, Y., Matsuo, T., Sakurai, T., & Matsuda, H., (1986). Mass distributions of negative cluster ions of copper, silver and gold, *Int. J. Mass Spectrom. Ion Processes, 74*, 33.

24. Heer, W. A. D., (1993). The physics of simple metal clusters: experimental aspects and simple models, *Rev. Mod. Phys, 65*, 611.

25. Gantefor, G., Gausa, M., Meiwes-Broer, K.-H., & Lutz, H. O., (1990). Photoelectron spectroscopy of silver and palladium cluster anions: electron delocalization versus localization, *J. Chem. Soc., Faraday Trans, 86*, 2483.

26. Leopold, D. G., Ho, J., & Lineberger, W. C., (1987). Photoelectron spectroscopy of mass-selected metal cluster anions. I. Cu_n^-, $n = 1–10$, *J. Chem. Phys*, *86*, 1715.

27. Lattes, A., Rico, I., Savignac, A. d., & Samii, A. A. Z., (1987). Formamide, a water substitute in micelles and microemulsions structural analysis using a diels-alder reaction as a chemical probe, *Tetrahedron, 43*, 1725.

28. Chen, F., Xu, G.-Q., & Hor, T. S.A., (2003). Preparation and assembly of colloidal gold nanoparticle in CTAB-stabilized reverse microemulsion, *Mater. Lett, 57*, 3282.

29. Taleb, A., Petit, C., & Pileni, M. P., (1998). Optical properties of self-assembled 2D and 3D superlattices of silver nanoparticles, *J. Phys. Chem. B, 102*, 2214.

30. Bishea, G. A., Marak, N., & Morse, M. D., (1991). Spectroscopic studies of jet-cooled CuAg, *J. Chem. Phys, 95*, 5618.

31. James, A. M., Lemire, G. W., & Langridge-Smith, P. R. R., (1994). Threshold photoionization spectroscopy of the CuAg molecule, *Chem. Phys. Lett., 227*, 503.

32. Barcaro, G., Fortunelli, A., Rossi, G., Nita, F., & Ferrando, R., (2006). Electronic and structural shell closure in AgCu and AuCu nanoclusters, *J. Phys. Chem. B, 110*, 23197.

33. Kilmis, D. A., & Papageorgiou, D. G., (2010). Structural and electronic properties of small bimetallic Ag-Cu clusters, *Eur. Phys. J. D, 56*, 189.

34. Ferrando, R., Fortunelli, A., & Rossi, G., (2005). Quantum effects on the structure and binary metallic nanoclusters, *Phys. Rev. B, 72*, 085449.

35. Ferrando, R., Fortunelli, A., & Johnston, R. L., (2008). *Phys. Chem.,* Searching for the optimum structures of alloy nanoclusters, *10*, 640.

36. Nunez, S., & Johnston, R. L., (2010). Structures and chemical ordering of small Cu-Ag clusters, *J. Phys. Chem. C, 114*, 13255.

37. Rapallo, A., Rossi, G., Ferrando, R., Fortunelli, A., Curley, B. C., Lloyd, L. D., et al., (2005). Global optimization of bimetallic cluster structures. I. Size-mismatched Ag-Cu, Ag-Ni, and Au-Cu systems, *J. Chem. Phys, 122*, 194308.

38. Yildirim, H., Kara, A., & Rahman, T. S., (2012). Tailoring electronic structure through alloying: The Ag_nCu_{34-n} ($n = 0–34$) nanoparticle family, *J. Phys. Chem. C, 116*, 281.

39. Li, W., & Chen, F., (2014). Effects of shape and dopant on structural, optical absorption, Raman and vibrational properties of silver and copper quantum clusters: A density functional theory study, *Chin. Phys. B, 23*, 117103.

40. Li, W., & Chen, F., (2014). Structural, electronic and optical properties of 7-atom Ag-Cu nanoclusters from density functional theory, *Eur. Phys. J. D, 68*, 91.

41. Li, W., & Chen, F., (2014). Ultraviolet-visible adsorption, Raman, vibration spectra of pure silver and Ag-Cu clusters: A density functional theory study, *Phys. B, 451*, 96.

42. Jiang, Z. Y., Lee, K. H., Li, S. T., & Chu, S. Y., (2006). Structures and charge distributions of cationic and neutral $Cu_{n-1}Ag$ clusters ($n = 2–8$), *Phys. Rev. B, 73*, 235423.

43. Ranjan, P., Dhail, S., Venigalla, S., Kumar, A., Ledwani, L., & Chakraborty, T., (2015). A theoretical analysis of bi-metallic $(Cu-Ag)_{n=1-7}$ nanoalloy clusters invoking DFT based descriptors, *Mat. Sci. Pol., 33, 719*.

44. Ranjan, P., Venigalla, S., Kumar, A., & Chakraborty, T., (2014). Theoretical study of bi-metallic Ag_mAu_n (m+n = 2-8) nano alloy clusters in terms of DFT based descriptors, *New Front. Chem., 23*, 111.

45. Ranjan, P., Venigalla, S., Kumar, A., & Chakraborty, T., (2016). A theoretical analysis of bimetallic $AgAu_n$ ($n = 1–7$) nano alloy clusters invoking DFT based descriptors, *Research Methodology in Chemical Sciences: Experimental and Theoretical Approaches,* Edited by Chakraborty, T., & Ledwani, L., Apple Academic Press, USA, p. 337.

46. Ranjan, P., Kumar, A., & Chakraborty, T., (2017). Theoretical analysis: Electronic and optical properties of small Cu-Ag nano alloy clusters, *Computational Chemistry Methodology in Structural Biology and Material Sciences,* Edited by Chakraborty, T., Ranjan P., Pandey, A., Apple Academic Press, USA, ISBN: 9781771885683, p. 259.

47. Ranjan, P., Chakraborty, T., & Kumar, A., (2017). A theoretical study of bimetallic CuAun (n = 1-7) nanoalloy clusters invoking conceptual DFT based descriptors, *Applied Chemsitry and Chemical Engineering*, Vol. 4, Edited by Haghi, A. K., Pogliani, L., Castro, E. A., Balkose, D., Mukbaniani, O. V., Chia, C. H., Apple Academic Press, USA, ISBN: 9781771885874.

48. Hafner, J., Wolverton, C., & Ceder, G., (2006). Toward computational materials design: the impact of density functional theory on materials research, *MRS Bulletin, 31,* 659.

49. Becke, A. D., (1993). Density-functional thermochemistry. III. The role of exact exchange, *J. Chem. Phys, 98,* 5648.

50. Lee, C., Yang, W., & Parr, R. G., (1988). Development of the Colle-Salvetti correlation-energy formula into a functional of the electron density, *Phys. Rev. B: Condens. Matter, 37,* 785.

51. Mielich, B., Savin, A., Stoll, H., & Preuss, H., (1989). Results obtained with the correlation energy density functionals of Becke and Lee, Yang and Parr, *Chem. Phys. Lett, 157,* 200.

52. Zhao, Y. R., Kuang, X. Y., Zheng, B. B., Li, Y. F., & Wang, S. J., (2011). Equilibrium geometries, stabilities, and electronic properties of the bimetallic M2-doped Aun (M = Ag, Cu; n = 1–10) clusters: Comparison with the pure gold clusters, *J. Phys. Chem. A, 115,* 569.

53. Gaussian, 03, Revision, C.02, Frisch, M. J., Trucks, G. W., Schlegel, H. B., Scuseria, G., et al., (2004). Gaussian, Inc., Wallingford CT.

54. Parr, R. G., & Yang, W., (1989). *Density Functional Theory of Atoms and Molecules*, Oxford, University Press: Oxford.

55. Xiao, H., Kheli, J. T., & Goddard, III W. A., (2011). Accurate band gaps for semiconductors from density functional theory, *J. Phys. Chem. Lett, 2,* 212

56. Li-Ping, D., Xiao-Yu, K., Peng, S., Ya-Ru, Z., & Yan-Fang, L., (2012). A comparative study on geometries, stabilities, and electronic properties between Ag_nX (X = Au, Cu; n = 1–8) and pure silver clusters, *Chin. Phys. B, 21,* 043601.

57. Hakkinen, H., & Landman, U., (2000). Gold clusters (AuN, 2 <~ N <~ 10) and their anions, *Phys. Rev, 62,* 2287.

58. Li, X. B., Wang, H. Y., Yang, X. D., Zhu, Z. H., & Tang, Y. J., (2007). Size dependence of the structures and energetic and electronic properties of gold clusters, *J. Chem. Phys., 126,* 084505.

59. Jain, P. K., (2005). A DFT-based study of the low-energy electronic structures and properties of small gold clusters, *Struct. Chem, 16,* 421.

60. Karelson, M., Lobanov, V. S., & Katritzky, A. R., (1996). Quantum-chemical descriptors in QSAR/QSPR studies, *Chem. Rev., 96,* 1027.

61. Beutel, V., Krämer, H. G., Bhale, G. L., Kuhn, M., Weyers, K., & Demtröder, W., (1993). High-resolution isotope selective laser spectroscopy of Ag^2 molecules, *J. Chem. Phys, 98,* 2699

62. Kabir, M., Mookerjee, A., & Bhattacharya, A. K., (2004). Structure and stability of copper clusters: a tight-binding molecular dynamics study, *Phys. Rev. A, 69,* 043203.

63. Balbuena, P. B., Derosa, P. A., & Seminario, J. M., (1999). Density functional theory study of copper clusters, *J. Phys. Chem. B, 103,* 2830

64. Spasov, V. A., Lee, T. H., & Ervin, K. M., (2000). Threshold collision-induced of anionic copper clusters and copper clusters monocarbonyls, *J. Chem. Phys, 112*, 1713.
65. Massobrio, C., Pasquarello, A., & Car, R., (1995). Structural and electronic properties of small copper clusters: a first principles study, *Chem. Phys. Lett, 238*, 215.
66. Kabir, M., Mookerjee, A., Datta, R. P., Banerjea, A., & Bhattacharya, A. K., (2003). Study of small metallic nanoparticles: an ab-initio full-potential muffin-tin orbitals based molecular dynamics study of small Cu clusters, *Int. J. Mod. Phys. B, 17*, 2061.
67. Tafoughalt, M. A., & Samah, M., (2014). Structural properties and relative stability of silver-doped gold clusters $AgAu_{n-1}$ ($n = 3–13$): Density functional calculations, *Comput. Theor. Chem., 1033*, 23.
68. Li, J., Liu, Y., Zhang, Y., Cai, H. L., & Xiong, R. G., (2013). Molecular ferroelectrics: where electronics meet biology, *Phys. Chem. Chem. Phys, 15*, 20786.
69. Heer, W. A. D., & Kresin, V. V., (2010). *Handbook of Nanophysics: Clusters and Fullerenes*, Edited by Klaus D. Slatter, Taylor & Francis, CRC Press.
70. Botti, S., Castro, A., Lathiotakis, N. N., Andrade, X., & Margues, M. A. L., (2009). Optical and magnetic properties of boron fullerenes, *Phys. Chem. Chem. Phys, 11*, 4523.

CHAPTER 2

SHAPE SIMILARITY MEASURES OF TRANSITION STRUCTURE ELECTRON DENSITIES AS TOOLS TO ASSESS SIMILARITIES OF REACTIONS AND CONFORMATIONAL CHANGES OF MOLECULES

PAUL G. MEZEY[1,2]

[1]*Distinguished Visiting Professor, Yukawa Institute for Theoretical Physics, Kyoto University, Kyoto 606-8502, Japan*

[2]*Canada Research Chair in Scientific Modeling and Simulation, Department of Chemistry, and Department of Physics and Physical Oceanography, Memorial University of Newfoundland 283 Prince Philip Drive, St. John's, NL, A1B 3X7, Canada, Tel.: +1-709-749 8768, Fax: +1-709-864-3702, E-mail: paul.mezey@gmail.com http://www.mun.ca/research/chairs/mezey.php*

CONTENTS

ABSTRACT

The studies of molecular shape-similarity are based on detailed shape analysis of molecular electron density clouds, which is typically applied to molecular models based on structural information from X-ray diffraction methods or on quantum-chemically energy-optimized structures. However, if the goal is to obtain comparisons and similarity measures between chemical reactions or conformational pathways, then alternative approaches are needed. One conceptually simple approach involves the shape comparisons of transition structures, and in this contribution, some of the fundamentals of the shape analysis of transition structure electron densities and the associated shape similarity measures will be discussed.

2.1   QUANTUM SIMILARITY MEASURES

Quantum similarity measures introduced and developed for a variety of applications by Carbó-Dorca and his coworkers [1–14] are at the foundations of all possible similarity measures among molecules. For an early summary of the approaches to molecular similarity studies, a collection of methodologies are described in a book edited by Johnson and Maggiora [15]. Quantum similarity measures are based on a quantum mechanical implementation of a natural requirement; similarity can be characterized by finding out how much is common between two entities, in our case, between two molecules. If in a quantum-mechanically well-founded framework the molecules are represented by wavefunctions, then a rigorous approach to the answer can be obtained by overlap-type computations, provided, that the two molecules fulfill the constraint that their mutual geometrical arrangement provides the "best" overlap. Whereas such a geometrical arrangement is not necessarily unique, for example, the existence of some symmetry for either one of the two molecules compared may imply that two or more equally "optimal" arrangements are possible, however, the actual numerical value for a measure of such an overlap is necessarily unique. Hence, a well-defined concept lies at the basis of such quantum similarity measures, which is well manifested by the large number and type of useful applications [3–14].

In the most frequently used implementation of quantum similarity measures the optimum mutual geometrical arrangement of the two molecules compared involves an optimization step implying that even if a large number of molecules is studied, for every pair comparison one needs a separate optimization for the best overlap. With the exception of small molecules of less than three nuclei, or systems which are constrained to be linear or planar, finding the best overlap is typically a six-dimensional optimization problem, involving three translations and three rotations [16]. Whereas such a task is computationally not very demanding, there are also advantages if not the molecules themselves are compared directly, but some shape descriptors which are characterizing the shapes of individual molecules instead of the relation between pairs of molecules. In such a case, no compatibility conditions on mutual orientations play any role, and the shape descriptors themselves can be compared. This task, typically, involves no optimization steps, such as the determination of the optimum overlap as needed in direct molecule-pair comparisons.

2.2 ELECTRON DENSITY SHAPE SIMILARITY MEASURES

One type of electron density shape similarity measure [17, 18], also used for the shapes of electrostatic potentials [17], is the *shape group method*, which provides intuitively straightforward interpretation and has been applied to a range of molecular shape problems [17–31]. The shape group method is based on the molecular electron density clouds, and it provides an individual shape code, a sequence of integer numbers for each molecule. If in a similarity study, two molecules are compared, this is not done directly on the molecules, instead, the two shape codes, that is, the corresponding two sets of these numbers representing the shape codes can be compared directly, and simple "match or no match" comparison along the sequence of numbers leads to a shape similarity measure between any two molecules. Consequently, the six-dimensional optimization step for finding the best mutual overlap between two molecules is avoided.

For the actual mathematical details of the shape group method, the reader may consult the original references [17–21], or possibly the relevant monograph on molecular shape and topology [28] that includes further developments.

Here only a brief summary will be given in order to serve as the background for the transition structure similarity approach for comparisons and similarity measures of chemical reactions.

An important concept in electron density shape analysis is that of the molecular isodensity contour, MIDCO, usually denoted by G(K,a). A

MIDCO is defined for the given nuclear configuration K and some electron density threshold value a as the collections of all those points $\mathbf{r}$ where the electron density value is equal to this threshold a:

$$G(K,a) = \{\mathbf{r}: \rho(K,\mathbf{r}) = a\}. \tag{1}$$

It is clear that, for each one of the infinitely many accessible electron density values for the given molecule, there is one such MIDCO, hence we need to deal collectively with all of them. If we want to describe the shape of the molecular electron density cloud, then we need a shape character-ization of the entire fuzzy object; that is, the complete electronic density cloud of the molecule. Consequently, one such isodensity contour MIDCO G(K,a) is insufficient, and ideally, we need a shape characterization of all of the infinitely many isocontour surfaces G(K,a) of the molecule. This would provide a complete shape characterization of the molecule, however, con-sidering infinitely many such G(K,a) contours individually is a tall order. Here is where topology can offer some help: some of these contours have shapes which can be regarded as equivalent by some topological criterion, and this fact allows one to consider only the topological equivalence classes of these MIDCO's G(K,a). As it can be shown [17–21, 28], there are only a finite number of such equivalence classes, if the topological conditions are chosen in a chemically relevant way. In turn, these equivalence classes can be characterized by some algebraic, group theoretical tools, leading to algebraic homology groups of various dimensions, among which the one-dimensional homology groups are called *shape groups*. The invariants, actually the ranks of these shape groups which are called the Betti numbers, provide a numerical shape code for each complete molecular electron den-sity cloud, relevant to the set of all MIDCO's G(K,a), for all values of the electron density threshold a.

The actual implementation of these ideas had some motivation originat-ing from a source not involved with the science of chemistry at all: some hints have been obtained from rather pure mathematics, and in some indi-rect way, from classical artistic achievements of famous sculptures, through the works of mathematician Felix Klein, the announcer of the Erlangen Program, who used differential geometry to analyze certain shapes with a special interest in mind.

Felix Klein was interested in the perception of beauty, especially, the beauty of shapes of ancient sculptures, and he tried to use some mathematical tools for describing and analyzing what geometrical features give us the impression of beauty.

In this context, we might recall the definition what is a parabolic point on a continuous and twice differentiable surface: if at some point **r** of such a surface one of the extreme local curvatures at a point **r** happens to be zero, then this point **r** is called a parabolic point.

If at such a point **r** one takes a tangent plane, then one straight line of the plane through **r** runs locally within the surface, along one of its extreme curvatures, and locally the rest of the surface falls on one side of the tangent plane, with the exception of some degenerate cases. Perpendicular to this straight line, the surface is, indeed, locally having parabolic shape.

Since such parabolic points are special, and occur on the boundaries of locally convex and locally concave regions of the surface, this fact was the reason why Felix Klein used them for local shape characterization. On replicas of some ancient and not so ancient statues, Felix Klein marked the parabolic points, forming continuous lines, and he hoped that they reveal something about the perception of beauty. He studied in detail the patterns obtained on the surfaces of the sculptures he generated by this method. Apparently, the markings of Felix Klein, as he produced them on a copy of the bust of the Belvedere Apollo, still can be found in the Mathematics Library of the University of Göttingen, Germany.

The patterns so obtained, however, were disappointing for him, and he did not find satisfactory correlation between the perceived beauty and these patterns. Felix Klein was not satisfied with the results of his approach, and never published anything about his findings. However, fortunately, some others did mention these efforts of Felix Klein, notably, Hilbert among other mathematicians, so his ideas still had the chance to motivate alternative studies.

In fact, if one is not focusing on beauty, not even on the beauty of molecules [31], but uses similar ideas for shape analysis, then, as it has happened, the studies of Felix Klein can serve as a strong motivation for an approach to molecular shape analysis.

Indeed, one can apply the ideas of Felix Klein to molecular surfaces, for example, to molecular isodensity contour surfaces, MIDCO's G(K,a), especially, if a useful, further generalization is introduced.

For this purpose, it is convenient to rephrase the partitioning used by Felix Klein by employing parabolic points in the following way:

For each point **r** of a MIDCO G(K,a), consider the local tangent plane P(**r**). In the close vicinity of point **r**, the tangent plane P(**r**) may be found locally in the inside of the MIDCO G(K,a), then, locally, the MIDCO G(K,a) is concave at this point **r**. If the tangent plane P(**r**) is found locally on the outside of the MIDCO G(K,a), then, locally, the MIDCO G(K,a) is convex at this point **r**. The third (non-degenerate) possibility is that the tangent plane P(**r**) is found locally to cut into the MIDCO G(K,a) in the immediate vicinity of point **r**, then, locally, the MIDCO G(K,a) is of the "saddle-surface type" at this point **r**.

An alternative, more formal, but less pictorial approach leads to equivalent results. Rotate the surface G(K,a), with tangent plane P(**r**) attached, to a position where the tangent plane P(**r**) is horizontal. Consider the MIDCO surface G(K,a) at this point as a function defined over the tangent plane P(**r**), where the function values correspond to the distance of the surface point from the tangent plane (the horizontal arrangement of the tangent plane P(**r**) is useful only for visualization purposes). Since the MIDCO surface G(K,a) is assumed to be twice differentiable (all MIDCOs are supposed to have this property), the Hessian matrix H(K,**r**) of second derivatives of this function can be determined. Consider the eigenvalues of this Hessian matrix H(K,**r**). In relation to the number of negative eigenvalues μ, there are three cases, leading to a convexity domain partitioning of the MIDCO surface G(K,a):

$$\mu = 0, \text{ locally concave region, denoted by } D_0, \tag{2}$$

$$\mu = 1, \text{ locally saddle-type region, denoted by } D_1, \tag{3}$$

$$\mu = 2, \text{ locally convex region, denoted by } D_2, \tag{4}$$

Indeed, this approach is equivalent to the one used by Felix Klein: points of the MIDCO surface G(K,a) where an eigenvalue of the Hessian matrix becomes zero, where a zero value is a necessary condition for an eventual sign change, are precisely the parabolic points of the MIDCO surface G(K,a), and indeed, the parabolic points form the boundary lines of these regions, concave regions D_0, saddle-type regions D_1, and convex regions D_2.

The above formulation, in terms of Hessian matrices and their eigenvalues allows one to progress beyond the approach of Felix Klein. What is the problem with his approach? There is no problem with this approach concerning the study of beauty, and someone may still discover a strong connection between his patterns and the perception of beauty [31]. However, for molecular shape characterization, or in general, for shape characterization, the above approach is rather limited.

As an example, consider the case of European football and American football. Both footballs are locally convex everywhere, in fact, both have just one single domain, a D_2 domain. Hence, according to the above approach, the shapes of the two footballs are not distinguishable. Evidently, one needs a finer characterization.

Fortunately, the above formulation in terms of tangent planes and local Hessian matrices allows for a useful generalization. In the original approach, the tangent plane is the reference, implying that the zero curvature, the curvature of the tangent plane, is also a reference: the question investigated is the following:

Is the MIDCO surface G(K,a) locally more curved or less curved along the eigenvectors of the Hessian matrix H(K,r) than the reference tangent plane?

We have seen that getting an answer to this question is equivalent to testing, what is the number of negative eigenvalues of the Hessian matrix H(K,**r**).

However, the concept of convexity, with reference to tangent planes P(r) can be easily generalized, in fact, the tangent plane approach discussed above can lead to a generalization that can be regarded as *convexity in a curved Universe.*

Indeed, simply by replacing the tangent planes P(**r**) by tangent spheres S(r,b) of curvature b, that is, by tangent spheres S(r,b) of radius 1/b, one can follow the same procedure as before, and test, whether locally this tangent sphere S(r,b) falls on the outside of the MIDCO surface G(K,a), or the tangent sphere S(r,b) falls on the inside of the MIDCO surface G(K,a), or the tangent sphere S(r,b) cuts into the MIDCO surface G(K,a) within the immediate vicinity of point **r**.

This is equivalent to testing the eigenvalues of the local Hessian matrix H(K,**r**), still defined as before, over the tangent plane P(r), and checking, how many eigenvalues of this local Hessian matrix H(K,**r**) has value that is less than the reference curvature b of the tangent sphere S(r,b).

One should note that in most of the recent applications one is using the convention adopted some time ago: a *positive b* value indicates that a tangent sphere S(r,b), as reference object, is placed on the *exterior side* of the MIDCO surface G(K,a), whereas a *negative b* value indicates that the tangent sphere S(**r**,b), is placed on the *interior side* of the MIDCO G(K,a), in which case the radius of sphere $S(\mathbf{r},b)$ is $-1/b$ (Note: a different convention was used in some of the early studies.).

In one aspect, the notation $S(r,b)$ for the tangent sphere might be misleading; one should note that point r is NOT the center of the tangent sphere, rather, it is a point of tangential contact between the MIDCO and the surface of the tangent sphere $S(r,b)$.

With reference to the number $\mu(b)$ of the eigenvalues of the Hessian matrix $H(K,\mathbf{r})$ which are less than reference curvature b, there are three cases, leading to a Relative Convexity Domain Partitioning of the MIDCO surface $G(K,a)$, where the term Relative Convexity refers to the fact that all curvature properties are interpreted relative to the reference curvature b of the tangent sphere $S(r,b)$:

$$\mu(b) = 0, \text{ locally concave region relative to curvature b,}$$
$$\text{denoted by } D_0(b), \qquad (5)$$

$$\mu(b) = 1, \text{ locally saddle-type region relative to curvature b,}$$
$$\text{denoted by } D_1(b), \qquad (6)$$

$$\mu(b) = 2, \text{ locally convex region relative to curvature b,}$$
$$\text{denoted by } D_2(b). \qquad (7)$$

One might recognize that the ordinary convexity approach that corresponds to the subdivision method by lines of parabolic points, as used by Felix Klein, is indeed only a special case of the more general, tangent sphere approach, since, if one takes a "tangent sphere of zero curvature," that is, a tangent sphere of infinite radius, then this case is equivalent to the case of a tangent plane of zero curvature. That is, the tangent sphere $S(r,b)$ of curvature $b = 0$ is equivalent to the tangent plane $P(r)$:

$$S(r,0) = P(r) \qquad (8)$$

Further generalizations have also been proposed [28], for example, if the goal is a shape analysis within some constrained environment, where some directional preferences exist, such as an enzyme cavity, then one may consider an Oriented Relative Convexity partitioning of MIDCO surface $G(K,a)$, as follows:

A detailed, orientation dependent shape description is obtained if the series of tangent spheres $S(r,b)$ for range of relative curvature values b is replaced by some other objects, for example, if a rotationally restricted

MIDCO surface G(K,a) is compared to a series of oriented tangent ellipsoids T, which are allowed translational motions only, but not rotations, to establish tangential contact with the MIDCO. A further generalization is obtained if some, but only limited rotations, say, a few degrees only, are allowed.

In the simplest case of oriented tangent ellipsoids, the directions of their axes are fixed, that is, they can be translated but not rotated as they are brought into tangential contact with the MIDCO surface G(K,a).

Another generalization is obtained if one chooses an oriented, otherwise general closed surface T, that can be chosen for some imposed condition, for example, T may be a specific MIDCO of another molecule, including the specific case of that of an enzyme cavity region, completed by some artificial surface area in order to obtain a closed surface T.

With reference to such a generalized tangent object T that may fall locally on the outside, or on the inside of the MIDCO, or it may cut into the given isodensity contour G(K,a) within the immediate neighborhood of the surface point r where the tangential contact occurs, the approach leads to a set of rather general oriented relative convexity domains labeled $D_2(T)$, $D_1(T)$, $D_0(T)$, respectively.

Returning to the examples of the two footballs, European football and American football, if the reference curvature value b of the tangent spheres $S(r,b)$ is small, that is, if the radius of the tangent spheres $S(r,b)$ is large, than even this generalized convexity approach will not distinguish the two footballs. If at every point r of the two football surfaces the tangent sphere $S(r,b)$ is curving less, than the surface of either one of the footballs, then the tangential contact between the tangent sphere $S(r,b)$ and both of the footballs will occur at every point in such a way that the tangent sphere will be located on the outside of the footballs, hence we get the same classification as before with the tangent plane: we obtain only one, single relative convexity domain of type $D_2(b)$, for both footballs. No discrimination is obtained.

If, however, a different, somewhat larger reference curvature b is chosen for the tangent sphere, then the European football, a sphere, will still have only a single domain, however, for a whole range of such b values, the American football will have two different types of relative convexity domains, relative to b, as a consequence of its "quasi-ellipsoidal" shape. In such a case, the relative convexity approach does already provide a shape discrimination.

This example illustrates an important consequence for molecular shape analysis: one needs to consider a whole range of relative curvature values b

for tangent spheres S(r,b), in order to make the approach truly discriminative concerning shape features.

As a consequence, we have to deal with two continua: the continuum range of the electron density threshold values a for all the relevant MIDCO surfaces G(K,a) of the molecule studied, and also a second continuum, the continuum range of relevant curvature values b occurring along all the relevant MIDCO surfaces. We have to consider two continua, a problem usually causing complications, however, topology helps to reduce these continua to a finite set of integer numbers, while still providing a somewhat simplified, but sufficiently detailed shape characterization.

The approach actually used to deal with these two continua a and b is based on a formal truncation of the MIDCOs G(K,a) with respect to one of types of convexity domains, by removing all domains from all MIDCOs with a specific convexity number $\mu(b)$, leading by this process to truncated MIDCOs denoted by G(K,a, $\mu(b)$):

the removal of all $D_\mu(b)$ relative convexity domains from the MIDCO G(K,a) for a specific $\mu(b)$ value results in G(K,a, $\mu(b)$).

In most applications the following choice has provided a practical, but already very useful shape characterization:

$$\mu(b) = 0 \qquad\qquad (10)$$

This truncation is, indeed, an important step in the shape characterization process, that links the geometrical, curvature characteristics of MIDCOs with some topological features, where the latter exhibit invariance to some, in our case unimportant, continuum dependence, and this very invariance allows the approach to ignore those unimportant features.

The primary tools used for shape characterization are the *homology groups of truncated surfaces. Such homology groups of algebraic topology* are topological invariants, expressing important features of the topological structure of bodies and surfaces.

The ranks of the one-dimensional homology groups, called shape groups, are the *Betti numbers*, themselves important topological invariants.

It is noteworthy that for the whole range of possible reference curvature values b and the whole range of isocontour density values a, there are only a *finite number of topologically different* truncated MIDCO's G(K,a,μ), resulting in a finite number of topological equivalence classes. In turn, these equivalence classes can be characterized by their one-dimensional homology groups, related, in the simplest case, to the number of holes created by the truncation. Although the actual topological treatment is more involved, as described in detail, for example, in the monograph on molecular shape and topology [28], nevertheless, the corresponding homology groups, the shape groups, are in fact finitely generated, with their ranks easily determined after truncation. These ranks, the so-called Betti numbers, are topological invariants and are the integers which generate the actual numerical shape codes for the complete electron density cloud.

In order to generate the numerical shape code, one can exploit the fact that only a finite number of topologically different truncated MIDCOs are produced. This implies that it is not necessary to consider all the infinitely many $G(K,a, \mu(b))$ truncated surfaces, and it is sufficient to sample the entire ranges of a and b values with a finite number of actual a and b values for the set of $G(K,a, \mu(b))$ truncated contours.

In most shape analysis applications, for the density threshold values a of the MIDCO surfaces $G(K,a)$ the range of [0.001, 0.1 au.] is taken, whereas for the reference curvature value b of the test spheres $S(\mathbf{r},b)$ the range of [0.1, 10] is taken. Since the relevant magnitudes of the a and b values can stretch over several orders of magnitude, a logarithmic scale has been chosen for both. On this logarithmic scale, 41 equidistant a values and 21 equidistant b values are used (where for negative b values their absolute values, |b| are considered), providing a rather detailed shape characterization of the electron density represented by the family of MIDCOs considered.

That is, for the complete, chemically relevant range of a and b values, a matrix of 41x21 entries is used, based on a logarithmic scale.

For each pair of a and b values of this grid, the shape groups of the molecule are determined, where their ranks, the Betti numbers represent the information for the next step, where these Betti numbers obtained for a given a,b pair of the grid are combined into a single number for each given a,b pair, that is taken as a numerical entry for a 41x21 matrix of integer numbers, called the (a,b)-parameter map. This map contains the shape information, that is used either in the 41x21 matrix form, or as a 41x21 component numerical shape code, giving a detailed shape description.

A shape similarity measure can be obtained using the above shape codes, for example, in the form of shape code matrices M(a,b) of 41x21 dimensions. If such shape code matrices, $M(a,b)_A$ and $M(a,b)_B$ have already been determined for two molecules, molecule A and molecule B, then, in the simplest approach, a comparison of entries can be made, and the number of matches is determined. Since 41x21 = 861, if the number of matches

$$m = m[M(a,b)_A, M(a,b)_B], \tag{11}$$

then the similarity measure for the two molecules a and b is

$$s(A,B) = m[M(a,b)_A, M(a,b)_B]/861 \tag{12}$$

A simple complementarity measure can also be determined based on such shape matrices $M(a,b)_A$ and $M(a,b)_B$. One may note that for complementarity, a convex domain matches well a concave domain, and similarly, high electron density region for one molecule matches well a low electron density region of another molecule.

Consequently, if one takes the original shape code matrix $M(a,b)_A$ for molecule A, and centrally invert the $M(a,b)_B$ shape code matrix for molecule b (where the choice for the actual center of the inversion of the shape code matrix $M(a,b)_B$ may require special considerations), leading to the centrally inverted matrix $M(a,b),''$ then the similarity measure

$$s(A,B'') = m[M(a,b)_A, M(a,b)''_B]/861 \tag{13}$$

obtained for the matrix pair of $M(a,b)_A$ and $M(a,b)''_B$ is in fact a complementarity measure for the two molecules A and B:

$$c(A,B) = m[M(a,b)_A, M(a,b)_B]/861 \tag{14}$$

where for both for the curvatures b, and for the electron density thresholds a, opposing values are matched up for the two molecules A and B.

These molecular shape analysis measures, the similarity measures

$$s(A,B) = m[M(a,b)_A, M(a,b)_B]/861 \tag{15}$$

and complementarity measures

$$c(A,B) = s(A,B'') = m[M(a,b)_A, M(a,b)''_B]/861, \tag{16}$$

based on the Betti numbers of the shape groups can be generated by suitable softwares.

Earlier applications show that such numerical shape codes can be generated for most molecules relevant to pharmaceutical chemistry, and drug design.

2.3 COMPARISONS OF SHAPES OF STABLE STRUCTURES AND TRANSITION STRUCTURES

For a discussion of some of the specific tools suggested for shape similarity studies of transition structures and an indirect similarity study of chemical reactions, as represented by some reaction paths or equivalence classes of reaction paths, we shall consider first some of the most general properties of molecular electron densities, as represented by density functional theory.

A fundamental theorem of density functional theory, the Hohenberg-Kohn theorem [32], usually applied to non-degenerate electronic ground states of molecules, states that *the molecular electron density determines the molecular energy and through the Hamiltonian, all other molecular properties.*

Consequently, the electron density actually stores the complete information about the molecule.

For artificial, bounded and finite Coulomb systems, Riess and Münch [33] have derived a result on the uniqueness of the local extension of electron densities from one finite and bounded region to a larger but still finite and bounded region. They have clearly stated that their proof of is not applicable for real molecules, and their result is limited to artificial, finite and bounded systems, since the analytic continuation theorem they have used works only for compact sets, and the full electron density cloud is not compact, hence their proof could not work for real molecules.

However, exploiting again the power of topology, in this case, some differential topology methods, topological tools are suitable to circumvent this limitation of the Riess and Münch [33] result. As a consequence, the holographic electron density theorem of Mezey [34], a strengthening of the Hohenberg-Kohn theorem has been proven, using another topological technique, the Alexandrov one-point compactification.

The holographic electron density theorem [34] states that *for any non-degenerate ground state, the complete molecular information is encoded in any small positive volume part of the electron density cloud.*

In the molecular world, each part of a molecule "knows" everything about the whole molecule. There is some importance in making a clear distinction between the above two theorems. The Hohenberg-Kohn Theorem refers to the *complete electron density*, that of the full molecule, by stating that all non-degenerate, ground state molecular properties are determined by this complete electron density.

On the other hand, the holographic electron density theorem asserts that the complete molecular information, in principle, must be available from the *part*, stating that any nonzero volume *part* of a molecular electron density in a non-degenerate ground state contains the *complete* information about all properties of the entire, boundaryless molecule. Of course, very accurate information is needed for the part to make reliable actual predictions for the hole, but, in some actual computational examples, excellent shape-activity correlations have been obtained for molecular series, relying only on a small molecular fragment from each molecule. In the actual applications, these fragments had the same formal chemical constitution, but of course, they had slightly different local shapes, due to the different influence of the surroundings provided by the rest of the actual molecule in each of these different molecules. Nevertheless, these local fragment electron densities contained enough global information about the complete molecules, to be able to correlate with over 90% accuracy in the predictions of some biochemical properties (including, for example, levels of fragrance intensities).

According the holographic electron density theorem, proven [34] and also confirmed by large numbers of independent studies involving both experimental and computational modeling results [39–54], the complete molecular information does not require the complete electron density, and local electron density ranges already fully determine all molecular properties. Additional applications of this theorem have been suggested for molecular design [35] and latent property studies [36], as well as in the context of a whole range of more general density functional approaches [37–54].

In particular, the combination of local molecular fragment approaches [38] with applications relying on the Holographic theorem, together with an adaptation of the shape group method to local shape analysis of molecular fragments appears promising. The efficient molecular fragment generation approach also used in *ab initio* quality, linear-scaling macromolecular quantum chemistry computational methods [55–58], has provided additional motivation for transition structure fragment shape analysis, and more recently, for a transition structure fragment shape similarity approach. Such fragments of transition structures provide a wider range of comparisons between various reactions, since smaller fragments which are approximately common for transition structures are more likely to be found for a wider family of different chemical reactions. Some of the natural compensation effects between molecular energy components [59], and general trends for molecular conformational processes and chemical reactions, for example, those involving the Hammond postulate [60, 61], can be influenced by fragment interactions [62].

Besides the natural applications of the holographic electron density theorem to developments [35] in molecular design research, a special, "latent property" aspect [36] of the holographic electron density theorem is important in extending some of the electron density shape analysis results to less easily modeled unstable structures. As it turns out, the comparison of transition structures is a useful tool for the detecting important differences of seemingly similar chemical mechanisms by comparing their respective transition structure electron densities.

A latent molecular property is one that is not manifested at the given stage, however, by some trigger, for example, by the arrival of a photon of the right energy, a sudden transition may occur to an excited state, and the molecule exhibits all the properties of this excited state. The amount of information in the photon is miniscule when compared to the information exhibited by the excited state molecule. Yet, after the arrival of the photon, the molecule is immediately able to exhibit all these properties. Much if not all this information must already be present in the ground state molecule's electron density cloud, as some latent property, not yet exhibited. Note that, for a ground state molecule, a property of one of its excited states is regarded a *latent property*.

Similar considerations apply for molecular deformations. For conformation A, a property of another conformation *b* of the same molecule is one of the latent properties of A, since the only obvious information in addition to those exhibited by conformation A is the set of new internal coordinates of the nuclei, and, again, this extra information is miniscule when compared to the richness of information on all the exhibited properties.

The holographic theorem as applied to latent properties is providing a theoretical framework for potential surface extrapolation. In practice, such an extrapolation is achieved by density matrices, in fact, by direct extrapolations of the density matrices themselves.

Using such extrapolations, it is possible to consider local ranges of closely related nuclear arrangements, for example, those near a potential energy minimum, or one near a formal transition structure of a reaction.

Extrapolations combined with shape analysis provides a new emphasis on a range of related structures, and on the shape analysis of such ranges of structures, as applied to transition structure modeling and similarity measures for chemical reactions.

One of the challenges of transition structure modeling is the difficulty of obtaining experimental information. For this reason, it is a useful approach to rely less on geometry optimization, and using instead ranges of structures, especially, since the precise "transition structure nuclear geometry" is in actual existence at most for only an instance, even within a semiclassical model. Nevertheless, using some of the transition structure search methods, such as saddle point determination, one can find it a valuable approach, however, here we shall be concerned with local families of nuclear arrangements, from a range including the ideal transition structure geometry.

It should be emphasized that there are further advantages in an approach that does not focus on a single point in nuclear configuration space, rather, it takes a range of nuclear arrangements in the vicinity of the assumed saddle point of the transition structure. In such an approach, the comparisons can be made not between individual electron densities of specific nuclear arrangements, but between families of electron densities, some possibly representing a range of reactant, another family representing a range of product nuclear arrangements.

Considering families of nuclear arrangements instead of individual nuclear geometries has a long tradition [63–66], leading to important concepts and methods of potential energy surface analysis [16].

The simplest tools for obtaining electron density representations for such ranges of families of nuclear arrangements is by extrapolation, that in one specific approach means density matrix extrapolation. Based on Löwdin's symmetric orthogonalization method [67, 68] for orbital sets, the same matrices which are used in this process, can also be used for other purposes, for example, in crystallographic applications as employed by Massa, Huang, and Karle [69, 70], and also for generating a good approximation to the density matrix $P(K)$ for some nuclear geometry K from an already computed density matrix $P(K')$ at some "not too different" nuclear geometry K'. For this purpose, the approach requires using additionally only the two overlap matrices S(K) and S(K') for the two nuclear geometries, K and K', respectively. Overlap matrices can be computed rapidly, hence this approach of density matrix extrapolation is rather inexpensive, yet, it generates a rather good quality approximation to the actual density matrix at the new nuclear arrangement. Of course, if density matrices (and the associated atomic orbitals) are available, the computation of a whole range of molecular properties is rather straightforward.

The method used is the Löwdin transform – inverse Löwdin transform (LIL) method, that generates an extrapolated approximate density matrix P(K,[K']) at conformation K from an already calculated density matrix P(K') at conformation K'.

The Löwdin transform – inverse Löwdin transform, in short, the LIL – transform, uses only a density matrix P(K') at some conformation K' and the two, inexpensive overlap matrices, S(K) and S(K') at both conformations, and it generates the following extrapolated, approximate density matrix P(K,[K']) at conformation K, where in the notation both old nuclear arrangement K', and the new one, K, are specified:

$$P(K,[K']) = S(K)^{-1/2} \, S(K')^{1/2} \, P(K') \, S(K')^{1/2} \, S(K)^{-1/2} \qquad (17)$$

Since the density matrix P of a molecule contains information equivalent to the molecular wavefunction, by this extrapolation, a whole range of molecular properties can be computed for the new nuclear geometry K, with an accuracy determined by the quality of P(K,[K']).

For actual applications, it is important that the electronic charge is preserved in such computations, that can often be diagnosed by testing the so-called "idempotency condition" or purity condition:

$$PSP = P \qquad (18)$$

where the rearranged form

$$PSP - P = 0 \qquad (19)$$

The above equation is more directly providing a numerical test, in terms of the magnitudes of the matrix elements o the right hand side.

It has been shown [20] that the LIL transform does not change the level of fulfillment of this condition, and the extrapolated density matrix P(K,[K']) has the same level of purity as the original density matrix P(K').

2.4   CHARACTERIZATION OF REACTIONS AND CONFORMATIONAL CHANGES BY TRANSITION STRUCTURES

The LIL transform approach is an efficient density matrix extrapolation that can be used for the generation of electron densities originating at one nuclear arrangement, but "transplanted" to a different nuclear arrangement.

This also opens the possibility for the following procedure: if density matrices

$$P(K_1), P(K_2), ..., P(K_i), ..., P(K_n), \tag{20}$$

and overlap matrices

$$S(K_1), S(K_2), ..., S(K_i), ..., S(K_n), \tag{21}$$

are available for a set of nuclear geometries

$$K_1, K_2, ..., K_i, ..., K_n, \tag{22}$$

where the K nuclear arrangements are the different conformations of the same molecule, then an artificial, averaged electron density can be computed as the average of all these densities expressed at some common nuclear geometry, that may well be different from any of the above listed nuclear geometries.

In fact, weighted averages may be of some interest, for example, weighted by the total energies of the individual electronic structures of the original entries, of density matrices $P(K_1)$, $P(K_2)$, ..., $P(K_i)$, ..., $P(K_n)$.

For example, if energies are available for a set of nuclear arrangements not very different from one another, then one can generate an "averaged electron density" for all these nuclear arrangements, at a common geometry K_{av}, and one can study the energy-shape correlations.

Such averages combine the features for these diverse solutions as obtained for different structures.

If similarity measures are determined between this averaged electron density and some or all of the individual densities, then correlations between similarities and energy deviations among the structures can provide insight, how electron density changes within the conformational range determined by the range of nuclear arrangements.

Such an approach is also likely to result in useful information, when a transition structure geometry is not exactly determined, but several nuclear configurations and the associated density and overlap matrices are known "somewhere" in the vicinity of the exact transition structure. In such a case, the averaged electron density, possibly a weighted average of the electron densities at some common nuclear geometry gives a "compromise" candidate for the study of the behavior of the associated conformational range.

2.5 HAMMOND POSTULATE AND ANALOGOUS SHAPE SIMILARITY POSTULATES

The original Hammond postulate describes the expectation, that in a chemical reaction or conformational change, the transition structure geometry is more similar to that of the reactant or product, if its energy value is more similar to the same participant in the reaction: reactant or product. Although the original postulate has been phrased in terms of exothermicity and endothermicity, the actual reasoning behind Hammond postulate is the expected positive correlation between the relative magnitudes of changes in energy and geometry.

The Hammond postulate has been also studied in the context of molecular shape variations along chemical reaction paths [60, 61], and some generalizations have been already suggested. In the present case, the LIL transform offers an alternative approach, where the comparisons of the various shape similarities can provide revealing details about the electron density changes during the reaction process.

The approach is equally valid if one applies it for a conformational change or for an actual chemical reaction between two reactants, but in the latter case, it is useful to consider several mutual arrangements of the reactants as well as the products as "supermolecules" that is, treating the combinations of two molecules as a single entity.

The conformational change is simpler to follow than an accrual bimolecular reaction, and this case will be reviewed in some more detail.

Consider a molecule with two stable conformations, separated by a formal transition structure along a conformational reaction path. If K_1 and K_2 are the nuclear geometries of the two conformers and K_T is that of the transition structure, let us assume that good quality density matrices are available for each, as well as the associated overlap matrices:

$$P(K_1), P(K_2), P(K_T), S(K_1), S(K_2), S(K_T). \tag{23}$$

Using these density matrices, and also all the possible extrapolated density matrices among these nuclear geometries, that is, using the following nine density matrices,

$$P(K_1), P(K_1,[K_2]), P(K_1,[K_T]), P(K_2), P(K_2,[K_1]),$$
$$P(K_2,[K_T]), P(K_T), P(K_T,[K_1]), P(K_T,[K_2]), \tag{24}$$

one can generate for each of the matrices the corresponding calculated electron densities:

$$\rho(P(K_1)), \rho(P(K_1,[K_2])), \rho(P(K_1,[K_T])), \rho(P(K_2)), \rho(P(K_2,[K_1])),$$
$$\rho(P(K_2,[K_T])), \rho(P(K_T)), \rho(P(K_T,[K_1])), \rho(P(K_T,[K_2])). \tag{25}$$

After generating shape codes for each of these electron densities using the shape group method, one can evaluate for each pair of the above electron densities the shape similarity measures.

In turn, these shape similarity measures, when combined with energy relations, are expected to lead to analogous results to the usual Hammond postulate: one expects that a larger energy difference between two densities, also corresponds to larger shape dissimilarities, that is, smaller numerical measures of their shape similarities. In addition, by testing he performance of extrapolated density matrices in comparison to the original density matrices, one can expect further insight, providing some finer details for this "shape-energy" Hammond postulate.

KEYWORDS

- density matrix extrapolation
- holographic electron density theorem
- molecular complementarity measures
- molecular fragments
- shape group method
- similarities of transition structures
- similarity measures for conformational changes
- similarity measures for reactions

REFERENCES

1. Carbó, R., Leyda. L., & Arnau, M., (1980). How similar is a molecule to another? An electron density measure of similarity between two molecular structures. *International Journal of Quantum Chemistry*, *17*(6), 1185–1189.
2. Carbó, R., & Arnau, M., (1981). Molecular Engineering: A General Approach to QSAR, in *Medicinal Chemistry Advances*, de las Heras, F. G., Vega, S. (eds.), Pergamon Press, Oxford.
3. Carbó, R., Calabuig, B., Vera, L., & Besalú, E., (1994). Molecular Quantum Similarity: Theoretical Framework, Ordering Principles, and Visualization Techniques, in *Advances in Quantum Chemistry, Vol. 25*, Löwdin, P-O., Sabin, R. J., Zerner, M. C. (eds.), Academic Press, New York.
4. Carbó, R. (Ed.)., (1995). *Molecular Similarity and Reactivity: From Quantum Chemical to Phenomenological Approaches*. Kluwer Academic Publishers, Dordrecht.
5. Besalú, E., Carbó, R., Mestres, J., & Solà, M., (1995). Foundations and recent developments on molecular quantum similarity, In *Topics in Current Chemistry, Vol. 173, Molecular Similarity*, Sen, K. (Ed.). Springer-Verlag, Heidelberg.
6. Amat. L. l., & Carbó-Dorca, R., (1997). Quantum similarity measures under atomic shell approximation: First-order density fitting using elementary Jacobi rotations. *J. Comput. Chem.*, *18*, 2023–2039.
7. Amat, L., & Carbó-Dorca, R., (1999). Ponec, R. Simple linear QSAR models based on quantum similarity measures. *J. Med. Chem*, *42*, 5169–5180.
8. Carbó-Dorca, R., Besalú, E., & Girones, X. (2000). Extended density functions. *Adv. Quantum Chem.*, *38*, 1–63.
9. Amat, L., Besalú, E., Carbó-Dorca, R., & Ponec, R. (2001). Identification of active molecular sites using quantum-self-similarity measures. *J. Chem. Inf. Model.*, *41*, 978–991.
10. Girones, X., Amat, L., & Carbó-Dorca, R., (2002). Modeling large macromolecular structures using promolecular densities. *J. Chem. Inf. Model*, *42*, 847–852.
11. Carbó-Dorca, R., & Besalú, E., (2010). Mathematical aspects of the LCAO MO first order density function (4): a discussion on the connection of Taylor series expansion of electronic density (TSED) function with the holographic electron density theorem (HEDT) and the Hohenberg–Kohn theorem (HKT). *J. Math. Chem*, *49*, 836–842.
12. Carbó-Dorca, R., & Besalú, E., (2010). Communications on quantum similarity (2): A geometric discussion on holographic electron density theorem and confined quantum similarity measures. *J. Comput. Chem*, *31*, 2452–2462.
13. Carbó-Dorca, R., & Besalú, E., (2010). A Gaussian holographic theorem and the projection of electronic density functions into the surface of a sphere. *J. Math. Chem*, *48*, 914–924.
14. Besalú, E., & Carbó-Dorca, R., (2012). Stereographic projection of density functions (DF) and the holographic electron density theorem (HEDT). *J. Chem. Theory Comput*, *8*, 854–861.
15. Johnson, M. A., & Maggiora, G. M., Eds. (1990). *Concepts and Applications of Molecular Similarity*, Wiley, New York, pp. 1–321.
16. Mezey, P. G., (1987). *Potential Energy Hypersurfaces*, Elsevier, Amsterdam, Chapter 5.
17. Mezey, P. G., (1986). Group theory of electrostatic potentials: a tool for quantum chemical drug design, *Int. J. Quant. Chem. Quant. Biol. Symp*, *12*, 113–122.

18. Mezey, P. G., (1987). Group theory of shapes of asymmetric biomolecules, *Int. J. Quantum Chem., Quant. Biol. Symp*, *14*, 127–132.

19. Mezey, P. G., (1987). The shape of molecular charge distributions: group theory without symmetry, *J. Comput. Chem*, *8*, 462–469.

20. Mezey, P. G., (2013). On the inherited 'purity' of certain extrapolated density matrices. *Computational and Theoretical Chemistry*, *1003*, 130–133.

21. Mezey, P. G., (1988). Shape group studies of molecular similarity: shape groups and shape graphs of molecular contour surfaces. *J. Math. Chem*, *2*, 299–323.

22. Arteca, G. A., & Mezey, P. G., (1988). Shape description of conformationally flexible molecules: application to two-dimensional conformational problems, *Internat. J. Quantum Chem., Quant. Biol. Symp. 15*, 33–54.

23. Mezey, P. G., (1990). Three-dimensional topological aspects of molecular similarity. In: *Concepts and Applications of Molecular Similarity;* Johnson, M. A., Maggiora, G. M., (eds.), Wiley, New York, pp. 321–368.

24. Mezey, P. G., (1990). Topological quantum chemistry. In: *Reports in Molecular Theory;* Náray-Szabó; G., Weinstein, H., (eds.), CRC Press: Boca Raton, FL, pp. 165–183.

25. Mezey, P. G., (1990). Molecular surfaces. In: *Reviews in Computational Chemistry;* Lipkowitz, K. B., Boyd, D. B., (eds.), VCH Publishers, New York, pp. 265–294.

26. Mezey, P. G., (1992). Dynamic shape analysis of biomolecules using topological shape codes. In *The Role of Computational Models and Theories in Biotechnology;* Bertran, J. (Ed.)., Kluwer Academic Publishers, Dordrecht, pp. 83–104.

27. Mezey, P. G., (1992). Shape-similarity measures for molecular bodies: A 3D topological approach to quantitative shape-activity relations, *J. Chem. Inf. Comput. Sci.*, *32*, 650–656.

28. Mezey, P. G., (1993). *Shape in Chemistry: Introduction to Molecular Shape and Topology.* VCH Publishers, New York.

29. Walker, P. D., Maggiora, G. M., Johnson, M. A., Petke, J. D., & Mezey, P. G., (1995). Shape Group Analysis of Molecular Similarity: Shape Similarity of Six-Membered Aromatic Ring Systems, *J. Chem. Inf. Comput. Sci.*, *35*, 568–578.

30. Walker, P. D., Mezey, P. G., Maggiora, G. M., Johnson, M. A., & Petke, J. D., (1995). Application of the Shape Group Method to Conformational Processes: Shape and Conjugation Changes in the Conformers of 2-Phenyl Pyrimidine, *J. Comput. Chem.*, *16*, 1474–1482.

31. Mezey, P. G., (2015). Topological Beauty and Molecular Shape. *Conference Proceedings of the XVIII. Generative Art Conference, University of Milano Press,* Milano, Italy, pp. 256–259. Available at http://www.generativeart.com/ga2015_WEB/topological-beauty_Mezey.pdf.

32. Hohenberg, P., & Kohn, W., (1964). Inhomogeneous electron gas. *Phys. Rev.*, *136*, B864–B871.

33. Riess, J., & Munch, W., (1981). The theorem of Hohenberg and Kohn for subdomains of a quantum system. *Theor. Chim. Acta*, *58*, 295–300.

34. Mezey, P. G., (1999). The Holographic Electron density theorem and quantum similarity measures. *Mol. Phys*, *96*, 169–178.

35. Mezey, P. G., (1999). Holographic electron density shape theorem and its role in drug design and toxicological risk assessment. *J. Chem. Inf. Comput. Sci*, *39*, 224–230.

36. Mezey, P. G., (2001). The Holographic principle for latent molecular properties. *J. Math. Chem*, *30*, 299–303.

37. Mezey, P. G., (2007). A Fundamental relation of molecular informatics: information carrying properties of density functions. *CCCC (Collection of Czechoslovak Chemical Communications), 11,* 153–163. (Volume dedicated to Prof Koutecky).

38. Mezey, P. G., (2012). Natural molecular fragments, functional groups, and holographic constraints on electron densities, *Phys. Chem. Chem. Phys. 14,* 8516–8522.

39. Ayers, P. W., (2000). Atoms in molecules, an axiomatic approach. I. Maximum transferability. *J. Chem. Phys., 113,* 10886–10899.

40. Ghosh, A., (2000). Local symmetry and chirality of molecular faces. *Theor. Chem. Acc., 104,* 157–159.

41. Geerlings, P., De Proft, F., & Langenaeker, W., (2003). Conceptual density functional theory. *Chem. Rev., 103,* 1793–1874.

42. Boon, G., Van Alsenoy, C., De Proft, F., Bultinck, P., & Geerlings, P., (2003). Similarity and chirality: quantum chemical study of dissimilarity of enantiomers. *J. Phys. Chem. A, 107,* 11120–11127.

43. De Proft, F., Ayers, P. W., Sen, K. D., & Geerlings, P., (2004). On the importance of the density per particle (shape function) in the density functional theory. *J. Chem. Phys., 120,* 9969–9974.

44. Geerlings, P., Boon, G., Van Alsenoy, C., & De Proft, F., (2005). Density functional theory and quantum similarity. *Int. J. Quantum Chem., 101,* 722–732.

45. Boon, G., Van Alsenoy, C., De Proft, F., Bultinck, P., & Geerlings, P., (2005). Molecular quantum similarity of enantiomers of amino acids: a case study. *J. Mol. Struct. (THEOCHEM), 727,* 49–56.

46. Boon, G., Van Alsenoy, C., De Proft, F., Bultinck, P., & Geerlings, P., (2006). Study of molecular quantum similarity of enantiomers of amino acids. *J. Phys. Chem. A, 110,* 5114–5120.

47. Janssens, S., Boon, G., & Geerlings, P., (2006). Molecular quantum similarity of enantiomers: substituted allenes. *J. Phys. Chem. A, 110,* 9267–9272.

48. Geerlings, P., De Proft, F., & Ayers, P. W., (2007). Chemical reactivity and the shape function. *Theor. Comput. Chem., 19,* 1–17.

49. Janssens, S., Van Alsenoy, C., & Geerlings, P., (2007). Molecular quantum similarity and chirality: enantiomers with two asymmetric centra. *J. Phys. Chem. A, 111,* 3143–3151.

50. Janssens, S., Borgoo, A., Van Alsenoy, C., & Geerlings, P., (2008). Information theoretical study of chirality: enantiomers with one and two asymmetric centra. *J. Phys. Chem. A, 112,* 10560–10569.

51. Debie, E., Jaspers, L., Bultinck, P., Herrebout W., & Van Der Veken, B., (2008). Induced solvent chirality: A VCD study of camphor in CDCl3. *Chem. Phys. Lett, 450,* 426–430.

52. Debie, E., Bultinck, P., Herrebout W., & Van Der Veken, B., (2008). Solvent effects on IR and VCD spectra of natural products: an experimental and theoretical VCD study of pulegone. *Phys. Chem. Chem. Phys, 10,* 3498–3508.

53. Janssens, S., Bultinck, P., Borgoo, A., Van Alsenoy, C., & Geerlings, P., (2010). Alternative Kullback–Leibler information entropy for enantiomers, *J. Phys. Chem. A, 114,* 640–645.

54. Geerlings, P., & Borgoo, A., (2011). Information carriers and (reading them through) information theory in quantum chemistry. *Phys. Chem. Chem. Phys, 13,* 911–922.

55. Mezey, P. G., (1995). Shape analysis of macromolecular electron densities, *Structural Chem., 6,* 261–270.

56. Mezey, P. G., (1995). Macromolecular density matrices and electron densities with adjustable nuclear geometries. *J. Math. Chem, 18*, 141–168.
57. Mezey, P. G., (1996). Local shape analysis of macromolecular electron densities. In *Computational Chemistry: Reviews and Current Trends,* Vol. 1, Leszczynski, J., (Ed.). World Scientific Publishers, Singapore, pp. 109–137.
58. Mezey, P. G., (1997). Quantum similarity measures and Löwdin's transform for approximate density matrices and macromolecular forces, *Int. J. Quantum Chem, 63*, 39–48.
59. Mezey, P. G., (2015). Compensation effects in molecular interactions and the quantum chemical *le Chatelier* principle. Invited paper to the Jacopo Tomasi Festschrift. *J. Phys. Chem. A, 119*, 5305–5312.
60. Arteca, G. A., & Mezey, P. G., (1988). Validity of the Hammond postulate and constraints on general one-dimensional barriers, *J. Comput. Chem, 9*, 728–744.
61. Arteca, G. A., & Mezey, P. G., (1989). Molecular similarity and molecular shape changes along reaction paths: a topological analysis and consequences on the Hammond postulate, *J. Phys. Chem, 83*, 4746–4751.
62. Mezey, P. G., (2014). Fuzzy electron density fragments in macromolecular quantum chemistry, combinatorial quantum chemistry, functional group analysis, and shape – activity relations, *Accounts of Chem. Research*, Invited paper, *47*, 2821–2827.
63. Tachibana, A., & Fukui, K., (1979). Intrinsic dynamism in chemically reacting systems. *Theoret. Chim. Acta, 51*,189–206.
64. Mezey, P. G., (1981). Catchment region partitioning of energy hypersurfaces, I. *Theor. Chim. Acta, 58*, 309–330.
65. Mezey, P. G., (1985). Catchment regions as "molecular loges" on potential energy hypersurfaces. *J. Mol. Structure, Theochem.* (volume dedicated to Prof. R. Daudel), *123*, 171–177.
66. Mezey, P. G., (1999). The topology of catchment regions of potential energy hypersurfaces. *Theor. Chem. Acc, 102*, 279–284.
67. Löwdin, P.-O., (1950). On the non- orthogonality problem connected with the use of atomic wave functions in the theory of molecules and crystals. *J. Chem. Phys., 18*, 365–375.
68. Löwdin, P.-O., (1970). On the orthogonality problem. *Adv. Quantum Chem., 5*, 185–199.
69. Massa, L., Huang, L., & Karle, J., (1995). Quantum crystallography and the use of kernel projector matrices. *Int. J. Quantum Chem. Quant. Chem. Symp, 29*, 371–384.
70. Huang, L., Massa, L., & Karle, J., (1996). Kernel projector matrices for Leu-Zervamicin, *Int. J. Quantum Chem. Quant. Chem. Symp., 30*, 1691–1699.
71. Mezey, P. G., (1998). Averaged electron densities for averaged conformations, *J. Comput. Chem., 19*, 1337–1344.
72. Exner, T. E., & Mezey, P. G., (2002). A comparison of nonlinear transformation methods for electron density approximation. *J. Phys. Chem. A, 106*, 5504–5509.
73. Mezey, P. G., (2005). Electron density extrapolation along reaction paths. *J. Mol. Struct.*, (special volume dedicated to Prof. Ramon Carbó-Dorca), *727*, 123–126.
74. Szekeres, Z., & Mezey, P. G., (2005). A one-step diophantine solution to the density matrix purification problem. *Mol. Phys.*, (special volume dedicated to Prof. N. Handy), *103*, 1013–1015.
75. Mezey, P. G., (2008). Charge-conserving electron density averaging for a set of nuclear configurations. *J. Math. Chem, 44*, 1023–1032.

CHAPTER 3

SUPERVISED DISTANCE METRIC LEARNING AND CURSE OF DIMENSIONALITY

FARNAZ HEIDAR-ZADEH[1-3]

[1]*Department of Chemistry and Chemical Biology, McMaster University, Hamilton, Ontario, Canada*

[2]*Department of Inorganic and Physical Chemistry, Ghent University, Krijgslaan 281 (S3), 9000 Gent, Belgium*

[3]*Center for Molecular Modeling, Ghent University, Technologiepark 903, 9052 Zwijnaarde, Belgium*

CONTENTS

3.1 PERSPECTIVE

The human brain is the world's best pattern recognition machine. Though machines are rapidly exceeding us in terms of raw computing power, they perform poorly when given a task requiring intelligence. By contrast, our brain excels at extracting a small number of characteristic features from vast amount of sensory data. Consider, for example, how quickly we discern a friendly smile from an unfriendly glare. In addition, we can also describe facial expressions compactly, with a few descriptive features. On the other hand, when computers are provided with digital images of faces, they are presented with hundreds of thousands of pixel intensities. Given such high-dimensional data, how a computer can identify the decisive modes of variability for a specific task? As a simple example, how a computer can classify images by facial expression? This task would be different from classifying faces based on gender. In other words, how a computer can learn to capture the task-specific descriptors needed to make accurate predictions about the class of an image [2–5]?

This problem arises frequently in the domain of computer vision, pattern recognition, and higher-level decision-making in artificial intelligence. One of the underlying challenges is making proper and accurate comparisons of data for a given task. For example, distance measures that compare images on a pixel-by-pixel basis, like mean-squared-difference between pixel intensities, do not support subtle aspects of similarities and differences. It is apparent that this distance is not ideal for tasks like facial-expression classification or gender classification. Distance metric learning algorithms address this problem by transforming raw data into a compact and task-specific representation. This is achieved by learning a proper measure of distance for comparing similarity of the data points. Most importantly, this approach allows incorporation of supervision (i.e., prior knowledge specifying which data points are similar and/or dissimilar), in developing a proper distance metric. The learned distance is then used in various machine-learning algorithms for purposes of classification, clustering or regression [3, 4].

These algorithms can be divided into two main categories: unsupervised and supervised distance metric learning. Figure 3.1 provides an overview of some of the available distance metric learning algorithms. The family of unsupervised distance metric learning methods aims to reduce the number of dimensions to simplify the representation of the high-dimensional data. This transformation not only increases the efficiency of various algorithms that process the data, but also reveals hidden modes of variability by embedding

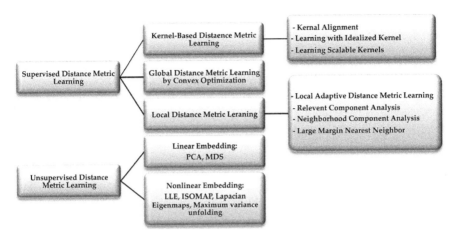

FIGURE 3.1 Some distance metric learning algorithms. The acronym PCA denotes principal component analysis, MDS refers to multidimensional scaling, and LLE stands for locally linear embedding.

the data into a subspace of lower dimension. The family of supervised algorithms takes advantage of prior knowledge to adapt the distance measure to the specific problem. In the same manner that humans learn, a computer is provided with a set of training examples to learn the proper measure of distance capturing task-specific features.

In the next sections, some (un)supervised distance metric learning methods are discussed. The ultimate goal is to apply these techniques to problems in cheminformatics, where the ability to make precise judgments about similarity and dissimilarity of compounds is of great importance. Specifically, the idea of learning problem-specific distances can be applied to classification and clustering of molecules, and predicting their chemical and biological properties. For example, one could make better models for identifying mutagenic compounds by learning a proper measure of distance between mutagens.

3.2   UNSUPERVISED DISTANCE METRIC LEARNING

The ubiquity of high-dimensional data like gene expression microarrays, digital images, and medical responses makes mining massive high-dimensional data an urgent problem of great practical importance. Usually, high dimensional data are redundant and impossible to visualize; they also require large storage and processing resources. In addition, the cost of many existing machine learning algorithms grows exponentially as the number of dimensions in the input space increases. It is often believed that there is a compact representation of data, i.e., that there is a low dimensional structure embedded in high dimensional data. This motivates statistical methods for dimensionality reduction.

Unsupervised distance metric learning algorithms address the curse of dimensionality by reducing the number of random variables representing the data. In other words, they map the initial high-dimensional data into a lower dimensional space where the axes are related to the intrinsic degrees of freedom of data. These include linear techniques, like principal component analysis (PCA), as well as non-linear techniques like Maximum Variance Unfolding discussed in the next section [5–13]. Here, no side-information in the form of labels or relative distance between sample points is available. It is basically assumed that the high-dimensional input data has intrinsic structure because it is being sampled from a lower dimensional manifold. Manifolds are mathematical spaces that are locally linear, so the Euclidean distances are only locally meaningful. For example, for a 2D manifold embedded in 3D space (e.g., a sphere, a torus, or a Swiss roll pictured in Figure 3.3), the Euclidean measure does not represent the distance on the manifold in the 3D input space; however, having the lower dimensional 2D representation of the data in terms of its intrinsic dimensions, the Euclidean distances become globally meaningful [3].

3.2.1 MAXIMUM VARIANCE UNFOLDING

Introduced by Weinberger et al. [14], maximum variance unfolding detects a low-dimensional embedding of high-dimensional data that preserves local isometry. Given n high-dimensional input points, $\{x_i\}_{i=1}^n$, where $x_i \in R^D$, this non-linear dimensionality reduction algorithm computes n output points, $\{y_i\}_{i=1}^n$, where $y_i \in R^d$ and $d<<D$, in one-to-one correspondence with the inputs. The local isometry constraints can be naturally represented by a graph with n nodes each corresponding to an input point. Each node is connected to its k nearest neighbors, where k is a free parameter of the algorithm. Now, the local isometry is imposed with constraints preserving the lengths of the edges as well as the angles between edges of the same node. Because, it is easier to handle the constraints on distances, as opposed to angles, additional edges between the neighbors of each node are placed and their lengths are preserved. Figure 3.2 illustrates how connecting the nodes preserves local isometry. The binary variable $h_{ij} = \{0,1\}$ denotes the connectivity between inputs x_i and x_j [3, 5, 9, 14].

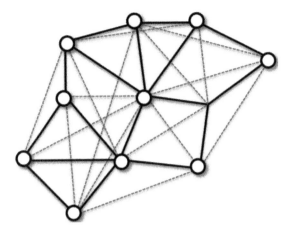

FIGURE 3.2 Preserving local distance and angle by connecting neighbors. The black lines connect each input point to its k nearest neighbors (distance-preserving edges), whereas the dashed red lines add edges between the neighbors of each input point (angle-preserving edges), if they don't already exist.

This algorithm attempts to pull the inputs apart without breaking the connection between connected nodes; this transformation is visualized in Figure 3.3. In other words, the algorithm outputs $\{\mathbf{y}_i\}_{i=1}^n$ that unfolds the

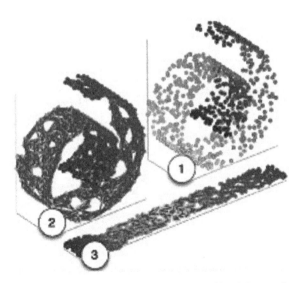

FIGURE 3.3 Maximum variance unfolding of a Swiss roll. (1) Data points sampled from the manifold in high-dimensional space; (2) Adding local isometry constraints by connecting nodes; (3) Data points in low-dimensional space preserving local isometry [1].

manifold represented by $\{x_i\}_{i=1}^n$, by maximizing the sum of pair-wise distances between outputs while nearby points remain close [9]:

$$\max \Sigma_{ij} \left\| y_i - y_j \right\|^2$$

$$\text{s.t. } \left\| y_i - y_j \right\|^2 = \left\| x_i - x_j \right\|^2 \quad \text{for all } (i, j) \text{ with } \eta_{i,j} = 1 \qquad (1)$$

$$\Sigma_i \, y_i = 0$$

This can be formulated as a quadratic optimization, Eq. (1), where the first set of constraints preserves local isometry, and the second constraints guarantees a unique solution (up to rotation) by centering the outputs. Unfortunately, optimizing over $\{y_i\}_{i=1}^n$, is not convex; however, by reformulating this algorithm in terms of the elements of inner product matrix, $K_{ij} = y_i \cdot y_j$, the optimization can be represented as a semidefinite program (SDP) that is convex. Using $\|y_i - y_j\|^2 = K_{ii} - 2K_{ij} + K_{jj}$, the SDP over the positive semidefinite K is written as a convex optimization:

$$\max_{K \succeq 0} \text{tr}(K)$$

$$\text{s.t. } \Sigma_{ij} K_{ij} = 0 \qquad (2)$$

$$K_{ii} - 2K_{ij} + K_{jj} = \left\| x_i - x_j \right\|^2 \quad \text{for all } (i,j) \text{ with } \eta_{ij} = 1$$

Having the solution K, the outputs $\{y_i\}_{i=1}^n$, satisfying $K_{ij} = y_i \cdot y_j$ are found by singular value decomposition. Also, the $y_i \in R^d$ for which $K_{ij} \approx y_i \cdot y_j$ can be obtained from the top d eigenvalues and eivenvectors of the K matrix. There are tricks to extend this approach to new data points that were not included in the original sampling that makes the out-of-sample mapping feasible [3, 5, 9].

3.3   SUPERVISED DISTANCE METRIC LEARNING

The performance of many algorithms critically depends on possessing a good distance metric over the input vector space that reflects important relationships between the data. As discussed, unsupervised metric learning algorithms learn the intrinsic dimensionality of a manifold embedded in the higher dimensional input space, and map the sample points into a space of reduced dimension where the Euclidean distance measure is globally proper. On the other hand, supervised metric learning algorithms specifically target the underlying metric, through a parameterized distance function, so that it is adapted to our specific purpose. In other words, based on the idea that manually tweaked distance metrics using expert knowledge improves efficiency, supervised metric learning algorithms systematically incorporate the prior knowledge to develop a task-specific distance. Specifically, given examples of (dis)similar pairs of points, supervised methods learn a distance metric over the input space that respects these relationships [4, 15].

Definition: A mapping $d: X \times X \rightarrow R_0^+$ over the vector space X is called a *metric*, if for all vectors $\mathbf{x}_i, \mathbf{x}_j, \mathbf{x}_k \in X$, it satisfies the properties below. If a metric satisfies only the first three properties, it is called a *pseudo-metric*.

1. Triangle Inequality: $d(\mathbf{x}_i, \mathbf{x}_j) + d(\mathbf{x}_j, \mathbf{x}_k) \geq d(\mathbf{x}_i, \mathbf{x}_k)$
2. Non-Negativity: $d(\mathbf{x}_i, \mathbf{x}_j) \geq 0$
3. Symmetry: $d(\mathbf{x}_i, \mathbf{x}_j) = d(\mathbf{x}_j, \mathbf{x}_i)$
4. Distinguishability: $d(\mathbf{x}_i, \mathbf{x}_j) = 0 \Leftrightarrow \mathbf{x}_i = \mathbf{x}_j$

Here, the focus is on the Mahalanobis distance, introduced by P. C. Mahalanobis in 1936, over the vector space X as defined in Eqs. (3) and (4); the matrix Σ denotes the covariance of matrix $\mathbf{X}$ in which $\mathbf{x}_i$ is the ith column. Note that when Σ is the identity matrix, the Euclidean distance is recovered.

$$d_{Mahalanobis}(\mathbf{x}_i, \mathbf{x}_j) = \sqrt{(\mathbf{x}_i - \mathbf{x}_j)^T \Sigma^{-1} (\mathbf{x}_i - \mathbf{x}_j)} \qquad (3)$$

$$\Sigma = E\left[(\mathbf{X} - E[\mathbf{X}])(\mathbf{X} - E[\mathbf{X}])^T \right] \qquad (4)$$

The Mahalanobis distance differs from Euclidean distance because it considers statistical regularities in the data. In other words, it can be interpreted as calculating the Euclidean distance of centered ($\mu = \mathbf{0}$) and whitened ($\Sigma = \mathbf{I}$) vectors. In the metric learning literature, the term "Mahalanobis distance" denotes any distance function of the form [4]:

$$d_{\mathbf{M}}(\mathbf{x}_i, \mathbf{x}_j) = \left\| \mathbf{x}_i - \mathbf{x}_j \right\|_{\mathbf{M}} = \sqrt{(\mathbf{x}_i - \mathbf{x}_j)^T \mathbf{M}(\mathbf{x}_i - \mathbf{x}_j)} \qquad (5)$$

Matrix **M** parameterizes a family of Mahalanobis distance over the input space $\mathcal{X}$. To ensure that Eq. (5) represents a pseudo-metric, the matrix **M** is required to be positive semi-definite, $\mathbf{M} \succeq \mathbf{0}$. Since **M** is a positive semi-definite matrix, it has a Cholesky factorization, $\mathbf{M} = \mathbf{L}^T\mathbf{L}$. Therefore,

$$d_{\mathbf{M}}(\mathbf{x}_i, \mathbf{x}_j) = \sqrt{(\mathbf{x}_i - \mathbf{x}_j)^T \mathbf{L}^T \mathbf{L}(\mathbf{x}_i - \mathbf{x}_j)} = \sqrt{(\mathbf{L}\mathbf{x}_i - \mathbf{L}\mathbf{x}_j)^T(\mathbf{L}\mathbf{x}_i - \mathbf{L}\mathbf{x}_j)} = \|\mathbf{L}(\mathbf{x}_i - \mathbf{x}_j)\| \quad (6)$$

In other words, the family of Mahalanobis distance metrics is obtained by computing the Euclidean distance after linear transformation of the data $\mathbf{L}\mathbf{x} = \mathbf{x}'$.

Now, suppose we are told that certain pairs of data points are similar or dissimilar through a set of pair-wise distance relationships, Eq. (7). These constraints can be provided as prior knowledge or derived from sample labels. Having this side-information, can we learn a distance metric, $d(\mathbf{x}_i, \mathbf{x}_j)$, over the input space $\mathcal{X}$ which satisfies these constraints?

$$\mathcal{S} = \{(\mathbf{x}_i, \mathbf{x}_j) | \mathbf{x}_i \text{ and } \mathbf{x}_j \text{ should be similar}\}$$
$$\mathcal{D} = \{(\mathbf{x}_i, \mathbf{x}_j) | \mathbf{x}_i \text{ and } \mathbf{x}_j \text{ should be dissimilar}\} \quad (7)$$

The goal of supervised metric learning is to adapt the distance metric to prior knowledge. As a result, we can obtain a problem-specific measure that incorporates the user's notion of distance. Considering the family of Mahalanobis distance matrices, the task of learning is equivalent to finding a linear transformation (scaling and rotation) of the data so that the Euclidean distance is a proper measure of similarity, Eq. (6). Note that Mahalanobis distance metric learning problem can be parameterized in terms of the positive semi-definite matrix **M** or the linear transformation **L**. Finally, the learned distance metric is incorporated into (un)supervised algorithms to make better predictions; this process is summarized in Figure 3.4.

FIGURE 3.4   Schematic representation of common practice in metric learning.

The first Mahalanobis distance metric learning algorithm was proposed in the seminal work of Xing et al. in 2002 [16]. Their goal was learning a Mahalanobis metric for clustering (MMC) respecting the side-information provided in Eq. (7) in the form of (in)equivalency constraints. They posed the following convex optimization algorithm to learn the proper distance metric:

$$\min_{\mathbf{M} \succeq 0} \sum_{(\mathbf{x}_i, \mathbf{x}_j) \in \mathcal{S}} d_{\mathbf{M}}^2(\mathbf{x}_i, \mathbf{x}_j)$$
$$\text{s.t.} \sum_{(\mathbf{x}_i, \mathbf{x}_j) \in \mathcal{D}} d_{\mathbf{M}}(\mathbf{x}_i, \mathbf{x}_j) \geq 1 \tag{8}$$

Simply, the metric was required to assign small squared distance between the pairs of points in the set $\mathcal{S}$ while keeping the pairs of points in set $\mathcal{D}$ distant. Following this work, there have been numerous attempts to learn global or local distance function in a supervised setting, as listed in Figure 3.1. This chapter will focus on only two algorithms: Large margin nearest neighbor (LMNN) classifier and kernel-based learning algorithms.

3.3.1 LARGE MARGIN NEAREST NEIGHBOR (LMNN) CLASSIFIER

Let $\{(\mathbf{x}_i, y_i)\}_{i=1}^N$, denote a training set of N labeled samples, each represented by an input feature vector $\mathbf{x}_i \in \Re^D$ and a discrete class label y_i. How can we determine the class label of a new sample? The traditional k-nearest neighbor (k-NN) algorithm classifies an unlabeled sample by the majority label of its k-nearest neighbors. It is a simple non-parametric pattern recognition technique, which performs surprisingly well in a wide range of applications. It is apparent that the performance of this decision rule is crucially dependent on the distance between samples. Using the Euclidean distance measure is a common practice; however, it has been demonstrated that learning an appropriate distance metric from labeled samples greatly improves the performance of k-NN classifier. This can be easily understood by a simple example. Suppose that k-NN is being used to classify face images by gender and age. It is hard to believe that the same distance metric (e.g., Euclidean distance between pixels) can optimally distinguish these two classification tasks. This motivates improving k-NN decision rule by learning problem-specific distance metrics [3–4, 17].

Inspired by recent work of Goldberger et al. [13] on neighborhood component analysis, Weinberger et al. [17] proposed the LMNN classification algorithm to improve the accuracy of k-NN predictions. Intuitively, it is based on the idea that the k-NN would correctly classify an unlabeled sample if all its k-nearest neighbors share the same label. Accordingly, the LMNN algorithm attempts to reinforce this property by learning a linear transformation of the input space prior to k-NN classification.

This supervised metric learning technique learns a Mahalanobis distance metric for the k-NN classification using a loss function, Eq. (9), composed of two competing terms: one that pulls target neighbors closer together and one that pushes samples with different labels far apart [3, 17].

$$\varepsilon(\mathbf{L}) = (1-\lambda)\varepsilon_{\text{pull}}(\mathbf{L}) + \lambda\,\varepsilon_{\text{push}}(\mathbf{L})$$

$$\varepsilon_{\text{pull}}(\mathbf{L}) = \sum_{i,j\rightsquigarrow i}\left\|\mathbf{L}(\mathbf{x}_i - \mathbf{x}_j)\right\|^2$$

$$\varepsilon_{\text{push}}(\mathbf{L}) = \sum_{i,j\rightsquigarrow i}\sum_{l}(1-y_{il})\left[1+\left\|\mathbf{L}(\mathbf{x}_i - \mathbf{x}_j)\right\|^2 - \left\|\mathbf{L}(\mathbf{x}_i - \mathbf{x}_l)\right\|^2\right]_{+} \qquad (9)$$

In this equation, $[z]_+ = \max(0,z)$ denotes the hinge loss, and $j \rightsquigarrow i$ means that input $\mathbf{x}_j$ is a target neighbor of input $\mathbf{x}_i$; that is, $\mathbf{x}_j$ is among similarly labeled k nearest neighbors of $\mathbf{x}_i$. The $\varepsilon_{\text{pull}}(\mathbf{L})$ term penalizes large distances between $\mathbf{x}_i$ and its target neighbors. By learning the linear transformation $\mathbf{L}$, the input $\mathbf{x}_i$ would become closer to its target neighbors than all imposters with different labels. In other words, we can assume that the target neighbors of each input establish a perimeter around it that should not include any differently labeled input point. To have a robust decision boundary, we can maintain a finite margin of safety between the imposters and the perimeter around $\mathbf{x}_i$. To capture this mathematically, for an input $\mathbf{x}_i$ labeled with y_i and any target neighbor $\mathbf{x}_j$, the input $\mathbf{x}_l$ with label $y_i \neq y_l$ satisfying Eq. (10) is called an imposter. The $e_{\text{push}}(\mathbf{L})$ term penalizes small distances between input $\mathbf{x}_i$ and the identified imposters in order to clearly separate them [3, 17].

$$\left\|\mathbf{L}(\mathbf{x}_i - \mathbf{x}_l)\right\|^2 \leq \left\|\mathbf{L}(\mathbf{x}_i - \mathbf{x}_j)\right\|^2 + 1 \qquad (10)$$

The auxiliary information regarding target inputs can be provided as prior knowledge or by determining the k nearest neighbors of $\mathbf{x}_i$ with the same label. For simplicity, target neighbors are considered fixed during the learning process. Figure 3.5 depicts the idea behind learning a LMNN classifier. Even though the LMNN's objective function is not convex, it can be cast as a convex optimization problem by reformulating it as an instance of semi-definite programming (SDP). This is obtained by working in terms of positive semi-definite matrix $\mathbf{M} = \mathbf{L}^T \mathbf{L}$, so the loss function is re-written as:

$$\varepsilon(\mathbf{M}) = (1-\lambda)\sum_{i,j \rightsquigarrow i} d_{\mathbf{M}}^2(\mathbf{x}_i, \mathbf{x}_j) + \lambda \sum_{i,j \rightsquigarrow i}\sum_{l}(1-y_{il})\left[1+d_{\mathbf{M}}^2(\mathbf{x}_i, \mathbf{x}_j)-d_{\mathbf{M}}^2(\mathbf{x}_i, \mathbf{x}_l)\right]_+ \quad (11)$$

Only the first term in Eq. (11) is linear in term of elements of $\mathbf{M}$, so in order to reformulate the optimization as a SDP, nonnegative slack variables $\{\xi_{ijl}\}$ for all triplets of two target neighbors and one imposter are introduced. These slack variables measure the violation of inequality expressed in Eq. (10). The result is the SDP formulation of LMNN:

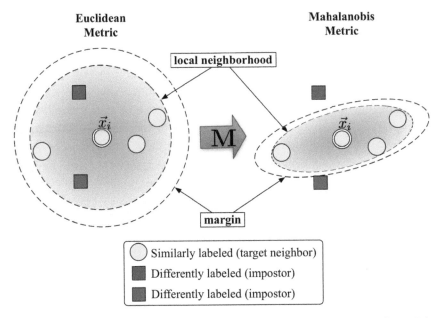

FIGURE 3.5   Schematic illustration of target neighborhood of x_i before and after training with $k = 3$ nearest neighbors [18].

$$\min_{\mathbf{M} \succeq 0} (1-\lambda) \sum_{i,j \leadsto i} (\mathbf{x}_i - \mathbf{x}_j)^T \mathbf{M}(\mathbf{x}_i - \mathbf{x}_j) + \lambda \sum_{i,j \leadsto i} \sum_l (1 - y_{il}) \xi_{ijl}$$

s.t. $\quad (\mathbf{x}_i - \mathbf{x}_l)^T \mathbf{M}(\mathbf{x}_i - \mathbf{x}_l) - (\mathbf{x}_i - \mathbf{x}_j)^T \mathbf{M}(\mathbf{x}_i - \mathbf{x}_j) \geq 1 - \xi_{ijl}$ $\quad$ (12)

$$\xi_{ijl} \geq 0$$

This optimization can be solved by many available SDP packages. Having the positive semidefinite matrix $\mathbf{M}$, the Mahalanobis distance metric is learned for the k-NN classifier. It is important to note that before classifying any new point, the same linear transformation encoded by the matrix $\mathbf{M}$ should be performed. Then, the k-nearest neighbors of the linearly transformed sample point are identified, and their majority vote assigns the class of the new sample point [3, 17].

3.3.2 KERNEL-BASED LEARNING

In the past decade, kernel-based algorithms have become a cornerstone of machine learning algorithms [19–22]. Briefly, these techniques generalize the linear methods to nonlinear domains by mapping the data from input space into a high-dimensional feature space where it is linearly separable; $\psi: \mathbf{x}_i \to \psi(\mathbf{x}_i)$. Note that in practice, this mapping can be applied implicitly, by introducing a kernel function that effectively represents the inner product between two samples in the feature space; $\kappa_0(\mathbf{x}_i, \mathbf{x}_j) = \psi(\mathbf{x}_i) \psi(\mathbf{x}_j)$. The kernel matrix $\mathbf{K}_{n \times n}$ is simply a square matrix whose $\mathbf{K}_{i,j}$ entry equals $\kappa_0(\mathbf{x}_i, \mathbf{x}_j)$. Once the kernel function is chosen, the kernel matrix can be constructed which contains all information required for learning in the feature space, so the kernel-based algorithm can be invoked; an example of kernelized algorithms is the kernel k-means clustering presented in Appendix I. It should be mentioned that the kernel function, κ_0, is valid (i.e., represents an inner product in some Hilbert space) if and only if the corresponding kernel matrix is positive semi-definite. Examples of kernel functions include the polynomial and Gaussian kernels [19, 20].

$$\kappa(\mathbf{x}_i, \mathbf{x}_j) = (\mathbf{x}_i \cdot \mathbf{x}_j + c)^d$$

$$\kappa(\mathbf{x}_i, \mathbf{x}_j) = \exp\left(-\frac{\|\mathbf{x}_i - \mathbf{x}_j\|^2}{2\alpha^2}\right) \qquad (13)$$

Despite their popularity, the cost of many kernel-based algorithms is prohibitive for large-scale learning tasks. They reveal poor scalability, $O(n^3)$, and require $O(n^2)$ memory overhead for n input data samples. In addition, they depend on the choice of a proper kernel function, which depends on the data. These drawbacks have raised interest in low-rank kernels and kernel learning methods [4, 23–26]. These methods, by assuming that the kernel matrix has rank $r << n$, achieve linear scalability to large problems because only the decomposition of the kernel matrix needs to be stored: $\mathbf{K} = \mathbf{G}\mathbf{G}^T$ where $\mathbf{G}$ is $n \times r$. Many kernel-based algorithms scale linearly with n after being reformulated in terms of $\mathbf{G}$. Unfortunately, standard kernel functions do not yield low-rank matrices. On the other hand, kernel learning algorithms attempt to learn a kernel matrix over the data from pair-wise (dis)similarity constraints. In the next section, a recent low-rank kernel learning method is presented which merges these two ideas; this can also be viewed as a metric learning problem with linear constraints on pairs of samples. The Bregman matrix divergences [23, 27] used in what follows are briefly introduced in Appendix II.

3.3.2.1 Low-Rank Kernel Learning

These learning algorithms find a low-rank kernel matrix by minimizing the divergence to an initial low-rank kernel matrix given side-information about the data. Two frameworks can be considered in formulating this algorithm: transductive learning and inductive learning. Transductive learning is simpler, but inductive learning is generalizable to new sample points, outside the original training set [4, 23, 26].

Transductive Learning [4, 22–23, 26]: Assume we are provided with both training and test sets, as well as the labels for the training samples. We will learn the kernel matrix over all the data using the constraints on the training set. Unfortunately, in this case it is not clear how to deal with new data points, as it requires re-learning the kernel matrix. Given an input kernel matrix, $\mathbf{K}_0$, using the Bregman matrix divergence D_φ, Kulis et al. [23] formulated the following algorithm to solve for $\mathbf{K}$:

$$\min_{\mathbf{K} \succeq 0} D_\phi(\mathbf{K}, \mathbf{K}_0)$$
$$\text{s.t.} \quad \text{tr}(\mathbf{K}\mathbf{A}_i) \leq b_i \quad 1 \leq i \leq c \tag{14}$$
$$\text{rank}(\mathbf{K}) \leq r$$

The first constraint denotes the general form of linear constraints that can be considered; note that $b_i \geq 0$, otherwise there cannot be a feasible solution. In kernel learning, we are interested in distance and kernel constraints. The squared Euclidean distance between j^{th} and k^{th} data points in feature space can be expressed as $\mathbf{K}_{jj} + \mathbf{K}_{kk} - 2\mathbf{K}_{jk}$, so the distance constraint $\mathbf{K}_{jj} + \mathbf{K}_{kk} - 2\mathbf{K}_{jk} \leq b_i$ can be written as $\mathrm{tr}(\mathbf{KA}_i) \leq b_i$, where $\mathbf{A}_i = (\mathbf{e}_j - \mathbf{e}_k)(\mathbf{e}_j - \mathbf{e}_k)^T$. Constraints with $\mathbf{K}_{jk} \leq b_i$ form can be represented as $\mathrm{tr}(\mathbf{KA}_i) \leq b_i$ having $\mathbf{A}_i = \frac{1}{2}(\mathbf{e}_j \mathbf{e}_k^T + \mathbf{e}_k \mathbf{e}_j^T)$. The optimization in Eq. (14) can be solved by the method of Bregman projections. Under mild conditions, this method of cyclic projections converges to the global optimum.

Because of the rank constraints, the problem in Eq. (14) is not generally convex. However, when the von Neumann or LogDet divergences are used, and $\mathrm{rank}(\mathbf{K}_0) \leq \mathrm{r}$, it is convex. This is because these divergences search for the optimal $\mathbf{K}$ in the linear subspace of matrices with the same range space as $\mathbf{K}_0$. In addition, as mentioned in Appendix II, the algorithm implicitly maintains positive semidefiniteness constraints when using the LogDet or von Neumann divergence measures.

Inductive Learning [4, 22–23, 26]: In inductive learning, the algorithm is generalizable to any query samples, so we must learn the kernel function explicitly. This is in contrast to transductive learning case, where the computationally simpler task of learning the kernel matrix subject to pairwise distance constraints between input points was discussed. Representing sample points in feature space by $\Psi = [\psi(\mathbf{x}_1),...,\psi(\mathbf{x}_n)]$, the inductive learning optimizes the kernel function of the form $\kappa(\mathbf{x}_i, \mathbf{x}_j) = \psi(\mathbf{x}_i)^T \mathbf{M}\psi(\mathbf{x}_j)$, where $\mathbf{M}$ is a positive (semi)definite matrix. This can be conceived as applying a linear transformation to the data in the feature space prior to building the kernel matrix. The parameterized kernel function is learned under the condition that we can only compute the original kernel function $\kappa_0(\mathbf{x}_i, \mathbf{x}_j) = \psi(\mathbf{x}_i)^T \psi(\mathbf{x}_j)$, i.e., having $\mathbf{M}_0 = \mathbf{I}$. Given a set of similarity $\mathcal{S}$ and dissimilarity constrains $\mathcal{D}$ over the learned distance, the basic approach to solving this problem, can be formulated as:

$$\min_{\mathbf{M} \succeq 0} D_{\ell d}(\mathbf{M}, \mathbf{M}_0)$$

$$\text{s.t.} \quad \mathrm{tr}\left(\mathbf{M}\left(\psi(\mathbf{x}_i) - \psi(\mathbf{x}_j)\right)\left(\psi(\mathbf{x}_i) - \psi(\mathbf{x}_j)\right)^T\right) \leq u \quad (i, j) \in \mathcal{S} \quad (15)$$

$$\mathrm{tr}\left(\mathbf{M}\left(\psi(\mathbf{x}_i) - \psi(\mathbf{x}_j)\right)\left(\psi(\mathbf{x}_i) - \psi(\mathbf{x}_j)\right)^T\right) \geq l \quad (i, j) \in \mathcal{D}$$

This algorithm was proposed by Davis et al. [28] in 2007 as an information-theoretic approach to Mahalanobis distance metric learning given pairwise distance constraints. It is important to note that the task of learning the kernel function can also be formulated in terms of Mahalanobis metric learning in feature space, i.e., kernelized Mahalanobis metric learning, because:

$$d_{\mathbf{M}}^2\left(\psi(\mathbf{x}_i),\psi(\mathbf{x}_j)\right)=\left(\psi(\mathbf{x}_i)-\psi(\mathbf{x}_j)\right)^T \mathbf{M}\left(\psi(\mathbf{x}_i)-\psi(\mathbf{x}_j)\right) \qquad (16)$$

The problem with learning the matrix $\mathbf{M}$ directly is that we don't have explicit knowledge of the mapping ψ: all we know is $\kappa_0(\mathbf{x}_i, \mathbf{x}_j) = \psi(\mathbf{x}_i)^T \psi(\mathbf{x}_j)$. So, how can we learn the kernel function? As discussed, the transductive kernel learning problem optimizes the kernel matrix $\mathbf{K}$ subject to the pairwise distance constraints by minimizing its divergence with respect to a specified kernel $\mathbf{K}_0$. That is,

$$\min_{\mathbf{K}\succeq 0} D_{\ell d}(\mathbf{K},\mathbf{K}_0)$$
$$\text{s.t.} \quad \mathbf{K}_{ii} + \mathbf{K}_{jj} - 2\mathbf{K}_{ij} \le u \quad (i,j)\in\mathcal{S} \qquad (17)$$
$$\mathbf{K}_{ii} + \mathbf{K}_{jj} - 2\mathbf{K}_{ij} \ge l \quad (i,j)\in\mathcal{D}$$

The lemma below establishes the coincidence of feasible solutions in transductive and inductive cases.

Lemma. Given $\mathbf{K} = \Psi^T\mathbf{M}\Psi$, the matrix $\mathbf{M}$ is feasible for Eq. (15) if and only if $\mathbf{K}$ is feasible for Eq. (17).

Proof: Considering that $\mathbf{K}_{ii} + \mathbf{K}_{jj} - 2\mathbf{K}_{ij} = (\mathbf{e}_i - \mathbf{e}_j)\mathbf{K}(\mathbf{e}_i - \mathbf{e}_j)^T = (\psi(\mathbf{x}_i) - \psi(\mathbf{x}_j))^T \mathbf{M}(\psi(\mathbf{x}_i) - \psi(\mathbf{x}_j))$, having a kernel matrix $\mathbf{K}$ satisfying $\mathbf{K}_{ii} + \mathbf{K}_{jj} - 2\mathbf{K}_{ij} \le u$ and $\mathbf{K}_{ii} + \mathbf{K}_{jj} - 2\mathbf{K}_{ij} \ge l$ constraints is equivalent to having a matrix $\mathbf{M}$ satisfying $(\psi(\mathbf{x}_i) - \psi(\mathbf{x}_j))^T \mathbf{M}(\psi(\mathbf{x}_i) - \psi(\mathbf{x}_j)) \le u$ and $(\psi(\mathbf{x}_i) - \psi(\mathbf{x}_j))^T \mathbf{M}(\psi(\mathbf{x}_i) - \psi(\mathbf{x}_j)) \ge l$.

An explicit relationship between the optimal solutions of algorithms in Eqs. (15) and (17) is presented in the theorem below. It can be easily proven by showing that, in each iteration of Bregman projections, the update coefficient of $\mathbf{K}$ and $\mathbf{M}$ are the same.

Theorem. Let $\mathbf{M}^*$ be the optimal solution to Eq. (15) and $\mathbf{K}^*$ be the optimal solution to Eq. (17). Then $\mathbf{K}^* = \Psi^T\mathbf{M}^*\Psi$.

As a result, the metric learning problem is equivalent to the low-rank transductive kernel learning problem. This connection allows generalization of the learned metric $\mathbf{M}^*$ to new points in kernel space using the learned kernel matrix $\mathbf{K}^*$. Having $\mathbf{M}^*$ or equivalently $\mathbf{K}^*$, the learned Mahalanobis distance in kernel space between any two points $\psi(\mathbf{x}_i)$ and $\psi(\mathbf{x}_j)$ in the training set can be calculated through:

$$d_{\mathbf{M}^*}^2 \left(\psi(\mathbf{x}_i), \psi(\mathbf{x}_j) \right) = \mathbf{K}_{ii}^* + \mathbf{K}_{jj}^* - 2\mathbf{K}_{ij}^* \tag{18}$$

In order to calculate this distance between any two points $\psi(\mathbf{z}_1)$ and $\psi(\mathbf{z}_2)$ which may or may not be in the training set, we need to calculate the inner product $\psi(\mathbf{z}_1)^T \mathbf{M}^* \psi(\mathbf{z}_2)$, so that Eq. (18) can be used to evaluate the learned kernelized Mahalanobis distance. This inner product can be computed via inner products between points as derived in Eq. (19). Note that the relation $\mathbf{M}^* = \mathbf{I} + \Upsilon\, \mathbf{S} \Upsilon^T$ has been used which relates the optimal solution $\mathbf{M}^*$ to $\mathbf{K}^*$ [4, 22–23, 26].

$$\psi(\mathbf{z}_1)^T \mathbf{M}^* \psi(\mathbf{z}_2) = \psi(\mathbf{z}_1)^T \left(\mathbf{I} + \Psi \mathbf{S} \Psi^T \right) \psi(\mathbf{z}_2)$$
$$= \psi(\mathbf{z}_1)^T \psi(\mathbf{z}_2) + \psi(\mathbf{z}_1)^T \Psi \mathbf{S} \Psi^T \psi(\mathbf{z}_2)$$
$$= \kappa_0(\mathbf{z}_1, \mathbf{z}_2) + \kappa_1^T \mathbf{S} \kappa_2 \tag{19}$$

where $\quad \Psi = [\psi(\mathbf{x}_1), \ldots, \psi(\mathbf{x}_n)]$

$\qquad \mathbf{S} = \mathbf{K}_0^{-1} \left(\mathbf{K}^* - \mathbf{K}_0 \right) \mathbf{K}_0^{-1} \quad$ having $\quad \mathbf{K}_0 = \Psi^T \Psi$

$\qquad \kappa_i = [\kappa_0(\mathbf{z}_i, \mathbf{x}_1), \ldots, \kappa_0(\mathbf{z}_i, \mathbf{x}_n)]^T \quad$ having $\quad \kappa_0(\mathbf{x}_i, \mathbf{x}_j) = \psi(\mathbf{x}_i)^T \psi(\mathbf{x}_j)$

3.4   PERSPECTIVE ON QUANTITATIVE STRUCTURE ACTIVITY RELATIONSHIPS (QSAR)

In the past decades, high-throughput screening (HTS) has offered large amounts of experimental data on biological activity of various chemical compounds. Even though these experimental screening methods have been very beneficial in drug discovery and chemical design, they are time-consuming and resource-intensive. On the other hand, data-mining and pattern-recognition algorithms, alongside advances in technology, have provided rapid, inexpensive, and efficient virtual screening techniques for selecting potent compounds for further experimental testing. These *in silico* techniques commonly guide decisions in environmental risk assessment of chemicals and structure-based drug design.

To improve these *in silico* methods, one needs to build robust statistical models for describing the relationship between a set of molecular features and their biological or physico-chemical properties. These quantitative structure activity relationship (QSAR) techniques search for patterns and signatures in the data sets, and provide insight about the chemical and/or biological characteristics of compounds. Because these types of mathematical models for predicting the target property are widely used prior to, or in lieu of, costly experimental measurements, finding reliable QSAR models is imperative. Such models should be statistically significant, robust, and generalizable with defined application boundaries [29, 30].

There are two main stages in conventional QSAR modeling: data preparation, and model generation. Data preparation includes collecting and cleaning a diverse data set of compounds with their corresponding acceptable target property values. Even though this sounds like a trivial task, it is extremely important to select accurate and consistent experimental data. In the next step, a set of numerical descriptors for each compound in the data set is calculated or compiled. These descriptors capture compositional, geometrical, electronic and steric features of the compounds under investigation. The information captured in the computed or measured descriptors (independent variables), is then related to the target property (dependent variable) of interest by generating a statistical model. The utility of a QSAR model is determined by its ability to accurately predict the target property for new chemicals that were not included in the training set. Therefore, after establishing the model, its predictive power and limitations need to be ascertained by assessing the statistical significance and robustness [31–33].

One goal of QSAR modeling is to interpret the model based on how descriptors mirror fundamental physico-chemical or mechanistic factors, which are presumably related to the target property. This favors models that use as few explanatory descriptors as possible. Some descriptors provide insight into the underlying molecular properties; e.g., the hydrophobicity, pK_a, and HOMO-LUMO gap. However, unfortunately, using a large set of descriptors, many of dubious chemical relevance, is a common practice in QSAR community; this has been facilitated by numerous (non-)commercial engines for computing descriptors. This prevents any simple interpretation of QSAR models, and necessitates incorporation of feature selection algorithms for picking the most significant descriptors in the process of model development.

Traditional QSAR approaches can be improved by using the state-of-the-art pattern recognition algorithms, a few of which were discussed in this chapter. The molecular descriptors compiled are commonly high-dimensional, which makes their analysis difficult, and obscure the proper measure of (dis)similarity. Reducing the dimensions of descriptor space, and/or systematically adjusting the (dis)similarity matrix to reflect our prior knowledge are key steps for developing task-specific descriptors. By transforming the raw data into a compact and task-specific representation through numerous metric learning algorithms, one selects a suitable feature space in a mathematically elegant manner. This potentially leads one toward a more relevant representation of chemical space, and enhances the performance of predictive (non)linear models, like multiple linear regression (MLR), partial least squares (PLS), artificial neural networks (ANNs), k-nearest neighbor (k-NN), support vector machines (SVM), kernel regression, and spectral learning, for chemical and biological properties [34–37].

3.4   CONCLUSION

Machine learning algorithms are commonly used in chemoinformatics for modeling the relationships between characteristic motifs of compounds and their physico-chemical properties. Many of these algorithms use a distance measure over the chemical space as a principle tool to compare molecules, and their performance critically depends on the quality of the distance metric used. This chapter presented the idea of learning a property-dependent distance measure; this is equivalent to adjusting the predictive model for our specific task. In doing so, various (un)supervised metric learning algorithms have been discussed to identify characteristic representations of input samples, that properly measures their distance for a special tasks, in a way that is consistent with prior knowledge.

For example, consider the problem of building a toxicity model using a dataset of aliphatic chemicals alongside their toxic potency. How should compounds be compared when they are being classified based on toxicity [38]? It is doubtful that a metric that was effective for classifying molecules based on carcinogenicity will also be effective for classifying them based on toxicity. Therefore, an accurate and context-sensitive assessment of the similarity and dissimilarity of the molecules in chemical space is crucial and dictates the success or failure of the predictive model (indeed, choosing the right assessment criteria can be more important that the specific machine-learning algorithm used). Conventional QSAR approaches try to select molecular features that best explain commonalities of compounds. This results in an expensive, but still not exhaustive, search in feature space, as well as possible loss of information that might be critical for proper modeling. For example, it could be that the key to correctly classifying toxic compounds is some nonlinear function of several features.

As distance measures are highly problem-specific, learning property-dependent features that respect our prior knowledge about the ideal distance metric seems to be the best approach. This justifies the utility of Mahalanobis distance metric learning algorithms for building accurate predictive models for various biological and chemical properties. It also motivates research into machine-learning methods that can learn from existing data and prior knowledge to build predictive and intelligent models for chemical properties. Such methods are particularly important for applications in human health and drug discovery, where poor models can have fatal consequences.

ACKNOWLEDGMENT

The author thanks Vanier Canada Graduate Scholarship by Natural Sciences and Engineering Research Council of Canada (NSERC), Compute Canada and Ghent University for funding. The author would like to sincerely thank Paul W. Ayers for fruitful discussions, continued guidance, and unsurpassed professional support. The author wishes to acknowledge the seminal contributions of Ramon Carbó-Dorca in the field of molecular similarity. The author first met Ramon Carbó-Dorca in autumn 2012 when attending a two-days meeting organized by the Ghent University Quantum Chemistry Group, and had the privilege of hearing his stimulating lecture on molecular similarity measures. Most of this work was inspired by his precedent endeavors, and extends trails that he started.

KEYWORDS

- (Un)supervised distance metric learning
- Bregman matrix divergences
- chemoinformatics
- Kernel-Based learning
- large margin nearest neighbor (LMNN) classifier

- **low-rank Kernel learning**
- **Mahalanobis distance metric learning**
- **maximum variance unfolding**
- **property-dependent distance measure**
- **quantitative structure activity relationships**

REFERENCES

1. Image borrowed from https://prateekvjoshi.com/2014/06/21/what-is-manifold-learning/.
2. Seung, H. S., & Lee, D. D., (2000). The manifold ways of perception. *Science, 290* (5500), 2268-2269.
3. Weinberger, K. Q., (2007). Metric Learning with Convex Optimization. Doctoral Dissertation at Computer Science Department of the University of Pennsylvania.
4. Kulis, B. J., (2008). Scalable Kernel Methods for Machine Learning. Doctoral Dissertation at Computer Science Department of The University of Texas at Austin.
5. Weinberger, K. Q., & Saul, L. S., (2006). An Introduction to Nonlinear Dimensionality Reduction by Maximum Variance Unfolding. In: *Proceedings of the 21st National Conference on Artificial Intelligence – Volume 2*, AAAI Press: Boston, Massachusetts, pp. 1683-1686.
6. von Luxburg, U., (2007). A tutorial on spectral clustering. *Statistics and Computing, 17*(4), 395–416.
7. Camastra, F., (2003). Data dimensionality estimation methods: a survey. *Pattern Recognition, 36*(12), 2945–2954.
8. Cao, L. J., Chua, K. S., Chong, W. K., Lee, H. P., & Gu, Q. M., (2003). A comparison of PCA, KPCA and ICA for dimensionality reduction in support vector machine. *Neurocomputing, 55*(1–2), 321–336.
9. Weinberger, K. Q., & Saul, L. K., (2006). Unsupervised learning of image manifolds by semidefinite programming. *International Journal of Computer Vision, 70*(1), 77–90.
10. Weinberger, K. Q., Sha, F., & Saul, L. K., (2010). Convex Optimizations for Distance Metric Learning and Pattern Classification. *IEEE Signal Processing Magazine, 27*(3), 146–158.
11. Roweis, S. T., & Saul, L. K., (2000). Nonlinear dimensionality reduction by locally linear embedding. *Science, 290*(5500), 2323–2326.
12. Hyvarinen, A., & Oja, E., (2000). Independent component analysis: algorithms and applications. *Neural Networks, 13* (4–5), 411–430.
13. Goldberger, J., Roweis, S., Hinton, & G., Salakhutdinov, R., (2004). Neighborhood Components Analysis. In *Proceedings of the 17th International Conference on Neural Information Processing Systems*, MIT Press: Vancouver, British Columbia, Canada, pp. 513-520.

14. Weinberger, K. Q., (2006). Unsupervised learning of image manifolds by semidefinite programming. *International Journal of Computer Vision, 70*(1), 77–90.
15. Kulis, B., (2013). Metric Learning: A Survey. *Foundations and Trends® in Machine Lerning, 5* (4), 287–364.
16. Xing, E. P., Ng, A. Y., Jordan, M. I., & Russell, S., (2002). Distance Metric Learning, with Application to Clustering with Side-Information. In *Proceedings of the 15th International Conference on Neural Information Processing Systems*, MIT Press, pp. 521–528.
17. Weinberger, K. Q., & Saul, L. K., (2009). Distance metric learning for large margin nearest neighbor classification. *Journal of Machine Learning Research, 10*, 207–244.
18. Image by Mlguy under CC-BY-SA-3.0 License, https://commons.wikimedia.org/w/index.php?curid = 10904310.
19. Müller, K. R., Mika, S., Rätsch, G., Tsuda, K., & Schölkopf, B., (2001). An introduction to kernel-based learning algorithms. *IEEE Transactions on Neural Networks, 12*(2), 181–201.
20. Schölkopf, B., (2000). The kernel trick for distances. In *Proceedings of the 13th International Conference on Neural Information Processing Systems*, MIT Press: Denver, CO, pp. 283–289.
21. Jäkel, F., Schölkopf, B., & Wichmann, F. A., (2007). A tutorial on kernel methods for categorization. *Journal of Mathematical Psychology, 51*(6), 343–358.
22. Weinberger, K. Q. & Tesauro, G., (2007). Metric Learning for Kernel Regression, In *Proceedings of the Eleventh International Conference on Artificial Intelligence and Statistics,* PMLR: Proceedings of Machine Learning Research; Vol. 2, pp. 612–619.
23. Kulis, B., Sustik, M. A., & Dhillon, I. S., (2009). Low-Rank Kernel Learning with Bregman Matrix Divergences. *Journal of Machine Learning Research, 10*, 341–376.
24. Weinberger, K. Q., & Saul, L. K. (2004). Learning a Kernel Matrix for Nonlinear Dimensionality Reduction, In *Proceedings of the Twenty-First International Conference on Machine Learning*, ACM: Banff, Alberta, Canada, pp. 106.
25. Kulis, B., Sustik, M., & Dhillon, I. (2006). Learning Low-Rank Kernel Matrices, In *Proceedings of the 23rd International Conference on Machine Learning*, ACM: Pittsburgh, Pennsylvania, USA, pp. 505–512.
26. Jain, P., Kulis, B., Davis, J. V., & Dhillon, I. S., (2012). Metric and Kernel Learning Using a Linear Transformation. *Journal of Machine Learning Research, 13*, 519–547.
27. Banerjee, A., Merugu, S., Dhillon, I. S., & Ghosh, J., (2005). Clustering with Bregman divergences. *Journal of Machine Learning Research, 6*, 1705–1749.
28. Davis, J. V., Kulis, B., Jain, P., Sra, S. & Dhillon, I. S., (2007). Information-Theoretic Metric Learning. In *Proceedings of the 24th International Conference on Machine Learning*, ACM: Corvalis, Oregon, USA, pp. 209–216.
29. Sushko, I., Novotarskyi, S., Körner, R., Pandey, A. K., Kovalishyn, V. V., Prokopenko, V. V., & Tetko, I. V., (2010). Applicability domain for in silico models to achieve accuracy of experimental measurements. *Journal of Chemometrics, 24* (3–4), 202–208.
30. Schneider, G., (2011). From hits to leads: challenges for the next phase of machine learning in medicinal chemistry. *Molecular Informatics, 30*(9), 759–763.
31. Tropsha, A., Gramatica, P., & Gombar, V. K., (2003). The importance of being earnest: Validation is the absolute essential for successful application and interpretation of QSPR models. *QSAR and Combinatorial Science, 22*(1), 69–77.

32. Schwaighofer, A., Schroeter, T., Mika, S., & Blanchard, G., (2009). How wrong can we get? A review of machine learning approaches and error bars. *Combinatorial Chemistry and High Throughput Screening, 12*(5), 453–468.

33. Golbraikh, A., & Tropsha, A., (2002). Beware of q(2)! *Journal of Molecular Graphics and Modelling, 20*(4), 269–276.

34. Hinselmann, G., Rosenbaum, L., Jahn, A., Fechner, N., Ostermann, C., & Zell, A., (2011). Large-Scale Learning of Structure-Activity Relationships Using a Linear Support Vector Machine and Problem-Specific Metrics. *Journal of Chemical Information and Modeling, 51*(2), 203–213.

35. Müller, K. R., Rätsch, G., Sonnenburg, S., Mika, S., Grimm, M., & Heinrich, N., (2005). Classifying 'drug-likeness' with kernel-based learning methods. *Journal of Chemical Information and Modeling, 45*(2), 249–253.

36. Heidar-Zadeh, F., & Ayers, P. W., Spectral Learning for Chemical Prediction (Chapter 4 of this book).

37. Heidar-Zadeh, F., Ayers, P. W., & Carbó-Dorca, R., (2017). A statistical perspective on molecular similarity. In *Conceptual Density Functional Theory* (in press).

38. Zadeh, F. H., & Ayers, P. W., (2013). Molecular alignment as a penalized permutation Procrustes problem. *J. Math. Chem., 51*(3), 927–936.

APPENDIX I: KERNEL *k*-MEANS CLUSTERING ALGORITHM [3, 4]

The *k*-means clustering algorithm partitions the n data points, $\{\mathbf{x}_i\}_{i=1}^n$, into k disjoint clusters with linear decision boundaries, $\{\pi_c\}_{c=1}^k$. Considering each data sample belonging to the cluster with nearest mean, the objective function for the *k*-means algorithm is:

$$\min_{\{\pi_c\}_{c=1}^k} \sum_{c=1}^k \sum_{\mathbf{x}_i \in \pi_c} \|\mathbf{x}_i - \mathbf{m}_c\|^2 \quad \text{where} \quad \mathbf{m}_c = \frac{\sum_{\mathbf{x}_i \in \pi_c} \mathbf{x}_i}{|\pi_c|} \tag{20}$$

The vector $\mathbf{m}_c$ represents the center of the cluster. Here, the sum of squared Euclidean distance is used to assign sample points to clusters. This algorithm can be generalized to partition the data along nonlinear decision boundaries using kernel methods. Mapping the input points, $\{\mathbf{x}_i\}_{i=1}^n$, into the feature space $\{\psi(\mathbf{x}_i)\}_{i=1}^n$, the generalized *k*-means algorithm is:

$$\min_{\{\pi_c\}_{c=1}^k} \sum_{c=1}^k \sum_{\mathbf{x}_i \in \pi_c} \|\psi(\mathbf{x}_i) - \mathbf{m}_c\|^2 \quad \text{where} \quad \mathbf{m}_c = \frac{\sum_{\mathbf{x}_i \in \pi_c} \psi(\mathbf{x}_i)}{|\pi_c|} \tag{21}$$

Using the kernel function κ, the Euclidean distance is expressed in terms of inner products.

$$\|\psi(\mathbf{x}_i) - \mathbf{m}_c\|^2 = \psi(\mathbf{x}_i) \cdot \psi(\mathbf{x}_i) - \frac{2\sum_{\mathbf{x}_j \in \pi_c} \psi(\mathbf{x}_i) \cdot \psi(\mathbf{x}_j)}{|\pi_c|} + \frac{\sum_{\mathbf{x}_j, \mathbf{x}_l \in \pi_c} \psi(\mathbf{x}_j) \cdot \psi(\mathbf{x}_l)}{|\pi_c|^2}$$

$$\|\psi(\mathbf{x}_i) - \mathbf{m}_c\|^2 = \kappa(\mathbf{x}_i, \mathbf{x}_i) - \frac{2\sum_{\mathbf{x}_j \in \pi_c} \kappa(\mathbf{x}_i, \mathbf{x}_j)}{|\pi_c|} + \frac{\sum_{\mathbf{x}_j, \mathbf{x}_l \in \pi_c} \kappa(\mathbf{x}_j, \mathbf{x}_l)}{|\pi_c|^2} \tag{22}$$

The only difference between the kernel *k*-means clustering algorithm and the traditional *k*-means clustering, Eqs. (21) and (20), is that the Euclidean distance computation is replaces by Eq. (22). Even though the kernel *k*-means clustering has linear decision boundaries in feature space, the corresponding boundaries in input space are non-linear.

APPENDIX II: BREGMAN MATRIX DIVERGENCES [4, 15, 23, 25–26, 28]

Bregman matrix divergences, measuring divergence between matrices, are generalization of Bregman vector divergence measures. Having a real-valued strictly convex function ϕ over a convex set $S = \text{dom}(\phi) \subseteq R^m$ such that ϕ is differentiable in the interior of its domain, the *Bregman vector divergence* with respect to ϕ is:

$$D_\varphi(\mathbf{x}, \mathbf{y}) = \varphi(\mathbf{x}) - \varphi(\mathbf{y}) - (\mathbf{x} - \mathbf{y})^T \nabla \varphi(\mathbf{y}) \tag{23}$$

Examples of Bregman vector divergence include:

$$\varphi(\mathbf{x}) = \mathbf{x}^T \mathbf{x} \qquad\qquad D_\varphi(\mathbf{x}, \mathbf{y}) = \|\mathbf{x} - \mathbf{y}\|_2^2$$

$$\varphi(\mathbf{x}) = \sum_i (x_i \log x_i - x_i) \qquad D_\varphi(\mathbf{x}, \mathbf{y}) = KL(\mathbf{x}, \mathbf{y}) = \sum_i \left(x_i \log \tfrac{x_i}{y_i} - x_i + y_i \right) \tag{24}$$

Now, given a strictly convex and differentiable function over real, symmetric, square matrices; $\varphi : S^n \to R$ the *Bregman matrix divergence* between two matrices is defined as:

$$D_\phi(\mathbf{X}, \mathbf{Y}) = \phi(\mathbf{X}) - \phi(\mathbf{Y}) - \text{tr}\left((\nabla\phi(\mathbf{Y}))^T (\mathbf{X} - \mathbf{Y}) \right) \tag{25}$$

Presented below are three examples of matrix divergences called the squared Frobenius norm, the von Neumann divergence, and the LogDet divergence, respectively.

$$\phi(\mathbf{X}) = \|\mathbf{X}\|_F^2 \qquad\qquad D_F(\mathbf{X}, \mathbf{Y}) = \|\mathbf{X} - \mathbf{Y}\|_F^2$$

$$\phi(\mathbf{X}) = \sum_i (\lambda_i \log \lambda_i - \lambda_i) = \text{tr}(\mathbf{X} \log \mathbf{X} - \mathbf{X}) \qquad D_{vN}(\mathbf{X}, \mathbf{Y}) = \text{tr}(\mathbf{X} \log \mathbf{X} - \mathbf{X} \log \mathbf{Y} + \mathbf{X} - \mathbf{Y})$$

$$\phi(\mathbf{X}) = -\sum_i \log \lambda_i = -\log|\mathbf{X}| \qquad D_{\ell d}(\mathbf{X}, \mathbf{Y}) = \text{tr}(\mathbf{X}\mathbf{Y}^{-1}) - \log|\mathbf{X}\mathbf{Y}^{-1}| - n \tag{26}$$

It is important to note that for all of these three divergences, the generating function can be represented as a composition $\varphi(\mathbf{X}) = (\phi \circ \lambda)(\mathbf{X})$, where function $\lambda(\mathbf{X})$ lists eigenvalues of $\mathbf{X}$ in decreasing order, and ϕ is the function defined over vectors when introducing Bregman vector divergences. In other words, every strictly convex function ϕ results in a Bregman matrix divergence over real, symmetric matrices. These divergence measures are called *spectral Bregman matrix divergences* for which $D_\varphi(\mathbf{X}, \mathbf{Y})$ can be expressed in terms of eigenvectors and eigenvalues of matrices $\mathbf{X}$ and $\mathbf{Y}$.

Lemma: Consider the eigendecomposition of $\mathbf{X} = \mathbf{V}\Lambda\mathbf{V}^T$ and $\mathbf{Y} = \mathbf{U}\Theta\mathbf{U}^T$, and assume ϕ is separable, that is, $\varphi(\mathbf{X}) = (\phi \circ \lambda)(\mathbf{X}) = \sum_i \phi_i(\lambda_i)$. Then

$$D_\phi(\mathbf{X},\mathbf{Y}) = \sum_{i,j}\left(\mathbf{v}_i^T\mathbf{u}_j\right)^2\left(\varphi_i(\lambda_i) - \varphi_j(\theta_j) - (\lambda_i - \theta_j)\nabla\varphi_j(\theta_j)\right) \qquad (27)$$

Corollary: Given $\mathbf{X} = \mathbf{V}\Lambda\mathbf{V}^T$ and $\mathbf{Y} = \mathbf{U}\Theta\mathbf{U}^T$, the von Neumann and LogDet divergences satisfy:

$$D_\phi(\mathbf{X},\mathbf{Y}) = \sum_{i,j}\left(\mathbf{v}_i^T\mathbf{u}_j\right)^2 D_\varphi(\lambda_i,\theta_j) \qquad (28)$$

The computational benefits as well as other important properties of the LogDet and von Neumann divergences justify their use for kernel learning. These include:

1. They are defined over positive definite matrices, so there is no need for an explicit constraint of the learned kernel matrix in the algorithm. Also, they have a range-space preserving property; as a result they can be generalized to the case of rank-deficient matrices.

2. The LogDet divergence is transformation-invariant. A corollary of this proposition would be the scale-invariance of the LogDet divergence; $D_{ld}(c\mathbf{X},c\mathbf{Y}) = D_{ld}(\mathbf{X},\mathbf{Y})$. This means that we can scale each feature in our data vector, and the learned kernel will be scaled by a diagonal transformation.

   - **Proposition 1:** Let $\mathbf{Q}$ be a square orthogonal matrix. Then, $D_\varphi(\mathbf{Q}^T\mathbf{X}\mathbf{Q}, \mathbf{Q}^T\mathbf{Y}\mathbf{Q}) = D_\varphi(\mathbf{X},\mathbf{Y})$ for all spectral Bregman matrix divergences.
   - **Proposition 2:** Let $\mathbf{M}$ be a square, non-singular matrix, and $\mathbf{X}$ and $\mathbf{Y}$ be square positive definite matrices. Then, for LogDet divergence, $D_{ld}(\mathbf{M}^T\mathbf{X}\mathbf{M},\mathbf{M}^T\mathbf{Y}\mathbf{M}) = D_{ld}(\mathbf{X},\mathbf{Y})$.

3. The generalization to out-of-sample points, i.e., to new data points, is feasible and does not require re-learning the kernel matrix.

The LogDet and von Neumann divergences are extendable to the case of positive semidefinite matrices; that is, low-rank matrices. The effective domain of the convex function generating the LogDet divergence is the set of positive definite matrices; i.e., $\varphi(\mathbf{X}) = -\log|\mathbf{X}|$ is infinite when $\mathbf{X}$ is singular. In the case of von Neumann divergence, it is possible to define $\varphi(\mathbf{X}) = \text{tr}(\mathbf{X}\log\mathbf{X}-\mathbf{X})$ for rank-deficient matrices via continuity. Briefly, in order to use these divergence measures for low-rank matrices, the convex function $\varphi(\mathbf{X})$ should be restricted to the range space of matrix φ. Here, we would motivate this using corollary 2; however, this observation can be formalized by showing that the low-rank problem can be mapped into a full-rank problem in a lower-dimensional space.

When $\mathbf{X}$ and $\mathbf{Y}$ are rank-deficient, some of their eigenvalues are zero, so if we could apply corollary 2, the $D_\phi(\lambda_i,\theta_j)$ terms involving zero eigenvalues can be infinite. In order to have finite matrix divergences, we require that $v_i^T u_j = 0$ whenever $D_\phi(\lambda_i,\theta_j)$ is infinite. So, in order to have a finite matrix divergence, the rank-deficient matrices should have the following properties.

- **Observation 1:** The von Neumann divergence $D_{vN}(\mathbf{X},\mathbf{Y})$ is finite if and only if range($\mathbf{X}$) $\subseteq$ range($\mathbf{Y}$)
- **Observation 2:** The LogDet divergence $D_{ld}(\mathbf{X},\mathbf{Y})$ is finite if and only if range($\mathbf{X}$) = range($\mathbf{Y}$).

Now, by assuming that the eigenvalues are listed in non-decreasing order, the low-rank equivalent of Eq. (28) simply becomes:

$$D_\phi(\mathbf{X},\mathbf{Y}) = \sum_{i,j \leq r} \left(v_i^T u_j\right)^2 D_\varphi(\lambda_i,\theta_j) \qquad (29)$$

CHAPTER 4

SPECTRAL LEARNING FOR CHEMICAL PREDICTION

FARNAZ HEIDAR-ZADEH[1-3] and PAUL W. AYERS[1]

[1]Department of Chemistry and Chemical Biology, McMaster University, Hamilton, Ontario, Canada

[2]Department of Inorganic and Physical Chemistry, Ghent University, Krijgslaan 281 (S3), 9000 Gent, Belgium

[3]Center for Molecular Modeling, Ghent University, Technologiepark 903, 9052 Zwijnaarde, Belgium

CONTENTS

4.1   MOTIVATION

Most chemical characterization involves the measurement of one or more spectrum, which are then interpreted to identify, characterize, and/or quantify a substance and its chemical transformations [1, 2]. Classic spectroscopic techniques are based on the way that molecules interact with light including radio waves (NMR), microwaves, infrared radiation, visual and ultraviolent light, and X-rays. However, there are also spectrometric methods based on neutron scattering. However, many different instrumental analysis methods provide spectrum-like output, for example, elution curves from chromatographic and electrophoretic methods, mass spectra (from mass spectrometry or other methods), cyclic voltammograms, differential scanning calorimetry, and various types of titration curves. More generally, any measured response to a control parameter (like wavelength, voltage, acidity, temperature, pressure, etc.) or to the time (transit time, reaction time, decay time, etc.) can be viewed as a type of "spectrum."

Skilled experimental chemists learn how to interpret spectra, and how to use key spectral features to elucidate molecular structure and chemical reactivity. Given the enormous amount of information that is contained in these measured spectra, can we use them to make *quantitative* predictions of (bio)chemical properties? For example, suppose we were given a group of compounds; could we use their infrared spectra and/or their mass spectra to predict their pK_a (acidity)? What about nonmolecular properties like toxicity? Insofar as one could probably infer the identity of the molecules from these spectra, this should be possible in theory. But is it possible in practice? The goal of this chapter is to develop a method for making property predictions from spectra. Our approach is based on the framework of molecular similarity [3–25] but, unlike traditional approaches, our approach is based on the similarity between molecular spectra.

4.2　BACKGROUND

For our purposes, a spectrum is any function, $R(\omega)$, where R is a response and ω is a control parameter. We further require that the spectrum be integrable,

$$-\infty < \int_{-\infty}^{\infty} |R(\omega)| d\omega < \infty \tag{1}$$

In spectroscopies based on light, R is signal (e.g., absorption or emission) and ω is frequency. For a chromatography experiment, ω is elution time. For a titration, ω is pH. (However, for a titration, one needs to differentiate the signal with respect to pH so that the "spectrum" is integrable, as required by our definition.)

In multidimensional spectroscopy, one combines several different spectra; typically these are spectra of the same basic type, measured at different times or under different conditions [26–28]. We extend the framework of multidimensional spectroscopy to address the case where multiple, possibly unrelated, spectra are available. Given T different types of spectra for M different molecules, $\left\{ R_{m;t}^{\mathrm{raw}}\left(\omega_t\right)\right\}_{m=1;t=1}^{M;T}$, it is convenient to "center" these spectra by subtracting a reference spectrum,

$$R_{m;t}\left(\omega_t\right) = R_{m;t}^{\mathrm{raw}}\left(\omega_t\right) - R_t^{\mathrm{ref}}\left(\omega_t\right) \tag{2}$$

In order to make firm statistical contact with statistical covariance analysis, it convenient to choose the reference spectrum as the average spectrum of the molecules in the data,

$$R_t^{\mathrm{ref}}\left(\omega_t\right) = R_t^{\mathrm{avg}}\left(\omega_t\right) = \frac{1}{M}\sum_{m=1}^{M} R_{m;t}^{\mathrm{raw}}\left(\omega_t\right) \tag{3}$$

In what follows, it is important that all the spectra have similar spectral range. That is, the highest-ω and lowest-ω features in each spectra should occur at comparable values. This can be achieved by simply scaling and shifting the spectra.

For each molecule, we then define the raw multidimensional spectrum as the product of the individual spectra,

$$S_m^{\mathrm{raw}}\left(\omega_1,\omega_2,\ldots,\omega_T\right) = R_{m;1}\left(\omega_1\right)R_{m;2}\left(\omega_2\right)\cdots R_{m;T}\left(\omega_T\right) \tag{4}$$

Again, it is useful to subtract a reference spectrum,

$$S_m\left(\omega\right) = S_m^{\mathrm{raw}}\left(\omega\right) - S^{\mathrm{ref}}\left(\omega\right) \tag{5}$$

and the average spectrum is a sensible choice for the reference,

$$S^{\mathrm{ref}}\left(\omega\right) = S^{\mathrm{avg}}\left(\omega\right) = \frac{1}{M}\sum_{m=1}^{M} S_m^{\mathrm{raw}}\left(\omega\right) \tag{6}$$

In Eqs. (5) and (6) we introduce the vector notation $\omega = (\omega_1,\omega_2,\ldots,\omega_T)$. These centered multidimensional spectra encapsulate a wealth of information. For simplicity, we will discuss the case where $R_{m;t}^{\text{raw}}(\omega_t) \geq 0$, that is, all the molecular responses are associated with positive features in the spectrum. (Spectra with both positive and negative features can be analyzed similarly.) The centered spectra $R_{m;t}(\omega_t)$ is positive where the value of the tth response of the mth molecule at ω_t is greater than it is for the average molecule (corresponding to strong/dominant spectral features of the mth molecule) and negative otherwise (corresponding to weak or missing spectral features of the mth molecule). The product spectrum, $S_m^{\text{raw}}(\omega)$, then captures correlations between features. For example, $S_m^{\text{raw}}(\omega_1, \omega_2)$, will be positive if the presence (absence) of a feature in the first spectrum at frequency ω_1 is associated with the presence (absence) of a feature in the second spectrum at frequency ω_2. These two spectral features are therefore positively correlated. $S_m^{\text{raw}}(\omega_1, \omega_2)$ will be negative if the presence (absence) of a feature in the first spectrum at frequency ω_1 is associated with the absence (presence) of a feature in the second spectrum at frequency ω_2. These two spectral features are therefore negatively correlated. The final, centered, multidimensional spectrum, $S_m(\omega)$, then reveals whether the (anti)correlations between molecule m's different types of spectra are larger, or smaller, than is typical for the reference set of molecules. That is, all the features in $S_m(\omega)$ correspond to deviations of the molecule m from the average behavior of the molecules in the data set.

The first step in our strategy for learning molecular properties from spectral input is to quantify the similarity between two molecules based on their spectra. At the simplest level, this can be achieved by computing their Gramian,

$$G_{mn}^{(0)} = \int S_m(\omega) S_n(\omega) d\omega \tag{7}$$

As shown in Sections 4.3 and 4.4, we can refine this zeroth-order approximation to the Gramian based on the properties we aim to predict. Given a suitable definition of the Gramian, G_{mn}, we can predict the properties of other molecules from their spectra. That is, if we know the properties of molecules within a training data set, $\{p_m\}_{m=1}^{M}$, then we can predict the property of a target molecule, p_{O}, by using its spectrum to compute the Gramian elements, $\{G_{m\text{O}}\}_{m=1}^{M}$, and then solving the following system of linear equations,

$$
\begin{bmatrix}
G_{11} & G_{12} & \cdots & G_{1M} & 1 \\
G_{21} & G_{22} & \cdots & G_{2M} & 1 \\
\vdots & \vdots & \ddots & \vdots & \vdots \\
G_{M1} & G_{M2} & \cdots & G_{MM} & 1 \\
1 & 1 & \cdots & 1 & 0
\end{bmatrix}
\begin{bmatrix}
w_1 \\ w_2 \\ \vdots \\ w_M \\ \mu
\end{bmatrix}
=
\begin{bmatrix}
G_{1\odot} \\ G_{2\odot} \\ \vdots \\ G_{M\odot} \\ 1
\end{bmatrix}
\tag{8}
$$

This lets one approximate the target molecule's property value as a weighted average of the properties of the molecules in the training data set,

$$
P_\odot = \sum_{m=1}^{M} w_m P_m
\tag{9}
$$

This approach to property prediction is called kriging, or Gaussian processes [29–32]. It can be derived by considering G_{mn} as a measure of the (spectral) similarity between the molecules m and n [33]. Kriging can also be modeled as a neural network with one hidden layer [34–39].

4.3 SIMILARITY MEASURES FOR MOLECULAR SPECTRA

The simple Gramian in Eq. (7) measures the similarity of the multidimensional spectra of molecules m and n. As discussed in the Appendix A, this matrix is proportional to a covariance matrix if one centers the spectra by choosing the reference spectrum to be the average spectrum. This is one motivation for centering the spectrum. Because the Gramian is proportional to a covariance matrix, the similarity-based kriging procedure in Eqs. (8) and (9) is justified.

The zeroth-order Gramian, however, implicitly assumes that all the spectra are equally useful for distinguishing between the molecules and that there is no correlation between the spectra. For example, in the extreme case where one spectrum is repeated multiple times in the product in Eq. (4), that spectrum will be overweighted in the spectral similarity measures based on the zeroth-order Gramian, Eq. (7). This problem is mitigated by using a more sophisticated Gramian based on a positive semidefinite integral kernel, $K(\omega, \omega')$,

$$
G_{mn}^{(K)} = \iint S_m(\omega) K(\omega, \omega') S_n(\omega') \, d\omega \, d\omega'
\tag{10}
$$

The zeroth-order Gramian corresponds to the delta-function kernel,

$$K(\omega,\omega') = \prod_{t=1}^{T} \delta(\omega_t - \omega_t')$$ (11)

The kernel-based Gramian in Eq. (10) is extremely general, and the rest of this paper focuses on developing suitable models for the kernel, $K(\omega,\omega')$. In the remainder of this section, we will focus on models based on the Gaussian kernel,

$$K(\omega,\omega') = \left(\frac{2|\mathbf{P}|}{(2\pi)^T}\right)^{\frac{1}{2}} \exp\left[-(\omega-\omega')^\dagger \mathbf{P}(\omega-\omega')\right]$$ (12)

Here $\mathbf{P}$ is a positive definite matrix. The normalization factor for the multivariate Gaussian kernel depends on the number of spectra, T. $\mathbf{v}^\dagger$ denotes the vector transpose.

To make a suitable choice for the Gaussian kernel, suppose we are given a validation set of molecules we can use to assess the quality of our kriging model, Eqs. (8)–(9). (If no validation set is available, we can use cross-validation on our training set of molecules.) Given the property values of the validation molecules, $\{p_l^{\text{actual}}\}_{l=1}^{N_{\text{Validation}}}$, we can minimize the prediction error as a function of the parameters in the Gaussian kernel, Eq. (12). A convenient way to do this is to write $\mathbf{P}$ in terms of its Cholesky decomposition,

$$\mathbf{P} = \mathbf{LDL}^T$$ (13)

where $\mathbf{L}$ is lower unitriangular,

$$l_{st} = \begin{cases} 0 & s < t \\ 1 & s = t \\ \text{arbitrary} & s > t \end{cases}$$ (14)

and $\mathbf{D}$ is diagonal and positive definite,

$$d_{st} = \alpha_t^2 \delta_{st}$$ (15)

One can then use the Gramian from Eq. (10) with the kernel from Eqs. (12)–(15) to predict the properties of the molecules in the validation set using the kriging equations, Eqs. (8)–(9). Minimizing the prediction errors with respect to the parameters in Eqs. (14) and (15),

$$\min_{\{\alpha_t ; l_{s<t}\}_{s,t=1}^{T}} \sum_{\ell=1}^{N_{\text{validation}}} \left(p_\ell^{\text{predicted}} - p_\ell^{\text{actual}} \right)^2 \tag{16}$$

allows one to determine the optimal choice of $\mathbf{P}$. While one can evaluate the gradient of the objective function in Eq. (16) analytically, the resulting expressions are mathematically complicated and computationally expensive to evaluate. It is probably more practical to use a gradient-free optimization technique like the covariance matrix adaptation evolution strategy, which has the additional advantage of finding (in principle) the global minimum [40]. Note that it is easy to extend this approach to a multiobjective optimization, where one wishes to find a Gramian, $G_{mn}^{(K)}$, that is suitable for optimizing several properties simultaneously. For example, given a vector of properties that one wishes to predict, one vectorizes the objective function in Eq. (16) to

$$\min_{\{\alpha_t ; l_{s<t}\}_{s,t=1}^{T}} \sum_{\ell=1}^{N_{\text{validation}}} \left| \mathbf{p}_\ell^{\text{predicted}} - \mathbf{p}_\ell^{\text{actual}} \right|^2 \tag{17}$$

If there is not enough data to fully optimize the kernel, one could choose to optimize only the variables corresponding to the diagonal elements, $\{\alpha_t\}_{t=1}^{T}$, i.e., set $\mathbf{L} = \mathbf{I}$, thereby removing $l_{s<t}$ from the variables to be optimized.

When one has a large number of spectra, it will not be practical to optimize the kernel directly, as in Eq. (16) (i.e., one cannot determine the parameters in the Gaussian kernel $\mathbf{P}$ unless $\frac{1}{2}T(T+1) \ll N_{\text{validation}}$. This means that one needs to estimate the kernel by another method. This is useful in any event, as the optimization in Eq. (16) is quite expensive, and is likely to have very many local minima, so having a good initial guess is critically important. It is also useful to have a nontrivial estimate of the kernel because for spectra with extremely high resolution (i.e., when the line shape approaches the delta-function limit), the overlap between molecular spectra may be negligible. This is likely to be particularly problematic for techniques, like mass spectrometry and chromatographic methods, where the "length" of the spectrum, $R_{m;t}(\omega_t)$, is vastly larger than the average peak width. In these cases, the zeroth-order Gramian is almost useless unless we broaden the spectra artificially using a simple diagonal Gaussian kernel, where

$$p_{st} = \alpha_t^2 \delta_{st} \tag{18}$$

To eliminate the problem where the feature width is vastly smaller than the spectral range, it is reasonable to choose α_t^2 to be inversely proportional to the square of the spectral range, where the spectral range is defined as the difference between the highest-ω feature and the lowest-ω features in the spectra of the molecules under consideration. For example, one might choose $\alpha_t^2 \sim N_{\text{lines}}(\omega_{t;\max} - \omega_{t;\min})^{-2}$, where N_{lines} is some measure of the number of lines (or features) that one typically finds in the tth spectrum (Obviously better estimates can be made when more information about the input spectra were available; this is merely a very simple model). One could use the diagonal kernel as an initial guess for the $\mathbf{P}$ matrix in Eq. (12). However, a better choice is to use the sample covariance matrix,

$$\Sigma_{st} = \frac{1}{M-1} \sum_{m=1}^{M} \int R_{m;s}(\omega) R_{m;t}(\omega) d\omega \tag{19}$$

to make the estimate

$$\mathbf{P} = \tfrac{1}{2} \Sigma^{-1} \tag{20}$$

If the spectra are too narrow for Eq. (19) to be meaningful, it would be reasonable to consider a Gaussian broadefining instead of the delta-function broadefining implicit in Eq. (19). For example, one could use,

$$\Sigma_{st} = \frac{1}{M-1} \sum_{m=1}^{M} \iint R_{m;s}(\omega_s) \left(\sqrt{\alpha_s \alpha_t} \exp\left[-\alpha_s \alpha_t (\omega_s - \omega_t)^2 \right] \right) R_{m;t}(\omega_t) d\omega_s d\omega_t \tag{21}$$

The sample covariance from Eq. (19) (or Eq. (21)) is positive semidefinite by construction. If it has any (nearly) zero eigenvalues, one should first eliminate the near linear dependencies in the spectra by dimensional reduction before proceeding further.

4.4 GRAMIAN LEARNING

4.4.1 SEMIDEFINITE PROGRAMMING APPROACH

In the previous section, we constructed a Gramian directly from the input spectra. Any practical optimization for the kernel in Eq. (10) can explore only an infinitesimal subspace of the possibilities. As discussed in the Chapter 3 of this book by Heidar-Zadeh [41], we can improve the Gramian, however, if we have (from some external source) information that certain molecules are (dis)similar. Then, owing to the link between the Gramian and the squared-distance between the two molecules,

$$d^2_{mn} = G_{mm} + G_{nn} - 2G_{mn} \qquad (22)$$

we can try to find a Gramian that satisfies the constraints,

$$
\begin{aligned}
G_{mm} + G_{nn} - 2G_{mn} \geq \lambda & \qquad m \text{ and } n \text{ are dissimilar molecules} \\
G_{mm} + G_{nn} - 2G_{mn} \leq \lambda & \qquad m \text{ and } n \text{ are similar molecules}
\end{aligned}
\qquad (23)
$$

The molecules that are similar in one context can be clearly dissimilar in another. For example, the refractive indices of water and methanol are very similar (differing by just 0.01%), but their freezing point and their toxicity are very dissimilar. It is clear, then, that we should define our Gramian in a problem-specific way.

While dissimilar molecules may have similar values of the property(ies), molecules with dissimilar property value(s) are certainly dissimilar for our purposes. Therefore, we can define

$$\left| \mathbf{p}_m - \mathbf{p}_n \right|^2 > \tau \rightarrow m \text{ and } n \text{ are dissimilar} \qquad (24)$$

We would like to use the information we have about the (dis)similarity between molecules to refine the Gramian we derived based on spectral data. One approach is to find the Gramian that is as close as possible to the original Gramian in an information-theoretic sense, but which satisfies the constraints in Eq. (23) [42–44]. This kernel learning problem gives a semi-definite optimization problem [43],

$$\min_{\left\{ G \succeq 0 \left| \substack{G_{mm}+G_{nn}-2G_{mn} \leq \eta \ \text{when} \ (m,n) \ \text{are similar} \\ G_{mm}+G_{nn}-2G_{mn} \geq \lambda \ \text{when} \ (m,n) \ \text{are dissimilar}} \right. \right\}} \left(\mathrm{Tr}\left[\mathbf{G}\left(\mathbf{G}^{(K)}\right)^{-1}\right] - \ln\left| \mathbf{G}\left(\mathbf{G}^{(K)}\right)^{-1}\right| - \mathrm{rank}\left(\mathbf{G}^{(K)}\right)\right) \tag{25}$$

Here $\mathbf{G}^{(K)}$ is the kernel Gramian constructed in Section 4.3, cf. Eq. (10). In the specific case of interest to us, this becomes

$$\min_{\left\{ G \succeq 0 \left| G_{mm}+G_{nn}-2G_{mn} \geq \lambda \ \text{when} \ |\mathbf{p}_m-\mathbf{p}_n|^2 > \tau \right. \right\}} \left(\mathrm{Tr}\left[\mathbf{G}\left(\mathbf{G}^{(K)}\right)^{-1}\right] - \ln\left| \mathbf{G}\left(\mathbf{G}^{(K)}\right)^{-1}\right| - \mathrm{rank}\left(\mathbf{G}^{(K)}\right)\right) \tag{26}$$

The model parameters λ (and even τ) can be optimized using the training or validation data, using a similar strategy to Eqs. (16) and (17).

There are several notable features of this procedure. First, if one reduces the rank of $\mathbf{G}^{(K)}$ by performing an eigenvector decomposition and then zeroing its lowest eigenvalues, the Gramian one learns in Eq. (26) will have the same reduced rank. This strategy is therefore useful for dimensionality reduction. Second, while the procedure in Eq. (26) does not distinguish between "very dissimilar" and "slightly dissimilar" molecules, this feature can be (partially) built in by first applying Eq. (26) with a very large value of τ, then iterating the procedure for gradually decreasing values of τ. Alternatively, one can employ a single-step procedure,

$$\min_{\left\{ G \succeq 0 \left| G_{mm}+G_{nn}-2G_{mn} \geq \lambda \cdot \max\left(|\mathbf{p}_m-\mathbf{p}_n|^2-\tau,0\right) \right. \right\}} \left(\mathrm{Tr}\left[\mathbf{G}\left(\mathbf{G}^{(K)}\right)^{-1}\right] - \ln\left| \mathbf{G}\left(\mathbf{G}^{(K)}\right)^{-1}\right| - \mathrm{rank}\left(\mathbf{G}^{(K)}\right)\right) \tag{27}$$

Finally, note that Eq. (26) does not affect the distance between molecules with similar properties: those molecules are unconstrained in this procedure, and can be either dissimilar or similar, depending on their spectral similarity and their (dis)similarity to other molecules.

4.4.2 PROJECTION APPROACH

The disadvantage of approaches based on Eqs. (25)–(27) is that semidefinite programming is computationally demanding, with many of the best methods scaling as $O(M^6)$. In this section we present a less elegant, but more pragmatic, approach inspired by methods for finding the closest positive semidefinite reduced density matrix in quantum chemistry [45]. The key theorem is that given an $M \times M$ matrix, $\tilde{\mathbf{V}}$, the closest positive semidefinite symmetric matrix with rank less than or equal to r is obtained by:

1. Symmetrize $\mathbf{V}$, $\tilde{\mathbf{V}} = \frac{1}{2}(\mathbf{V} + \mathbf{V}^\dagger)$.
2. Diagonalize $\tilde{\mathbf{V}}$, obtaining a list of eigenvalues $\varpi_1 \geq \varpi_2 \geq \cdots \geq \varpi_M$ and the corresponding eigenvectors, $\{\mathbf{v}_m\}_{m=1}^M$.
3. Construct the closest rank r positive semidefinite approximation to $\mathbf{V}$ as

$$\mathbf{V}^{(r)} = \sum_{m=1}^{r} \max(\varpi_m, 0) \mathbf{v}_m \mathbf{v}_m^\dagger \tag{28}$$

This allows us to develop analogues of the procedures in Eqs. (25)–(27) that require finding only a subset of the eigenvalues and eigenvectors of the Gramian, with a maximum cost of $O(M^3)$; however, iterative diagonalization methods will be much faster if $\mathbf{V}$ is sparse or if $r = M$.

To derive an analogue to Eq. (25), suppose we have a list of (dis)similar molecules, as in Eq. (23), then:

a.   Initialize the Gramian to the kernel-based Gramian, $\tilde{\mathbf{G}}^{(0)} = \mathbf{G}^{(K)}$. Find the closest rank r approximation to this matrix using Step 3 above (Eq. (28)).

b.   For every molecule pair in the (dis)similarity lists, update the Gramian according to

$$G_{mn}^{(i+1)} = \begin{cases} \min\left(\tilde{G}_{mn}^{(i)}, \frac{1}{2}\left(\tilde{G}_{mm}^{(i)} + \tilde{G}_{nn}^{(i)} - \lambda\right)\right) & m \text{ and } n \text{ are dissimilar} \\ \max\left(\tilde{G}_{mn}^{(i)}, \frac{1}{2}\left(\tilde{G}_{mm}^{(i)} + \tilde{G}_{nn}^{(i)} - \eta\right)\right) & m \text{ and } n \text{ are similar} \\ \tilde{G}_{mn}^{(i)} & \text{otherwise} \end{cases} \tag{29}$$

If this does not cause any update to the matrix (i.e., $G_{mn}^{(i+1)} - \tilde{G}_{mn}^{(i)} = 0$), then the constraints are satisfied and the Gramian learning procedure is finished.

c.   The new Gramian is symmetric, but it may not be positive semidefinite and may have a rank that is larger than the target rank, r. Diagonalize the Gramian and find the closest positive semidefinite rank r approximation to it using Eq. (28), obtaining $\tilde{G}_{mn}^{(i+1)}$. This revision may cause the (dis)similarity constraints to be violated, so one needs to go back and repeat step a. Alternatively, it is possible that the (dis)similarity constraints are incompatible with the rank restriction requirement and the positive semidefinite constraints. In this case, one should exit the procedure when $\|\tilde{G}_{mn}^{(i+1)} - \tilde{G}_{mn}^{(i)}\|$ is sufficiently small.

The analogues of Eqs. (26) and (27) correspond to alternative definitions of Eq. (29). Specifically, in step a one uses (analogous to Eq. (26))

$$G_{mn}^{(i+1)} = \begin{cases} \min\left(\tilde{G}_{mn}^{(i)}, \frac{1}{2}\left(\tilde{G}_{mm}^{(i)} + \tilde{G}_{nn}^{(i)} - \lambda\right)\right) & |\mathbf{p}_m - \mathbf{p}_n|^2 > \tau \\ \tilde{G}_{mn}^{(i)} & |\mathbf{p}_m - \mathbf{p}_n|^2 \leq \tau \end{cases} \qquad (30)$$

or (analogous to Eq. (27))

$$G_{mn}^{(i+1)} = \min\left(\tilde{G}_{mn}^{(i)}, \frac{1}{2}\left(\tilde{G}_{mm}^{(i)} + \tilde{G}_{nn}^{(i)} - \lambda \max\left(|\mathbf{p}_m - \mathbf{p}_n|^2 - \tau, 0\right)\right)\right) \qquad (31)$$

As before, the parameters that control this procedure, λ and τ, can be optimized using the training and/or validation data.

4.4.3 OUT OF SAMPLE EXTENSION FOR GRAMIAN LEARNING

The machine learning procedures Sections 4.4.1 and 4.4.2 produce a refined Gramian, $\mathbf{G}$, which improves on the Gramian-based kernel in Eq. (10). At the end of this procedure, we have a suitable measure of the similarity between all the molecules in the training set, $\{G_{mn}\}_{m,n=1}^{M}$. However, in order to predict the properties of a target molecule that is not in the training set, we also need to know the similarity between that molecule and the molecules in the training set. To achieve this, we divide the spectrum of the target molecule, $S_o(\omega)$, into two pieces: a piece that can be expressed as a linear combination of the spectra of the molecules in the training set, and a piece that is orthogonal to the spectral subspace defined by the training molecules. We then use the machine-learned Gramian for the piece of the spectrum that can be described using the training molecules, and use the kernel Gramian to describe the similarity of the remnant that cannot be expressed in this way.

This results in the following formula for the similarity of a target molecule to a molecule m in the training set [43],

$$G_{m\odot} = G_{m\odot}^{(K)} + \sum_{j=1}^{M} \sum_{n=1}^{M} \left(G_{mj} - G_{mj}^{(K)} \right) \left(G_{jn}^{(K)} \right)^{-1} G_{n\odot}^{(K)} \qquad (32)$$

$$G_{m\odot}^{(K)} = \iint S_m(\omega) K(\omega, \omega') S_\odot(\omega') d\omega d\omega' \qquad (33)$$

Appendix B provides a justification for this strategy and extends this approach so that the similarity between two molecules, neither of which are in the training set, can be computed.

4.4.4 INTERPRETATION OF LEARNED GRAMIANS

In traditional nonstatistical scientific modeling approaches, one does not only wish to predict property values, but also to gain insight into what molecular features are responsible for the observed values of the properties. This allows one to formulate hypotheses (e.g., adding an electronegative group on the α-carbon will increase the pK_a) to be tested in subsequent experimental studies. Similarly, in traditional spectroscopic assignment, one identifies which spectral signatures are associated with which molecular properties or structural features. In statistical prediction approaches like linear regression, the procedure by which one makes property predictions is opaque, and does not lend itself to conceptual insight and hypothesis formulation. This problem is exacerbated by the additional complexity of machine-learning approaches, where the link between the input data and the output prediction is usually undecipherable.

One advantage of the spectral learning approach we propose, however, is that one can, *post facto*, determine which input spectral features are most important for predicting the output properties. Consider the eigendecomposition of the learned Gramian,

$$\boldsymbol{G} = \sum_{m=1}^{M} \gamma_m \boldsymbol{g}_m \boldsymbol{g}_m^\dagger \qquad (34)$$

$$\gamma_1 \geq \gamma_2 \geq \cdots \geq \gamma_M$$

Recall that the Gramian is proportional to a covariance matrix (Appendix A). The eigenvectors of the Gramian, therefore, provide a list of key spectral features,

$$\phi_m(\omega) = \sum_{n=1}^{M} g_{mn} S_n(\omega) \tag{35}$$

These are the spectral features that effectively distinguish between the molecules in the training data set, listed in decreasing order of importance in the context of the properties whose values we are striving to predict. Therefore, the spectral machine learning method we are proposing here not only predicts molecular properties, but also identifies, which special spectral features are most strongly associated with the property(ies) we are interested in.

If one wishes to refine this analysis further, one can correlate the key spectral features to the property values, constructing the property covariance,

$$\sigma_{mn} = \sum_{m=1}^{M} P_m C_{mn}$$
$$C_{mn} = \int S_m(\omega)\phi_n(\omega)d\omega \tag{36}$$

This lets one discern the extent to which different spectral features are (anti)correlated with the property value we are striving to predict.

4.5   DISCUSSION

This research began with a question: can we leverage the enormous quantity of experimental spectral data to make chemical predictions? The simplest such predictions would be the (nonetheless challenging) task of interpreting the spectra (e.g., given a collection of spectra for a single molecule, could one predict the structure of the molecule's most stable conformer?). A bolder quest would be to predict properties, like toxicity or mutagenicity, that lack a clear molecular basis.

Our strategy for achieving this goal is to use a training set of molecular spectra to first train a kernel (Section 4.3). This kernel defines a Gramian, which measures the spectral similarity between molecules. This Gramian can be refined by using context-dependent prior knowledge about the (dis) similarity of molecules in the training and validation sets. This procedure, as described in Section 4.4, allows us to adapt the Gramian to the problem of interest. After the Gramian has been learned, ordinary kriging can be used to make predictions of molecular properties. While we do not claim that the individual detailed steps proposed here are optimal, we believe the strategy is nonetheless enticing, partly because it represents a computational realization of the unquestionably useful "human learning" approach by which chemists use experimental measurements to make chemical inferences.

ACKNOWLEDGMENTS

The authors thank Natural Sciences and Engineering Research Council of Canada (NSERC) and Compute Canada for funding. We also wish to acknowledge the seminal role of Ramon Carbó-Dorca in this work. PWA's first exposure to molecular similarity was through his interactions with Ramon, and it was on a fruitful visit to Girona in the summer of 2010 that the utility of kriging as a tool for making predictions based on molecular similarity measures was first conceived. Ramon's realization of the power of molecular similarity measures, three decades before either of us began to contemplate the subject, was truly prescient and his work remains genuinely inspiring.

KEYWORDS

- Gaussian processes
- Gramian learning
- kriging
- machine learning
- metric learning
- molecular similarity
- molecular spectrum

REFERENCES

1. McHale, J. L., (1999). *Molecular Spectroscopy*. 1st edn., Prentice-Hall: Upper Saddle River, New Jersey.
2. Atkins, P., & Friedman, R., (2005). *Molecular Quantum Mechanics*. Oxford UP: Oxford.
3. Ballester, P. J., (2011). Ultrafast shape recognition: method and applications. *Future Medicinal Chemistry*, *3*(1), 65–78.
4. Sukumar, N., & Das, S., (2011). Current Trends in Virtual High Throughput Screening Using Ligand-Based and Structure-Based Methods. *Combinatorial Chemistry & High Throughput Screening*, *14*(10), 872–888.
5. Rupp, M., & Schneider, G., (2010). Graph Kernels for Molecular Similarity. *Molecular Informatics*, *29*(4), 266–273.
6. Sukumar, N., Krein, M., & Breneman, C. M., (2008). Bioinformatics and cheminformatics: Where do the twain meet? *Current Opinion in Drug Discovery & Development*, *11*(3), 311–319.
7. Ballester, P. J., & Richards, W. G., (2007). Ultrafast shape recognition to search compound databases for similar molecular shapes. *J. Comput. Chem.*, *28*(10), 1711–1723.
8. Ralaivola, L., Swamidass, S. J., Saigo, H., & Baldi, P., (2005). Graph kernels for chemical informatics. *Neural Networks*, *18*(8), 1093–1110.
9. Bender, A., & Glen, R. C., (2004). Molecular similarity: a key technique in molecular informatics. *Organic & Biomolecular Chemistry*, *2*(22), 3204–3218.
10. Klebe, G., (2000). Recent developments in structure-based drug design. *Journal of Molecular Medicine*, *78* (5), 269–281.
11. Maggiora, G. M., & Shanmugasundaram, V., (2010). Molecular similarity measures. *Methods in Molecular Biology*, *272*, 39–100.
12. Bultinck, P., Girones, X., & Carbó-Dorca, R., (2005). Molecular quantum similarity: theory and applications. *Rev. Comput. Chem.*, *21*, 127–207.
13. Bultinck, P., & Carbó-Dorca, R., (2005). Molecular quantum similarity using conceptual DFT descriptors. *J. Chem. Sci.*, *117*, 425–435.
14. Geerlings, P., Boon, G., Van Alsenoy, C., & De Proft, F., (2005). Density functional theory and quantum similarity. *Int. J. Quantum Chem.*, *101*(6), 722–732.
15. Besalú, E., Girones, X., Amat, L., & Carbó-Dorca, R., (2002). Molecular quantum similarity and the fundamentals of QSAR. *Acc. Chem. Res.*, *35*, 289–295.
16. Carbó-Dorca, R., Amat, L., Besalú, E., Girones, X., & Robert, D., (2000). Quantum mechanical origin of QSAR: theory and applications. *J. Mol. Struct. theochem*, *504*, 181–228.
17. Carbó-Dorca, R., Amat, L., Besalú, E., & Lobato, M., (1998). Quantum similarity. In: *Advances in Molecular Similarity*, Carbo Dorca, R., Mezey, P. G., (eds.), Vol. *2*, pp. 1–42.
18. Carbó-Dorca, R., & Besalú, E., (1998). A general survey of molecular quantum similarity. *Journal of Molecular Structure*, *451*(1–2), 11–23.
19. Carbó, R., Besalú, E., Amat, L., & Fradera, X., (1996). On quantum molecular similarity measures (QMSM) and indices (QMSI). *J. Math. Chem.*, *19*, 47–56.
20. Carbó, R., & Calabuig, B., (1992). Molecular quantum similarity measures and N-dimensional representation of quantum objects. 1. Theoretical foundations. *Int. J. Quantum Chem.*, *42*, 1681–1693.

21. Carbó, R., & Calabuig, B., (1992). Molecular quantum similarity measures and N-dimensional representation of quantum objects. 2. Practical applications. *Int. J. Quantum Chem.*, *42*, 1695–1709.

22. Carbó, R., Leyda, L., & Arnau, M., (1980). How similar is a molecule to another: an electron-density measure of similarity between two molecular structures. *Int. J. Quantum Chem.*, *17*, 1185–1189.

23. Miranda-Quintana, R. A., Cruz-Rodes, R., Codorniu-Hernandez, E., & Batista-Leyva, A. J., (2010). Formal theory of the comparative relations: its application to the study of quantum similarity and dissimilarity measures and indices. *J. Math. Chem.*, *47*(4), 1344–1365.

24. Zadeh, F. H., & Ayers, P. W., (2013). Molecular alignment as a penalized permutation Procrustes problem. *J. Math. Chem.*, *51*(3), 927–936.

25. Carbó-Dorca, R., (2016). Aromaticity, quantum multimolecular polyhedra, and quantum QSPR fundamental equation. *J. Comput. Chem.*, *37*(1), 78–82.

26. Harrington, P. D., Urbas, A., & Tandler, P. J., (2000). Two-dimensional correlation analysis. *Chemometrics and Intelligent Laboratory Systems*, *50*(2), 149–174.

27. Noda, I., (1993). Generalized 2-dimensional correlation method applicable to infrared, raman, and other types of spectroscopy. *Appl. Spectrosc.*, *47*(9), 1329–1336.

28. Noda, I., & Ozaki, Y., (2004). *Two-Dimensional Correlation Spectroscopy: Applications in Vibrational and Optical Spectroscopy.* Wiley: West Sussex.

29. Wackernagel, H., (2003). *Multivariate Geostatistics: An Introduction with Applications.* Springer-Verlag: New York.

30. Kitanidis, P. K., (1993). Generalized covariance functions in estimation. *Mathematical Geology*, *25*, 525–540.

31. Isaaks, E. H., & Srivastava, R. M., (1989). *An Introduction to Applied Geostatistics.* Oxford University Press: New York.

32. Clark, I., (1979). *Practical Geostatistics.* Applied Science Publishers: London.

33. Heidar-Zadeh, F., Ayers, P. W., & Carbó-Dorca R. "A Statistical Perspective on Molecular Similarity" In: *Conceptual Density Functional Theory Apple Academic* (ed. Nazmul Islam) (in press).

34. Sarle, W. S., (1994). In: *Neural Networks and Statistical Models*, Nineteenth Annual SAS Users Group International Conference, SAS Institute.

35. Couvreur, C., & Couvreur, P., (1996). Neural networks and statistics: A naive comparison. *Belgian Journal of Operations Research, Statistics, and Computer Science*, *36*, 217–225.

36. Resop, J. P., (2006). A comparison of artificial neural networks and statistical regression with biological resources applications. University of Maryland at College Park, 2006.

37. Cheng, B., & Titterington, D. M., (1994). Neural networks: a review from a statistical perspective. *Statistical Science*, *9*(1), 2–30.

38. Warner, B., & Misra, M., (1996). Understanding neural networks as statistical tools. *American Statistician*, *50*(4), 284–293.

39. Ciampi, A., & Lechevallier, Y., (1997). Statistical models as building blocks of neural networks. *Communications in Statistics-Theory and Methods*, *26*(4), 991–1009.

40. Hansen, N., (2006). The CMA evolution strategy: A comparing review. In: *Towards a New Evolutionary Computation: Advances in Estimation of Distribution Algorithms*, Lozano, J. A., Larranaga, P., Inza, I., Bengoetxea, E., (eds.), Springer: Berlin, pp. 75–102.

41. Heidar-Zadeh, F., Supervised distance metric learning and the curse of dimensionality, In *Theoretical and Quantum Chemistry at the Dawn of 21st Century,* pp. 57–82.
42. Xing, E. P., Ng, A. Y., Jordan, M. I., & Russel, S., (2003). Distance metric learning, with application to clustering with side-information, *Advances in Neural Information Processing Systems, 15,* 505–512.
43. Davis, J. V., Kulis, B., Jain, P., Sra, S., & Dhillon, I. S., (2007). Information-Theoretic Metric Learning, *Proceedings of the 24th International Conference on Machine Learning,* ACM, pp. 209–216.
44. Muller, K. R., Mika, S., Ratsch, G., Tsuda, K., & Scholkopf, B., (2001). An introduction to kernel-based learning algorithms. *IEEE Transactions on Neural Networks, 12*(2), 181–201.
45. Lanssens, C., Ayers, P. W., Bultinck, P., & Van Neck, D. *J. Chem. Phys.* (submitted).

APPENDIX A

For completeness, we show that the Gramian,

$$G_{mn}^{(0)} = \int S_m(\omega) S_n(\omega) d\omega \tag{37}$$

is proportional to a unitary transformation of the covariance matrix, albeit only if the spectra are centered first (cf. Eqs. (5) and (6)), so that

$$0 = \sum_{n=1}^{M} S_n(\omega) \tag{38}$$

To show this, we perform a symmetric (Löwdin) orthogonalization of the centered spectra, obtaining the feature vectors

$$\phi_k(\omega) = \sum_{m=1}^{M} \left(G_{km}^{(0)} \right)^{-1/2} S_m(\omega) \tag{39}$$

Any unitary transformation of this feature space is also acceptable, i.e.,

$$\tilde{\phi}_j(\omega) = \sum_{k=1}^{M} \sum_{m=1}^{M} U_{jk} \left(G_{km}^{(0)} \right)^{-1/2} S_m(\omega) \tag{40}$$

is also acceptable. The expansion coefficients can be interpreted as "property values" for the spectra. So,

$$\tilde{c}_{jn} = \int S_n(\omega) \tilde{\phi}_j(\omega) d\omega = \int S_n(\omega) \left(\sum_{k=1}^{M} \sum_{m=1}^{M} U_{jk} \left(G_{km}^{(0)} \right)^{-1/2} S_m(\omega) \right) d\omega$$

$$= \sum_{k=1}^{M} \sum_{m=1}^{M} U_{jk} \left(G_{km}^{(0)} \right)^{-1/2} G_{mn}^{(0)} = \sum_{k=1}^{M} U_{jk} \left(G_{kn}^{(0)} \right)^{1/2} \tag{41}$$

The Gramian for two vectors of properties is defined as:

$$\Gamma_{mn} = \sum_{j=1}^{M} \tilde{c}_{jm} \tilde{c}_{jn} = \sum_{j=1}^{M} \sum_{k=1}^{M} \sum_{l=1}^{M} \left(U_{jk} \left(G_{km}^{(0)} \right)^{1/2} \right) U_{jl} \left(G_{ln}^{(0)} \right)^{1/2}$$

$$= \sum_{k=1}^{M} \sum_{l=1}^{M} \left(G_{km}^{(0)} \right)^{1/2} \left(G_{ln}^{(0)} \right)^{1/2} \sum_{j=1}^{M} U_{jk} U_{jl} = \sum_{k=1}^{M} \sum_{l=1}^{M} \left(G_{km}^{(0)} \right)^{1/2} \left(G_{ln}^{(0)} \right)^{1/2} \delta_{kl} \tag{42}$$

$$= G_{mn}^{(0)}$$

The covariance is given by the similar formula,

$$\Sigma_{ij} = \frac{1}{M-1}\sum_{m=1}^{M}\left(\tilde{c}_{im} - \frac{1}{M}\sum_{n=1}^{M}\tilde{c}_{in}\right)\left(\tilde{c}_{jm} - \frac{1}{M}\sum_{n=1}^{M}\tilde{c}_{jn}\right) \tag{43}$$

To simplify the covariance matrix, notice that

$$\begin{aligned}
\sum_{n=1}^{M}\tilde{c}_{jn} &= \sum_{n=1}^{M}\sum_{k=1}^{M}U_{jk}\left(G_{kn}^{(0)}\right)^{1/2} = \sum_{n=1}^{M}\int S_n(\omega)\tilde{\phi}_k(\omega)d\omega \\
&= \sum_{n=1}^{M}\int\left(S_n(\omega) - \frac{1}{M}\sum_{n'=1}^{M}S_{n'}(\omega)\right)\tilde{\phi}_k(\omega)d\omega \\
&= \sum_{n=1}^{M}\int S_n(\omega)\tilde{\phi}_k(\omega)d\omega - \frac{1}{M}\sum_{n=1}^{M}\sum_{n'=1}^{M}\int S_{n'}(\omega)\tilde{\phi}_k(\omega)d\omega \\
&= \sum_{n=1}^{M}\int S_n(\omega)\tilde{\phi}_k(\omega)d\omega - \frac{1}{M}\left(\sum_{n'=1}^{M}\int S_{n'}(\omega)\tilde{\phi}_k(\omega)d\omega\right)\left(\sum_{n=1}^{M}1\right) \\
&= \sum_{n=1}^{M}\int S_n(\omega)\tilde{\phi}_k(\omega)d\omega - \sum_{n'=1}^{M}\int S_{n'}(\omega)\tilde{\phi}_k(\omega)d\omega \\
&= 0
\end{aligned} \tag{44}$$

In the second line of this expression we use the fact that the spectra are centered, cf. Eq. (38). The inner summations in Eq. (43) therefore vanish, so that the covariance matrix has the simple form,

$$\begin{aligned}
\Sigma_{kl} &= \frac{1}{M-1}\sum_{m=1}^{M}\tilde{c}_{im}\tilde{c}_{jm} \\
&= \frac{1}{M-1}\sum_{m=1}^{M}\sum_{k=1}^{M}\sum_{l=1}^{M}U_{ik}\left(G_{km}^{(0)}\right)^{1/2}U_{jl}\left(G_{lm}^{(0)}\right)^{1/2} \\
&= \frac{1}{M-1}\sum_{k=1}^{M}\sum_{l=1}^{M}U_{ik}U_{jl}\sum_{m=1}^{M}\left(G_{km}^{(0)}\right)^{1/2}\left(G_{lm}^{(0)}\right)^{1/2} \\
&= \frac{1}{M-1}\sum_{k=1}^{M}\sum_{l=1}^{M}U_{ik}G_{kl}^{(0)}U_{jl}
\end{aligned}$$

This is to say, the covariance matrix is proportional to the Gramian if the Löwdin orthogonalization (corresponding to $U_{ik} = \delta_{ik}$) is used to define the "properties". For more general choices, the covariance is proportional to a unitary transformation of the Gramian. This unitary transformation is irrelevant as it merely reflects on the fact that one can replace one set of properties with different linear combinations of the same properties.

APPENDIX B

Suppose that after the Gramian is trained using the methods in Section 4.3, we are given new molecules to consider. These could be additional molecules whose properties are known, which could be used as training data in the ordinary kriging procedure. Alternatively, these could be molecules whose properties we wish to predict. Suppose we are given some new molecules and their raw spectra, $\{R_{p;t}(\omega_t)\}$. We form the centered multidimensional spectra using the procedure in Eqs. (2)–(6),

$$S_p(\omega) = \prod_{t=1}^{T}\left(R_{p;t}^{\mathrm{raw}}(\omega_t) - R_t^{\mathrm{ref}}(\omega_t)\right) - S^{\mathrm{ref}}(\omega) \qquad (45)$$

Note that the reference spectra are computed using only the M molecules in the original training dataset. We can compute the kernel approximation to the Gramian from Eq. (10),

$$G_{pq}^{(K)} = \iint S_p(\omega) K(\omega,\omega') S_q(\omega') d\omega d\omega' \qquad (46)$$

This expression is true regardless of whether $S_p(\omega)$ refers to a spectrum from the original training set or to the spectrum of one of the new molecules. We propose that the learned kernel between two arbitrary molecules can be approximated by

$$G_{pq} = G_{pq}^{(K)} + \sum_{m=1}^{M}\sum_{i=1}^{M}\sum_{j=1}^{M}\sum_{n=1}^{M} G_{pm}^{(K)}\left(G_{mi}^{(K)}\right)^{-1}\left(G_{ij} - G_{ij}^{(K)}\right)\left(G_{jn}^{(K)}\right)^{-1}G_{nq}^{(K)} \qquad (47)$$

Notice that if p and q are both in the initial M-molecule training set, then this equation is simply the identity,

$$G_{pq} = G_{pq}^{(K)} + \left(G_{pq} - G_{pq}^{(K)}\right) \qquad\qquad 1 \le p,q \le M \qquad (48)$$

To understand the interpretation of Eq. (47) for molecules outside the original training set, it is convenient to expand the spectrum using the Löwdin orthogonalized basis of the (kernel) Gramian,

$$\phi_k(\omega) = \sum_{m=1}^{M} \left(G_{km}^{(K)}\right)^{-1/2} S_m(\omega) \tag{49}$$

This is an orthonormal basis in the sense that

$$\iint \phi_j(\omega) K(\omega,\omega') \phi_k(\omega') d\omega d\omega'$$

$$= \sum_{m=1}^{M} \sum_{n=1}^{M} \left(G_{jm}^{(K)}\right)^{-1/2} \left(\iint S_m(\omega) K(\omega,\omega') S_n(\omega') d\omega d\omega'\right) \left(G_{kn}^{(K)}\right)^{-1/2}$$

$$= \sum_{m=1}^{M} \sum_{n=1}^{M} \left(G_{jm}^{(K)}\right)^{-1/2} G_{mn}^{(K)} \left(G_{kn}^{(K)}\right)^{-1/2} \tag{50}$$

$$= \delta_{jk}$$

The expansion coefficients for an arbitrary spectrum in this basis are

$$c_{pk} = \iint S_p(\omega) K(\omega,\omega') \phi_k(\omega') d\omega d\omega'$$

$$= \sum_{m=1}^{M} \iint S_p(\omega) K(\omega,\omega') \left(G_{km}^{(K)}\right)^{-1} S_m(\omega) d\omega d\omega'$$

$$= \sum_{m=1}^{M} G_{pm}^{(K)} \left(G_{mk}^{(K)}\right)^{-1/2} \tag{51}$$

Substituting this result into Eq. (47), we have

$$G_{pq} = G_{pq}^{(K)} + \sum_{m=1}^{M} \sum_{i=1}^{M} \sum_{j=1}^{M} \sum_{n=1}^{M} c_{pm} \left(G_{mi}^{(K)}\right)^{-1/2} \left(G_{ij} - G_{ij}^{(K)}\right) \left(G_{jn}^{(K)}\right)^{-1/2} c_{qn}$$

$$= \left[G_{pq}^{(K)} - \sum_{m=1}^{M} c_{pm} c_{qm} \right] + \sum_{m=1}^{M} \sum_{i=1}^{M} \sum_{j=1}^{M} \sum_{n=1}^{M} c_{pm} \left(G_{mi}^{(K)}\right)^{-1/2} G_{ij} \left(G_{jn}^{(K)}\right)^{-1/2} c_{qn} \tag{52}$$

The term in square brackets is the Gramian from Eq. (46) minus the contribution that can be expressed as a linear combination of the spectra in the training set. Notice, for example, that this term is zero if the spectra of the molecules being considered can be exactly written as a linear combination of existing spectra,

$$0 = S_p(\omega) - \sum_{m=1}^{M} c_{pm}\phi_m(\omega) = S_q(\omega) - \sum_{m=1}^{M} c_{qm}\phi_m(\omega) \qquad (53)$$

This equation therefore amounts to projecting the new spectrum onto the basis of the existing spectrum, using the learned Gramian, $\mathbf{G}$, from Section 4.3 for those components of the spectrum, and using the kernel Gramian $\mathbf{G}^{(K)}$ for the remaining components (for which there is no "learned" information).

CHAPTER 5

A POLYNOMIAL-SCALING ALGORITHM FOR COMPUTING THE PROBABILITY OF OBSERVING SPECIFIED NUMBERS OF ELECTRONS IN MULTIPLE DOMAINS USING CORRELATION FUNCTIONS

PAUL W. AYERS,[1] GUILLAUME ACKE,[2] STIJN FIAS,[1] DEBAJIT CHAKRABORTY,[1] and PATRICK BULTINCK[2]

[1]*Department of Chemistry and Chemical Biology, McMaster University, Hamilton, Ontario, L8P 4Z2, Canada*

[2]*Department of Inorganic and Physical Chemistry, Ghent University, Krijgslaan 281 (S3), 9000 Gent,Belgium*

CONTENTS

5.1   MOTIVATION

The vocabulary chemists use to interpret molecular structure (atoms, bonds, functional groups, atomic charges, etc.) and even electronic structure (electron pairs, electronegativity, electrophilicity, nucleophilicity, oxidation states, etc.) often predates the Schrödinger equation, and has no clear basis in quantum mechanics. Finding ways to express the output of quantum mechanical calculations in this vocabulary to obtain qualitative insights remains one of the salient challenges of theoretical chemistry [1–4]. In this chapter, we focus on one specific facet of this general problem: how are Lewis electron pairs evinced by the N-electron wavefunction?

The precept that covalent chemical bonds are associated with electron pairs and, more generally, that electron pairs are the fundamental building block of molecules' electronic structure, predates the Schrödinger equation [5]. After the advent of computational quantum chemistry, the electron-pair concept was approached primarily through valence-bond theory and the theory of resonance [6, 7], as well as through (closely related) methods based on geminals (2-electron wavefunctions) or more-general electron-group functions [8–13]. If the electron-pair picture of molecular electronic structure is valid, however, it should not depend on the choice of computational approach: one should be able to *find* electron pairs directly from any accurate electronic wavefunction. For example, Daudel and his coworkers asserted that one could divide a molecule into regions of space containing electron pairs, called loges [14–18]. Their approach inspired other attempts along the same lines based, for example, on the topological structure of the Laplacian of the electron density, the electron localization function, or other similar indicators [19–46]. Many of these descriptions are computationally unwieldy or theoretically shaky. However, the proposal by Savin and his coworkers to interpret loges as maximum probability domains is theoretically sound and, at least for wavefunctions that can be expressed as a linear combination of a (very) few Slater determinants, computationally tractable [47–52].

A maximum probability domain is defined as a region in space, Ω, where the probability of observing exactly N_Ω electrons is locally maximum [50]. If $N_\Omega = 2$, this region can be identified with an electron pair, as it indicates a molecular region where an electron pair is most likely to be localized [40, 42, 47–55]. (For now we do not restrict the spin of the electrons in the pair.) Even without the shape optimization step required to find the optimal region Ω, computing the probability of observing N_Ω electrons in a specified region,

$p_\Omega(N_\Omega)$, is interesting, and can lead to significant insights into molecular electronic structure and bonding. For example, by observing the probability of finding various numbers of electrons in the atoms composing a chemical bond, Pendás et al. [55–65] have been able to characterize chemical bond order.

For wavefunctions that are Slater determinants, it is reasonably easy to compute $p_W(N_W)$ when the molecule is divided into only two regions [49]. For molecules divided into three or more regions and quantitatively accurate wavefunctions (which necessarily have contributions from myriad Slater determinants), the cost of computing $p_W(N_W)$ grows exponentially with both the number of regions and the number of determinants [51, 60, 62, 64, 66–77]. In this chapter, we will present a (necessarily approximate) method for computing $p_W(N_W)$ that has polynomial cost.

5.2   ELECTRON NUMBER PROBABILITIES

Suppose we wish to divide our molecule into D regions, $\left\{\Omega^{(d)}\right\}_{d=1}^{D} \subset \mathbb{R}^3$, each containing $\left\{N_s^{(d)}\right\}_{d=1;s=\alpha,\beta}^{D}$ electrons with spin $s \in \{\alpha, \beta\}$. (We can choose to not specify the spin of the regions. However, an Lewis-electron-pair corresponds to a region with $N_\alpha^{(a)} = N_\beta^{(a)} = 1$.

For convenience and generality, we will represent the regions themselves as a partition of unity. That is, we select a set of weighting functions, $\left\{w^{(d)}(\mathbf{r})\right\}_{d=1}^{D}$, that satisfy the constraints

$$0 \le w^{(d)}(\mathbf{r}) \le 1 \tag{1}$$

$$\sum_{d=1}^{D} w^{(d)}(\mathbf{r}) = 1 \tag{2}$$

If we further require that, $w^{(d)}(\mathbf{r}) \in \{0,1\}$ then every point in space belongs to one and only one region, and we have a strict partitioning into spatial regions $\left\{\Omega^{(d)}\right\}_{d=1}^{D}$. In the more general case, the regions are fuzzy and overlapping. We will assume that the weighting functions themselves are not spin-dependent, but this assumption is inessential.

To make our notation clearer, we will often consider the special case of N = 5, where the five electrons are partitioned into three regions, two regions containing an α–β electron pair, and one region containing a solitary α-spin electron. In this case, the probability of observing the desired number of electrons in the regions is

$$
p\left(N_\alpha^{(1)} = N_\beta^{(1)} = N_\alpha^{(2)} = N_\beta^{(2)} = N_\alpha^{(3)} = 1; N_\beta^{(3)} = 0\right)
$$

$$
= 5! \iint \iint \int \left[\begin{array}{l} w^{(1)}\left(\mathbf{r}_1\right) w^{(1)}\left(\mathbf{r}_2\right) w^{(2)}\left(\mathbf{r}_3\right) w^{(2)}\left(\mathbf{r}_4\right) w^{(3)}\left(\mathbf{r}_5\right) \\ \times \delta_{s_1\alpha} \delta_{s_2\beta} \delta_{s_3\alpha} \delta_{s_4\beta} \delta_{s_5\alpha} \\ \times \left| \Psi\left(\mathbf{x}_1, \mathbf{x}_2, \ldots, \mathbf{x}_5\right) \right|^2 \end{array} \right] dx_1 dx_2 \ldots dx_5 \tag{3}
$$

Here we have used $\left\{\mathbf{x}_i\right\}_{i=1}^N = \left\{\mathbf{r}_i, s_i\right\}_{i=1}^N$ to denote the spatial and spin coordinates of the electrons, and the shorthand $\int dx$ to denote integration over all space and summation over spin. The factor of 5! arises because all possible permutations of the electronic coordinates are allowed. If we perform an explicit summation over the spin degrees of freedom, we obtain a more compact and explicit expression,

$$
p\left(N_\alpha^{(1)} = N_\beta^{(1)} = N_\alpha^{(2)} = N_\beta^{(2)} = N_\alpha^{(3)} = 1; N_\beta^{(3)} = 0\right)
$$

$$
= 3! 2! \iint \iint \int \left[\begin{array}{l} w^{(1)}\left(\mathbf{r}_1\right) w^{(2)}\left(\mathbf{r}_2\right) w^{(3)}\left(\mathbf{r}_3\right) w^{(1)}\left(\mathbf{r}_4\right) w^{(2)}\left(\mathbf{r}_5\right) \\ \times \left| \Psi\left(\mathbf{r}_1, \alpha; \mathbf{r}_2, \alpha; \mathbf{r}_3, \alpha; \mathbf{r}_4, \beta; \mathbf{r}_5, \beta\right) \right|^2 \end{array} \right] dr_1 dr_2 \ldots dr_5 \tag{4}
$$

In the general case of D partitions specified by the weighting functions $\left\{w^{(d)}(\mathbf{r})\right\}_{d=1}^D$, the probability of observing exactly $\left\{N_s^{(d)}\right\}_{d=1; s=\alpha,\beta}^D$ in each partition can be expressed as:

$$
p\left(\left\{N_s^{(d)}\right\}_{d=1}^D\right) = \frac{N_\alpha! N_\beta!}{\displaystyle\prod_{d=1}^D N_\alpha^{(d)}! N_\beta^{(d)}!}
$$

$$
\times \iint \cdots \int \left[\begin{array}{l} w^{(1)}\left(\mathbf{r}_1\right) w^{(1)}\left(\mathbf{r}_2\right) \ldots w^{(1)}\left(\mathbf{r}_{N_\alpha^{(1)}}\right) w^{(2)}\left(\mathbf{r}_{N_\alpha^{(1)}+1}\right) \ldots w^{(2)}\left(\mathbf{r}_{N_\alpha^{(1)}+N_\alpha^{(2)}}\right) \ldots w^{(D)}\left(\mathbf{r}_{N_\alpha}\right) \\ \times w^{(1)}\left(\mathbf{r}_{N_\alpha+1}\right) \ldots w^{(1)}\left(\mathbf{r}_{N_\alpha+N_\beta^{(1)}}\right) w^{(2)}\left(\mathbf{r}_{N_\alpha+N_\beta^{(1)}+1}\right) \ldots w^{(2)}\left(\mathbf{r}_{N_\alpha+N_\beta^{(1)}+N_\beta^{(2)}}\right) \ldots w^{(D)}\left(\mathbf{r}_N\right) \\ \times \sigma_N^{\alpha\alpha\ldots\alpha\beta\beta\ldots\beta}\left(\mathbf{r}_1, \mathbf{r}_2, \ldots, \mathbf{r}_N\right) dr_1 dr_2 \ldots dr_N \end{array} \right] \tag{5}
$$

The combinatorial factor on the first line is the number of permutations of electronic coordinates one can make without putting the wrong number of electrons in one or more of the partitions. For compactness, Eq. (5) also uses the definition of the N-electron shape distribution function,

$$\sigma_N^{\alpha\alpha...\alpha\beta\beta...\beta}\left(\mathbf{r}_1,\mathbf{r}_2,\ldots,\mathbf{r}_N\right)=\left|\Psi\left(\mathbf{r}_1,\alpha;\mathbf{r}_2,\alpha;\ldots;\mathbf{r}_{N_\alpha},\alpha;\mathbf{r}_{N_\alpha+1},\beta;\mathbf{r}_{N_\alpha+2},\beta;\ldots;\mathbf{r}_N,\beta\right)\right|^2 \quad (6)$$

where

$$N_\alpha = \sum_{d=1}^{D} N_\alpha^{(d)} \quad (7)$$

$$N_\beta = \sum_{d=1}^{D} N_\beta^{(d)} \quad (8)$$

and

$$N = N_\alpha + N_\beta \quad (9)$$

are the number of α-spin electrons, the number of β-spin electrons, and the total number of electrons, respectively. The choice to list the α-spin electrons before the β-spin electrons is only a notational convenience; the general formula for the N-electron shape distribution function is

$$\sigma_N^{s_1 s_2 ... s_N}\left(\mathbf{r}_1,\mathbf{r}_2,\ldots,\mathbf{r}_N\right)=\left|\Psi\left(\mathbf{r}_1,s_1;\mathbf{r}_2,s_2;\ldots;\mathbf{r}_N,s_N\right)\right|^2 \quad (10)$$

The many-electron shape distribution functions are related to the many-electron distribution functions in the same way that the shape function is related to the electron density [68–72], so they satisfy the obvious normalization condition

$$1=\iint\cdots\int\sigma_N^{s_1 s_2 ... s_N}\left(\mathbf{r}_1,\mathbf{r}_2,\ldots,\mathbf{r}_N\right)d\mathbf{r}_1 d\mathbf{r}_2 \ldots d\mathbf{r}_N \quad (11)$$

To compute, much less maximize, the probability of observing $\left\{N_s^{(d)}\right\}_{d=1;s=\alpha,\beta}^{D}$ electrons in a partitioned system, one needs to be able to efficiently numerically evaluate expressions like Eq. (5). This requires building appropriate approximations to the N-electron shape distribution function. Alternatively, one can build a suitable approximation for a quantity, like the N-electron density matrix or the N-electron distribution function, from which the N-electron shape distribution function can be efficiently evaluated [73–78].

5.3 CORRELATION FUNCTION APPROXIMATIONS TO THE
N-ELECTRON DISTRIBUTION FUNCTION

One can envision several strategies for approximating the N-electron shape distribution function in terms of simpler quantities. For example, one could imagine using a cumulant expansion to approximate the N-electron density matrix (and then take its diagonal element) or to approximate the N-electron distribution function. (These methods are not the same; in general it is better to reconstruct the N-electron density matrix and then take its diagonal element, as the reconstruction of the N-electron density matrix—but not the reconstruction of the N-electron distribution function—respects anti-symmetry [79].) Cumulant expansions only give computationally tractable approaches if the cumulant expansion is truncated at some order, which corresponds to neglecting many-electron correlations. This might be a sensible strategy for partitions into two parts [74], but it does not seem helpful for general multidomain partitioning. There is also a disadvantage: the N-electron quantities constructed by truncated cumulant expansions may not be correctly normalized.

An approach that leads to normalized many-electron distribution functions is called the (generalized) convolution approximation [80–91]. Convolution approximations satisfy the sequential relations between the k-electron shape distribution functions,

$$\sigma_k^{s_1 s_2 \ldots s_k}\left(\mathbf{r}_1, \mathbf{r}_2, \ldots, \mathbf{r}_k\right) = \iint \cdots \int \sigma_N^{s_1 s_2 \ldots s_N}\left(\mathbf{r}_1, \mathbf{r}_2, \ldots, \mathbf{r}_N\right) d\mathbf{r}_{k+1} d\mathbf{r}_{k+2} \ldots d\mathbf{r}_N \quad (12)$$

and therefore also the normalization constraints,

$$1 = \int \sigma_1^s\left(\mathbf{r}\right) d\mathbf{r}$$

$$1 = \iint \sigma_2^{s_1 s_2}\left(\mathbf{r}_1, \mathbf{r}_2\right) d\mathbf{r}_1 d\mathbf{r}_2$$

$$\vdots \qquad\qquad\qquad\qquad\qquad\qquad\qquad (13)$$

$$1 = \iint \cdots \int \sigma_k^{s_1 s_2 \ldots s_k}\left(\mathbf{r}_1, \mathbf{r}_2, \ldots, \mathbf{r}_k\right) d\mathbf{r}_1 d\mathbf{r}_2 \ldots d\mathbf{r}_k$$

$$\vdots$$

$$1 = \iint \cdots \int \sigma_N^{s_1 s_2 \ldots s_N}\left(\mathbf{r}_1, \mathbf{r}_2, \ldots, \mathbf{r}_N\right) d\mathbf{r}_1 d\mathbf{r}_2 \ldots d\mathbf{r}_N$$

Unfortunately, the convolution approximations typically give N-electron shape distribution functions that are negative for certain electron configurations. Moreover, many-particle convolution approximations seem complicated, and therefore computationally unwieldy.

In general, it seems very difficult to practically construct N-electron shape distribution functions that satisfy the sequential relations (Eq. (12)), the normalization condition (Eq. (11)), and the nonnegativity constraint

$$\sigma_N^{s_1 s_2 \cdots s_N}(\mathbf{r}_1, \mathbf{r}_2, \ldots \mathbf{r}_N) \geq 0 \tag{14}$$

Imposing more general N-representability conditions is even more daunting [75, 79, 82–92]. Recognizing that the nonnegativity constraint is of primal importance (because there is no sensible interpretation for negative probabilities), we choose to consider an approximation for the N-electron distribution function based on the (generalized) Kirkwood superposition principle [79, 93–96].

The generalized Kirkwood superposition principle approximates the higher-order shape distribution functions in terms of lower-order electron correlation functions, which are defined as

$$g_k^{s_1 s_2 \cdots s_k}(\mathbf{r}_1, \mathbf{r}_2, \ldots, \mathbf{r}_k) = \frac{\sigma_k^{s_1 s_2 \cdots s_k}(\mathbf{r}_1, \mathbf{r}_2, \ldots, \mathbf{r}_k)}{\sigma_1^{s_1}(\mathbf{r}_1)\sigma_2^{s_2}(\mathbf{r}_2)\cdots\sigma_1^{s_k}(\mathbf{r}_k)} \tag{15}$$

To the extent that electrons that are far apart move independently, the correlation functions approach unity when electrons are far apart.

The generalized Kirkwood superposition formula for the N-electron correlation function is

$$g_N^{s_1 s_2 \cdots s_N}(\mathbf{r}_1, \mathbf{r}_2 \ldots, \mathbf{r}_N) = \Delta^{s_1 s_2 \cdots s_N}(\mathbf{r}_1, \mathbf{r}_2 \ldots, \mathbf{r}_N) \prod_{k=2}^{N-1}\left(\prod_{1 \leq l_1 < l_2 < \cdots < l_k \leq N} g_k^{s_{l_1} s_{l_2} \cdots s_{l_k}}(\mathbf{r}_{l_1}, \mathbf{r}_{l_2} \ldots, \mathbf{r}_{l_k})\right)^{-1^{(k+N-1)}} \tag{16}$$

The function Δ is defined so that Eq. (16) is exact; if one wishes to neglect all the N-electron correlations that cannot be described as simple products and quotients of N–1 and lower-order electron configurations, then one makes the Kirkwood superposition approximation, $\Delta = 1$. The generalized Kirkwood superposition principle can be derived in numerous ways, including by maximum entropy and minimum free-energy approaches [94–96].

Equation (16) still requires the (prohibitively expensive) N–1-electron correlation function. However, we can approximate this correlation function in terms of lower-order correlation functions by repeatedly applying the superposition approximation. If one does not wish to consider any correlation functions beyond K^{th} order, one obtains the formula

$$g_N^{s_1 s_2 \cdots s_N}\left(\mathbf{r}_1, \ldots, \mathbf{r}_N\right) \approx \prod_{k=2}^{K}\left(\prod_{1 \leq i_1 < i_2 < \cdots < i_k \leq N} g_k^{s_{i_1} s_{i_2} \cdots s_{i_k}}\left(\mathbf{r}_{i_1}, \mathbf{r}_{i_2}, \ldots \mathbf{r}_{i_k}\right)\right)^{-1^{(K-k)}} \tag{17}$$

This superposition approximation is accurate except for electron configurations where more than K electrons are close together. The sequential relations of the electron shape distribution functions are not exactly satisfied, however, because the integration regions in Eq. (13) include electron configurations where the superposition approximation is inaccurate.

It is helpful to rewrite Eq. (16) so that no terms appear in the denominator. We consider the alternative expression,

$$g_N^{s_1 s_2 \cdots s_N}\left(\mathbf{r}_1 \cdots \mathbf{r}_N\right) = \prod_{1 \leq i_1 < i_2 \leq N} \eta_2^{s_{i_1} s_{i_2}}\left(\mathbf{r}_{i_1}, \mathbf{r}_{i_2}\right) \prod_{1 \leq i_1 < i_2 < i_3 \leq N} \eta_3^{s_{i_1} s_{i_2} s_{i_3}}\left(\mathbf{r}_{i_1}, \mathbf{r}_{i_2}, \mathbf{r}_{i_3}\right)$$

$$\times \prod_{1 \leq i_1 < i_2 < i_3 < i_4 \leq N} \eta_4^{s_{i_1} s_{i_2} s_{i_3} s_{i_4}}\left(\mathbf{r}_{i_1}, \mathbf{r}_{i_2}, \mathbf{r}_{i_3}, \mathbf{r}_{i_4}\right) \tag{18}$$

$$\vdots$$

$$\times \eta_N^{s_1 s_2 \cdots s_N}\left(\mathbf{r}_1, \mathbf{r}_2, \ldots, \mathbf{r}_N\right)$$

where we have defined

$$\eta_2^{s_1 s_2}\left(\mathbf{r}_1, \mathbf{r}_2\right) = g_2^{s_1 s_2}\left(\mathbf{r}_1, \mathbf{r}_2\right) \tag{19}$$

and, for $n > 2$,

$$\eta_n^{s_1 s_2 \cdots s_n}\left(\mathbf{r}_1, \mathbf{r}_2, \ldots, \mathbf{r}_n\right) \equiv \frac{g_n^{s_1 s_2 \cdots s_n}\left(\mathbf{r}_1, \mathbf{r}_2, \ldots, \mathbf{r}_n\right)}{\prod_{k=2}^{n-1}\left(\prod_{1 \leq i_1 < i_2 < \cdots < i_k \leq n} g_k^{s_{i_1} s_{i_2} \cdots s_{i_k}}\left(\mathbf{r}_{i_1}, \mathbf{r}_{i_2}, \ldots, \mathbf{r}_{i_k}\right)\right)^{-1^{(k+n-1)}}}$$

$$= \frac{g_n^{s_1 s_2 \cdots s_n}\left(\mathbf{r}_1, \mathbf{r}_2, \ldots, \mathbf{r}_n\right)}{\prod_{k=2}^{n-1}\left(\prod_{1 \leq i_1 < i_2 < \cdots < i_k \leq n} \eta_k^{s_{i_1} s_{i_2} \cdots s_{i_k}}\left(\mathbf{r}_{i_1}, \mathbf{r}_{i_2}, \ldots, \mathbf{r}_{i_k}\right)\right)} \tag{20}$$

The Kirkwood superposition approximation is then to set $\eta_n^{s_1 s_2 \ldots s_n} = 1$ for all $n > K$. Because evaluating the n-electron correlation function, $g_n^{s_1 s_2 \ldots s_n}$, requires evaluating the n-electron reduced density matrix, it will probably be infeasible to consider any truncation of this expression at an order $K > 4$.

To the extent that electrons that are far apart move independently, the value of $\eta_n^{s_1 s_2 \ldots s_n}$ tends towards unity for all electron configurations except those where n or more electrons are close together. (More precisely, $\eta_n^{s_1 s_2 \ldots s_n} \sim 1$ whenever the Kirkwood superposition approximation to $g_n^{s_1 s_2 \ldots s_n}$ is accurate.) It is therefore sensible to define the (generalized) hole correlation functions, which are defined as

$$h_n^{s_1 s_2 \ldots s_n}(\mathbf{r}_1, \mathbf{r}_2, \ldots, \mathbf{r}_n) = \eta_n^{s_1 s_2 \ldots s_n}(\mathbf{r}_1, \mathbf{r}_2, \ldots, \mathbf{r}_n) - 1 \tag{21}$$

The n-electron hole correlation functions approach zero for far-apart electrons. A digression: while the asymptotic decay rates $\eta_n^{s_1 s_2 \ldots s_n} \sim 1$ and $h_n^{s_1 s_2 \ldots s_n}$ are "usually" valid in the limit of large interelectronic separation, this asymptotic decay is not universal, and will not hold for certain special electron configurations. For example, if one considers the value of $\eta_2^{s_1 s_2}(\mathbf{r}_1, \mathbf{r}_1 + \mathbf{u})$ when $|\mathbf{u}| \to \infty$, averaged over the probability of observing an electron at $\mathbf{r}_1$, one will observe $\eta_2^{s_1 s_2} \sim 1$. (Recall that $\eta_2^{s_1 s_2} = g_2^{s_1 s_2}$.) However, if $\mathbf{r}_1$ is very far from a molecule, the probability of observing another electron very far from the molecule is negligible, so $\eta_2^{s_1 s_2} \sim 0$ in that case. As a specific pernicious example, if one considers the asymptotic decay of $\eta_2^{s_1 s_2}(\mathbf{R} + \mathbf{u}, \mathbf{R} - \mathbf{u})$ when $\mathbf{R}$ is chosen to be a position of high electron density and $|\mathbf{u}| \to \infty$, one will find $\eta_2^{s_1 s_2} \sim 0$.

In terms of the hole correlation functions, the N-electron correlation function is:

$$g_N^{s_1 s_2 \ldots s_N}(\mathbf{r}_1 \ldots \mathbf{r}_N) = \prod_{1 \le i_1 < i_2 \le N} \left(1 + h_2^{s_{i_1} s_{i_2}}\left(\mathbf{r}_{i_1}, \mathbf{r}_{i_2}\right)\right) \prod_{1 \le i_1 < i_2 < i_3 \le N} \left(1 + h_3^{s_{i_1} s_{i_2} s_{i_3}}\left(\mathbf{r}_{i_1}, \mathbf{r}_{i_2}, \mathbf{r}_{i_3}\right)\right)$$

$$\times \prod_{1 \le i_1 < i_2 < i_3 < i_4 \le N} \left(1 + h_4^{s_{i_1} s_{i_2} s_{i_3} s_{i_4}}\left(\mathbf{r}_{i_1}, \mathbf{r}_{i_2}, \mathbf{r}_{i_3}, \mathbf{r}_{i_4}\right)\right) \tag{22}$$

$$\vdots$$

$$\times \left(1 + h_N^{s_1 s_2 \ldots s_N}\left(\mathbf{r}_1, \mathbf{r}_2, \ldots, \mathbf{r}_N\right)\right)$$

Notice that this expansion converges very quickly when electrons are sufficiently well-separated, and therefore for all of the most probable electron distributions.

5.4   ALGORITHM FOR COMPUTING THE PROBABILITY OF OBSERVING SPECIFIED NUMBERS OF ELECTRONS IN MULTIPLE DOMAINS

The generalized Kirkwood superposition principle in the form of Eq. (22) is the key ingredient in our algorithm. Truncating Eq. (22) by setting $h_k^{s_1 s_2 \ldots s_k} = 0$ for all k greater than some maximum order, K, and then inserting it into the expression for the electron probability (Eq. (5)) gives the expression,

$$
p\left(\left\{N_s^{(d)}\right\}_{\substack{d=1 \\ s=\alpha,\beta}}^{D}\right) = \frac{N_\alpha! N_\beta!}{\prod\limits_{d=1}^{D} N_\alpha^{(d)}! N_\beta^{(d)}!}
$$

$$
\times \iint \cdots \int
\begin{bmatrix}
w^{(1)}\left(\mathbf{r}_1\right) w^{(1)}\left(\mathbf{r}_2\right) \ldots w^{(1)}\left(\mathbf{r}_{N_\alpha^{(1)}}\right) w^{(2)}\left(\mathbf{r}_{N_\alpha^{(1)}+1}\right) \ldots w^{(2)}\left(\mathbf{r}_{N_\alpha^{(1)}+N_\alpha^{(2)}}\right) \ldots w^{(D)}\left(\mathbf{r}_{N_\alpha}\right) \\
\times w^{(1)}\left(\mathbf{r}_{N_\alpha+1}\right) \ldots w^{(1)}\left(\mathbf{r}_{N_\alpha+N_\beta^{(1)}}\right) w^{(2)}\left(\mathbf{r}_{N_\alpha+N_\beta^{(1)}+1}\right) \ldots w^{(2)}\left(\mathbf{r}_{N_\alpha+N_\beta^{(1)}+N_\beta^{(2)}}\right) \ldots w^{(D)}\left(\mathbf{r}_N\right) \\
\times \prod_{n=1}^{N} \sigma^{s_n}\left(\mathbf{r}_n\right) \prod_{1 \le i_1 < i_2 \le N}\left(1 + h_2^{s_{i_1} s_{i_2}}\left(\mathbf{r}_{i_1}, \mathbf{r}_{i_2}\right)\right) \prod_{1 \le i_1 < i_2 < i_3 \le N}\left(1 + h_3^{s_{i_1} s_{i_2} s_{i_3}}\left(\mathbf{r}_{i_1}, \mathbf{r}_{i_2}, \mathbf{r}_{i_3}\right)\right) \\
\times \cdots \times \prod_{1 \le i_1 < \cdots < i_K \le N}\left(1 + h_K^{s_{i_1} s_{i_2} \ldots s_{i_K}}\left(\mathbf{r}_{i_1}, \mathbf{r}_{i_2}, \ldots, \mathbf{r}_{i_K}\right)\right) d\mathbf{r}_1 d\mathbf{r}_2 \ldots d\mathbf{r}_N
\end{bmatrix} \quad (23)
$$

   To clarify the notation, we present the result for the special case with three partitions into two electron-pair regions and one single-electron region. Using the hole correlation functions in Eq. (4) and truncating at $K = 3$,

$$
p\left(N_\alpha^{(1)} = N_\beta^{(1)} = N_\alpha^{(2)} = N_\beta^{(2)} = N_\alpha^{(3)} = 1; N_\beta^{(3)} = 0\right)
$$

$$
= 3!2! \iint \iint \int
\begin{bmatrix}
w^{(1)}\left(\mathbf{r}_1\right) w^{(1)}\left(\mathbf{r}_2\right) w^{(2)}\left(\mathbf{r}_3\right) w^{(2)}\left(\mathbf{r}_4\right) w^{(3)}\left(\mathbf{r}_5\right) \\
\times \sigma^{s_1}\left(\mathbf{r}_1\right) \sigma^{s_2}\left(\mathbf{r}_2\right) \cdots \sigma^{s_5}\left(\mathbf{r}_5\right) \prod_{1 \le i < j \le 5}\left(1 + h_2^{s_i s_j}\left(\mathbf{r}_i, \mathbf{r}_j\right)\right) \\
\prod_{1 \le i < j < k \le 5}\left(1 + h_3^{s_i s_j s_k}\left(\mathbf{r}_i, \mathbf{r}_j, \mathbf{r}_k\right)\right) d\mathbf{r}_1 d\mathbf{r}_2 \ldots d\mathbf{r}_5
\end{bmatrix} \quad (24)
$$

   These equations are still computationally intractable. For example, the expansion in Eq. (23) has an exponential number of terms,

$$N_{\text{terms}} = 2^{\frac{N^K (N-1)^{K-1} \cdots (N-K+1)^2 (N-K)}{2^K 3^{K-1} \cdots (K-1)^2 K}} \tag{25}$$

and this exponential cost persists even true even if we truncate at the lowest possible level, $K = 2$. However, we expect that because $h_k^{s_1 s_2 .. s_k} \approx 0$ except when k electrons of the specified spins are close together, many of the terms in Eq. (23) that include products of hole correlations functions are entirely negligible. This suggests that we there may be only a polynomial number of significant contributions to Eq. (23). The key is to design an algorithm that focuses on the significant contributions. To do this, we use an approach inspired by the recent work of Holmes and Umrigar, et al. [97, 98]. The basic strategy is to precompute all the types of contributions involving one partition, two partitions, three partitions, etc. Then we consider only the higher-order combinations of these terms that exceed a prespecified threshold, ε.

The fundamental one-body contributions have the form,

$$n_s^{(d)} = \int w^{(d)} (\mathbf{r}) \sigma_1^s (\mathbf{r}) d\mathbf{r} \tag{26}$$

There are no contributions of this type if $N_s^{(d)} = 0$, which slightly reduces the number of numerical integrations that need to be performed. The fundamental two-body contributions have the form,

$$m_{s_1 s_2}^{(d_1, d_2)} = \frac{\iint w^{(d_1)} (\mathbf{r}_1) w^{(d_2)} (\mathbf{r}_2) \sigma_1^{s_1} (\mathbf{r}_1) \sigma_1^{s_2} (\mathbf{r}_2) h_2^{s_1 s_2} (\mathbf{r}_1, \mathbf{r}_2) d\mathbf{r}_1 d\mathbf{r}_2}{n_{s_1}^{(d_1)} n_{s_2}^{(d_2)}} \tag{27}$$

This term does not contribute to Eq. (23) when $N_{s_1}^{(d_1)} N_{s_2}^{(d_2)} = 0$. Also, there is no "diagonal" term unless the region has two electrons of the same spin. That is, there is no contribution from $m_{ss}^{(d,d)}$ unless $N_s^{(d)} \geq 2$. Without loss of generality, one can assume that the electronic coordinates occur in the same order they did in Eq. (23). This means that α-spin electrons always appear before β-spin electrons (so one never has terms like $m_{\beta\alpha}^{(d_1, d_2)}$) and for electrons with the same spin, the domains appear in increasing order (so one never has terms like $m_{ss}^{(d_1, d_2)}$ where $d_1 > d_2$). Therefore, the only types of terms that can appear are $m_{\alpha\alpha}^{(d_1 \leq d_2)}$, $m_{\alpha\beta}^{(d_1, d_2)}$, and $m_{\beta\beta}^{(d_1 \leq d_2)}$.

In cases where $N_{s_1}^{(d_1)}N_{s_2}^{(d_2)}N_{s_3}^{(d_3)} \neq 0$, we need to compute the three-partition contributions, which we write as:

$$m_{s_1 s_2 s_3}^{(d_1,d_2,d_3)} = \frac{\iiint w^{(d_1)}(\mathbf{r}_1) w^{(d_2)}(\mathbf{r}_2) w^{(d_3)}(\mathbf{r}_3) \sigma_1^{s_1}(\mathbf{r}_1) \sigma_1^{s_2}(\mathbf{r}_2) \sigma_1^{s_3}(\mathbf{r}_3) f_3^{s_1 s_2 s_3}(\mathbf{r}_1,\mathbf{r}_2,\mathbf{r}_3) d\mathbf{r}_1 d\mathbf{r}_2 d\mathbf{r}_3}{n_{s_1}^{(d_1)} n_{s_2}^{(d_2)} n_{s_3}^{(d_3)}} \qquad (28)$$

where we have defined

$$f_3^{s_1 s_2 s_3}(\mathbf{r}_1,\mathbf{r}_2,\mathbf{r}_3) = h_3^{s_1 s_2 s_3}(\mathbf{r}_1,\mathbf{r}_2,\mathbf{r}_3)\left[1+U_1^{s_1 s_2 s_3}(\mathbf{r}_1,\mathbf{r}_2,\mathbf{r}_3)+U_2^{s_1 s_2 s_3}(\mathbf{r}_1,\mathbf{r}_2,\mathbf{r}_3)\right]+U_1^{s_1 s_2 s_3}(\mathbf{r}_1,\mathbf{r}_2,\mathbf{r}_3) \quad (29)$$

The intermediate functions appearing in this expression are:

$$\begin{aligned} U_1^{s_1 s_2 s_3}(\mathbf{r}_1,\mathbf{r}_2,\mathbf{r}_3) &= h_2^{s_1 s_2}(\mathbf{r}_1,\mathbf{r}_2) h_2^{s_1 s_3}(\mathbf{r}_1,\mathbf{r}_3) + h_2^{s_1 s_2}(\mathbf{r}_1,\mathbf{r}_2) h_2^{s_2 s_3}(\mathbf{r}_2,\mathbf{r}_3) \\ &+ h_2^{s_1 s_3}(\mathbf{r}_1,\mathbf{r}_3) h_2^{s_2 s_3}(\mathbf{r}_2,\mathbf{r}_3) + h_2^{s_1 s_2}(\mathbf{r}_1,\mathbf{r}_2) h_2^{s_1 s_3}(\mathbf{r}_1,\mathbf{r}_3) h_2^{s_2 s_3}(\mathbf{r}_2,\mathbf{r}_3) \end{aligned} \qquad (30)$$

$$U_2^{s_1 s_2 s_3}(\mathbf{r}_1,\mathbf{r}_2,\mathbf{r}_3) = \sum_{1 \leq i < j \leq 3} h_2^{s_i s_j}(\mathbf{r}_i,\mathbf{r}_j) = h_2^{s_1 s_2}(\mathbf{r}_1,\mathbf{r}_2) + h_2^{s_1 s_3}(\mathbf{r}_1,\mathbf{r}_3) + h_2^{s_2 s_3}(\mathbf{r}_2,\mathbf{r}_3) \qquad (31)$$

Similar to the 2-partition case in Eq. (27), repeated indices are only allowed if the number of electrons in a partition is sufficiently large. Terms with two repeated indices (i.e., terms with the form $m_{\bar{s}\bar{s}s_3}^{(\bar{d},\bar{d},d_3)}$ and $m_{s_1 \bar{s}\bar{s}}^{(d_1,\bar{d},\bar{d})}$) appear in Eq. (23) only if $N_{\bar{s}}^{(\bar{d})} \geq 2$. Terms with three repeated indices (i.e., $m_{\bar{s}\bar{s}\bar{s}}^{(\bar{d},\bar{d},\bar{d})}$) appear only if $N_{\bar{s}}^{(\bar{d})} \geq 3$. As before, we may assume that one never has β-spin electrons appearing before α-spin electrons (so one never encounters terms with the form $m_{\beta\alpha\alpha}^{(d_1,d_2,d_3)}$, $m_{\beta\beta\alpha}^{(d_1,d_2,d_3)}$, or $m_{\beta\alpha\beta}^{(d_1,d_2,d_3)}$) and that partitions with the same spin never appear in nondecreasing order. That is, the only types of terms that can appear are $m_{\alpha\alpha\alpha}^{(d_1 \leq d_2 \leq d_3)}$, $m_{\alpha\alpha\beta}^{(d_1 \leq d_2,d_3)}$, $m_{\alpha\beta\beta}^{(d_1,d_2 \leq d_3)}$, and $m_{\bar{s}\bar{s}\bar{s}}^{(\bar{d},\bar{d},\bar{d})}$.

Contributions from even more partitions can be computed by the same strategy, namely examining Eq. (23) and determine which terms include hole correlation functions coupling exactly K partitions. To give one final example, if $N_{s_1}^{(d_1)}N_{s_2}^{(d_2)}N_{s_3}^{(d_3)}N_{s_4}^{(d_4)} \neq 0$, there is a contribution from four coupled partitions which has the complicated expression,

$$m_{s_1 s_2 s_3 s_4}^{(d_1,d_2,d_3,d_4)} = \frac{\iiint \begin{bmatrix} w^{(d_1)}(\mathbf{r}_1) w^{(d_2)}(\mathbf{r}_2) w^{(d_3)}(\mathbf{r}_3) w^{(d_4)}(\mathbf{r}_4) \sigma_1^{s_1}(\mathbf{r}_1) \sigma_1^{s_2}(\mathbf{r}_2) \sigma_1^{s_3}(\mathbf{r}_3) \sigma_1^{s_4}(\mathbf{r}_4) \\ \times f_4^{s_1 s_2 s_3 s_4}(\mathbf{r}_1,\mathbf{r}_2,\mathbf{r}_3,\mathbf{r}_4) d\mathbf{r}_1 d\mathbf{r}_2 d\mathbf{r}_3 d\mathbf{r}_4 \end{bmatrix}}{n_{s_1}^{(d_1)} n_{s_2}^{(d_2)} n_{s_3}^{(d_3)} n_{s_4}^{(d_4)}} \qquad (32)$$

with

$$f_4^{s_1s_2s_3s_4}\left(\mathbf{r}_1,\mathbf{r}_2,\mathbf{r}_3,\mathbf{r}_4\right)=h_4^{s_1s_2s_3s_4}\left(\mathbf{r}_1,\mathbf{r}_2,\mathbf{r}_3,\mathbf{r}_4\right)\left(\begin{array}{c}1+V_1^{s_1s_2s_3s_4}\left(\mathbf{r}_1,\mathbf{r}_2,\mathbf{r}_3,\mathbf{r}_4\right)+V_2^{s_1s_2s_3s_4}\left(\mathbf{r}_1,\mathbf{r}_2,\mathbf{r}_3,\mathbf{r}_4\right)\\+V_3^{s_1s_2s_3s_4}\left(\mathbf{r}_1,\mathbf{r}_2,\mathbf{r}_3,\mathbf{r}_4\right)\end{array}\right) \quad (33)$$
$$+V_1^{s_1s_2s_3s_4}\left(\mathbf{r}_1,\mathbf{r}_2,\mathbf{r}_3,\mathbf{r}_4\right)$$

$$V_1^{s_1s_2s_3s_4}\left(\mathbf{r}_1,\mathbf{r}_2,\mathbf{r}_3,\mathbf{r}_4\right)=T_1^{s_1s_2s_3s_4}\left(\mathbf{r}_1,\mathbf{r}_2,\mathbf{r}_3,\mathbf{r}_4\right)+T_2^{s_1s_2s_3s_4}\left(\mathbf{r}_1,\mathbf{r}_2,\mathbf{r}_3,\mathbf{r}_4\right)+T_3^{s_1s_2s_3s_4}\left(\mathbf{r}_1,\mathbf{r}_2,\mathbf{r}_3,\mathbf{r}_4\right) \quad (34)$$

$$V_2^{s_1s_2s_3s_4}\left(\mathbf{r}_1,\mathbf{r}_2,\mathbf{r}_3,\mathbf{r}_4\right)=\sum_{1\le i<j<k\le4}f_3^{s_is_js_k}\left(\mathbf{r}_i,\mathbf{r}_j,\mathbf{r}_k\right) \quad (35)$$

$$V_3^{s_1s_2s_3s_4}\left(\mathbf{r}_1,\mathbf{r}_2,\mathbf{r}_3,\mathbf{r}_4\right)=\sum_{1\le i<j\le4}h_2^{s_is_j}\left(\mathbf{r}_i,\mathbf{r}_j\right) \quad (36)$$

$$T_1^{s_1s_2s_3s_4}\left(\mathbf{r}_1,\mathbf{r}_2,\mathbf{r}_3,\mathbf{r}_4\right)=\prod_{1\le i<j\le4}h_2^{s_is_j}\left(\mathbf{r}_i,\mathbf{r}_j\right)$$
$$+\prod_{1\le i<j\le4}h_2^{s_is_j}\left(\mathbf{r}_i,\mathbf{r}_j\right)\left(\sum_{1\le k<l\le4}\frac{1}{h_2^{s_ks_l}\left(\mathbf{r}_k,\mathbf{r}_l\right)}\right)$$
$$+\prod_{1\le i<j\le4}h_2^{s_is_j}\left(\mathbf{r}_i,\mathbf{r}_j\right)\left(\sum_{\substack{1\le k<l\le4\\1\le m<n\le4}}\frac{1-\delta_{km}\delta_{ln}}{h_2^{s_ks_l}\left(\mathbf{r}_k,\mathbf{r}_l\right)h_2^{s_ms_n}\left(\mathbf{r}_m,\mathbf{r}_n\right)}\right) \quad (37)$$
$$+\sum_{\{i,j,k\}=\text{permutations of }\{2,3,4\}}h_2^{s_1s_i}\left(\mathbf{r}_1,\mathbf{r}_i\right)h_2^{s_is_j}\left(\mathbf{r}_i,\mathbf{r}_j\right)h_2^{s_js_k}\left(\mathbf{r}_j,\mathbf{r}_k\right)$$

The first term in T_1 includes all possible 2-electron hole correlations; the second (third) term includes all possible ways to remove one (two) of these 2-electron correlations; the final term includes all possible ways to remove three of these 2-electron correlations without decoupling one of the electronic coordinates from the others.

$$T_2^{s_1s_2s_3s_4}\left(\mathbf{r}_1,\mathbf{r}_2,\mathbf{r}_3,\mathbf{r}_4\right)=\left[h_3^{s_1s_2s_3}\left(\mathbf{r}_1,\mathbf{r}_2,\mathbf{r}_3\right)\left(\sum_{i=1}^{3}h_2^{s_is_4}\left(\mathbf{r}_i,\mathbf{r}_4\right)+\sum_{1\le i<j\le3}f_3^{s_is_js_4}\left(\mathbf{r}_i,\mathbf{r}_j,\mathbf{r}_4\right)\right)\right]$$
$$+\left(\text{term in }[\ ]\text{ but interchange indices 1 and 4}\right)$$
$$+\left(\text{term in }[\ ]\text{ but interchange indices 2 and 4}\right) \quad (38)$$
$$+\left(\text{term in }[\ ]\text{ but interchange indices 3 and 4}\right)$$
$$+\left(\sum_{1\le i<j<k\le4}h_3^{s_is_js_k}\left(\mathbf{r}_i,\mathbf{r}_j,\mathbf{r}_k\right)\right)T_1^{s_1s_2s_3s_4}\left(\mathbf{r}_1,\mathbf{r}_2,\mathbf{r}_3,\mathbf{r}_4\right)$$

$$T_3^{s_1 s_2 s_3 s_4}(\mathbf{r}_1,\mathbf{r}_2,\mathbf{r}_3,\mathbf{r}_4) = \left(\begin{array}{c} \prod_{1 \le i < j < k \le 4} h_3^{s_i s_j s_k}(\mathbf{r}_i,\mathbf{r}_j,\mathbf{r}_k) \\[2mm] + \prod_{1 \le i < j < k \le 4} h_3^{s_i s_j s_k}(\mathbf{r}_i,\mathbf{r}_j,\mathbf{r}_k) \left(\sum_{1 \le m < n \le 4} \frac{1}{h_3^{s_j s_m s_n}(\mathbf{r}_l,\mathbf{r}_m,\mathbf{r}_n)} \right) \\[2mm] + \prod_{1 \le i < j < k \le 4} h_3^{s_i s_j s_k}(\mathbf{r}_i,\mathbf{r}_j,\mathbf{r}_k) \left(\sum_{\substack{1 \le m < n \le 4 \\ 1 \le o < p \le 4}} \frac{1 - \delta_{lo}\delta_{mp}\delta_{nq}}{h_3^{s_j s_m s_n}(\mathbf{r}_l,\mathbf{r}_m,\mathbf{r}_n) h_3^{s_o s_p s_q}(\mathbf{r}_o,\mathbf{r}_p,\mathbf{r}_q)} \right) \end{array} \right)$$

$$\times \left(1 + T_1^{s_1 s_2 s_3 s_4}(\mathbf{r}_1,\mathbf{r}_2,\mathbf{r}_3,\mathbf{r}_4) + V_2^{s_1 s_2 s_3 s_4}(\mathbf{r}_1,\mathbf{r}_2,\mathbf{r}_3,\mathbf{r}_4) + V_3^{s_1 s_2 s_3 s_4}(\mathbf{r}_1,\mathbf{r}_2,\mathbf{r}_3,\mathbf{r}_4)\right) \tag{39}$$

The first term in T_3 includes all possible 3-electron hole correlation functions; the second (third) term includes all possible ways to remove one (two) of these 3-electron correlations.

As before, terms with two repeated indices (i.e., terms with the form $m_{\bar{s}\bar{s}s_3 s_4}^{(\bar{d},\bar{d},d_3,d_4)}$, $m_{s_1 \bar{s}\bar{s}s_4}^{(d_1,\bar{d},\bar{d},d_4)}$, and $m_{s_1 s_2 \bar{s}\bar{s}}^{(d_1,d_2,\bar{d},\bar{d})}$) contribute only if $N_{\bar{s}}^{(\bar{d})} \ge 2$. Terms with three repeated indices (i.e., $m_{\bar{s}\bar{s}\bar{s}s_4}^{(\bar{d},\bar{d},\bar{d},d_4)}$ and $m_{s_1 \bar{s}\bar{s}\bar{s}}^{(d_1,\bar{d},\bar{d},\bar{d})}$) contribute only if $N_{\bar{s}}^{(\bar{d})} \ge 3$. Terms with four repeated indices, $m_{\bar{s}\bar{s}\bar{s}\bar{s}}^{(\bar{d},\bar{d},\bar{d},\bar{d})}$ appears only if $N_{\bar{s}}^{(\bar{d})} \ge 4$. As before, we may assume that for electrons with the same spin, the domains appear in non-decreasing order. Similarly, one never has β-spin electrons appearing before α-spin electrons (i.e., the only types of terms that appear are $m_{\alpha\alpha\alpha\alpha}^{(d_1 \le d_2 \le d_3 \le d_4)}$, $m_{\alpha\alpha\alpha\beta}^{(d_1 \le d_2 \le d_3,d_4)}$, $m_{\alpha\alpha\beta\beta}^{(d_1 \le d_2,d_3 \le d_4)}$, $m_{\alpha\beta\beta\beta}^{(d_1,d_2 \le d_3 \le d_4)}$, and $m_{\beta\beta\beta\beta}^{(d_1 \le d_2 \le d_3 \le d_4)}$).

It seems intractable to manually go beyond four-partition coefficients; one could use symbolic programming to compute higher-order results.

The algorithm we present below does not fail even if one includes all possible terms with repeated indices, even "nonsensical" terms like $k_{\bar{s}\bar{s}s_3 s_4}^{(\bar{d},\bar{d},d_3,d_4)}$ *when* $N_{\bar{s}}^{(\bar{d})} < 2$*. However, the algorithm assumes that the terms that are provided respect the electron ordering from Eq. (24).*

An algorithm for approximating the probability integral in Eq. (23), neglecting only terms that are smaller than ε, follows:

1. Compute the population of each partition, $n_s^{(d)}$, and calculate the independent particle probability,

$$P_0 = \prod_{i=1}^{N} n_{s_i}^{(d_i)} \tag{40}$$

Here $\{d_i\}_{i=1}^{N}$ and $\{s_i\}_{i=1}^{N}$ denote the partition and spin that the ith electron is assigned to in the probability integral, Eq. (23). Because of the way we have defined the contributions in Eqs. (27), (28), and (32), we should neglect contributions whose absolute value is smaller than $\varepsilon' = \varepsilon/p_0$. This precomputation is trivial to parallelize.

2. Precompute the contributions corresponding to two partitions (Eq. (27)), three partitions (Eq. (28)), four partitions (Eq. (32)), etc. Assume the ordering of terms from Eq. (23). This means that we "only" need to compute terms that are of the form, $m_{\alpha...\alpha,\beta...\beta}^{d_1 \leq \cdots \leq d_k, d_l \leq \cdots \leq d_n}$. For example, if we consider all terms up to 4-electron configurations, we have to compute $m_{\alpha\alpha}^{(d_1 \leq d_2)}$, $m_{\alpha\beta}^{(d_1, d_2)}$, $m_{\beta\beta}^{(d_1 \leq d_2)}$, $m_{\alpha\alpha\alpha}^{(d_1 \leq d_2 \leq d_3)}$, $m_{\alpha\alpha\beta}^{(d_1 \leq d_2, d_3)}$, $m_{\alpha\beta\beta}^{(d_1, d_2 \leq d_3)}$, $m_{\beta\beta\beta}^{(d_1 \leq d_2 \leq d_3)}$, $m_{\alpha\alpha\alpha\alpha}^{(d_1 \leq d_2 \leq d_3 \leq d_4)}$, $m_{\alpha\alpha\alpha\beta}^{(d_1 \leq d_2 \leq d_3, d_4)}$, $m_{\alpha\alpha\beta\beta}^{(d_1 \leq d_2, d_3 \leq d_4)}$, $m_{\alpha\beta\beta\beta}^{(d_1, d_2 \leq d_3 \leq d_4)}$, and $m_{\beta\beta\beta\beta}^{(d_1 \leq d_2 \leq d_3 \leq d_4)}$ for $d_j = 1, 2, ...D$. This precomputation is trivial to parallelize.

3. The terms we computed in step two can contribute to the integral not at all, once, or even multiple times. To compute the number of times an integral contributes, we need to know how many times a given partition-spin pair, (d_k, s_k), appears in the integral. To introduce a concrete notation, let $\mathbf{z}$ denote the partition-spin pairs inside an integral. i.e.,

$$m_{s_1 s_2}^{d_1 d_2} \rightarrow \mathbf{z} \equiv \left[(d_1, s_1), (d_2, s_2) \right]$$
$$m_{s_1 s_2 s_3}^{d_1 d_2 d_3} \rightarrow \mathbf{z} \equiv \left[(d_1, s_1), (d_2, s_2), (d_3, s_3) \right] \qquad (41)$$
$$m_{s_1 s_2 s_3 s_4}^{d_1 d_2 d_3 d_4} \rightarrow \mathbf{z} \equiv \left[(d_1, s_1), (d_2, s_2), (d_3, s_3), (d_4, s_4) \right]$$

For $\mathbf{Z} = [(d_1, s_2), (d_2, s_2), ...]$, let $m(\mathbf{z})$ denote the value of the integral $m_{s_1 s_2 ...}^{(d1, d2, ...)}$ and $q_s^{(d)}(\mathbf{z})$ denote the number of electrons of spin s in partition d in $\mathbf{z}$. E.g., for $\mathbf{Z} = [(1, \alpha), (4, \alpha), (4, \alpha), (2, \beta)]$, $m(\mathbf{z}) = m_{\alpha\alpha\alpha\beta}^{(1,4,4,2)}$, $q_\alpha^{(1)}(\mathbf{z}) = q_\beta^{(2)}(\mathbf{z}) = 1$, $q_\alpha^{(4)}(\mathbf{z}) = 2$, and all of the other values of $q_s^{(d)}(\mathbf{z})$ are zero (e.g., $q_\alpha^{(2)}(\mathbf{z}) = q_\beta^{(4)}(\mathbf{z}) = 0$). The combinatoric factor is defined as:

$$\chi(\mathbf{z}) = \prod_{\substack{d=1 \\ s=\alpha, \beta}}^{D} \binom{N_s^{(d)}}{q_s^{(d)}(\mathbf{z})} \qquad (42)$$

Recall that the definition of the "number of combinations" is:

$$\binom{N_s^{(d)}}{q_s^{(d)}} = \begin{cases} \dfrac{N_s^{(d)}!}{\left(N_s^{(d)} - q_s^{(d)}\right)! q_s^{(d)}!} & N_s^{(d)} \geq q_s^{(d)} \\ 0 & N_s^{(d)} < q_s^{(d)} \end{cases} \qquad (43)$$

The total contribution of the integrals is the product of the integral's value and a combinatoric factor, $c(\mathbf{z}) = \chi(\mathbf{Z}) \cdot m(\mathbf{z})$. We will now make a master list of contributions, C, which contains all the integrals that can possibly contribute more than ε' to the electron number probability expression in Eq. (23).

For all the integrals computed in step 2, if $c(\mathbf{z}) = \chi(\mathbf{Z}) \cdot m(\mathbf{z}) \geq \varepsilon'$, add this contribution to the master list, $C = [(c_1, m_1, \mathbf{z}_1), (c_2, m_2, \mathbf{z}_2),...]$.

4. Sort the master list, C, so that the contributions in the list occur in decreasing order of magnitude, $|c_1| \geq |c_2| \geq |c_3| \geq$. Thus c_j denotes the jth largest-magnitude contribution to Eq. (23) (so far).

5. Starting with the largest contribution in the list, try to make a composite contribution by combining it with subsequent terms. To make a composite contribution between two terms, $m_{s_1 s_2 \ldots}^{d_1 d_2 \ldots}$ and $m_{s_1' s_2' \ldots}^{d_1' d_2' \ldots}$ one first merges the lists of their partition/spin labels,

$$\mathbf{z}_{new} = [\mathbf{z}, \mathbf{z}'] = \left[(d_1, s_1), (d_2, s_2), \ldots (d_1', s_1'), (d_2', s_2'), \ldots \right] \quad (44)$$

The amount this composite term contributes to the expression is

$$c(\mathbf{z}_{new}) = \chi(\mathbf{z}_{new}) m(\mathbf{z}_{new}) \quad (45)$$

where $\chi(\mathbf{z}_{new})$ is a combinatoric factor (which will be zero if the term is disallowed, cf. Eqs. (42)–(43)) $m(\mathbf{z}_{new})$ is defined as the product of the composing integrals,

$$m(\mathbf{z}_{new}) = m(\mathbf{z}) m(\mathbf{z}') \quad (46)$$

Explicitly,

a. Initialize $j = 1$.

b. Initialize $k = j + 1$. Compute the quantities associated with the composite contribution $\mathbf{z}_{new} = [\mathbf{z}_j, \mathbf{z}_k]$, specifically $m(\mathbf{z}_{new}) = m(\mathbf{z}_j) m(\mathbf{z}_k)$ and $c(\mathbf{z}_{new}) = \chi(\mathbf{z}_{new}) m(\mathbf{z}_{new})$.

c. If $c(\mathbf{z}_{new}) < \varepsilon'$ then there can be no further composite contributions including $\mathbf{z}_j$.

   1. Increment j (set $j: = j + 1$) go to Step b.
   2. If j corresponded to the last entry in the list, go to Step f.

d. If $c(\mathbf{z}_{new}) \geq \varepsilon'$, then store $(c(\mathbf{z}_{new}), m(\mathbf{z}_{new}), (\mathbf{z}_{new})$ in the master list, C, placing it at the appropriate position so that the list remains sorted in order of decreasing magnitude of contributions.

e. Prepare to look for the next composite contribution by incrementing k (set $k: = k+1$) and go to step c. If k corresponded to the last value entry of the list, increment j (set $j: = j+1$) and go to step b. If j cannot be incremented because it at the end of the list, go to step f.

f. Compute the probability by adding up all the simple and composite contributions in the list,

$$p\left(\left\{N_s^{(d)}\right\}_{\substack{d=1 \\ s=\alpha,\beta}}^{D}\right) = \frac{N_\alpha!N_\beta!}{\prod\limits_{d=1}^{D} N_\alpha^{(d)}!N_\beta^{(d)}!} \times P_0\left(1+\mathrm{sum}\left(\mathcal{C}\right)\right) \tag{47}$$

This algorithm will have polynomial scaling if the final length of the master list C grows polynomially with the size of the problem, which should be true for sufficiently large ε. The algorithm rests on three key approximations. (1)The use of a parameter, ε, to reduce the number of terms in the sum is a controlled approximation. Truncation of the sum over terms is also very well justified: there is no reason to include composite contributions that are much smaller in magnitude than the contributions that were neglected due to (2) the truncation of the superposition approximation (in Eq. (23)) or (3) the omission of terms that require higher-dimensional numerical integration than one wishes to perform. These latter two approximations are uncontrolled, and are based on the *assumption* that contributions from many-electron correlations becomes increasingly small as the number of electrons increases. The error inherent in these approximations can be assessed by performing a series of calculations at different orders, e.g., by comparing the results obtained when the superposition approximation is truncated after two-body, three-body, and four-body terms. Similarly, if one discovers that all the n-body correlations beyond some order are negligible, e.g., that

$$\varepsilon' > \frac{\int \cdots \int \left[w^{(d_1)}\left(\mathbf{r}_1\right) \cdots w^{(d_n)}\left(\mathbf{r}_n\right) \sigma_1^{s_1}\left(\mathbf{r}_1\right) \cdots \sigma_n^{s_n}\left(\mathbf{r}_n\right) f_n^{s_1 s_2 \cdots s_n}\left(\mathbf{r}_1,\mathbf{r}_2,\ldots,\mathbf{r}_n\right) \right] d\mathbf{r}_1 \ldots d\mathbf{r}_n}{n_{s_1}^{(d_1)} \cdots n_{d_n}^{(n)}} \tag{48}$$

then one may safely neglect those contributions.

5.5   NUANCES

While we have not implemented the above algorithm, we have thought about it thoroughly enough to have some appreciation for its nuances. We first discuss a few key challenges:

1. Determining the correlation functions from the output of a quantum chemistry calculation is challenging. Computing $h_k^{s_1 s_2 \ldots s_k}(\mathbf{r}_1, \mathbf{r}_2, \ldots, \mathbf{r}_k)$ requires the k-electron reduced density matrix, and it is rarely practical to compute reduced density matrices for $k > 4$. (Even $k = 3$ is quite challenging.) It is also true, however, that most conventional electronic structure methods are inaccurate for high-order reduced density matrices (due, for example, to cumulant approximations) and also for describing situations where many electrons are close together (due to their inadequacy in describing electron-electron cusps). This means that low-order truncation of the superposition approximation is inherently consistent with the approximations that are most commonly made in quantum chemistry.

2. Numerically evaluating the many-electron integrals in Eqs. (26), (27), (28), and (32) is time-consuming, not only because many-electron integrals are daunting, but also because there are an intimidatingly large number of integrals to perform. However, each integral is localized on a few partitions, which means that much of the molecular integration grid is irrelevant for these integrals. By considering only the grid points that are near the partitions under scrutiny, one can drastically reduce the integration region, to the point where each integral may have a cost similar to the numerical integrations in atomic calculations. Also, the integration step is a pre-processing step and is perfectly parallelizable.

3. If one wishes to compute the probability of not just one, but several choices of electronic populations, one does not need to redo all the integrals in Eqs. (26), (27), (28), and (32). One only needs to restart the algorithm in Step 3.

So far we have discussed only the problem of computing the probability of observing a specified number of electrons, $\{N_s^{(d)}\}_{d=1; s=\alpha,\beta}$, in partitions defined by the weighting functions $\{W^{(d)}(\mathbf{r})\}_{d=1}$. If we would like to optimize these domains, we will need an initial guess. We will also need a method for computing the change in the probability with respect to a change in the partitioning weight function.

As an initial guess, one could use the localized orbitals corresponding to a "guide determinant" that corresponds (one hopes) to a Lewis structure of interest. For example, if one assigns different orbitals to different domains, then one might define an initial guess as,

$$w^{(d)}(\mathbf{r}) = \frac{\sum_{k \in \text{domain}} |\varphi_k(\mathbf{r})|^2}{\rho(\mathbf{r})} \tag{49}$$

A similar formula could be constructed using the geminals from a geminal-based wavefunction. In that case one could use the extracule density [99] of the individual geminals to define the electron-pair regions,

$$w^{(d)}(\mathbf{R}) = \frac{\sum_{k \in \text{domain}} E_k(\mathbf{R})}{\sum_{k=1}^{N/2} E_k(\mathbf{R})} \tag{50}$$

Different initializations could be constructed by using different terms in a linear combination of geminal products to construct the guiding extracules.

There are many different ways to optimize the weighting functions, but the simplest approach is to consider the values of the weighting functions on a molecular numerical integration grid as variables. We denote the numerical integration grid as,

$$\int f(\mathbf{r}) d\mathbf{r} \approx \sum_{g=1}^{G} \varpi_g f(\mathbf{r}_g) \tag{51}$$

One maximizes the probability expression in Eq. (47) are $\left\{ w^{(d)}(\mathbf{r}_g) \right\}_{g=1; d=1}^{G;D}$, subject to the constraints

$$0 \le w^{(d)}(\mathbf{r}_g) \le 1$$
$$\sum_{d=1}^{D} w^{(d)}(\mathbf{r}_g) = 1 \tag{52}$$

We can recast this as an unconstrained optimization by using the transformation of coordinates,

$$w^{(1)}\left(\mathbf{r}_g\right) = f\left(x_g^{(1)}\right)$$

$$w^{(2)}\left(\mathbf{r}_g\right) = \left(1 - f\left(x_g^{(1)}\right)\right) f\left(x_g^{(2)}\right)$$

$$\vdots$$ (53)

$$w^{(D-1)}\left(\mathbf{r}_g\right) = \left(1 - f\left(x_g^{(1)}\right)\right)\left(1 - f\left(x_g^{(2)}\right)\right)\cdots\left(1 - f\left(x_g^{(D-2)}\right)\right) f\left(x_g^{(D-1)}\right)$$

$$w^{(D)}\left(\mathbf{r}_g\right) = \left(1 - f\left(x_g^{(1)}\right)\right)\left(1 - f\left(x_g^{(2)}\right)\right)\cdots\left(1 - f\left(x_g^{(D-2)}\right)\right)\left(1 - f\left(x_g^{(D-1)}\right)\right)$$

where $f(x)$ is any function from the real numbers to the unit interval. Commonly one chooses $f(x) = \cos^2(x)$, $f(x) = \frac{1}{2}(1+\cos(x))$, or $f(x) = (x^2+1)^{-1}$, but one could also use a single-valued function like $f(x) = (e^x+1)^{-1}$. Notice that in the special case where $D = 2$, only integrals over the first partition need to be evaluated (i.e., one only needs to evaluate the integrals $m_{ss\ldots}^{(1,1,\ldots)}$).

Unconstrained maximization of the probability expression with respect to $\left\{x_g^{(d)}\right\}_{g=1;d=1}^{G,D}$ will give a fuzzy (weighted) partitioning of space. If one wishes to obtain a strict partitioning into disjoint maximum probability domains, then one can add a penalty to the objective function being maximized, e.g.,

$$\sum_{d=1}^{D}\sum_{g=1}^{G} \varpi_g \rho\left(\mathbf{r}_g\right)\left[a\left(2w^{(d)}\left(\mathbf{r}_g\right) - 1\right)\right]^{2n} \quad a>1; n\to\infty$$ (54)

In the $n\to\infty$ limit, this term forces the maximum of the objective function to correspond to a strict partitioning.

An efficient optimization method usually requires information not only about the objective function, but about its gradients. The gradients are easy to evaluate by combining the gradients of the individual terms in the sum, Eq. (47). For example, corresponding to Eq. (27) one has:

$$m_{s_1 s_2}^{(a,b)} = \frac{\displaystyle\sum_{g_1=1}^{G}\sum_{g_2=1}^{G} \varpi_{g_1}\varpi_{g_2} w^{(a)}\left(\mathbf{r}_{g_1}\right) w^{(b)}\left(\mathbf{r}_{g_2}\right) \sigma_1^{s_1}\left(\mathbf{r}_{g_1}\right) \sigma_1^{s_2}\left(\mathbf{r}_{g_2}\right) h_2^{s_1 s_2}\left(\mathbf{r}_{g_1}, \mathbf{r}_{g_2}\right)}{n_{s_1}^{(a)} n_{s_2}^{(b)}}$$ (55)

with gradients

$$\frac{\partial m_{s_1 s_2}^{(a,b)}}{\partial w^{(a)}\left(\mathbf{r}_{g_1}\right)} = \frac{\varpi_{g_1} \sigma_1^{s_1}\left(\mathbf{r}_{g_1}\right) \displaystyle\sum_{g_2=1}^{G} \varpi_{g_2} w^{(b)}\left(\mathbf{r}_{g_2}\right) \sigma_1^{s_2}\left(\mathbf{r}_{g_2}\right) h_2^{s_1 s_2}\left(\mathbf{r}_{g_1},\mathbf{r}_{g_2}\right)}{n_{s_1}^{(a)} n_{s_2}^{(b)}} - \frac{m_{s_1 s_2}^{(a,b)}}{n_{s_1}^{(a)}} \frac{\partial n_{s_1}^{(a)}}{\partial w^{(a)}\left(\mathbf{r}_{g_1}\right)}$$

$$\text{(56)}$$

$$\frac{\partial m_{s_1 s_2}^{(a,b)}}{\partial w^{(b)}\left(\mathbf{r}_{g_2}\right)} = \frac{\varpi_{g_2} \sigma_1^{s_s}\left(\mathbf{r}_{g_2}\right) \displaystyle\sum_{g_1=1}^{G} \varpi_{g_1} w^{(a)}\left(\mathbf{r}_{g_1}\right) \sigma_1^{s_1}\left(\mathbf{r}_{g_1}\right) h_2^{s_1 s_2}\left(\mathbf{r}_{g_1},\mathbf{r}_{g_2}\right)}{n_{s_1}^{(a)} n_{s_2}^{(b)}} - \frac{m_{s_1 s_2}^{(a,b)}}{n_{s_2}^{(b)}} \frac{\partial n_{s_2}^{(b)}}{\partial w^{(b)}\left(\mathbf{r}_{g_2}\right)}$$

where the derivative in the second term is domain independent,

$$\frac{\partial n_s^{(a)}}{\partial w_s^{(a)}\left(\mathbf{r}_g\right)} = \varpi_g \sigma_1^s\left(\mathbf{r}_g\right) \tag{57}$$

Because computing and storing the gradient is quite expensive, it is probably advisable to do a preliminary optimization for a short master list (corresponding to a large value of ε in the algorithm proposed in Section 5.4).

It seems probable that the maximization of the probability with respect to the partitioning weights/domains is difficult not only because of the expense of evaluating the objective function and its gradient, but also because the optimization surface may be extremely flat. The maximum probability will be very small in most cases, because it corresponds to the joint probability of finding exactly $\left(N_\alpha^{(1)}, N_\beta^{(1)}\right)$ electrons in partition 1 *and* exactly $\left(N_\alpha^{(2)}, N_\beta^{(2)}\right)$ electrons in partition 2 *and* ... *and* exactly $\left(N_\alpha^{(D)}, N_\beta^{(D)}\right)$ electrons in partition D. Even in a favorable case where the probability that a domain has the desired population is high, the joint probability will be extremely small anytime D is large. This means that the optimization procedure is being asked to decide between several competing (local) maximum probability partitions with nearly equal numerical values. Numerically, this problem can be (partially) fixed by using the logarithm of the probability as the objective function. Ignoring constant terms, the log-probability is then

$$\text{logp}\left(\left\{N_s^{(d)}\right\}_{\substack{d=1 \\ s=\alpha,\beta}}^{D}\right) = \ln P_0 + \ln\left(1 + \text{sum}(\mathcal{C})\right) \tag{58}$$

This numerical improvement does not solve the *interpretative* difficulty: if $p\left(\{N_s^{(d)}\}_{\substack{d=1 \\ s=\alpha,\beta}}^{D}\right)$ is vanishingly small, then the relative importance of the electron configuration it describes may be doubted. A more robust descriptor might be the total probability associated with the topological attractor associated with $p\left(\{N_s^{(d)}\}_{\substack{d=1 \\ s=\alpha,\beta}}^{D}\right)$ (that is, the total probability of every partitioning that can be used as an initial guess for a (local) maximization that leads to the targeted maximum-probability-partitioning). However, it seems practically impossible to implement this approach. For example, even a very approximate numerical approach would require evaluating the second derivative of the electron number probability function, which seems intractable.

5.6   CONCLUSION

This paper grew out of our fascination with the problem of computing and optimizing maximum probability domains. The first problem we identified was that maximum probability domains were computed primarily for single-Slater-determinant wavefunctions, and rarely for the types of modern wavefunctions that have quantitative accuracy. One may be skeptical of any interpretation based on a wavefunction that one would not be willing to use for quantitative prediction. Moreover, molecules where the identity of the most appropriate Lewis structure(s) is uncertain tend to be multireference, and in these cases the results from single-Slater-determinant wavefunctions are *qualitatively* incorrect. Therefore maximum probability domains are most useful exactly in the case where they are most challenging to apply.

One advantage of the algorithm we propose is that the key inputs—the hole-correlation integrals in Eqs. (27), (28), and (32)—are the same for all wavefunction forms. We only presuppose the ability to compute the low-order electron distribution functions of the wavefunction form of interest (e.g., via the reduced density matrices). A disadvantage of our algorithm is that it will not be as efficient as other approaches for "simple" cases (a few Slater determinants partitioned into very few regions). Another disadvantage of our approach is that it is inherently limited by the accuracy of the generalized Kirkwood superposition principle, truncated at low order (probably $K = 3$ or 4, in practice). We assume that the superposition approximation is accurate enough (and, as discussed earlier, the approximations inherent in the superposition approximation are closely aligned to the approximations commonly made in quantum chemistry).

The second problem we identified is that maximum probability domains are usually used only for division of a molecule into two regions. Occasionally three or more regions are considered, but the methods for doing so have exponential computational cost. Our approach allows us to directly compute the probability of a specific partitioning of electrons into regions, $p\left(\left\{N_s^{(d)}\right\}_{d=1;s=\alpha,\beta}^{D}\right)$. The cost of our approach grows as D^K, where D is the number of partitions and K is the maximum number of distinct partitions over which we integrate a correlated probability. (This scaling is determined by the number of numerical integrals that have to be performed. It is favorable, however, that as the number of partitions increases, the integration volume of each region decreases.) Our algorithm remains practical even when one attempts to partition the system into Lewis pairs, where the total number of regions is half the total number of electrons.

ACKNOWLEDGMENTS

The authors Paul W. Ayers, Stijn Fias, and Debajit Chakraborty would like to thank NSERC (Canada) and Compute Canada for funding. SF acknowledges support from a Marie Curie postdoctoral fellowship. The authors Guillaume Acke and Patrick Bultinck acknowledge support from the FWO (Belgium). All of us wish to acknowledge the inspiration that Prof. Ramon Carbó-Dorca has provided to us over the years. Three of us (Paul W. Ayers, Stijn Fias, and Patrick Bultinck) have had the privilege of collaborating with Ramon. His spark of creativity, his intellectual fecundity, his propensity for creative outside-the-box (and sometimes even outside-the-building!) thinking, his breadth of knowledge and skills, his humor, his thoughtfulness, and his generosity are *sui generis*.

KEYWORDS

- **electron number probabilities**
- **Kirkwood superposition principle**
- **Lewis pairs**
- **maximum probability domains**
- **multidomain partitioning**
- **N-electron distribution function**

REFERENCES

1. Ayers, P. W., Boyd, R. J., Bultinck, P., Caffarel, M., Carbó-Dorca, R., & Causà, M., (2015). Six questions on topology in theoretical chemistry. *Comput. Theor. Chem.*, *1053*, 2–16.
2. Mulliken, R. S., (1965). *J. Chem. Phys.*, *43*, S2.
3. Head-Gordon, M., (1996). Quantum Chemistry and Molecular Processes. *J. Phys. Chem.*, *100*, 13213–13225.
4. Ayers, P. W., (2007). The physical basis of the hard/soft acid/base principle. *Faraday Discuss*, *135*, 161–190.
5. Lewis, G. N., (1916). The atom and the molecule. *J. Am. Chem. Soc.*, *38*, 762–785.
6. Wheland, G., (1955). Resonance in Organic Chemistry. Wiley: New York.
7. Gallup, G. A., (2002). *Valence Bond Methods: Theory and Applications*. Cambridge UP: Cambridge.
8. Hurley, A. C., Lennard-Jones, J., & Pople, J. A., (1953). The molecular orbital theory of chemical valency XVI. A theory of paired-electrons in polyatomic molecules *Proc. R. Soc. London, Ser. A*, *220*, 446–455.
9. Parr, R. G., Ellison, F. O., & Lykos, P. G., (1956). Generalized antisymmetric product wave functions for atoms and molecules. *J. Chem. Phys.*, *24*, 1106.
10. Parks, J. M., & Parr, R. G., (1958). Theory of separated electron pairs. *J. Chem. Phys.*, *28*, 335–345.
11. Zoboki, T., Jeszenszki, P., & Surjan, P. R., (2013). Composite particles in quantum chemistry: From two-electron bonds to cold atoms. *Int. J. Quantum Chem.*, *113*(3), 185–189.
12. Surjan, P. R., (1999). An introduction to the theory of geminals. In: *Correlation and Localization*, Vol. 203, Surjan, P. R.(ed.), pp. 63–88.
13. Tecmer, P., Boguslawski, K., Johnson, P. A., Limacher, P. A., Chan, M., Verstraelen, T., et al., (2014). Assessing the Accuracy of New Geminal-Based Approaches. *J. Phys. Chem. A*, *118*(39), 9058–9068.
14. Daudel, R., Brion, H., & Odiot, S., (1955). Localizability of electrons in atoms and molecules: Application to the study of the notion of shell and the nature of chemical bonds. *J. Chem. Phys.*, *23*, 2080–2083.
15. Aslangul, R., Constanciel, R., Daudel, R., & Kottis, P., (1972). Aspects of the localizability of electrons in atoms and molecules: Loge theory and related methods. *Adv. Quantum Chem.*, *6*, 93–141.
16. Daudel, R., Bader, R. F. W., Stephens, M. E., & Borett, D. S., (1974). The electron pair in chemistry. *Can. J. Chem.*, *52* 1310–1320.
17. Daudel, R., Bader, R. F. W., Stephens, M. E., & Borett, D. S., (1974). The electron pair in chemistry. *Can. J. Chem.*, *52*, 3077.
18. Daudel, R., (1976). Quantum mechanical facets of chemical bonds. In:*The New World of Quantum Chemistry: Proceedings of the Second International Congress of Quantum Chemistry*, Pullman, B., Parr, R. G., (eds.). Reidl: Dordrecht, Vol. 2, pp. 33–56.
19. Ayers, P. W., (2005). Electron localization functions and local measures of the covariance. *J. Chem. Sci.*, *117*, 441–454.
20. Kohout, M., (2004). A measure of electron localizability. *Int. J. Quantum Chem.*, *97*, 651–658.

21. Becke, A. D., & Edgecombe, K. E., (1990). A simple measure of electron localization in atomic and molecular systems. *J. Chem. Phys.*, *92*, 5397–5403.
22. Savin, A., (2005). On the significance of ELF basins. *J. Chem. Sci.*, *117* (5), 473–475.
23. Savin, A., (2005). The electron localization function (ELF) and its relatives: interpretations and difficulties. *J. Mol. Struct. theochem*, *727* (1–2), 127–131.
24. Chamorro, E., Fuentealba, P., & Savin, A., (2003). Electron probability distribution in AIM and ELF basins. *J. Comput. Chem.*, *24*, 496–504.
25. Savin, A., Nesper, R., Wengert, S., & Fassler, T. F., (1997). ELF: The electron localization function. *Angew. Chem.*, *36*, 1809–1832.
26. Silvi, B., & Savin, A., (1994). Classification of chemical bonds based on topological analysis of electron localization functions. *Nature*, *371*, 683–686.
27. Ayers, P. W., Parr, R. G., & Nagy, A., (2002). Local kinetic energy and local temperature in the density-functional theory of electronic structure. *Int. J. Quantum Chem.*, *90*, 309–326.
28. Schmider, H. L., & Becke, A. D., (2002). Two functions of the density matrix and their relation to the chemical bond. *J. Chem. Phys.*, *116* (8), 3184–3193.
29. Schmider, H. L., & Becke, A. D., (2000). Chemical content of the kinetic energy density. *Theochem-Journal of Molecular Structure*, *527*, 51–61.
30. de Silva, P., & Korchowiec, J., & Wesolowski, T. A., (2014). Atomic shell structure from the Single-Exponential Decay Detector. *J. Chem. Phys.*, *140*, 164301.
31. Bohorquez, H. J., & Boyd, R. J., (2010). A localized electrons detector for atomic and molecular systems. *Theor. Chem. Acc.*, *127*, 393–400.
32. Bohorquez, H. J., Matta, C. F., & Boyd, R. J., (2010). The localized electrons detector as an Ab initio representation of molecular structures. *Int. J. Quantum Chem.*, *110*, 2418–2425.
33. Bader, R. F. W., Gillespie, R. J., & Macdougall, P. J., (1988). A physical basis for the VSEPR model of molecular geometry. *J. Am. Chem. Soc.*, *110*, 7329–7336.
34. Sperber, G., (1971). Analysis of Reduced Density Matrices in the Coordinate Representation. II. The Structure of Closed-Shell Atoms in the Restricted Hartree-Fock Approximation. *Int. J. Quantum Chem.*, *5*, 189–214.
35. de Silva, P., & Corminboeuf, C., (2014). Simultaneous visualization of covalent and noncovalent interactions using regions of density overlap. *J. Chem. Theory Comp.*, *10* (9), 3745–3756.
36. de Silva, P., Korchowiec, J., & Wesolowski, T. A., (2012). Revealing the Bonding Pattern from the Molecular Electron Density Using Single Exponential Decay Detector: An Orbital-Free Alternative to the Electron Localization Function. *Chem Phys Chem*, *13*, 3462–3465.
37. de Silva, P., Korchowiec, J., Ram, J. S. N., & Wesolowski, T. A., (2013). Extracting Information about Chemical Bonding from Molecular Electron Densities via Single Exponential Decay Detector (SEDD). *Chimia*, *67*, 253–256.
38. Sagar, R. P., & Guevara, N. L., (2007). Local correlation measures and atomic shell structure. *Chem. Phys. Lett.*, *438* (4–2), 330–335.
39. Rahm, M., & Christe, K. O., (2013). Quantifying the nature of lone pair domains. *Chemphyschem*, *14* (16), 3714–3725.
40. Causà, M., Savin, A., & Silvi, B., (2014). Atoms and bonds in molecules and chemical explanations. *Foundations of Chemistry*, *16* (1), 3–26.
41. Astakhov, A. A., & Tsirelson, V. G., (2014). Spatial localization of electron pairs in molecules using the Fisher information density. *Chem. Phys.*, *435*, 49–56.

42. Acke, G., De Baerdemacker, S., Claeys, P. W., Van Raemdonck, M., Poelmans, W., Van Neck, D., et al., (2016). Maximum probability domains for Hubbard models. *Mol. Phys.*, *114* (7–8), 1392–1405.

43. Jacobsen, H., (2008). Localized-orbital locator (LOL) profiles of chemical bonding. *Canadian Journal of Chemistry-Revue Canadienne De Chimie*, *86*, 695–702.

44. Urbina, A. S., Torres, F. J., & Rincon, L., (2016). The electron localization as the information content of the conditional pair density. *J. Chem. Phys.*, *144* (24), 244104.

45. Francisco, E., Martín Pendás, A., Garcia-Revilla, M., & Boto, R. A., (2013). A hierarchy of chemical bonding indices in real space from reduced density matrices and cumulants. *Comput. Theor. Chem.*, *1003*. 71–78.

46. Angyan, J. G., (2009). Electron Localization and the Second Moment of the Exchange Hole. *Int. J. Quantum Chem.*, *109* (11), 2340–2347.

47. Scemama, A., Caffarel, M., & Savin, A., (2007). Maximum probability domains from quantum Monte Carlo calculations. *J. Comput. Chem.*, *28*, 442–454.

48. Gallegos, A., Carbó-Dorca, R., Lodier, F., Cances, E., & Savin, A., (2005). Maximal probability domains in linear molecules. *J. Comput. Chem.*, *26*, 455–460.

49. Cances, E., Keriven, R., Lodier, F., & Savin, A., (2004). How electrons guard the space: shape optimization with probability distribution criteria. *Theor. Chem. Acc.*, *111* (2–2), 373–380.

50. Savin, A., (2002). Probability distributions and valence shells in atoms. In: *Reviews of Modern Quantum Chemistry*, Sen, K. D., Ed. World Scientific: Singapore, Vol. 1, pp. 43–62.

51. Acke, G., (2016). Maximum probability domains: Theoretical foundations and computational algorithms. Ph.D., Ghent University.

52. Scemama, A., (2005). Investigating the volume maximizing the probability of finding v electrons from Variational Monte Carlo data. *Journal of Theoretical & Computational Chemistry*, *4* (2), 397–409.

53. Causà, M., & Savin, A., (2011). Maximum Probability Domains in the Solid-State Structures of the Elements: the Diamond Structure. *Z. Anorg. Allg. Chem.*, *637* (7–8), 882–884.

54. Causà, M., & Savin, A., (2011). Maximum Probability Domains in Crystals: The Rock-Salt Structure. *J. Phys. Chem. A*, *115* (45), 13139–13148.

55. Menéndez, M., Martín Pendás, A., Braida, B., & Savin, A., (2015). A view of covalent and ionic bonding from Maximum Probability Domains. *Comput. Theor. Chem.*, (*1053*), 142–149.

56. Martín Pendás, A., Francisco, E., & Blanco, M. A., (2007). An electron number distribution view of chemical bonds in real space. *PCCP*, *9* (9), 1087–1092.

57. Martín Pendás, A., Francisco, E., & Blanco, M. A., (2007). Pauling resonant structures in real space through electron number probability distributions. *J. Phys. Chem. A*, *111* (6), 1084–1090.

58. Martín Pendás, A., Francisco, E., & Blanco, M. A., (2007). Spatial localization, correlation, and statistical dependence of electrons in atomic domains: The X-1 S(+)(g) and b(3)S (+) (u) states of H-2. *Chem. Phys. Lett.*, *437* (4–2), 287–292.

59. Martín Pendás, A., Francisco, E., &Blanco, M. A., (2007). Spin resolved electron number distribution functions: How spins couple in real space. *J. Chem. Phys.*, *127* (14), 144103.

60. Francisco, E., Martín Pendás, A., & Blanco, M. A., (2008). EDF: Computing electron number probability distribution functions in real space from molecular wave functions. *Comput. Phys. Commun.*, *178* (8), 621–634.

61. Francisco, E., Martín Pendás, A., & Blanco, M. A., (2009). A connection between domain-averaged Fermi hole orbitals and electron number distribution functions in real space. *J. Chem. Phys.*, *131* (12), 124125.

62. Francisco, E., Martín Pendás, A., & Blanco, M. A., (2011). Generalized electron number distribution functions: real space versus orbital space descriptions. *Theor. Chem. Acc.*, *128* (4–2), 433–444.

63. Francisco, E., Martín Pendás, A., & Costales, A., & Garcia-Revilla, M., (2011). Electron number distribution functions with iterative Hirshfeld atoms. *Comput. Theor. Chem.*, *975* (1–3), 2.

64. Francisco, E., & Martín Pendás, A., (2014). Electron number distribution functions from molecular wavefunctions. Version 2. *Comput. Phys. Commun.*, *185* (10), 2663–2682.

65. Ferro-Costas, D., Francisco, E., Martín Pendás, A., & Mosquera, R. A., (2015). Revisiting the carbonyl n -> p(star) electronic excitation through topological eyes: expanding, enriching and enhancing the chemical language using electron number distribution functions and domain averaged Fermi holes. *PCCP*, *17* (39), 26059–26071.

66. Martín Pendás, A., Francisco, E., & Blanco, M. A., (2007). Spin resolved electron number distribution functions: how spins couple in real space. *The Journal of Chemical Physics*, *127* (14), 144103–144103.

67. Francisco, E., Martín Pendás, A., & Blanco, M. A., (2007). Electron number probability distributions for correlated wave functions. *J. Chem. Phys.*, *126* (9), 094102.

68. Parr, R. G., & Bartolotti, L. J., (1983). Some remarks on the density functional theory of few-electron systems. *J.Phys.Chem.*, *87*, 2810–2815.

69. De Proft, F., Ayers, P. W., Sen, K. D., & Geerlings, P., (2004). On the importance of the "density per particle" (shape function) in the density functional theory. *J. Chem. Phys.*, *120*, 9969–9973.

70. Ayers, P. W., (2000). Density per particle as a descriptor of Coulombic systems. *Proc. Natl. Acad. Sci.*, *97*, 1959–1964.

71. Ayers, P. W., & Cedillo, A., (2009). The shape function. In: *Chemical Reactivity Theory: A Density Functional View*, Chattaraj, P. K., Ed. Taylor and Francis: Boca Raton, p. 269.

72. Bultinck, P., & Carbó-Dorca, R., (2004). A mathematical discussion on density and shape functions, vector semispaces and related questions. *J. Math. Chem.*, *36* (2), 191–200.

73. Ziesche, P., (2000). On relations between correlation, fluctuation and localization. *J. Mol. Struct. Theochem*, *527*, 35–50.

74. Ziesche, P., (2000). Cumulant expansions of reduced densities, reduced density matrices, and Green's functions. In *Many-electron densities and reduced density matrices*, Cioslowski, J., Ed. Kluwer: New York, pp. 33–56.

75. Ayers, P. W., & Davidson, E. R., (2007). Linear inequalities for diagonal elements of density matrices. *Adv. Chem. Phys.*, *134*, 443–483.

76. Davidson, E. R., (1976). *Reduced Density Matrices in Quantum Chemistry*. Academic Press: New York.

77. Coleman, A. J., & Yukalov, V. I., (2000). *Reduced Density Matrices: Coulson's Challenge*. Springer: Berlin.

78. Mazziotti, D. A., (2006). Quantum chemistry without wave functions: Two electron reduced density matrices. *Acc. Chem. Res.*, *39*, 207–215.

79. Ayers, P. W., (2005). Generalized density functional theories using the k-electron densities: Development of kinetic energy functionals. *J. Math. Phys.*, *46*, 062107.

80. Lee, D. K., Jackson, H. W., & Feenberg, E., (1967). *Ann. Phys. (Amsterdam, Neth.)*, *44*, 84–94.

81. Wu, F. Y., & Chien, M. K., (1970). Convolution approximation for the n-particle distribution function. *J. Math. Phys.*, *11*, 1912–1916.

82. Ayers, P. W., (2008). Constraints for hierarchies of many electron distribution functions. *J. Math. Chem.*, *44*, 311–323.

83. Ayers, P. W., & Davidson, E. R., (2006). Necessary conditions for the N-representability of pair distribution functions. *Int. J. Quantum Chem.*, *106*, 1487–1498.

84. Ayers, P. W., (2006). Using classical many-body structure to determine electronic structure: An approach using k-electron distribution functions. *Phys. Rev. A*, *74*, 042502.

85. Davidson, E. R., (1970). Uncertainty Principle for Ensembles. *Phys. Rev. A*, *1*, 30–32.

86. Garrod, C., & Percus, J. K., (1964). Reduction of the N-Particle Variational Problem. *J. Math. Phys.*, *5* 1756–1776.

87. Pistol, M. E., (2007). Investigations of random pair densities and the application to the N-representability problem. *Chem. Phys. Lett.*, *449*, 208–211.

88. Pistol, M. E., (2006). Characterization of N-representable n-particle densities when N is infinite. *Chem. Phys. Lett.*, *417*, 521–523.

89. Pistol, M. E., (2006). Relations between N-representable n-particle densities. *Chem. Phys. Lett.*, *422*, 363–366.

90. Pistol, M. E., (2006). N-representable distance densities have positive Fourier transforms. *Chem. Phys. Lett.*, *431*, 216–218.

91. Pistol, M. E., (2004). N-representability of two-electron densities and density matrices and the application to the few-body problem. *Chem. Phys. Lett.*, *400*, 548–552.

92. Mazo, R. M., & Kirkwood, J. G., (1958). Statistical Thermodynamics of Quantum Fluids. *J. Chem. Phys.*, *28* (4), 644–647.

93. Kirkwood, J. G., (1935). Statistical mechanics of fluid mixtures. *J. Chem. Phys.*, *3*, 300–313.

94. Reiss, H., (1972). Superposition Approximations from a Variation Principle. *Journal of Statistical Physics*, *6* (1), 39–47.

95. Singer, A., (2004). Maximum entropy formulation of the Kirkwood superposition approximation. *J. Chem. Phys.*, *121*, 3657–3666.

96. Attard, P., Jepps, O. G., & Marcelja, S., (1997). Information content of signals using correlation function expansions of the entropy. *Phys. Rev. E*, *56*, 4052–4067.

97. Holmes, A. A., Tubman, N. M., & Umrigar, C. J., (2016). Heat-bath configuration interaction: an efficient selected configuration interaction algorithm inspired by heat-bath sampling. *J. Chem. Theory Comp.*, *12* (8), 3674–3680.

98. Holmes, A. A., Changlani, H. J., & Umrigar, C. J., (2016). Efficient heat-bath sampling in Fock space. *J. Chem. Theory Comp.*, *12* (4), 1561–1571.

99. Coleman, A. J., (1967). Density matrices in the quantum theory of matter: Energy, intracules, and extracules. *Int. J. Quantum Chem.*, *1* (S1), 457–464.

CHAPTER 6

SOME USEFUL PROCEDURES AND CONCEPTS IN QSAR/QSPR

EMILI BESALÚ,[1] LIONELLO POGLIANI,[2,3] and
J. VICENTE DE JULIÁN-ORTIZ[2,4]

[1]Department of Chemistry and Institute of Computational Chemistry and Catalysis, University of Girona, C/Maria Aurélia Capmany 69, 17003 Girona, Spain

[2]Drug Design and Molecular Connectivity Investigation Unit, Department of Physical Chemistry, Pharmacy Faculty, University of Valencia, Av V. Andrés Estellés 0, 46100 Burjassot, Valencia, Spain

[3]MOLware SL, C/Burriana 36-3, 46005 Valencia, Spain

[4]ProtoQSAR SL, Scientific Park of the University of Valencia, C/Catedrático Agustín Escardino 9, 46980 Paterna, Valencia, Spain

CONTENTS

The QSAR/QSPR field relies on several concepts like parameter definitions, algorithms for prediction, plots and data fitting. All these notions help to get good fitting or classification models and to assess their predictive ability. Here, some of these concepts are discussed to highlight their usefulness and misconceptions.

6.1   TRUE PREDICTION CONCEPT: INTERNAL TEST SETS ALGORITHM

In almost all the modern publications in the QSAR field it is compulsory to present reliable and robust models. A way to test for the predictive abilities of those models consist into validate them. Certainly, the validation via an external set is paradigmatic. This step usually consists to simulate a known molecular set to be predicted *after* the model has been build and tested. But it is recognized that this check is expensive in the sense that more molecules are to be available. In many cases researches try to simulate by means of predictions obtained via cross-validation (CV) procedures, for instance leave-one-out (L1O), leave-two-out (L2O), leave-many-out (LMO) or n-fold CV. It is admitted that, in general, a CV is not as demanding as a pure external validation test. Nevertheless, the need to incorporate in training (i.e., model construction) as many molecules as possible renders the CV algorithms a routine procedure.

Several years ago, one of us described the Internal Test Sets (ITS) algorithm (1). The ITS procedure is based on the concept of CV but preserving the idea to generate molecular predictions exactly in the same manner as if these were generated *a posteriori*, i.e., if the compounds were released *after* the model was build. The concept is simple, but the conditions are not met in many modern QSAR studies. For instance, prior to model construction a variable selection is done. This is a necessary step for QSAR modeling, but normally in the selection process *all* the compounds are included, even the ones belonging to the external set. This is not realistic. In the same way, when CV procedures are followed, a previous global variable selection conditioned the data. ITS algorithm focuses on the idea to do any kind of CV but from scratch. This means that, at every CV cycle, all (absolutely all) the model design steps are to be started from the very beginning. This implies, for instance, that all the variables are to be kept and at every CV step the variable selection is to be redone without taking into account the cross-validated compound/s. This serves to simulate that the left out struc-tures will be 'known' *a posteriori*, when the variable selection is already done. ITS procedure is more time consuming than other CV protocols, but it constitutes a coherent algorithm. One of its main advantages relies on the fact that the predictions made are risky and hence more realistic. This helps to test the model robustness in a very direct way, and for to avoid overparametrization. Additionally, the procedure serves to identify outliers as these compounds will normally show very bad predictions. After the CV, the simple counting of variables entering in models serves to select them if a final and unique model is to be constructed.

The ITS concept is to be considered in any kind of model building or test by CV procedures. It has been applied to select linear QSAR models. In Ref. [1], a set of 21 previously published COX-2 inhibitors was analyzed in order to obtain true predictions of the activity. Overfitting in the models has effectively been prevented even for small training sets. The most significant variables for the prediction of the property were selected, and some outli-ers were easily identified with this approach. In another study, Topological Indices (TI) were applied to build QSAR models for a set of 20 antima-larial cyclic peroxy ketals [2]. In order to evaluate the reliability of the proposed linear models leave-n-out and Internal Test Sets (ITS) approaches were considered. The proposed procedure resulted in a robust and valid pre-diction equation that is superior to the employed standard cross-validation algorithms involving multilinear regression (MLR) models. For instance, the classical MLR-CV models involving three variables gave r^2 values of 0.88 (fitting) and 0.83 for L1O and L2O. Some models were coincident respect to the variables selected. This seemed promising and robust. But the application of the ITS protocol gave very distinct results. The prediction performance varied irregularly and the results for 3 descriptors worsened dramatically: r^2 turned to be 0.0! All in all, the ITS footage helped to detect that all the models involving 1–4 descriptors were spurious. Only the mod-els involving five descriptors showed to be robust and predictive (six or more descriptors were not tested): $r^2 = 0.97$ for classical MLR fit and 0.95 for L1O and L2O, and the respective ITS counterparts gave 0.66 (L1O) and 0.68 (L2O).

ITS protocol forces to obtain diverse models (one for each CV cycle) opening the possibility to performing a statistical study over the selected descriptors, and even the construction of a consensus model. In Ref. [3] two outliers were detected and, hence, 19 simple linear models were obtained.

This helped to detect how some indices were used most the time, even pre-serving their coefficient sign.

ITS paradigm was also considered to model anti-tuberculosis compounds [4] involving 64 benzoxazine and phenylquinazoline derivatives. Predictions were made *in silico* for 512 new molecules, proposing new structures as especially promising candidates for active anti-tuberculotic drugs. The pro-tocol helped to rely only on stable models, avoiding again overparameter-ization and overfitting. Curiously, again the best model resulted to have five descriptors.

In Ref. [5], ITS was considered to model a database of structurally het-erogeneous chemical structures. In this case, the molecular property was the lowest observed adverse effect levels (LOAELs). ITS was used here to select the variables entering in both MLR and linear discriminant (LD) models involving graph theoretical descriptors.

6.2 PROBABILISTIC ANALYSIS: DISCERNING BETWEEN THE CONCEPTS OF DIFFICULTY AND USEFULNESS

Discerning between the concepts of difficulty and usefulness of a molecular ranking classification is of capital importance in virtual design chemistry. Here, both concepts are viewed from the statistical and practical point of view according to the standard definitions of enrichment and statistical sig-nificance p-values. These parameters are useful not only to compare dis-tinct rankings obtained for the same molecular database, but also in order to compare from an objective point of view the ones established in distinct molecular sets [6]. Simple formulations are able to quantify the concepts of *difficulty* and *utility* of a molecular classification ranking.

6.2.1 DIFFICULTY: MEASURED BY P-VALUES

One way to quantify the difficulty to rank a molecular set is to rely on sta-tistical p-values which, ultimately, are based on probabilistic evaluations. For instance, let us consider a series of m molecules, a database, for which $n<m$ of them have a property of interest (being a drug, being active, not being hemolytic, being easily synthesizable, ...). Then, consider the situa-tion where s of these m molecules are randomly selected. Then, one may ask what is the probability that r of the selected molecules are also of interest

(i.e., belong to the subset of n). Under an equal basis, this probability is given by the hypergeometrical combinatorial number [4, 7]:

$$P(r,n;s,m) = \frac{\binom{n}{r}\binom{m-n}{s-r}}{\binom{m}{s}} \quad \text{with } r \leq s \leq m \text{ and } r \leq n \leq m \quad (1)$$

From this, the statistical significance levels (p-values) are obtained from cumulated probabilities, that is, given some valid fixed values for m, n and s, the p significance level corresponds to the probability to select r or more (denoted $r+$) structures of interest:

$$p(r+,n;s,m) = \sum_{i=r}^{\min(n,s)} P(i,n;s,m) = 1 - \sum_{i=\max(0,s+n-m)}^{r-1} P(i,n;s,m) \quad (2)$$

Clearly, this quantity informs about the *difficulty* to reproduce the situation of embracing r or more active compounds. This formulation has been considered to design the superposing significant interaction rules (SSIR) method [8, 9] or in other fields [4, 7].

6.2.2 UTILITY: MEASURED BY ENRICHMENT FACTORS

Despite a molecular ranking obtained by a particular QSAR technique may give apparently impressive p-values (i.e., very small values as, for instance, 10^{-8}), this has to be faced against the utility or usefulness of the ranking. Many times, for medicinal chemistry scientists and QSAR practitioners, the ranking is as useful as the enrichment of compounds of interests it collects at the beginning of the sorted list. And this enrichment sometimes is attached to quite big p-values or, some other times, is inherent to a small p-value. In other words, despite a correlation exists between the obtained enrichment and the attached p-values (the higher the first, the lesser the last), a simple look at p-values does not suffice to know about the practical help given by the ranked family.

The basic definition for the natural enrichment factor corresponds to the actual ratio of molecules of interest found in a subset (r/s) divided by the overall ratio of target molecules within the whole database (n/m). Following the notation employed in previous section, this is

$$e = \frac{r/s}{n/m} = \frac{r}{\frac{n}{m}s} \tag{3}$$

The last equality explicitly shows an equivalent and practical definition: the enrichment is the same as the quotient between the number of active compounds found in the subset and the proportional expected number of active structures in a random selection of s items.

6.3 SOME PARTICULARITIES RELATED TO THE ORDINARY OR MULTILINEAR LINEAR REGRESSIONS

For many scientists, an inherent characteristic of the ordinary Least Squares (LS) method has been overlooked. This feature is also found in the multilinear linear regression (MLR) context and it consists on a subtle asymmetry found when representing the fitted values by the linear method (LS or MLR) against the experimental ones [10]. An example will clarify the concepts. Let us assume that a particular dependent variable y is to be adjusted by a single independent variable x (as said, the result is also applicable if several independent variables are being considered, as in the MLR case). For instance take the data in Table 6.1 where 100 pairs of (x,y) samples are shown. The reader can consider another table (changing the points, the number of them, the correlation among variables, taking only a single column or spread points, and so on) and perform a parallel simulation getting the same results, as these are general and sustained by a theorem.

The regression equation for the 100 points is $y_{adj} = 2.025886\,x + 1.477874$ and the coefficient of determination is $r^2 = 0.831254$. In Figure 6.1a, it has been represented the 'experimental' values ($y = y_{exp}$) against the adjusted ones (y_{adj}), whereas in Figure 6.1b, it has been represented the 'inverted' graph, i.e., the y_{adj} values against the y ones. In each graph a new one-variable ordinary linear regression equation has been depicted. Hence, in Figure 6.1a the experimental values have been fitted according to a simple LS equation as a function of the adjusted values. Conversely, in Figure 6.1b the adjusted values have been fitted according to a simple LS equation as a function of the experimental ones. Both fittings correspond to the depicted diagonal solid lines.

TABLE 6.1   A Set of 100 Data Pairs of Correlated Variables, One Considered Dependent (y) and the Other Independent (x)

x	y	x	y	x	y	x	y	x	y
0.10	1.52	0.08	1.39	0.25	2.44	0.74	3.33	0.40	2.42
0.50	2.34	0.12	1.65	0.28	2.26	0.60	2.90	0.71	2.80
0.81	3.44	0.14	2.07	0.68	3.03	0.98	3.61	0.63	2.83
0.78	3.26	0.01	1.04	0.41	2.54	0.43	2.29	0.97	3.90
0.51	2.72	0.63	3.19	0.89	3.01	0.28	2.48	0.43	2.20
0.32	1.88	0.12	1.78	0.45	1.96	0.99	3.85	0.58	2.70
0.27	2.10	0.38	1.94	0.23	2.23	0.65	2.97	0.79	3.32
0.27	1.82	0.38	1.83	0.05	1.87	0.00	1.30	0.11	1.23
0.45	2.76	0.95	3.03	0.02	1.79	0.48	2.38	0.51	2.51
0.76	2.55	0.70	2.93	0.34	2.63	0.06	1.73	0.54	2.26
0.51	2.62	0.35	1.73	0.91	3.01	0.51	2.32	0.42	2.15
0.22	1.87	0.03	1.40	0.65	2.43	0.19	2.10	0.70	2.93
0.11	1.98	0.51	2.66	0.23	1.75	0.89	3.51	0.41	2.25
0.77	2.91	0.78	2.85	0.95	3.05	0.93	3.71	0.49	2.95
0.97	3.14	0.02	1.19	0.79	2.76	0.11	1.54	0.46	2.33
0.15	1.40	0.29	2.09	0.22	2.16	0.32	2.47	0.95	3.29
0.24	1.88	0.10	1.98	0.70	2.45	0.25	2.49	0.45	2.01
0.94	3.39	0.95	3.80	0.76	2.75	0.84	3.24	0.63	2.92
0.48	2.59	0.18	1.68	0.94	3.65	0.99	3.38	0.81	2.92
0.25	2.00	0.22	1.93	0.13	1.90	0.76	2.52	0.17	1.78

6.3.1 THEOREM ON SLOPE

The interesting feature is that the ordinary regression equation depicted in Figure 6.1a is *always* $y_{exp} = y_{adj}$, the straight line of slope 1 and ordinate at the origin zero. In turn, the ordinary regression equation depicted in Figure 6.1b (solid line) is

$$y_{adj} = 0.831254\, y_{exp} + 0.414811 \qquad (4)$$

Is now the reader able to relate this last slope with another parameter? Note that this slope is *always* equal to the original coefficient of determination cited above. As said, this result is general, even for data coming from a MLR fitting. The theorem [10–12] states that the above slopes are always coincident with 1 and r^2, respectively. In this context,

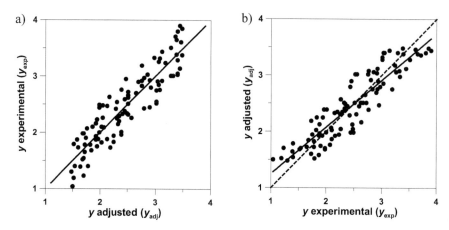

FIGURE 6.1A,B   Graphical representation of experimental and adjusted values by the linear method. (a) experimental *vs.* adjusted. (b) adjusted vs. experimental.

this constitutes a more specific result than the general one which states the relationship between the slopes of 'inverted' regressions: the product of slopes is always r^2 (in any case, the slopes are not both 1, as some people could anticipate).

So, a word of caution must be stated here: when QSAR results coming from multilinear equations are reported (direct MLR, CoMFA, ...) it is erroneous to represent the adjusted property values *vs.* the experimental ones as in Figure 6.1b, and claim that the points seem to follow the *ideal* bisector line $y = x$. This 'optical' effect is ensured by the cited theorem. This behavior is a consequence of the so-called regression towards the mean effect [13], for which fitted y values tend to be greater than the original ones if the corresponding original value is lesser than the y_{exp} mean, whereas on the other side, i.e., for y values greater than its mean, the fitted values tend to be smaller. This feature becomes visually evident when the determination coefficient is far away from the unit (e.g., 0.8 as in the example above): note the asymmetric tendency for rotation of the spread points in Figure 6.1b around the dashed line (the quadrant bisector). This feature is normally overlooked when the r^2 value approaches the unit. This is the case found in many precision fittings, as those found in the analytical chemistry field. These concepts have brought us to clarify the relationship between the so-called direct and inverse calibration [14].

6.3.2 REGRESSION TOWARDS THE MEAN EFFECTS IN MLR-L1O CROSS-VALIDATION

The leave-one-out (L1O) approach is a commonly used technique. In particular, this procedure is many times considered when MLR models are being constructed. This is so because the L1O results can be obtained rapidly without explicitly reproducing all the L1O steps [14]. Related with previous results, it was shown how MLR-L1O predictions will always present systematic deviations which magnify the regression towards the mean effects. Values predicted by the MLR-L1O technique are obtained from the following equation [12]:

$$y_i' = \frac{h_{ii} y_i - \hat{y}_i}{h_{ii} - 1}, \quad i = 1, 2, \ldots, n, \tag{5}$$

where y_i are experimental values, $\hat{y}_i$ are the values adjusted by the overall MLR data fitting, and each h_{ii} term is a diagonal element of the so-called hat matrix, $H = D(D^T D)^{-1} D^T$. Here D is the matrix of QSAR descriptors. From the last equation it can be seen how to relate the numerical differences between experimental, fitted and cross-validated values:

$$\hat{y}_i - y_i = (1 - h_{ii})(y_i' - y_i) \tag{6}$$

On the other hand, it is well known that the h_{ii} terms are bounded, $m/n \leq h_{ii} \leq 1$, m being here the number of descriptors and n the number of equations. This condition implies $0 \leq 1 - h_{ii} < 1$, because $n > m$. Then, it is immediate to see that the differences $\hat{y}_i - y_i$ and $y_i' - y_i$ appearing above bear the same sign and, additionally, the second difference increases, *in absolute value*, relatively to the first one. This leads to the following general interpretation: fitted points that where overestimated when fitting the original data are attached to L1O cross-validated values, which in turn are more overestimated. Conversely, fitted points originally underestimated will always give even more underestimated L1O predictions once they are cross-validated. Another consequence being that the r^2_{LOO} parameter will always decrease relatively to the fitting one, as expected. Due to the regression effects reviewed above, the procedure will tend to generate many cross-validated points with a regression towards the mean effect (raising predicted values) on the left of the bidimensional graph. In the same way, due to regression towards the mean effect decreasing predicted values are expected to be normally found on the right part of the graph.

6.4   ORTHOGONAL REGRESSION

In many fields of chemistry, the ordinary least-squares method is preferentially used to fit data. Nevertheless, univariate linear regression by classical least-squares (CLS) analysis has some drawbacks that are usually overlooked in experimental science courses and even in many chemical research papers. Mainly, some users, and especially students, tend to "forget" the basic assumptions of the method and apply it over series of data pairs having errors in both variables. Evenmore concerning, as said above, some people still believe that the classical univariate linear regression performs a symmetric treatment of the data series x and y. Then, at the end of the treatment when the fitting equation is obtained, the mathematical expression is erroneously and freely manipulated, ignoring that the regression equation of y over x is not the same as the equation for x over y fit. Orthogonal least-squares (OLS) fitting is a good method to avoid the unsymmetrical treatment of the data [16]. This is because the CLS method (in the y over x case, for instance) only minimizes the squared distance parallel to the y axis between the experimental points and the fitting line, whereas the distance parallel to the x axis is not considered because it is understood or assumed to be error or uncertainty free. The OLS regression is an alternative for obtaining symmetric treatments because it minimizes the sum of quadratic orthogonal distances to the equation line, as can be seen in Figure 6.2. It is an intuitive method that deals with errors in both variables.

The details on this OLS method can be seen in the cited reference 16. A critical examination of four different least squares orthogonal methods and of the ordinary least squares (LS) method, which is normally used in QSAR and QSPR studies, and in many scientific and chemistry-related fields, reveals that not always the orthogonal regression methods perform better than LS in the aforementioned fields. Nonetheless the OR methods, whose use in statistics and economics are considered superior in most cases to LS, relying on the minimization of the sum or quadratic orthogonal distances offer an interesting alternative method for obtaining a graphical 'symmetric' representation, which is not rendered by the LS method [13, 17].

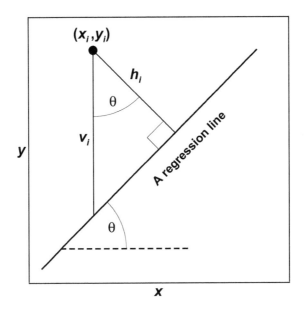

FIGURE 6.2   The two distances which are minimized by the two methods: the vertical, v_i, and the orthogonal, h_i, distances.

6.5   ROC CURVES

Many QSAR/QSPR models are intended to rank a series of compounds having a dichotomic property. In order to evaluate the model rank performance along all the series receiver operating characteristic (ROC) curves can be considered [18]. ROC curves [19–21] were developed after the World War II in the context of radar signals research in order to treat noise. In the sixties ROC curves were used in psychophysics, then in medicine and more recently for the evaluation of machine learning results. Table 6.2 shows the typical counts of positive and negative results (the true ones and the ones obtained by a classifier at a certain threshold cut level).

A ROC curve constitutes a graphical representation of the global efficiency of a classifier. It is a graphical representation for the frequencies of true positives and negatives and also for false positives and negatives along all the series of ranked compounds. Table 6.2 shows a typical confounding matrix collecting the following frequencies: true positives fraction (TPF), true negatives fraction (TNF), false positives fraction (FPF) and the false negatives fraction (FNF). The relations FNF + TPF = 1 and FPF + TNF = 1 hold. The TPF is also called sensitivity and measures how good the

TABLE 6.2 Contingency Table for Positive and Negative Cases and the Predictions Made by a Classifier at a Certain Cut-Off Value

		Classifier result	
		Positive	**Negative**
Real cases	**Positive**	TPF	FNF
	Negative	FPF	TNF

TPF: True positive fraction, FPF: False positive fraction, FNF: False negative fraction, TNF: True negative fraction.

classification method at detecting positive cases is. The sensitivity is the probability that the classification method gives a true positive result knowing that the case is a real positive one. This is the conditional probability $P(C+|+)$. On the other hand, specificity measures the ability of the classifier to pick out real negative cases and corresponds to the TNF or the $P(C-|-)$ conditional probability. The FNF is the probability of the classifier to set a case as negative when it is a real positive one, $P(C-|+)$. The FPF is the same as the $P(C+|-)$ probability. All these terms can be graphically displayed as being the respective areas under two Gaussian normalized density distribution functions. This is shown in Figure 6.3, where the vertical line stands for the threshold value the classifier uses as frontier for the two classes.

Ideally, a good classification method is able to generate two non-overlapping distributions. It has to be noted that for a non-ideal model, if the distributions are maintained and the vertical cut value line is moved to the

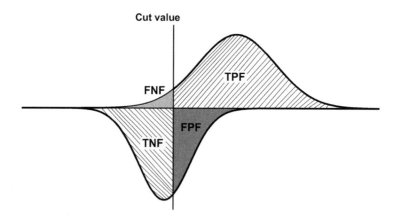

FIGURE 6.3 The idealized areas of the FNF, TPF, TNF and FPF parameters under Gaussian curves.

left (to the right), the proportion of false negatives decreases (increases) and, accordingly, the false positives fraction increases (decreases). The ROC curve is obtained moving the threshold cut value along all the ranked series and representing the (1-specificity, sensitivity) points being generated. This is equivalent to depict the (1-TNF, TPF) = (FPF, TPF) values along the cut line translation. Figure 6.4 shows an example of a ROC curve.

Ideally, a random or a neutral classifier will produce a ROC curve equal to the diagonal depicted dashed line. The total squared area depicted in Figure 6.4 integrates the unit. The shaded area is often used to quantify the ROC performance. It is named area under the ROC curve (AUROC or AU-ROC). The nearer this value to the unit the better the classifier performance. It has to be said that a neutral classifier shows an AU-ROC equal to 0.5. If the ROC curve (or a part of it) goes below the diagonal, it means that the classifier must have its criteria reversed, and the area can result to be lesser than 0.5. An AU-ROC greater than 0.75 is normally considered a god result if one is dealing with realistic prediction values.

The AU-ROC corresponds to the classifier probability to correctly classify a couple of training random cases, one being labeled as positive and the other

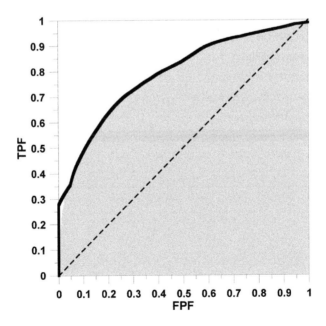

FIGURE 6.4   A ROC curve example.

as negative. Specifically, the AU-ROC corresponds to the "a posteriori" probability that the classifier correctly sorts the pair [21]. Recently, this concept has been used together with the definition of the balanced leave-2-out (BL2O) procedure [8]. This CV process is applicable for a dichotomized property (i.e., items labeled as positive and negative, of interest or not of interest, active or not active, etc.). The algorithm consists into exhaustively generate all the possible molecular pairs, one compound being positively labeled and the other being negatively tagged. For each left out pair the classifier model is build (rebuild from scratch, following the ITS paradigm) and a prediction is made for each member of the couple. During the cycles (pairs generation) it is counted how many times the relative classification inside the pair is correct (e.g., the positive item receives a higher vote), is incorrect (inverse classification) or if a tie is reproduced (both structures are predicted to belong to the same binary classification group). The proportion of correct classifications respect all the generated pairs (or respect all the non-ties reproduced) is a parameter lying between 0 and 1 which can be interpreted as a sort of AU-ROC area.

KEYWORDS

- **cross-validation**
- **enrichment factor**
- **internal test set**
- **multilinear linear regression**
- **orthogonal regression**
- **regression towards the mean effect**
- **ROC curves**

REFERENCES

1. Besalú, E., & Vera, L., (2008). Internal test sets (ITS) method: A new cross-validation technique to assess the predictive capability of QSAR models: application to a benchmark set of steroids *J. Chil. Chem. Soc*, *53* (3), 1576–1580.
2. De Julián-Ortiz, J. V., Besalú, E., & García-Domenech, R., (2003). True prediction by consensus for small sets of cyclooxigenase-2 inhibitors. *Indian Journal of Chemistry*, *42*(6). A, 1392–1404.

3. De Julián-Ortiz, J. V., & Besalú, E. (2006). Internal test sets studies in a group of anti-malarials. *Int. J. Mol. Sci.*, *7* (10), 456–468.

4. Besalú, E., Ponec, R., & De Julián-Ortiz, J. V., (2003). Virtual generation of agents against mycobacterium tuberculosis. A QSAR study. *Molecular Diversity*, *6* (2), 107–120.

5. García-Domenech, R., De Julián-Ortiz, J. V., & Besalú, E., (2006). True prediction of lowest observed adverse effect levels. *Molecular Diversity*, *10* (2), 159–168.

6. De Julián-Ortiz, J. V., & Besalú, E., Pogliani, L., (2014). A probabilistic analysis about the concepts of difficulty and usefulness of a molecular ranking classification. *Curr. Comp. Aid. Drug Des*, *10*, 107–114.

7. Barroso, J. M., & Besalú, E., (2005). Design of experiments applied to QSAR: ranking a set of compounds and establishing a statistical significance test. *Theochem*, *727*(1–3), 89–96.

8. Besalú, E., (2016). Fast modeling of binding affinities by means of superposing significant interaction rules (SSIR) method. *Int. J. Mol. Sci.*, *17*, 827.

9. Besalú, E., Pogliani, L., & De Julián-Ortiz, J. V., The superposing significant interaction rules (SSIR) method. Chapter 7 in: *Applied Chemistry and Chemical Engineering*, Volume 4. Haghi, A. K. et al., Ed., Apple Academic Press (AAP), Waretown, New Jersey, 2017.

10. Besalú, E., De Julián-Ortiz, J. V., & Pogliani, L., (2006). Some Plots Are not that Equivalent, *MATCH Commun. Math. Comput. Chem.*, *55*, 281–286.

11. Besalú, E., De Julián-Ortiz, J. V., Iglesias, M., & Pogliani, L., (2006). An overlooked property of plot methods. *.J Math. Chem.*, *39*, 475–484.

12. Besalú, E., De Julián-Ortiz, J. V., & Pogliani, L., (2007). Trends and plot methods in MLR studies, *J. Chem. Inf. Model.*, *47*, 751–760.

13. Besalú, E., De Julián-Ortiz, J. V., & Pogliani, L., (2010). Ordinary and orthogonal regressions in SAR/QSPR and chemistry-related studies, *MATCH Commun. Math. Comput. Chem.*, *63*, 573–583.

14. Besalú, E., (2013). The connection between inverse and classical calibration. *Talanta*, *116*, 45–49.

15. Besalú, E., (2001). Fast computation of cross-validated properties in full linear leave-many-out procedures. *J. Math. Chem.*, *29* (3), 191–204.

16. De Julián-Ortiz, J. V., Pogliani, L., & Besalú, E., (2010). Two-variable linear regression: modeling with orthogonal least-squares analysis, *J. Chem. Educ.*, *87*, 994–995.

17. Besalú, E., De Julián-Ortiz, J. V., & Pogliani, L., (2012). Orthogonal regression methods in chemical modeling. In: *Chemical Information and Computational Challenges in the 21st Century*, Putz, M., Ed. Nova Science Pub., New York, Chapter 10.

18. Besalú, E., De Julián-Ortiz, J. V., & Pogliani, L., (2010). On plots in QSAR/QSPR methodologies. In: *Quantum Frontiers of Atoms and Molecules in Physics, Chemistry, and Biology*; Putz, M. (Ed.), Nova Publishing Inc., New York, Chapter 22, pp. 581–598.

19. Egan, J. P., (1975). *Signal Detection Theory and ROC Analysis*; Academic Press, New York, Chapter 2.6, p. 37.

20. Forlay-Frick, P., Van Gyseghem, E., Héberger, K., & Vander Heyden, Y., (2005). Selection of orthogonal chromatographic systems based on parametric and non-parametric statistical tests. *Anal. Chim. Acta*, *539*, 1–10.

21. Mason, S. J., & Graham, N. E., (2002). Areas beneath the relative operating characteristics (ROC) and relative operating levels (ROL) curves: Statistical significance and interpretation. *Q. J. R. Meteorol. Soc.*, *128*, 2145–2166.

CHAPTER 7

TIME-DEPENDENT DENSITY FUNCTIONAL THEORY

TAKAO TSUNEDA[1] and KIMIHIKO HIRAO[2]

[1]*Professor, Fuel Cell Nanomaterials Center, University of Yamanashi, Kofu – 4000021, Japan, Tel./Fax: +81-55-254-7139, E-mail: ttsuneda@yamanashi.ac.jp*

[2]*Director, Computational Chemistry Unit, RIKEN Advanced Institute for Computational Science, Kobe, Hyogo 650-0047, Japan, Tel./Fax: +81-78-940-5555, E-mail: hirao@riken.jp*

CONTENTS

ABSTRACT

Time-dependent density functional theory (TDDFT) based on linear response theory has been frequently used to perform excited-state calculations in quantum chemistry since the 21th century's beginning. TDDFT has become the most widely-used theory in the calculations of electronic spectra and excited-state potential energy surfaces of large systems because it needs much less computational time than that of other reliable excited-state calculation methods such as *ab initio* multi reference methods. However, original TDDFT calculations have been reported to cause many problems, for example, the underestimation of charge transfer, Rydberg and core excitation energies and the neglect of multi electron excitations, relativistic spin-orbit interactions, and nonadiabatic couplings. These problems have been solved or drastically reduced by various types of physical corrections. For example, the long-range correction of exchange functionals has clearly solved the underestimation of charge transfer excitation energies. Despite the remarkable successes of physical corrections, several problems remain in TDDFT calculations, which are chiefly attributed to the single configuration characteristics of TDDFT. However, TDDFT will be mainly used in excited-state calculations of large systems for decades to come because it has become one of the most advanced excited-state theories.

7.1   FUNDAMENTALS

7.1.1   BACKGROUND

Time-dependent density functional theory (TDDFT) had hardly ever been used in the 20th century in excited-state calculations of molecules, but its frequency of usage has gradually increased in the 21st century to become the main theory used in more than 60% of excited-state calculations (currently in 2015 in chemical journals on the Web of Science: Thomson Reuters). *Ab initio* multiconfigurational self-consistent-field (MCSCF) [1] and multireference wavefunction theories [2] had nearly been the only theories until the 1990s to be used to quantitatively investigate the excited states of molecules. Multireference theories use MCSCF wavefunctions as the reference functions of configuration interaction (CI) theories [3], perturbation theories [4], and coupled-cluster theories [5]. The complete-active-space (CAS) self-consistent field (SCF) wavefunction [6] is frequently used as the reference function in major multireference theories, for example, CAS second-order perturbation theory (CASPT2) [7], multireference Møller-Plesset (MRMP) perturbation theory [8], and MC quasi-degenerate perturbation theory (MCQDPT) [9]. Since excited states generally have very different electronic states from the ground state, they essentially require highly-correlated theories involving many-body electron correlations. By definition, electron correlation is the difference between the exact and Hartree-Fock

(HF) energies [10]. Electron correlation is classified into two types [11]: the first is "dynamical correlation" emerging from the correlation hole (correlation cusp [12]) of wavefunctions, which is generated by the repulsion of short-range opposite-spin electron pairs, and the second is "nondynamical correlation" attributed to kinetic energy loss by mixing near-degenerate electronic states. Both these two electron correlations should be incorporated in a balanced manner to quantitatively investigate excited states. That is why only multireference theories have dominated the excited-state calculations of molecules. These theories actually provide even better accuracies for excitation energies than experimental ones. However, multireference theories have some serious problems: high computational order (more than $O(N^5)$ for the number of electrons N), immense amounts of computational time, and poor convergence and arbitrariness in selecting electron configurations in MCSCF calculations. The high computational order, in particular, is the most critical problem in the calculations of large molecules. TDDFT [13] has become the main theory in the excited-state calculations of large systems due to its low computational order ($O(N^4)$ or less). TDDFT has, therefore, been applied to studies on the excited states of large systems, for example, the mechanism elucidation of photochemical reactions [14, 15] and excited-state dynamics simulation [16].

We briefly explain the fundamentals and physical corrections of TDDFT in excited-state calculations in this chapter. The high applicability of TDDFT has increased excited-state calculations for two decades. However, many problems have been reported in conventional TDDFT calculations using standard functionals like generalized-gradient-approximation (GGA) and hybrid GGA functional [17], except for the lowest-lying excitations of small molecules. As will be explained in Section 7.2, various physical corrections have been proposed to solve these problems. Consequently, TDDFT has increased the accuracies so that they are comparable to multireference ones in electronic spectrum calculations. These corrections have, however, often been neglected in many TDDFT studies on photochemical reactions. For example, the band calculations of solid states usually use standard GGA functionals with no physical corrections. The calculated highest occupied molecule orbital-lowest unoccupied molecular orbital (HOMO-LUMO) gaps are taken as the (optical) band gaps, i.e., excitation energies, in the band calculations. However, semi-conductors are known to have considerable differences between HOMO-LUMO and band gaps, even though DFT calculations using GGA functionals often provide HOMO-LUMO gaps close to the experimental optical band gaps due to error cancellation. This discrepancy is eliminated by long-range correction for exchange functionals [18, 19] because it results from the lack of long-range exchange interactions [20]. We will first introduce the history and formulations of TDDFT in Section 7.1 of this chapter, and then detail the physical corrections for TDDFT in Section 7.2.

7.1.2 TIME-DEPENDENT HARTREE-FOCK METHOD

Let us first consider the time-dependent HF (TDHF) method, which forms the present formulation of TDDFT, to understand the fundamental concept of TDDFT. Even though the original TDHF method was derived by Dirac [21] in 1930, it had hardly been used until the 1950s when computers appeared. This method was redeveloped by Bohm and Pines in 1951 as a random phase approximation (RPA) for the collective motion calculations of electron gas and has been applied since then to various calculations of electronic properties [22]: e.g., the dielectronic constant of electrons in metals [23] and the rotational and vibrational levels of nuclei [24]. McLachlan and Ball first applied the TDHF method to the calculations of molecules in 1964 [25]. The fundamental equation for the TDHF method is the equation-of-motion for molecular orbitals $\{\phi_i\}$ under a small perturbation, $-f(t)$:

$$i\frac{\partial \phi_i}{\partial t} = \left[\hat{F}(t) - f(t)\right]\phi_i,$$

(1)

where $F(t)$ is the time-dependent Fock operator. Atomic units ($\hbar = e^2 = m = 1$, energies are in Hartree, and distances are in Bohr) are used in this review. When this perturbation acts on electrons in the ground state, the wavefunction, Ψ, becomes time-dependent:

$$\Psi = \Psi(t)\exp(-iE_0 t),$$

(2)

in which E_0 is the energy of the time-independent ground state Ψ_0 and

$$\Psi(t) = \Psi_0 + \sum_{i,a} C_{ia}(t)\Psi(i \rightarrow a),$$

(3)

where $\Psi(i \rightarrow a)$ is the excited state electron configuration in which one electron jumps from ϕ_i to ϕ_a. Consider the equation of motion based on the HF method,

$$i\frac{\partial \phi_i}{\partial t} = \hat{F}(t)\phi_i = \epsilon_i\phi_i + \sum_a (ia \| jb) P_{jb}\phi_a \,, \tag{4}$$

where P_{jb} is the two-electron density matrix, and

$$(ia \| jb) = \int\int d^3\mathbf{r}_1 d^3\mathbf{r}_2 \phi_i^*(\mathbf{r}_1)\phi_a(\mathbf{r}_1)r_{12}^{-1}\left(1-\hat{P}_{12}\right)\phi_j(\mathbf{r}_2)\phi_b^*(\mathbf{r}_2) \tag{5}$$

where $\hat{P}_{12}$ is the replacement operator for electrons. Equation (4) is provided for coefficients $\{C_{ia}\}$ in Eq. (3) as

$$i\frac{dC_{ia}}{dt} = \left(\epsilon_a - \epsilon_i\right)C_{ia} + \left(ia \| jb\right)C_{jb} + \left(ia \| bj\right)C_{jb}^* - f_{ai}(t), \tag{6}$$

and

$$-i\frac{dC_{ia}^*}{dt} = \left(\epsilon_a - \epsilon_i\right)C_{ia}^* + \left(ai \| jb\right)C_{jb} + \left(ai \| bj\right)C_{jb}^* - f_{ia}(t). \tag{7}$$

Under the force having the periodic time factor, $\exp(-i\omega t)$, the coefficients are represented as

$$C_{ia} = X_{ia}\exp(-i\omega t) + Y_{ia}^*\exp(+i\omega t), \tag{8}$$

and

$$C_{ia}^* = X_{ia}^*\exp(+i\omega t) + Y_{ia}\exp(-i\omega t). \tag{9}$$

Using these coefficients, Eqs. (6) and (7) are given as

$$\omega X_{ia} = \left(\epsilon_a - \epsilon_i\right)X_{ia} + \left(ia \| jb\right)X_{jb} + \left(ia \| bj\right)Y_{jb} - f_{ai}(\omega), \tag{10}$$

and

$$-\omega Y_{ia} = \left(\epsilon_a - \epsilon_i\right) Y_{ia} + \left(ai \| jb\right) X_{jb} + \left(ai \| bj\right) Y_{jb} - f_{ia}(\omega). \qquad (11)$$

For no forced oscillation, $f = 0$, Eqs. (10) and (11) are represented using the matrix of coefficients X and Y as

$$\omega \begin{pmatrix} \mathbf{X} \\ -\mathbf{Y} \end{pmatrix} = \begin{pmatrix} \left(\epsilon_a - \epsilon_i\right)\delta_{ia,jb} + \left(ia \| jb\right) & \left(ia \| bj\right) \\ \left(ai \| jb\right) & \left(\epsilon_a - \epsilon_i\right)\delta_{ia,jb} + \left(ai \| bj\right) \end{pmatrix} \begin{pmatrix} \mathbf{X} \\ \mathbf{Y} \end{pmatrix} \equiv \begin{pmatrix} \mathbf{A} & \mathbf{B} \\ \mathbf{B}^* & \mathbf{A}^* \end{pmatrix} \begin{pmatrix} \mathbf{X} \\ \mathbf{Y} \end{pmatrix}. \qquad (12)$$

This equation is called the "time-dependent HF (TDHF) equation." The TDHF method has rarely been applied to excited-state calculations because this method lacks electron correlations, which are required in excited-state calculations. Several multiconfigurational TDHF methods [26–28] have been suggested to incorporate electron correlation. There is also a TDHF method that combines the coupled-cluster wavefunctions, which is called the "symmetry-adapted-cluster CI (SAC-CI) method" [29] or the "equation of motion coupled-cluster (EOM-CC) method" [30, 31]. Even a multiref-erence TDHF method [32] involving both dynamical and nondynamical electron correlations has very recently been proposed. However, all these methods need a great deal of computational time that is even comparable to those of multiconfigurational and multireference theories. TDDFT was developed to incorporate electron correlations in the TDHF method without incurring additional computational cost.

7.1.3   FUNDAMENTAL THEOREMS OF TIME-DEPENDENT DENSITY FUNCTIONAL THEORY

Similarly to the Hohenberg-Kohn theorems [33], i.e., the fundamental theo-rems of DFT, there are also fundamental theorems of TDDFT. Runge and Gross proposed the fundamental theorems of TDDFT for periodically time-dependent electronic states [34]. The following two conditions are assumed in these theorems for the external potential, V_{ext}:

1. $V_{ext}(\mathbf{r}, t)$ depends on time periodically; and
2. $V_{ext}(\mathbf{r}, t)$ consists of a time-independent stationary part, V_{stat}, and a slightly time-dependent perturbation part, V_{pert}.

The following four theorems are derived based on these assumptions:

1. For V_{ext} expansible in terms of time, define that the $V_{\text{ext}}(\mathbf{r}, t) \rightarrow \rho(\mathbf{r}, t)$ transformation corresponds to solving the time-dependent Schrödinger equation. Based on this definition, the inverse transformation $\rho \rightarrow V_{\text{ext}}$ can be performed in the case of the second assumption above. This is regarded as the first time-dependent Hohenberg-Kohn theorem.

2. The time-derivative of the current density j is represented as a density functional, $\Omega[\rho](\mathbf{r}, t)$:

$$\frac{\partial \mathbf{j}(\mathbf{r},t)}{\partial t} = \Omega[\rho](\mathbf{r},t), \tag{13}$$

where $\mathbf{j}$ is defined as

$$\frac{\partial \rho(\mathbf{r},t)}{\partial t} = -\nabla \cdot \mathbf{j}(\mathbf{r},t). \tag{14}$$

3. The action integral in an arbitrary time interval from t_0 to t_1:

$$S = \int_{t_0}^{t_1} dt \Psi^*[\rho] \left(i\frac{\partial}{\partial t} - \hat{H} \right) \Psi[\rho], \tag{15}$$

is representable as a density functional $S[\rho]$ and is decomposable to:

$$S[\rho] = \int_{t_0}^{t_1} dt \Psi^*[\rho] \left[i\frac{\partial}{\partial t} - \left(\hat{T} + \hat{V}_{ee}\right) \right] \Psi[\rho] - \int_{t_0}^{t_1} dt \int d^3\mathbf{r}\rho(\mathbf{r},t)V_{\text{ext}}(\mathbf{r},t), \tag{16}$$

where $\Psi[\rho]$ is a wavefunction giving the electron density, ρ. The first term on the right-hand side is a universal functional, because it does not depend on the external potential, V_{ext}. This action density functional $S[\rho]$ satisfies the variational principle to give the stationary value at the exact density. This is interpreted as the second time-dependent Hohenberg-Kohn theorem.

4. Time-dependent orbitals $\{\phi_i(\mathbf{r}, t)\}$ satisfy the time-dependent Schrödinger equation:

$$i\frac{\partial}{\partial t}\phi_i(\mathbf{r},t) = \left(-\frac{1}{2}\nabla^2 + V_{\text{eff}}[\mathbf{r},t;\rho(\mathbf{r},t)]\right)\phi_i(\mathbf{r},t),\qquad(17)$$

where the effective potential, V_{eff}, is represented as

$$V_{\text{eff}}[\mathbf{r},t;\rho(\mathbf{r},t)] = V_{\text{ext}}[\mathbf{r},t] + \int d^3\mathbf{r}\,\frac{\rho(\mathbf{r}',t)}{|\mathbf{r}-\mathbf{r}'|} - \frac{\delta S_{\text{xc}}[\rho]}{\delta\rho(\mathbf{r},t)}.\qquad(18)$$

Note that S_{xc} is the exchange-correlation part of the action integral and is approximated on the basis of the adiabatic approximation as

$$\frac{\delta S_{\text{xc}}[\rho]}{\delta\rho(\mathbf{r},t)} \approx V_{\text{xc}}[\rho](\mathbf{r},t) = \frac{\delta E_{\text{xc}}[\rho]}{\delta\rho(\mathbf{r},t)}.\qquad(19)$$

These theorems are called "Runge-Gross theorems." Equation (17) using the effective potential in Eq. (18) is broadly called "the time-dependent Kohn-Sham (TDKS) equation" [34].

The Runge-Gross theorem has a severe problem in the use of the wavefunction giving electron density in the third theorem [35]. Since time-dependent wavefunctions contain time-dependent phase terms, the one-to-one correspondence of the wavefunction and electron density is only established for a specific phase. This problem can be avoided by representing the wavefunction as a functional of the external potential $\Psi[V_{\text{ext}}]$, for which the one-to-one correspondence with electron density is established. However, this approach leads to the denial of the concept of the universal functional. Even if it would be accepted, the one-to-one correspondence of the time-derivative of the wavefunction $\delta\Psi$ and that of the potential δV_{ext} is not assured. Therefore, different to that of the time-independent case, the variational principle is not strictly established in the time-dependent case [36]. Even though the one-to-one correspondence can be proven to exist if the external potential is Taylor-expandable in time [37], it can also be proven that no such Taylor-expandable potential exists [38].

This dilemma is resolved by taking relativistic effects into account because there is a theorem interpreted as the time-dependent Hohenberg-Kohn theorem, which is the relativistic expansion of the Hohenberg-Kohn theorem. Rajagopal and Callaway [39] proved so-called relativistic Hohenberg-Kohn theorems using the Hamiltonian operator of quantum electrodynamics (QED) and four-component current density j^μ containing electron density ρ and current density j:

$$\mathbf{j}^\mu(\mathbf{r}) = (\rho(\mathbf{r}), \mathbf{j}(\mathbf{r})/c). \qquad (20)$$

The theorems consist of two theorems, which are almost the same as the Hohenberg-Kohn theorems [33], for non-degenerate ground states:

1. The four-component external potential, which includes nuclear-electron electrostatic potential and vector potential, is determined by the four-component current density.

2. For any four-component current density, the variational principle is always satisfied.

The first Rajagopal-Callaway theorem establishes that four-component external potential,

$$\mathbf{V}_0^\mu(\mathbf{r}) = (V_0(\mathbf{r}), -\mathbf{A}(\mathbf{r})), \qquad (21)$$

where V_0 is nonrelativistic external potential and $\mathbf{A}$ is vector potential, is uniquely determined by the four-component current density. This theorem indicates that the relativistic Hamiltonian operator is uniquely represented by the four-component current density. Similar to the first Hohenberg-Kohn theorem, this theorem is proven by reductio ad absurdum [40]. Assuming that two different four-component external potentials exist for the same four-component current density, this theorem is proven by showing that this assumption is inconsistent with the variational principle for wavefunctions. What is different from the Hohenberg-Kohn theorem is that this theorem has a problem attributed to magnetic field B for the four-component external potential. That is, this theorem is not established to provide the same ground state in the case that the magnetic field eliminates the difference between two four-component external potentials. However, this problem is fortunately not so serious because this theorem does not need the one-to-one correspondence between the four-component external potential and the ground state. What is more problematic is that the one-to-one correspondence between the four-component current density and the four-component wavefunction is not established similarly to the Hohenberg-Kohn theorem. That is, the four-component external potential also has the V-representability problem [17]. The one-to-one correspondence between the four-component wavefunction, which is derived from the four-component external potential, and the four-component current density is proven by excluding the degree of freedom in the gauge transformation of the four-component external potential [17]. It is known that the degree of freedom of the transformation is eliminated for electrostatic potentials for gauge transformation due to the pinning effect based on the Helmholtz theorem ($V(\mathbf{r}) \xrightarrow{r \to \infty} 0$). Consequently, the first Rajagopal-Callaway theorem is established. The second Rajagopal-Callaway theorem is the variational principle establishing that the Hamiltonian operator has the minimum energy solution for the four-component current density. There is no rigorous proof for this theorem because the variational principle is not proven even in the framework of QED without the renormalization [40]. However, if the renormalization is assumed to be correct as usual, this proposition is proven for nondegenerate ground states by *reductio ad absurdum* similarly to the Hohenberg-Kohn theorem [40]. Moreover, the existence of the energy derivative, which is the prerequisite for the variational principle, is also not proven for the four-component current density. The redefinition of energy based on the Legendre transformation is required to solve this problem [41].

The four-component relativistic TDDFT based on the Rajagopal-Callaway theorems is needed, as was explained above, to investigate the time propagation of electrons for the external potential depending on time nonperiodically. Note, however, that there are no sophisticated exchange-correlation functionals based on the four-component current density, as far as we know. The current density-independent TDDFT based on the Runge-Gross theorems, on the other hand, is applicable to the calculations of excitation spectra and excited-state dynamics simulations, which are targeted in this chapter because these calculations do not need to perform the explicit time propagation of electrons.

7.1.4 TIME-DEPENDENT KOHN-SHAM METHOD

Applying the TDKS equation to linear response theory makes it possible to calculate excitation energies [42]. The excitation energies in the linear response theory appear to be poles of the response to the electrostatic field. Since this linear response to the TDKS method is now widely-used in excitation energy calculations, the TDKS method for excited state calculations usually indicates this linear response. Note that there is a propagator TDKS method, which does not use the linear response theory but instead uses Eq. (17) for analyzing the time propagation of electronic states. This method is often used in the calculations of excitation spectra [43]. What is significant is that the linear response TDKS method is only applicable to single excitation calculations. Although it is possible to explicitly incorporate double and more excitation effects in the TDKS method, this is presently too time-consuming to be used in excited-state calculations of large systems. A simple method of implicitly incorporating double-excitation effects in the TDKS method is described in Subsection 7.2.3.

Let us first consider the linear response of the electron density to the external potential. Assuming that only a weak perturbation, δV_{ext}, is added to the external potential following the Runge-Gross theorems, the electron density is interpreted to undergo an infinitesimal change, $\delta\rho(\mathbf{r}, t)$, in the stationary part, ρ_{stat}. The exchange-correlation potential, V_{xc}, is therefore given as:

$$V_{xc}[\rho](\mathbf{r}_1,t_1) = V_{xc}^{stat}[\rho](\mathbf{r}_1) + \int\int dt_2 d^3\mathbf{r}_2 f_{xc}[\rho_{stat}](\mathbf{r}_1,\mathbf{r}_2,t_2-t_1)\delta\rho(\mathbf{r}_2,t_2) \quad (22)$$

where f_{xc} is the exchange-correlation integral kernel,

$$f_{xc}[\rho_{stat}](\mathbf{r}_1,\mathbf{r}_2,t_2-t_1) = \left.\frac{\delta V_{xc}(\mathbf{r}_1,t_1)}{\delta\rho(\mathbf{r}_2,t_2)}\right|_{\rho=\rho_{stat}} \quad (23)$$

The derivative of the exchange-correlation potential in terms of electron density f_{xc} is called the "exchange-correlation integral kernel." Define the response function of the electron density, χ_{KS}, for the infinitesimal change in the Kohn-Sham potential, δV_{KS}, as:

$$\delta\rho(\mathbf{r}_1,t_1) = \int\int dt_2 d^3\mathbf{r}_2 \chi_{KS}[\rho_{stat}](\mathbf{r}_1,\mathbf{r}_2,t_2-t_1)\delta V_{KS}(\mathbf{r}_2,t_2) \quad (24)$$

The response function in this definition is given by Green function theory as:

$$\chi_{KS}(\mathbf{r}_1,\mathbf{r}_2,\omega) = 2\lim_{\eta\to 0+}\sum_i^{n_{occ}}\sum_a^{n_{vir}}\left[\frac{\phi_i^*(\mathbf{r}_1)\phi_a(\mathbf{r}_1)\phi_i(\mathbf{r}_2)\phi_a^*(\mathbf{r}_2)}{\omega-(\epsilon_a-\epsilon_i)+i\eta} - \frac{\phi_i(\mathbf{r}_1)\phi_a^*(\mathbf{r}_1)\phi_i^*(\mathbf{r}_2)\phi_a(\mathbf{r}_2)}{\omega+(\epsilon_a-\epsilon_i)-i\eta}\right] \quad (25)$$

Note that this response function is Fourier-transformed ($t \to \omega$).

Based on this linear response of density, Casida transformed the TDHF equation in Eq. (12) to a DFT-based formulation [13]:

$$\omega \begin{pmatrix} \mathbf{X} \\ -\mathbf{Y} \end{pmatrix} = \begin{pmatrix} \mathbf{A} & \mathbf{B} \\ \mathbf{B}^* & \mathbf{A}^* \end{pmatrix} \begin{pmatrix} \mathbf{X} \\ \mathbf{Y} \end{pmatrix} \tag{26}$$

where A and B matrices have the elements of:

$$A_{ia\sigma,jb\tau} = \delta_{\sigma\tau}\delta_{ij}\delta_{ab}\left(\epsilon_{a\sigma}-\epsilon_{i\sigma}\right)+K_{ia\sigma,jb\tau} \tag{27}$$

where ε_{is} is the i-th σ-spin orbital energy, and

$$B_{ia\sigma,jb\tau} = K_{ia\sigma,jb\tau}, \tag{28}$$

respectively, where $K_{ias,jbt}$ is provided as:

$$K_{ia\sigma,jb\tau} = \left(ia\sigma \mid jb\tau\right)+\int\int d^3\mathbf{r}_1 d^3\mathbf{r}_2 \phi_{i\sigma}^*(\mathbf{r}_1)\phi_{a\sigma}(\mathbf{r}_1)f_{xc}(\mathbf{r}_1,\mathbf{r}_2)\phi_{j\tau}(\mathbf{r}_2)\phi_{b\tau}^*(\mathbf{r}_2) \tag{29}$$

Note that the spins of the orbitals (σ, τ, $\sigma' = \sigma$) are explicitly given in Eqs. (27) and (28) for purposes of accuracy. In Eq. (29), the first term of the right-hand side, which is called the Hartree integral, is provided as:

$$\left(ia\sigma \mid jb\tau\right) = \int\int d^3\mathbf{r}_1 d^3\mathbf{r}_2 \phi_{i\sigma}^*(\mathbf{r}_1)\phi_{a\sigma}(\mathbf{r}_1)r_{12}^{-1}\phi_{j\tau}(\mathbf{r}_2)\phi_{b\tau}^*(\mathbf{r}_2) \tag{30}$$

The exchange-correlation integral kernel, f_{xc}, in Eq. (29) is assumed to have the local form:

$$f_{xc}(\mathbf{r}_1,\mathbf{r}_2) = \frac{\delta^2 E_{xc}}{\delta\rho^2(\mathbf{r}_1)}\delta(\mathbf{r}_1-\mathbf{r}_2) \tag{31}$$

Equation (26) is called the "time-dependent response Kohn-Sham equation" and the method using this equation is called the "time-dependent response Kohn-Sham method." The "time-dependent Kohn-Sham (TDKS) method" in excitation energy calculations usually indicates this method.

Since Eq. (26) needs the asymmetric matrix to be diagonalized, this equation is usually transformed into a symmetric matrix form [44]:

$$\Omega F = \omega^2 F, \tag{32}$$

where for singlet excitations:

$$\Omega_{ia\sigma,jb\tau}^{singlet} = \delta_{\sigma\tau}\delta_{ij}\delta_{ab}\left(\epsilon_{a\sigma} - \epsilon_{i\sigma}\right)^2 + 2\left(\epsilon_{a\sigma} - \epsilon_{i\sigma}\right)^{1/2}\left(K_{ia,jb}^{\sigma\sigma} + K_{ia,jb}^{\sigma\sigma'}\right)\left(\epsilon_{b\tau} - \epsilon_{j\tau}\right)^{1/2} \tag{33}$$

and for triplet excitations:

$$\Omega_{ia\sigma,jb\tau}^{triplet} = \delta_{\sigma\tau}\delta_{ij}\delta_{ab}\left(\epsilon_{a\sigma} - \epsilon_{i\sigma}\right)^2 + 2\left(\epsilon_{a\sigma} - \epsilon_{i\sigma}\right)^{1/2}\left(K_{ia,jb}^{\sigma\sigma} - K_{ia,jb}^{\sigma\sigma'}\right)\left(\epsilon_{b\tau} - \epsilon_{j\tau}\right)^{1/2} \tag{34}$$

The F in Eq. (32) is the response coefficient matrix consisting of:

$$F_{ia\sigma} = \left(\epsilon_{a\sigma} - \epsilon_{i\sigma}\right)^{-1/2}\left(X_{ia\sigma} - X_{ai\sigma}\right) \tag{35}$$

where

$$X_{ia\sigma}(\omega) = \frac{-1}{\omega + \left(\epsilon_{a\sigma} - \epsilon_{i\sigma}\right)}\int d^3r \phi_{i\sigma}^*(\mathbf{r})\delta\left(2\sum_i^n \hat{J}_i + V_{xc}\right)(\mathbf{r},\omega)\phi_{a\sigma}(\mathbf{r}) \tag{36}$$

Excitation energies $\{\omega_{ia}\}$ and the corresponding response coefficient vectors $\{F_{ias}\}$ in usual TDKS calculations are given by solving Eq. (35). The TDKS calculations often use the Tamm-Dancoff approximation [45], which only takes the diagonal terms of the TDKS matrix, to reduce the computational time in large-system calculations.

The TDKS method also provides the peak strengths of electronic excitation spectra, which correspond to transition probabilities from the ground to excited states. The transition probabilities are proportional to the oscillator strengths. The oscillator strength for the excitation to the I-th excited state is provided by:

$$f_I = \frac{2}{3}\left(E_I - E_0\right)\sum_{v=x,y,z}\left|\langle\Psi_0|\mathbf{r}_v|\Psi_I\rangle\right|^2 \tag{37}$$

where Ψ_0 and Ψ_I correspond to the ground and I-th excited-state wavefunctions having total energies of E_0 and E_I. The oscillator strength in the framework of the TDKS method is represented as [13]:

$$f_I = \frac{2}{3}\left[\sum_{v=x,y,z}\sum_{ia\sigma}\left\{d_{ia\sigma}^v\left(\epsilon_{a\sigma}-\epsilon_{i\sigma}\right)^{1/2}F_{ia\sigma}^I\right\}^2\right]\bigg/\sum_{ia\sigma}\left(F_{ia\sigma}^I\right)^2 \qquad (38)$$

where $d_{ia\sigma}^v$ is the transition dipole moment:

$$d_{ia\sigma}^v = \int\psi_{i\sigma}(\mathbf{r})r_v\psi_{a\sigma}(\mathbf{r})d^3\mathbf{r} \qquad (39)$$

and $F_{ia\sigma}^I$ is the response function coefficient of the I-th excited state, which are obtained by solving Eq. (35). Electronic excitation spectra are usually illustrated by plotting the calculated oscillator strengths at the corresponding excitation energies.

Since the TDKS method has given reasonable valence excitation energies of small molecules despite this, it has become the most-used excited-state calculation method (current in 2016) as mentioned in Subsection 7.1.1. It has, however, been reported that the TDKS method using standard functionals has many problems and various physical corrections have been proposed to solve these. The problems and the physical corrections are detailed in Section 7.2.

7.1.5 TDKS EXCITATION ENERGY GRADIENTS

The potential energy gradients, besides excitation energies, in terms of nuclear coordinates are required to perform excited-state dynamics simulations of photochemical reactions as was explained in Subsection 7.2.5. Several methods have been suggested to evaluate TDKS excitation energy gradients [46–48]. Since the potential energy gradient is essentially a response property to nuclear coordinates [49], it needs the coupled-perturbed Kohn-Sham (CPKS) equation to be solved for all the nuclear coordinates [17]. It is, however, impractical to solve all of them for large systems containing huge numbers of nuclear coordinates due to the immense amounts of computational time required.

Furche and Ahlrichs proposed an efficient way to solve this problem by using the Z-vector equation [48]. The TDKS energy gradient, ω^ξ, (ξ denotes the derivative for the nuclear coordinate) in their formulation is provided as:

$$\omega^\xi = \sum_{ia\sigma}\left\{h_{ia}^\xi P_{ia\sigma} - S_{ia}^\xi W_{ia\sigma} + \left(V_{xc}^\xi\right)_{ia\sigma} P_{ia\sigma} + \sum_{jb\tau}(ia\,|\,bj)^\xi \Gamma_{ia\sigma\,jb\tau} + \sum_{jb\tau}\left(f_{xc}^\xi\right)_{ia\sigma\,jb\tau}(\mathbf{X}+\mathbf{Y})_{ia\sigma}(\mathbf{X}+\mathbf{Y})_{jb\tau}\right\} \quad (40)$$

where h_{ia} and S_{ia} correspond to the core Hamiltonian and overlapping integral matrix elements, and exchange-correlation potential integral $\left(V_{xc}\right)_{ia\sigma}$ and exchange-correlation integral kernel integral $\left(f_{xc}\right)_{ia\sigma,jb\tau}$ are respectively given as:

$$\left(V_{xc}\right)_{ia\sigma} = \int d^3\mathbf{r}\,\phi_{i\sigma}^*(\mathbf{r})\frac{\delta E_{xc}}{\delta\rho_\sigma(\mathbf{r})}\phi_{a\sigma}(\mathbf{r}). \quad (41)$$

and

$$\left(f_{xc}\right)_{ia\sigma,jb\tau} = \int\int d^3\mathbf{r}_1 d^3\mathbf{r}_2\,\phi_{i\sigma}^*(\mathbf{r}_1)\phi_{a\sigma}(\mathbf{r}_1)\frac{\delta^2 E_{xc}}{\delta\rho_\sigma(\mathbf{r}_1)\delta\rho_\tau(\mathbf{r}_1)}\delta(\mathbf{r}_1-\mathbf{r}_2)\phi_{j\tau}(\mathbf{r}_2)\phi_{b\tau}^*(\mathbf{r}_2) \quad (42)$$

respectively. The $P_{\mu\nu\sigma}$ in Eq. (40) is the element of relaxed one-particle difference density matrix $\mathbf{P}$:

$$\mathbf{P} = \mathbf{T} + \mathbf{Z} \quad (43)$$

where $\mathbf{T}$ is the unrelaxed difference density matrix containing three types of elements,

$$T_{ab\sigma} = \frac{1}{2}\sum_i\left\{(\mathbf{X}+\mathbf{Y})_{ia\sigma}(\mathbf{X}+\mathbf{Y})_{ib\sigma} + (\mathbf{X}+\mathbf{Y})_{ia\sigma}(\mathbf{X}-\mathbf{Y})_{ib\sigma}\right\}. \quad (44)$$

$$T_{ij\sigma} = -\frac{1}{2}\sum_a\left\{(\mathbf{X}+\mathbf{Y})_{ia\sigma}(\mathbf{X}+\mathbf{Y})_{ja\sigma} + (\mathbf{X}-\mathbf{Y})_{ia\sigma}(\mathbf{X}-\mathbf{Y})_{ja\sigma}\right\}. \quad (45)$$

and

$$T_{ia\sigma} = T_{ai\sigma} = 0 \quad (46)$$

What is most important is that the remaining $\mathbf{Z}$ matrix is determined by solving the following Z-vector equation:

$$\sum_{jb\tau} (A+B)_{ia\sigma, jb\tau} Z_{jb\tau} = -R_{ia\sigma} \tag{47}$$

The **R** matrix element in the right-hand side is given as

$$R_{ia\sigma} = \sum_{b} \left\{ (X+Y)_{ib\sigma} H_{ab\sigma}^{+} [X+Y] + (X-Y)_{ib\sigma} H_{ab\sigma}^{-} [X-Y] \right\}$$

$$- \sum_{j} \left\{ (X+Y)_{ja\sigma} H_{ji\sigma}^{+} [X+Y] + (X-Y)_{ja\sigma} H_{ji\sigma}^{-} [X-Y] \right\} \tag{48}$$

$$+ H_{ia\sigma}^{+} [T] + 2 \sum_{jb\tau, kc\lambda} (g_{xc})_{ia\sigma, jb\tau, kc\lambda} (X+Y)_{jb\tau} (X+Y)_{kc\lambda},$$

where for arbitrary vector **V**,

$$H_{ia\sigma}^{+} [V] = \sum_{jb\tau} \left\{ 2(ia\sigma \mid jb\tau) + 2(f_{xc})_{ia\sigma, bj\tau} - c_{x}\delta_{\sigma\tau} \left[(ij\sigma \mid ab\sigma) + (ib\sigma \mid aj\sigma) \right] \right\} V_{bj\sigma} \tag{49}$$

$$H_{ia\sigma}^{-} [V] = \sum_{jb\tau} \left\{ c_{x}\delta_{\sigma\tau} \left[(ij\sigma \mid ba\sigma) - (ib\sigma \mid ja\sigma) \right] \right\} V_{bj\sigma} \tag{50}$$

and $(g_{xc})_{ia\sigma, jb\tau, kc\lambda}$ is the third-order derivative matrix element,

$$(g_{xc})_{ia\sigma, jb\tau, kc\lambda} = \int\int\int d^3 r_1 d^3 r_2 d^3 r_3 \phi_{i\sigma}^{*}(r_1)\phi_{a\sigma}(r_1)\phi_{j\tau}(r_2)\phi_{b\tau}^{*}(r_2)\phi_{k\lambda}^{*}(r_3)\phi_{c\lambda}(r_3)$$

$$\times \frac{\delta^3 E_{xc}}{\delta\rho_{\sigma}(r_1)\delta\rho_{\tau}(r_1)\delta\rho_{\lambda}(r_1)} \delta(r_1 - r_2)\delta(r_1 - r_3). \tag{51}$$

The $W_{\mu\nu\sigma}$ in Eq. (40) is the element of the Lagrangian multiplier matrix **W**:

$$W_{ij\sigma} = \left(1 - \frac{1}{2}\delta_{ij}\right) \left[\sum_{a} \omega \left\{ (X+Y)_{ia\sigma} (X-Y)_{ja\sigma} + (X-Y)_{ia\sigma} (X+Y)_{ja\sigma} \right\} \right.$$

$$+ \sum_{a} \epsilon_{a\sigma} \left\{ (X+Y)_{ia\sigma} (X+Y)_{ja\sigma} + (X-Y)_{ia\sigma} (X-Y)_{ja\sigma} \right\} \tag{52}$$

$$\left. + H_{ia\sigma}^{+} [P] + 2 \sum_{kc\tau, ld\lambda} (g_{xc})_{ij\sigma, kc\tau, ld\lambda} (X+Y)_{kc\tau} (X+Y)_{ld\lambda} \right]$$

$$W_{ab\sigma} = \left(1 - \frac{1}{2}\delta_{ab}\right)\left[\sum_i \omega\left\{(\mathbf{X}+\mathbf{Y})_{ia\sigma}(\mathbf{X}-\mathbf{Y})_{ib\sigma} + (\mathbf{X}-\mathbf{Y})_{ia\sigma}(\mathbf{X}+\mathbf{Y})_{ib\sigma}\right\}\right.$$
$$\left. + \sum_a \epsilon_{i\sigma}\left\{(\mathbf{X}+\mathbf{Y})_{ia\sigma}(\mathbf{X}+\mathbf{Y})_{ib\sigma} + (\mathbf{X}-\mathbf{Y})_{ia\sigma}(\mathbf{X}-\mathbf{Y})_{ib\sigma}\right\}\right] \tag{53}$$

and

$$W_{ia\sigma} = \sum_j \left\{(\mathbf{X}-\mathbf{Y})_{ja\sigma}H_{ji\sigma}^+[\mathbf{X}+\mathbf{Y}] + (\mathbf{X}-\mathbf{Y})_{ia\sigma}H_{ji\sigma}^-[\mathbf{X}-\mathbf{Y}]\right\} + \epsilon_{i\sigma}Z_{ia\sigma} \tag{54}$$

The remaining Γ matrix in Eq. (40) is a two-particle difference density matrix, which contains:

$$\Gamma_{ia\sigma,jb\tau} = P_{ia\sigma}D_{jb\tau} + (\mathbf{X}+\mathbf{Y})_{ia\sigma}(\mathbf{X}+\mathbf{Y})_{jb\tau} - c_x\delta_{\sigma\tau}\left[P_{ia\sigma}D_{ja\tau} + P_{ij\sigma}D_{ba\tau} + (\mathbf{X}+\mathbf{Y})_{ib\sigma}(\mathbf{X}+\mathbf{Y})_{ja\tau}\right.$$
$$\left. + (\mathbf{X}+\mathbf{Y})_{ij\sigma}(\mathbf{X}+\mathbf{Y})_{ba\tau} - (\mathbf{X}-\mathbf{Y})_{ib\sigma}(\mathbf{X}-\mathbf{Y})_{ja\tau} - (\mathbf{X}-\mathbf{Y})_{ij\sigma}(\mathbf{X}-\mathbf{Y})_{ba\tau}\right] \tag{55}$$

Since Eq. (40) provides TDKS excited-state energy gradients, it not only enables us to perform excited-state geometry optimizations but also excited-state dynamics simulations. Our program for calculating the TDKS excited-state energy gradients [50] has been released in the official version of the free *ab initio* quantum chemistry program package General Atomic and Molecular Electronic Structure System (GAMESS) (http://www.msg. ameslab.gov/gamess/). Note, however, that the TDKS gradients require sophisticated functionals including especially long-range exchange interactions and well-balanced electron correlations in investigating photochemical reactions. Section 7.2 provides details on physical corrections for solving these problems.

7.2 PHYSICAL CORRECTIONS FOR TDKS METHOD

This section reviews physical corrections to solve various problems or to extend applicabilities in TDKS excited-state calculations. The standard TDKS method using GGA or hybrid GGA functionals presents five major difficulties:

- Charge transfer and Rydberg excitation energies are seriously under-estimated [51, 52].

- Core excitation energies are also significantly underestimated [53, 54].
- No multielectron excitation is taken into consideration [49].
- No relativistic spin-orbit effect is incorporated [55].
- No nonadiabatic coupling is included.

Various physical corrections have been proposed to resolve these difficulties and some corrections have succeeded in solving them. It should also be emphasized that other difficulties in TDKS calculations remain, for example, poor excited-state potential energy surface calculations around conical intersections. This section only introduces major physical corrections to solve these difficulties. Note that all these corrections have clear physical meanings in common that are different from artificial corrections simply to improve the calculated excitation energies. Basically, these corrections, therefore, can simply be combined without changing adjustable parameters in the corrections.

7.2.1 LONG-RANGE CORRECTION

The long-range correction for exchange functionals [18] solves the underestimation of charge transfer (CT) excitation energies in TDKS calculations. Since CT excitations are the precursory process of most photochemical reactions such as photosynthesis, these are required to investigate photochemical reactions to reproduce accurate CT excitation energies. The underestimation of CT excitation energies has been the most serious problem in TDKS calculations [51, 56] for many years. This problem is clearly solved by long-range correction [19], where the two electron operator, r^{-1}, is divided into short- and long-range parts and adopts exchange functionals to the short-range part and the HF exchange integral to the long-range part to supplement long-range exchange interactions into exchange functionals [18, 57] (for the history and applicabilities of the long-range correction, see Ref. [58]). Figure 7.1 illustrates the CT excitation energy of the tetrafluoroethylene-ethylene dimer in terms of the intermolecular distance [56], for which the CT excitation energy is seriously underestimated when using standard functionals, such as GGA and hybrid GGA functionals [56]. As has clearly been shown in the figure, the TDKS method using an LC functional provides accurate CT excitation energy very close to the *ab initio* SAC-CI results, while the TDKS method seriously underestimates the CT excitation energy when using the standard functionals. The long-range correction was also found

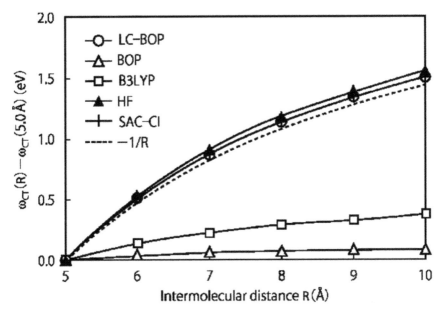

FIGURE 7.1 The lowest charge transfer excitation energies of ethylene-tetrafluoroethylene dimer for the long intermolecular distance calculated by the TDKS method using various functionals and the TDHF method in eV. The excitation energies in the figure are set at zero at 5 Å for each method [19].

to significantly improve the underestimation of Rydberg excitation energies [59] and oscillator strengths [60] simultaneously while maintaining the accuracy of valence excitation energies [19]. Long-range correction, which is also currently called "range-separation" (even though it is confusing due to the presence of short-range corrected functionals with very different characteristics such as the Heyd-Scuseria-Ernzerhof (HSE) hybrid functional [61]) has been established as the tool for reproducing accurate CT excitations in TDKS calculations. Thus far, long-range corrected (LC) TDKS studies have discovered various unknown mechanisms, for example, the initial process of TiO_2 photocatalytic reactions [14] and the photo-induced phase transition of tetrathiafulvalene-p-chloranil (TTF-CA) [15]. The LC-TDKS method in excited-state geometry optimizations provides very accurate excited-state geometries comparable to the multireference results [50], and it also succeeds in explaining several photochemical reaction mechanisms such as the dual fluorescence mechanism of dimethylaminobenzonitrile (DMABN) [62] and the ultrafast photoisomerization of cis-stilbene in solution [16]. Due to its accuracy in CT excitation calculations, the LC-TDKS method has recently been applied to various approximations of the TDKS method such as the tight-binding approximation [63, 64]. The elements of **A** and **B** matrices of the TDKS equation in Eq. (26) in the LC-TDKS method are represented as [19]:

$$A_{ia\sigma,jb\tau} = \delta_{ij}\delta_{ab}\delta_{\sigma\tau}\left(\epsilon_{a\sigma} - \epsilon_{i\sigma}\right) + K_{ia\sigma,jb\tau}^{LC} \tag{56}$$

and

$$B_{ia\sigma,jb\tau} = K_{ia\sigma,bj\tau}^{LC} \tag{57}$$

where $\varepsilon_{i\sigma}$ is the i-th s-spin orbital energy and $K_{ia\sigma,jb\tau}^{LC}$ is given by

$$K_{ia\sigma,jb\tau}^{LC} = (ia\sigma \mid jb\tau) + \int\int d^3r_1 d^3r_2 \phi_{i\sigma}^*(\mathbf{r}_1)\phi_{a\sigma}(\mathbf{r}_1) f_{xc}^{LC(sr)}(\mathbf{r}_1,\mathbf{r}_2)\phi_{j\tau}(\mathbf{r}_2)\phi_{b\tau}^*(\mathbf{r}_2) + K_{ia\sigma,jb\tau}^{LC(lr)} \tag{58}$$

where $f_{xc}^{LC(sr)}(\mathbf{r}_1,\mathbf{r}_2)$ is the local LC exchange-correlation integral kernel,

$$f_{xc}^{LC(sr)}(\mathbf{r}_1,\mathbf{r}_2) = \frac{\delta^2\left[E_x^{LC(sr)} + E_c\right]}{\delta\rho_\sigma(\mathbf{r}_1)\delta\rho_\tau(\mathbf{r}_1)}\delta(\mathbf{r}_1 - \mathbf{r}_2) \tag{59}$$

The first term on the right side of Eq. (58) is the Hartree integral in Eq. (30) and $K_{ia\sigma,bj\tau}^{LC(lr)}$ is the long-range exchange term:

$$K_{ia\sigma,jb\tau}^{LC(lr)} = -\delta_{\sigma\tau}\int\int d^3r_1 d^3r_2 \phi_{i\sigma}^*(\mathbf{r}_1)\phi_{j\sigma}(\mathbf{r}_1)r_{12}^{-1}\mathrm{erf}(\mu r_{12})\phi_{a\tau}^*(\mathbf{r}_2)\phi_{b\tau}(\mathbf{r}_2) \tag{60}$$

Note that the orbitals and orbital energies of the LC-KS calculations must be used in the LC-TDKS calculations to obtain accurate charge transfer excitation energies.

The remarkable accuracy of LC-TDKS excitation energies results from the very accurate exchange-correlation integral kernel, f_{xc}, of LC functionals [65]. It is noteworthy that LC functionals have been established to provide accurate orbital energies of both occupied and unoccupied orbitals for the first time ever [65]. The accuracy of calculated orbital energies is proven to depend on self-interaction error (Subsection 7.2.2) through the exchange-correlation integral kernel in Eq. (31) [65]. Since the K terms consist of the exchange-correlation integral kernel in Eq. (58), all the **A** and **B** matrix elements in Eqs. (56) and (57) are dependent on the exchange integral kernel of the functional used. It has been revealed that the exchange-correlation integral kernel of LC-DFT is usually dominated by the long-range exchange part even for the self-interactions [65]. Note, however, that plausible excitation energies are often obtained even without long-range correction due to the error cancellation of small orbital energy gaps and small exchange integral kernels. The most prominent example is the valence excitations of small molecules, for which accurate excitation energies are obtained due to the similar shapes of orbitals before and after excitations. In contrast, the orbital shapes in CT and Rydberg excitations greatly differ before and after the excitations to yield a negligible **K** matrix in Eqs. (27) and (28). The underestimation of orbital energy gaps consequently causes severe errors in calculated excitation energies.

7.2.2   SELF-INTERACTION CORRECTION

Core excitation energies have attracted attention in catalytic reaction analyses in recent years because these energies are available to investigate changes in the electronic structures of materials and surface adsorption species by way of X-ray absorption hyper-fine spectrum (XAFS) analyses. These XAFS analyses are frequently used to specify reaction intermediates in catalytic reactions. However, it has been reported that TDKS calculations using conventional functionals including even LC functionals seriously underestimate core excitation energies [66]. Even though several corrections have been proposed to improve core excitation energies in the TDKS method [67, 68], there have been no corrections for simultaneously providing accurate valence and core excitation energies, until very recently.

It has been suggested that the underestimation of core excitation energies results from the self-interaction error of exchange functionals [53, 54]. Self-interaction error is the sum of Coulomb and exchange self-interactions, which remains due to the use of exchange functionals as a substitute for the HF exchange integral in the exchange part of the Kohn-Sham equation:

$$\Delta E^{\text{SIE}} = \sum_i^n \left(J_{ii} + E_{\text{xc}}[\rho_i] \right) \tag{61}$$

where J_{ii} is the Coulomb self-interaction of the i-th orbital electron, E_{xc} is the exchange-correlation energy functional, ρi is the electron density of the i-th orbital, and n is the number of electrons. Various types of self-interaction corrections have been developed to remove this error. (For the history and various methods of the self-interaction corrections in DFT, see Tsuneda and Hirao [69]). Perdew-Zunger correction (PZ-SIC) [70], which simply eliminates the self-interaction error from the energy and potential, is the most widely-used SIC. However, it has been reported that this correction seriously deteriorates Kohn-Sham orbital energies [71] and consequently worsens TDKS excitation energies.

Regional self-interaction correction (RSIC) is an SIC, which only removes self-interaction error for electrons with no two-electron interaction. Tsuneda et al. proposed RSIC, which only corrects for exchange functionals in self-interaction regions, where the ratio of the von Weizsäcker kinetic energy density, τ^W, to the total one, τ^{total}:

$$t_\sigma = \tau_\sigma^W / \tau_\sigma^{total} \tag{62}$$

It is close to one [72]. This is based on a relationship between the density matrix of self-interacting electrons and the kinetic energy density [73]. For σ-spin self-interacting electrons, the density matrix is represented as:

$$P_\sigma^{SI}(\mathbf{r}_1,\mathbf{r}_2) = \rho_\sigma^{1/2}(\mathbf{r}_1)\rho_\sigma^{1/2}(\mathbf{r}_2) \tag{63}$$

Using this density matrix, the kinetic energy density, τ, in the kinetic energy, $T \equiv \int d^3\mathbf{r}\tau$, becomes the von Weizsäcker one, τ^W, such as [73]:

$$\tau = \sum_\sigma \tau_\sigma = -\frac{1}{2}\sum_\sigma \nabla^2 P_\sigma^{SI}(\mathbf{r}_1,\mathbf{r}_2)\Bigg|_{\mathbf{r}_1=\mathbf{r}_2} \rightarrow \tau^W = \sum_\sigma \tau_\sigma^W = \sum_\sigma \frac{1}{8}\frac{|\nabla\rho_\sigma|^2}{\rho_\sigma} \tag{64}$$

The RSIC cuts out the self-interaction regions by using a region-separation function, f_{RS}, to only select these self-interacting electrons and replaces the exchange integral kernel, f_{xs}, in the self-interaction regions with the exchange integral kernel of the exacted exchange self-interactions, i.e.,

$$f_{x\sigma}^{RSIC} = [1 - f_{RS}(t)]f_{x\sigma}^{DF} + f_{RS}(t)f_{x\sigma}^{SI} \tag{65}$$

That is, the RSIC-TDKS method has similar **A** and **B** matrices to the LC-TDKS one in Eqs. (56) and (57):

$$A_{ia\sigma,jb\tau} = \delta_{ij}\delta_{ab}\delta_{\sigma\tau}\left(\epsilon_{a\sigma} - \epsilon_{i\sigma}\right) + K_{ia\sigma,jb\tau}^{\text{RSIC}} \tag{66}$$

and

$$B_{ia\sigma,jb\tau} = K_{ia\sigma,bj\tau}^{\text{RSIC}} \tag{67}$$

The $K_{ia\sigma,jb\tau}^{\text{RSIC}}$ in these equations is analogous to the LC-TDKS one such as:

$$K_{ia\sigma,jb\tau}^{\text{RSIC}} = (ia\sigma \mid jb\tau) + \iint d^3\mathbf{r}_1 d^3\mathbf{r}_2 \phi_{i\sigma}^*(\mathbf{r}_1)\phi_{a\sigma}(\mathbf{r}_1) f_{\text{xc}}^{\text{non-SI}}(\mathbf{r}_1,\mathbf{r}_2)\phi_{j\tau}(\mathbf{r}_2)\phi_{b\tau}^*(\mathbf{r}_2) + K_{ia\sigma,jb\tau}^{\text{SI}} \tag{68}$$

where $f_{\text{xc}}^{\text{non-SI}}$ is the non-self-interaction-part of the exchange-correlation integral kernel:

$$f_{\text{xc}}^{\text{non-SI}}(\mathbf{r}_1,\mathbf{r}_2) = \frac{\delta^2\left([1 - f_{\text{RS}}(t)]E_{\text{x}}^{\text{non-SI}} + E_{\text{c}}\right)}{\delta\rho_\sigma(\mathbf{r}_1)\delta\rho_\tau(\mathbf{r}_1)}\delta(\mathbf{r}_1 - \mathbf{r}_2) \tag{69}$$

The exact exchange of the self-interaction integral kernel is the best way to use the pseudospectral exchange integral kernel [54]. In this case, $K_{ia\sigma,bj\tau}^{\text{SI}}$ is represented as:

$$K_{ia\sigma,jb\tau}^{\text{SI}} = -\frac{1}{4}\sum_{\mu\nu\lambda\kappa}\int d^3\mathbf{r}_1 P_{\mu\nu}P_{\lambda\kappa}\chi_\nu^*(\mathbf{r}_1)\chi_\lambda(\mathbf{r}_1)\int d^3\mathbf{r}_2 \frac{\chi_\kappa^*(\mathbf{r}_2)\chi_\mu(\mathbf{r}_2)}{|\mathbf{r}_2 - \mathbf{r}_1|} \tag{70}$$

where $P_{\mu\nu}$ is the density matrix based on atomic orbitals and χ_μ is the μ-th atomic orbital, which can be used with the LC-DFT [54, 74]. The region-separation function of the PSRSIC method f_{RS} is given by a step function such as:

$$f_{\text{RS}}^{\text{PSRSIC}} = \begin{cases} 0 & (t_\sigma < a) \\ 1 & (t_\sigma \geq a) \end{cases} \tag{71}$$

where a is the number of t_s s needed to cut out the self-interaction regions, $a = -0.0204\ Z + 1.0728$ (Z is the nuclear charge of the atom) in PSRSIC.

The RSIC using the pseudospectral exchange energy density is called "pseudospectral RSIC (PSRSIC)." Note that similarly to the LC-TDKS method, the orbitals and orbital energies of the RSIC-KS calculations should be used in the RSIC-TDKS calculations. The PSRSIC solves the underestimation of TDKS core excitation energies [54, 74]. Figure 7.2 compares the mean absolute errors in the calculated core excitation energies. As shown in the figure, TDKS calculations using standard functionals including LC functional seriously underestimate core excitation energies, while the PSRSIC drastically improves these core excitation energies to the accuracy of valence excitation energies. This remarkable accuracy is found to be attributed to the significant improvement of core orbital energies [54]. Note that the PSRSIC maintains or improves the accuracy of LC-TDKS valence excitation energies [74]. This clearly indicates that the underestimation of core excitation energies results from self-interaction error.

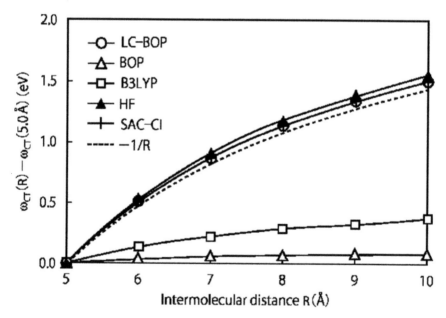

FIGURE 7.2 The mean absolute deviations (MADs) of the occupied and unoccupied orbital energies from the corresponding minus ionization potentials (IPs) and electron affinities (EAs), respectively, in eV: (a) HOMOs and LUMOs of hydrogen and rare gas atoms, (b) core 1s orbitals, and (c) HOMOs and LUMOs of typical molecules. "LC-PR-BOP" indicates the long-range and pseudospectral regional self-interaction corrected BOP functional [54].

7.2.3 DOUBLE-EXCITATION CORRECTION

The usual TDKS method only takes single excitations into account [75]. However, *ab initio* excited-state studies have revealed that double-excitation configurations are frequently mixed in highly-excited states and appear even in the ground state of ozone molecules [76] and the lowest excited state of benzene molecules [77]. Double-excitation effects are, therefore, required in the TDKS method to perform photochemical reaction simulations through excited state transitions. Even though there are *ab initio* wavefunction theories incorporating double-excitations such as multiconfiguration and multireference theories, these theories need too much computational time to calculate the excited-states of large systems, as was explained in Subsection 7.1.1. There is a method incorporating the double-excitation effect of the perturbation theory into the TDKS method [78]. However, this method also needs too much computational time because it performs the perturbation energy calculations for each ground or excited state. Note that it is possible to explicitly incorporate double-excitation effects in the TDKS method [49]. However, it also requires excessive computational time to calculate quite a lot of the additional terms. These theories have, therefore, only been applied to the calculations of small molecules.

There is a method of easily incorporating double-excitation effects in the TDKS method: the spin-flip (SF) TDKS method [79, 80]. The SF-TDKS method generates singlet ground and excited states incorporating double-excitation configurations by the spin-flip excitations of a triplet reference configuration. This method only supplements the exchange term into the TDKS matrix through the exchange-correlation integral kernel, f_{xc} [80]:

$$K^{SF}_{ia\sigma,jb\tau} = \int\int d^3\mathbf{r}_1 d^3\mathbf{r}_2 \phi^*_{i\sigma}(\mathbf{r}_1)\phi_{a\sigma}(\mathbf{r}_1) f_{xc}(\mathbf{r}_1,\mathbf{r}_2)\phi_{j\tau}(\mathbf{r}_2)\phi^*_{b\tau}(\mathbf{r}_2) \quad (72)$$

This term only provides a nonzero value for the HF part of the exchange integral kernel. When using GGA functionals with no HF exchange integral, the SF-TDKS method, therefore, yields the same results as the usual TDKS method in form, though these results are generally different due to the discrepancy in the orbitals of the reference configurations. In contrast, LC and hybrid GGA functionals provide clear differences between TDKS and SF-TDKS excitation energies.

Note that the standard SF-TDKS results partly contain triplet excitations as well as singlet excitations. These triplet excitations are attributed to the spin-asymmetric triplet configuration used as the reference of the SF-TDKS calculations. The singlet and triplet excitations can be classified in the SF-TDKS calculations on the basis of the expectation values of the squared total spin operator, $\langle S^2 \rangle$. Although triplet wavefunctions have three types of spin functions, i.e., $\alpha\alpha$, $\beta\beta$, and $\alpha\beta + \beta\alpha$ functions for singly occupied molecular orbitals (SOMOs), the usual SF-TDKS calculations only adopt the spin-asymmetric $\alpha\alpha$-spin configuration as the triplet reference function. Note that the generated triplet excitations do not cause the elimination of the singlet excitations because they are assigned to the triplet excitations of the TDKS method. The spin-asymmetric reference also generates the excitations of $\langle S^2 \rangle \approx 1$ in the SF-TDKS calculations. To avoid spin-contamination, the spin-symmetric configurations should be generated by using a spin-symmetric triplet reference configuration, which consists of both $\alpha\alpha$ and $\beta\beta$ states, or by supplementing spin-complement configurations [81, 82]. It has, however, been reported that the spin-contamination effect is not so large in the calculated excitation energies of the $\langle S^2 \rangle \approx 1$ excitations [82].

By combining the long-range correction in Subsection 7.2.1, the SF-TDKS method can provide accurate excitation energies [20]. It has been reported that the SF-TDKS method significantly underestimates even valence excitation energies. The BHHLYP functional, in which the Becke 1988 exchange functional [83] is combined with the HF exchange integral in a one-to-one ratio for the exchange part, has been used to avoid the underestimation of SF-TDKS excitation energies. However, even when using the BHHLYP functional, the calculated SF-TDKS excitation energies are considerably higher than the TDKS ones for excitations with negligible double excitations. This indicates that the SF-TDKS method violates the correspondence principle that the SF-TDKS excitation energies should be close to the TDKS ones for excitations with no double-excitation configurations, even though this principle is required to validate the SF-TDKS excitation energies. It has recently been found that long-range correction makes the SF-TDKS method meet the correspondence principle [20]. The SF-TDKS method using LC functionals provides excitation energies that are very close to the TDKS ones for single excitations with no significant double excitations and it reproduces very accurate excitation energies. The calculated excitation energies of long-chain polyacetylenes are plotted in Figure 7.3. The figure shows that the SF-TDKS method gives very accurate excitation energies for both systems when using the LC-BLYP functional, while it gradually increases the errors as the chains are extended when using the B3LYP hybrid and BLYP GGA functionals. SF-LC-TDBLYP particularly succeeds in reproducing the *ab initio* multireference results in that the lowest excitations of polyacetylenes are alternated from the $1B_{2u}$ to $1B_{3u}$ excited states as the chain lengthens. Note from the figure that the differences between these excitation energies are considerable in the LC-TDBLYP calculations. This indicates that SF-LC-TDBLYP incorporates double-excitation effects that match the accuracy of multireference theories.

7.2.4   RELATIVISTIC SPIN-ORBIT CORRECTION

Relativistic effects are required in the TDKS method to perform the calculations of electronic spectra and photochemical reactions of heavy atom compounds containing sixth and seventh-row atoms in the periodic table. Relativistic spin-orbit interaction plays a particularly significant role in photochemical reactions because it causes forbidden intersystem crossings (spin-orbit transitions) as seen in the Jablonski diagram (Figure 7.4). Incorporating the spin-orbit interactions in the TDKS method also makes it possible to investigate the spin-orbit splitting of the excited states [55, 84]. For the relativistic TDKS method, the Rajagopal-Callaway theorems [39], which are the application of QED to DFT, form the foundation as mentioned in Subsection 7.1.3, while QED is constructed on the basis of the time-dependent Dirac equation. Even though there is the four-component time-dependent Dirac-Kohn-Sham method [85], the applicability of this method is limited to the calculations of small molecules due to the immense computational time needed. This section focuses on the relativistic TDKS method [86–90] based on linear response theory.

In the relativistic spin-orbit TDKS method based on the two-component zeroth-order regular approximation (ZORA), relativistic effects including spin-orbit couplings are incorporated on the basis of the ZORA Hamiltonian operator $\hat{H}^{ZORA}$ such as [91]:

$$\hat{H}^{ZORA}\Phi = \left(\hat{T}^{ZORA} + V\right)\Phi \qquad (73)$$

where Φ is the two-component spinor wavefunction:

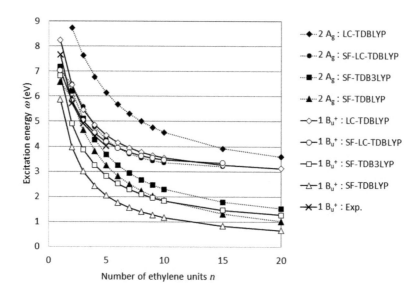

FIGURE 7.3   The SF-TDKS excitation energies (ω) of the $1^1B^-_{2u}$(HOMO→LUMO) and 2^1A_g(HOMO-1→LUMO and HOMO→LUMO+1) excitations of linear polyacetylenes [20].

$$\hat{T}^{ZORA} = \sigma\cdot\mathbf{p}\,\frac{c^2}{2c^2-V}\,\sigma\cdot\mathbf{p} \tag{74}$$

and σ, $\hat{\mathbf{p}}$, and c correspond to a 2×2 Pauli spin matrix, the momentum operator, and the speed of light. The spin-orbit couplings are incorporated through
the kinetic energy in Eq. (79). The spinor wavefunction is represented
by orbital spinors $\{\phi_i\}$, which are the linear combinations of α and β spin
orbitals:

$$\phi_i = \sum_\mu \left(c^\alpha_{\mu i}\chi_\mu\alpha + c^\beta_{\mu i}\chi_\mu\beta \right) \tag{75}$$

where χ_m is the μ-th atomic orbital. The density matrix, **P**, in the spin-orbit
ZORA is given as:

$$\mathbf{P} = \begin{pmatrix} \mathbf{P}^{\alpha\alpha} & \mathbf{P}^{\alpha\beta} \\ \mathbf{P}^{\beta\alpha} & \mathbf{P}^{\beta\beta} \end{pmatrix} \tag{76}$$

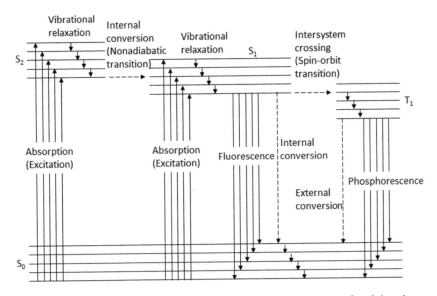

FIGURE 7.4   Jablonski diagram, which indicates possible consequences of applying photons.

where $\mathbf{P}^{\zeta\tau}$ has the elements:

$$P_{\mu\nu}^{\zeta\tau} = \sum_i c_{\mu i}^{\zeta*} c_{\nu i}^{\tau} \quad (\zeta, \tau = \alpha \text{ or } \beta) \tag{77}$$

The difference between the relativistic and nonrelativistic TDKS methods is found in the exchange-correlation part. Similar to the TDKS equation, the relativistic TDKS equation has the following **A** and **B** matrix elements for closed-shell systems:

$$A_{ia,jb} = \delta_{ij}\delta_{ab}(\epsilon_a - \epsilon_a) + \frac{\partial F_{ia}}{\partial P_{jb}} \tag{78}$$

and

$$B_{ia,jb} = \frac{\partial F_{ia}}{\partial P_{jb}} \tag{79}$$

where F_{ia} and P_{jb} are the elements of the Fock matrix and density matrix. The difference from the TDKS matrix elements is given in the noncollinearity of the exchange-correlation part [88]. That is, since the wavefunction becomes a non-eigenfunction of the spin angular momentum operator, $\hat{S}_z$, by incorporating the spin-orbit interaction terms, the derivatives in terms of spin appear in the exchange-correlation potential:

$$V_{xc} = \frac{\delta E_{xc}}{\delta \rho} + \frac{\delta E_{xc}}{\delta s} \frac{\mathbf{m} \cdot \boldsymbol{\sigma}}{s} \tag{80}$$

where $\mathbf{m}$ is magnetic vector:

$$\mathbf{m} = \sum_i \phi_i^* \, \boldsymbol{\sigma} \phi_i \tag{81}$$

It is interesting to exemplify the formulation of the relativistic spin-orbit TDKS method using LC functionals. The $\mathbf{A}$ and $\mathbf{B}$ matrix elements in the spin-orbit TDKS method in Eqs. (78) and (79) are expressed as [55]:

$$A_{ia,jb} = \delta_{ij}\delta_{ab}(\epsilon_a - \epsilon_a) + (ia \mid jb) - (ij \mid ab)^{lr} + \left(f_x^{sr}\right)_{ia,jb} + \left(f_c\right)_{ia,jb} \tag{82}$$

and

$$B_{ia,jb} = (ia \mid bj) - (ib \mid aj)^{lr} + \left(f_x^{sr}\right)_{ia,bj} + \left(f_c\right)_{ia,bj} \tag{83}$$

The long-range exchange integral term in these matrices is given by:

$$(ij \mid ab)^{lr} = \int d^3\mathbf{r}_1 d^3\mathbf{r}_2 \phi_i^*(\mathbf{r}_1)\phi_j(\mathbf{r}_1)\frac{\mathrm{erf}(\mu r_{12})}{r_{12}}\phi_a^*(\mathbf{r}_2)\phi_b(\mathbf{r}_2) \tag{84}$$

and $\left(f_x^{sr}\right)_{ia,jb}$ and $\left(f_c\right)_{ia,jb}$ are the integral terms for the short-range exchange and correlation integral kernels, respectively. Considering the energy derivatives in terms of spin, the short-range exchange integral kernel term gives a complicated form [55]:

$$\left(f_x^{sr}\right)_{ia,jb} = \frac{\partial^2 E_x^{sr}}{\partial P_{ia}\partial P_{jb}}$$

$$= \int d^3\mathbf{r}\phi_i^*(\mathbf{r})\phi_a(\mathbf{r})\frac{\partial^2 E_x^{sr}}{\partial\rho^2}\phi_j^*(\mathbf{r})\phi_b(\mathbf{r}) + \int d^3\mathbf{r}\phi_i^*(\mathbf{r})\phi_a(\mathbf{r})\left(2\frac{\partial^2 E_x^{sr}}{\partial\rho\partial(|\nabla\rho|^2)}\right)(\nabla\rho(\mathbf{r})\cdot(\phi_j^*\phi_b)(\mathbf{r}))$$

$$+ \int d^3\mathbf{r}(\phi_i^*(\mathbf{r})\sigma\phi_a(\mathbf{r}))\frac{\partial^2 E_x^{sr}}{\partial\sigma^2}(\phi_j^*(\mathbf{r})\sigma\phi_b(\mathbf{r})) + \int d^3\mathbf{r}(\phi_i^*(\mathbf{r})\sigma\phi_a(\mathbf{r}))\frac{\partial^2 E_x^{sr}}{\partial s\partial(\nabla\rho\cdot\sigma)}(\nabla\rho(\mathbf{r})\cdot(\phi_j^*(\mathbf{r})\sigma\phi_b(\mathbf{r})))$$

$$+ \int d^3\mathbf{r}\nabla(\phi_i^*(\mathbf{r})\phi_a(\mathbf{r}))\left(2\frac{\partial^2 E_x^{sr}}{\partial\rho\partial(|\nabla\rho|^2)}\right)\phi_j^*(\mathbf{r})\phi_b(\mathbf{r})\nabla\rho(\mathbf{r})$$

$$+ \int d^3\mathbf{r}\nabla(\phi_i^*(\mathbf{r})\phi_a(\mathbf{r}))\left(4\frac{\partial^2 E_x^{sr}}{\partial(|\nabla\rho|^2)^2}\right)(\nabla\rho\cdot(\phi_j^*(\mathbf{r})\phi_b(\mathbf{r})))\nabla\rho(\mathbf{r})$$

$$+ \int d^3\mathbf{r}\nabla(\phi_i^*(\mathbf{r})\phi_a(\mathbf{r}))\left(2\frac{\partial E_x^{sr}}{\partial(|\nabla\rho|^2)}\right)\nabla(\phi_j^*(\mathbf{r})\phi_b(\mathbf{r}))$$

$$+ \int d^3\mathbf{r}\nabla(\phi_i^*(\mathbf{r})\sigma\phi_a(\mathbf{r}))\frac{\partial^2 E_x^{sr}}{\partial s\partial(\nabla\rho\cdot\sigma)}(\phi_j^*(\mathbf{r})\sigma\phi_b(\mathbf{r}))\nabla\rho(\mathbf{r})$$

$$+ \int d^3\mathbf{r}\nabla(\phi_i^*(\mathbf{r})\sigma\phi_a(\mathbf{r}))\frac{\partial^2 E_x^{sr}}{\partial s\partial(\nabla\rho\nabla\sigma)}(\phi_j^*(\mathbf{r})\sigma\phi_b(\mathbf{r}))\nabla\rho(\mathbf{r})$$

$$+ \int d^3\mathbf{r}\nabla(\phi_i^*(\mathbf{r})\sigma\phi_a(\mathbf{r}))\frac{\partial^2 E_x^{sr}}{\partial(\nabla\rho\cdot\sigma)^2}(\nabla\rho(\mathbf{r})\nabla(\phi_j^*(\mathbf{r})\sigma\phi_b(\mathbf{r})))\nabla\rho(\mathbf{r})$$

$$+ \int d^3\mathbf{r}\nabla(\phi_i^*(\mathbf{r})\sigma\phi_a(\mathbf{r}))\left(2\frac{\partial E_x^{sr}}{\partial(\nabla\sigma\cdot\sigma)}\right)\nabla(\phi_j^*(\mathbf{r})\sigma\phi_b(\mathbf{r}))$$

$$(85)$$

The correlation integral kernel term, $\left(f_c\right)_{ia,jb}$, is only given by replacing E^{sr} of Eq. (85) with E_c when using a pure GGA exchange functional. The spin-orbit electric dipole transition moment in the spin-orbit TDKS calculations, which is proportional to the spin-orbit transition rate and the phosphorescence lifetime, is calculated by determining the singlet and triplet wavefunctions with the response function coefficients. The a-axis projection of the electric dipole transition moment between the ground state, S_0, and the k-th substrate of the lowest triplet states T^k is represented as [92]:

$$M_a^k = \sum_{n=0}^{\infty}\frac{\langle S_0|\hat{\mu}_a|S_n\rangle\langle S_n|\hat{\mu}_a|T_1^k\rangle}{E(S_n)-E(T_1)} + \sum_{n=1}^{\infty}\frac{\langle S_n|\hat{H}_{SO}|T_1^k\rangle\langle S_0|\hat{\mu}_a|S_n\rangle}{E(T_n)-E(S_0)} \quad (86)$$

where $\hat{H}_{SO}$ is the spin-orbit operator and μ_a is an a-axis dipole moment operator. Using this transition moment, the phosphorescence lifetime from $T_1^k(\tau_k)$ is calculated by:

$$\frac{1}{\tau_k} = \frac{4}{3}\alpha_0^3 \left[E(T_1) - E(S_0)\right]^3 \sum_{a=\{x,y,z\}} \left|M_a^k\right|^2 \tag{87}$$

where α_0 is the fine-structure constant.

Table 7.1 summarizes the mean absolute errors in the calculated excitation energies and spin-orbit splittings of the 12-th group transition metal atoms

TABLE 7.1   Mean Absolute Errors of the Calculated Excitation Energies and Spin-Orbit Splittings (SOSs) of Two-Component ZORA-TDKS from the Experimental Values in eV [55]

Excitation	LDA	BLYP	B3LYP	LC-BLYP
The $ns^2 \rightarrow ns^1 np^1$ excitation energies of the 12-th group transition metal atoms				
$^3P_0$	0.36	0.26	0.08	0.25
$^3P_1$	0.35	0.25	0.07	0.25
$^3P_2$	0.32	0.19	0.07	0.26
$^1P_1$	0.29	0.19	0.14	0.17
Total	0.33	0.22	0.09	0.23
The $np^6 \rightarrow np^5 (n+1)s^1$ excitation energies of the 12-th group transition metal atoms				
$^3P_2$	0.93	1.34	0.78	0.37
$^3P_1$	0.94	1.34	0.77	0.35
$^3P_0$	1.03	1.47	0.87	0.37
$^1P_1$	1.05	1.46	0.87	0.33
Total	0.99	1.40	0.82	0.35
The SOSs of the $ns^1 np^1$ configurations of the 12-th group transition metal atoms				
$^3P_1$	0.01	0.02	0.01	0.01
$^3P_2$	0.05	0.07	0.06	0.03
$^1P_1$	0.09	0.12	0.12	0.17
Total	0.05	0.07	0.07	0.07
The SOSs of the $np^5(n+1)s^1$ configurations of rare gas atoms				
$^3P_1$	0.01	0.01	0.01	0.02
$^3P_0$	0.13	0.15	0.13	0.10
$^1P_1$	0.16	0.19	0.18	0.15
Total	0.10	0.12	0.11	0.09

The Tamm-Dancoff approximation is used. Three 12-th group transition metal atoms, Zn, Cd, and Hg, and five rare gas atoms, Ne, Kr, Xe, and Rn, are used.

and rare gas atoms [55]. As listed in the table, the long-range correction significantly improves the excitation energies of rare gas atoms, even though it hardly contributes to the excitation energies of the 12-th group transition metal atoms. The B3LYP hybrid GGA functional, on the other hand, provides the best excitation energies for the excitation energies of the 12-th group transition metal atoms. This results from the difference in the contributing exchange interaction ranges of these excitations. In contrast, the table indicates that the spin-orbit splittings hardly depend on the functionals used. This may be due to the small long-range exchange effect on the spin-orbit splittings, for which similar electron configurations are compared. The long-range correction for spin-orbit transitions is supposed to be efficient because spin-orbit interactions are essentially large between the electronic states of very different spin multiplicities or angular momenta. Since these states have vastly different electron distributions, the long-range exchange interactions are supposed to be significant to enable spin-orbit transitions to be investigated.

7.2.5   NONADIABATIC CORRECTION

Thus far, the TDKS method has been discussed on the basis of the adiabatic approximation (Born-Oppenheimer approximation [93]), which neglects the explicit motion of atomic nuclei. However, electronic and nuclear motions in actual systems interact with one another to proceed with nonadiabatic transitions, which are the transitions between electronic states due to electron-nuclear interactions. As can be seen from the Jablonski diagram (Figure 7.4), internal conversions (nonadiabatic transitions) frequently take place in photochemical reactions and develop useful functions in many optical functional materials. Let us consider the total wavefunction incorporating both electronic and nuclear motions:

$$\Psi(\mathbf{r},\mathbf{R},t) = \sum_{I,A} \Phi_I(\mathbf{r},\mathbf{R})\chi_{IA}(\mathbf{R}) \tag{88}$$

where Φ_I is the wavefunction of the I-th electron motion and χ_{IA} is the wavefunction of the A-th atom, to which the I-th electron belongs. The total Hamiltonian operator, $\hat{H}$ for this wavefunction is given by:

$$\hat{H} = \hat{H}_{\text{elec}} + \hat{T}_{\text{nuc}} \tag{89}$$

where $\hat{H}_{\text{elec}}$ is the electronic Hamiltonian operator and $\hat{T}_{\text{nuc}}$ is the nuclear kinetic energy operator. Assuming that electronic motion obeys the Born-Oppenheimer approximation, the eigenequation of the electronic motion is expressed as:

$$\hat{H}_{\text{elec}}(\mathbf{R})\Phi_I = U(\mathbf{R})\Phi_I \tag{90}$$

where U is the eigenenergy of electronic motion, and the equation of nuclear motion is obtained as:

$$\left[i\frac{\partial}{\partial t} - \left(\hat{T}_{\text{nuc}} + U_I(\mathbf{R})\right)\right]\chi_{IA} + \sum_J \left[i\left\langle\Phi_I(\mathbf{R})\left|\frac{\partial\Phi_J(\mathbf{R})}{\partial t}\right.\right\rangle - i\left\langle\Phi_I(\mathbf{R})\left|\hat{T}_{\text{nuc}}\right|\Phi_J(\mathbf{R})\right\rangle\right]\chi_{JA} = 0 \tag{91}$$

In this equation,

$$\left\langle\Phi_I(\mathbf{R})\left|\frac{\partial\Phi_J(\mathbf{R})}{\partial t}\right.\right\rangle = \int d^3\mathbf{r}\,\Phi_I(\mathbf{R},\mathbf{r})\frac{\partial\Phi_J(\mathbf{R},\mathbf{r})}{\partial t}, \tag{92}$$

is the nonadiabatic coupling term, which causes nonadiabatic transitions. Even though the nuclear wavefunction is determined by solving Eq. (91) by using, for example, the wave packet method, it requires an immense amount of computational time to calculate the potential energies and nonadiabatic couplings.

Tully's surface hopping method [94] is the most widely-used to perform nonadiabatic transitions based on the TDKS method. Nonadiabatic transition probability in the surface hopping method is momentarily calculated at each snapshot of excited-state (on the fly) dynamics simulations. That is, this method stochastically determines the adiabatic potential energy surface that the nuclear motion passes through. After the nonadiabatic transitions, the generated de-excitation energies are distributed to each velocity component based on the velocity scaling. The fewest switch method [95, 96] is used to calculate the nonadiabatic transition probability, $P_{I \to J}$, in the surface hopping method:

$$P_{I \to J} = -2\int_t^{t+\Delta t} d\tau \frac{\text{Re}\left[C_I^*(\tau)\,\sigma_{\text{nonad}}(\tau)C_J^*(\tau)\right]}{\left|C_I(\tau)\right|^2} \tag{93}$$

where C_I is the expansion coefficient of the nonadiabatic wavefunction corresponding to the combined I-th adiabatic wavefunction and the element of the σ_{nonad} matrix is expressed by:

$$\left(\sigma_{\text{nonad}}\right)_{IJ}(t) = \left\langle \Phi_I(t) \middle| \dot{\Phi}_J(t) \right\rangle \tag{94}$$

Using the nonadiabatic transition probability in Eq. (93), the surface hopping method is performed in four steps [94]:

1. Calculate the TDKS excitation energy gradients for the initial structure of a system, $\{\mathbf{R}_0\}$, by solving the TDKS equation in Eq. (26) and its excitation energy gradients in Eq. (40).

2. Solve the Newton equation of motion to obtain the structure of the system after a constant time by using the calculated excited-state potential energy gradient, $\nabla_A V(\{\mathbf{R}_0\})$, which is the sum of the gradient of the ground state energy and excitation energy in terms of the nuclear coordinate of the A-th atom, $\mathbf{R}_A$, such as:

$$M_A \frac{d^2 \mathbf{R}_A}{dt^2}(t) = -\nabla_A V(\{\mathbf{R}_0\}) \tag{95}$$

where M_A is the mass of the A-th atom in the system.

3. Calculate the expansion coefficients, and the σ_{nonad} matrix is obtained by solving the time-dependent Schrödinger equation for the updated structure $\{\mathbf{R}_I\}$:

$$i\frac{d\mathbf{C}_I}{dt}(t) = \sum_J \left(E_I \delta_{IJ} - \left(\sigma_{\text{nonad}}\right)_{IJ}(t) \right) \mathbf{C}_J(t) \tag{96}$$

and then determine the nonadiabatic transition probability in Eq. (93).

4. Return to step 1 after setting the updated structure, $\{\mathbf{R}_I\}$, as the initial one $\{\mathbf{R}_0\}$ and repeat the above steps.

This method can easily be carried out once the TDKS excitation energy gradients are obtained, and has therefore been recently applied to theoretical investigations into various photochemical reactions. However, this method has several problems from both theoretical and practical points of view. The nonadiabatic transition probability in this method is calculated using the expansion coefficients of the excited-state configurations, which are determined using the TDKS response function coefficients in step 3. Since this is based on the concept of multiconfiguration theories, it has not been established to be coherent with the TDKS method in the other steps. This step also requires a great deal of computational time.

Several TDKS-based methods have been suggested to calculate the non-adiabatic transition probability [97–99]. A representative example is the Chernyak-Mukamel density matrix-based transition probability method [97]. In this method, the TDKS-based polarizability:

$$\alpha_{kl}(\omega) = \sum_I \frac{2\mathbf{h}_k^\dagger \left(\mathbf{A}-\mathbf{B}\right)^{-1/2} \mathbf{F}_I \mathbf{F}_I^\dagger \left(\mathbf{A}-\mathbf{B}\right)^{-1/2} \mathbf{h}_l}{\omega_I^2 - \omega^2} \tag{97}$$

where $\mathbf{h}_k$ is the derivative of the Hamiltonian matrix in terms of the k-th nuclear coordinate space, is compared with the wavefunction-based one:

$$\alpha_{kl}(\omega) = \sum_I \frac{2\omega_I \left\langle \Phi_0 | \hat{h}_k | \Phi_I \right\rangle \left\langle \Phi_I | \hat{h}_k | \Phi_0 \right\rangle}{\omega_I^2 - \omega^2} \tag{98}$$

Based on this comparison, Chernyak and Mukamel derived a nonadia-batic coupling form, in which the ground-state density matrix is used to cal-culate h_k [97]. Instead, Hu et al. [99] proposed a simple form to calculate this nonadiabatic coupling:

$$A_{I,k}(\mathbf{R}) = -\frac{\mathbf{h}_k^\dagger \left(\mathbf{A}-\mathbf{B}\right)^{-1/2} \left(\mathbf{X}+\mathbf{Y}\right)_I}{\omega_I^{3/2}} \tag{99}$$

This nonadiabatic coupling form is very useful due to the use of only the excitation energies and response functions given in the TDKS calcula-tions. This form actually has been applied to the nonadiabatic coupling cal-culations of small molecules such as LiH molecules [100]. However, this form has several problems in applying it to the investigation of nonadia-batic transitions. First, it has no rigorous form for the transitions between excited states and gives infinite values for the transitions of degenerate states, which are normal to proceed through conical intersections. This nonadiabatic coupling form, therefore, needs further improvement in the applications.

7.3   CONCLUSIONS

TDDFT has become the most widely-used excited-state calculation method because it accurately reproduces excitation energies with much less computational time than those of *ab initio* excited-state calculation methods such as multireference theories. This chapter has focused on TDDFT based on linear response theory, which is mainly applied to excited-state calculations and is well established by the Runge-Gross theorem [34]. TDDFT [13] has much more applicability to excited-state calculations than the TDHF method because excited-state calculations need to incorporate well-balanced electron correlations due to the vastly different electronic structures of the excited states. Due to its high applicability, even TDDFT excited-state energy gradients, which are required to perform excited-state dynamics simulations like photochemical reactions, are now available in many quantum chemistry calculation packages.

However, TDDFT calculations have many critical problems, for example, the underestimations of charge transfer, Rydberg and core excitation energies, disregard of multielectron excitations, relativistic spin-orbit interactions, and nonadiabatic couplings. Various types of physical corrections have been suggested to solve these problems. For example, the long-range correction (LC) for exchange functionals [18] clearly solves the underestimation of charge transfer excitation energies and improves the underestimated Rydberg excitation energies. The serious underestimation of core excitation energies, which results from the self-interaction error of exchange functionals, is clearly improved by pseudospectral regional self-interaction correction (PRSIC) [54]. Double-excitation effects are accurately incorporated in ground and excited states by the spin-flip TDKS method [80]. Relativistic spin-orbit interactions are included in TDDFT on the basis of ZORA incorporating the derivative of exchange-correlation functionals in terms of spin. Nonadiabatic transition dynamics simulations are carried out using the TDDFT excited-state potential gradients in the surface hopping method [94].

As was previously explained, many TDDFT problems have so far been solved or drastically reduced. There are, however, several outstanding problems in TDKS calculations. For example, it is difficult to reproduce nonadiabatic transitions through conical intersections or bond dissociation potential energy surfaces. These problems are mainly attributable to the single configurational characteristics of the standard TDKS method. There are still no established multiconfigurational TDDFTs for solving these problems as far as we know. Moreover, the computational time and computational order of TDDFT should be drastically reduced to perform the excited-state calculations of large-scale systems such as the band calculations of solids. We, however, expect that these problems will be solved in the near future because TDDFT is one of the most advanced methods of excited-state calculation. TDDFT will be the main theory for the excited-state calculations of large-scale systems for some time to come.

ACKNOWLEDGMENTS

We would like to acknowledge the contributions made thus far by Professor Ramon Carbó-Dorca for his kind hospitality and the productive discussions we had with him. This review was supported by the Japanese Ministry of Education, Culture, Sports, Science and Technology (MEXT) (Grant nos.: 23225001, 24350005, and 17H01188).

KEYWORDS

- double-excitation correction
- excited state calculations
- long-range correction
- nonadiabatic correction
- relativistic correction
- self-interaction correction

REFERENCES

1. Frenkel, J., (1934). Wave mechanics, *Advanced Genera Theory*. Clarendon Press, Oxford, pp. 1-525.
2. Whitten, J. L., & Hackmeyer, M., (1969). Configuration interaction studies of ground and excited states of polyatomic molecules. I. The CI formulation and studies of formaldehyde, *J. Chem. Phys.*, *51*, 5584–5596.
3. Condon, E. U., (1930). The theory of complex spectra, *Phys. Rev.*, *36*, 1121–1133.
4. Møller, C., & Plesset, M. S., (1934). Note on an approximation treatment for many-electron systems, *Phys. Rev.*, *46*, 618–622.

5. Čížek, J., (1966). On the Correlation Problem in Atomic and Molecular Systems. Calculation of Wavefunction Components in Ursell-Type Expansion Using Quantum-Field Theoretical Methods, *J. Chem. Phys.*, *45*, 4256–4266.

6. Roos, B. O., Taylor, P. R., & Siegbahn, P. E. M., (1980). A complete active space SCF method (CASSCF) using a density matrix formulated super-CI approach, *Chem. Phys.*, *48*, 157–173.

7. Andersson, K., Malmqvist, P. A., Roos, B. O., Sadlej, A. J., & Wolinski, K., (1990). Second-order perturbation theory with a CASSCF reference function, *J. Phys. Chem.*, *94*, 5483–5488.

8. Hirao, K., (1992). Multireference Møller-Plesset method, *Chem. Phys. Lett.*, *190*, 374–380.

9. Nakano, H., (1993). Quasidegenerate perturbation theory with multiconfigurational self-consistent-field reference functions, *J. Chem. Phys.*, *99*, 7983–7992.

10. L̈owdin, P.-O., (1955). Quantum Theory of Many-Particle Systems. III. Extension of the Hartree-Fock Scheme to Include Degenerate Systems and Correlation Effects, *Phys. Rev.*, *97*, 1509–1520.

11. Sinanoğlu, O., (1964). Many-Electron Theory of Atoms, Molecules and Their Interactions, *Adv. Chem. Phys.*, *6*, 315–412.

12. Kato, T., (1957). On the eigenfunctions of many-particle systems in quantum mechanics, *Commun. Pure Appl. Math.*, *10*, 151–177.

13. Casida, M. E., (1996). Time-Depedent Density Functional Response Theory of Molecular Systems: Theory, Computational Methods and Functionals, *Recent Developments and Applications of Modern Density Functional Theory*, edited by Seminario, J. J. Elsevier, Amsterdam, pp. 391–439.

14. Suzuki, S., Tsuneda, T., & Hirao, K., (2012). A theoretical investigation on photocatalytic oxidation on the TiO_2 surface, *J. Chem. Phys.*, *136*, 024706(1–6).

15. Nakatsuka, Y., Tsuneda, T., Sato, T., & Hirao, K., (2011). Theoretical Investigations on the Photoinduced Phase Transition Mechanism of Tetrathiafulvalene-*p*-chloranil, *J. Chem. Theor. Comput.* 7, 2233–2239.

16. Takeuchi, S., Ruhman, S., Tsuneda, T., Chiba, M., Taketsugu, T., & Tahara, T., (2008). Spectroscopic tracking of structural evolution in ultrafast stilbene photoisomerization, *Science*, *322*, 1073–1077.

17. Tsuneda, T., (2014). *Density Functional Theory in Quantum Chemistry*, Springer, Tokyo, Chap. 4, pp. 79–99.

18. Iikura, H., Tsuneda, T., Yanai, T., & Hirao, K., (2001). A long-range correction scheme for generalized-gradient-approximation exchange functionals, *J. Chem. Phys.*, *115*, 3540–3544.

19. Tawada, Y., Tsuneda, T., Yanagisawa, S., Yanai, T., & Hirao, K., (2004). A long-range-corrected time-dependent density functional theory, *J. Chem. Phys.*, *120*, 8425–8433.

20. Tsuneda, T., Singh, R. K., & Nakata, A., (2016). Relationship Between Orbital Energy Gaps and Excitation Energies for Long-Chain Systems, *J. Comput. Chem.*, *37*, 1451–1462.

21. Dirac, P. A. M., (1930). Note on exchange phenomena in the Thomas atom, *Camb. Phil. Soc.*, *26*, 376–385.

22. Bohm, D., & Pines, D., (1951). A collective description of electron interactions. I. Magnetic interactions, *Phys. Rev.*, *82*, 625–634.

23. Ehrenreich, H., & Cohen, M. H., (1959). Self-consistent field approach to the many-electron problem, *Phys. Rev.*, *115*, 786–790.

24. Thouless, D. J., (1960). Stability conditions and nuclear rotations in the Hartree-Fock theory, *Nucl. Phys.*, *21*, 225–232.

25. Mclachlan, A. D., & Ball, M. A., (1964). Time-Dependent Hartree-Fock Theory for Molecules, *Rev. Mod. Phys.*, *36*, 844–855.

26. Moccia, R., (1974). Static and dynamic first- and second-order properties by variational wave functions, *Int. J. Quantum Chem.*, *8*, 293–314.

27. Yeager, D. L., & Jorgensen, P., (1979). A multiconfigurational time-dependent Hartree-Fock approach, *Chem. Phys. Lett.*, *65*, 77–80.

28. McWeeny, R., (1992). *Methods of Molecular Quantum Mechanics,* 2nd Ed. Academic Press, San Diego, Chap. 12.7, pp. 442-446.

29. Nakatsuji, H., & Hirao, K., (1978). Cluster expansion of the wavefunction. Symmetry-adapted-cluster expansion, its variational determination, and extension of open-shell orbital theory, *J. Chem. Phys.*, *68*, 2053–2065.

30. Koch, H., & Jørgensen, P., (1990). Coupled cluster response functions, *J. Chem. Phys.*, *93*, 3333–3344.

31. Stanton, J. F., & Bartlett, R. J., (1993). A coupled-cluster based effective Hamiltonian method for dynamic electric polarizabilities, *J. Chem. Phys.*, *99*, 5178–5183.

32. Samanta, P. K., Mukherjee, D., Hanauer, M., & Kohn, A., (2014). Excited states with internally contracted multireference coupled-cluster linear response theory, *J. Chem. Phys.*, *140*, 134108(1–14).

33. Hohenberg, P., & Kohn, W., (1964). Inhomogeneous electron gas, *Phys. Rev. B*, *136*, 864–871.

34. Runge, E., & Gross, E. K. U., (1984). Density-functional theory for time-dependent systems, *Phys. Rev. Lett.*, *52*, 997–1000.

35. Gross, E. K. U., Ullrich, C. A., & Gossmann, U. A., (1995). Density functional theory of time-dependent systems, *Density Functional Theory, NATO ASI Series B*, edited by Dreizler, R; Gross, E. K. U., Plenum, New York, pp. 149–171.

36. van Leeuwen, R., (2006). Beyond the Runge-Gross theorem, *Lect. Notes Phys.*, *706*, 17–31.

37. Ullrich, C. A., (2012). *Time-Dependent Density-Functional Theory,* Oxford University Press, New York, Chap. 10, pp. 213-251.

38. Yang, Z., & Burke, K., (2013). Nonexistence of a Taylor expansion in time due to cusps, *Phys. Rev. A*, *88*, 042514(1–14).

39. Rajagopal, A. K., & Callaway, J., (1972). Inhomogeneous electron gas, *Phys. Rev. B*, *7*, 1912–1919.

40. Engel, E., & Dreizler, R. M., (2011). *Density Functional Theory an Advanced Course,* Springer, Berlin Heidelberg, Chap. 8, pp. 351-400.

41. Eschrig, H., (2003). *The Fundamentals of Density Functional Theory, 2nd Ed.* EAGLE, Leipzig, pp. 99-126.

42. Gross, E. K. U., & Burke, K., (2006). Basics, *Lect. Notes Phys.*, *706*, 1–17.

43. Yabana, K., & Bertsch, G. F., (1996). Time-dependent local-density approximation in real time, *Phys. Rev. B*, *54*, 4484–4487.

44. Bauernschmitt, R., & Ahlrichs, R. (1996). Treatment of electronic excitations within the adiabatic approximation of time dependent density functional theory, *Chem. Phys. Lett.*, *256*, 454–464.

45. Hirata S., & Head-Gordon, M., (1999). Time-dependent density functional theory within the Tamm-Dancoff approximation, *Chem. Phys. Lett.*, *314*, 291–299.

46. Caillie, C. V., & Amos, R. D., (1999). Geometric derivatives of excitation energies using SCF and DFT, *Chem. Phys. Lett. 308*, 249–255.

47. Caillie, C. V., & Amos, R. D., (2000). *Chem. Phys. Lett., 317*, 159–164.

48. Furche, F., & Ahlrichs, R., (2002). Adiabatic time-dependent density functional methods for excited state properties, *J. Chem. Phys., 117*, 7433–7447.

49. Jensen, F., (2017). *Introduction to Computational Chemistry,* Wiley, Chichester, Chap. 11.6, pp. 353-357.

50. Chiba, M., Tsuneda, T., & Hirao, K., (2006). Excited state geometry optimizations by analytical energy gradient of long-range corrected time-dependent density functional theory, *J. Chem. Phys., 124*, 144106(1–11).

51. Dreuw, A., & Head-Gordon, M., (2004). Failure of Time-Dependent Density Functional Theory for Long-Range Charge-Transfer Excited States: The Zincbacteriochlorin-Bacteriochlorin and Bacteriochlorophyll-Spheroidene Complexes, *J. Am. Chem. Soc., 126*, 4007–4016.

52. Gritsenko, O., & Baerends, E. J., (2004). Asymptotic correction of the exchange-correlation kernel of time-dependent density functional theory for long-range charge-transfer excitations, *J. Chem. Phys., 121*, 655–660.

53. Nakata, A., Tsuneda, T., & Hirao, K., (2009). Modified regional self-interaction corrected time-dependent density functional theory for core excited-state calculations, *J. Comput. Chem., 30*, 2583–2593.

54. Nakata, A., & Tsuneda, T., (2013). Density functional theory for comprehensive orbital energy calculations, *J. Chem. Phys., 139*, 064102(1–9).

55. Nakata, A., Tsuneda, T., & Hirao, K., (2011). Spin-orbit relativistic long-range corrected time-dependent density functional theory for investigating spin-forbidden transitions in photochemical reactions, *J. Chem. Phys., 135*, 224106(1–9).

56. Dreuw, A., Weisman, J. L., & Head-Gordon, M., (2003). Long-range charge-transfer excited states in time-dependent density functional theory require non-local exchange, *J. Chem. Phys., 119*, 2943–2946.

57. Savin, A., (1996). On Degeneracy, Near-degeneracy and Density Functional Theory, *Recent Developments and Applications of Modern Density Functional Theory*, edited by J. J. Seminario, Elsevier, Amsterdam, pp. 327–357.

58. Tsuneda, T., & Hirao, K., (2014). Long-range correction for density functional theory, *WIREs Comput. Mol. Sci., 4*, 375–390.

59. Tozer, D. J., & Handy, N. C., (1998). Improving virtual Kohn-Sham orbitals and eigenvalues: Application to excitation energies and static polarizabilities, *J. Chem. Phys., 109*, 10180–10189.

60. van Gisbergen, S. J. A., Kootstra, F., Schipper, P. R. T., Gritsenko, O. V., Snijders, J. G., & Baerends, E. J., (1998). Density-functional-theory response-property calculations with accurate exchange-correlation potentials, *Phys. Rev. A, 57*, 2556–2571.

61. Heyd, J., Scuseria, G. E., & Ernzerhof, M., (2003). Hybrid functionals based on a screened Coulomb potential, *J. Chem. Phys., 118*, 8207–8215.

62. Chiba, M., Tsuneda, T., & Hirao, K., (2007). Long-range corrected time-dependent density functional study on fluorescence of 4,4'-dimethylaminobenzonitrile, *J. Chem. Phys., 126*, 034504(1–10).

63. Humeniuk, A., & Mitrić, R., (2015). Long-range correction for tight-binding TD-DFT, *J. Chem. Phys., 143*, 134120(1–21).

64. Lutsker, V., Aradi, B., & Niehaus, T. A., (2015). Implementation and benchmark of a long-range corrected functional in the density functional based tight-binding method, *J. Chem. Phys., 143*, 184107(1–14).

65. Tsuneda, T., Song, J.-W., Suzuki, S., & Hirao, K., (2010). On Koopmans' theorem in density functional theory, *J. Chem. Phys., 133*, 174101(1–9).

66. Imamura, Y., & Nakai, H., (2006). Time-dependent density functional theory (TDDFT) calculations for core-excited states: Assessment of an exchange functional combining the Becke88 and van Leeuwen-Baerends-type functionals, *Chem. Phys. Lett.*, *419*, 297–303.

67. Stener, M., Fronzoni, G., & de Simone, M., (2003). Time dependent density functional theory of core electrons excitations, *Chem. Phys. Lett.*, *373*, 115–123.

68. Tu, G., Carravetta, V., Vahtras, O., & Agren, H., (2007). Core ionization potentials from self-interaction corrected Kohn-Sham orbital energies, *J. Chem. Phys.*, *127*, 174110(1–11).

69. Tsuneda, T., & Hirao, K., (2014). Self-interaction corrections in density functional theory, *J. Chem. Phys.*, *140*, 18A513(1–13).

70. Perdew, J. P., & Zunger, A., (1981). Self-interaction correction to density-functional approximations for many-electron systems, *Phys. Rev. B*, *23*, 5048–5079.

71. Vydrov, O. A., Heyd, J., Krukau, A., & Scuseria, G. E., (2006). Importance of short-range versus long-range Hartree-Fock exchange for the performance of hybrid density functionals, *J. Chem. Phys.*, *125*, 074106(1–9).

72. Tsuneda, T., Kamiya, M., & Hirao, K., (2003). Regional self-interaction correction of density functional theory, *J. Comput. Chem.*, *24*, 1592–1598.

73. Dreizler, R. M., & Gross, E. K. U., (1990). *Density-Functional Theory An Approach to the Quantum Many-Body Problem*, Springer, Berlin, pp. 75-137.

74. Nakata, A., Tsuneda, T., & Hirao, K., (2010). Modified Regional Self-Interaction Correction Method Based on the Pseudospectral Method, *J. Phys. Chem. A*, *114*, 8521–8528.

75. Maitra, N. T., Burke, K., & Woodward, C., (2002). Memory in time-dependent density functional theory, *Phys. Rev. Lett.*, *89*, 023002 (1–4).

76. Tsuneda, T., Nakano, H., & Hirao, K., (1995). Study of low-lying electronic states of ozone by multireference Møller-Plesset perturbation method, *J. Chem. Phys.*, *103*, 6520–6528.

77. Hashimoto, T., Nakano, H., & Hirao, K., (1996). Theoretical study of the valence $\pi \rightarrow \pi^*$ excited states of polyacenes: benzene and naphthalene, *J. Chem. Phys.*, *104*, 6244–6258.

78. Maitra, N. T., Zhang, F., Cave, R. J., & Burke, K., (2004). Double excitations within time-dependent density functional theory linear response, *J. Chem. Phys.*, *120*, 5932–5937.

79. Krylov, A. I., (2001). Size-consistent wave functions for bond-breaking: the equation-of-motion spin-flip model, *Chem. Phys. Lett.*, *338*, 375–384.

80. Shao, Y., Head-Gordon, M., & Krylov, A. I., (2003). The spin-flip approach within time-dependent density functional theory: Theory and applications to diradicals, *J. Chem. Phys.*, *118*, 4807–4818.

81. Sears, J. S., Sherill, C. D., & Krylov, A. I., (2003). A spin-complete version of the spin-flip approach to bond breaking: What is the impact of obtaining spin eigenfunctions?, *J. Chem. Phys.*, *118*, 9084–9094.

82. Li, Z., Liu, W., Zhang, Y., & Suo, B., (2011). Spin-adapted open-shell time-dependent density functional theory. II. Theory and pilot application, *J. Chem. Phys.*, *134*, 134101(1–22).

83. Becke, A. D., (1988). Density-functional exchange-energy approximation with correct asymptotic behavior, *Phys. Rev. A*, *38*, 3098–3100.

84. Kuühn, M., & Weigend, F., (2015). Two-component hybrid time-dependent density functional theory within the Tamm-Dancoff approximation, *J. Chem. Phys.*, *142*, 034116(1–8).

85. Bast, R., Jensen, H. J. A., & Saue, T., (2009). Relativistic adiabatic time-dependent density functional theory using hybrid functionals and noncollinear spin magnetization, *Int. J. Quantum Chem.*, *109*, 2091–2112.

86. Parpia, F. A., & Johnson, W. R., (1984). The relativistic time-dependent local-density approximation, *J. Phys. B*, *17*, 531–540.

87. Toffoli, D., Stener, M., & Decleva, P., (2002). Photoionization of mercury: A relativistic time-dependent density-functional-theory approach, *Phys. Rev. A*, *66*, 012501(1–16).

88. Wang, F., & Ziegler, T., (2004). Time-dependent density functional theory based on a noncollinear formulation of the exchange-correlation potential, *J. Chem. Phys.*, *121*, 12191–12196.

89. Wang, F., Ziegler, T., van Lenthe, E., van Gisbergen, S., & Baerends, E. J., (2005). The calculation of excitation energies based on the relativistic two-component zeroth-order regular approximation and time-dependent density-functional with full use of symmetry, *J. Chem. Phys.*, *122*, 204103(1–12).

90. Peng, D., Zou, W., & Liu, W., (2005). Time-dependent quasirelativistic density-functional theory based on the zeroth-order regular approximation, *J. Chem. Phys.*, *123*, 144101(1–13).

91. van Lenthe, E., Baerends, E. J., & Snijders, J. G., (1993). Relativistic regular two-component Hamiltonians, *J. Chem. Phys.*, *99*, 4597–4610.

92. Jansson, E., Minaev, B., Schrader, S., & Agren, H., (2007). Time-dependent density functional calculations of phosphorescence parameters for fac-tris (2-phenylpyridine) iridium, *Chem. Phys.*, *333*, 157–167.

93. Born, M., (1927). Oppenheimer, R. Zur quantentheorie der molekeln, *Ann. Phys.*, *389*, 457–484.

94. Tully, J. C., (1990). Molecular dynamics with electronic transitions, *J. Chem. Phys.*, *93*, 1061–1071.

95. Hammes-Schiffer, S., & Tully, J. C., (1994). Proton transfer in solution: Molecular dynamics with quantum transitions, *J. Chem. Phys.*, *101*, 4657–4667.

96. Tapavicza, E., Tavernelli, I., & Rothlisberger, U., (2007). Trajectory surface hopping within linear response time-dependent density-functional theory, *Phys. Rev. Lett.*, *98*, 023001(1–4).

97. Chernyak, V., & Mukamel, S., (2000). Density-matrix representation of nonadiabatic couplings in time-dependent density functional (TDDFT) theories, *J. Chem. Phys.*, *112*, 3572–3579.

98. Baer, R., (2002). Non-adiabatic couplings by time-dependent density functional theory, *Chem. Phys. Lett.*, *364*, 75–79.

99. Hu, C., Hirai, H., & Sugino, O., (2007). Nonadiabatic couplings from time-dependent density functional theory: Formulation in the Casida formalism and practical scheme within modified linear response, *J. Chem. Phys.*, *127*, 064103(1–9).

100. Hu, C., Sugino, O., Hirai, H., & Tateyama, Y., (2010). Nonadiabatic couplings from the Kohn-Sham derivative matrix: Formulation by time-dependent density-functional theory and evaluation in the pseudopotential framework, *Phys. Rev. A*, *82*, 062508(1–9).

CHAPTER 8

APPLICATIONS OF LEVELING METHODS TO PROPERTIES OF SMALL MOLECULES AND PROTEIN SYSTEMS

LAURENCE LEHERTE

Laboratory of Computational Physical Chemistry, Unit of Theoretical and Structural Physico-Chemistry, Department of Chemistry, Namur MEdicine and Drug Innovation Center (NAMEDIC), University of Namur, Rue de Bruxelles 61, B-5000 Namur, Belgium, Tel.: +32-81-72-45-60, E-mail: laurence.leherte@unamur.be

CONTENTS

ABSTRACT

Despite the advent of high performance computing resources, the calculations applied to the large systems may remain intractable. Methods to reduce the level of details are therefore essential to allow fast calculations, but also to provide new insights into the systems under study. In this chapter, various techniques and application domains related to the leveling of molecular properties through low-resolution, smoothing, denoising, or coarse-graining approaches, are presented. A focus is done on Gaussian smoothing, wavelet multi-resolution analysis, crystallography-based methods, as well as discretization methods. An emphasis is given on the use of smoothed charge density distribution functions and their extrema to generate reduced point charge models (RPCM) of proteins. Molecular dynamics simulations based on RPCMs are reported for three ubiquitin complexes. Results are discussed based on the ability of such models to generate stable protein-ligand conformations.

8.1 INTRODUCTION

Computer resources have now become sufficiently powerful to enable simulations of large systems at a classical level. Therefore, biological macromolecules and supramolecular complexes, for example, can be modeled with atomic details. However, when the systems include huge numbers of degrees of freedom and/or environment considerations, or when they contain unnecessary details like noise, calculations may remain too long. Low-resolution and smoothing techniques can thus bring an aid to the modeling of large systems. On the experimental point of view, low-resolution representations are also extremely useful in the refinement or interpretation of images generated by experimental approaches such as electron microscopy (EM) or X-ray diffraction. For example, a challenge in structural biology is to establish the structure of complex systems, which require high-resolution structural determination methods to generate atomic models of the individual components. Complexes created from the well-resolved individual components are imaged at a lower resolution to validate the interpretation of the experimental low-resolution image.

In this chapter, some methods that are used to level out molecular properties and fields, or to generate reduced discrete molecular representations of biomolecules, are presented. Applications are then given in various fields, with an emphasis on the use of smoothed charge density (CD) distribution functions and their extrema to generate reduced point charge models (RPCM) of proteins. Specific calculations based on RPCMs are reported for ubiquitin-ligand complexes modeled through Molecular Dynamics (MD) simulations. Results are discussed based on the ability of such models to generate stable protein-ligand conformations.

8.2 METHODS

8.2.1 SPLINE APPROXIMATION

The approximation of mathematical functions within a given interval defined by control points is among the numerous applications of, e.g., B-splines (basis splines) in science. Such piecewise polynomial functions (Eq. 1), whose shape is determined by the control points, are characterized by continuity conditions at their junctions, and are for example often used to smooth experimental data.

$$p(x) = a_0 + a_1 x + a_2 x^2 + \ldots + a_{k-1} x^{k-1} \qquad (1)$$

In Eq. (1), k and $k-1$ are the order and the maximum degree of the B-spline function, respectively. The resolution of the smoothed function thus depends on the number of polynomials used to approximate the initial function and on their order k. The use of B-splines does not require any a priori knowledge regarding the trend followed by the data. In a relatively recent paper, Klasson details how to construct spline functions in spreadsheets to smooth experimental data [1]. Earlier, Oberlin and Scheraga [2] used B-spline functions to approximate, through a pre-calculated potential energy, the interaction energy between rigid and fixed parts of a molecular system. A well-known application of splines is the ribbon representation of molecules like proteins and DNA, which allows a clear picture of the secondary structure and fold of the macromolecules (Figure 8.1). Contributions to such representations were brought by Carson [4] who also used B-splines to model molecular surfaces [5]. In his work, Carson applied B-spline filters to represent protein backbones, folds, as well as surfaces, with a suggested implication in structure-based drug design. Additional references regarding the approximation of molecular surfaces can be found in the work of Bajaj et al. [6].

8.2.2 GAUSSIAN TRANSFORMATION

Gaussian transformation, also known as Gaussian smoothing or blurring, belongs to the so-called kernel-based techniques. A function $f(x)$ is smoothed through a convolution product with a Gaussian smoothing kernel $G(x\text{-}y,t)$:

$$F(x,t) = \int G(x-y,t) f(x) dx \qquad (2)$$

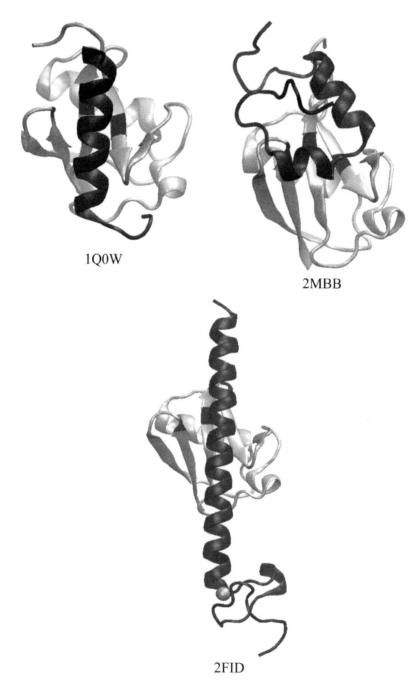

1Q0W

2MBB

2FID

FIGURE 8.1 Catmull-Rom spline representation of Ubiquitin complexes obtained using VMD [3]. The ligand is displayed in black. Residues Leu8, Ile44, and Val70 of Ubiquitin are shown in black. The zinc ion in structure 2FID is shown as a gray sphere.

where:

$$G(x-y,t) = \frac{1}{2\sqrt{\pi t}} e^{-\frac{(x-y)^2}{4t}} \tag{3}$$

The convolution product leads to mathematical equations that involve a smoothing parameter t, also called deformation or smoothing parameter, that is modulated to smooth (t is increased) or unsmooth (t is decreased) $f(x)$. Various transformations of elementary functions are reported by Moré et al. [7]. Such a smoothing technique is easily applicable to three-dimensional (3D) molecular properties represented themselves by Gaussian functions. Indeed, the convolution products can be calculated using analytical formula as illustrated later in the paper for the treatment of electron density (ED) and CD fields. Smoothing the function $f(x)$ through a convolution product with a Gaussian is equivalent to define $f(x,t)$ as its deformed version that is directly expressed as the solution of the diffusion equation according to the formalism presented by Kostrowicki et al. [8]. The method is thus known as diffusion equation method (DEM). In their paper, the authors used the procedure to the smoothing of interaction potentials in order to facilitate the global optimization of atom clusters. The technique was also applied to the prediction of a crystal structures, like S_6 [9]. An example of a one-dimensional (1D) smoothed potential energy function $f(x,t)$ is illustrated in Figure 8.2. As the smoothing factor t increases, the two initial potential energy wells progressively disappear to eventually lead to a single minima.

In the following sub-sections, 3D molecular fields like ED and CD are given as particular application cases.

8.2.2.1 Application to Promolecular Electron Density Distribution Functions

Promolecular models, i.e., molecular models built with non-interacting atoms, have often turned out to lead to very good approximated representations of ED distributions for the purpose of a number of applications as varied as chemical bond analysis or molecular similarity applications [10–17]. In the promolecular atomic shell approximation (PASA) approach developed by Carbó-Dorca and co-workers, a promolecular ED distribution r_M is calculated as a weighted summation over atomic ED distributions r_i, i.e., $\rho_M = \sum_i^{No.atoms} Z_i \rho_i$, where Z_i is the atomic number of atom i. r_i is described in

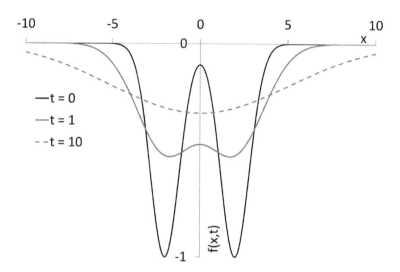

FIGURE 8.2 Gaussian smoothing of a hypothetical 1D potential energy function
$f(x,t) = -\left((1+4b)^{-1/2} e^{-b*(x-2)^2/(1+4b)} + (1+4b)^{-1/2} e^{-b*(x+2)^2/(1+4b)}\right)$ with $b = 0.75$ (arbitrary units).
The original unsmoothed signal is obtained when $t = 0$.

terms of series of squared 1s-type Gaussian functions fitted to atomic basis
set representations [18,19]:

$$\rho_i(\mathbf{r} - \mathbf{R}_i) = \sum_{j=1}^{5} w_{i,j} \left[\left(\frac{2\varsigma_{i,j}}{\pi}\right)^{3/4} e^{-\varsigma_{i,j}|\mathbf{r}-\mathbf{R}_i|^2} \right]^2 \tag{4}$$

where $\mathbf{R}_i$ is the position vector of atom i, and $w_{i,j}$ and $z_{i,j}$ are the fitted parameters, respectively. The number of 1s-type functions used to approximate
the ED of an atom may vary depending on the model. When applied to r_i as
given in Eq. (4), the Gaussian smoothing approach leads to:

$$\rho_{i,t}(\mathbf{r} - \mathbf{R}_i) = \sum_{j=1}^{5} a_{i,j}(1+4b_{i,j}t)^{-3/2} e^{\frac{-b_{i,j}|\mathbf{r}-\mathbf{R}_i|^2}{1+4b_{i,j}t}} \tag{5}$$

where:

$$b_{i,j} = 2\varsigma_{i,j} \qquad a_{i,j} = w_{i,j}\left(\frac{b_{i,j}}{\eth}\right)^{6/4} \tag{6}$$

In this context, the smoothing parameter t is seen as the product of a diffusion coefficient with time. Figure 8.3 shows the evolution of the promolecular ED distribution of Piroxicam, an anti-inflammatory drug molecule, as t increases from 0.0 to 2.5 bohr^2. Coordinates were retrieved from the crystallographic structural database (CSD) [20,21]. The smoothing involves a decrease in the number of maxima, initially located on the atoms, and eventually leads to a single maximum located on the SO_2 group of the 1,2-thiazine dioxide ring of the molecule.

8.2.2.2   Application to Electrostatic Potential Functions

The electrostatic potential function $F_M(\mathbf{r})$ generated by a molecule M can be approximated by a summation over its atomic contributions using the Coulomb equation:

$$\Phi_M(\mathbf{r}) = \sum_{i \in M}^{No.\,atoms} \frac{q_i}{|\mathbf{r} - \mathbf{R}_i|} \tag{7}$$

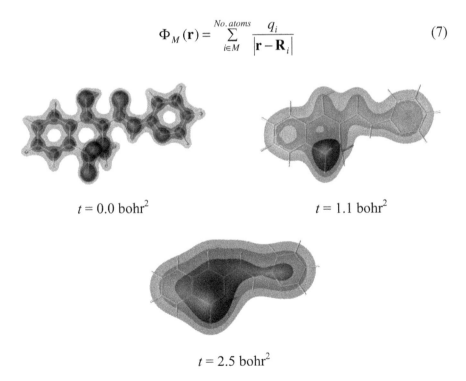

$t = 0.0$ bohr^2 $t = 1.1$ bohr^2

$t = 2.5$ bohr^2

FIGURE 8.3 Iso-contours of the PASA ED distributions of Piroxicam (CSD code: BIYSEH05) calculated at $t = 0.0$ bohr^2 (iso = 0.1, 0.2, 0.3 e$^-$/bohr^3), $t = 1.1$ bohr^2 (iso = 0.1, 0.15, 0.2 e$^-$/bohr^3), and $t = 2.5$ bohr^2 (iso = 0.05, 0.075, 0.10 e$^-$/bohr^3).

where q_i being the net charge of atom i. A smoothed version of the potential generated by atom i, $F_{i,t}(\mathbf{r} - \mathbf{R}_i)$ can be written as:[22]

$$\Phi_{i,t}(\mathbf{r} - \mathbf{R}_i) = \frac{q_i}{|\mathbf{r} - \mathbf{R}_i|} erf\left(\frac{|\mathbf{r} - \mathbf{R}_i|}{2\sqrt{t}}\right) \tag{8}$$

where erf stands for the error function. An analytical expression for the corresponding CD function $r_{i,t}(\mathbf{r} - \mathbf{R}_i)$ can be obtained from the Poisson equation:

$$-\nabla^2 \Phi_{i,t}(\mathbf{r} - \mathbf{R}_i) = \rho_{i,t}(\mathbf{r} - \mathbf{R}_i) \tag{9}$$

and is expressed as:

$$\rho_{i,t}(\mathbf{r} - \mathbf{R}_i) = \frac{q_i}{(4\pi)^{3/2}} e^{-|\mathbf{r}-\mathbf{R}_i|^2/4t} \tag{10}$$

In such a formalism, $r_{i,t}(\mathbf{r} - \mathbf{R}_i)$ cannot be calculated at $t = 0$. Indeed, that situation corresponds to the original Coulomb potential for which the solution of the Poisson equation is zero. Figure 8.4 shows the evolution of the CD distribution of Piroxicam as t increases from 0.05 to 2.0 bohr². Atomic charges q_i were calculated using the restrained electrostatic potential (RESP) method applied at the Hartree-Fock (HF) 6–31G* level and calculated with the program Gaussian09 [23]. From a situation where the maxima and minima of the electrostatic potential are located at the level of the atoms, one evolves toward a decrease in the number of extrema which can be located away from the molecular structure.

8.2.3 CRYSTALLOGRAPHY-BASED METHOD APPLIED TO THE ELECTRON DENSITY

Within the crystallographic approach, an ED distribution function $r(\mathbf{r})$ is written as the Fourier transform of the structure factors $F(\mathbf{h})$:

$$\rho(\mathbf{r}) = \frac{1}{V} \sum_{\{\mathbf{h}\}} F(\mathbf{h}) e^{-2\pi i \mathbf{h}.\mathbf{r}} \tag{11}$$

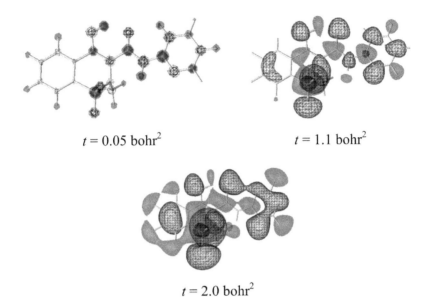

$t = 0.05$ bohr^2 $t = 1.1$ bohr^2

$t = 2.0$ bohr^2

FIGURE 8.4 Iso-contours of the CD distributions of Piroxicam (CSD code: BIYSEH05) calculated using RESP charge values q_a obtained using the program Gaussian09 [22] (HF 6–31G* level) at $t = 0.05$ bohr^2 (iso = –0.25, –0.10, 0.10, 0.25 e$^-$/bohr^3), $t = 1.1$ bohr^2 (iso = –0.0025, 0.0025, 0.01 e$^-$/bohr^3), and $t = 2.0$ bohr^2 (iso = –0.0005, 0.0005, 0.003 e$^-$/bohr^3). Negative and positive iso-contours are displayed using meshes and plain surfaces, respectively.

where V is the volume of the unit cell and **h** is a reciprocal space vector. The structure factors $F(\mathbf{h})$ are mathematically expressed as:

$$F(\mathbf{h}) = \sum_{i=1}^{No.\,atoms} f_i e^{-B_i \left(\frac{\sin\theta}{\lambda}\right)^2} e^{2\pi i \mathbf{h}.\mathbf{R}_i} \qquad (12)$$

where f_i is the atomic form factor of atom i, and B_i is the corresponding isotropic temperature factor. Such ED maps can be calculated at various resolution levels using crystallography programs such as XTAL [24]. In practice, the number of known structure factors occurring in Eq. (11) is not infinite and varies with the resolution.

In crystallography, the resolution d_{min} is a well-known concept which is defined using Bragg's law:

$$\left(\frac{\sin\theta}{\lambda}\right)_{max} = \frac{1}{2d_{min}} \qquad (13)$$

where 2θ is the angle between the diffracted and the primary beams of wavelength l, and d_{min} depends on different parameters including the quality of the crystal, the chemical composition, the radiation used, and the temperature of the experiment. Figure 8.5 depicts the crystallographic ED distribution of the Piroxicam molecule calculated from tabulated f_j factors for independent atoms using the program XTAL [24] at two resolution levels.

If one considers that the so-called overall isotropic temperature factor B is equivalent to $8p^2u^2$, where u^2 is the mean square atomic displacement, it is found that $u^2 = 2t$ [25]. The crystallographic resolution d_{min} thus differs from the smoothing parameter t which is here related to the dynamical quantity u^2.

8.2.4   WAVELET-BASED MULTI-RESOLUTION ANALYSIS

Wavelet-based techniques do not require the treated signal to be a Gaussian function as needed with the blurring method described above. In practice, wavelet theory is commonly applied to the treatment of signals and images, e.g., in analytical chemistry [26–29], or in spectroscopy [30, 31], but applications to bioinformatics and computational biology [32] as well as to chemistry [33, 34] and chemometrics [34–36] have been reported.

8.2.4.1   Wavelet Transforms

A wavelet transform (WT) is a localized transform in both space (or time) and frequency which uses integration kernels called wavelets. A basis set of wavelet functions $\{Y_{ab}(x)\}$ is built on translated and dilated versions of a so-called mother wavelet $Y(x)$:

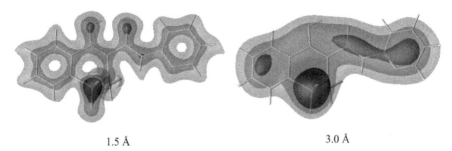

1.5 Å   3.0 Å

FIGURE 8.5   Iso-contours of the crystallography-based promolecular ED distributions of Piroxicam (CSD code: BIYSEH05) calculated using the program XTAL [24] at a resolution of (left) 1.5 Å (iso = 1.0, 3.0, 5.0 e /Å³) and (right) 3.0 Å (iso = 1.0, 1.5, 2.0 e /Å³).

$$\Psi_{ab}(x) = \frac{1}{\sqrt{a}}\, \Psi\!\left(\frac{x-b}{a}\right) \text{ with } a \in R_0, b \in R \qquad (14)$$

where a is the scaling parameter which allows to capture changes in frequency, and b is the shift along the x axis applied to analyze space (time)-dependent variations of a signal. The projection $\langle f, \Psi_{ab} \rangle$ of a square integrable signal $f(x)$ onto this basis according to:

$$\langle f, \Psi_{ab} \rangle = \int\limits_{-\infty}^{+\infty} f(x)\Psi_{ab}^{*}(x)\, dx \qquad (15)$$

is the result from a continuous wavelet transform (CWT). Y is often required to have a certain number p of vanishing moments:

$$\int\limits_{-\infty}^{+\infty} x^{n}\, \Psi(x)\, dx = 0 \text{ with } n = 0, 1, \ldots, p\text{-}1 \qquad (16)$$

where p is also known as the order of Y. In Figure 8.6, one illustrates the absolute values of the CWT coefficients obtained from the analysis of a-helix propensity descriptors [38] of the amino acids that constitute three ubiquitin ligands (Figure 8.1). The descriptors are reported under the name BLAM930101 in the amino acid database AA index that is publically available [39]. BLAM930101 is found to be at the center of a cluster of a-propensity descriptors of AA index [40]. The initial signal is progressively smoothed up to the point where a limited number of minima are obtained. For Vps27 UIM-1, a single minimum appears at scale $a = 16$, around Ile267 (residue 13), and is further stabilized at Glu268 (residue 14) at larger scales. For iota UBM1, four minima occur at scale $a = 13$, at the level of Leu63, Asp71, Asp81, and Lys94 (residues 2, 10, 20, and 33). For the Rabex-5 fragment, Cys19-Lys20, Gly27, Cys38-Trp39, Gln50, and Gln59-Glu67 (residues 5–6, 14, 25–26, 37, and 46–54), correspond to minima in the signal transformed at scale $a = 13$. It is a scale value that corresponds to a pronounced shift of the coefficient minima along the residue axis. These specific locations will be discussed in the Section 8.3.4.2.

For numerical purposes, the CWT can be discretized, by restricting the parameters a and b to the points of a dyadic lattice. Thus, if a and b/a are set equal to 2^{-j} and 1, respectively, Eqs. (14) and (15) are written as:

$$\Psi_{jk} = 2^{j/2}\Psi(2^{j} - k) \text{ with } j,k \in Z \qquad (17)$$

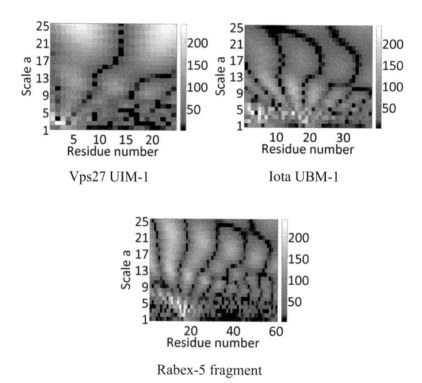

Vps27 UIM-1

Iota UBM-1

Rabex-5 fragment

FIGURE 8.6 Absolute value of the wavelet coefficients obtained from the CWT using Y = D15 applied to α-helix propensity descriptors of three Ubiquitin ligands. (Top left) Vps27 UIM-1 (PDB code: 1Q0W), (top right) iota UBM1 (PDB code: 2MBB), and (bottom) a bovine Rabex-5 fragment (PDB code: 2FID). Calculations were achieved using MATLAB [37].

$$f(x) = \sum_{j,k} \langle f, \Psi_{jk} \rangle \Psi_{jk}(x) \qquad (18)$$

The discrete wavelet transform (DWT) of $f(x)$ is calculated by passing the signal through two filters, i.e., a low-pass filter F to obtain the convolution of $f(x)$ with F, and a high-pass filter Y to generate the details, or wavelet coefficients. The procedure can be iteratively repeated by applying the decomposition to the first convolution product, and so on.

8.2.4.2 Multi-Resolution Analysis

A wavelet multi-resolution analysis (WMRA) is a mathematical construction used to express an arbitrary function $f \square L^2(R)$ at various levels of detail.

The function $f(x)$ is developed as in Eq. (18) where $<f, Y_{jk}>$, also written d_{jk}, are called the wavelet coefficients. In practice, the wavelet expansion is truncated at a scale J:

$$f(x) = \sum_k c_{Jk} \Phi_{Jk}(x) + \sum_{j=J}^{J_0-1} \sum_k d_{jk} \Psi_{jk}(x) \text{ with } c_{Jk} = \langle f, \Phi_{Jk} \rangle \quad (19)$$

where it is chosen here to set the resolution of the original signal J_0 equal to zero. Coefficients c_{Jk} are projections of the function f onto a space built on the basis set $\{F_{Jk}(x)\}$. Thus, lower resolution signals are characterized by negative values for J. In Eq. (19), the first sum is a coarse representation of f, where f is replaced by a linear combination of a finite number of translations of the scaling functions $F_{j0}(x)$. The remaining terms are refinements (details) determined at each scale j by translations of the wavelet $\Psi_{j0}(x)$ that are added to obtain a successively more detailed approximation of $f(x)$. For example, B-spline functions can be considered as scaling functions as illustrated by Stollnitz et al. [41,42] and applied by Carson [5] to model protein backbones and DNA surfaces at various levels of resolution.

A fast and accurate algorithm due to Mallat, named the 'pyramid algorithm' or the fast wavelet transform (FWT) [43], is applicable to signals consisting of 2^n data points. Its aim is to derive a mapping between the sequence $\{c_j\}$ and the sequences $\{c_{j-1}\}$ and $\{d_{j-1}\}$ through the following identities [44,45]:

$$c_{j-1,l} = \sum_k h_k c_{j,2l+k} \quad (20)$$

$$d_{j-1,l} = \sum_k g_k d_{j,2l+k} \quad (21)$$

where the numbers h_k are called the filter coefficients, and the wavelet coefficients g_k are obtained directly from the filter coefficients h_k [46]:

$$g_k = (-1)^k h_{k_{max}-k}, \text{ where } k = 0, 1, \ldots, k_{max} \quad (22)$$

Equations (20) and (21) are further applied to the sequence $\{c_{j-1}\}$ in order to obtain the new sequences $\{c_{j-2}\}$ and $\{d_{j-2}\}$. This procedure is repeated until the full FWT is achieved. The full procedure is named the cascade algorithm. Equation (21) shows that the calculation of coefficients $\{c_{j-1}\}$ from coefficients $\{c_j\}$ implies a downsampling, i.e., the number of coefficients is reduced by 2. It is also known as a decimated wavelet analysis. On the contrary, reconstruction implies an upsampling procedure, i.e., the number of data point is multiplied by 2 at each level of resolution.

The inverse mapping can be derived according to:

$$c_{jl} = \sum_k h_{l-2k} c_{j-1,k} + \sum_k g_{l-2k} d_{j-1,k} \tag{23}$$

The inverse FWT is obtained by repeated application of this equation for $j = J+1, J+2, \ldots$, up to J_0.

For example, Main and Wilson used an inverse WT approach to increase the resolution of ED maps [47]. The method proposed by the authors is based on a preliminary design of histograms of wavelet coefficients obtained from one-level WT decomposition of identified ED maps. Then, their procedure consists, as briefly summarized, of the following steps: (i) a one-level WT decomposition of a low-resolution ED map, (ii) the creation of an ordered list of the wavelet coefficients, (iii) a match of the wavelet coefficients with the preliminary obtained histograms, (iv) an inverse WT to generate the higher resolution ED map.

8.2.4.3   Multidimensional Cases and Smoothing

A simple way to obtain wavelet coefficients in dimensions higher than one is to carry out a 1D wavelet decomposition for each variable separately. The standard decomposition, described by Stollnitz et al., consists in the application of a 1D FWT to each row of data values [41,42]. The operation gives, for each of them, an average signal along with detail coefficients. Next, these transformed rows are treated as if they formed an image, and a 1D FWT is applied to each column. In the non-standard decomposition scheme, operations in rows and columns alternate. One applies a one-level decomposition to each row, followed by a one-level decomposition to each column. Then, one repeats the process on the resulting filtered image, and so on.

To obtain an image at various levels of decomposition, a smoothing procedure is required, which is applied before reconstructing the original signal as follows: all details generated after a given number of decomposition levels using a FWT are set equal to zero before a full reconstruction procedure is applied to restore the original number of data points. An example applied to the 3D promolecular ED of Piroxicam using the Daubechies' wavelet of order 10, D10, is displayed in Figure 8.7. Similarly to the crystallography-based approach, smoothing the PASA ED grid involves a decrease in the number of maxima which are initially located on the atoms, then on the rings and heteroatoms. The procedure eventually leads to a single maximum.

In the so-called 'à trous' algorithm, the smoothing procedure is implemented as a convolution product with a symmetric filter. The corresponding WT is achieved by inserting zeroes between the h_k coefficients [48]. The algorithm is illustrated by Gonzalèz-Audicana et al. in a paper comparing the Mallat and the 'à trous' algorithms [49], and relations between the two approaches are discussed by Shensa [50]. While the Mallat algorithm leads, at each resolution level, to a decrease in the number of points in an image, that number remains constant with the 'à trous' algorithm. As already mentioned for the Mallat algorithm, a reconstruction is necessary to preserve the number of data points in an image. The application of the 'à trous' algorithm to the promolecular ED grid of Piroxicam using the Daubechies' wavelet of order 10, D10, is illustrated in Figure 8.8. As in Figure 8.7, at the wavelet transformation level $J = -5$, there is only one maximum left, which is slightly displaced away from the SO_2 group. That single maximum was located on the sulfur atom with the Gaussian blurring and crystallography-based approaches, which thus appear to be more sensitive to the atomic number of the heaviest atom.

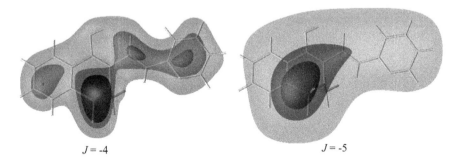

$J = -4$ $J = -5$

FIGURE 8.7 Iso-contours of the smoothed PASA ED distributions of Piroxicam (CSD code: BIYSEH05) calculated using the FWT approach with $\Phi = D10$. (Left) $J = -4$ (iso = 0.1, 0.2, 0.25 e⁻/Å³) and (right) $J = -5$ (iso = 0.03, 0.1, 0.125 e⁻/Å³). The grid size is $2^8 \times 2^8 \times 2^8$ and the grid interval is 0.125 Å.

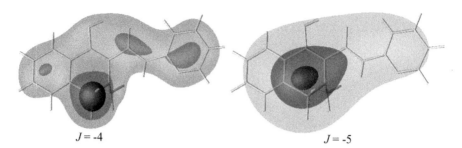

$J = -4$ $J = -5$

FIGURE 8.8 Iso-contours of the smoothed PASA ED distributions of Piroxinam (CSD code: BIYSEH05) calculated using the 'à trous' approach with the Lagrangian-based D10 filter. (Left) $J = -4$ (iso = 0.1, 0.2, 0.3 e/Å³) and (right) $J = -5$ (iso = 0.05, 0.1, 0.125 e/Å³). The grid size is 2^8 x 2^8 x 2^8 and the grid interval is 0.125 Å.

8.2.5 MOLECULAR DYNAMICS-RELATED APPROACHES

In addition to direct smoothing procedures described, e.g., by Eqs. (2) and (19), specific approaches are also currently applied to artificially smooth functions such as complex potential energy hyper-surfaces (PES), to allow a molecular system to visit less probable energy wells in a faster way during a MD simulation. Rather than modifying the force field (FF) itself, like it is done in potential smoothing approaches, the FF is biased by an extra term. This can be done either by reducing the energy barriers, through the hyper-dynamics approach, or by progressively filling the already visited energy wells, through the metadynamics and flooding approaches.

8.2.5.1 Hyperdynamics

The aim of hyperdynamics is to build an auxiliary system, which actually is the original system with a faster dynamics [51–53]. A bias potential $\Delta V(\mathbf{r})$ is added to the original potential energy function $V(\mathbf{r})$, which allows a reduction of the height of the energy barriers:

$$\Delta V(\mathbf{r}) = \begin{cases} 0 & \text{if } V(\mathbf{r}) \geq E \\ \dfrac{(E - V(\mathbf{r}))^2}{\alpha + E - V(\mathbf{r})} & \text{if } V(\mathbf{r}) < E \end{cases} \tag{24}$$

The approach thus requires a bias potential characterized by E and a that control the depth and flatness of the biased potential wells, respectively. At each time step of the simulation, the system undergoes forces that correspond to an increased potential energy value, allowing it to more easily cross energy barriers. This is illustrated in Figure 8.9 for a hypothetical 1D potential energy curve. The figure shows the effect of decreasing the value α on the energy barrier occurring between two potential wells. With respect to the Gaussian blurring method (Figure 8.2), the two potential wells stay preserved but are leveled out in the hyperdynamics approach.

8.2.5.2 Metadynamics

As for hyperdynamics, an auxiliary system is built as being the original system driven by collective variables $s(\mathbf{r})$. Such variables can be distances, torsion angles, a coordination number, lattice vectors, a solvation energy, a root mean square deviation (RMSD), etc. depending upon the system under consideration. A time-dependent bias potential $\Delta V(s(\mathbf{r}), time)$ is added to the original potential function $V(\mathbf{r})$ to avoid the system visiting already explored regions in the collective variable space [54, 55]. Doing so, at each time step, the potential energy involves a sum of extra Gaussian-type contributions, each depending on a previously visited state:

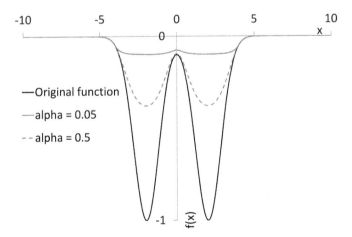

FIGURE 8.9 Effect of the application of a hyperdynamics bias to a hypothetical 1D potential energy function $V(x) = -(e^{-0.75*(x-2)^2} + e^{-0.75*(x+2)^2})$ with $E = -0.05$ (arbitrary units).

$$\Delta V_i(s(\mathbf{r}), time) = \overset{\substack{\text{all previous MD steps}}}{\underset{i=1}{\sum}} w\exp\left(-\frac{\|s(\mathbf{r}) - s(\mathbf{r}_i)\|^2}{2\delta s^2}\right) \qquad (25)$$

This can be visualized as a well filling procedure, as nicely illustrated in Figure 1 of the paper by Barducci et al. [55].

To limit the total number of PES minima, or to avoid having to visit several times the minima as done, for example, when using Monte Carlo (MC) or MD approaches, the principle of flooding energy minima can be applied until the system finds a way towards the global minima. When carrying out MD simulations, the kinetic energy can be adjusted [56]. In a MC procedure, the so-called basin hopping method was proposed which consists in the transformation of the PES such that all the potential energy values characterizing an energy well are replaced by the single minimal energy value [57]. The PES thus adopts a staircase shape where local energy barriers are neglected.

8.2.6 DISCRETIZATION TECHNIQUES

In this section, procedures used to replace a discrete or continuous molecular property by a limited number of discrete data points are briefly presented.

The so-called vector quantization (VQ) algorithms are well-known in data compression processes. They approximate an initial distribution probability of data points by a set of representative vectors. The methodology consists in the partitioning of the initial space into compact and well-separated clusters of points in such a way that all data points in one group are replaced by a single representative data point for that particular group. Each representative point is named a code vector. Clusters can, for example, be estimated from distance criteria, by a tessellation technique like Delaunay triangulation, … or by convolution with a Gaussian function [58]. De-Alarcón et al. applied their technique to the lowering of resolution of van der Waals surfaces of proteins and to cryo-EM maps [58]. Starting from an initial low-resolution map, they obtain a limited number of pseudo-atoms (code vectors), each characterized by a probability distribution function. Volumetric elements, named alpha shapes, are associated with the pseudo-atoms of the map and are used for the detection of deep clefts and channels in the protein system. Wriggers et al. [59, 60] proposed VQ procedures, implemented in the package Situs, to quantize two EM grids, at an atomic and a lower resolution, to rapidly enable the search for the best match between the two so-obtained VQ representations. In partial relation with VQ, Vorobjev proposed a method to locate low-resolution binding sites of a protein from its solvent accessible surface (SAS) represented by a set of discrete dots [61]. Binding sites result from the clustering of SAS-related dots followed by the location of centers of dense clusters.

In their approach, Glick et al. [62, 63] represented small molecules through a limited number of points obtained using a clustering of atoms based on their separating distances. The authors developed a method for ligand binding site identification on a protein. A hierarchy of models generated using a k-mean clustering algorithm for the ligand under consideration is established starting from the lowest resolution representation of the ligand, i.e., one single point located at the mean position of the ligand atoms. The resulting graphs were used in ligand-docking applications and illustrated the decrease in the possible number of conformers.

An inverse procedure, starting from all the atoms of a molecule, was implemented by Leherte et al. to locate critical points (CP), i.e., points where the gradient of the 3D field vanishes, in smoothed molecular fields [64]. It is based on the work of Leung et al. whose algorithm was originally established to cluster data by modeling the blurring effect of lateral retinal interconnections based on scale space theory [65]. The various steps of the resulting algorithm are as follow. (i) At scale $t = 0$, each atom of a molecular structure is considered as a local extremum of the molecular field to be analyzed. All atoms are then considered as the starting points of trajectories whose merging procedure is described hereafter. (ii) As t increases from 0.0 to a given maximum value, each extremum moves continuously along a path to reach a location in the 3D space where the gradient of the molecular field is zero. As t increases, trajectories progressively merge to CP locations. It is thus possible to assign, to each CP, a number of atoms which correspond to the starting points of the merged trajectories. (iii) The procedure can be carried out until the whole set of extrema becomes one single point. This is the ultimate stopping criteria of the merging procedure. Various applications can be found in previous publications [25, 66–69].

To analyze low-resolution ED grids obtained using crystallography-based approaches, we used the program ORCRIT, that was developed by Johnson [70]. The information generated by the topological analysis method implemented in ORCRIT allowed, for example, as sign ellipsoids centered at the CPs of ED grids of biomolecular systems to probe the interaction potential between a ligand and a DNA fragment [71], and between protein-DNA partners [72, 73], or to generate descriptors for small molecules [74, 75].

8.3 APPLICATION FIELDS

In this section, wherein some application domains are presented, a distinction is made between the smoothing of continuous or pseudo-continuous functions as in global optimization, denoising, and molecular similarity applications, and discrete representations, for example, in coarse-graining studies. In that latter sub-section, specific applications of RPCMs to MD simulations of proteins of ubiquitin complexes are also reported.

8.3.1 GLOBAL GEOMETRY OPTIMIZATION

Global optimization is one of the major fields of research that may require the use of smoothed or low-resolution molecular properties [76]. Its aim is to optimize a function considering some constraints, e.g., finding the global minimum of a potential energy hypersurface. Constraining degrees of freedom and/or reducing the level of detail are helpful ways to tackle multiple minima problems. Since they facilitate the overcome of energy barriers by reducing the number and depth of energy wells, global geometry optimization techniques are widely used, e.g., to generate atom clusters [56, 77]. A method to smooth interaction potential functions like well-known FFs is based on the results of the DEM [78], as for example implemented in the program package TINKER [79]. It consists in calculating a convolution product of the original energy function with a Gaussian as described earlier in the papers [8, 22, 77, 80–82]. As energy terms of conventional FFs are not always suitable for direct applications of the diffusion equation, they need to be replaced by, e.g., Gaussian approximations [8]. Mathematical formalisms and applications to Ar clusters, small molecules, and docking of α-helices were treated by Pappu et al. [78], and a study regarding a short peptide was proposed by Hart et al. [22]. Other mathematical approaches used to reduce the number of energy minima were reviewed by Schelstrate et al. [80] and different smoothing functions were proposed by Grossfield and Ponder [81] and Shao et al. [83].

In molecular distance geometry problems, which consist in the determination of a molecular structure from a set of interatomic distances, global optimization techniques can be based on a smoothing of the objective function. Various smoothing procedures are proposed in literature [7, 84–86]. The Gaussian transformation however appears to be mostly used [7, 84, 85].

The case of molecular docking applications, i.e., the search for an optimal arrangement of molecular partners, asks for a scoring function, which is often selected to be the potential energy of the system. For example, in the so-called 'stochastic approximation with smoothing', the mathematical aspects of the implementation differ from the DEM, but the main philosophy is similar in the sense that one initially looks for a single minimum in a smoothed energy hypersurface and then one iteratively recovers the initial resolution of the energy function while performing a minimization calculation at each step [87].

8.3.2 *MOLECULAR SIMILARITY*

Global optimization algorithms are often sought to tackle molecular similarity problems, which also involve many local solutions. A recent review of similarity-based methods was, for example, proposed by Cai et al. [88]. A way to reduce the number of possible alignments is to lower the level of detail of the molecular field under consideration [89–92]. To compare molecular structures or surfaces, it is useful to model the molecular properties to be compared using mathematical functions that allow fast calculations involving the maximization of a similarity score or the minimization of a distance-based score. The use of Gaussian functions for the evaluation of the molecular similarity has been an attractive strategy as it both allows short calculation times and it is easy to implement [93]. Indeed, using such functions, similarity measures are directly related to distances between the atoms that constitute the molecular structures to be compared [94, 95].

In our works about the superposition of small drug molecules [91, 92], we used well-known similarity indices and applied them to smoothed ED and CD distribution functions. We also evaluated molecular similarity by representing molecular systems as graphs of CPs obtained from smoothed ED distributions [74, 96].

Carbó-Dorca and co-workers were the first to report the now widely used quantum molecular similarity measure (QMSM) Z_{AB} definition between two chemical systems characterized by ED distribution functions ρ_A and ρ_B [97–99]:

$$Z_{AB} = \int\int \rho_A(\mathbf{r}_1)\, O(\mathbf{r}_1,\mathbf{r}_2)\, \rho_B(\mathbf{r}_2)\, d\mathbf{r}_1 d\mathbf{r}_2 \tag{26}$$

Depending upon the nature of the operator $O(r_1,r_2)$, overlap-, Coulomb-like, ... similarities are obtained [94, 100]. The similarity measures can be combined to lead to similarity indices, such as the well-known Cosine-like (also known as Carbó), Tanimoto, or Hodgkin-Richards indices [93, 101–103]:

$$S_{AB,Carbó} = \frac{Z_{AB}}{\sqrt{Z_{AA} Z_{BB}}} \tag{27}$$

$$S_{AB,Tanimoto} = \frac{Z_{AB}}{Z_{AA} + Z_{BB} - Z_{AB}} \tag{28}$$

$$S_{AB,Hodgkin-Richards} = \frac{2Z_{AB}}{Z_{AA} + Z_{BB}} \tag{29}$$

As already mentioned in the Section 8.2, Carbó-Dorca and coworkers developed the so-called atomic shell approximation (ASA) method, in which atomic or molecular ED are expressed as linear combinations of 1s Gaussian-type functions centered at atomic positions [104]. Such approximations were shown to be useful in QMSM calculations, especially to model promolecular ED distribution functions [105] where the EDs are expressed as sums over the contributions of independent atoms [18, 106, 107]. The use of smoothed PASA models obtained using Eq. (5) allows to significantly reduce the number of local solutions as illustrated in Figure 8.10 for the alignment of $C_2(CN)_4$ (CSD code: TCYETY) onto acridine (CSD code: ACRDIN01), generated using $S_{AB,Tanimoto}$ together with the overlap integral:

$$Z_{AB} = \int\int \rho_A(\mathbf{r}_1)\, \rho_B(\mathbf{r}_2)\, d\mathbf{r}_1 d\mathbf{r}_2 \tag{30}$$

For the unsmoothed case, i.e., at $t = 0$ bohr^2, multiple maxima are obtained. Their number is drastically reduced to four at $t = 0.1$ bohr^2, and a

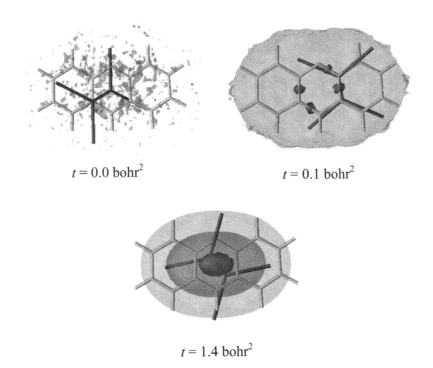

$t = 0.0$ bohr^2 $t = 0.1$ bohr^2

$t = 1.4$ bohr^2

FIGURE 8.10 Iso-similarity contour maps calculated from the alignment of the PASA ED of $C_2(CN)_4$ (CSD code: TCYETY) (black sticks) on Acridine (CSD code: ACRDIN01) (light gray sticks) using the overlap integral measure combined with the Tanimoto index, obtained at $t = 0.0$ bohr^2 (iso = 0.04, 0.07), 0.1 bohr^2 (iso = 0.1, 0.3), and 1.4 bohr^2 (iso = 0.3, 0.5, 0.7). To generate the maps, a grid is defined around the largest ligand. The center of mass of $C_2(CN)_4$ is placed at each grid point and its optimal orientation is determined as corresponding to the maximum value of $S_{AB,Tanimoto}$ at that point. The example is taken from the test cases studied in Constans et al. work [108].

single maximum $S_{AB,Tanimoto}$ value of 0.74 is obtained at $t = 1.4$ bohr^2. At $t = 0$, the atoms tends to be superimposed while at larger smoothing values, the global shape of the molecules are aligned.

Molecular surfaces can also be approximated using low-resolution functions, e.g., spherical harmonics [109]. Ritchie et al. suggested that surface representations which contain too high-resolution details may not be particularly convenient to search for regions of similarity or complementarity between two molecules [109]. The authors therefore proposed the use of low-resolution real spherical harmonics to represent and compare macromolecular surface shapes. Rotations of a molecular surface can thus be simulated by rotating only the harmonic expansion coefficients.

Hakkoymaz et al. adapted well-known molecular similarity indices, like the Carbó-Dorca and Hodgkin indices, to wavelet coefficients [110]. They applied the revised similarity indices to a multi-resolution analysis (MRA) decomposition of electrostatic potential grid points. Rather than working with modified similarity indices, Beck and Schindler used the original indices but applied them to a MRA decomposition of a 3D molecular field, like the ED [111]. Martin et al. applied a similar MRA analysis, but used graph descriptions of the MRA molecular images in molecular alignment applications [112].

The search for similarity degrees through wavelet coefficients is a well-known technique to also compare protein sequences, represented either by their 3D $C\alpha$ coordinates [113], or by their amino acid sequence [114]. Transmembrane proteins are good study cases since their membrane and non-membrane regions are constituted by sequences of contiguous amino acid residues. In Fisher's work, the smoothing of a hydropathy profile to predict the location of helices in transmembrane proteins is achieved by setting to zero wavelet coefficients associated with high frequencies of the hydropathy signals [114]. The sequences are submitted to a wavelet-based filtering algorithm, that provides smoothed profiles whose reduced number of extrema are identified to transmembrane helices, as also achieved using a CWT by Qiu et al. [115] and by Vannuci and Liò [116]. de Trad et al. proposed a method that is based on the MRA decomposition of a protein sequence signal built from the Fourier transform of amino acid properties like the electron-ion interaction potential [117]. A DWT is then applied to the transformed sequences in order to decompose the data into a number of levels. Two protein sequences can be compared at each level through the calculation of a cross-correlation coefficient, which is seen as a similarity score. Sabarish and Thomas applied a similar approach to protein sequence similarity and functional classification [118]. They used a DWT to decompose profiles of amino acid properties. It was followed by a correlation analysis that allows the classification of protein sets into functional classes. The conservation of physico-chemical properties in a functionally similar family of proteins was also studied by MRA [119]. Wen et al. applied 46 kinds of wavelets to a set of protein profiles, at various levels of decomposition, and defined a metrics to quantitatively evaluate the similarity degree between any two protein sequences with low identity [120]. In addition to the calculation of cross-correlation analysis, the comparison of protein profiles reconstructed after zeroing detail coefficients can be quantified using a distance-based criteria, such as in the work of Krishnan et al. [121].

The MRA technique was also applied to DNA strands, transformed to a sequence of integers (Adenine = 1, Thymine = 2, Cytosine = 3, Guanine = 4), by Tsonis et al. who detected localized periodic patterns and suggested DNA construction rules [122]. Machado et al. used complex numbers (Adenine = $1 + i0$, Thymine = $0 + i1$, Cytosine = $-1 + i0$, Guanine = $0 - i1$) to encode human DNA and showed that the Shannon continuous wavelet led to interpretable patterns [123], while Mena-Chalco et al. decomposed DNA strands into four binary sequences, one for each nucleic base, so as to consider a single descriptor for each base [124].

8.3.3 DENOISING OF SIGNALS AND IMAGES

Smoothing and denoising are related by the fact that the methods aim at separating useful and useless information in a data set or a signal [34]. In a smoothing approach, high-frequency components are removed while, in a denoising approach, low-amplitude components are removed [125].

In a DWT approach, smoothing of a signal is achieved through the following four steps: (i) transform the signal up to a selected level, (ii) recognize the wavelet coefficients associated with the highest frequencies, i.e., the detail coefficients, (iii) cancel those coefficients, and (iv) apply an inverse WT to the resulting signal. It has, for example, been achieved to smooth PASA ED or QM ED distribution functions of drug molecules in order to further generate a limited amount of CPs used in molecular alignments [74].

Denoising is achieved by canceling wavelet detail coefficients that are lower than a threshold value. This is called hard tresholding. One can distinguishes between hard and soft tresholding where values slightly below the threshold are not set to zero but attenuated so as to obtain smoother transitions between the original and the deleted values. These two approaches were tested by Pilard and Epelboin in their work about the restoration of noisy X-ray topographs [126]. Several threshold values and methods are presented by Ergen [127] to denoise heart sounds, and by Jeena et al. [128]. Non-reconstructed signals can also be useful, as shown by Chen et al. in their work about the parametrization of CG potential energy functions of DNA [129]. The authors decompose all-atom distance, angle, ... probability distributions calculated from all-atom MD simulations to generate the corresponding stretching, bending, and non-bonded coarse-grained versions.

Simulating proteins at low-resolution is a way to overcome structural inaccuracies issued from NMR data or from an approximate model. In their paper, Vakser et al. digitized a protein image onto a 3D grid [130]. Any grid point outside the molecular volume was set equal to 0, otherwise it was set equal to a numerical value corresponding to the protein surface or to the protein core, as also applied by Katchalski-Katzir et al. in a docking procedure [131]. The structural elements smaller than the grid interval are thus eliminated from the initial protein structure. The approach is implemented in the program GRAMM (Global RAnge Molecular Matching) reviewed together with other docking techniques by Russell et al. [132]. Vakser and colleagues studied large sets of protein complexes, at various resolution levels, and showed that low-resolution docking can provide gross structural features of protein-protein organization [133, 134].

Denoising procedures are also useful to locate water molecules in experimental ED maps. For example, Nittinger et al. [135] first generate a Gaussian expression for the ED associated with water molecules from a set of PDB structures and analyze it to classify the water molecules. They suggest that the procedure could be extended to detect misleadingly placed water molecules through a difference of Gaussian filters characterized each by a different width.

8.3.4   DISCRETE REPRESENTATIONS OF BIOMOLECULAR STRUCTURES AND PROPERTIES

8.3.4.1   Coarse-Grained Representations

For several years, much effort has been put into accelerating computational techniques such as MD and normal mode analysis for simulating large biological systems [136–138]. Enhancements to these well-known algorithmic procedures are based, notably, on a spatial coarse-graining of the molecular structures [139, 140]. Techniques that are relevant to coarse-graining of molecular structures are not necessarily linked to the smoothing of molecular properties, but they are nevertheless based on a decrease in the number of degrees of freedom and in the level of details. Rather than simulating the molecules at their atomic level, one reduces their description to a limited set of points, either centered on selected sites/atoms such the $C\alpha$ atoms of a protein backbone [136, 141, 142], on the center of mass of specific groups of atoms like amino acid residues [143], or on a set of merged atoms [144]. A shape-based coarse-graining approach [145, 146], now implemented in the program VMD [3], was proposed to generate highly coarse-grained descriptions of biomolecular complexes like viral capsids, proteins, and membranes. The authors used a reduced set of point masses, determined from a Voronoï-based partitioning of the molecules into domains of atoms, to reproduce the overall shape of the systems while respecting the mass distribution.[145,146] Reviews on the progresses on coarse grain (CG) models can be found in several other references [147–154].

The development of CG interaction potential functions is generally made either from atomistic interaction potential [155] or MD results [156–159], via experimental data such as B-factors [160], or through the fitting of a potential function achieved by matching CG and atomistic distributions [156]. For example, Lyman et al. presented a new method for fitting spring constants to mean square CG-CG distance fluctuations computed from atomistic MD [161]. Orellana et al. also developed an approach to design robust and transferable elastic network models that fit MD simulation data so as to best approximate local and global protein flexibility [162]. CG interaction potential can also be designed by fitting energy functions to smoothed all-atom energy profiles [129], or by applying the Inverse Monte Carlo approach [163], used for iteratively adjusting a CG potential function until it matches a target radial distribution function. Another example is the parametrization of the MARTINI FF, designed to reproduce partitioning free energies between polar and a polar phases of a large number of chemical systems [164, 165]. The model is based on a four-to-one mapping, i.e., four heavy atoms are represented by a single interaction center, except for small ring-like fragments. In the UNRES model, a peptidic chain is represented by a sequence of backbone beads located at peptide bonds, while side chains are modeled as single beads attached to the $C\alpha$ atoms, which are considered only to define the molecular geometry [166]. In the so-called SimFold CG description and energy function, a mixed representation is used. Residues of aqueous proteins are represented by backbone atoms N, $C\alpha$, C, O, and H, and one side chain centroid [167, 168]. The more recent SIRAH force field, applicable to DNA and proteins, is intended to capture temperature and solvent effects without the need of any structural constraints as required in MARTINI. [142]. Multiscale methods, that combine several levels of description, are also appealing since they allow to model limited regions of space with details while representing the outer regions by coarser models [149, 169, 170]. Multiscaling is not only used at the level of molecular structure representation, but is also applied to solve the atom equations of motion in MD simulations. Particularly, He et al. described the so-called smoothed molecular dynamics (SMD) method wherein the equations of motion are solved for nodes of a grid to generate nodal velocities and accelerations [171]. Atomic velocities and positions are then updated using the properties of the node they are associated with. In such a way, the SMD time step can be much larger than the conventional MD time step, depending upon the grid element size.

The advantage of using non-atomic representations is not limited to the increase of the speed of computation. Simplified representations of protein geometry have also been used by many groups to reduce sensitivity to small perturbations in conformation, e.g., when docking a ligand versus a receptor [172, 173]. Sternberg et al. replaced amino acid residues with spheres of varying size and performed docking to maximize the buried surface area [173].

In DNA, structural elements such as chains and bases can be modeled as rigid segments connected through energetic terms [174–176]. Particularly, von Kitzing and coworkers described an approach which reduces the number of degrees of freedom by assembling certain groups of atoms into configurational structures with less degrees of freedom [174, 175]. Corresponding potential energy functions were constructed with respect to these new variables using methods from the theory of wavelets, splines, and radial basis functions. DNA is, under such a description, represented by two kinds of rigid substructures: the bases and the phosphate groups, and one rotational group: the ribose subunit. In a recent paper, Naômé et al. developed a CG model and interaction potential to accurately reproduce the structural features of the underlying atomistic DNA system [159]. The authors used a one-site representation of the DNA nucleotides together with a limited number of intramolecular and intermolecular pair interaction potentials.

To model small molecules, the grouping of atoms to form pseudo-atoms is often achieved to accelerate the search of substructures in large databases [177, 178], in molecular similarity applications [74], and in docking calculations [62, 63]. Among the most recent review papers in the field of reduced graph representation of small molecules, one can cite, for example, the work of Birchall and Gillet [179].

8.3.4.2   Reduced Point Charge Models of Proteins

In previous works, we described an approach to model, through MD simulations, protein systems using an hybrid Amber99SB FF mixing all-atom bonded and van der Waals terms with reduced point charge sets [67–69]. Such a model allows to preserve the information regarding the atomic positions and does not require any back-mapping procedure to recover the all-atom structure of the system. It also eliminates some drawback related to the use of CG models, such as structure collapsing [180].

The RPCMs involve, for each amino acid, a limited amount of point charges located at the extrema of their smoothed positive and negative CD distribution function (Figure 8.11). Cao and Voth also used a CG generation procedure wherein they treat separately positive and negative charges to represent effective dipole moments [181]. Figure 8.11 illustrates positive and negative iso-contours of the CD of the structure Gly-Hisδ-Gly built from Amber99SB atomic charges according to Eq. (10). The CD is smoothed at a level of $t = 1.7$ bohr^2 and is characterized by two extrema located on the amino acid backbones and three extrema located on the Hisδ side chain.

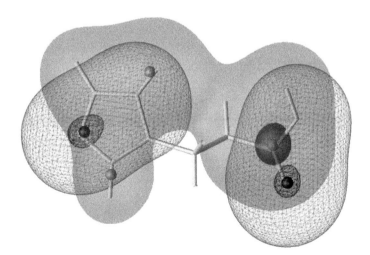

FIGURE 8.11   Iso-contours of the smoothed CD distributions ($t = 1.7$ bohr^2) built from the positive (plain surface; iso = 0.001, 0.0055) and the negative Amber99SB charges (mesh; iso = −0.001, −0.0055 e /bohr) for Gly-Hisδ-Gly. Point charges are located at the extrema of the negative (black spheres) and positive CD (gray spheres).

In our last study [69], it was shown that the RPCMs built with charges values fitted to electrostatic forces, and mostly located on selected atoms, led to MD trajectories that are, to some extent, similar to the all-atom ones. Here, we perform MD simulations of three ubiquitin complexes over longer time scales than previously reported to verify the stability of the complexes generated under RPCM conditions. Applications are given for the three ubiquitin complexes Vps27 UIM-1–ubiquitin [182], Iota UBM1–ubiquitin [183], and a bovine Rabex-5 fragment complexed with ubiquitin [184]. Representations of the PDB structures are shown in Figure 8.1. The three ligands interact with a hydrophobic patch of ubiquitin centered around its Leu8, Ile 44, and Val70 residues. The Vps27 UIM-1 is a helix with contiguous hydrophobic residues [182], Iota UBM-1 is characterized by a helix-turn-helix motif [183], and Rabex-5 binds to ubiquitin very similarly to Vps27 UIM-1, in a reverse orientation [184]. The Vps27 UIM-1 and Rabex-5 are helices whose amino acid sequence interacting with the hydrophobic patch of ubiquitin is constituted by alternating charged and nonpolar residues. The helix fold is such that the nonpolar residues face the receptor while the charged residues are oriented towards the solvent. Particularly, hydrophobic residues Leu262, Ile 263, Ala266, Ile267, and Leu271 of Vps27 UIM-1, and Ile51, Trp55, Leu57, Ala58, and Leu61 of Rabex-5are involved in hydrophobic ligand-protein 'contacts'. The ligand iota UBM1 of complex 2MBB contains less charged residues, with a nonpolar sequence, i.e., Leu78-Pro79-Val80, that is located at the level of its turn. The sequence is surrounded by other nonpolar residues facing the receptor, like Pro67, Val70, Val74, Phe75, Ile82 and Ile86.

Each of the Leu8, Ile44, and Val70 residues of ubiquitin belongs to a β-strand, i.e., $\beta 1$, $\beta 3$, and $\beta 5$, respectively. Minimum distance maps between each ligand and its receptor (Figure 8.12), obtained from the analysis of the 100 ns all-atom MD trajectories described below, clearly show three regions of minimal distances involving those three strands. In complexes 1Q0W and 2MBB, about the whole sequence of the ligand closely interact with strands $\beta 1$ (residues 1 to 8), $\beta 3$ (residues 40 to 45), and $\beta 5$ (residues 66 to 72) of ubiquitin.

MD simulations carried out with GROMACS 4.5.5 [185,186] and the Amber99SB FF [187] were first applied to model all-atom and RPCM structures, under conditions described in our last work, i.e., an equilibration stage of 40.1 ns followed by a 20 ns production stage [69]. All crystal water molecules

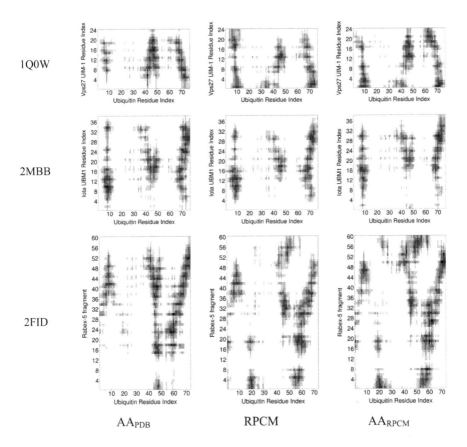

FIGURE 8.12 Ligand-receptor minimum distance maps calculated from the 100 ns production stages of the AA_{PDB} and AA_{RPCM} MD simulations and from the 20 ns production stage of the RPCM MD simulation. (Top) Vps27 UIM-1–Ubiquitin (PDB code: 1Q0W), (center) iota UBM1–Ubiquitin (PDB code: 2MBB), and (bottom) a bovine Rabex-5 fragment complexed with Ubiquitin (PDB code: 2FID). Scale goes from 0 to 1.5 nm (black to white) using a distance increment of 0.30 nm.

and the Zn ion of structure 2FID were removed. Structures were solvated in TIP4P-Ew water [188], and sodium ions were used to cancel the total electric charge. The final conformations of the 20 ns RPCM MD trajectories were considered as starting points for additional all-atom simulations, named here AA_{RPCM}, carried out with a 20 ns equilibration stage and a 100 ns production stage (50×10^6 steps with a time step of 0.002 ps) in the NPT ensemble at 1 bar and 300 K. Frames were saved every 20,000 steps. In addition, the initial all-atom MD trajectory was extended by a 100 ns long calculation. Thus, for each of the three protein systems, three MD trajectories were analyzed, i.e., a 100 ns long all-atom MD started from the PDB structure of the complex (AA_{PDB}), a 20 ns long RPCM MD, and a 100 ns long all-atom MD started from the final RPCM frame (AA_{RPCM}). In addition, all-atom 100 ns long MD simulations of the unbound ligands in water were also carried out at 1 bar and 300 K.

The minimum ligand-ubiquitin distance maps established from the 20 ns RPCM and the 100 ns AA_{RPCM}MD trajectories show that the three β-strands of ubiquitin mentioned above are still involved in the interaction with the ligands like they are during the AA_{PDB} simulations (Figure 8.12). Particularly, RPCM and AA_{RPCM} maps of the iota UBM1 ligand are significantly similar to the corresponding AA_{PDB} map. It is due to structural changes that are weaker in complex 2 MBB than in the two other protein complexes, as illustrated by the last conformation of the MD trajectories (Figure 8.13). In the case of Vps27 UIM-1, the appearance of a turn during the RPCM-based simulation induces a break in the first region of the corresponding minimum distance map, i.e., below residue 15 of ubiquitin. The turn involves residues Ala266-Ile267-Glu268-Leu269 of the ligand (residues 12 to 15), and also appears in 100 ns all-atom MD trajectories of the unbound ligand of structure 1Q0W simulated in water (Figure 8.14). This is thus a weak point in the ligand sequence whose regularity is easily perturbed when the RPCM model is used. The all-atom MD simulation of the isolated 2MBB ligand preserves the turn involving residues Leu78-Pro79-Val80 (residues 17 to 19 in Figure 8.14), and two end chain bents appear at the level of residues Gly89-Lys90 (residues 28 and 29) and Gly69 (residue 8). Regarding the Rabex-5 ligand, the structural changes that occur during the AA_{RPCM} MD simulation of the complex lead to a minimum distance map that is similar to the one obtained from the 20 ns RPCM simulation (Figure 8.12). The initial helix structure of the unbound ligand of 2FID simulated at the all-atom level is now destructed at residues Ser36, Ile51-Glu52, and Glu65 (residues 23, 38–39, and 52 in Figure 8.14).

Very interestingly, there is a rather good match between the residues involved in non-helix regions of the ligands and those previously identified through a CWT analysis (Figure 8.6). Indeed, low absolute values of the wavelet coefficients actually correspond either to deconstructed regions of the initial a-helix structure of the three ligands, or turns, or characterize residues that precede preserved α-helix segments (Figure 8.14). It may involve that an α-propensity set of descriptors such as BLAM930101 [38] is useful in

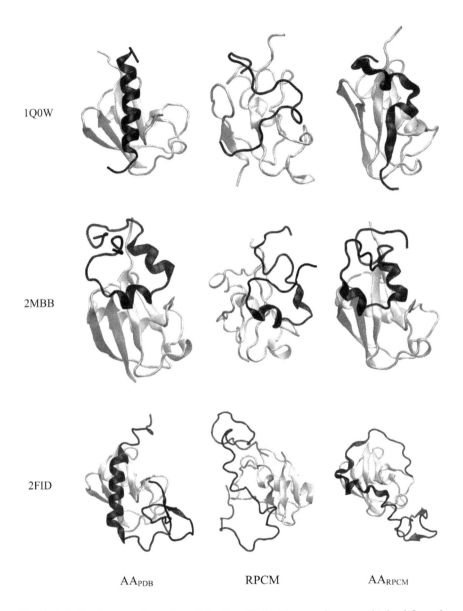

FIGURE 8.13 Last conformation of the three Ubiquitin complexes as obtained from the 100 ns production stages of the AA_{PDB} and AA_{RPCM} MD simulations and from the 20 ns RPCM simulation. (Top) Vps27 UIM-1–Ubiquitin (PDB code: 1Q0W), (center) iota UBM1–Ubiquitin (PDB code: 2MBB), and (bottom) a bovine Rabex-5 fragment complexed with Ubiquitin (PDB code: 2FID). The ligand is displayed in black.

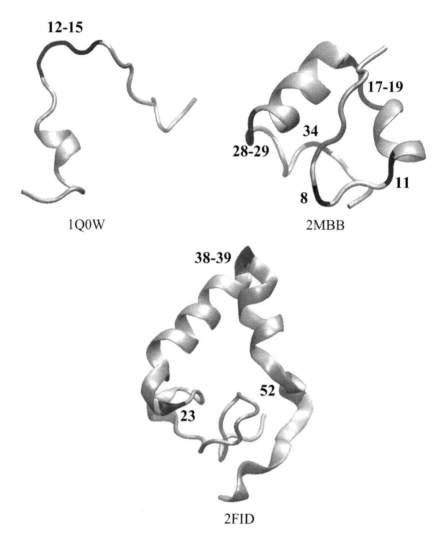

FIGURE 8.14 Last conformations of the three unbound Ubiquitin ligands simulated in water as obtained from 100 ns all-atom MD production stages. Residues occurring in destructed helix regions and turns are numbered and shown in black.

wavelet-based analysis to detect amino acid regions that are likely to lose their α-helical structure.

Figure 8.13 also illustrates that the AA_{PDB} conformations stay close to the PDB structure (Figure 8.1), except for 2FID whose long helix structure is partly deconstructed. In the complex 2FID, major structural changes appear during the AA_{PDB} simulation due to the selection of a 1:1 ligand-ubiquitin complex while, in the crystal structure, the ligand is actually in close interaction with two ubiquitin molecules, the second one also interacting with a Zn ion. As mentioned by Chakrabartty et al., isolated helices derived from proteins are unstable due to the lack of side chain interactions [189], which initially occurred in the present case between the receptor and the ligand. A loss of secondary structure elements is also seen in RPCM conformations (Figure 8.13). The trend of 1Q0W and 2FID ligands to be deconstructed during RPCM simulations can be explained by the presence, in these two peptides, of numerous charged residues, i.e., 11 in 1Q0W and 19 in 2FID. Particularly, for 1Q0W and 2FID ligands, a cluster of five contiguous charged residues are present, at the level of the N-terminal and C-terminal ends, respectively. Thus, a modification in the point charge description is likely to more strongly affect their dynamical behavior than for the 2MBB ligand.

The return to all-atom interactions in AA_{RPCM} simulations allows to recover some regular secondary features in the structures. As emphasized earlier [67], RPCMs can lead to deformed conformations of the systems due to a lack of short-range electrostatic descriptions, which affect, notably, the existence of intra- and inter-molecular H-bonds. However, such conformations may appear to remain stable under all-atom MD simulation conditions. The RMSD versus the initially optimized PDB conformations of the complexes as well as the number of clusters detected during the production stages are presented in Figure 8.15 and Table 8.1, respectively. The low variation of the RMSD functions and the small number of clusters emphasize the stability of the AA_{RPCM} trajectories, thus providing clues that RPCM simulations can lead to the sampling of diverse and stable conformations. For complex 2FID, the AA_{RPCM} RMSD function is even lower than the AA_{PBD} one (Figure 8.15). This higher stability comes with an increased number of ligand-ubiquitin H-bonds, i.e., 15 rather than 11 (Table 8.1).

The number of clusters is determined using the approach called 'Gromos' method [190]. It consists, for each conformation in a trajectory, in counting the number of similar frames (called neighbors) considering a cut-off. The structure with the largest number of neighbors, and all its neighbors, are assumed to form a cluster, which is eliminated from the pool of already existing clusters. The procedure is repeated for the remaining structures in the pool. Cut-off values of 0.3 and 0.45 nm were selected to probe the receptor and ligand conformations, respectively. These values were chosen so as to keep small numbers of clusters for the ligand and ubiquitin when

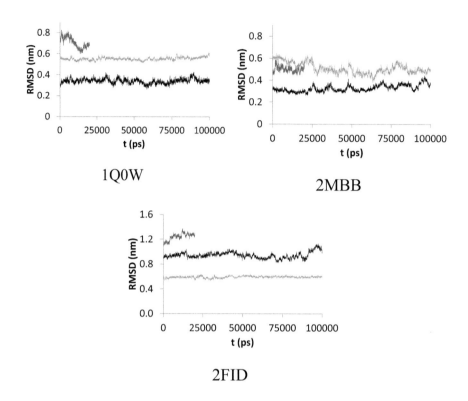

FIGURE 8.15 Time dependence of the protein RMSD calculated versus the initially optimized solvated structure from the 100 ns production stages of the AA_{PDB} and AA_{RPCM} MD simulations and from the 20 ns production stage of the RPCM MD simulation. Plain line = AA_{PDB}, gray line = RPCM, light gray line = AA_{RPCM}.

simulated at the AA_{PDB} level. Indeed, smaller cut-off values would involve additional clusters. Models RPCM lead to high numbers of clusters observed for the ligands, denoting a more flexible structure. The ligand iota UBM1 involved in complex 2MBB is an exception where the limited number of clusters reflects the high stability of the complex even at the RPCM level of representation. Despite a large number of AA_{RPCM} clusters, i.e., three, the good conformational conservation of the 2MBB ligand might be explained by the longer amino acid sequence of the ligand interacting with ubiquitin, i.e., 38 residues rather than 24 for 1Q0W. In all cases, the ubiquitin structure stays rather stable during all production stages.

A close examination of the H-bonds formed between the ligand and ubiquitin (Table 8.1) shows that the ligand forms less numerous H-bonds

TABLE 8.1   Properties (Mean and Standard Deviation Values) of the Solvent Layers of Thickness 0.35 nm Around the Protein Systems*

	AA_{PDB}	RPCM	AA_{RPCM}
MD production stage (ns)	100	20	100
1Q0W			
No. of molecules	1953 ± 46	2264 ± 64	2063 ± 65
No. of water-water H-bonds	436 ± 20	537 ± 21	464 ± 25
No. of protein-water H-bonds	282 ± 9	291 ± 10	312 ± 12
No. of ligand-Ubiquitin H-bonds	5 ± 1	1 ± 1	5 ± 2
No. of ligand/Ubiquitin clusters	1/1	13/1	1/1
2MBB			
No. of molecules	2084 ± 60	2444 ± 67	2155 ± 81
No. of water-water H-bonds	465 ± 25	582 ± 26	486 ± 30
No. of protein-water H-bonds	306 ± 10	294 ± 7	310 ± 13
No. of ligand-Ubiquitin H-bonds	5 ± 2	2 ± 1	3 ± 2
No. of ligand/Ubiquitin clusters	1/1	2/1	3/1
2FID			
No. of molecules	2522 ± 63	3047 ± 68	2579 ± 78
No. of water-water H-bonds	572 ± 26	737 ± 25	590 ± 28
No. of protein-water H-bonds	364 ± 10	375 ± 10	388 ± 15
No. of ligand-Ubiquitin H-bonds	11 ± 2	3 ± 2	15 ± 2
No. of ligand/Ubiquitin clusters	4/1	6/1	2/1

*The number of conformational clusters observed during the MD production stages is also given for the ligand and Ubiquitin.

at the RPCM level than at the all-atom level. For example, one H-bond rather than five is formed in structure 1Q0W, while only two and three are present in the 2MBB and 2FID complexes, respectively. One also observes that for complexes 1Q0W and 2FID, a high number of ligand-ubiquitin H-bonds are recovered at the AA_{RPCM} level versus their corresponding value obtained during the AA_{PDB} simulations. Values of 5 ± 2 and 15 ± 2 rather than 5 ± 1 and 11 ± 2 are indeed obtained (Table 8.1). Contrarily, for structure 2MBB, a lower number of H-bonds at the AA_{RPCM} level, i.e., 3 ± 2 rather than 5 ± 2, is observed, but this does not affect the orientation of the ligand versus the receptor as conformations are rather similar (Figure 8.13).

The analysis of the protein-solvent interactions is also carried out by focusing on the hydrogen bonds. A water layer of thickness 0.35 nm was determined around the protein complex and the mean numbers of water-water and water-protein H bonds were calculated considering that layer of solvent molecules. Analyses of the geometrical properties of the H-bonds were already reported previously [67, 68]. They showed that the usually expected first layer of water molecules is not structured any longer and intramolecular H-bonds strongly lose their orientation preference as reflected by the H-Donor-Acceptor angle values. The donor–acceptor distance distribution is however rather well preserved. Regarding the protein-water H-bonds, both distance and angle distributions stay similar to the all-atom results. From Table 8.1, one observes a larger number of water molecules in the layer close to the protein structure when the RPCM is used. For example, in structure 1Q0W, one gets 2264 molecules rather than 1953 and 2063 in the AA_{PDB} and AA_{RPCM} cases, respectively. It comes with an increase in the number of water-water H-bonds, i.e., 266 versus 217 and 233. Contrarily, the number of protein-water H-bonds is rather similar between the three charge models. For example, in structure 1Q0W, the number of such H-bonds is 282, 291, and 312, for trajectories AA_{PDB}, RPCM, and AA_{RPCM}, respectively. The AA_{RPCM} and AA_{PDB} trajectories behave similarly in terms of water-water H-bonds. It was actually shown previously that the use of a RPCM make the system very sensitive to the choice of the water force field [68]. The solvent is largely over-structuring, in opposition to what is known at the all-atom level [191, 192].

A statistical analysis of the potential energy terms calculated from the MD trajectories is reported in Table 8.2. Regarding the RPCM simulations, a post-processing calculation of the energy terms was carried out at the all-atom level. A good agreement is seen between the ligand-ubiquitin interaction energy terms calculated for the RPCM conformations using the RPCM and all-atom FFs. For example, values of –310.00 and –309.96 kJ.mol^{-1} are obtained for the system 1Q0W. The ligand-ubiquitin energy contributions calculated at the all-atom level for the RPCM conformations are systematically higher than those obtained from the AA_{PDB} and AA_{RPCM} trajectories. Indeed, the latters involve additional stabilizing interactions like H-bonds. As an example, the system 1Q0W is characterized by intermolecular ubiquitin-ligand energy values of –309.96 kJ.mol^{-1} rather than –474.22 and –571.81 kJ.mol^{-1}. The all-atom version of the RPCM intramolecular potential energy term is also always larger than the corresponding AA_{PDB} and AA_{RPCM} values. For example, one observes values of 20109.40, 18391.47, and 18984.40 kJ.mol^{-1}, for the RPCM, AA_{PDB}, and AA_{RPCM} simulations, respectively. It illustrates that a change in the point charge model involves short-range constraints

TABLE 8.2 All-Atom Mean Potential Energy Components (kJ.mol^{-1}) and Standard Deviation Calculated From the Production Stages of the MD Simulations**

MD production stage (ns)	AA$_{PDB}$	RPCM	AA$_{RPCM}$
	100	20	100
1Q0W			
Intramolecular ubiquitin and ligand	18391.47 ± 195.57	20109.40 ± 191.14* *16378.33 ± 180.78*	18984.40 ± 236.71
Intermolecular ligand-solvent	−4238.10 ± 200.47	−4826.66 ± 230.72* *−4893.84 ± 227.79*	−4521.04 ± 210.44
Intermolecular ubiquitin-solvent	−8913.81 ± 258.05	−9556.34 ± 259.89* *−9834.53 ± 270.38*	−9517.71 ± 345.16
Intermolecular ligand-ubiquitin	−474.22 ± 72.52	−309.96 ± 88.11* *−310.00 ± 90.95*	−517.81 ± 88.58
2MBB			
Intramolecular ubiquitin and ligand	22890.77 ± 199.86	24099.64 ± 234.47* *18311.47 ± 206.65*	22672.63 ± 245.78
Intermolecular ligand-solvent	−5427.15 ± 210.47	−5782.59 ± 195.94* *−5945.99 ± 197.60*	−5377.71 ± 263.47
Intermolecular ubiquitin-solvent	−8658.39 ± 266.12	−8694.14 ± 262.15* *−8987.46 ± 262.91*	−8788.56 ± 322.74
Intermolecular ligand-ubiquitin	−603.10 ± 102.54	−307.30 ± 66.47* *−304.33 ± 61.62*	−358.22 ± 92.73
2FID			
Intramolecular ubiquitin and ligand	25861.16 ± 226.01	28035.89 ± 208.74* *22115.24 ± 181.82*	26576.76 ± 306.60
Intermolecular ligand-solvent	−8826.55 ± 290.96	−9966.49 ± 241.66* *−10203.90 ± 244.34*	−9631.28 ± 293.92
Intermolecular ubiquitin-solvent	−7843.12± 237.08	−8373.19 ± 228.68* *−8655.84 ± 234.09*	−7833.00 ± 387.20
Intermolecular ligand-ubiquitin	−868.46 ± 93.99	−604.86 ± 137.12* *−602.52 ± 137.77*	−1176.00 ± 116.91

*Values obtained with the all-atom AMBER99SB FF applied to the conformations generated during the RPCM MD simulations.

**Long-range electrostatic contributions, calculated in the reciprocal space, are not considered in the reported values. Values in italics are energy terms calculated with the RPCM charges.

to the proteins. Those constraints are obviously less affecting the intermolecular energy terms. Table 8.2 also shows that the protein-solvent contributions are systematically lower for the RPCM models versus its all-atom counterpart. The case of the complex 1Q0W is again given here as an example, with values of -4893.84 and -9834.53 versus -4826.66 and -9556.34 kJ.mol^{-1}, for the RPCM and all-atom energy terms, respectively. It can be an explanation to the increased influence of the solvent when a RPCM is used [68].

An interesting point to mention is related to the stability of the AA_{RPCM} conformations, which may be characterized by more stabilizing ligand-solvent and ubiquitin-solvent interaction energy values than in the AA_{PDB} case. This is verified for complex 1Q0W, ant partly for systems 2MBB and 2FID where the ubiquitin-solvent and ligand-solvent energy values are more stabilizing than in the AA_{PDB} cases with -8788.56 and -9631.28 kJ.mol^{-1}, respectively. However, the ligand-solvent and ubiquitin-solvent counterparts are only slightly less stabilizing with -5377.71 and -7833.00 kJ.mol^{-1}, respectively. Also, in the cases of 1Q0W and 2FID, the intermolecular ligand-ubiquitin energy values calculated from the AA_{RPCM} trajectories are lower than in the AA_{PDB} cases. This is not observed at all for 2MBB, a nevertheless well-preserved complex, with a value of -358.22 versus -603.10 kJ.mol^{-1}.

8.4 CONCLUSIONS AND PERSPECTIVES

The current development of computer resources allows to study more and more complex systems over extended time scales. Consequently, more and more data are generated for storage and analysis. The need for simple models thus remains crucial to decrease the complexity of a problem, to get rid of useless data, or to reduce calculation time. Leveling and coarse-graining procedures remain widely used and still present vivid perspectives in the modeling of complex molecular systems and in the interpretation of experimental low-resolution data.

Two points of views can be adopted when using reduced molecular descriptions. Either one focuses on the generation of results that are similar to those obtained at a higher level of detail, but at a lower cost and with simpler algorithms, or one wishes to get results that differ from those obtained at a higher level of detail, thus providing new and/or different insights to a problem. In literature, various approaches are reported which, when applied to a same problem, may lead to different results.

Well-known methods such as spline approximation, Gaussian smoothing, wavelet multi-resolution analysis, and crystallography reconstruction, allow to level molecular properties. Discretization approaches such as vector quantization, critical point analysis, and coarse-graining procedures, generate models that replace a molecular property or representation by a limited number of data points. Leveling techniques are very common, e.g., in molecular graphics representations. Applications are found in many other research fields like global structure optimization, denoising of modeled or experimental data, molecular structure elucidation, similarity analysis, and molecular simulations.

We present selected results obtained from the implementation of smoothed electron density and charge density distribution functions in similarity of small molecules and molecular dynamics applications of proteins. Smoothing in similarity applications affects the number of possible solutions but also provides new or different solutions to problem. For example, it tends to align molecules in terms of global shape rather than in terms of atoms. On the other hand, the use of reduced point charge models of protein complexes favors ligand conformations that are not particularly stabilized at the all-atom level. Such conformations can however appear to be stable when returning to the all-atom description. A change in the level of detail of a protein complex is associated with a modification in the protein-solvent interactions. Thus, a focus on the influence of the solvent, especially in regards to the balance of protein-solvent interactions, and on the properties of protein systems as a function of the coarse-graining degree of the solvent, are planned as perspectives to the results reported in the present chapter.

ACKNOWLEDGMENTS

This research used resources of the 'Plateforme Technologique de Calcul Intensif (PTCI)' (http://www.ptci.unamur.be) located at the University of Namur, Belgium, which is supported by the F.R.S.-FNRS. The PTCI is member of the 'Consortium des Équipements de Calcul Intensif (CÉCI)' (http://www.ceci-hpc.be). The author gratefully acknowledges Daniel Vercauteren, Director of the 'Laboratory of Computational Physical Chemistry' at the University of Namur.

KEYWORDS

- charge density
- coarse graining
- denoising
- discretization
- electron density
- Gaussian convolution
- molecular dynamics
- molecular similarity
- multiresolution analysis
- point charge
- protein
- smoothing
- spline
- ubiquitin
- vector quantization
- wavelets

REFERENCES

1. Klasson, K. T., (2008). Construction of spline functions in spreadsheets to smooth experimental data. *Adv. Eng. Softw.*, *39*, 422–429.
2. Oberlin, D., Jr., & Scheraga, H. A., (1998). B-spline method for energy minimization in grid-based molecular mechanics calculations. *J. Comput. Chem.*, *19*, 71–85.
3. Humphrey, W., Dalk, A., & Schulten, K., (1996). VMD – Visual Molecular Dynamics. *J. Mol. Graphics*, *14*, 33–38.
4. Carson, M., (1987). Ribbon models of macromolecules. *J. Mol. Graphics*, *5*, 103–106.
5. Carson, M., (1996). Wavelets and molecular structure. *J. Comput. Aided Mol. Des.*, *10*, 273–283.
6. Bajaj, C. L., Pascucci, V., Shamir, A., Holt, R. J., & Netravali, A. N., (2003). Dynamic maintenance and visualization of molecular surfaces. *Discrete Appl. Math.*, *127*, 23–51.
7. Moré, J. J., & Wu, Z., (1997). Global continuation for distance geometry problems. *SIAM J. Optim.*, *7*, 814–836.
8. Kostrowicki, J., Piela, L., Cherayil, B. J., & Scheraga, H. A., (1991). Performance of the diffusion equation method in searches for optimum structures of clusters of Lennard-Jones atoms. *J. Phys. Chem.*, *95*, 4113–4119.

9. Wawak, R. J., Gibson, K. D., Liwo, A., & Scheraga, H. A., (1996). Theoretical prediction of a crystal structure. *Proc. Natl. Acad. Sci. USA, 93*, 1743–1746.

10. Gironés, X., Amat, L., & Carbó-Dorca, R., (1998). A comparative study of isodensity surfaces using ab initio and ASA density functions. *J. Mol. Graph. Model., 16*, 190–196.

11. Tsirelson, V. G., Avilov, A. S., Abramov, Y. A., Belokoneva, E. L., & Kitaneh, R., (1998). Feil, D. X-ray and electron diffraction study of MgO. *Acta Crystallogr. B, 54*, 8–17.

12. Tsirelson, V. G., Abramov, Y., Zavodnik, V., Stash, A., Belokoneva, E., & Stahn, J., (1998). Critical points in a crystal and procrystal. *Struct. Chem., 9*, 249–254.

13. Mitchell, A. S., & Spackman, M. A., (2000). Molecular surfaces from the promolecule: A comparison with Hartree-Fock ab initio electron density surfaces. *J. Comput. Chem., 21*, 933–942.

14. Gironés, X., Carbó-Dorca, R., & Mezey, P. G., (2001). Application of promolecular ASA densities to graphical representation of density functions of macromolecular systems. *J. Mol. Graph. Model., 19*, 343–348.

15. Downs, R. T., Gibbs, G. V., Boisen, M. B., Jr., & Rosso, K. M., (2002). A comparison of procrystal and ab initio model representations of the electron-density distributions of minerals. *Phys. Chem. Miner., 29*, 369–385.

16. Gironés, X., Amat, L., Carbó-Dorca, R., (2002). Modeling large macromolecular structures using promolecular densities. *J. Chem. Inf. Comput. Sci., 42*, 847–852.

17. Bultinck, P., Carbó-Dorca, R., (2003). Van Alsenoy Ch. Quality of approximate electron densities and internal consistency of molecular alignment algorithms in molecular quantum similarity. *J. Chem. Inf. Comput. Sci., 43*, 1208–1217.

18. Amat, L., & Carbó-Dorca, R., (1997). Quantum similarity measures under atomic shell approximation: First order density fitting using elementary Jacobi rotations. *J. Comput. Chem., 18*, 2023–2039.

19. Amat, L., & Carbó-Dorca, R., (2016). Quantum similarity measures under atomic shell approximation: First order density fitting using elementary Jacobi rotations, http://iqc.udg.es/cat/similarity/ASA/funcset.html (accessed May 18).

20. Allen, F. H., (2002). The Cambridge Structural Database: A quarter of a million crystal structures and rising. *Acta Crystallogr. B, 58*, 380–388.

21. Groom, C. R., Bruno, I. J., Lightfoot, M. P., & Ward, S. C., (2016). The Cambridge Structural Database, *Acta Crystallogr. B, 72*, 171–179.

22. Hart, R. K., Pappu, R. V., & Ponder, J. W., (2000). Exploring the similarities between potential smoothing and simulated annealing. *J. Comput. Chem., 21*, 531–552.

23. Frisch, M. J., Trucks, G. W., Schlegel, H. B., Scuseria, G. E., Robb, M. A., Cheeseman, J. R., et al., (2009). Gaussian 09, Revision E.01, Gaussian Inc., Wallingford, CT, USA.

24. Hall, S. R., du Boulay, D. J., & Olthof-Hazekamp, R., Eds., (2000). Xtal3.7 System, University of Western Australia. The source code is available at http://xtal.sourceforge.net/ (accessed May 18, 2016).

25. Leherte, L., (2004). Hierarchical analysis of promolecular full electron density distributions: Description of protein structure fragments. *Acta Crystallogr. D, 60*, 1254–1265.

26. Walczak, B., & Massart, D. L., (1997). Wavelets – Something for analytical chemistry? *Trends Anal. Chem., 16*, 451–463.

27. Leung, A. K.-M., Chau, F.-T., & Gao, J.-B., (1998). A review on applications of wavelet transform techniques in chemical analysis: 1989–1997. *Chemom. Intell. Lab. Syst., 43*, 165–184.

28. Ehrentreich, F., (2002). Wavelet transform applications in analytical chemistry. *Anal. Bioanal. Chem., 372*, 115–121.

29. Dinç, E., & Baleanu, D., (2007). A review on the wavelet transform applications in analytical chemistry. In: *Mathematical Methods in Engineering*; Tas, K., Tenreiro Machado, J. A., Baleanu, D., Eds., Springer: Dordrecht, The Netherlands, pp. 265-284.

30. Chau, F.-T., & Leung, A. K.-M., (2000). Applications of wavelet transform in spectroscopic studies. In: *Wavelets in Chemistry*; vol.22; Walczak, B., Ed., Elsevier. Amsterdam, The Netherlands, pp. 241-261.

31. Teitelbaum, H., (2000). Application of wavelet analysis to physical chemistry. In: *Wavelets in Chemistry*; vol. 22; Walczak, B., Ed., Elsevier: Amsterdam, The Netherlands, pp. 263-289.

32. Liò, P., (2003). Wavelets in bioinformatics and computational biology: State of art and perspectives. *Bioinformatics, 19*, 2–9.

33. Shao, X.-G., Leung, A. K.-M., & Chau, F.-T., (2003). Wavelet: A new trend in chemistry. *Acc. Chem. Res., 36*, 276–283.

34. Sundling, C. M., Sukumar, N., Zhang, H., Embrechts, M. J., & Breneman, C. M., (2006). Wavelets in chemistry and cheminformatics. *Rev. Comput. Chem., 22*, 295–329.

35. Alsberg, B. K., Woodward, A. M., & Kell, D. B., (1997). An introduction to wavelet transforms for chemometricians: A time-frequency approach. *Chemom. Intell. Lab. Syst., 37*, 215–239.

36. Jetter, K., Depczynski, U., Molt, K., & Niemöller, A., (2000). Principles and applications of wavelet transformation to chemometrics. *Anal. Chim. Acta, 420*, 169–180.

37. Matlab and Statistics Toolbox, Release 2012b; The MathWorks, Inc.: Natick, MA, 2012.

38. Blaber, M., Zhang, X. J., & Matthews, B. W., (1993). Structural basis of amino acid alpha helix propensity. *Science, 260*, 1637–1640.

39. Kawashima, S., Pokarowski, P., Pokarowska, M., Kolinski, A., Katayama, T., & Kanehisa, M., (2008). AAindex: Amino acid index database, progress report 2008. *Nucl. Acids Res., 36*, D202–D205.

40. Saha, I., Maulik, U., Bandyopadhyay, S., & Plewczynski, D., (2012). Fuzzy clustering of physicochemical and biochemical properties of amino acids. *Amino Acids 43*, 583–594.

41. Stollnitz, E. J., DeRose, T. D., & Salesin, D. H., (1995). Wavelets for computer graphics: A primer 1. *IEEE Comput. Graphics Appl., 15*(3), 76–84.

42. Stollnitz, E. J., DeRose, T. D., & Salesin, D. H., (1995). Wavelets for computer graphics: A primer 2, *IEEE Comput. Graphics Appl., 15*(4), 75–85.

43. Mallat, S. G. A., (1989). theory for multiresolution signal decomposition: The wavelet representation. *IEEE Trans. Pattern Anal. Machine Intell., 11*, 674–693.

44. Jawerth, B., & Sweldens, W., (1993). An overview of wavelet based multiresolution analyses; Technical Report, University of South Carolina, USA. http://citeseerx.ist.psu.edu/viewdoc/summary?doi=10.1.1.45.5864 (accessed May 18, 2016).

45. Nielsen, O. M., (1998). Wavelets in scientific computing, Ph.D. Dissertation, Technical University of Denmark, Lyngby, http://orbit.dtu.dk/fedora/objects/orbit:83353/datastreams/file_5265681/content (accessed May 18, 2016).

46. Daubechies, I., (1988). Orthonormal bases of compactly supported wavelets. *Commun. Pure Appl. Math., 41*, 909–996.

47. Main, P., & Wilson, J., (2000). Wavelet analysis of electron density maps. *Acta Crystallogr. D, 56*, 618–624.

48. Starck, J.-L., Murtagh, F., & Bijaoui, A., (1998). *Image Processing and Data Analysis: The Multiscale Approach*; Cambridge University Press: Cambridge, United Kingdom, pp. 287.

49. González-Audícana, M., Otazu, X., Fors, O., & Seco, A., (2005). Comparison between Mallat's and the 'à trous' discrete wavelet transform based algorithms for

the fusion of multispectral and panchromatic images. *Int. J. Remote Sensing, 26,* 595–614.

50. Shensa, M. J., (1992). The discrete wavelet transform: Wedding the à trous and the Mallat algorithms. *IEEE Trans. Signal Process., 40,* 2464–2482.

51. Voter, A. F., (1997). A method for accelerating the molecular dynamics simulation of infrequent events. *J. Chem. Phys., 106,* 4665–4677.

52. Hamelberg, D., Mongan, J., & McCammon, J. A., (2004). Accelerated molecular dynamics: A promising and efficient simulation method for biomolecules. *J. Chem. Phys., 120,* 11919–11929.

53. Perez, D., Uberuaga, B. P., Shim, Y., Amar, J. G., & Voth, A. F., (2009). Accelerated molecular dynamics methods: Introduction and recent developments. *Annu. Rep. Comput. Chem., 5,* 79–98.

54. Leone, V., Marinelli, F., Carloni, P., & Parrinello, M., (2010). Targeting biomolecular flexibility with metadynamics. *Curr. Opin. Struct. Biol., 20,* 148–154.

55. Barducci, A., Bonomi, M., & Parrinello, M., (2011). *WIREs Comput. Mol. Sci., 1,* 826–843.

56. Goedecker, S., (2004). Minima hopping: An efficient search method for the global minimum of the potential energy surface of complex molecular systems. *J. Chem. Phys., 120,* 9911–9917.

57. Wales, D. J., & Doye, J. P. K., (1997). Global optimization by basin-hopping and the lowest energy structures of Lennard-Jones clusters containing up to 110 atoms. *J. Phys. Chem. A, 101,* 5111–5116.

58. De-Alarcón, P. A., Pascual-Montano, A., Gupta, A., & Carazo, J. M., (2002). Modeling shape and topology of low-resolution density maps of biological macromolecules. *Biophys. J., 83,* 619–632.

59. Wriggers, W., Milligan, R. A., & McCammon, J. A., (1999). Situs: A package for docking crystal structures into low-resolution maps from electron microscopy. *J. Struct. Biol., 125,* 185–195.

60. Wriggers, W., & Birmanns, S., (2001). Using Situs for flexible and rigid-body fitting of multiresolution single-molecule data. *J. Struct. Biol., 133,* 193–202.

61. Vorojbev, Y. N., (2010). Blind docking method combining search of low-resolution binding sites with ligand pose refinement by molecular dynamics-based global optimization. *J. Comput. Chem., 31,* 1080–1092.

62. Glick, M., Robinson, D. D., Grant, G. H., & Richards, W. G., (2002). Identification of ligand binding sites on proteins using a multi-scale approach. *J. Amer. Chem. Soc., 124,* 2337–2344.

63. Glick, M., Grant, G. H., & Richards, W. G., (2002). Docking of flexible molecules using multiscale ligand representations. *J. Med. Chem., 45,* 4639–4646.

64. Leherte, L., Dury, L., & Vercauteren D. P., (2003). Structural identification of local maxima in low-resolution promolecular electron density distributions. *J. Phys. Chem. A, 107,* 9875–9886.

65. Leung, Y., Zhang, J.-S., & Xu, Z.-B., (2000). Clustering by scale-space filtering. *IEEE Trans. Pattern Anal. Mach. Intell., 22,* 1396–1410.

66. Leherte, L., & Vercauteren, D. P., (2009). Coarse point charge models for proteins from smoothed molecular electrostatic potentials. *J. Chem. Theory Comput., 5,* 3279–3298.

67. Leherte, L., & Vercauteren, D. P., (2014). Evaluation of reduced point charge models of proteins through Molecular Dynamics simulations: Application to the Vps27 UIM-1–ubiquitin complex. *J. Mol. Graphics Model., 47,* 44–61.

68. Leherte, L., & Vercauteren, D. P., (2014). Comparison of reduced point charge models of proteins: Molecular dynamics simulations of ubiquitin. *Sci. China Chem.*, *57*, 1340–1354.

69. Leherte, L., (2016). Reduced point charge models of proteins: Assessment based on molecular dynamics simulations. *Mol. Simul.*, *42*, 289–304.

70. Johnson, C. K., (1977). Orcrit. The Oak Ridge Critical Point Network Program; Technical report; Oak Ridge National Laboratory: Oak Ridge, TN.

71. Leherte, L., & Allen, F. H., (1994). Shape information from a critical point analysis of calculated electron density maps: Application to DNA-drug Systems. *J. Comput. Aided Mol. Des.*, *8*, 257–272.

72. Becue, A., Meurice, N., Leherte, L., & Vercauteren, D. P., (2003). Description of protein-DNA complexes in terms of electron density topological features. *Acta Crystallogr. D*, *59*, 2150–2162.

73. Becue, A., Meurice, N., Leherte, L., & Vercauteren, D. P., (2004). Evaluation of the protein solvent-accessible surface using reduced representations in terms of critical points of the electron density. *J. Comput. Chem.*, *25*, 1117–1126.

74. Leherte, L., (2001)Applications of multiresolution analyses to electron density maps of small molecules: Critical point representations for molecular superposition. *J. Math. Chem.*, *29*, 47–83.

75. Burton, J., Meurice, N., Leherte, L., & Vercauteren, D. P., (2008). Can descriptors of the electron density distribution help to distinguish functional groups. *J. Chem. Inf. Model.*, *48*, 1974–1983.

76. Troyer, J. M., & Cohen, F. E., (1990). Simplified models for understanding and predicting protein structure, *Rev. Comput. Chem.*, *2*, 57–80.

77. Piela, L., (1998). Search for the most stable structures on potential energy surfaces. *Collect. Czech. Chem. Commun.*, *63*, 1368–1380.

78. Pappu, R. V., Hart, R. K., & Ponder, J. W., (1998). Analysis and applications of potential energy smoothing and search methods for global optimization. *J. Phys. Chem. B*, *102*, 9725–9742.

79. TINKER – Software Tools for Molecular Design, Jay Ponder Lab, Department of Chemistry, Washington University, Saint Louis (MI) USA, (2015). http://dasher.wustl.edu/tinker/ (accessed 30 June 2016).

80. Schelstraete, S., Schepens, W., & Verschelde, H., (1999). Energy minimization by smoothing techniques: A survey. In: *Molecular Dynamics: From Classical to Quantum Methods*; vol. 7; Balbuena, P. B., Seminario. J. M., Eds., Elsevier: Amsterdam, The Netherlands, pp. 129-185.

81. Grossfield, A., & Ponder, J. W., (2016). Global optimization via a modified potential smoothing kernel; CCB Report 2002-01; Washington University School of Medicine: St Louis, MO, (2002). http://dasher.wustl.edu/ponder/papers/ccb-report-2002-01.pdf (accessed May 18).

82. Goldstein, M., Fredj, E., & Gerber, R. B., (2011). A new hybrid algorithm for finding the lowest minima of potential surfaces: Approach and application to peptides. *J. Comput. Chem.*, *32*, 1785–1800.

83. Shao, C.-S., Byrd, R., Eshow, E., & Schnabel, R. B., (2011). Global optimization for molecular clusters applied to molecular structure determination. *Oper. Res. Lett.*, *39*, 461–465.

84. Moré, J. J., & Wu, Z., (1999). Distance geometry optimization for protein structures. *J. Global Optim.*, *15*, 219–234.

85. Liberti, L., Lavor, C., & Macula, N., (2009). Double variable neighborhood search with smoothing for the molecular distance geometry problem. *J. Global Optim., 43,* 207–218.

86. Souza, M., Xavier, A. E., Lavor, C., & Maculan, N., (2011). Hyperbolic smoothing and penalty techniques applied to molecular structure determination. *Oper. Res. Lett., 39,* 461–465.

87. Diller, D. J., Verlinde, & Ch, M. L. J., (1999). A critical evaluation of several optimization algorithms for the purpose of molecular docking. *J. Comput. Chem., 20,* 1740–1751.

88. Cai, C., Gong, J., Liu, X., & Gao, D., Li, H., (2013). Molecular similarity: Methods and performances. *Chin. J. Chem., 31,* 1123–1132.

89. Duncan, B. S., & Olson, A. J., (1993). Shape analysis of molecular surfaces. *Biopolymers, 33,* 231–238.

90. Maggiora, G. M., Rohrer, D. C., & Mestres, J., (2001). Comparing protein structures: A Gaussian-based approach to the three-dimensional structural similarity of proteins. *J. Mol. Graph. Model., 19,* 168–178.

91. Leherte, L., (2006). Similarity measures based on Gaussian-type promolecular electron density models: Alignment of small rigid molecules. *J. Comput. Chem., 27,* 1800–1816.

92. Leherte, L., & Vercauteren, D. P., (2012). Smoothed Gaussian molecular fields: An evaluation of molecular alignment problems, *Theor. Chem. Acc., 131,* 1259/1–1259/16.

93. Maggiora, G. M., & Shanmugasundaram, V., (2004). Molecular similarity measures. In: *Chemoinformatics: concepts, methods, and tools for drug discovery,* vol. 275 of series: *Methods in Molecular Biology;* Bajorath, J., Ed., Humana Press: Totowa, NJ, USA, pp. 1-50.

94. Bultinck, P., Gironés, X., & Carbó-Dorca, R., (2005). Molecular quantum similarity: Theory and applications. *Rev. Comput. Chem., 21,* 127–207.

95. Carbó-Dorca, R., Besalú, E., & Mercado, L. D., (2011). Communications on quantum similarity, Part 3: A geometric-quantum similarity molecular superposition algorithm. *J. Comput. Chem., 32* 582–599.

96. Leherte, L., Meurice, N., & Vercauteren, D. P., (2005). Influence of conformation on the representation of small flexible molecules at low resolution: Alignment of endothiapepsin ligands. *J. Comput. Aided Mol. Des., 19,* 525–549.

97. Carbó, R., Leyda, L., & Arnau, M., (1980). How similar is a molecule to another? An electron density measure of similarity between two molecular structures. *Int. J. Quantum Chem., 17,* 1185–1189.

98. Carbó, R., & Calabuig, B., (1992). Molecular quantum similarity measures and N-dimensional representation of quantum objects. I. Theoretical foundations. *Int. J. Quantum Chem., 42,* 1681–1693.

99. Carbó, R., & Besalú, E., (1995). Theoretical foundations of quantum molecular similarity. In: *Molecular Similarity and Reactivity – From Quantum Chemical to Phenomenological Approaches;* vol. 14 of series: *Understanding Chemical Reactivity;* Carbó-Dorca, R., Ed., Kluwer: Dordrecht, The Netherlands, pp. 3–28.

100. Carbó-Dorca, R., & Besalú, E., (1998). A general survey of molecular quantum similarity. *J. Mol. Struct. (Theochem), 451,* 11–23.

101. Robert, D., & Carbó-Dorca, R., (1998). A formal comparison between molecular quantum similarity measures and indices. *J. Chem. Inf. Comput. Sci., 38,* 469–475.

102. Maggiora, G. M., Petke, J. D., & Mestres, J., (2002). A general analysis of field-based molecular similarity indices. *J. Math. Chem., 31,* 251–270.

103. Monev, V., (2004). Introduction to similarity searching in chemistry. *Commun. Math. Comput. Chem., 51,* 7–38.

104. Constans, P., & Carbó, R., (1995). Atomic Shell Approximation: Electron density fitting algorithm restricting coefficients to positive values. *J. Chem. Inf. Comput. Sci., 35,* 1046–1053.

105. Constans, P., Amat, L., & Fradera, X., & Carbó-Dorca, R., (1996). Quantum molecular similarity measures (QMSM) and the atomic shell approximation (ASA). *Adv. Mol. Simil., 1,* 187–211.

106. Amat, L., & Carbó-Dorca, R., (1999). Fitted electronic density functions from H to Rn for use in quantum similarity measures: cis-diammine-dichloroplatinum(II) complex as an application example. *J. Comput. Chem., 20,* 911–920.

107. Amat, L., & Carbó-Dorca, R., (2000). Molecular electronic density fitting using elementary Jacobi rotations under atomic shell approximation. *J. Chem. Inf. Comput. Sci., 40,* 1188–1198.

108. Constans, P., Amat, L., & Carbó-Dorca, R., (1997). Toward a global maximization of the molecular similarity function: Superposition of two molecules. *J. Comput. Chem., 18,* 826–846.

109. Ritchie, D. W., & Kemp, G. J. L., (1999). Fast computation, rotation and comparison of low resolution spherical harmonic molecular surfaces. *J. Comput. Chem., 20,* 383–395.

110. Hakkoymaz, H., Kieslich, Ch.A., Gorham, R. D., Jr., Gunopulos, D., & Morikis, D., (2011). Electrostatic similarity determination using multiresolution analysis. *Mol. Inf., 30,* 733–746.

111. Beck, M. E., & Schindler, M., (2009). Quantitative structure-activity relations based on quantum theory and wavelet transformations. *Chem. Phys., 356,* 121–130.

112. Martin, R. L., Gardiner, E. J., Senger, S., & Gillet, V. J., (2012). Compression of molecular interaction fields using wavelet thumbnails: Application to molecular alignment. *J. Chem. Inf. Model., 52,* 757–769.

113. Crippen, G. M., (2003). Series approximation of protein structure and constructing conformation space. *Polymer, 44,* 4373–4379.

114. Fischer, P., Baudoux, G., & Wouters, J., (2003). Wavpred: A wavelet-based algorithm for the prediction of transmembrane proteins. *Comm. Math. Sci., 1,* 44–56.

115. Qiu, J., Liang, R., Zou, X., & Mo, J., (2003). Prediction of protein secondary structure based on continuous wavelet transform. *Talanta, 61,* 285–293.

116. Vannuci, M., & Liò, P., (2001). Non-decimated wavelet analysis of biological sequences: Applications to protein structure and genomics. *Indian J. Stat., 63,* 218–233.

117. de Trad, C. H., Fang, Q., & Cosic, I., (2002). Protein sequence comparison based on the wavelet transform approach. *Prot. Eng., 15,* 193–203.

118. Sabarish, R. A., & Thomas, T., (2011). A frequency domain approach to protein sequence similarity analysis and functional classification. *Signal Image Process, 2,* 36–48.

119. Chua, G.-H., Krishnan, A., Li, K.-B., & Tomita, M., (2006). Multiresolution analysis uncovers hidden conservation of properties in structurally and functionally similar proteins. *J. Bioinform. Comput. Biol., 4,* 1245–1267.

120. Wen, Z.-N., Wang, K.-L., Li, M.-L., Nie, F.-S., & Yang, Y., (2005). Analyzing functional similarity of protein sequences with discrete wavelet transform. *Comput. Biol. Chem., 29,* 220–228.

121. Krishnan, A., Li, K.-B., & Isaac, P., (2004). Rapid detection of conserved regions in protein sequences using wavelets. *In Silico Biol.*, *4*, 133–148.

122. Tsonis, A. A., Kumar, P., Elsner, J. B., & Tsonis, P. A., (1996). Wavelet analysis of DNA sequences. *Phys. Rev. E*, *53*, 1828–1834.

123. Machado, J.-A., Costa, A., & Quelhas, M. D., (2011). Wavelet analysis of human DNA. *Genomics*, *98*, 155–163.

124. Mena-Chalco, J., Carrer, H., Zana, Y., & Cesar, R. M., Jr., (2008). Identification of protein coding regions using the modified Gabor-wavelet transform. *IEEE/ACM Trans. Comput. Biol. Bioinform.*, *5*, 198–207.

125. Barclay, V. J., & Bonner, R. F., (1997). Application of wavelet transforms to experimental spectra: Smoothing, denoising, and data set compression. *Anal. Chem.*, *69*, 78–90.

126. Pilard, M., & Epelboin, Y., (1998). Multiresolution analysis for the restoration of noisy X-ray topographs. *J. Appl. Crystallogr.*, *31*, 36–46.

127. Ergen, B., (2013). Comparison of wavelet types and thresholding methods on wavelet based denoising of heart sounds. *J. Signal Inf. Process*, *4*, 164–167.

128. Jeena, J., Salice, P., & Neetha, J., (2013). Denoising using soft thresholding, *Int. J. Adv. Res. Elec. Electron. Instrum. Eng.*, *2*, 1027–1032.

129. Chen, J.-S., Teng, H., & Nakano, A., (2007). Wavelet-based multi-scale coarse graining approach for DNA molecules. *Finite Elem. Anal. Des.*, *43*, 346–360.

130. Vakser, I. A., Matar, O. G., & Lam, C. F., (1999). A systematic study of low-resolution recognition in protein protein complexes. *Proc. Natl. Acad. Sci. USA*, *96*, 8477–8482.

131. Katchalski-Katzir, E., Shariv, I., Eisenstein, M., Friesem, A. A., Aflalo, C., & Vakser, I. A., (1992). Molecular surface recognition: Determination of geometric fit between proteins and their ligands by correlation techniques. *Proc. Natl. Acad. Sci. USA*, *89*, 2195–2199.

132. Russell, R. B., Alber, F., Aloy, P., Davis, F. P., Korkin, D., Pichaud, M., et al., (2004). A structural perspective on protein-protein interactions. *Curr. Op. Struct. Biol.*, *14*, 313–324.

133. Vakser, I. A., (1996). Low-resolution docking: Prediction of complexes for underdetermined structures. *Biopolymers*, *39*, 455–464.

134. Tovchigrechko, A., Wells, Ch. A., & Vakser, I. A., (2002). Docking of protein models. *Prot. Sci.*, *11*, 1888–1896.

135. Nittinger, E., Schneider, N., Lange, G., & Rarey, M., (2015). Evidence of water molecules: A statistical evaluation of water molecules based on electron density. *J. Chem. Inf. Model.*, *55*, 771–783.

136. Emperador, A., Carrillo, O., Rueda, M., & Orozco, M., (2008). Exploring the suitability of coarse-grained techniques for the representation of protein dynamics. *Biophys. J.*, *95*, 2127–2138.

137. Hinsen, K., (2008). Structural flexibility in proteins: Impact of the crystal environment. *Bioinform.*, *24*, 521–528.

138. Moritsugu, K., & Smith, J. C., (2008). REACH Coarse-grained biomolecular simulation: Transferability between different protein structural classes. *Biophys. J.*, *95*, 1639–1648.

139. Hills, R. D., Jr., Lu, L., & Voth, G. A., (2010). Multiscale coarse-graining of the protein energy landscape. *PLoS Comput. Biol.*, *6*, e1000827/1–e1000827/12.

140. Spijker, P., van Hoof, B., Debertrand, M., Markvoort, A. J., Vaidehi, N., & Hilbers, A. J., (2010). Coarse-grained molecular dynamics simulations of transmembrane protein-lipid systems. *Int. J. Mol. Sci.*, *11*, 2393–2420.

141. Doruker, P., Jernigan, R. L., & Bahar, I., (2002). Dynamics of large proteins through hierarchical levels of coarse-grained structures. *J. Comput. Chem.*, *23*, 119–127.

142. Darré, L., Machado, M. R., Brandner, A. F., González, H. C., Ferreira, S., & Pantano, S., (2014). SIRAH: A structurally unbiased coarse-grained force field for proteins with aqueous solvation and long-range electrostatics. *J. Chem. Theory. Comput.*, *11*,723–739.

143. Basdevant, N., Ha-Duong, T., & Borgis, D., (2007). A coarse-grained protein-protein potential derived from an all-atom force field. *J. Phys. Chem. B*, *111*, 9390–9399.

144. Gohlke, H., & Thorpe, M. F., (2006). A natural coarse-graining for simulating large biomolecular motion. *Biophys. J.*, *91*, 2115–2120.

145. Arkhipov, A., Freddolino, P. L., & Schulten, K., (2006). Stability and dynamics of virus capsids described by coarse-grained modeling. *Structure*, *14*, 1767–1777.

146. Arkhipov, A., Yin, Y., & Schulten, K., (2008). Four-scale description of membrane sculpting by BAR domains. *Biophys. J.*, *95*, 2806–2821.

147. Chng, Ch.-P., & Yang, L.-W., (2008). Coarse-grained models reveal functional dynamics – II. Molecular dynamics simulation at the coarse-grained level – Theories and biological applications. *Bioinform. Biol. Insights*, *2*, 171–185.

148. Yang, L.-W., & Chng, Ch.-P., (2008). Coarse-grained models reveal functional dynamics – I. Elastic network models – Theories, comparisons and perspectives. *Bioinf. Biol. Insights*, *2*, 25–45.

149. Kamerlin, S. C. L., Vicatos, S., Dryga, A., & Warshel, A., (2011). Coarse-grained (multiscale) simulations in studies of biophysical and chemical systems. *Annu. Rev. Phys. Chem.*, *62*, 41–64.

150. Naganathan, A. N., (2013). Coarse-grained models of protein folding as detailed tools to connect with experiments. *WIREs Comput. Mol. Sci.*, *3*, 504–514.

151. Baaden, M., & Marrink, S. J., (2013). Coarse-grain modelling of protein-protein interactions. *Curr. Opin. Struct. Biol.*, *23*, 878–886.

152. Brini, E., Algaer, E. A., Ganguly, P., Li, C., Rodríguez-Ropero, F., & van der Vegt, N. F.A., (2013). Systematic coarse-graining methods for soft matter simulations – A review. *Soft Matter*, *9*, 2108–2119.

153. Meier, K., Choutko, A., Dolenc, J., Eichenberger, A. P., Riniker, S., & van Gunsteren, W. F., (2013). Multi-resolution simulation of biomolecular systems: A review of methodological issues. *Angew. Chem. Int. Ed.*, *52*, 2820–2834.

154. Saunders, M., & Voth, G. A., (2013). Coarse-graining methods for computational biology. *Annu. Rev. Biophys.*, *42*, 73–93.

155. Paramonov, L., & Yaliraki, S. N., (2005). The directional contact distance of two ellipsoids: Coarse-grained potentials for anisotropic interactions. *J. Chem. Phys.*, *123*, 194111/1–194111/11.

156. Izvekov, S., & Voth, G. A., (2005). A Multiscale coarse-graining method for biomolecular systems. *J. Phys. Chem. B*, *109*, 2469–2473.

157. Liu, P., Izvekov, S., & Voth, G. A., (2007). Multiscale coarse-graining of monosaccharides. *J. Phys. Chem. B*, *111*, 11566–11575.

158. Carbone, P., Varnazeh, H. A.K., Chen, X., & Müller-Plathe, F., (2008). Transferability of coarse-grained force fields: The polymer case. *J. Chem. Phys.*, *128*, 064904/1–064904/11.

159. Naômé, A., Laaksonen, A., & Vercauteren, D. P., (2014). A solvent-mediated coarse-grained model of DNA derived with the systematic Newton inversion method. *J. Chem. Theory Comput.*, *10*, 3541–3549.

160. Kondrashov, D. A., Cui, Q., & Phillips, G. N., Jr., (2006). Optimization and evaluation of a coarse-grained model of protein motion using X-ray crystal data. *Biophys. J.*, *91*, 2760–2767.

161. Lyman, E., Pfaendtner, J., & Voth, G. A., (2008). Systematic multiscale parametrization of heterogeneous elastic network models of proteins. *Biophys. J.*, *95*, 4183–4192.

162. Orellana, L., Rueda, M., Ferrer-Costa, C., Lopez-Blanco, J. R., Chacón, P., & Orozca, M., (2010). Approaching elastic network models to molecular dynamics flexibility. *J. Chem. Theory Comput.*, *6*, 2910–2923.

163. Lyubartsev, A. P., & Laaksonen, A., (1995). Calculation of effective interaction potentials from radial distribution functions: A reverse Monte Carlo approach. *Phys. Rev. E*, *52*, 3730–3737.

164. Marrink, S. J., Risselada, H. J., Yefimov, S., Tieleman, D. P., & de Vries, A. H., (2007). The MARTINI forcefield: Coarse-grained model for biomolecular simulations. *J. Phys. Chem. B*, *111*, 7812–7824.

165. Monticelli, L., Kandasamy, S. K., Periole, X., Larson, R. G., Tieleman, D. P., & Marrink, S. J., (2008). The MARTINI coarse-grained forcefield: Extension to proteins. *J. Chem. Theory Comput.*, *4*, 819–834.

166. Liwo, A., Czaplewski, C., Oldziej, S., Rojas, A. V., Kazmierkiewicz, R., Makowski, M., et al., (2009). In Coarse-graining of condensed phase and biomolecular systems; Voth, G. A., Ed., CRC Press: Boca Raton, FL.

167. Fujitsuka, Y., Takada, S., Luthey-Schulten, Z., & Wolynes, P. G., (2004). Optimizing physical energy functions for protein folding. *Proteins*, *54*, 88–103.

168. Hori, N., Chikenji, G., Berry, R. S., & Takada, S., (2009). Folding energy landscape and network dynamics of small globular proteins. *Proc. Natl. Acad. Sci. USA*, *106*, 73–78.

169. Clementi, C., (2008). Coarse-grained models of protein folding: Toy models or predictive tools? *Curr. Opin. Struct. Biol.*, *18*, 10–15.

170. Sherwood, P., Brooks, B. R., & Sansom, M. S. P., (2008). Multiscale methods for macromolecular simulations. *Curr. Opin. Struct. Biol.*, *18*, 630–640.

171. He, N., Liu, Y., & Zhang, X., (2016). Molecular dynamics - Smoothed molecular dynamics (MD-SMD) adaptive coupling method with seamless transition. *Int. J. Numer. Meth. Engng.* doi:10.1002/nme.5224.

172. Cherfils, J., Duquerroy, S., & Janin, J., (1991). Protein–protein recognition analyzed by docking simulation. *Proteins*, *11*, 271–280.

173. Sternberg, M. J. E., & Gabb, H. A., & Jackson, R. M., (1998). Predictive docking of protein–protein and protein–DNA complexes. *Curr. Opin. Struct. Biol.*, *8*, 250–256.

174. Butzlaff, M., Dahmen, W., Diekmann, S., Dress, A., Schmitt, E., & von Kitzing, E., (1994). A hierarchical approach to force field calculations through spline approximations. *J. Math. Chem.*, *15*, 77–92.

175. von Kitzing, E., & Schmitt, E., (1995). Configurational space of biological macromolecules as seen by semiempirical force fields: Inherent problems for molecular design and strategies to solve them by means of hierarchical force fields. *J. Mol. Struct. (Theochem)*, *336*, 245–259.

176. Olson, W. K., (1996). Simulating DNA at low resolution. *Curr. Opin. Struct. Biol.*, *6*, 242–256.

177. Dury, L., Latour, Th., Leherte, L., Barberis, F., & Vercauteren, D. P., (2001). A new graph descriptor for molecules containing cycles. Application as screening criterion for searching molecular structures within large databases of organic compounds. *J. Chem. Inf. Comput. Sci.*, *41*, 1437–1445.

178. Fischer, J. R., Lessel, U., & Rarey, M., (2011). Improving similarity-driven library design: Customized matching and regioselective feature trees. *J. Chem. Inf. Model.*, *51*, 2156–2163.

179. Birchall, K., & Gillet, V. J., (2011). Reduced Graphs and Their Applications in Che-moinformatics. In: *Cheminformatics and Computational Chemical Biology*, vol.672 of series: *Methods in Molecular Biology*; Bajorath, J., Ed., Humana Press: New York, NY, USA, pp. 197-212.

180. Jia, Z., & Chen, J., (2016). Necessity of high-resolution for coarse-grained modeling of flexible proteins. *J. Comput. Chem.*, *37*, 1725–1733.

181. Cao, Z., & Voth, G. A., (2015). The multiscale coarse-graining method. XI. Accurate interaction based on the centers of charge of coarse-grained sites. *J. Chem. Phys.*, *143*, 243116/1–243116/11.

182. Swanson, K. A., Kang, R. S., Stamenova, S. D., Hicke, L., & Radhakrishnan, I., (2003). Solution structure of Vps27 UIM-ubiquitin complex important for endosomal sorting and receptor downregulation. *EMBO J.*, *22*, 4597–4606.

183. Wang, S., & Zhou, P., (2014). Sparsely-sampled, high-resolution 4D omit spectra for detection and assignment of intermolecular NOEs of protein complexes. *J. Biomol. NMR*, *59*, 51–56.

184. Lee, S., Tsai, Y. C., Mattera, R., Smith, W. J., Kostelansky, M. S., Weissman, A. M., et al., (2006). Structural basis for ubiquitin recognition and autoubiquitination by Rabex-5. *Nat. Struct. Mol. Biol.*, *13*, 264–271.

185. Hess, B., Kutzner, C., van der Spoel, D., & Lindahl, E., (2008). GROMACS 4: Algo-rithms for highly efficient, load-balanced, and scalable molecular simulation. *J. Chem. Theory Comput.*, *4*, 435–447.

186. Pronk, S., Páll, S., Schulz, R., Larsson, P., Bjelkmar, P., Apostolov, R., et al., (2013). GROMACS 4.5: A high-throughput and highly parallel open source molecular simula-tion toolkit. *Bioinformatics*, *29*, 845–854.

187. Showalter, S. A., & Brüschweiler, R., (2007). Validation of molecular dynamics simu-lations of biomolecules using NMR spin relaxation as benchmarks: Application to the AMBER99SB force field. *J. Chem. Theory Comput.*, *3*, 961–975.

188. Horn, H. W., Swope, W. C., Pitera, J. W., Madura, J. D., Dick, T. J., Hura, G. L., & Head-Gordon, T., (2004). Development of an improved four-site water model for bio-molecular simulations: TIP4P-Ew. *J. Chem. Phys.*, *120*, 9665–9678.

189. Chakrabartty, A., Kortemme, T., & Baldwinn, R. L., (1994). Helix propensity of the amino acids measured in alanine-based peptides without helix-stabilizing side-chain interactions. *Prot. Sci.*, *3*, 843–852.

190. Daura, X., Gademann, K., Jaun, B., Seebach, D., Gunsteren, W. F., & Mark, A. E., (1999). Peptide folding: When simulation meets experiment. *Angew. Chem. Int. Ed.*, *38*, 236–240.

191. Best, R. B., Zheng, W., & Mittal, J., (2014). Balanced protein-water interactions improve properties of disordered proteins and non-specific protein association. *J. Chem. Theory Comput.*, *10*, 5113–5124.

192. Henriques, J., Gragnell, C., & Skepö, M., (2015). Molecular Dynamics simulations of intrinsically disordered proteins: Force field evaluation and comparison with experi-ment. *J. Chem. Theory Comput.*, *11*, 3420–3431.

CHAPTER 9

THE NETWORK REPRESENTATION OF CHEMICAL SPACE: A NEW PARADIGM

ALFONSO NIÑO, CAMELIA MUÑOZ-CARO, and SEBASTIÁN REYES

SciCom Research Group, Escuela Superior de Informática, Universidad de Castilla-La Mancha, Paseo de la Universidad 4, 13004 Ciudad Real, Spain

CONTENTS

9.1 GENERAL CONCEPTS

More or less explicitly the concept of chemical space has been around for decades being a recurrent topic in nowadays cheminformatics and medicinal chemistry. As it is now understood the chemical space is defined as the set of all possible small organic molecules [35]. Of course, the meaning of "small" is absolutely arbitrary but, in the present context, it is related to the considerations of Bohacek et al. [8]. These authors, considered the "universe of organic molecules" from the point of view of structure-based drug design (i.e., what we now call chemical space). In such space, they focused in molecules up to 30 atoms containing just C, N, O, S as heavy atoms (this amounts roughly to a mass of 500 Daltons) allowing for branching and rings of three or four atoms. In these conditions, they made a rough estimate of 10^{60} possible compounds. This value is frequently quoted as the size of chemical space. However, it must be taken into account that just by allowing additional heteroatoms and larger rings the number of compounds would increase exponentially. The size of the chemical space is so large that in practice, the studies are restricted to specific sets of molecules, for instance, to specific kinds of bioactive compounds. So, if we consider the chemical space as the whole set, in practice we always deal with chemical subspaces defining a subset of the former. The key point when dealing when chemical subspaces are the representation we use to make the most of the information contained in them. Different approaches have been developed over the years, as shown in the following section of this work. An appropriate representation is needed to facilitate the application of pattern recognition/machine learning [66] algorithms. In this context, especially promising is the network representation of chemical space, a new approach arising from the new field of complex networks.

Complex network, or network science, is a modern interdisciplinary field involving computer science, physics, mathematics, biology, social sciences and, of course chemistry, among almost every current discipline. The field focuses in the structural and dynamic study of complex systems under the prism of a networked set of interrelated elements. In fact, the field has experienced a great development since the beginning of this century [2, 11]. From the network point of view, any system can be considered as a set of entities, describing the nodes of the network, interrelated by a set of interactions, describing the edges. Clearly, the network approach is closely related to graph theory. However, despite the similarities of networks and graphs, there is a key difference: in a network, in addition to data, nodes and edges can have an associated behavior, see Figure 9.1. This is a fundamental difference when analyzing and modeling the dynamics of a problem.

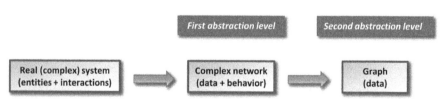

FIGURE 9.1 Schematic representation of the double abstraction process relating real systems to complex networks and graphs.

In a general sense, network studies can be traced back to the introduction of graph theory by Leonard Euler in the famous Königsberg bridges problem [23]. After that, the next key milestone in the field is found in the late 1950s, when Erdös and Rényi made a breakthrough in classical mathematical graph theory by describing a network with complex topology as a random graph [21, 22]. The main characteristic of an Erdös-Rényi (ER) network is that the connectivity of the nodes follows a Poisson distribution. Therefore, almost all nodes have the same number of neighbors. The ER model was the dominant model for half a century. From a more applied point of view, networks were mainly associated to the social sciences field until the end of the 20th century. Thus, in the early 1930s Jacob Moreno introduced the sociogram [43]. The sociogram is a network where nodes represent individuals and edges social relationships. Moreno used sociograms to obtain specific information about the structure and behavior of social systems. In this same context, Mark Granovetter introduced the concept of weak ties in the 1970s [30]. Weak ties are links between different social clusters (groups of closely interrelated nodes) that are essential for information dissemination. Weak ties are related to the small-world characteristic introduced in the social sciences by Milgram in the late 1960s [42]. The small-world behavior implies that the distance (measured in number of edges) between any pair of nodes is small with respect to the number of nodes in the network.

However, in the late 1990s, technological advances made possible to collect and process information regarding large real networks. Examples of them are communication and collaboration networks, the Internet or the World Wide Web. In sharp contrast to the ER model, in these networks the nodes connectivity follows approximately a power law distribution. This implies that the networks self-organize into a scale-free structure, a feature unpredicted by the ER model. Thus, for instance, the Internet [25], the World Wide Web [1], and the citation patterns of scientific publications [56,58] exhibit this characteristic. In fact, this behavior has been observed in systems as diverse as communication, social, economic, technological, biological, or transportation networks, to name just a few. A detailed survey can be found in [15]. In this context, two seminal works started the current interest in the study of complex networks. The first one was the work by Watts and Strogatz on small-world networks [74], where the authors showed how a regular lattice network is transformed into a random network, exhibiting the small-world property, by randomly reconnecting a small fraction of edges. The second is the work by Barabási and Albert on the generation of scale-free networks by a process of preferential attachment [3]. The observation of a common behavior pattern in networks, independently of the network nature, has led to the establishment of a series of general principles and to a unified treatment of networked systems defining what is now considered the new field of complex networks [2, 48]. Its current ubiquity is due to its capability of describing virtually any kind of system of scientific or technological interest such as computer and communication networks, organizations, coupled chemical reactions, or interacting biological systems, just to name a few. In addition, the complex networks approach offers a theoretical and practical way to deal with complexity, breaking the limitations of the reductionist approach traditional in science. The huge amount of data currently available on any real system makes complex networks an interesting way of accessing to the large amount of information contained in these data. There are points of contact, therefore, with the new world of big data.

As stated above, the complex networks approach represents a powerful way to make sense of the information contained in the chemical space. Here, we review the work done so far in this field considering the different approaches used to generate and process chemical networks. Anyhow, the starting point is to determine how to represent the chemical space.

9.2 CHEMICAL SPACE REPRESENTATIONS

As stated above, the chemical space is defined as the set of all possible small organic molecules [35]. More specifically [17], each chemical compound is assumed to be fully characterized by a set of n descriptors (think about the set of descriptors used in QSAR/QSPR studies such as lipophilicity, dipole moment, molecular refractivity or topological indexes, among many other). This set of n descriptors characterize the molecule defining a feature vector, $(d_1, d_2, ..., d_n)$, which, mathematically, corresponds to an n-tuple. In addition, the descriptors define a hyper dimensional space where each molecule is represented as a single point, see Figure 9.2. The figure represents a simplified view of the chemical space concept. Here, we have just three descriptors (d_1 to d_3) and three molecules (m_a to m_c) defined on that space. Actually, the chemical space resembles the concept of phase space in mechanics. A crucial difference exists, though. Whereas in mechanics the finite phase space defines the state of the system, there is no finite set of descriptors that define univocally a molecule. Actually, in practice we use

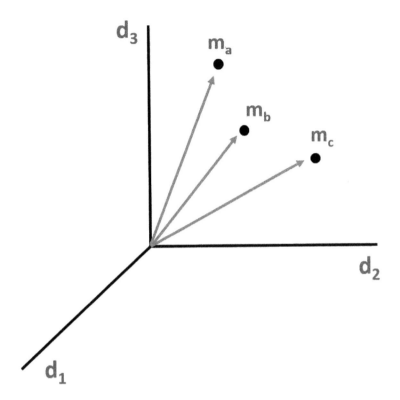

FIGURE 9.2   Simplified representation of the chemical space concept. Here, we have three different molecular descriptors, d_1 to d_3, defining the coordinates, and three molecules, m_a to m_c, characterized by them.

a given set of descriptors and usually specific sets of molecules (i.e., bioactive compounds) in the studies. In other words, in practice we use chemical subspaces. Anyway, the main point when dealing with chemical space is to use an appropriate representation approach. This topic is considered below.

9.2.1   COORDINATE-BASED REPRESENTATION

Using the above definitions, the chemical space is a large but finite set of points on a hyperdimensional space, usually compared to stars in the universe [17]. This is the most direct representation of chemical space: the coordinate-based representation. Here, each molecule is defined by its feature vector. However, this approach presents several drawbacks. The main ones are the following [39]:

- The coordinate space (descriptors) is continuous, but the set of molecules in this space is finite. So, we have a continuous representation of a large but finite set of objects.
- Relationships in the chemical space are not invariant to the representation used to describe the molecular set. In other words, different representations could modify the relationships (i.e., proximity) of molecules in the space.
- Due to the usually high number of descriptors needed to characterize molecular sets, the coordinate-based representation experiences the curse of dimensionality as introduced by Bellman in dynamic optimization problems [4]. Thus, the volume of the coordinate space for high number of dimensions is large when compared to the number of discrete points (molecules) defined in it. Therefore, we have a sparse set of objects, which makes hard the interpretation of many procedures. For instance, in these high dimensional spaces, all the points are far from each other, which makes difficult to obtain significant results in clustering studies. The solution is to resort to some dimension reduction technique such as the principal components analysis. However, this procedure can lead to a loss of information.
- The different axes of the coordinate space correspond to different magnitudes with different physical units. Thus, some kind of scaling or normalization is necessary for the sake of compatibility.

The previous problems (and others) yield the coordinate-based representation complicated to use. Thus, different coordinate-free representations have arisen. The first one was the cell-based representation.

9.2.2   CELL-BASED REPRESENTATION

The cell-based representation of chemical space was introduced by Pearlman and Smith [53]. These authors focused on the description of chemical diversity by selecting an assorted subset of compounds from a larger population of compounds defined in chemical space. Pearlman and Smith remarked that distance-based algorithms use relative distances in chemical space. Therefore, these algorithms are limited since no absolute distances of the compounds are used. For instance, it is virtually impossible to identify diversity voids in chemical space. In the cell-based approach, each axis of the hyperdimensional chemical space is divided into sections that define non-overlapping hypercubic cells. This allows not only for the use of relative, but also of absolute distances. Although information is lost, the cell-based representation allows many procedures such as comparing compound collections, selecting diverse subsets of compounds, and compound acquisition. These tasks can be more difficult to carry out in a coordinate-based framework. However, some problems arise in this representation model [39]:

- The location of cell boundaries can be problematic for nearby molecules since small changes in position close to cell boundaries can lead to significant changes in cell occupancies.
- The cellular location of a molecule depends on its coordinates. Thus, the continuous nature of coordinate-based representations can also influence the distribution of molecules over discrete partitionings.
- Cell-based spaces must typically be orthogonal and of low dimension to ensure meaningful analysis of compound distributions. Therefore, dimensionality reduction is usually required. This implies a loss of information, which influences the final results.

A third approach is represented by the application of the network paradigm to chemical space. This approach is considered independently in this work. However, before introducing the network paradigm to the description of chemical space, it is necessary to present some basic notions of complex networks.

9.3 COMPLEX NETWORKS BASICS

There are a few basic concepts needed to understand any complex networks processing activity. We introduce them here.

9.3.1 BASIC DEFINITIONS

Due to the intimate relationship between networks and graphs, network studies need some key concepts related to graph theory, the first one being the notion of graph (network). Consider a finite set of N nodes or vertices, $V = \{n_1, n_2, ..., n_N\}$, and a set of M relationships or links between those nodes called edges, E, where $E \subseteq \{(n_i, n_j) \mid n_i, n_j \in V\}$. A graph, G, is an ordered pair defined as $G = (V, E)$. In simpler terms, a graph is a set of nodes and their associated edges. When the edges between nodes have no direction, the graph is said undirected, otherwise it is said directed (or digraph), see Figures 9.3(a) and 9.3(b), respectively. In addition, the edges of the graph can have

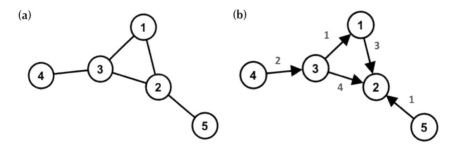

FIGURE 9.3 Schematic illustration of (a) an undirected and unweighted graph and (b) a directed and weighted graph.

an associated value (representing distances or similarities, for instance). In this case, the graph is said weighted, otherwise it is unweighted. Figure 9.3(b) shows an example of a graph which is both directed and weighted.

Of key importance when dealing with graphs is the concept of walk. Given a graph G, a walk between two nodes is a sequence of alternating nodes and edges where each edge is adjacent in the sequence to its endpoints. Thus, for instance, in Figure 9.3(a) a walk between nodes 1 and 5 is: 1, (1,2), 2, (2,5), 5. In a graph without multiple edges between the same nodes nor edges from a node to itself (a simple graph) a path is unambiguously specified just by stating the nodes in it. A path is defined as a walk without repeated edges nor vertices (except, maybe, the first and last vertices).

Now, we can define the concept of connected graph. A graph is connected if a path connects every pair of nodes. Otherwise, it is said to be disconnected or not connected. In this last case, the graph is formed at least by two isolated subgraphs without any edge between them. In directed graphs, we have the concept of strongly connected graph. A directed graph is strongly connected if every vertex is reachable from every other vertex. In addition, a strongly connected component is a subgraph of the original graph, which is strongly connected.

For representing graphs, we have different techniques. From the pure mathematical standpoint, the simplest and more powerful representation is the adjacency matrix. The adjacency matrix, A, is a squared matrix with as many rows and columns as vertices in the graph. The elements of A are defined as,

$$a_{ij} = \begin{cases} 1 \; iff \; (n_i, n_j) \in E \\ 0 \; otherwise \end{cases} \tag{1}$$

In other works, a_{ij} equals 1 only if there is an edge from vertex i to vertex j. For an undirected graph $\mathbf{A}$ is symmetric. However, it is not so for a directed one since an edge form vertex i to vertex j does not imply the existence of an edge from vertex j to vertex i.

9.3.2 NETWORK MEASURES

Different measures, indexes or metrics are associated to networks, which are used to characterize them topologically. Here, we collect the most common ones [2, 48], which are needed later to discuss the job done on chemical space networks.

9.3.2.1 Node Degree

The degree of a node n_i, denoted as k_i, is the number of nodes adjacent to n_i or, equivalently, the number of edges incident on that node. Thus, in Figure 9.3(a) the degree of node 3 is three (it is connected to three neighbors), and the degree of node 4 is one. In directed networks, we can distinguish between the in-degree, k^{in}, defined as the number of incoming edges, and the out-degree, k^{out}, which is the number of outcoming edges. As an example, in Figure 9.3(b) the in-degree of node 3 is one, and its out-degree is two. In the directed case, the total degree of a node is obtained as $k_i = k^{in}_i + k^{out}_i$. A global index of connectivity in the network is the average node degree defined as,

$$\langle k \rangle = \frac{1}{N}\sum_{i=1}^{N} k_i = 2M/N \tag{2}$$

In the previous result, we make use of the fact that the sum of the degrees is twice the number of edges.

9.3.2.2 Degree Probability Distribution

Related to the node degree is the key concept of degree probability distribution. The degree probability distribution, $P(k)$, is a function giving the probability that a randomly chosen node has degree k. So, for instance, $P(3)$, gives the probability of finding a node with degree 3 when a node of the graph is randomly selected. Using the frequency interpretation of probability, and for a large enough network, $P(k)$ can be visualized as the quotient between the number of nodes with degree k with respect to N, the total number of nodes.

9.3.2.3 Clustering Coefficient

Another basic index is the clustering coefficient. Given a node or vertex i, its clustering coefficient, c_i, is the quotient between, e_i, the number of edges between the k_i nodes adjacent to i (i.e., connected by edges to i) and the maximum possible number of edges between the nodes adjacent to i. The meaning of e_i can be easily apprehended considering a social network of friends. In this case, k_i represents the number of friends of node i and e_i represents the number of friends of i who are friends among themselves. Since the maximum number of edges between the k_i nodes adjacent to i is $k_i(k_i - 1)/2$ we have,

$$c_i = \frac{2e_i}{k_i(k_i-1)} \tag{3}$$

It is possible to define an average clustering coefficient for the whole graph, which is used as one of the basic graph indexes,

$$\langle c \rangle = \frac{1}{N}\sum_{i=1}^{N} c_i \tag{4}$$

9.3.2.4 Path Length

The length of a given path between nodes i and j is the number of edges in the path. Special importance has the shortest path between nodes i and j, d_{ij}. This path is the shortest one among all possible paths between nodes i and j. Usually, it is known as the geodesic path, or just geodesic, between i and j. Determining the shortest path between nodes in a graph is a key algorithmic problem. The Breath-First-Search (BFS) and Disjkstra's algorithms solve the problem in the unweighted and weighted cases, respectively [13]. As in the previous cases, it is possible to define an overall index for the whole graph: the average (shortest) path length, l. For an undirected graph of N nodes l is given by,

$$l = \frac{1}{N(N-1)}\sum_{j \neq i}^{N} d_{ij} \tag{5}$$

9.3.2.5   Edge Density

The total number of possible edges between N nodes or vertices is given by the number of combinations without repetition of two elements out of N,

$$\binom{N}{2} = \frac{N(N-1)}{2} \tag{6}$$

Therefore, the quotient between this quantity and the actual number of edges, $M = |E|$ is an indication of sparseness or edge density (i.e., how many edges there are with respect to all possible),

$$edge\ density = D = \frac{2M}{N(N-1)} \tag{7}$$

This metric is defined in the range [0, 1] and measures how close to a fully connected network our network is.

9.3.2.6   Degree-Degree Correlation

As pointed out by Newman more than a decade ago [47] networks exhibit a certain pattern, a correlation, among the degrees of nodes connected together. This degree–degree correlation is not accounted for by the degree distribution. To analyze such behavior, Newman [46] introduced the use of the Pearson correlation coefficient between degrees at both ends of edges, r, to measure of degree-degree correlation,

$$r = \frac{\sum_{j,i}^{N} k_i k_j a_{ij} - (1/2M) \sum_{j,i}^{N} (k_i k_j)^2}{\sum_{j,i}^{N} k_i \delta_{ij} - (1/2M) \sum_{j,i}^{N} (k_i k_j)^2} \tag{8}$$

where a_{ij} represents the elements of the adjacency matrix and δ_{ij} is the Kronecker delta. Clearly, $-1 \leq r \leq 1$. If $r < 0$ nodes of different degree tend to be connected, and the network is said to be dissasortative. When $r > 0$, nodes of similar degree tend to connect, and the network is called assortative. In the case $r = 0$ the network is called non-assortative and there no special relationship between the degree of the nodes. However, the situation is much more complex than expressed by a single parameter since different zones of a network can have different behavior, see Refs. [44, 65].

9.3.2.7 Modularity

Modularity, Q, in a network is a measure of how much a given distribution of nodes in groups (communities) differs from an equivalent random network, which is expected to have no community structure. The concept was introduced by Newman and Girvan [46], and in generalized form can be defined as [48]:

$$Q = \frac{1}{2M} \sum_{i,j}^{N} \left(a_{ij} - \frac{k_i k_j}{2M} \right) \delta(n_i, n_j) \qquad (9)$$

where the summation runs on the nodes of the network, and a_{ij} is the corresponding element of the adjacency matrix. In addition, the second term within the parenthesis corresponds to the expected number of edges between nodes i and j if the network were randomly connected. Finally, $\delta(n_i, n_j)$ equals 1 if nodes i and j belong to the same community and 0 otherwise. Modularity is strictly smaller than 1. It is positive if the number of edges between nodes of the same community is greater than expected by chance and negative otherwise. If Q is large, nodes in communities are much more strongly connected than between communities.

9.3.3 COMPLEX NETWORKS MODELS

When dealing with network problems one of the main issues is how to model their formation. Different models of network formation have been proposed able to reproduce the properties observed in real networks [2, 48]. Here, we introduce the two most widely considered network models: the random one and the scale-free Barabási-Albert models. The interested reader can find useful the review by Prettejohn et al. [55] which presents methods, and the corresponding algorithms, for generating complex networks.

9.3.3.1 The Erdös and Rényi (ER) Random Graph Model

The random model, usually called the Erdös and Rényi (ER) model, takes its name from the seminal work by these two authors [22]. Maybe, the simplest definition of an ER network is a graph of N nodes where each pair of nodes has a probability p of having an edge. Curiously, this is not the approach taken by Erdös and Rényi but the one presented simultaneously by Gilbert [28]. This model is known as the G (N, p) model. It is simply to show [2, 21, 28, 48] that the $G(N, p)$ degree probability distribution is given by

$$P(k) = \binom{N-1}{k} p^k (1-p)^{N-1-k} \tag{10}$$

which is a binomial distribution. For large networks, $N \to \infty$, equation (10) simplifies to

$$P(k) = e^{-\langle k \rangle} \frac{\langle k \rangle^k}{k!} \tag{11}$$

a Poisson distribution. Equation (11) tells us that in a random, ER, network most of the nodes have average degree, $<k>$, and just a few have smaller or greater values than $<k>$, see Figure 9.4(a). In addition, it can be shown [2] that the average path length, l, and average clustering coefficient, c, in a random network are given by

$$l \approx \frac{\ln N}{\ln \langle k \rangle}$$
$$c = \frac{\langle k \rangle}{N} = p \tag{12}$$

So, l depends logarithmically on the network size (small-world property), and the clustering coefficient is given by p, the probability of finding and edge between two randomly selected nodes.

9.3.3.2 The Scale-Free Barabási-Albert Model

In contrast with the ER model, initial studies regarding large networks such as author collaboration networks, communication networks, the World Wide Web or the Internet led to the discovery that many natural and artificial networks follow, at least approximately, a power law distribution, $P(k) \square \ k^{-\gamma}$ where usually $2 < \gamma < 3$ [2]. Since $P(k)$ follows a power law distribution, we have a few nodes with high degree and a lot of nodes with small degrees, see Figure 9.4(b). The nodes of high degree, accumulating large number of edges, are known as hubs.

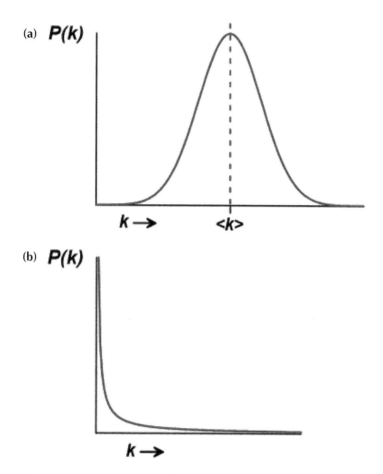

FIGURE 9.4 Graphical representation of degree probability distributions. (a) Poisson distribution. (b) Power law distribution.

In the seminal paper by Barabási and Albert scale-free networks are generated by a process of growth and preferential attachment [3]. The process begins with an initial network of N_0 nodes. Then, new nodes are added and connected to the previous ones with a probability proportional to the degree of each node. Thus, the probability of a new node being connected to a node i with degree k_i is

$$p_i = \frac{k_i}{\Sigma_j k_j} \tag{13}$$

The net effect is a "the rich get richer" behavior where nodes with higher degrees tend to accumulate more edges. A network so built is known as a Barabási-Albert, BA, network and its degree probability function is given by $P(\mathrm{k}) \propto k^{-3}$. In practice, the term BA network has become synonymous of scale-free network. For a BA network the average path length, l, and average clustering coefficient, c, are [2]

$$l \approx \frac{\ln N}{\ln \ln N}$$
$$c \approx \frac{(\ln N)^2}{N} \tag{14}$$

Again, the logarithmic dependence of l on N tells us that the BA model exhibits the small-world characteristic.

9.3.4 COMMUNITY DETECTION IN NETWORKS

In many networks, nodes group together forming sections of the network where nodes are more related to each other than to other nodes in the network. So, networks usually are organized, at what can be called mesoscale, in communities of nodes with similar characteristics.

Community structure is a consequence of inhomogeneities in the network. Thus, some nodes are more "strongly" connected, or more related, among them than with others. This fact defines communities of nodes [26] with probably similar state or behavior. The problem is how to characterize this "strength" of the connection since no single approach exists. The definition usually depends on the problem considered, and it is frequently related to edge density [26]. Therefore, many different types of methods exist such as hierarchical, partitional or spectral algorithms, among others [2, 14, 26, 64, 77].

Community detection in networks presents a great interest due to the large number of applications. The key point is that nodes in the same community represent entities with similar characteristics. So, in a social network, communities represent interrelated individuals such as families, friend circles, or co-workers. In communications networks, they correspond to systems strongly interrelated where ease of communication would represent an increase of the total efficiency of the system. In a sales system, communities can represent sets of customers with similar preferences. In terrorist or criminal networks (dark networks), they can reveal the organizational structure [24]. In gene or metabolic networks, communities identify functional blocks [57].

Many different network-clustering methods have been developed. However, no single classification does exist. Here, we use as a guide the excellent reviews by Fortunato [26], Schaeffer [64], Coscia et al. [14], Xie el al. [77] and Barabási [2]. Thus, we can classify the basic community detection methods as:

- *Hierarchical clustering methods*: designed to reveal, specifically, the multilevel, nested, structure of a network. As canonical examples, we have the divisive Girvan-Newman [29] and the agglomerative Ravasz et al. [57] algorithms.
- *Modularity based methods*: those making use of the concept of modularity introduced by Newman and Girvan [46] as a clustering quality index to maximize. As an example, we have the greedy algorithm introduced by Newman [46] and also the greedy Louvain algorithm [6].
- *Overlapping communities methods*: which assign nodes to more than one community, such as the 'CFinder clique percolation algorithm' [51]. In this group, it is interesting to highlight the so-called fuzzy methods. They represent a generalization of the overlapping concept. Here, each node belongs, in different proportion, to each network community. The most known algorithm in this context is fuzzy c-means developed by Dunn [19] and Bezdek [5].
- *Partitional clustering methods*: here, the number of clusters is defined a priori, such as in the k-means method [38].
- *Spectral methods*: based in the eigenvalues of some similarity matrix defined among the nodes of the network or the Laplacian matrix of the associated graph, such as the method by Donetty and Muñoz [18].
- *Dynamic methods*: based on the simulation of processes in the network, such as spin coupling [59], random walks [80], or synchronization [7].
- *Information-based methods*: which uses information theory to determine the optimal partition of the network. These methods seek to determine the community structure by minimizing some information theory magnitude. As examples we have, Infomod [62] that maximizes the mutual information contained in the partition and Infomap [63], which seeks to minimize the amount of information needed to represent a random walk in the network.

9.4 CHEMICAL SPACE NETWORKS

The concept of chemical space networks (CSNs) is fairly new being a powerful way to represent, interpret and use the more general concept of chemical space. First of all, the network paradigm allows for a coordinate-free representation of chemical space. In addition, it permits an easy visualization of chemical space and opens the possibility of using the arsenal of methods and techniques developed by network scientists. Thus, for instance, Figure 9.5 shows the network representation of 15 nodes (representing hypothetical molecules) connected by 18 edges with random weights (representing similarity coefficients). As it can be seen the network approach permits a simple, clear and direct representation of all the relationships among the molecules. Thus, as seen in Figure 9.5, the strength of the interactions can be visualized by making the edges thickness proportional to the, in this case, similarity coefficients. In addition, Figure 9.5 is organized according to the three communities identified by the Louvain algorithm [6] making this information directly available in the representation.

9.4.1 STATE OF THE ART

To the best of our knowledge, the first work specifically involving some kind of chemical (in this case pharmacological) space network is the one by

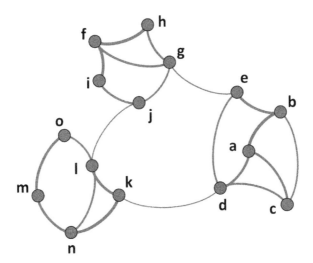

FIGURE 9.5 Network representation of 15 hypothetical molecules (labeled *a* to *o*) and 18 edges with random similarity values as weights. The edge width in the representation is proportional to the similarity value. The network is organized around the three communities identified by the Louvain algorithm (*a* to *e*, *f* to *j*, and *k* to *o*). The hypothetical data have been analyzed and represented using the Gephi network visualization and manipulation software [27].

Paolini et al. [52]. Here, the authors build an interaction network between proteins in chemical space. Two proteins are considered connected if both bind one or more compounds within a defined difference in binding energy threshold. This defines a polypharmacology network. From this information, the authors were able to navigate polypharmacology relationships between protein targets.

A few months later, a second work [41] used shannon entropy descriptors (SHED) to build a nuclear receptor network relating 25 nuclear receptors to 2033 ligands. A link is established between a nuclear receptor and a ligand if its minimum SHED Euclidean distance (as defined in the work) is greater than 0.6. The results permit to determine that phylogenetic relationships between nuclear receptors are preserved and apparent in the network. In addition, the network permitted to identify several ligand molecules connecting different sets of nuclear receptors.

Later, some works used indirectly the notion of CSNs to connect macromolecular targets through the similarities of their ligands [3, 31].

None of the previous works relate directly one ligand to another in order to build a true CSN. The first work where this is accomplished is the one by Waver et al. [75]. Here, similarity-based molecular networks are built to identify multiple structure-activity relationship components in sets of active compounds. In a similar way, Tanaka et al. [68] build similarity molecular networks to analyze the structure of chemical libraries. On the other hand, Krein and Sukumar [36] study the network topology and scaling relationships of chemical space networks independently of any biological activity. In addition, Ripphausen et al. [60] analyze the dependence of the success of ligand-based virtual screening on the structure-activity tolerance of screening targets. To such an end, the authors use similarity-based CSNs to model activity landscapes. Finally, Stumpfe at al. [67] have considered compounds active on human targets with high-confidence activity data. The network is built using the concept of activity cliff (i.e., pairs of structurally similar active compounds having a large difference in activity). In this form, the modular structure of activity cliffs is determined.

All these works generate the networks in an ad-hoc basis. However, in a five paper series [39, 76, 79, 78, 81] a serious effort has been made to systematize the generation of CSNs from a set of molecular structures. A comprehensive summary of the results is available in [72].

9.4.2 NETWORK-BASED CHEMICAL SPACE WORKFLOW

Considering the studies reported so far, we can formulate a general protocol for the non-trivial problem of the generation and treatment of CSNs. Thus, to such an end, it is possible to develop a scientific workflow, see Figure 9.6, for tackling in a general way the problem of the generation and use of CSNs. Figure 9.6 shows that the whole process can be organized in three sequential stages: first, the generation of the molecular space (actually, subspace) defined by a given set of molecules; second, the creation of the CSNs, i.e., the network generation; third, the processing of the network using the methods and techniques of network science. Let us consider each of these three stages independently.

9.4.2.1 First Stage: Chemical Space Generation

As shown in Figure 9.6, the first step in the workflow is the definition of the requirements that the set of molecules of interest must fulfill. This set of requirements translate into appropriate queries to one or more chemical databases. Nowadays, there are several large chemical databases available where specific sets of compounds can be recovered, see Table 9.1. These databases admit on-line queries and some of them, in particular, ChEMBL, can be downloaded and processed with database management systems such as MySQL [45]. Therefore, from the present standpoint, the databases

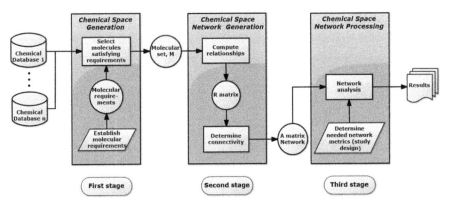

FIGURE 9.6   Scientific workflow representing the activities involved in the generation and interpretation of chemical space networks.

TABLE 9.1 Main Chemical Databases Currently Used in CSNs Studies of Bioactive Compounds

Database	Molecules contained	Curation	URL
ChEMBL	Drug-like	European Bioinformatics Institute	www.ebi.ac.uk/chembl
PubChem	Bioactive	National Center for Biotechnology Information (USA)	www.pubchem.ncbi.nlm.nih.gov
ZINC	Commercially available	John Irwin. University of California, San Francisco	www.zinc.docking.org
DrugBank	Drugs and drug targets	University of Alberta and The Metabolomics Innovation Centre	www.drugbank.ca
BindingDB	Drug-like and bioactive	Michael Gilson. University of California. San Diego	www.bindingdb.org/bind/index.jsp

represent the chemical space molecular set, CS, from which, we select the subspace, $M \subseteq CS$, relevant for our study. This first step is important since the design of chemical spaces for the analysis of compound sets, compound classification or activity prediction is one of the central tasks in cheminformatics [76]. As shown in Figure 9.6, the result of this first stage is the CS subset of molecules.

9.4.2.2 Second Stage: Chemical Space Network Generation

The second stage of the workflow, see Figure 9.6, is the generation of the CSN. According to the CSN definition, we start from a set of unique molecules $M = \{m_1, m_2, ..., m_N\}$. Thus, for the graph describing the chemical network, G_{CSNs}, we have:

$$G_{CSNs} = (V, E), \text{ with } V = M, \text{ and } E \subseteq \{(m_i, m_j) \mid m_i, m_j \in M\} \quad (15)$$

From the previous workflow stage, we have $M = CS$, but we still need the definition of an appropriate criterion to determine set E. So, the problem is how to answer the following question: when are two molecules connected? The answer depends on what is the relevant relationship for the problem at hand. Thus, assuming the existence of a given relationship criterion, c, involving an arbitrary relationship function, $r(m_i, m_j)$, so that when satisfied implies the existence of a connection between molecules, we have

$$(m_i, m_j) \in E \quad iff \quad i \neq j \land c(r(m_i, m_j)) = true \tag{16}$$

With the previous definition, no self-loops are allowed in a chemical network. Equation (16) defines a relationship matrix, $\mathbf{R}$, as shown in the second stage of the workflow in Figure 9.6.

In the study of bioactive molecules is quite frequent to be interested in the action mechanism of the molecules comprising the set M. Here, we consider the structure-activity relationship hypothesis, which states that structurally similar molecules should exhibit similar behavior [32]. Therefore, the key aspect is the molecular similarity of the molecules in the M molecular set. Actually, molecular similarity is the most used relationship function in CSNs. Assuming we have a similarity measure, s, for molecules i and j, s_{ij}, we can define a similarity matrix $\mathbf{S}$ as:

$$\mathbf{S} = \begin{bmatrix} s_{1,1} & \cdots & s_{1,N} \\ \vdots & \ddots & \vdots \\ s_{N,1} & \cdots & s_{N,N} \end{bmatrix} \ni s_{ij} = s(m_i, m_j) \tag{17}$$

In the present case, the similarity function s corresponds to the r function in Eq. (16). On the other hand, $\mathbf{S}$ is generally a symmetric matrix but this is not necessarily the case as exemplified by asymmetric Tversky index-based similarities [69]. In the current similarity-based representation, matrix $\mathbf{S}$ contains the information needed to define the network through the similarity functions. So, in this case $\mathbf{S} = \mathbf{R}$, the matrix of relationship. As shown in Figure 9.6, we determine the connectivity in the network from the $\mathbf{R}$ (in this case $\mathbf{S}$) matrix data. We do that by applying a transformation, T, depending on the connectivity relationship defined in equation (16), to the $\mathbf{S}$ matrix in order to obtain the adjacency matrix, $\mathbf{A}$:

$$\mathbf{A} = T(\mathbf{S}) = \begin{bmatrix} a_{1,1} & \cdots & a_{1,N} \\ \vdots & \ddots & \vdots \\ a_{N,1} & \cdots & a_{N,N} \end{bmatrix} \ni a_{ij} = \begin{cases} 1 \; iff \; i \neq j \land c(s_{ij}(m_i, m_j)) = true \\ 0 \; otherwise \end{cases} \tag{18}$$

Equation (18) generates a binary, unweighted adjacency matrix. Alternatively, rather than as a direct $\mathbf{A}$ matrix representation, Eq. (18) can be used to build the network one edge at a time storing them in an appropriate data structure [44, 49]. The previous presentation highlights the two main problems in the construction of CSNs: the definition of the functions r and c.

Function r defines the closeness between molecules. Usually, in CSNs, some kind of similarity between molecules is used. Table 9.2 collects the diverse r functions used in the CSNs jobs available to date. It can be seen that only four different r functions have been used. All of them are some kind of similarity measure. Table 9.2 shows that so far only one of the jobs uses the concept of maximum common substructure (MCS). The MCS of two molecules is the largest substructure shared by them. Thus, a similarity index is generated from this common substructure rather than from the whole molecules. On the other hand, Table 9.2 shows that just two works resort to the matched molecular pairs approach (MMP). An MMP is defined as a pair of compounds that only differ by a structural change at a single point [34]. So, in essence, an edge is considered to connect two molecules in set M if they conform an MMP. Finally, Table 9.2 also shows that most of the jobs apply some kind of fingerprint-based similarity.

Fingerprints are a common measure of similarity in cheminformatics. Molecular similarity is a ubiquitous concept admitting different definitions. The most formal one was introduced by Carbó et al. [10] as a measure of molecular electron density overlap. From this basic similarity measure, similarity indexes, normalized in the interval (0, 1], can be constructed by determining the cosine of the angle subtended by the two electron densities in its Hilbert semispace [9]. This defines the Carbó similarity index. Despite its sound theoretical foundation, calculation of Carbó similarity is time consuming due to the computation of the overlap of molecular electron density functions. Thus, less rigorous but faster techniques have been developed to search large chemical databases. Fingerprints represent the most used

TABLE 9.2 Nature of the, $r(m_i, m_j)$, Function Used in the Definition of CSNs in the Jobs Available to Date

$r(m_i, m_j)$	References
Maximum common substructures-based Tanimoto similarity	[79]
Matched molecular pairs-based similarity	[67, 78]
Fingerprints-based Tanimoto similarity	[31, 36, 60, 68, 75 78, 81]
Fingerprints-based Tversky similarity	[76]

technique for such a goal. Fingerprints are bitstrings representing the presence (1) or not (0) of specific substructural fragments in a molecule [37]. Thus, a bitstring with N bits is associated with a set of 2^N elements defined in an N-dimensional hypercubic space where each vertex corresponds to one of the members of the set. Fingerprints are essentially 2D descriptors and can be divided in two broad categories. The first corresponds to structural fingerprints where each bit in the bitstring corresponds to a given structural feature (key). These are exemplified by the MACCS fingerprint [20], which exists in two versions, 166 and 960 keys. The other category corresponds to hashed fingerprints where the bits in the bitstring do not correspond directly to structural features. Rather, the bits are set as the result of applying a hash function (for more information on hashing check [13] to some metric derived from the environment of each atom. The paradigmatic example is the Daylight fingerprint [16]. Special mention within hashed fingerprints deserves the so-called circular fingerprints. Here, the environment of each atom up to a given radius is used. The most common fingerprints in this class are the extended-connectivity fingerprints (ECFP), see [61]. ECFP is a very popular choice in any of its two variants, with diameter 4 (ECFP4) and diameter 6 (ECFP6). For a detailed review of chemical fingerprints refer Ref. [11].

The result of fingerprint calculations on a set of molecular compounds is a set of bitstrings from which we must produce a measure of similarity or distance. Different similarity/distance metrics are possible between two bitstrings, A and B. Table 9.3 collects some of the more common metrics [11,37]. For a deeper account of similarity/distance metrics the interested reader can consult [12] where the topic is broadly considered applied to density probability functions. Of those listed in Table 9.3, the Tanimoto coefficient is the most popular measure of fingerprint-based chemical similarity, see for instance, Table 9.2.

Definition of a similarity coefficient solves the problem of determining the r function in Eq. (16). The next step is to define the function c, see Eq. (16), which determines the existence of an edge between molecules m_i and m_j depending on the value of $r (m_i, m_j)$. As shown in Eq. (18), the final goal is to transform the similarity matrix $\mathbf{S}$ in the adjacency matrix $\mathbf{A}$. The simplest approach is to consider a direct transformation

$$\mathbf{A} = T(\mathbf{S}) = \begin{bmatrix} a_{1,1} & \cdots & a_{1,N} \\ \vdots & \ddots & \vdots \\ a_{N,1} & \cdots & a_{n,N} \end{bmatrix} \ni a_{ij} = \begin{cases} 1 \; iff \; s_{ij}(m_i, m_j) > 0 \\ 0 \; otherwise \end{cases} \quad (19)$$

33333
f333333

TABLE 9.3 Common Similarity Coefficients and Distance Measures for Two Sets (A and B) of N Binary (Dichotomous) Variables (Bitstrings of Length N)**

	Coefficient/Measure	Definition	Range
Similarity	Tanimoto or Jaccard	$S_{A,B} = c/(a+b-c)$	[0, 1]
	Dice (Hodgkin index)	$S_{A,B} = 2c/(a+b)$	[0, 1]
	Cosine (Carbó index)	$S_{A,B} = c/\sqrt{a \cdot b}$	[0, 1]
	Tversky*	$S_{A,B} = c/[\alpha \cdot (a-c) + \beta \cdot (b-c) + c]$	[0, 1]
Distance	Euclidean	$S_{A,B} = \sqrt{a+b-2c}$	[0, N]
	Hamming, Manhattan or City-block	$S_{A,B} = a+b-2c$	[0, N]
	Soergel	$S_{A,B} = (a+b-2c)/(a+b-c)$	[0, 1]

* The Tversky similarity coefficient is asymmetric so $S_{A,B} \neq S_{B,A}$. α and β are user-defined constants obeying the condition $\alpha, \beta \geq 0$. For $\alpha = \beta = 1$ the Tversky coefficient becomes the Tanimoto one. For $\alpha = \beta = 1/2$ it reduces to the Dice coefficient.

**As usual, a is the number of bits set to 1 in A, b is the number of bits set to 1 in B, and c is the number of bits set to 1 in both A and B.

This approach leads to a fully connected network where lots of edges correspond to very weakly connected nodes (not very similar molecules). The $O(N^2)$ space complexity of fully connected networks substantially complicates the storage of the network and the application of graph-based algorithms. Thus, the goal is to select the relevant edges to build the network or, as shown in Figure 9.6, to determine the "significant" connectivity of the network. Essentially what is needed is to edge-prune the complete network generated by Eq. (19). However, the meaning of "significant" is, though, ambiguous. This problem is general to the complex networks world and it has been clearly addressed in the field of brain complex networks generated from functional magnetic resonance imaging (fMRi) [70]. The general technique consists in using a given threshold value, t_h, such that an edge is considered to exist between molecules m_i and m_j if $s_{ij}(m_i, m_j) \geq t_h$. The problem now is how to construct a meaningful **A** matrix using t_h. Van Wijk et al. [70] indicate three different ways to tackle this problem:

- Use of a fixed threshold: In this case, we must select an appropriate th value. This value would depend on the problem, and we face here an "educated guess" situation. Several approaches are available.

Thus, we can select a threshold value corresponding to a statistically significant level of similarity. Furthermore, we can select a t_h value as large as possible while maintaining the network as a single connected component. The main problem of any of these approaches relays in the comparison of networks. The same t_h value applied to different networks would lead to very different connectivity patterns. A useful technique here is to analyze the network for different t_h values. Then, the most appropriate one, according to some problem-specific criterion, is selected. Another possibility is to analyze the variation of the network metrics relevant for the study at hand as a function of the t_h value.

- Use of a fixed average degree: Here, a given average degree, <k>, is selected and the t_h value is adjusted until it reproduces the <k> value. However, this approach also exhibits a clear drawback. A given <k> value can be too large for a network with low average connectivity and too small for a network with high connectivity. Thus, in a network with low connectivity, a lot of irrelevant edges could be included whereas in a highly connected network it would be possible to exclude some important ones. Again, a solution is to analyze the behavior of the network as a function of <k> (which indirectly means as a function of t_h).

- Use of a fixed edge density: This approach is similar to the previous one. Now, we fix the edge density value, D (see Eq. (7)), and we select the t_h reproducing that value. This approach has a sound theoretical ground since in an Erdös-Rényi network, fixing the edge density is equivalent to fixing the probability, p, of having an edge between two nodes. However, as in the previous cases, different D values can modify significantly the topology of the network.

As it can be seen, in all cases, it is necessary to select, somehow, a reference value for some parameter, t_h, <k>, or D. In the absence of a sound theoretical technique to select the optimal value, we must resort to an empirical approach. Thus, a problem-specific value is usually selected or the variation of the network behavior is analyzed in a range of values.

In the available CSNs studies, different approaches have been considered for building the networks, as collected in Table 9.4. In all cases, the networks are generated, in last instance, selecting a specific similarity threshold, t_h using it to build the network through Eq. (18). However, as shown in

TABLE 9.4 Similarity-Based Threshold Values Used in the Available CSNs Studies

Fingerprint	Similarity coefficient	Threshold (t_h)	Paper
MDLPublicKeys	Tanimoto	0.80	[68]
MDLPublicKeys	Tanimoto	[0.70, 0.90] ($\Delta = 0.05$)	
ECFP4	Tanimoto	[0.60, 0.80] ($\Delta = 0.05$)	
See footnote*	Tanimoto	[0.75, 0.85] ($\Delta = 0.05$)	[36]
ECFP4	Tanimoto	0.40	[60]
MACCS	Tanimoto	[0.85, 0.90] ($\Delta = 0.05$)	[81]
MACCS	Tanimoto	$D = 0.050$	
ECFP4	Tanimoto	$D = 0.025$	
MACCS	Tanimoto	$D \in [0, 1]$	
MMP-based	—	$D = 0.046, 0.052, 0.025$	[78]
ECFP4	Tanimoto	$D = 0.046, 0.052, 0.025$	
ECFP4	Tanimoto	$D = 0.025$	[79]
MMP-basedMCS-based	—	$D = 0.021$	
	Tanimoto-like	$D = 0.021$	
ECFP4	Tversky	$D = 0.001$	[76]
ECFP4	Tanimoto	$D = 0.001$	

* In this chapter, the authors used the CDK, CDK extended, GraphOnly, MACCS, EState, and PubChem fingerprints.

Table 9.4 two approaches have been used. The first one consists in applying directly a t_h value or interval. The second approach is to focus in a given edge density selecting t_h as the threshold value reproducing that density. Thus, in Refs. [36, 60, 68], the first approach was used. In particular, in the two first jobs, the variation of network behavior was analyzed in an interval of threshold values. However, the selection of t_h values and intervals is somewhat arbitrary, a fact that reflects the difficulty of selecting an appropriate threshold value.

In Ref. [81], the authors introduce the construction and analysis of CSNs as a function of edge density, D, rather than as a direct function of t_h. For the construction of networks, the authors select fixed D values that lead to a clear community structure in the network. As it would be expected, low D values are found to lead to meaningful community structures. From the results of the study, the authors proposed an edge density value of 0.025. From an operational point of view, the network is built as a similarity-based network using a t_h value reproducing the desired edge density. This approach is interesting due to the relationship between edge density and the probability of having an edge between randomly selected nodes. Therefore, the edge density is expected to remain constant in subsets of networks obtained by random sampling of nodes. In fact, in this work the authors show that building networks of the same edge density allows a direct and meaningful comparison of network measures and topology for different CSNs.

Despite its advantages, the empirical selection of edge density has a subjective component. In Ref. [79], see Table 9.4, a new approach is proposed. The authors build CSNs considering two molecules connected if they form a MMP. This leads directly to MMP-CSNs without any subjective intermediate step. However, this produces to "binary" networks where the only information is if two given molecules are or not similar (connected). On the other hand, similarity-based CSNs include a higher wealth of information. Thus, in this study, the authors build MMP-CSNs, compute their edge density and generate similarity-based CSNs using a t_h value reproducing the same edge density. Three similarity-based CSNs were considered. These were inhibitors/antagonists of matrix metalloproteinase, orexin receptor and bradykinin B1 receptors. The authors found that in average a $t_h = 0.55$ using ECFP4 fingerprints and Tanimoto coefficients seems to be a reasonable choice for building similarity-based CSNs.

As a consequence of the previous results, Zhang et al. [78], Table 9.4, consider how to combine the strengths of fingerprint and substructure-based similarity measures. Thus, these authors generate a Tanimoto-like coefficient quantifying pairwise similarity relationships on the basis of MCS. Using ChEMBL and DrugBank the authors select bioactive compounds generating a total of 39 combined activity classes with 15255 bioactive compounds and 1489 approved drugs. With these data, three different CSNs were built. The first is an ECFP4 fingerprint-based network with Tanimoto coefficients using an edge density of 0.025 (as suggested by the results of [81]. The second corresponds to an MMP-based network which yields an edge density of 0.021. Finally, a MCS-based network is built, using the Tanimoto-like coefficient, with a t_h that reproduces the former 0.021 edge density. The results show that the three CSNs exhibit well-resolved and similar topologies.

In Ref. [76], the authors build CSNs selecting from ChEMBL a total of 46,837 bioactive compounds exhibiting direct interaction with human targets. As shown in Table 9.4, the ECFP4 fingerprint is used to compute similarities through the asymmetric Tversky index. Clearly, the asymmetry of the index leads to a directed network where two edges in opposite directions are possible between each molecular pair. This asymmetric approach allows to tune the weight given to compounds that are substructures of others. To build the CSN the authors select an edge density of 0.01 setting as t_h a threshold value reproducing that density. In order to keep one edge between each pair of related compounds the authors only keep the edge with higher similarity if both exceed t_h. If both edges were exactly the same, no edge is kept.

Applying the appropriate approach, the result of this workflow stage, see Figure 9.6, is a network represented as an adjacency matrix, **A**, or as a data structure. The next step is to extract the information contained in that chemical space representation.

9.4.2.3   Third Stage: Chemical Space Network Processing

After generating a network representation of the desired chemical (sub) space, the use of different network metrics and techniques permits to obtain new information from the wealth of data comprising the network nodes and edges. We consider here the approaches used in the studies available so far, including the insight provided by them.

Thus Ref. [75] apply a hierarchical agglomerative clustering algorithm due to Ward [73] to similarity-based chemical networks for six different enzyme inhibitors. This makes possible to identify, within each network, groups corresponding to structurally related compounds with local structure-activity relationships. In a later study, Tanaka et al. [68], with an interest in fragment-based drug discovery, consider several fragment libraries. For these libraries, similarity-based chemical networks are built, and their degree distribution is analyzed. The authors find that their fragment library networks are scale-free, therefore, exhibiting the small-world property. On the basis of this finding, they propose a method of compound-prioritization for fragment-linking based on the hubs of the network. In a similar way, Krein and Sukumar [36] built CSNs using molecular sets taken from the ZINC database and from the PubChem AID361 bioassay, see Table 9.1. The authors analyze the degree distribution, clustering coefficient and degree-degree correlation. The results show that the networks follow power laws, although the noisy nature of the AID361 bioassay data does not give great confidence to the results. The networks exhibit positive clustering coefficients and assortative behavior (large and positive degree-degree correlation coefficients). In the same year, Ripphausen et al. [60] use active compounds taken from the ChEMBL and BindingDB databases, see Table

9.1, to build similarity-based networks. The authors apply a network representation where the nodes are color coded according to biological potency and scaled according to the structure-activity relationship index (SARI), see [54]. The inspection of the networks permits to identify activity cliffs and zones of local SAR continuity. In turn, Stumpfe at al. [67] use compounds with specified equilibrium constants for human targets taken from ChEMBL, see Table 9.1, to build an activity cliff network. Here, the authors determine activity cliff clusters by identifying the disconnected subgraphs within the activity cliff network.

In the series of interrelated papers on CSNs published between 2015 and 2016, a consistent pattern is applied to the analysis of networks. Thus, Zwierzyna et al. [81] use several network metrics such as edge density, clustering coefficient, degree-degree correlation or modularity to compare the different similarity-based chemical networks they built from a random set of compounds taken from the ZINC database, see Table 9.1. In addition, they make use of Newman [46] greedy clustering algorithm. They find, first, that the networks are assortative and second, the existence of a clear modular structure. In the next paper of the series, Zhang et al. [79] use ChEMBL data to build and compare MMP-based CSNs and similarity-based CSNs. To such an end, they apply, as in the previous work, basic network metrics: edge density, clustering coefficient, degree-degree correlation or modularity. Again, the structure of the network is determined by applying the Newman [46] greedy clustering algorithm. The authors use a mutual information measure [40, 71] to compare community distributions in different networks. The results show that MMP-based CSNs and similarity-based CSNs have similar topology and community structure. Thus, they provide similar global views of chemical space. However, a virtually non-existent correlation between assortativity in both networks is found. Therefore, for structure-activity relationship analysis both networks should be considered. As a continuation of this work, Zhang et al. [78] introduced MCS-based networks and perform a comparison with MMP-based CSNs and similarity-based CSNs. The comparison follows the same lines as in the previous paper. Here, though, information entropy is used to determine the network with higher number of compounds in the communities detected. The results show that all the CSNs exhibit similar topologies. However, the clustering coefficients of the MCS-CSNs are in general higher than for the other networks. This results in an improved separation of similar molecules into distinct communities.

Finally Ref. [76] consider the effect of the asymmetric Tversky similarity index in the behavior of CSNs. Again, as in the previous jobs, basic network metrics are considered to compare Tversky-CSNs with regular Tanimoto-CSNs. In addition, the degree distribution is considered and the cluster structure of networks is once more determined through the Newman [46] clustering algorithm. The results show that when asymmetry is increased in the Tversky coefficients, the number of hubs in the networks also increases. This leads to a clear power-law behavior in the out-degree distribution of the Tversky-CSN, which is associated to a decrease of assortativity. This fact leads to communities often centered on hubs, which are specially interesting for structure-activity relationship studies.

9.5 CONCLUDING REMARKS

The network paradigm represents a powerful tool for the analysis and visualization of chemical space. It allows for a new and useful way to derive information from the molecular data resorting to the techniques and tools of network science. To date, only the surface of this new approach has been scratched. Thus, in the available jobs, basic network metrics and clustering methods have been applied. Further studies can be carried out on the unambiguous generation of CSNs, the use of centrality measures different from the degree (i.e., hubs location) to identify key compounds in CSNs or the effect of different clustering approaches. Despite that, a lot of useful information regarding the organization of bioactive compounds in different chemical (sub)spaces has already been generated. Recalling the introductory speech of the classical Start Trek TV series, it is time to use the powerful network approach to explore (chemical) *Space, the final frontier....*

KEYWORDS

- **chemical space**
- **chemical space networks**
- **chemoinformatics**
- **complex networks**
- **complex systems**
- **network analysis**

REFERENCES

1. Albert, R., Jeong, H., & Barabási, A. L., (1999). Internet: Diameter of the World-Wide Web. *Nature, 401*, 130–131.
2. Barabási, A. L., (2016). *Network Science*; Cambridge University Press.
3. Barabási, A. L., & Albert, R., (1999). Emergence of Scaling in Random Networks. *Science, 286*, 509–512.
4. Bellman, R. E., (1961). Computational aspects of dynamic programming, *Adaptive Control Processes*; Princeton University Press: Princeton, pp 94..
5. Bezdek, J. C., (1981). Objective function clustering, *Pattern Recognition with Fuzzy Objective Function Algorithms*; New York: Plenum Press, pp 65.
6. Blondel, V. D., Guillaume, J. L., Lambiotte, R., & Lefebvre, E., (2008). Fast unfolding of communities in large networks. *Journal of Statistical Mechanics: Theory and Experiment, 10*, 10008.
7. Boccaletti, S., Ivanchenko, M., Latora, V., Pluchino, A., & Rapisarda, A., (2007). Detecting complex network modularity by dynamical clustering. *Phys. Rev. E., 75*, 45102–45105.
8. Bohacek, R. S., McMartin, C., & Wayne C. G., (1996). The art and practice of structure-based drug design: A molecular modeling perspective. *Med. Res. Rev., 16*, 3–50.
9. Carbó, R., Leyda, L., & Arnau, M., (1980). How similar is a molecule to another? An electron density measure of similarity between two molecular structures. *Int. J. Quantum Chem., 17*(6), 1185–1189.
10. Carbó, R., & Gironés, X., (2005). Foundation of Quantum Similarity Measures and Their Relationship to QSPR: Density Function Structure, Approximations, and Application Examples. *Int. J. Quantum Chem., 101*, 8–20.
11. Cereto-Massagué, A., Ojeda, M. J., Valls, C., Mulero, M., Garcia-Vallvé, S., & Pujadas, G., (2015). Molecular fingerprint similarity search in virtual screening. *Methods., 71*, 58–63.
12. Cha, S. H., (2007). Comprehensive Survey on Distance Similarity Measures between Probability Density Functions. *International Journal of Mathematical Models and Methods in Applied Sciences., 4*, 300–307.
13. Cormen, T. H., Leiserson, C. E., Rivest, R. L., & Stein, C., (2009). *Introduction to Algorithms, 3rd Edition*; The MIT Press.
14. Coscia, M., Giannotti, F., & Pedreschi, D., (2012). A Classification for Community Discovery Methods in Complex Networks. *Statistical Analysis and Data Mining, 4*, 512–546.
15. Costa, L. F., Oliveira Jr. O. N., Travieso, G., Rodrigues, F. A., Boas, P. R. V., Antiqueira, L., (2011). Analyzing and modeling real-world phenomena with complex networks: A survey of applications. *Adv. Phys., 60*(3), 329–412.
16. Daylight. Chemical Information Systems. http://www.daylight.com/ (accessed October, 2016).
17. Dobson, C. M., (2004). Chemical space and biology. *Nature, 432*, 824–828.
18. Donetti, L., & Muñoz, M. A., (2004). Detecting network communities: A new systematic and efficient algorithm. *Journal of Statistical Mechanics: Theory and Experiment, 10*, 10012.
19. Dunn, J. C., (1973). A fuzzy relative of the ISODATA process and its use in detecting compact well-separated clusters. *J. Cybernetics, 3*, 32–57.

20. Durant, J. L., Leland, B. A., Henry, D. R., & Nourse, J. G., (2002). Reoptimization of MDL keys for use in drug discovery. *J. Chem. Inf. Comput. Sci., 42,* 273–1280.

21. Erdős, P., & Rényi, A., (1959). On Random Graphs. I. *Publ. Math., 6,* 290–297.

22. Erdös, P., & Rényi, A., (1960). On the evolution of random graphs, *Publ. Math. Inst. Hung. Acad. Sci., 5,* 17–61.

23. Euler, L., (1736). Solutio problematis ad geometriam situs pertinentis. *Commentarii Academiae Scientarum Imperialis Petropolitanae., 8,* 128–140.

24. Everton, S. F., (2013). Cohesion and Clustering, Structural Analysis in the Social Sciences Series; *Disrupting Dark Networks*; Edition 34; Cambridge University Press, pp 170.

25. Faloutsos, M., Faloutsos, P., & Faloutsos, C., (1999). On power-law relationships of the internet topology. *Comput. Commun., 29,* 251–262.

26. Fortunato, S., (2010). Community detection in graphs. *Phys. Rep, 486,* 75–174.

27. Gephi. The Open Graph Viz Platform. https://gephi.org/ (accessed October 2016).

28. Gilbert, E. N., (1959). Random Graphs. *Annals of Mathematical Statistics, 30* (4), 1141–1144.

29. Girvan, M., & Newman, M. E. J., (2002). Community structure in social and biological networks. *Proceedings of the National Academy of Sciences of the United States of America, 99,* 7821–7826.

30. Granovetter. M. S., (1973). The Strength of Weak Ties, *Am. J. Sociol., 78,* 1360–1380.

31. Hert, J., Keiser, M. J., Irwin, J. J., Oprea, T., & Shoichet, B. K., (2008). Quantifying the Relationships among Drug Classes. *J. Chem. Inf. Model., 48,* 755–765.

32. Johnson, M. A., & Maggiora, G. M., (1990). *Concepts and Applications of Molecular Similarity*; John Wiley and Sons: New York.

33. Keiser, M. J., Roth, B. L., Armbruster, B. N., Ernsberger, P., Irwin, J. J., & Shoichet, B. K., (2007). Relating protein pharmacology by ligand chemistry. *Nat. Biotechnol., 25*(2), 197–206.

34. Kenny, P. W., & Sadowski, J., (2005). Structure Modification in Chemical Databases. In: *Chemoinformatics in Drug Discovery*; Oprea, T. I., Ed., Wiley-VCH: Weinheim, Germany, pp. 271–285.

35. Kirkpatrick, P., & Ellis, C., (2004). Chemical space. *Nature, 432,* 823.

36. Krein, M. P., & Sukumar, N., (2011). Exploration of the topology of chemical spaces with network measures. *J. Phys. Chem. A., 115,* 12905–12918.

37. Leach, A. R., & Gillet, V. J., (2007). Molecular Descriptors, *An Introduction to Chemoinformatics, Revised Edition*; Springer, pp.62.

38. Macqueen, J. B., (1967). Some Methods for classification and Analysis of Multivariate Observations. *Proceedings of the 5th Berkeley Symposium on Mathematical Statistics and Probability*; University of California Press, Vol. 1, pp. 281–297.

39. Maggiora, G. M., & Bajorath, J., (2014). Chemical space networks: a powerful new paradigm for the description of chemical space. *J. Comput. Aid. Mol. Des., 28,* 795–802.

40. Maggiora, G. M., & Shanmugasundaram, V., (2005). An information theoretic characterization of partitioned property spaces. *J. Math. Chem., 38,* 1–20.

41. Mestres, J., Martin-Couce, L., Gregori-Puigjane, E., Cases, M., & Boyer, S., (2006). Ligand-Based Approaches to in Silico Pharmacology: Nuclear Receptor Profiling. *J. Chem. Inf. Model., 46,* 2725–2736.

42. Milgram, S., (1967). The small world problem. *Psychol. Today., 2,* 60–67.

43. Moreno, J. L., (1934). Evolution of groups, *Who Shall Survive?* Beacon House Inc. New York, pp.26.

44. Muñoz-Caro, C., Niño, A., Reyes, S., & Castillo, M., (2016). APINetworks Java. A Java approach to the efficient treatment of large-scale complex networks. *Comput. Phys. Commun. 207*, 549–552.

45. MySQL: Open-source relational database management system. http://www.mysql.com/ (accessed October, 2016).

46. Newman, M. E. J., (2002). Assortative mixing in networks. *Phys. Rev. Lett., 89*, 208701–4.

47. Newman, M. E. J., (2010). *Networks: An Introduction*; Oxford University Press.

48. Newman, M. E. J., & Girvan, M., (2004). Finding and evaluating community structure in networks. *Phys. Rev. E, 69*, 026113–026127.

49. Niño, A., & Muñoz-Caro, C., (2013). *Quantitative Modeling of Degree-Degree Correlation in Complex Networks. Phys. Rev. E., 88*, 32805–32813.

50. Niño, A., Muñoz-Caro, C., & Reyes, S., (2015). APINetworks: A general API for the treatment of complex networks in arbitrary computational environments. *Comput. Phys. Commun., 196*, 446–454.

51. Palla, G., Derényi, I., Farkas, I., & Vicsek, T., (2005). Uncovering the overlapping community structure of complex networks in nature and society. *Nature, 435*, 814–818.

52. Paolini, G. V., Shapland, R. H. B., van Hoorn, W. P., Mason, J. S., & Hopkins, A. L., (2006). Global mapping of pharmacological space. *Nat. Biotechnol., 24*, 805–815.

53. Pearlman, R., & Smith, K., (2002). Novel software tools for chemical diversity. *3D QSAR Drug Design, 2*, 339–353.

54. Peltason, L., & Bajorath, J., (2007). SAR Index: Quantifying the Nature of Structure-Activity Relationships. *J. Med. Chem., 50*, 5571–5578.

55. Prettejohn, B. J., Berryman, M. J., & McDonnell, M. D., (2011). Methods for generating complex networks with selected structural properties for simulations: A review and tutorial for neuroscientists. *Front. Comput. Neurosci., 5*, 11.

56. Price, D. J., (1965). Networks of scientific papers. *Science, 149*, 510–515.

57. Ravasz, E., Somera, A. L., Mongru, D. A., Oltvai, Z. N., & Barabási, A. L., (2002). Hierarchical organization of modularity in metabolic networks. *Science, 297*, 1551–1555.

58. Redner, S., (1998). How popular is your paper? An empirical study of the citation distribution. *Eur. Phys. J. B., 4*, 131–134.

59. Reichardt, J., & Bornholdt, S., (2004). Detecting fuzzy community structures in complex networks with a Potts model. *Phys. Rev. Lett., 93*, 218701–218704.

60. Ripphausen, P., Nisius, B., Wawer, M., & Bajorath, J., (2011). Rationalizing the role of SAR tolerance for ligand-based virtual screening. *J. Chem. Inf. Model, 51*, 837–842.

61. Rogers, D., & Mathew, H., (2010). Extended-Connectivity Fingerprints. *J. Chem. Inf. Model, 50* (5), 742–754.

62. Rosvall, M., & Bergstrom, C., (2007). An information-theoretic framework for resolving community structure in complex networks. *P. Natl. Acad. Sci., 104*, 7327–7331.

63. Rosvall, M., & Bergstrom, M., (2008). Maps of random walks on complex networks reveal community structure. *P. Natl. Acad. Sci., 105*, 1118–1123.

64. Serrano, M. A., Boguñá, M., Pastor-Satorras, R., & Vespignani, A., (2007). *Large Scale Structure and Dynamics of Complex Networks: From Information Technology to Finance and Natural Sciences*; Caldarelli, G. and Vespignani, A., Eds., World Scientific Publishing: Singapore, pp. 35–66.

65. Schaeffer, S. E., (2007). Graph clustering. *Computer Science Review*, *1*, 27–64.

66. Shalev-Shwartz, S., & Ben-David, S., (2014). A gentle start, *Understanding Machine Learning: From Theory to Algorithms, 1st. Ed.,* Cambridge University Press, New York, pp. 13.

67. Stumpfe, D., Dimova, D., & Bajorath, J., (2014). Composition and topology of activity cliff clusters formed by bioactive compounds. *J. Chem. Inf. Model.*, *54*, 451–461.

68. Tanaka, N., Ohno, K., Niimi, T., Moritomo, A., Mori, K., & Orita, M., (2009). Small-world phenomena in chemical library networks: application to fragment-based drug discovery. *J. Chem. Inf. Model.*, *49*, 2677–2686.

69. Tversky, A., (1977). Features of similarity. *Psychol. Rev.*, *84*, 327–352

70. Van Wijk, B. C., Stam, C. J., & Daffertshofer, A., (2010). Comparing brain networks of different size and connectivity density using graph theory. *PLoS One.*, *5*(10), e13701.

71. Vinh, N. X., Epps, J., & Bailey, J., (2010). Information theoretic measures for clusterings comparison: variants, properties, normalization and correction for chance. *J. Mach. Learn. Res.*, *11*, 2837–2854.

72. Vogt, M., Stumpfe, D., Maggiora, G. M., & Bajorath, J., (2016). Lessons learned from the design of chemical space networks and opportunities for new applications. *J. Comput. Aid. Mol. Des.*, *30*, 191–208.

73. Ward, J. H., (1963). Hierarchical Grouping to Optimize an Objective Function. *J. Am. Stat. Assoc.*, *58*, 236–244.

74. Watts, D. J., & Strogatz, S. H., (1998). Collective dynamics of 'small-world' networks. *Nature*, *393*, 440–442.

75. Wawer, M., Peltason, L., Weskamp, N., Teckentrup, A., & Bajorath, J., (2008). Structure-activity relationship anatomy by network-like similarity graphs and local structure-activity relationship indices. *J. Med. Chem*, *51*, 6075–6084.

76. Wu, M., Vogt, M., Maggiora, G. M., & Bajorath, J., (2016). Design of chemical space networks on the basis of Tversky similarity. *J. Comput. Aid. Mol. Des.*, *30*, 1–12.

77. Xie, J., Kelly, S., & Szymanski, B. K., (2013). Overlapping Community Detection in Networks: The State of the Art and Comparative Study. *ACM Computing Surveys*, *45*, 43.

78. Zhang, B., Vogt, M., Maggiora, G. M., & Bajorath, J., (2015a). Comparison of bioactive chemical space networks generated using substructure-and fingerprint-based measures of molecular similarity. *J. Comput. Aid. Mol. Des*, *29*, 595–608.

79. Zhang, B., Vogt, M., Maggiora, G. M., & Bajorath, J., (2015b). Design of chemical space networks using a Tanimoto similarity variant based upon maximum common substructures. *J. Comput. Aid. Mol. Des*, *29*, 937–950.

80. Zhou, H., (2003). Distance, dissimilarity index, and network community structure. *Phys. Rev. E.*, *67*, 61901–61908.

81. Zwierzyna, M., Vogt, M., Maggiora, G. M., & Bajorath, J., (2015). Design and characterization of chemical space networks for different compound data sets. *J. Comput. Aid. Mol. Des.*, *29*, 113–125.

CHAPTER 10

AN APPLICATION OF THE MAXIMUM PRINCIPLE IN CHEMISTRY: A METHOD TO LOCATE TRANSITION STATES

JOSEP MARIA BOFILL[1] and WOLFGANG QUAPP[2]

[1]*Universitat de Barcelona, Departament de Química Inorgànica i Orgànica, Secció de Química Orgànica, Universitat de Barcelona, and Institut de Química Teòrica i Computacional, Universitat de Barcelona, (IQTCUB), Martí i Franquès, 1, 08028 Barcelona, Spain, E-mail: jmbofill@ub.edu*

[2]*Mathematisches Institut, Universität Leipzig, PF 100920, D-04009 Leipzig, Germany, E-mail: quapp@uni-leipzig.de*

CONTENTS

ABSTRACT

The solution curves of the gentlest ascent dynamics (GAD) follow a maximum principle. Under application of the Pontryagin maximum principle, it is demonstrated that the optimal control vector is exactly given by the second GAD equation. The variational nature of GAD curves is discussed, as well as that of some other known reaction pathways.

10.1 INTRODUCTION

An important goal in chemistry is the control of processes, if possible in an optimal manner, for the quantitative conversion of a molecule to a desired product. Normally, the control in chemistry is achieved modifying the thermodynamics of the process or the reaction through the external parameters like temperature, pressure, concentration, or solvent. This is the first and the most widely used control in chemistry. A second way is to manipulate the kinetics of the process by adding or modifying an appropriate catalyst. The third way is the most recent and now under strong development, namely, the control of the reactions through special light sources offering the opportunity to control quantum systems coherently. Within the third option were first the theoretical proposals due to Brumer and Shapiro [1], and to Tannor-Kosloff-Rice [2]. The latter authors proposed a pump–pump scheme where the laser light is used to create and steer nuclear wave packets to control the molecular reaction. The first experimental realization of the theoretical proposal was demonstrated by Zewail and coworkers [3, 4]. From a theoretical point of view, optimal pulses steering a reaction coherently from the given reactant to a desired or predefined product: that can be found in a more direct way by utilizing the approach of optimal control theory (OCT) [5, 6].

Other important applications of optimal control theory are given in the area of chemical engineering. The chemical industry moves towards the field of life sciences in which fed-batch processes are predominant (e.g., production of food), optimization and control of fed-batch bioreactors have become more challenging than ever. By programming substrate feeding, one can control important phenomena such as substrate inhibition, glucose effect, and catabolite repression. From the control engineering point of view, fed-batch processes are quite challenging, since the optimization of the substrate feed rate is a dynamic problem [7].

Finally, workers in theoretical chemistry have proposed algorithms to find transition states that are based in the optimal control theory [8]. This theory is contained therein to find the set of rules of a system in a way that a certain optimality criteria are achieved. It is the part of mathematics that formalizes and solves the problem to choose the best way of realizing a controlled process in a prescribed sense. Depending of the parameters, or control parameters, which usually are subject to some constraints, the optimal control process is described through differential or integral functionals. According to the formulation of the problem, the search of the controls and the realization of the process are chosen in accord to certain prescribed constraints. In fact the term "mathematical theory of optimal control" is applied to the part of mathematical science dealing with the solution of non-classical variational problems of optimal control. The type of problems permits the search of non-smooth functionals and arbitrary constraints related with the control parameters or on the dependent variables. The term covers mathematical methods involving a statistical process or the dynamic optimization, and its interpretation is given in terms of applied procedures for adopting optimal solutions. Taking this into account, the mathematical theory of optimal control contains elements of operations research, mathematical programming, game theory and machine learning. The set of problems studied in the mathematical theory of optimal control have arisen from practical demands like automatic control theory. In methods and in applications, the mathematical theory of optimal control is closely related with analytic mechanics, in the areas relating to the variational principles of classical mechanics.

The variational theory is applied extensively in theoretical chemistry especially in quantum chemistry, see, e.g., Refs. [9–11].

The reaction path (RP) concept [12] is one of the most widely used models in theoretical chemistry. The nature of many types of curves representing RPs has been proved to be variational [8, 13–17]. The recently proposed curve for a RP model, the gentlest ascent dynamics (GAD) [18] is an example of a curve that falls in the group of a variational problem of optimal control [8]. We describe such a reaction path in this chapter. Its nature is based on the Maximum Principle, the basis of the OCT, and it can be used to locate transition states on a potential energy surfaces (PES). Additionally we review the variational nature of some other types of paths. We treat steepest descent, gradient extremal, and distinguished coordinate or its modern version, the Newton trajectory. An extension of the model and its behavior is also discussed.

10.2   THE MAXIMUM PRINCIPLE AS A BASIS OF THE GENTLEST ASCENT DYNAMICS MODEL

In this section we report a proof of the optimal control character of the curves of the GAD. For this purpose we consider a system with N degrees of freedom represented by a point vector $\mathbf{x} \in \mathfrak{R}^N$. Curves in the $\mathfrak{R}^N$ are usually characterized by $\mathbf{x}(t)$ with a parameter t. The potential energy is described by the PES function, $V(\mathbf{x})$. The concept of the GAD model is that of a dynamical system. The solution curves of the GAD equations [18] evolve from a point close to a minimum to a stationary point (SP) on the PES. The GAD model is based on the gradient field of the PES, $\mathbf{g}(\mathbf{x}) = \nabla_x V(\mathbf{x})$, and a normalized control vector, $\mathbf{w}$. The Hessian matrix is also used, $\mathbf{H}(\mathbf{x}) = \nabla_x \mathbf{g}^T(\mathbf{x})$. The control vector itself is generated on the path, point by point, by a continuous version of the power method for finding the eigenvector of the Hessian matrix, which belongs to the smallest eigenvalue. The first GAD equation for the tangent or velocity vector $\dot{\mathbf{x}}$ of a GAD curve is the sum of the reverse (negative) gradient plus two times an effect of the control vector, $\mathbf{w}$, shorten with the projection on the gradient

$$\dot{\mathbf{x}} = -\,[\mathbf{I} - 2\mathbf{w}\mathbf{w}^T]\,\mathbf{g}(\mathbf{x}), \tag{1}$$

where we assume that the w-vector is normalized. Geometrically, the matrix $[\mathbf{I} - 2\mathbf{w}\mathbf{w}^T]$ is a mirror transformation at the mirror line through the control vector $\mathbf{w}$, the Householder orthogonal transformation [19]. Note that $(\mathbf{w}\mathbf{w}^T)$ is a dyadic product matrix. The control vector $\mathbf{w}(t)$ depends on the curve parameter, t. Finding the variational bases of this model is, in general, difficult. It was proved to be an optimal control problem by Bofill and Quapp [8] based on Zermelo's navigation problem [20], see also Zermelo's navigation problem in Carathéodory's book [21] and Carathéodory's 1926 article [22], which can be seen as precursors of the maximum principle and OCT. They are the foundations attributed to the field of variational studies, which are realized during the last 50 years of the 20th century. In the GAD model is (realistically) assumed that the gradient vector field cannot be controlled and that the control is to execute by the normalized vector, $\mathbf{w}$, which is here generated by the power method to find the eigenvector with the lowest eigenvalue of the Hessian, $\mathbf{H}$; thus

$$\dot{\mathbf{w}} = -[\mathbf{I} - \mathbf{w}\mathbf{w}^T]\,\mathbf{H}\mathbf{w}, \tag{2}$$

the matrix $[I - \mathbf{w}\mathbf{w}^T]$ is the projection orthogonal to the control vector, $\mathbf{w}$.

The equivalence between Zermelo's navigation problem and the GAD system can be seen as follows: in the navigation problem the central question is the present location of a ship in the sea, with a given current distribution characterized by a local dependent vector field. The current is assumed to be independent of a time; it only depends on the position. One desires to find the optimal control of the ship so to reach the destination in the shortest possible time. In the GAD model the gradient vector field of the PES function can be thought of as representing the current of the sea, which we cannot change, whereas the normalized vector w determines the control. The destination is the next SP of the PES. We recall that the set of coupled first-order ordinary differential equations, Eqs. (1) and (2), constitute the fundamental expressions of the GAD model [18, 23]. In the recent reference [8], we use both, a device due to Zermelo as well as the Lagrange multipliers method [20], see also Carathéodory [21]. There the variational nature of the GAD model was proved. Now we will proof that GAD is an example of OCT based on a Legendre construction.

10.2.1   VARIATIONAL NECESSARY CONDITIONS

Let us consider a controlled object, which is represented by a point $\mathbf{x} = (x_1, ..., x_N)^T$ in the N-dimensional configuration space, and we use the system of N non-autonomous differential equations (1) with the control parameters $\mathbf{w} = (w_1, ..., w_N)^T$ which are the components of the $\mathbf{w}$-vector. The values of the $\mathbf{w}$-vector are assumed to be on the unit sphere of $\mathfrak{R}^N$ $\mathbf{w}^T\mathbf{w} = 1$. For the reason we have $N - 1$ control parameters in the GAD model. Now, we have given an initial state that is supposed to be a minimum of the PES, $\mathbf{x}_0 = \mathbf{x}(t_0)$, and a final state, a stationary point of index one of this PES, $\mathbf{x}_{TS} = \mathbf{x}(t_f)$ with a variable t_f depending from the pathway to the TS. We will find a control $\mathbf{w}(t)$-vector on the unit sphere of $\mathfrak{R}^N$ for $t_0 \leq t \leq t_f$ such that it minimizes the transition of the state point $\mathbf{x}$ moving on a GAD path from $\mathbf{x}_0$ to $\mathbf{x}_{TS}$ according to the non-autonomous system of Eq. (1).

In a more precise way, the GAD model consists in the determination of the minimum of the t-parameter, $J[\mathbf{x}_{TS}(\mathbf{w}(t_f))] = t_f - t_0$. A controlled point can be evolved from a given minimum point of the PES, $\mathbf{x}_0 = \mathbf{x}(t_0)$, to a final transition state of this PES, $\mathbf{x}_{TS} = \mathbf{x}(t_f)$. The evolution of the test point is described by the system of ordinary differential equations (1). Note that $\mathbf{x}(t)$

is a N-dimensional vector of the configuration space, where $\mathbf{w}(t)$ is a normalized N-dimension vector of the control parameters. Due to the normalization condition of the $\mathbf{w}(t)$-vector we have $N-1$ control parameters, and for every t these control parameters belong to the unit sphere.

Since t_0 is fixed, the required minimum t_f is merely the minimization of the functional $J[\mathbf{x}_{TS}(\mathbf{w}(t_f))]$ that depends on the chosen $\mathbf{w}(t)$-control normalized vector. Thus the GAD model is a t-parameter-optimal control problem and can be considered as a particular instance of the Mayer problem of the Theory of calculus of variations, and is obtained from these problems by the special form of the functional to be optimized. The GAD, as a case of an optimal control problem, must satisfy the Pontryagin Maximum Principle, which is a necessary condition that generalizes the necessary conditions of Euler and Weierstrass, used in the classical theory of calculus of variations [21, 24], see here the Mayer problem.

From this formulation of the GAD model and following Pontryagin [25] we can formulate the GAD problem as follows: if the pair of vectors $\mathbf{x}(t)$ and $\mathbf{w}(t)$ for $t_0 \le t \le t_f$ is an optimal solution then there exists a nonzero covector-function $\mathbf{y}(t)$ such that $\mathbf{x}(t)$, $\mathbf{w}(t)$ and $\mathbf{y}(t)$ for $t_0 \le t \le t_f$ is a solution to the system of differential equations Eq. (1) and the equation

$$\dot{\mathbf{y}}(t) = \frac{d}{dt}\mathbf{y}(t) = -\left[\nabla_{\mathbf{x}}\dot{\mathbf{x}}(t)^T\right]\mathbf{y}(t) = H(\mathbf{x}(t))\left[\mathbf{I} - 2\mathbf{w}(t)\mathbf{w}(t)^T\right]\mathbf{y}(t) \qquad (3)$$

where $\dot{\mathbf{x}}(t)$ is that given by Eq. (1), and along the solution, for every t a type of maximization with respect to the normalized $\mathbf{w}(t)$-vector is satisfied. The type of maximization will be treated below.

We suppose that the set of admissible $(w_1, ..., w_N)$ values of the control belongs to the unit sphere with $\mathbf{w}^T\mathbf{w} = 1$. Since the curve, $\mathbf{x}(t)$, for $t_0 \le t \le t_f$ is optimal, it traverses at each value of t a plane. The covector-function, $\mathbf{y}(t)$, is orthogonal to this plane, and Eq. (3) represents the "transportation" of this plane along the optimal curve $\mathbf{x}(t)$, for $t_0 \le t \le t_f$. The first order variation on $\mathbf{w}(t_0)$ produces a first order variation on the difference vector, $\mathbf{x}_v(t_f) - \mathbf{x}(t_f)$, being $\mathbf{x}_v(t)$ the curve that starts at $\mathbf{x}_0 = \mathbf{x}_v(t_0) = \mathbf{x}(t_0)$ but with a different $\mathbf{w}(t_0)$-vector of the control. According to Pontryagin et al., [25], the above difference vector $\mathbf{x}_v(t_f) - \mathbf{x}(t_f)$ is orthogonal to the $\mathbf{y}(t_f)$-vector and all the difference vectors generated in this way are orthogonal to the $\mathbf{y}(t_f)$-vector. The full set of difference vectors are in a plane whose normal is the $\mathbf{y}(t_f)$-vector and which contains the point $\mathbf{x}(t_f)$. The plane formula is $\mathbf{y}(t_f)^T[\mathbf{x}_v(t_f) - \mathbf{x}(t_f)] = 0$

and the dimension of the subspace formed by the set of difference vectors, $\{[\mathbf{x}_v(t_f) - \mathbf{x}(t_f)]\}$, is $N - 1$.

Now a problem is to solve. We have only the initial point $\mathbf{x}_0 = \mathbf{x}(t_0)$, and we take an arbitrary initial value $\mathbf{y}_0 = \mathbf{y}(t_0) \neq \mathbf{0}$, and we attempt to solve the system of $2N$ equations, namely, Eqs. (1) and (3) with $3N$ unknowns, $\mathbf{x}(t)$, $\mathbf{y}(t)$, and $\mathbf{w}(t)$ proceeding along an arbitrary extremal GAD curve passing through $\mathbf{x}_0$. Clearly we need another equation for $\mathbf{w}(t)$ such that the above problem can be solved uniquely. If this is possible then the $2N$ unknown parameters are left, $\mathbf{x}(t)$ and $\mathbf{y}(t)$, subject to the system of $2N$ differential Eqs. (1) and (3) and the initial conditions, $\mathbf{x}_0 = \mathbf{x}(t_0)$ and $\mathbf{y}_0 = \mathbf{y}(t_0)$. Because the adjoint Eq. (3) is linear in $\mathbf{y}$ the function $\mathbf{y}(t)$ is defined up to a nonzero constant factor. This property we will use later.

Collecting the two necessary conditions (1) and (3) given above we can put a certain combination of symbols in the scalar-valued function of three arguments

$$\mathbb{H}(\mathbf{x}(t), \mathbf{y}(t), \mathbf{w}(t)) = \dot{\mathbf{x}}(t)^T \mathbf{y}(t) = -\mathbf{g}(\mathbf{x}(t))^T [\mathbf{I} - 2\mathbf{w}(t)\mathbf{w}(t)^T]\, \mathbf{y}(t) \qquad (4)$$

where Eq. (1) has been used. The function enables us to rewrite the system of Eqs. (1) and (3) as a Hamiltonian system and the function of Eq. (4) as a Hamiltonian function,

$$\begin{aligned}\dot{\mathbf{x}}(t) &= \nabla_y \mathbb{H}(\mathbf{x}(t), \mathbf{y}(t), \mathbf{w}(t)),\\ \dot{\mathbf{y}}(t) &= -\nabla_x \mathbb{H}(\mathbf{x}(t), \mathbf{y}(t), \mathbf{w}(t)).\end{aligned} \qquad (5)$$

We have used that the Hamiltonian function is a scalar function, being a product of a matrix that is multiplied by two different vectors from the left and from the right. For this reason this scalar function can be written as

$$\begin{aligned}\mathbb{H}(\mathbf{x}(t), \mathbf{y}(t), \mathbf{w}(t)) &= -\mathbf{y}(t)^T [\mathbf{I} - 2\mathbf{w}(t)\mathbf{w}(t)^T]\, \mathbf{g}(\mathbf{x}(t))\\ &= -\mathbf{g}(\mathbf{x}(t))^T [\mathbf{I} - 2\mathbf{w}(t)\mathbf{w}(t)^T]\, \mathbf{y}(t).\end{aligned} \qquad (5b)$$

So, $\mathbb{H}(\mathbf{x}(t), \mathbf{y}(t), \mathbf{w}(t))$ is a function of $\mathbf{x}$ through the gradient vector, $\mathbf{g}(\mathbf{x}(t))$. Applying Eq. (5b) to this Hamiltonian function we get Eq. (3).

The Hamiltonian function of Eq. (4) is the GAD Hamiltonian, see also Eqs. (18) and (41) of Ref. [8]. Because the optimal GAD curve, $\mathbf{x}(t)$, traverses at this point a plane, (it is not tangent to this plane) to which the $\mathbf{y}(t)$-covector is the normal vector, we obtain

$$\mathbb{H}(\mathbf{x}(t), \mathbf{y}(t), \mathbf{w}(t)) = \mathbf{y}(t)^T \dot{\mathbf{x}}(t) > 0 \tag{6}$$

Hence the plane divides the configuration space, $\mathfrak{R}^N$, in two distinguishable half-spaces, $\mathfrak{R}^N_-$ before the optimal GAD curve, $\mathbf{x}(t)$, intersects the plane, and $\mathfrak{R}^N_+$, after the intersection. In particular every initial control vector, $\mathbf{w}_v(t_0)$, different with respect to the initial optimal control vector $\mathbf{w}(t_0)$, displaces the endpoint of the optimal curve $\mathbf{x}(t_f)$ to another point. The real displacement $\Delta\mathbf{x}(t_f) = \mathbf{x}_v(t_f) - \mathbf{x}(t_f)$ is certainly a nonlinear function. Generally, it stays off the plane, namely, $\mathbf{x}(t_f) + \Delta\mathbf{x}(t_f) \in \mathfrak{R}^N_-$ or $\mathbf{x}(t_f) + \mathbf{x}(t_f) \in \mathfrak{R}^N_+$. From a geometric point of view, the optimal GAD curve, $\mathbf{x}(t)$, $t_0 \le t \le t_f$ consists in the assertion that the displacement of its endpoint due to the differences of the control vector, $\mathbf{w}_v(t)$ and $\mathbf{w}(t)$ falls into the half-space $\mathfrak{R}^N_-$, in other words, $\mathbf{x}(t_f) + \Delta\mathbf{x}(t_f) \in \mathfrak{R}^N_-$ for any variation of the control vector.

Along the above reasoning we can conclude that additional to the necessary conditions on a GAD curve given by the Eqs. (1) and (3) another necessary condition emerges. The scalar product $\mathbf{y}(t_f)^T \Delta\mathbf{x}(t_f)$, where $\mathbf{x}(t_f)$ is obtained from any variation of the initial control vector is nonpositive provided that the covector $\mathbf{y}(t_f)$, which is the normal to the plane, is correctly

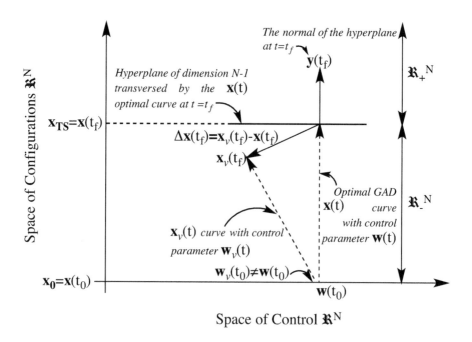

FIGURE 10.1   Scheme of the Pontryagin Maximum Principle applied to the GAD model.

normalized (directed toward the half-space $\mathfrak{R}^N_+$). In other words, $\mathbf{y}(t_f)^T \Delta \mathbf{x}(t_f)$ ≤ 0, for any variation of the initial control vector, see Figure 10.1.

In Figure 10.1, we show the optimal GAD curve, $\mathbf{x}(t)$, which starts at $\mathbf{x}_0 = \mathbf{x}(t_0)$ with the initial control vector, $\mathbf{w}(t_0)$. At $t = t_f$ the curve transverses the hyperplane of dimension $N - 1$. The normal of this hyperplane is the $\mathbf{y}(t_f)$-vector. The other curve $\mathbf{x}_v(t)$ with initial control vector $\mathbf{w}_v(t_0)$ does not achieve the $\mathbf{x}_{TS}$ position at the time $t = t_f$. It is not the optimal curve and for this reason $\mathbf{y}(t_f)^T \Delta \mathbf{x}(t_f) = \mathbf{y}(t_f)^T [\mathbf{x}_v(t_f) - \mathbf{x}(t_f)] \leq 0$, which is the additional necessary condition supporting the Maximum Principle. In the GAD model this additional necessary condition is achieved by maximization of the Hamiltonian given in Eq. (4) with respect to any direction $\mathbf{y}(t)$ under the normalized control vector $\mathbf{w}(t)$.

For the GAD model this necessary condition can be written as

$$\mathbb{H}(\mathbf{x}(t), \mathbf{y}(t), \mathbf{w}(t)) = \max_{\{\mathbf{y}(t)|\mathbf{y}^T\mathbf{y}=1\}} \mathbb{H}(\mathbf{x}(t), \mathbf{y}(t), \mathbf{w}(t))$$
$$= \max_{\{\mathbf{y}(t)|\mathbf{y}^T\mathbf{y}=1\}} \{-\mathbf{y}(t)^T[\mathbf{I} - 2\mathbf{w}(t)\mathbf{w}(t)^T]\mathbf{g}(\mathbf{x}(t))\}, \tag{7}$$

where Eq. (4) has been used. The direction $\mathbf{y}(t)$ that maximizes the Hamiltonian is attained as follows:

We observe that $\mathbf{w}^T(t)$ is a left eigenvector of the matrix $[-\mathbf{I}+2\mathbf{w}(t)\mathbf{w}(t)^T]$ with eigenvalue $+1$. This is the only one positive eigenvalue. All the possible $(N - 1)$ linear independent eigenvectors orthogonal to $\mathbf{w}(t)$ have the $(N-1)$-fold degenerate eigenvalue -1. Thus to get a maximum in Eq. (7), we must chose the direction $\mathbf{y}(t) = \mathbf{w}(t)$. Taking this direction we would get in Eq. (4) the value $\mathbf{w}(t)^T \mathbf{g}(\mathbf{x}(t))$ which still depends on the length of the gradient, the $\mathbf{g}(\mathbf{x}(t))$-vector. To avoid this and to ensure that for any $t_0 \leq t \leq t_f$ the Hamiltonian function of Eq. (4) achieves a constant maximum value, we take

$$\mathbf{y}(t) = \frac{\mathbf{w}(t)}{\mathbf{w}(t)^T \mathbf{g}(\mathbf{x}(t))} \tag{8}$$

for a solution of the maximum direction in Eq. (7). With this solution we get $\mathbb{H}(\mathbf{x}(t), \mathbf{y}(t), \mathbf{w}(t)) = 1$. This is the maximum value, which the GAD Hamiltonian achieves along the optimal GAD curve. Substituting solution (8) in Eq.(3) we obtain a new expression for the $\mathbf{y}(t)$-vector,

$$\dot{\mathbf{y}}(t) = -\frac{\mathbf{H}(\mathbf{x}(t))\mathbf{w}(t)}{\mathbf{w}(t)^T \mathbf{g}(\mathbf{x}(t))}. \tag{9}$$

If we differentiate Eq. (8) with respect to t and if we equating the resulting expression to Eq. (9) we obtain

$$\frac{\mathbf{w}(t)^T \mathbf{g}(\mathbf{x}(t)) \dot{\mathbf{w}}(t) - \mathbf{w}(t) \left[\dot{\mathbf{w}}(t)^T \mathbf{g}(\mathbf{x}(t)) + \mathbf{w}(t)^T \dot{\mathbf{g}}(\mathbf{x}(t)) \right]}{(\mathbf{w}(t)^T \mathbf{g}(\mathbf{x}(t)))^2} = -\frac{\mathbf{H}(\mathbf{x}(t)) \mathbf{w}(t)}{\mathbf{w}(t)^T \mathbf{g}(\mathbf{x}(t))} \quad (10)$$

Multiplying Eq. (10) from the left by $[\mathbf{I} - \mathbf{w}(t)\mathbf{w}(t)^T]$ and using that $\mathbf{w}(t)^T \dot{\mathbf{w}}(t) = 0$ and $[\mathbf{I} - \mathbf{w}(t)\mathbf{w}(t)^T] \mathbf{w}(t) = 0$ we obtain the Eq. (2) which is the searched second GAD equation for $t_0 \leq t \leq t_f$.

With these results we can paraphrase the Pontryagin Maximum Principle [25] applied to the GAD curve model. We assert that the GAD extremal curves are solution of the Hamiltonian system of Eq. (5) or equivalently, Eqs. (1)–(2) and according to the Eq. (7) their points maximize the Hamiltonian function (4) with respect to the control vector $\mathbf{w}(t)$; furthermore according to the Eq. (7), along the GAD extremal curves the control vector for which the Hamiltonian function attains its local maximum is given by Eq.(8). Along the optimal GAD curve the Hamiltonian $\mathbb{H}(\mathbf{x}(t), \mathbf{y}(t), \mathbf{w}(t))$ achieves its maximum value being equal to 1.

10.2.2 EXTENSIONS OF GAD MODEL

The results of the previous section motivate us to propose new curves based on the GAD model and the requirements of the Pontryagin Maximum Principle. With this consideration a curve was proposed such that the tangent is given by the general expression.

$$\dot{\mathbf{x}} = -\mathbf{g}(\mathbf{x}) + f(\Phi, \mathbf{x}, \mathbf{w}) \, \mathbf{w}, \quad (11)$$

where the control vector, $\mathbf{w}$, is assumed to be normalized. The function $f(\Phi, \mathbf{x}, \mathbf{w})$ is a continuous and differentiable function with respect to $\mathbf{x}$, and Φ is a constant. In this case the expression for $\dot{\mathbf{w}}$ is

$$\dot{\mathbf{w}} = -[\mathbf{I} - \mathbf{w}\mathbf{w}^T][\nabla_{\mathbf{x}} f(\Phi, \mathbf{x}, \mathbf{w}) - \mathbf{H}(\mathbf{x}) \, \mathbf{w}]. \quad (12)$$

To delimit the function $f(\Phi, \mathbf{x}, \mathbf{w})$, we should organize that if the control vector, $\mathbf{w}$, is associated with the uphill direction of the evolution of the curve, then the general expression should be one that minimizes the potential energy in the subspace orthogonal to the **w**-vector and maximizes it along the **w** direction, like in the GAD case. An interesting special case is

$$\dot{\mathbf{x}} = -\mathbf{g}(\mathbf{x}) + \Phi(\mathbf{w}^T \mathbf{g}(\mathbf{x}))\mathbf{w}. \tag{13}$$

Here it is $f(\Phi, \mathbf{x}, \mathbf{w}) = (\mathbf{w}^T \mathbf{g}(\mathbf{x}))$ with Φ is greater than one. Taking $\Phi = 2$ we obtain the standard GAD model. The behavior of this kind of curves given in Eq. (13) is reported in Ref. [8]. Notice that if one takes $f(\Phi, \mathbf{x}, \mathbf{w}) = \Phi$ being Φ a positive constant and changing $-\mathbf{g}(\mathbf{x})$ by $\mathbf{g}(\mathbf{x})$ we obtain the original Zermelo problem, namely, $\dot{\mathbf{x}} = \mathbf{g}(\mathbf{x}) + \Phi\mathbf{w}$, where as before it is $\mathbf{w}^T\mathbf{w} = 1$. It is easy to prove that in this case the expression for $\dot{\mathbf{w}}$ coincides with that given in Eq. (2). This possibility has not been yet investigated as a possible curve model to locate transition states, as well as for the representation of reaction paths.

10.3 THE VARIATIONAL NATURE OF OTHERS REACTION PATHS

The reaction path concept is based on the definition of a curve located on the PES, which is monotonely increasing in the potential energy from the reactant minimum to the SP and monotonely decreasing from this point downhill to the product minimum. Many geodesic curves satisfy the definition, and for this reason, there is a large set of curves proposed as a model of the reaction path. The most widely used curves for this purpose are the steepest descent [26–28], the gradient extremals [29–32], and the distinguished reaction coordinate [33] and in its newer version, the Newton trajectory (NT) [34, 35].

The steepest-descent/ascent curve is the curve that at each point follows the gradient of the PES. This curve is variational and extremalizes the integral functional [13, 14]

$$I(\mathbf{x}) = \int_{t_0}^{t} F(\mathbf{x}, \dot{\mathbf{x}}) = \int_{t_0}^{t} \sqrt{\mathbf{g}^T\mathbf{g}}\sqrt{\dot{\mathbf{x}}^T\dot{\mathbf{x}}}dt' = \int_{s_0}^{s} \sqrt{\mathbf{g}^T\mathbf{g}}ds', \tag{14}$$

where s is the arc length and, $F(\mathbf{x}, \dot{\mathbf{x}})$, is a functional homogeneous of degree one with respect to the argument $\dot{\mathbf{x}}$. The second variation indicates that the steepest-descent/ascent curve that joints two minimums of the PES through an SP only minimizes the integral functional of Eq. (14). The special curve, the so-called intrinsic reaction coordinate (IRC) [26, 27], always satisfies the reaction path definition, but the second variation [13] does not imply that the curve is fully located in a deep valley; see a counter example elsewhere [36]. The second variation is not related to the Minimum Energy Path (MEP) condition. The gradient curve of the type IRC is always a reaction path but it can be or not be a MEP. This depends on the shape of the PES [8].

Another type of curve is the so-called gradient extremal (GE) [37]. A GE curve was proposed as a reaction path some time ago [29, 30]. Its definition can be described assuming first that we are on a "valley ground" of the PES with respect to the variations of $\mathbf{x}$ within the equipotential hypersurface, $V(\mathbf{x}) = v = const$. The functional integral to be extremalized is

$$I(\mathbf{x}) = \frac{1}{2} \int_{t_0}^{t} \mathbf{g} \left(\mathbf{x}(t')\right)^T \mathbf{g} \left(\mathbf{x}(t')\right) - \lambda(t')[V(\mathbf{x}(t')) - \nu(t')]dt' . \quad (15)$$

The curve that extremalizes the integral functional of Eq. (15) corresponds to the curve that also satisfies at each point the eigenvalue equation

$$\mathbf{H}(\mathbf{x}(t)) \, \mathbf{g}(\mathbf{x}(t)) = \lambda(t)\mathbf{g}(\mathbf{x}(t)) \quad (16)$$

This curve is the GE. In Eq. (16) the λ is the eigenvalue and the gradient $\mathbf{g}$ is the corresponding eigenvector. The demonstration of the variational nature of this type of curves was formulated in Ref. [17] (see also references therein). Within the theory of the calculus of variations, the GE problem is classified as a Bolza variational problem, which is related to a Lagrangian multipliers problem [21, 38]. As mentioned above the GE curve extremalizes the integral functional given in Eq. (15), and in its evolution it transverses the set of equipotential surfaces, $V(\mathbf{x}) - v = 0$. From a variational point of view it is important to consider that if the GE curve joins two minimums of the PES and if it does not have on this subarc turning points, then this GE curve describes a reaction path belonging to the category of the MEP [17]. A GE curve joining two minimums of the PES minimizes the integral functional of Eq. (15) if and only if it does not have on this subarc turning points. Otherwise other arbitrary curves joining the same two minimums lower the value of the integral functional of Eq. (15) [17]. We conclude that GE curves which join two minimums of the PES and which do not have turning points in between, minimize the functional given in Eq. (15), and satisfy the reaction path definition and the MEP requirement. In this case, the model of GE curves has a direct relation between variationality, reaction path and MEP definition. Unfortunately, the GE curves do not cover all the PES; in other words, a GE curve does not exist at the most points of the PES. Additionally, there can be parts of a reaction valley of a PES where no continuous GE exists [39, 40].

The distinguished reaction coordinate [33] or its new reformulation, the NT [34, 35], is a model curve, which is often used to locate SPs. The curve can be used as representation of a reaction path, again if no turning point emerges. The variational nature of the curves was studied in Ref. [16]. It corresponds to a problem where the functional only depends on the arguments, coordinates, and the parameter that characterizes the curve

$$I(\mathbf{x}_v) = \int_{x_{rc}^0}^{x_{rc}} V(\mathbf{x}_v, x_{rc}')dx_{rc}' , \qquad (17)$$

where the $\mathbf{x}_v$-vector is the coordinate vector $\mathbf{x}$ without the x_{rc} component. It can be shown [16, 41] that the curve which extremalizes the functional integral of Eq. (17) is the curve which satisfies the Branin equation [42]

$$\dot{\mathbf{x}} = \frac{d\mathbf{x}}{dt} = \pm\mathbf{A}(\mathbf{x})\mathbf{g}(\mathbf{x}) , \qquad (18)$$

where $\mathbf{A}(\mathbf{x})$ is the adjoint matrix of the Hessian matrix, $\mathbf{H}(\mathbf{x})$ [41], and the parameter t plays the role of x_{rc}. If $\det(\mathbf{A}(\mathbf{x}))$ is positive definite along the whole NT curve joining two minimums of the PES, then this curve is a reaction path and it has the MEP category, because for this model curve both reaction path and MEP formulation coincide. In addition, in Ref. [16], it is shown that the second variation of the integral functional given in Eq.(17) is positive definite if the NT curve satisfies the inequality $\mathbf{g}(\mathbf{x})^T\mathbf{A}(\mathbf{x})\mathbf{g}(\mathbf{x}) > 0$ which is noting more than the MEP requirement. Thus, for the NT model coincides the minimum variational condition with the MEP condition. However, if the NT has a turning point or a valley-ridged inflection point (VRI), then the minimum variational character is lost [16] and the reaction path and the MEP conditions are not satisfied. An NT curve can start at any point of the PES.

10.4   THE BRANCHING OF A REACTION PATH: VALLEY-RIDGE-INFLECTION POINTS

The analysis of PESs remains an important basis for classifying and understanding the reasons of the mechanisms of chemical reactions as well as their dynamics. It is associated to the concept of the reaction path or to the definition of the minimum energy path on a PES. The chemical reaction may be composed by a number of elementary processes characterizing the mechanism of the reaction. Reaction path bifurcations are omnipresent on PESs; they happen at VRI points already on the PES of very small molecules like H_2O [43], H_2O, H_2S, H_2Se, H_2CO [44], HCN [45, 46], the ethyl cation [47], H_3CO, C_2H_3F [48], and many others. The importance of VRI points for the chemical reactivity is described in the reviews of Ess et al. [49, 50] and Refs. [51, 52]. The intrinsic reaction coordinate (IRC) is a type of a reaction path, which is widely used. However, in "skew," non-symmetric cases this curve usually does not meet a VRI point being nearby [53, 54]. As mentioned in Section 10.3, there is a variety of types of curves that can be used as reaction path models, (a former ansatz was coordinate driving) which can be used in many cases to characterize the reaction path [16, 34]. Sometimes the GE curves [17, 29–31, 55, 56] also appear to form a suitable ansatz for such purposes. Certain NTs describe the valley or cirque structures of a PES, as well as their complements of ridges or cliffs (for the definition of such structures see Ref. [57]. The structures are related to important chemical properties of the PES of the reaction under study [16, 52]. The use of NTs opens the possibility to find and to study VRI points and, in succession, the bifurcation points or the branching points of reaction channels, because the reaction channel-branching is related to the existence of a special class of points of the PES, the VRI points [58. 59]. A VRI point is that point in the configuration space where, orthogonally to the gradient, at least one main curvature of the PES becomes zero [60]. This definition implies that the gradient vector is orthogonal to an eigenvector of the Hessian matrix where its eigenvalue is zero. Thus, at least one PES-direction orthogonally to the gradient is flat. Usually, VRI points represent nonstationary points of the PES. Note that the VRI points are independent of the RP curve model used. They are related to the nature of the PES topography. Normally the VRI points are not related to the branching point of the reaction path curve except for NT curves [16, 59]. So to say, a geometrical indicator of a VRI point is the bifurcation of a singular NT.

The IRC curve is mathematically expressed through an autonomous system of differential equations for the tangent vector describing its evolution [13]. Its solution is unique; due to this fact no bifurcations can occur before reaching the next stationary point after the SP. No branching of PES valleys will be truly described or located by using the IRC curve as an RP type model [60, 61]. It orthogonally traverses the family of levels, the equipotential energy surfaces [13]. Hirsch and Quapp [36] gave an example of a two-dimensional PES where the IRC is going over a skew ridge, however, it does not follow the valley ground nearby, which is here characterized by a GE. The IRC or any other SD curve does not take into account the curvature of the traversed contours in its evolution. In other words it does not give information on the valley floor or ridge character of its pathway. After a change of levels from convex to concave form the IRC curve ceases to be a valley pathway and is actually a merely RP. An early visualization of such an unstable minimum energy path was given by Mezey in Ref. [62], see also Ref. [63]. As explained, the IRC curve traverses in its evolution a family of equipotential energy surfaces. At any point of an SD curve we can define a tangential plane to the equipotential energy surface orthogonally traversed by the SD curve at the point, and the normal of the tangential plane is the gradient vector of the point. All direction vectors contained in the tangential plane are orthogonal to the gradient vector. If at least one of these direction vectors is connected with the curvature zero then we say that the steepest descent curve crosses a Valley-Ridge transition (VRT) point. The curve leaves a valley and enters a ridge region of the PES or vice-versa. The VRT points are the border between valley- and ridge-regions. The concept of a VRT point is much more general than the VRI point concept. In fact a VRI point is a special case of a VRT point. In the general VRT situation, the gradient vector is not orthogonal to the set of eigenvectors of the Hessian matrix. This is the most general behavior. The zero curvature of the PES along the level line or equipotential energy surface at the VRT point comes from a suitable linear combination of the eigenvectors with their eigenvalues of the Hessian matrix. A manifold of points with these features exist on a PES. They are border points between quasi-convex valley regions and ridges. NTs there have a turning point. So to say, a turning point of an NT is the geometrical indicator of a VRT point, see Ref. [41]. An example is given in Figure 10.2. This modified BQC-PES is near to the original reported in Ref. [17]. Its equations is

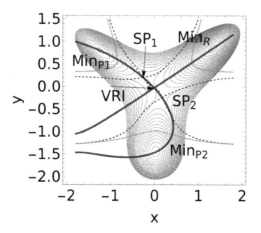

FIGURE 10.2   A modified BQC-PES with three minimums and two saddle points of index one. The green line corresponds to the condition $Det(\mathbf{H}(\mathbf{x})) = 0$. It indicates a change of the sign of an eigenvalue. The dashed line is the convexity border of the PES, namely, $Det[\mathbf{S}^T(\mathbf{x}) \mathbf{H}(\mathbf{x})\mathbf{S}(\mathbf{x})] = 0$, where the matrix, $\mathbf{S}(\mathbf{x})$, is formed by the set of all orthonormalized vectors all of them are orthogonal to the $\mathbf{g}(\mathbf{x})$ vector. The fat line corresponds to a singular NT curve, which crosses the VRI point.

$$V(x, y) = 1000/3(y^3 - 3yx^2) - 30(y + 3x) + 250((y + 0.7/4)^4 + x^4). \quad (19)$$

ACKNOWLEDGMENTS

Financial support from the Spanish Ministerio de Economía y Competitividad, Project CTQ2016-76423-P, is acknowledged.

KEYWORDS

- maximum principle
- minimum energy path
- optimal control
- reaction path
- saddle point
- transition state

REFERENCES

1. Brumer, P., & Shapiro, M., (1986). *Chem. Phys. Lett., 126,* 541.
2. Tannor, D. J, Kosloff, R., & Rice, S. A., (1986). *J. Chem. Phys., 85,* 5805.
3. Potter, E. D., Herek, J. L., Pedersen, S., Liu, Q., & Zewail, A. H., (1992). *Nature, 355,* 66.
4. Zewail, A. H., (2000). *J. Phys. Chem. A, 104,* 6660.
5. Peirce, A. P., Dahleh, v M. A., & Rabitz, H., (1988). *Phys. Rev. A: At., Mol., Opt. Phys., 37,* 4950.
6. Zhu, W., & Rabitz, H., (1998). *J. Chem. Phys., 109,* 385.
7. Rani, K., & Rao, V., (1999). *Bioprocess Eng., 21,* 77.
8. Bofill, J. M., & Quapp, W., (2016). *Theor. Chem. Acc., 135,* 11.
9. McWeeny, R., (1992). *Methods of Molecular Quantum Mechanics, 2nd ed.*, Academic Press, Oxford.
10. Carbó, R., & Riera, J. M., (1976). A General SCF Theory. *Lecture Notes in Chemistry,* Vol. 5, Springer, New York.
11. Bofill, J. M., Bono, H., & Rubio, J., (1998). *J. Comput. Chem., 19,* 368.
12. Heidrich, D., Kliesch, W., & Quapp, W., (1991). *Properties of Chemically Interesting Potential Energy Surfaces*, Springer, Berlin, Heidelberg.
13. Crehuet, R., & Bofill, J. M., (2005). *J. Chem. Phys., 122,* 234105.
14. Aguilar-Mogas, A., Crehuet, R., Giménez, X., & Bofill, J. M., (2007). *Mol. Phys., 105,* 2475.
15. Aguilar-Mogas, A., Giménez, X., & Bofill, J. M., (2008). *J. Chem. Phys., 128,* 104102.
16. Bofill, J. M., & Quapp, W., (2011). *J. Chem. Phys., 134,* 074101.
17. Bofill, J. M., Quapp, W., & Caballero, M., (2012). *J. Chem. Theory Comput., 8,* 927.
18. Zhou, W. E, X., (2011). *Nonlinearity, 24,* 1831.
19. Householder, A. S., (1958). *J. ACM, 5,* 339.
20. Zermelo, E., (1931). *Z. Angew. Math. Mech., 11,* 114.
21. Carathéodory, C., (1935). *Variationsrechnung und Partielle Differentialgleichungen erster Ordnung* B. G. Teubner, Leipzig und Berlin.
22. Carathéodory, C., (1926). *Acta Math., 47,* 199.
23. Bofill, J. M., Quapp, W., & Caballero, M., (2013). *Chem. Phys. Lett., 583*(0), 203.
24. Vapnyarskii, I. B., *Encyclopedia of Mathematics.*
25. Pontryagin, L. S., Boltyanski, V.G., Gamkrelidze., R.V., & Mishechenko, E. F., (1962). *The Mathematical Theory of Optimal Processes* John Wiley & Sons, NY.
26. Fukui, K., (1970). *J. Phys. Chem., 74,* 4161.
27. Quapp, W., & Heidrich, D., (1984). *Theor. Chim. Acta., 66,* 245.
28. Hratchian, H., Frisch, M. J., & Schlegel, H. B., (2010). *J. Chem. Phys., 133,* 224101.
29. Basilevsky, M., & Shamov, A., (1981). *Chem. Phys., 60,* 347.
30. Hoffmann, D. K., Nord, R. S., & Ruedenberg, K., (1986). *Theor. Chim. Acta., 69,* 265.
31. Quapp, W., (1989). *Theoret. Chim. Acta., 75,* 447.
32. Schlegel, H. B., (1992). *Theor. Chim. Acta., 83,* 15.
33. Rothman, M. J., & Lohr, L. L., (1980). *Chem. Phys. Lett., 70,* 405, 18.
34. Quapp, W., Hirsch, M., Imig, O., & Heidrich, D., (1998). *J. Comput. Chem., 19,* 1087.
35. Anglada, J. M., Besalú, E., Bofill, J. M., & Crehuet, R., (2001). *J. Comput. Chem., 22,* 387.
36. Hirsch, M., & Quapp, W., (2004). *Chem. Phys. Lett., 395*(1–3), 150.

37. Rowe, D. J., & Ryman, A., (1982). *J. Math. Phys., 23,* 732.
38. Bliss, G. A., (1946). *Lectures on the Calculus of Variations,* Chicago University Press, Chicago, IL.
39. Quapp, W., (2003). *J. Theoret. Comput. Chem., 2,* 385.
40. Quapp, W., (1994). *J. Chem. Soc., Faraday Trans., 90,* 1607.
41. Hirsch, M., & Quapp, W., (2004). *J. Math. Chem., 36,* 307.
42. Branin, F. H., (1972). *IBM J. Res. Develop., 16,* 504.
43. Hirsch, M., Quapp, W., & Heidrich, D., (1999). *Phys. Chem. Chem. Phys., 1,* 5291.
44. Quapp, W., & Melnikov, V., (2001). *Phys. Chem. Chem. Phys., 3,* 2735.
45. Quapp, W., & Schmidt, B., (2011). *Theor. Chem. Acc., 128,* 47.
46. Schmidt, B., & Quapp, W., (2012). *Theor. Chem. Acc., 132,* 1305.
47. Quapp, W., & Heidrich, D., (2002). *J. Mol. Struct., (THEOCHEM), 585*(1–3), 105. DOI 10.1016/S0166-1280(02)00037-4.
48. Kumeda, Y., & Taketsugu, T., (2000). *J. Chem. Phys., 113,* 477.
49. Ess, D. H., Wheeler, S. E., Iafe, R., Xu, L., Çelebi-Ölçüm, N., & Houk, K., (2008). *Angew. Chem. Int. Ed., 47,* 7592.
50. Bakken, V., Danovich, D., Shaik, S., & Schlegel, H. B., (2001). *J. Am. Chem. Soc., 123,* 130.
51. Quapp, W., Bofill, J. M., & Aguilar-Mogas, A., (2011). *Theor. Chem. Acc., 129,* 803.
52. Quapp, W., & Bofill, J. M., (2012). *J. Math. Chem., 50*(5), 2061, 19.
53. Quapp, W., (2015). *J. Chem. Phys., 143,* 177101.
54. Harabuchi, Y., Ono, Y., Maeda, S., & Taketsugu, T., (2015). *J. Chem. Phys., 143,* 014301.
55. Sun, J. Q., & Ruedenberg, K., (1993). *J. Chem. Phys., 98,* 9707.
56. Bondensgård, K., & Jensen, F., (1996). *J. Chem. Phys., 104,* 8025.
57. Atchity, G. J., Xantheas, S. S., & Ruedenberg, K., (1991). *J. Chem. Phys., 95,* 1862.
58. Valtazanos, P., & Ruedenberg, K., (1986). *Theor. Chim. Acta., 69,* 281.
59. Quapp, W., Hirsch, M., & Heidrich, D., (1998). *Theor. Chem. Acc., 100*(5–6), 285.
60. Quapp, W., (2004). *J. Mol. Struct., (THEOCHEM), 95,* 695–696.
61. Quapp, W., Hirsch, M., & Heidrich, D., (2004). *Theor. Chem. Acc., 112,* 40.
62. Mezey, P., (1987). *Potential Energy Hypersurfaces,* Elsevier, Amsterdam.
63. Mezey, P., (1980). *Theor. Chem. Acc., 54,* 95.

CHAPTER 11

KINETIC STABILITY OF NOBLE GAS ATOMS WITHIN SINGLE-WALLED ALN AND GAN NANOTUBES

DEBDUTTA CHAKRABORTY and PRATIM KUMAR CHATTARAJ

*Department of Chemistry and Center for Theoretical Studies,
Indian Institute of Technology, Kharagpur – 721302, West Bengal,
India, Tel.: +91-3222-283304, Fax: +91-3222-255303,
E-mail: pkc@chem.iitkgp.ernet.in*

CONTENTS

11.1 INTRODUCTION

Noble gases (Ng), members of the group 18 of the periodic table, constitute an enigmatic family of elements. Due to the closed shell electronic configuration of these elements, they exhibit remarkable apathy towards showing chemical reactivity. Extreme physicochemical conditions can, however, facilitate the reactions of noble gas atoms with different ligands. Particularly noteworthy in this direction are the heavier noble gas atoms

[1] whose substantially high polarizability and low ionization energy allow them to display some chemical reactivity as compared to their lighter congeners. Of late, inspired by successful experimental syntheses [2–9] of several Ng-containing compounds, noble gas chemistry has received a major thrust from experimentalists and theoreticians alike [10–15]. From a theoretical point of view, the pertinent question is of course to characterize and rationalize the nature of binding of Ng-atoms with different chemical moieties. One probable physical condition that has emerged as a veritable testing ground of the reactivity of noble gas atoms is the effect of geometrical confinement. It has been demonstrated experimentally that cage-like compounds like fullerene [16–18] or dodecahedrane [19] can encapsulate Ng-atoms. Following this idea, theoretical analysis of analogous situations has been carried out in detail in the cases of different cage-like moieties [20–25]. Apart from the obvious goal to achieve some enhanced degree of reactivity in the cases of different Ng-atoms in an encapsulated state and thereby revealing important physical insights into the behavior of Ng-atoms, such a study bears important consequences from a practical point of view. Noble gases are industrially important elements and they have been identified as having importance in different domains such as anesthetics [26, 27], insulation [28], carrier gases [29], excimer lasers [30, 31], refrigerants [32], illuminating agents [33, 34], etc. On the other hand, radioactive isotopes of noble gas atoms such as [85]Kr and [133]Xe [35], which are generated during nuclear fission and may enter the atmosphere during reprocessing of the spent fuel possess great danger. All the aforementioned cases have one important prerequisite and that is to find a suitable host that may potentially sequestrate noble gas atoms at physically relevant temperature and pressure. This task is, however, far from being a trivial one. An inherent problem in dealing with such a problem is related to the chemical inertness of the Ng atoms and as a result obtaining a reasonably strong binding in between Ng atoms and the host moiety may not be feasible. The situation is further aggravated by the low natural abundance of Ng atoms in earth's atmosphere. Except Ar which constitutes roughly 1% of air, all other noble gases are scarcely present. As a result the quest for obtaining a viable host moiety that may potentially encapsulate different Ng atoms has gained a major recent research interest.

Ever since the experimental synthesis of carbon nanotube [36], carbon and non-carbon based nanostructures have attained wide spread attention. Particularly noteworthy in this direction are the group III–V compounds, group III nitrides being a special case [37, 38]. Herein, we seek to investigate the noble gas (Ng = He_{1-2}, Ne_{1-2}, Ar_{1-2}, Kr_{1-2}, Xe_{1-2}) encapsulating ability of AlN and GaN single-walled nanotubes. For this purpose we have considered the aforementioned noble gas atoms in an endohedrally confined geometry inside the nanotubes and carried out density functional theory (DFT) based calculations. The choice of the considered nanotubes is dictated by the fact that these geometries allow us to 'compress' the confined Ng atoms to a significant extent than their bigger analogues. It is expected that such a consideration would enable the Ng atoms to be more reactive and thereby initiating the process of binding in between themselves as well as with the inner walls of the nanotubes. The observed results obtained from theoretical calculations have been rationalized by carrying out detailed thermochemical analysis in concomitance with global reactivity descriptors within the purview of conceptual density functional theory (CDFT) [39–42]. To provide further insights into the nature of interaction in between Ng atoms as well as Ng-AlN/GaN surface, electron density analysis within the realms of atoms-in a-molecule (AIM) theory and energy decomposition analysis (EDA) have been performed. In order to confirm the dynamic stability of the observed minimum energy geometries of the concerned Ng-encapsulated nanotubes, an ab initio molecular dynamics simulation at 298 K temperature has been carried out.

11.2   COMPUTATIONAL DETAILS

The reported geometries have been modeled by making use of graphical software GaussView 5.0.8 [43]. Finite chunks of the AlN and GaN single walled nanotubes with molecular formulae $Al_{36}N_{36}H_{12}$ and $Ga_{30}N_{30}H_{12}$, respectively have been considered herein. The edge atoms have been saturated with H atoms in order to alleviate any boundary problem. Because of the larger atomic radius of the Ga atom than that of Al, the GaN nanotube is larger in diameter than AlN nanotube. Keeping in mind a reasonable balance in between computational cost and accuracy, we have employed dispersion-corrected hybrid DFT level of theory for the present study. Unconstrained geometry optimization in conjunction with harmonic vibrational frequency calculation have been done at wB97X-D/6–311G(d, p) level of theory for He_{1-2}/Ne_{1-2}/Ar_{1-2}/Kr_{1-2},@AlN/GaN. For the cases of Xe_{1-2}@AlN/GaN, in order to take into account the relativistic effects for Xe atoms, an effective core potential as described by a LANL2DZ basis set has been used for all the Xe atoms. Rest of the atoms has been treated by the 6–311G(d, p) basis set and geometry optimization followed by harmonic vibrational frequency calculation performed using the wB97X-D functional. All the structures reported herein correspond to minima on the concerned potential energy surfaces. In order to determine the extent of electronic charge accumulated/depleted on each atomic site, natural population analysis (NPA) has been carried out. We have also determined the Wiberg bond index (WBI) in order to ascertain the bond order existing in between Ng-Ng and Ng-AlN/GaN surface. Global reactivity descriptors such as hardness (h) and mean polarizability (α) have been computed according to the following expressions:

$$\eta = I - A \tag{1}$$

$$\alpha = \frac{1}{3}\left(\alpha_{xx} + \alpha_{yy} + \alpha_{zz}\right) \tag{2}$$

Here I and A represent ionization potential and electron affinity, respectively, whereas $\alpha_{xx/yy/zz}$ are the diagonal components of the polarizability tensor. Applying Koopmans' theorem I and A have been evaluated as $I = -E_{HOMO}$ and $A = -E_{LUMO}$. All the above-mentioned calculations have been performed using Gaussian 09 software package [44]. We have characterized the nature of interaction in between the concerned moieties by making use of AIM theory [45]. For this purpose, Multiwfn [46] software has been used. The electron density ($\rho(r_c)$) at the bond critical point (BCP) and the Laplacian of electron density ($\nabla^2 r(r_c)$) along the respective BCP are the two important descriptors which can categorize the observed nature of interaction into different forms viz. covalent or non-covalent. Additionally local kinetic energy density ($G(r_c)$), local potential energy density ($V(r_c)$) and electron localization function (ELF) at the BCP have been computed. The local electron energy density ($H(r_c)$) has been evaluated by adding $G(r_c)$ and $V(r_c)$ terms at the concerned BCP. The EDA has been performed with ADF software [47, 48] at the revPBE-D3/TZ2P level of theory by importing the above stated minimum energy geometries. For the cases of Xe_{1-2}, zero order regular approximation (ZORA) has been used in order to take into account the relativistic effect at the revPBE-D3/TZ2P level. The ab initio molecular dynamics study has been performed through an atom-centered density matrix propagation (ADMP) method [49] as implemented in Gaussian 09 software. The above-mentioned minimum energy geometries have been simulated up to 500 fs time scale at 298 K temperature at the B3LYP-D2/6-31G(d, p) level for the cases of $He_2/Ne_2/Ar_2/Kr_2@AlN/GaN$ whereas for $Xe_2@AlN/GaN$ the employed level is B3LYP-D2/LANL2DZ. Boltzmann distribution at 298 K has been used in order to assign the initial kinetic energies. Velocity scaling thermostat has been employed to preserve the temperature throughout the temporal evolution. In order to assign initial mass weighted Cartesian velocities, default random number generator seed has been used.

11.3   RESULTS AND DISCUSSION

Let us firstly consider the minimum energy geometries of calculated $Ng_{1-2}@$ AlN nanotubes (Figure 11.1). The AlN nanotube in its pristine form has an approximate diameter of 5.40 Å whereas upon encapsulation of Ng_2, it undergoes minute expansion in order to accommodate the guest atoms. The extent of increase in diameter of the nanotube is clearly dictated by the atomic radii of the guest atoms and the values are approximately 5.41,

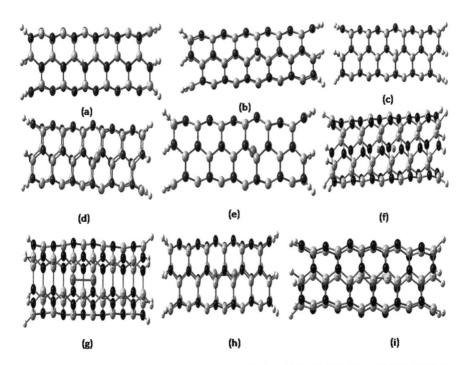

FIGURE 11.1   Optimized geometries of (a) He@AlN, (b) Ne@AlN, (c) Ar@AlN, (d) Kr@ AlN, (e) Xe@AlN, (f) $He_2@AlN$, (g) $Ar_2@AlN$, (h) $Kr_2@AlN$ and (i) $Xe_2@AlN$, respectively.

5.42, 5.44, 5.45 Å for the cases of $He_2/Ar_2/Kr_2/Xe_2@AlN$, respectively. It should be noted that the nanotube diameter is not uniform in case of the Ng_2 encapsulated state and the middle portion is slightly more elongated than the edge. The Ng atoms align themselves along the tube axis in order to minimize the extent of Coulomb repulsion with the constituent atoms of the nanotube. A different orientation viz. perpendicular to the tube axis, seems to be prohibitively costly since that will increase the repulsive forces manifold. The Ng-Ng distances at the encapsulated state have been presented in Table 11.1. Clearly, this particular geometrical alignment forces the Ng atoms to stay in close proximity of each other. For the corresponding

TABLE 11.1   Free Energy Change (ΔG, kcal/mol) and Reaction Enthalpy (ΔH, kcal/mol) at 298 K for the Process: $Ng_{1-2} + AlN/GaN \rightarrow Ng_{1-2}@AlN/GaN$; Ng-Ng Distance in the Encapsulated State (Å); NPA (Q_K) Charge (a.u.) on Ng Atoms at the Encapsulated State; Wiberg Bond Index (WBI_{tot}) for Ng Atoms at the Encapsulated State; Hardness (h), Polarizability (α) (in au^3) for the $Ng_{1-2}@AlN/GaN$ Moieties [52]

System	ΔG	ΔH	R_{Ng-Ng}	Q_K	WBI_{tot}	h	a
He@AlN	20.22	12.86	–	0.01	0.12	8.13	888.77
Ne@AlN	26.87	18.13	–	0.10	0.19	8.09	889.46
Ar@AlN	76.06	66.25	–	0.23	0.47	8.02	897.96
Kr@AlN	99.00	89.30	–	0.31	0.68	8.03	905.70
Xe@AlN	141.20	130.16	–	0.42	0.81	7.60	894.29
He@GaN	17.03	8.27	–	0.05	0.11	7.53	861.58
Ne@GaN	20.97	10.98	–	0.10	0.16	7.47	862.07
Ar@GaN	60.64	51.17	–	0.20	0.42	7.36	870.65
Kr@GaN	81.59	71.74	–	0.28	0.60	7.29	877.68
Xe@GaN	132.19	122.36	–	0.47	0.88	7.18	884.47
He_2@AlN	39.40	26.09	2.47	0.06, 0.06	0.12, 0.13	8.20	889.83
Ar_2@AlN	148.86	132.31	3.23	0.23, 0.23	0.47, 0.48	8.06	907.49
Kr_2@AlN	195.70	179.38	3.51	0.32, 0.31	0.69, 0.69	7.99	922.50
Xe_2@AlN	309.29	293.13	3.70	0.51, 0.51	0.96, 0.96	7.63	940.17
He_2@GaN	30.44	18.11	2.47	0.05, 0.05	0.10, 0.11	7.55	861.52
Ne_2@GaN	36.67	23.72	3.02	0.08, 0.08	0.17, 0.17	7.46	864.06
Ar_2@GaN	119.93	103.95	3.31	0.20, 0.20	0.42, 0.42	7.29	879.94
Kr_2@GaN	162.90	145.99	3.52	0.28, 0.28	0.60, 0.61	7.16	892.96
Xe_2@GaN	261.48	244.89	3.81	0.46, 0.46	0.90, 0.89	6.98	908.81

situation in the cases of Ng@AlN species, we note that except for the case of Xe@AlN where the nanotube undergoes some distortion (diameter of 5.42 Å), all other noble gases accommodate themselves inside the host moiety without causing any noticeable distortion to the AlN nanotube. The corresponding changes in Gibbs free energy (ΔG) and reaction enthalpy (ΔH) associated with the formation of endohedrally confined Ng_{1-2}@AlN moieties at 298 K temperature and 1 atmosphere pressure have been provided in Table 11.1. It becomes clear that all the Ng_{1-2} encapsulation processes are thermodynamically unfavorable. The noble gas atoms, upon encapsulation inside the nanotubes, loose significant extent of entropy as they are forced to stay in a particular geometrical alignment thereby curtailing their translational, rotational and vibrational degrees of freedom. The thermochemical situation is further aggravated by the fact that all the Ng_{1-2} encapsulation processes are endothermic in nature. One probable cause behind such an observation could be that some amount of energy needs to be provided in order to account for the structural deformation of the nanotubes, necessary to accommodate the guest atoms. The fact that this steric factor could be an important reason behind our observed thermochemistry data is validated by the fact that as we go from He_{1-2} to Kr_{1-2}, the associated process of Ng_{1-2} encapsulation becomes energetically more demanding. We preclude ourselves from a quantitative comparison in between the cases of Xe_{1-2}@AlN and other Ng_{1-2}@AlN moieties since Xe atoms have been represented by a different basis set. However, we seek a qualitative understanding of the basic physical picture that is occurring and henceforth state some qualitative information. It is useful to consider the global reactivity descriptors at this point in order to shed some light on the aforementioned thermochemical data. The Ng_{1-2}@AlN moieties become more polarizable and less hard as we move from He_2 to Xe_2. Therefore, taking note of the popular electronic structure principles [39–42] such as maximum hardness principle (MHP) and minimum polarizability principle (MPP) we may state that as the size of the encapsulated Ng atoms increases, the AlN nanotubes become more reactive and thus susceptible to further physical or chemical transformation. In such a situation, the constituent atoms of the nanotube should certainly accumulate a significantly different electronic charge distribution as compared to its pristine form. To verify this point we consider the NPA charges of Ng atoms as well as Al and N atoms. Maximum influence of this asymmetric charge concentration/depletion should be present near the vicinity of the encapsulated atoms. Therefore, we take a note of the NPA charges of the Al and N atoms which are present in close proximity of the Ng atoms (NPA values are presented in Table 11.1). Firstly it becomes clear that all the Ng atoms act as Lewis bases and donate some electron density to the nanotube. The extent of such electron density transfer becomes more prominent as we move from He_{1-2} to Xe_{1-2}. In the pristine form of AlN nanotube, the NPA charges on Al and N atoms towards the middle of the tube are 1.97 and -1.97, respectively. For He@AlN, the corresponding values are 1.97 and -1.98 for Al and N atoms, respectively. Similar values are noted for Ne@AlN. For Ar/Kr@AlN, the concerned values do change (varying in the range of 1.95–1.98 and -1.99 for Al and N atoms, respectively). For Xe@AlN, the corresponding values are 2.04 to 2.06 and -2.08 to -2.09, respectively. In case of He_2@AlN, the corresponding values are 1.98 and -1.99. For Ar_2/Kr_2@AlN, the concerned values vary in the range of 1.98 to 1.95 and -1.99 to -2.00; 1.94 to 1.98 and -1.99 to -2.00, respectively. For the case of Xe_2@AlN the observed values are 1.91 to 1.97 and -2.01 to -2.00. It becomes clear that the heavier noble gas atoms particularly, can bring about significant changes in the charge concentration around the nanotube constituents. In order to characterize this interaction we invoke the WBI values. The individual Ng-Ng WBI values are small and of the order of 10^{-1} to 10^{-2} signifying a closed shell type inter-action. However, a marked increase is observed upon considering the Ng-N and Ng-Al interactions. Individual Ng atoms interact with multiple centers. The total WBI values for each Ng atom have been presented in Table 11.1. It shows that all the Ng atoms interact strongly with the nanotube thereby rendering them reactive than usual. In order to have an idea of the orbitals that get involved in these interaction, the natural electronic configurations of the Ng atoms in the encapsulated state have been presented in Table 11.2. Particularly noteworthy are the heavier Ng atoms that transfer electron den-sity principally from their valence shell s and p orbitals to the nanotube.

We now turn our attention to the cases of Ng_{1-2}@GaN (Figure 11.2). Here the nanotube is having an approximate diameter of 5.58 Å in its pris-tine form therefore rendering a little more space for the Ng_2 atoms to reside. The approximate diameters of the tube in presence of noble gas atoms are: 5.58, 5.59, 5.59, 5.60 and 5.61 Å, respectively, for $He_2/Ne_2/Ar_2/Kr_2/Xe_2$@ GaN. The same values in the cases of singly occupied noble gas atoms are: 5.58, 5.58, 5.59, 5.59, 5.59 Å, respectively, for He/Ne/Ar/Kr/Xe@GaN. As in the previous case, here also the Ng_2 atoms orient themselves in a parallel direction with respect to the tube axis. Gibbs free energy (ΔG) and reaction enthalpy (ΔH) changes associated with the complex formation processes are

TABLE 11.2 The Valence Shell Orbital Populations of Ng Atoms in the Ng-Bound Nanotubes

System	Natural electron configuration
He@AlN	$1s^{(1.94)}$
Ne@AlN	$2s^{(1.97)}2p^{(5.94)}$
Ar@AlN	$3s^{(1.91)}3p^{(5.85)}3d^{(0.01)}$
Kr@AlN	$4s^{(1.89)}4p^{(5.78)}4d^{(0.02)}$
Xe@AlN	$5s^{(1.84)}5p^{(5.74)}$
He@GaN	$1s^{(1.95)}$
Ne@GaN	$2s^{(1.97)}2p^{(5.95)}$
Ar@GaN	$3s^{(1.92)}3p^{(5.87)}3d^{(0.01)}$
Kr@GaN	$4s^{(1.90)}4p^{(5.80)}4d^{(0.01)}$
Xe@GaN	$5s^{(1.86)}5p^{(5.68)}$
He$_2$@AlN	$1s^{(1.94)}$ and $1S^{(1.94)}$
Ar$_2$@AlN	$3s^{(0.37)}3p^{(0.64)}3d^{(0.02)}5p^{(0.01)}$ and $3s^{(0.55)}3p^{(0.72)}3d^{(0.02)}$
Kr$_2$@AlN	$4s^{(1.88)}4p^{(5.78)}4d^{(0.02)}$ and $4s^{(1.88)}4p^{(5.78)}4d^{(0.02)}$
Xe$_2$@AlN	$5s^{(1.83)}5p^{(5.66)}$ and $5s^{(1.83)}5p^{(5.66)}$
He$_2$@GaN	$1s^{(1.95)}$ and $1s^{(1.94)}$
Ne$_2$@GaN	$2s^{(1.97)}2p^{(5.94)}$ and $2s^{(1.97)}2p^{(5.94)}$
Ar$_2$@GaN	$3s^{(1.92)}3p^{(5.87)}3d^{(0.01)}$ and $3s^{(1.92)}3p^{(5.87)}3d^{(0.01)}$
Kr$_2$@GaN	$4s^{(1.90)}4p^{(5.80)}4d^{(0.01)}$ and $4s^{(1.90)}4p^{(5.80)}4d^{(0.01)}$
Xe$_2$@GaN	$5s^{(1.85)}5p^{(5.68)}$ and $5s^{(1.85)}5p^{(5.68)}$

thermodynamically unfavorable (Table 11.1). The concerned global reactivity descriptors reveal that as the atomic radii of the Ng atoms increase, the Ng$_2$@GaN moieties become more reactive. The Ng atoms upon insertion into the GaN nanotube, alters the atomic charge distribution to a significant extent as evidenced from NPA values. In its pristine form, Ga and N atoms towards the middle of the tube are having NPA charges of 1.80 and −1.80, respectively. The same quantities in Ng$_2$ inserted states acquire the values: 1.80 to 1.82 and −1.83 to −1.82; 1.81 to 1.80 and −1.81; 1.78 and −1.83 to

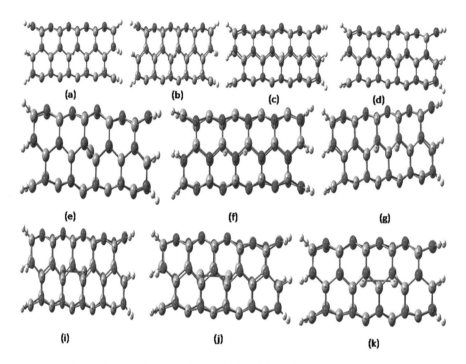

FIGURE 11.2 Optimized geometries of (a) He@GaN, (b) Ne@GaN, (c) Ar@GaN, (d) Kr@GaN, (e) Xe@GaN, (f) He$_2$@GaN, (g) Ne$_2$@GaN, (i) Ar$_2$@GaN, (j) Kr$_2$@GaN and (k) Xe$_2$@GaN, respectively.

−1.82; 1.81 to 1.79 and −1.84 to −1.82; 1.81 to 1.77 and −1.85 to −1.82, respectively, for He$_2$/Ne$_2$/Ar$_2$/Kr$_2$/Xe$_2$@GaN moieties. Corresponding values in the cases of Ng@GaN are as follows: 1.81 and −1.81; 1.80 and −1.81; 1.78 and −1.82; 1.77 to 1.81 and −1.81 to −1.82; 1.75 and −1.81, respectively, for He/Ne/Ar/Kr/Xe@GaN moieties. It becomes clear that a significant extent of electron density gets transferred to the nanotube from the Ng atoms (from the s and p orbitals). As a result the Ng atoms attain unusually high total WBI values as compared to their free analogues.

In order to characterize the nature of interaction prevalent in between the Ng atoms themselves and with the nanotube constituents, we now present the AIM analysis (Table 11.3). It should be noted that all the Ng atoms remain multiply coordinated and we present the topological descriptors at the concerned BCPs for some representative ones. It is prudent at this point to mention the well-known conditions governing the AIM analysis. Generally if $\nabla^2 r(r_c) > 0$ and $H(r_c) < 0$, then the bonding is of partially covalent nature

TABLE 11.3 Electron Density Descriptors (in a.u.) at the Bond Critical Points (BCP)

Systems	BCP	$\rho(r_c)$	$\nabla^2 r(r_c)$	$H(r_c)$	ELF
He@AlN	He-N	0.01	0.04	0.00	0.03
Ne@AlN	Ne-N	0.02	0.06	0.00	0.04
Ar@AlN	Ar-N	0.02	0.09	0.00	0.07
Kr@AlN	Kr-N	0.02	0.08	0.00	0.09
Xe@AlN	Xe-N	0.03	0.10	−0.01	0.10
He@GaN	He-N	0.01	0.03	0.00	0.03
Ne@GaN	Ne-N	0.01	0.05	0.00	0.03
Ar@GaN	Ar-N	0.02	0.07	0.00	0.06
Kr@GaN	Kr-N	0.02	0.07	0.00	0.06
Xe@GaN	Xe-N	0.02	0.07	0.00	0.09
He$_2$@AlN	He-He,	0.01	0.02	0.00	0.01
	He-N	0.00	0.04	0.00	0.03
Ar$_2$@AlN	Ar-Ar,	0.01	0.04	0.00	0.02
	Ar-N	0.03	0.09	0.00	0.07
Kr$_2$@AlN	Kr-Kr, Kr-N, Kr-Al	0.01	0.03	0.00	0.03
		0.02	0.05	−0.01	0.11
		0.01	0.04	0.00	0.02
Xe$_2$@AlN	Xe-Xe, Xe-N, Xe-Al	0.01	0.03	0.00	0.05
		0.03	0.05	−0.01	0.14
		0.03	0.05	−0.01	0.14
He$_2$@GaN	He-He,	0.01	0.02	0.00	0.01
	He-N	0.01	0.04	0.00	0.03
Ne$_2$@GaN	Ne-Ne,	0.00	0.02	0.00	0.00
	Ne-N	0.02	0.06	0.00	0.04
Ar$_2$@GaN	Ar-Ar, Ar-N, Ar-Ga	0.01	0.04	0.00	0.02
		0.02	0.07	0.00	0.07
		0.02	0.08	0.00	0.06

TABLE 11.3 (Continued)

Systems	BCP	$\rho(r_c)$	$\nabla^2 r(r_c)$	$H(r_c)$	ELF
Kr$_2$@GaN	Kr-Kr, Kr-N, Kr-Ga	0.01	0.03	0.00	0.02
		0.02	0.07	0.00	0.09
		0.03	0.08	0.00	0.08
Xe$_2$@GaN	Xe-Xe, Xe-N, Xe-Ga	0.01	0.03	0.00	0.04
		0.03	0.08	0.00	0.09
		0.03	0.08	−0.01	0.11

[50, 51]. On the other hand the ELF represents a reliable descriptor, significantly high value of which quantifies the tendency of electron density to remain localized in between the intervening atoms. All the Ng atoms in case of AlN nanotube interact with the other Ng atom in a non-covalent manner. A similar trend is observed in case of GaN nanotube as well. As the atomic number of the Ng atoms increase, the Ng-Ng BCPs show some enhancement in interaction as slightly more electron density gets localized there than the lighter congeners. Significant interaction is, however, observed when we consider the Ng-Al/N BCPs. Particularly noteworthy are the Kr-N, Xe-N, Xe-Al BCPs in cases of Ng$_{1-2}$@AlN. Since for these BCPs $\nabla^2 r(r_c) > 0$ and $H(r_c) < 0$, we may state that a partially covalent character is present therein. Therefore, Kr$_2$ and Xe$_{1-2}$ can be classified as being chemisorbed on the host moiety. He$_{1-2}$, Ne and Ar$_{1-2}$, on the other hand remain physisorbed on AlN surface. In order to provide a graphical illustration of the scenario we have plotted the contour line diagrams of $\nabla^2 r(r_c)$ on the xz plane for the Ng$_{1-2}$@AlN moieties (Figure 11.3). It becomes clear that as we move from He through Xe, the extent of deformation of the scalar field of $\nabla^2 r(r_c)$ becomes gradually more distorted along both Ng-Ng axis as well as along Ng-N/Al BCPs. Therefore, more electron density gets shifted towards the concerned BCPs thereby rendering some form of covalent character. One may note that the heavier Ng analogues viz. Kr and in particular Xe can redistribute significant amount of their electron densities from their valence orbitals towards the Ng-N/Al BCPs. Such a behavior is rather unusual given the inherent 'inertness' of the Ng atoms. The observed binding is clearly dictated by the 'squeezing' effect of the nanotube which is sufficient to polarize the Ng atoms thereby initiating the process of binding. Kr and Xe being

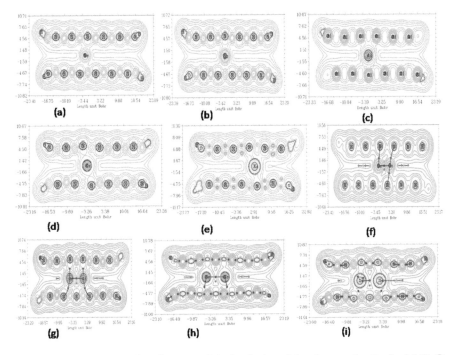

FIGURE 11.3   Contour line diagrams of the Laplacian of the electron density for (a) He@ AlN, (b) Ne@AlN, (c) Ar@AlN, (d) Kr@AlN, (e) Xe@AlN, (f) He$_2$@AlN, (g) Ar$_2$@AlN, (h) Kr$_2$@AlN and (i) Xe$_2$@AlN, respectively at the xz plane.

more polarizable than their lighter congeners, feel this effect most prominently. In the confined state, having been chemically 'activated' by the effect of confinement, the individual Ng atoms have several options to interact with. Instead of interacting with the other Ng atom present in the vicinity, all the Ng atoms interact strongly with N and Al atoms of the nanotube where superior electron accepting power of nanotube constituents as compared to noble gas atoms plays a significant role. Therefore, we only obtain van der Walls type dimers of Ng$_2$ inside AlN nanotube. In the case of GaN nanotube, we obtain a qualitatively similar picture. We note that because of the larger volume available inside GaN nanotube, the effect of confinement and therefore subsequent polarization of the Ng atoms inside GaN should be qualitatively slightly smaller as compared to AlN. We observe that all Ng atoms from He through Kr, interact in a non-covalent fashion among themselves as well as with Ga and N atoms. For the case of Xe$_2$@GaN, however, we observe a partially covalent character along the Xe-Ga BCP. Plots of $\nabla^2 r(r_c)$ on the xz plane support the above-stated observations (Figure 11.4).

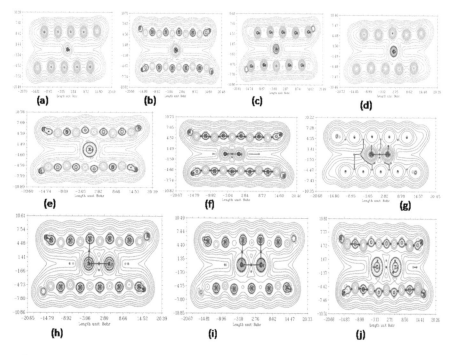

FIGURE 11.4   Contour line diagrams of the Laplacian of the electron density for (a) He@ GaN, (b) Ne@GaN, (c) Ar@GaN, (d) Kr@GaN, (e) Xe@GaN, (f) He$_2$@GaN, (g) Ne$_2$@GaN, (i) Ar$_2$@GaN, (j) Kr$_2$@GaN and (k) Xe$_2$@GaN, respectively at the xz plane.

In order to provide further information on the nature of interaction of the studied systems, we now consider the EDA data (Table 11.4). The total inter-action energy (ΔE_{total}) in between Ng$_{1-2}$ and AlN/GaN nanotubes has been decomposed into various fragments such as electrostatic interaction (ΔE_{elest}), orbital interaction (ΔE_{orb}), Pauli repulsion (ΔE_{Pauli}) and dispersion energy (ΔE_{disp}). The ΔE_{total} is positive in all cases considered here thereby reaffirm-ing the earlier thermochemical analysis. The chief destabilizing factor is the ΔE_{Pauli}, contribution from which increases as we go down the Ng series. As the atomic radii of the encapsulated Ng atoms increase, they are forced to come in close distance with the nanotube thereby enhancing the Coulomb repulsive forces. On the other hand, ΔE_{elest}, ΔE_{orb} and ΔE_{disp} are attractive in nature. For the cases of Ng$_{1-2}$@AlN, ΔE_{elest} and ΔE_{orb} provide the major con-tributions towards the total attractive interaction. We note the decrease in the contribution of ΔE_{disp} towards total attractive interaction as we go down the Ng series. Had the nature of interaction been purely of weak non-covalent

TABLE 11.4 EDA Results of the Ng_{1-2}@AlN/GaN Moieties

Systems	Fragments	ΔE_{total}	ΔE_{elest}	ΔE_{orb}	ΔE_{Pauli}	ΔE_{disp}
He@AlN	[He] + [AlN]	11.62	−7.37	−6.13	28.97	−3.86
Ne@AlN	[Ne] + [AlN]	20.40	−25.26	−5.66	59.10	−7.78
Ar@AlN	[Ar] + [AlN]	50.81	−74.74	−28.11	168.14	−14.48
Kr@AlN	[Kr] + [AlN]	65.63	−113.68	−43.29	239.91	−17.30
Xe@AlN	[Xe] + [AlN]	78.15	−146.50	−59.18	302.89	−19.06
He@GaN	[He] + [GaN]	8.51	−5.88	−4.47	23.02	−4.15
Ne@GaN	[Ne] + [GaN]	15.14	−20.55	−4.08	48.00	−8.23
Ar@GaN	[Ar] + [GaN]	39.91	−62.91	−22.27	140.88	−15.79
Kr@GaN	[Kr] + [GaN]	53.80	−98.19	−34.95	205.95	−19.01
Xe@GaN	[Xe] + [GaN]	72.64	−153.84	−58.82	306.28	−20.98
He_2@AlN	[He_2] + [AlN]	22.93	−14.47	−11.92	57.05	−7.72
Ar_2@AlN	[Ar_2] + [AlN]	100.05	−145.22	−54.85	328.87	−28.75
Kr_2@AlN	[Kr_2] + [AlN]	137.58	−222.23	−80.03	473.86	−34.00
Xe_2@AlN	[Xe_2] + [AlN]	181.67	−337.09	−130.47	686.65	−37.42
He_2@GaN	[He_2] + [GaN]	17.91	−11.48	−8.07	45.82	−8.36
Ne_2@GaN	[Ne_2] + [GaN]	31.77	−39.48	−8.22	96.09	−16.62
Ar_2@GaN	[Ar_2] + [GaN]	91.13	−123.85	−39.67	286.01	−31.37
Kr_2@GaN	[Kr_2] + [GaN]	129.19	−191.16	−57.26	414.86	−37.26
Xe_2@GaN	[Xe_2] + [GaN]	181.68	−301.35	−97.03	621.47	−41.40

type, one would have expected a major share of attractive terms to come from ΔE_{disp}. Here, instead we observe a different scenario and polarization of Ng atoms by the effect of confinement and subsequent creation of charge centers play a major role in stabilizing the Ng_{1-2}@AlN moieties. Also significant are the interactions among the frontier orbitals of Ng and nanotube surface. Since there exists a nice correlation in between the size, shape and symmetry of the frontier orbitals of Ng (principally s and p orbitals) to that present in Al and N atoms, the intervening orbitals get into a favorable condition to overlap. As a result a significant extent of charge transfer as well as polarization effect persists in between the Ng atoms and the inner surface of the nanotube. Therefore, our earlier observation gained from NBO and AIM analyses gets further support with regard to the nature of interaction of the host and guest atoms. For Ng_{1-2}@GaN cases, we observe a slightly different situation. Here for the lighter atoms such as He and Ne, contribution from ΔE_{disp} term is significant. One probable reason could stem from the fact that because of larger radii of GaN as opposed to AlN nanotube, the guest atoms are slightly less 'confined' thereby enabling particularly the lighter Ng atoms to maintain some distance from the inner wall of the nanotube. Therefore, the intervening orbitals are probably not close enough spatially to have a favorable overlap. However, as we consider the heavier Ng analogues Ar through Xe, contribution from ΔE_{orb} gets a gradual boost along with a concomitant fall in the contribution from ΔE_{disp}.

In order to check whether the observed Ng_{1-2} encapsulated moieties are kinetically stable or not, we now present results obtained from ADMP simulation (Figures 11.5 and 11.6) study at 298 K temperature. Firstly let us consider the case of He_2@AlN. Herein, the two He atoms exhibit significant extent of translational drift away from its equilibrium position situated near the middle of the ring albeit remaining inside the nanotube. All along their translational motion, the two He atoms maintain a close distance and during the said motion they move in a correlated manner. The He-He distance, however, fluctuates to a certain extent, at 100, 200, 300, 400 and 500 fs the distance stands at 2.25, 2.35, 2.16, 2.32 and 2.16 Å, respectively. Therefore, the He atoms come even closer to each other in a dynamical situation as compared to its equilibrium structure. The He_2 dimer tries to perform some extent of rotational motion and aligns itself in an inclined orientation with respect to the tube axis as opposed to its equilibrium structure. The host moiety remains intact up to the studied timescale although it undergoes minute expansion and contraction mimicking

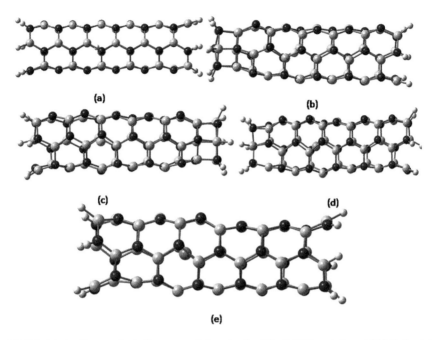

FIGURE 11.5   Geometrical alignment of the adsorbed Ng@AlN moieties at 500 fs for (a) He@AlN, (b) Ne@AlN, (c) Ar@AlN, (d) Kr@AlN and (e) Xe@AlN, respectively.

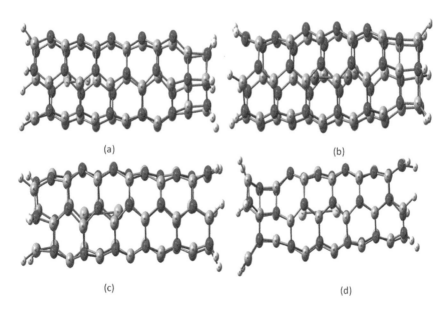

FIGURE 11.6   Geometrical alignment of the adsorbed Ng_2@AlN moieties at 500 fs for (a) He_2@AlN, (b) Ar_2@AlN, (c) Kr_2@AlN and (d) Xe_2@AlN, respectively.

vibrational mode of oscillation. For the case of He@AlN, the guest atom remains inside the host till 500 fs. Very little translational drift is noted till that time although the guest undergoes some movement vertically. The dynamical situation changes significantly as we consider the heavier Ng analogues. Because of the structural rigidity as well as the small radius of the host moiety, very little space is available to the Ng atoms (in the doubly occupied case) to perform rotational motion. The translational motion of the Ng_{1-2} species is also curtailed because of the powerful interaction in between the Ng atoms and the nanotube. Two opposing factors should be operative in the present context. The kinetic energy of the Ng atoms should encourage them to fly apart from each other. On the other hand large gain in entropic degree of freedom should also assist the Ng_2 species to get desorbed from the host surface and ultimately facilitate them to escape the nanotube. However, the Ng-nanotube interactions should prohibit such an outcome and hold the guest species inside the nanotube. In cases of Ne, Ar_{1-2}, Kr_{1-2} and Xe_{1-2} the latter force seems to be the determining factor. As a result Ne, Ar_{1-2}, Kr_{1-2} and Xe_{1-2} not only remain inside the nanotube up to 500 fs time, they also exhibit negligible extent of translational drift in sharp contrast to He_2@AlN. Since in case of He_2@AlN, the He-nanotube interactions are considerably weaker as compared to other cases, the He atoms experience a lot more freedom to perform its dynamical evolution. The Ng-Ng distances show very little (for $Ar_2/Kr_2/Xe_2$@AlN) change during the course of temporal evolution and the Ar-Ar, Kr-Kr and Xe-Xe distances at 500 fs are 3.37, 3.44 and 3.86 Å, respectively. It should be noted though that the amplitude of vibration of the host moiety increases slightly as we go from Ar_2 to Xe_2 cases and the increasing bulk of the guest seem to be the chief factor for such an observation. The host relaxes some of the induced strain generated due to the encapsulation of Ng_2 atoms by performing more pronounced vibration.

In case of He_2@GaN (Figure 11.7), the encapsulated He atoms exhibit much greater translational movement with sharp fluctuation in the He-He distance. The said distance at 50, 130, 389 and 500 fs are 1.93, 3.59, 2.43 and 2.47 Å, respectively, representing the extent of fluctuation. The greater translational maneuver of He atoms is clearly dictated by the fact that unlike the case of AlN nanotube, inside GaN a greater extent of space is available thus rendering the He atoms slightly more mobile. Despite their vigorous motion, the He atoms reside well inside the GaN nanotube up to 500 fs. No significant nanotube distortion is noted. For the case of He@

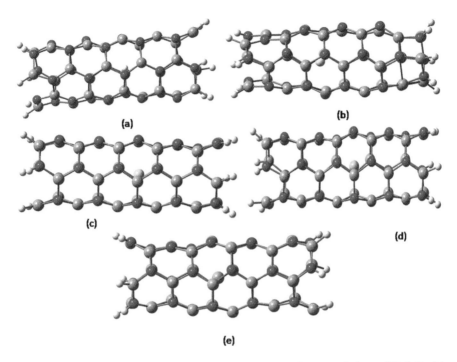

FIGURE 11.7   Geometrical alignment of the adsorbed Ng@GaN moieties at 500 fs for (a) He@GaN, (b) Ne@GaN, (c) Ar@GaN, (d) Kr@GaN and (e) Xe@GaN, respectively.

GaN (Figure 11.8), we note minute vertical movement of the He atom inside the nanotube up to 500 fs time scale. The guest remains inside the nanotube till the studied time scale. All other Ng atoms in the singly occupied cases, show similar dynamical behavior and stay inside the nanotube till 500 fs. In case of Ne_2@GaN (Despite repeated attempts for the case of Ne_2@AlN, the geometry optimization calculation could not be converged within the prescribed tolerance limit. Therefore, we have excluded this case from our present study.), we observe a lesser extent of motion and the Ne-Ne distance at 500 fs is 2.88 Å. As we consider the heavier Ng congeners such as Ar_2 and Kr_2, the translational or rotational motion is severely curtailed and one obtains a stable geometry of the noble gas encapsulated GaN nanotube up to 500 fs. In case of Xe_2 similar situation persists albeit the fact that here the GaN nanotube expands and contracts to a slightly more extent. As a result we may conclude that all the Ng_2 encapsulated nanotubes considered in this work are dynamically stable at 298 K temperature up to 500 fs.

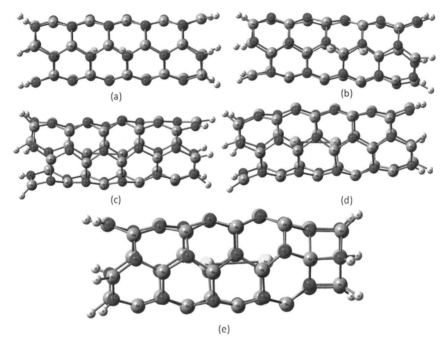

(a)

(b)

(c)

(d)

(e)

FIGURE 11.8 Geometrical alignment of the adsorbed Ng_2@GaN moieties at 500 fs for (a) He_2@GaN, (b) Ne_2@GaN, (c) Ar_2@GaN, (d) Kr_2@GaN and (e) Xe_2@GaN, respectively.

11.4 CONCLUSION

Herein, we have studied the feasibility of having Ng_{1-2} encapsulated AlN and GaN nanotubes, through DFT calculations. Formation process of all the studied systems from their respective parent moieties is thermodynamically unfavorable at 298 K temperature and one atmosphere pressure as revealed by detailed thermochemical analyses. The stability of the Ng_{1-2} encapsulated nanotubes decreases as we move from He through Xe as illustrated by CDFT based global reactivity descriptors. However, the studied systems were found to be kinetically stable up to 500 fs timescale at the same temperature as demonstrated by ADMP simulation results. The Ng-Ng binding is of closed-shell type in case of both AlN and GaN nanotubes. However, the Ng-AlN/GaN interaction showed some degree of covalent character particularly in the cases of Kr_2@AlN, Xe_{1-2}@AlN and Xe_2@GaN whereas the others were bound in a non-covalent manner. Therefore, one can state that adsorption process of Kr_2@AlN, Xe_{1-2}@AlN and $Xe_2@GaN$ constitutes an example of chemisorption whereas other cases considered herein reveal a situation where the Ng atoms remain physisorbed on the nanotubes. The Ng atoms act as Lewis bases and donate electron density to the concerned nanotubes thereby attaining a significant extent of positive charge. The orbital and electrostatic energies in between the Ng atoms and nanotube play a significant role in stabilizing the Ng_{1-2} encapsulated host moieties. The dispersion interaction plays an important role in case of lighter elements viz., He or Ne encapsulated in the nanotubes. It is well appreciated in scientific discourse that both thermodynamic as well as kinetic stability determine the overall durability of the concerned physical processes. In that deliberation it becomes evident that the present study sheds some light on the viability of having Ng_{1-2} encapsulated AlN and GaN nanotubes. The fact that $He@C_{20}H_{20}$ was synthesized [19] despite being thermodynamically unfavorable is worth to note in this context. Unlike the cases of cage molecules where majority of Ng encapsulation studies have been devoted, tubular geometries such as nanotubes as considered in this study present a viable alternative choice for entrapping Ng atoms. For a host to be an effective Ng-carrier both adsorption as well as desorption should play a subtle role. In open face host moieties, no covalent bond breaking is required unlike the cage molecules in order to inject or release the Ng atoms. Therefore, nanotubes such as AlN or GaN could probably be considered as a reasonable choice as far as Ng-encapsulation process is concerned.

ACKNOWLEDGMENTS

DC thanks CSIR, New Delhi for the financial assistance. PKC would like to thank DST, New Delhi for the J. C. Bose National Fellowship.

KEYWORDS

- **noble gas**
- **partially covalent bonding**
- **kinetic stability**

- **encapsulation**
- **nanotube**
- **energy decomposition analysis**

REFERENCES

1. Pauling, L., (1933). The formulas of antimonic acid and the antimonates. *J. Am. Chem. Soc.*, *55*, 1895–1900.
2. Bartlett, N., (1962). Xenon hexafluoroplatinate (V) $Xe^+[PtF_6]^-$. *Proc. Chem. Soc.*, 218.
3. Graham, L., Graudejus, O., Jha, N. K., & Bartlett, N., (2000). Concerning the nature of $XePtF_6$. *Coord. Chem. Rev.*, *197*, 321–324.
4. Fields, P. R., Stein, L. H., & Zirin, M. H., (1962). Radon fluoride. *J. Am. Chem. Soc.*, *84*, 4164–4165.
5. Turner, J. J., & Pimentel, G. C., (1963). Krypton fluoride: preparation by the matrix isolation technique. *Science*, *140*, 974–975.
6. Nelson, L. Y., & Pimentel G. C., (1967). Infrared detection of xenon dichloride. *Inorg. Chem.*, *6*, 1758–1759.
7. Pettersson, M., Lundell, J., & Räsänen, M., (1995). Neutral rare-gas containing charge-transfer molecules in solid matrices. II: HXeH, HXeD, and DXeD in Xe. *J. Chem. Phys.*, *103*, 205–210.
8. Pettersson, M., Nieminen, J., Khriachtchev, L., & Räsänen, M., (1997). The mechanism of formation and IR-induced decomposition of HXeI in solid Xe. *J. Chem. Phys.*, *107*, 8423–8431.
9. Khriachtchev, L., Pettersson, M., Lundell, J., Tanskanen, H., Kiviniemi, T., Runeberg, et al., (2003). Neutral xenon-containing radical, HXeO. *J. Am. Chem. Soc.*, *125*, 1454–1455.
10. Koch, W., Frenking, G., Gauss, J., Cremer, D., & Collins, J. R., (1987). Helium chemistry: theoretical predictions and experimental challenge. *J. Am. Chem. Soc.*, *109*, 5917–5934.
11. Frenking, G., & Cremer, D., (1990). The chemistry of the noble gas elements helium, neon, and argon—experimental facts and theoretical predictions. *Struct. Bond.*, *73*, 17–95.
12. Rzepa, H. S., (2010). The rational design of helium bonds. *Nature Chemistry*, *2*, 390–393.
13. Grochala, W., (2007). Atypical compounds of gases, which have been called 'Noble'. *Chem. Soc. Rev.*, *36*, 1632–1655.
14. Borocci, S., Bronzolino, N., & Grandinetti, F., (2006). Neutral helium compounds: theoretical evidence for a large class of polynuclear complexes. *Chem. Eur. J.*, *12*, 5033–5042.
15. Chakraborty, D., & Chattaraj, P. K., (2015). In Quest of a Superhalogen Supported Covalent Bond Involving a Noble Gas Atom, *J. Phys. Chem. A*, *119*, 3064–3074.
16. Saunders, M., Jiménez-Vázquez, H. A., Cross, R. J., & Poreda, R. J., (1993). Stable compounds of Helium and Neon: He@C60 and Ne@C60. *Science*, *259*, 1428–1430.

17. Saunders, M., Jiménez-Vázquez, H. A., Cross, R. J., Mroczkowski, S., Gross, M. L., Giblin, D. E. et al., (1994). Incorportion of Helium, Neon, Argon, Krypton, and Xenon into Fullerenes using High Pressure. *J. Am. Chem. Soc.*, *116*, 2193–2194.
18. Saunders, M., Cross, R. J., Jiménez-Vázquez, H. A., Shimshi, R., & Khong, A., (1996). Noble gas atoms inside fullerenes. *Science*, *271*, 1693–1697.
19. Cross, R. J., Saunders, M., & Prinzbach, H., (1999). Putting helium inside dodecahedrane. *Org. Letts.*, *1*, 1479–1481.
20. Krapp, A., & Frenking, G., (2007). Is this a chemical bond? A theoretical study of $Ng_2@C_{60}$ (Ng = He, Ne, Ar, Kr, Xe). *Chem. Eur. J.*, *13*, 8256–8270.
21. Cerpa, E., Krapp, A., Flores-Moreno, R., Donald, K. J., & Merino, G., (2009). Influence of endohedral confinement on the electronic interaction between He atoms: A $He_2@C_{20}H_{20}$ Case Study. *Chem. Eur. J.*, *15*, 1985–1990.
22. Khatua, M., Pan, S., & Chattaraj, P. K., (2014). Confinement Induced Binding of Noble Gas Atoms. *J. Chem. Phys.*, *140*, 164306–164311.
23. Perry IV, J. J., Teich-McGoldrick, S. L., Meek, S. T., Greathouse, J. A., & Haranczyk, M., Allendorf, M. D., (2014). Noble gas adsorption in metal–organic frameworks containing open metal sites. *J. Phys. Chem. C*, *118*, 11685–11698.
24. Chakraborty, D., & Chattaraj, P. K., (2015). Confinement induced binding in noble gas atoms within a BN-doped carbon nanotube. *Chem. Phys. Lett.*, *621*, 29–34.
25. Chakraborty, D., Pan, S., & Chattaraj, P. K., (2016). Encapsulation of small gas molecules and rare gas atoms inside the octa acid cavitand. *Theor. Chem. Acc.*, *135*, 119.
26. Preckel, B., Weber, N. C., Sanders, R. D., Maze, M., & Schlack, W., (2006). Molecular mechanisms transducing the anesthetic, analgesic, and organ protective actions of xenon. *Anesthesiology*, *105*, 187–197.
27. Kennedy, R. R., Stokes, J. W., & Downing, P., (1992). Anesthesia and the inert-gases with special reference to xenon. *Anaesthesiol. Intensive Care*, *20*, 66–70.
28. Kebabian, P. L., Romano, R. R., & Freedman, A., (2003). Determination of argon-filled insulated glass window seal failure by spectroscopic detection of oxygen. *Meas. Sci. Technol.*, *14*, 983–988.
29. Kuo, D. H., Cheung, B. Y., & Wu, R. J., (2001). Growth and properties of alumina films obtained by low-pressure metalorganic chemical vapor deposition. *Thin Solid Films*, *398*, 35–40.
30. Hutchinson, M. H. R., (1980). Excimers and excimer lasers. *Appl. Phys.*, *21*, 95–114.
31. McIntyre, I. A., & Rhodes, C. K., (1991). High-power ultrafast excimer lasers. *J. Appl. Phys.*, *69*, R1–R19.
32. Stephan, K., & Abdelsalam, M., (1980). Heat-transfer correlations for natural-convection boiling. *Int. J. Heat Mass Transfer*, *23*, 73–87.
33. Lomaev, M. I., Sosnin, E. A., Tarasenko, V. F., Shits, D. V., Skakun, V. S., Erofeev, M. V. A., et al., (2006). Capacitive and barrier discharge excilamps and their applications. *Instrum. Exp. Technol.*, *49*, 595–616.
34. Justel, T., Krupa, J. C., & Wiechert, D. U., (2001). VUV spectroscopy of luminescent materials for plasma display panels and Xe discharge lamps. *J. Lumin.*, *93*, 179–189.
35. Soelberg, N. R., Garn, T. G., Greenhalgh, M. R., Law, J. D., Jubin, R., Strachan, D. M., et al., (2013). Radioactive iodine and krypton control for nuclear fuel reprocessing facilities. *Sci. Technol. Nucl. Inst.*, *2013*, 1-12.
36. Iijima, S., (1991). Helical microtubules of graphitic carbon. *Nature*, *354*, 56–58.

37. Wu, Q., Hu, Z., Wang, X., Lu, Y., Chen, X., Xu, H., et al., (2003). Synthesis and characterization of faceted hexagonal aluminum nitride nanotubes. *J. Am. Chem. Soc.*, *125*, 10176–10177.

38. Goldberger, J., He, R., Zhang, Y., Lee, S., Yan, H., Choi, H., et al., (2003). Single-crystal gallium nitride nanotubes. *Nature*, *422*, 599–602.

39. Parr, R. G., & Yang, W., (1989). *Density Functional Theory of Atoms and Molecules*, Oxford University Press: New York.

40. Chattaraj P. K., (2009). Ed., *Chemical Reactivity Theory: A Density Functional View*, Taylor & Francis, CRC Press: Florida.

41. Geerlings, P., DeProft, F., & Langenaeker, W., (2003). Conceptual density functional theory. *Chem. Rev.*, *103*, 1793–1873.

42. Chattaraj, P. K., Sarkar, U., & Roy, D. R., (2006). Electrophilicity index. *Chem. Rev.*, *106*, 2065–2091.

43. Dennington, R., Keith, T., & Millam, J., (2009). GaussView, version 5; Semichem, Inc. Shawnee Mission, KS.

44. Frisch, M. J., Trucks, G. W., Schlegel, H. B., Scuseria, G. E., Robb, M. A., & Cheeseman, J., (2010). Gaussian 09, Revision C.01; Gaussian, Inc., Wallingford City.

45. Bader, R. F. W., (1990). *Atoms in Molecules: A Quantum Theory*; Clarendon Press: Oxford, UK.

46. Lu, T., & Chen, F. W., (2012). Multiwfn: a multifunctional wavefunction analyzer. *J. Comput. Chem.*, *33*, 580–592.

47. Bickelhaupt, F. M., & Baerends, E. J., (2000). In: *Reviews of Computational Chemistry*; Boyd, D. B., Lipkowitz, K. B., Eds., Wiley-VCH: New York.Kohn-Sham density functional theory: predicting and understanding chemistry, *15*, 1-86.

48. te Velde, G., Bickelhaupt, F. M., Baerends, E. J., van Gisbergen, S. J. A., Fonseca Guerra, C., Snijders, J. G., et al., (2001). Chemistry with ADF. *J. Comput. Chem.*, *22*, 931–967.

49. Schlegel, H. B., (2011). Ab initio molecular dynamics: propagating the density matrix with gaussian orbitals. *J. Chem. Phys.*, *114*, 9758–9763.

50. Macchi, P., Garlaschelli, L., Martinengo, S., & Sironi, A., (1999). Charge Density in Transition Metal Clusters: Supported vs Unsupported Metal–Metal Interactions. *J. Am. Chem. Soc.*, *121*, 10428–10429.

51. Cremer, D., & Kraka, E., (1984). Chemical bonds without bonding electron density—does the difference electron-density analysis suffice for a description of the chemical bond? *Angew. Chem., Int. Ed.*, *23*, 627–628.

CHAPTER 12

ONE-ELECTRON DENSITIES OF HARMONIUM ATOMS

JERZY CIOSLOWSKI

Institute of Physics, University of Szczecin, Wielkopolska 15, 70-451 Szczecin, Poland

CONTENTS

12.1   INTRODUCTION

Among nonrelativistic observables, the one-electron density $\rho(\vec{r})$ undoubtedly plays the most important role in the quantum-mechanical description of atoms and molecules. Not only does it allow calculations of all the one-electron properties described by multiplicative operators but it also enters expressions for measures of molecular similarity [1] and electron localization [2]. Moreover, its critical points and gradient paths enable delineation of atoms in molecules and the major interactions among them [3]. Last but certainly not least, the one-electron density is the fundamental variable quantity of the density functional theory (DFT) [4]. In light of this importance, the availability of benchmarks for $\rho(\vec{r})$ is of great interest to quantum chemists.

In general, the benchmarks of quantum chemistry are derived from either experimental data or exactly solvable/quasi-solvable model systems. Although the agreement of numerical predictions with experimental data constitutes the ultimate validation of theoretical methods, experiment-based benchmarks suffer from several disadvantages. First of all, they usually require adjustments of the measured quantities in order to account for the phenomena neglected at the commonly employed levels of quantum-mechanical description. These adjustments (such as the removal of relativistic contributions to electronic energies and anharmonic contributions to vibrational frequencies) carry own uncertainties that add to possible experimental errors. In the case of the one-electron densities, this problem is even more acute as they are actually obtained by fitting of *ad hoc* models [5] rather than from direct measurement. Second, properties of chemical species with unusual bonding situations or uncommon electronic structures are often not amenable to experimental determination. Third, benchmarking against experimental data is not suitable for analysis of errors in the computed electronic properties in terms of their components due to different types of electron correlation (dynamical vs. nondynamical, short- vs. long-range, in finite vs. in infinite systems, etc.). Fourth, such comparisons do not result in formulation of explicit physically-motivated constraints upon the key theoretical quantities, such as effective Hamiltonians and approximate functionals.

All of these problems are eliminated upon replacing experimental data with those obtained from exactly solvable/quasi-solvable model systems. In order to be successful in this role, such systems have to satisfy several conditions. In particular, their Hamiltonians have to depend on at least one parameter, variation of which has to result in a continuous tuning of the extent of the electron correlation effects and the relative magnitudes of their different types for each combination of the number of particles and the electronic state. Moreover, the pertinent exact wavefunctions have to be computable by analytic means for an infinite number of values of the parameter or at least (in the form of asymptotic expressions) for some of its limits. The Hamiltonians should permit evaluation of approximate yet highly accurate wavefunctions at reasonable computational cost when their exact counterparts are not available.

There is an often-studied system, called the harmonium atom, that conforms to the above requirements. The properties of the one-electron densities of its diverse electronic states are the subject of this chapter.

12.2   THE HARMONIUM ATOM

The two-electron harmonium atom, i.e. the species described by the nonrelativistic Hamiltonian

$$\hat{H} = \frac{1}{2}\sum_{i=1}^{N}\left(-\hat{\nabla}_i^2 + \omega^2 r_i^2\right) + \sum_{i>j=1}^{N}\frac{1}{r_{ij}} \tag{1}$$

with $N = 2$, has been thoroughly investigated with both rigorous mathematical analysis and numerical simulations [6–10]. As described in detail in the following sections of this chapter, its general properties are representative of those of the harmonium atoms comprising arbitrary numbers of electrons in the sense that they transition smoothly between the weak- and strong-correlation regimes. Thus, for large values of the confinement strength ω, the two-electron harmonium atom closely resembles helium-like systems, whereas at the strong-correlation limit of $\omega \to 0$ it becomes an archetype of species with exclusively nondynamical correlation and complete spatial localization of electrons (i.e. the Wigner crystallization). Consequently, combinations of dynamical and nondynamical electron correlation effects with arbitrary relative magnitudes can be modeled with a proper choice of ω. One should be reminded that ordinary atoms do not permit such an arbitrary tuning of electron correlation as they undergo spontaneous ionization upon their nuclear charges falling below certain critical values, which occurs well before the Wigner crystallization sets in.

Together with the availability of exact wavefunctions and energies for select values of ω (such as $\frac{1}{2}$, $\frac{1}{10}$, etc.), the continuous tunability of electron correlation effects makes the two-electron harmonium atom a convenient benchmark that has been already applied to a multitude of electronic structure methods, including those based on DFT [11–17] and other formalisms [18–20]. Unfortunately, its general usefulness is severely limited by its failure to provide any new information for those approximate electron correlation methods that are already exact for two-electron systems.

The harmonium atoms with $N > 2$, upon which this chapter focuses, are free of this limitation. Moreover, due to the abundance of electronic states with diverse spins and character (single- or multideterminantal) [21–24], they provide a much richer testing ground for approximate methods of quantum chemistry. In order to appreciate this fact, one has first to elucidate their properties within the weak- and strong-correlation regimes.

12.2.1 THE WEAK-CORRELATION REGIME

Within the weak-correlation (or strong-confinement) regime of large values of ω, the harmonium atoms are essentially systems of three-dimensional harmonic oscillators weakly perturbed by the interelectron repulsion. Consequently, their energies $E(\omega)$ are given by the power series [10, 25–28, 30]

$$E(\omega) = \sum_{k=0}^{\infty} E_k \, \omega^{(2-k)/2}. \tag{2}$$

The zeroth- and first-order energy coefficients, E_0 and E_1, are trivial to compute. In contrast, evaluation of E_2 and its higher-order counterparts involves either solution of respective differential equations (which is practical only for $N = 2$ [25, 26]) or infinite summations that arise from the sum-over-states expressions of the second- and higher-order perturbation theory. Unlike in fully Coulombic systems, the continuum states do not potentially contribute to these sums, evaluation of which is fairly straightforward for $N = 2$ [27]. For $N = 3$, the computations of E_2 are aided by intermediate evaluations of electron-pair contributions (with nonadditivity corrections) [28]. A general approach employing specialized algebraic methods [29] that yields closed-form expressions for E_2 of arbitrary states of many-electron harmonium atoms has been published [30].

In the case of the three-electron species, the two lowest-energy states are of the single-determinantal nature, the $^2P_-$ doublet originating from the $|s\bar{s}p_z|$ Slater determinant and the $^4P_+$ quartet stemming from the $|sp_x p_y|$ one [22]. For $N = 4$, low-lying states with both the single- and multideterminantal zeroth-order wavefunctions are encountered. In particular, the $^3P_+$ triplet and the $^5S_-$ quintet arise from the $|s\bar{s}p_x p_y|$ and $|sp_x p_y p_z|$ configurations, respectively, whereas the wavefunctions of both the $^1S_-$ and $^1D_-$ singlets converge at the $\omega \to \infty$ limit to linear combinations of the $\{|s\bar{s}p_m \bar{p}_m|,\ m = -1, 0, 1\}$ Slater determinants [23, 31]. Similarly, one assigns the $|s\bar{s}p_x p_y p_z|$ configuration to the $^4S_-$ state of the five-electron harmonium atom, whereas its $^2P_-$ and $^2D_-$ counterparts require the zeroth-order descriptions involving the $\{|s\bar{s}p_m p_{m'} \bar{p}_{m''}|,\ m,\ m',\ m'' = -1, 0, 1,\ m + m' + m'' = 0\}$ determinants [24].

The nature of these limiting wavefunctions is reflected in the values of the coefficients E_0 and E_1 compiled in Table 12.1. It also has to be taken into account in computations of the respective second-order energy coefficients for the aforementioned states, which are given by the general expression [30]

$$E_2 = \mathcal{E}_1 + \frac{1}{\pi}\mathcal{E}_2 + \frac{\sqrt{3}}{\pi}\mathcal{E}_3 + \frac{\ln 2}{\pi}\mathcal{E}_4 + \frac{\ln(1+\sqrt{3})}{\pi}\mathcal{E}_5. \tag{3}$$

TABLE 12.1   The Zeroth- and First-Order Energy Coefficients of Several States of the Harmonium Atoms [24, 28, 30]

N	State	E_0	E_1
3	$^2P_-$	$\frac{11}{2}$	$\frac{5}{2}\sqrt{\frac{2}{\pi}}$
	$^4P_+$	$\frac{13}{2}$	$2\sqrt{\frac{2}{\pi}}$
4	$^3P_+$	8	$\frac{14}{3}\sqrt{\frac{2}{\pi}}$
	$^1S_+$	8	$\frac{59}{12}\sqrt{\frac{2}{\pi}}$
	$^1D_+$	8	$\frac{143}{30}\sqrt{\frac{2}{\pi}}$
	$^5S_-$	9	$4\sqrt{\frac{2}{\pi}}$
5	$^4S_-$	$\frac{21}{2}$	$\frac{15}{2}\sqrt{\frac{2}{\pi}}$
	$^2P_-$	$\frac{21}{2}$	$\frac{31}{4}\sqrt{\frac{2}{\pi}}$
	$^2D_-$	$\frac{21}{2}$	$\frac{153}{20}\sqrt{\frac{2}{\pi}}$

The coefficients $\{\varepsilon_1, \varepsilon_2, \varepsilon_3, \varepsilon_4, \varepsilon_5\}$ and the resulting numerical values of E_2 are listed in Table 12.2.

The reciprocal of the square root of ω is the natural coupling strength within the weak-correlation regime. It appears as the expansion variable in the power series (2), provides the length scale, and governs the behavior of the natural spinorbitals (NSOs). In particular, as clearly demonstrated with the example of the NSOs of the four-electron harmonium atom in the $^5S_-$ state (Figure 12.1), their occupancies $\{\nu_n(\omega)\}$ follow the expansion analogous to Eq. (2), namely (note that $\nu_{n1} = 0$) [10]

$$\nu_n(\omega) = \sum_{k=0}^{\infty} \nu_{nk}\, \omega^{-k/2}, \tag{4}$$

where ν_{n0} equals either one or zero depending on whether the nth natural spinorbital is strongly or weakly occupied. At the $\omega \to \infty$ limit, only the former NSOs contribute to the individual (nonvanishing) spin components $\rho_s(R,\theta;\omega)$ of the one-electron density in terms of the spherical coordinates [32, 33], allowing derivation of the asymptotic expression [31]

TABLE 12.2   The Second-Order Energy Coefficients of Several States of the Harmonium Atoms [24, 28, 30]

N	State	$\mathcal{E}_1$	$\mathcal{E}_2$	$\mathcal{E}_3$	$\mathcal{E}_4$	$\mathcal{E}_5$	E_2
3	$^2P_-$	$\frac{49}{9}$	$-\frac{44}{3}$	$\frac{1}{3}$	$-\frac{173}{6}$	$\frac{49}{3}$	-0.176 654 125
	$^4P_+$	$\frac{23}{9}$	$-\frac{32}{3}$	$\frac{8}{3}$	$-\frac{56}{3}$	$\frac{32}{3}$	-0.075 610 347
4	$^3P_+$	$\frac{370}{27}$	$-\frac{1204}{27}$	$\frac{40}{9}$	$-\frac{848}{9}$	$\frac{520}{9}$	-0.344 824 960
	$^1S_+$	$\frac{488}{27}$	$-\frac{20603}{432}$	$-\frac{49}{36}$	$-\frac{7439}{72}$	$\frac{2279}{36}$	-0.400 572 997
	$^1D_+$	$\frac{10274}{675}$	$-\frac{300647}{6750}$	$\frac{407}{225}$	$-\frac{43751}{450}$	$\frac{13391}{225}$	-0.370 649 862
	$^5S_-$	$\frac{62}{9}$	$-\frac{320}{9}$	$\frac{32}{3}$	$-\frac{208}{3}$	$\frac{128}{3}$	-0.195 514 553
5	$^4S_-$	$\frac{80}{3}$	$-\frac{890}{9}$	15	$-\frac{1285}{6}$	135	-0.604 550 408
	$^2P_-$	$\frac{488}{15}$	$-\frac{18329}{180}$	$\frac{129}{20}$	$-\frac{27217}{120}$	$\frac{2857}{20}$	-0.665 067 347
	$^2D_-$	$\frac{1348}{45}$	$-\frac{894241}{9000}$	$\frac{2867}{300}$	$-\frac{44223}{200}$	$\frac{41771}{300}$	-0.644 386 259

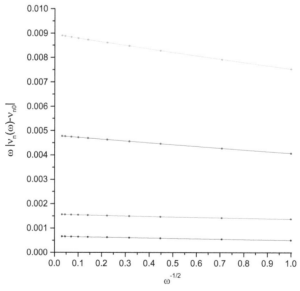

FIGURE 12.1   The reduced occupancies $\omega\,|\nu_n(\omega) - \nu_{n0}|$ of the strongly occupied (blue: s, green: p) and the dominant weakly occupied (red: d, purple: s) natural spinorbitals of the ^5S state of the four-electron harmonium atom vs. $\omega^{-1/2}$ [23].

$$\rho_s(R, \theta; \omega) \xrightarrow[\omega \to \infty]{} \pi^{-3/2} \omega^{3/2} \left[1 + \omega R^2 \kappa_s(\theta)\right] \exp(-\omega R^2), \qquad (5)$$

where s stands for either α or β. The functions $\kappa_s(\theta)$ for several states of the harmonium atoms are compiled in Table 12.3. When $\kappa_s(\theta) > 1$, the asymptotic one-electron densities (5) exhibit radial maxima at which $\partial\rho_s(R,\theta;\omega)/\partial R = 0$

TABLE 12.3   The Functions $\kappa_s(\theta)$ for Several States of the Harmonium Atoms

N	State	m_L	$\kappa_\alpha(\theta)$	$\kappa_\beta(\theta)$
3	$^2P_-$	0	$2\cos^2\theta$	0
	$^4P_+$	0	$2 - 2\cos^2\theta$	n/a
4	$^3P_+$	0	$2 - 2\cos^2\theta$	0
	$^1S_+$	0	$\frac{2}{3}$	$\frac{2}{3}$
	$^1D_+$	2	$1 - \cos^2\theta$	$1 - \cos^2\theta$
	$^5S_-$	0	2	n/a
5	$^4S_-$	0	2	0

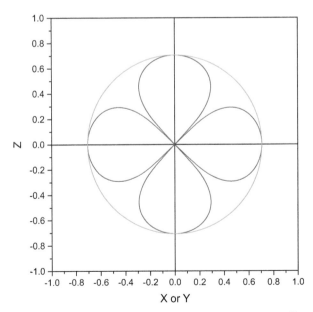

FIGURE 12.2   The xz (or yz) cross sections (in the length units of $\omega^{-1/2}$) of the surfaces delineated by the condition $\partial\rho_\alpha(R,\theta;\omega)/\partial R = 0$ at $R \neq 0$ at the limit of $\omega \to \infty$ for the $^2P_-$ (red), $^4P_+$ and $^3P_+$ (blue), as well as $^5S_-$ and $^4S_-$ (green) states of the three-, four-, and five-electron harmonium atoms.

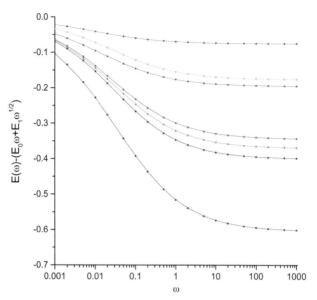

FIGURE 12.3   The differences between the actual energies $E(\omega)$ and the sums $E_0\,\omega+E_1\,\omega^{1/2}$ vs. ω for the $^2P_-$ (green), $^4P_+$ (blue), $^3P_+$ (red), $^1S_+$ (purple), $^1D_+$ (orange), $^5S_-$ (magenta), and $^4S_-$ (navy blue) states of the three-, four-, and five-electron harmonium atoms [22–24].

at $R \neq 0$. This condition is satisfied for certain values of θ in the cases of the α electrons of the $^2P_-$ and $^4P_+$ states ($N = 3$), the $^3P_+$ and $^5S_-$ states ($N = 4$), and the $^4S_-$ state ($N = 5$). The xz (or yz) cross sections of the surfaces delineated by these maxima at the limit of $\omega\to\infty$ are displayed in Figure 12.2. However, only in the instance of the $^5S_-$ quintet, the resulting topology of the spin-summed one-electron density can be regarded as that exhibiting the so-called fat attractor consisting of the cage point at the center of the coordinate system surrounded by an attractor surface [10].

The weak-correlation regime extends down to the magnitudes of the confinement strength as low as 100, below which the errors in the energy predictions provided by the power series (2) truncated after the first three terms start to grow significantly (Figure 12.3). Upon a further decrease in ω, the second limiting regime, namely that of strong correlation, is eventually entered through a transitional region.

12.2.2   THE STRONG-CORRELATION REGIME

Once the regime of the strongly correlated species is entered, the classical Coulomb crystals [34–36] with energies proportional to $\omega^{2/3}$ and spatial extents of the order of $\omega^{-2/3}$ become the unperturbed systems while the kinetic energy with the relative coupling constant scaling like $\omega^{2/3}$ assumes the role of the perturbation. At the first order, this perturbation gives rise to the zero-point energy proportional to ω and fluctuations in particles' position vectors $\vec{r}_j$ with magnitudes that carry over from the other regime, i.e. scale like $\omega^{-1/2}$. Thus, weakening of the confinement leads to delocalization of particles in the absolute sense that, on the other hand, amounts to localization relative to the overall size of the system in question. The resulting freely rotating species comprising particles oscillating about their equilibrium positions determined by the minima (global or local) of the potential energy are called Wigner molecules [37].

In accordance with the above observations, the energies of the electronic states of the Wigner molecules are given by the power series [6, 10, 38]

$$E(\omega) = \sum_{k=0}^{\infty} \tilde{E}_k \, \omega^{(2+k)/3} \, , \tag{6}$$

where each of the energy coefficients $\tilde{E}_k$ originates from different physical phenomena. Thus, $\tilde{E}_0 \omega^{2/3}$ is the classical energy of the Coulomb crystal (i.e. the Wigner molecule with clamped constituting electrons), which is obtained by minimizing the potential energy

$$V(\vec{r}_1, \dots, \vec{r}_N) = \frac{1}{2} \omega^2 \sum_{i=1}^{N} r_i^2 + \sum_{i>j=1}^{N} \frac{1}{r_{ij}} \tag{7}$$

with respect to the position vectors $\{\vec{r}_1, \dots, \vec{r}_N\}$ of the electrons. As the lowest-energy states correspond to the global minimum of $V(\vec{r}_1, \dots, \vec{r}_N)$, they share the same value of $\tilde{E}_0$. The second term, namely $\tilde{E}_1 \omega$, is the energy of harmonic oscillations about the minimum, which for $N > 2$ depends on the set of $3N - 3$ vibrational numbers. The third term, i.e. $\tilde{E}_2 \omega^{4/3}$, arises from rotations of the Wigner molecule (described in general with three quantum numbers) and the lowest-order anharmonic contributions from the vibrations.

The wavefunctions that correspond to the energies yielded by the expansion (6) are available from the recently developed formalism based upon antisymmetrization of the eigenfunctions of a rovibrational Hamiltonian [38]. In the case of the Wigner molecule derived from the three-electron harmonium atom, in which the electrons vibrate about vertices of an equilateral triangle, the $^2P_-$ and $^4P_+$ states turn out to be described by the wavefunctions (note that here and in the following, the standard notation $j \equiv \{\vec{r}_j, s_j\}$ is used)

$$\Psi(1,2,3) = 2^{-5/4} \, 3^{-11/24} \, \pi^{-5/2} \, \omega^{19/6} \left|1\bar{1}z\right| \, \Phi(\vec{r}_1, \vec{r}_2, \vec{r}_3) \tag{8}$$

and

$$\Psi(1,2,3) = 2^{-1/4} \, 3^{-19/24} \, \pi^{-5/2} \, \omega^{23/6} \left|1xy\right| \, \Phi(\vec{r}_1, \vec{r}_2, \vec{r}_3), \tag{9}$$

respectively, that are products of normalization factors, unnormalized Slater determinants, and the correlated bosonic function

$$\Phi(\vec{r}_1, \vec{r}_2, \vec{r}_3)$$
$$= \exp\left[-\omega\left(\frac{3}{2}r_{CM}^2 + \frac{\sqrt{3}}{6}\left[(r_{12}-r_0)^2 + (r_{23}-r_0)^2 + (r_{31}-r_0)^2\right]\right.\right.$$
$$+ \frac{\sqrt{6}-\sqrt{3}}{9}\left[(r_{12}-r_{23})(r_{12}-r_{31}) + (r_{23}-r_{12})(r_{23}-r_{31})\right.$$
$$\left.\left.+ (r_{31}-r_{12})(r_{31}-r_{23})\right]\right)\right], \tag{10}$$

where $\vec{r}_{CM} = \frac{1}{3}(\vec{r}_1 + \vec{r}_2 + \vec{r}_3)$ and $r_0 = 3^{1/3} \, \omega^{-2/3}$. These wavefunctions, which are asymptotically exact at the limit of $\omega \to 0$, pertain to the $m_L = 0$ components of the states in question (for both of which $\tilde{E}_0 = \frac{1}{2}3^{5/3}$ and $\tilde{E}_1 = \frac{1}{2}(3+\sqrt{3}+\sqrt{6})$ [38]).

At the limit of a vanishing confinement strength, the four-electron harmonium atom turns into a Wigner molecule with the equilibrium geometry of a regular tetrahedron. Several states of this molecule are described by wavefunctions having the same form as their three-electron counterparts. For example, the expressions for the asymptotically exact wavefunctions of the $m_L = 0$ components of the $^5S_-$ quintet, $^3P_+$ triplet, and $^1D_+$ singlet read [38]

$$\Psi(1,2,3,4) = 2^{-43/8} \, 3^{1/4} \, \pi^{-13/4} \, \omega^{21/4} \left|1xyz\right| \, \Phi(\vec{r}_1, \vec{r}_2, \vec{r}_3, \vec{r}_4), \tag{11}$$

$$\Psi(1,2,3,4) = 2^{-125/24} \, 3^{3/4} \, \pi^{-13/4} \, \omega^{55/12} \left(\left|1\bar{1}x\bar{y}\right| + \left|1\bar{1}\bar{x}y\right|\right) \, \Phi(\vec{r}_1, \vec{r}_2, \vec{r}_3, \vec{r}_4), \tag{12}$$

and

$$\Psi(1,2,3,4) = 2^{-125/24}\, 3^{-1/4}\, 5^{1/2}\, \pi^{-13/4}\, \omega^{55/12} \left(\left|1\bar{1}x\bar{x}\right| + \left|1\bar{1}y\bar{y}\right| - 2\left|1\bar{1}z\bar{z}\right| \right)$$
$$\times \Phi(\vec{r}_1, \vec{r}_2, \vec{r}_3, \vec{r}_4), \quad (13)$$

respectively, where the bosonic function is given by the expression

$$\Phi(\vec{r}_1, \vec{r}_2, \vec{r}_3, \vec{r}_4)$$
$$= \exp\left[-\omega \left(2r_{CM}^2 + \frac{\sqrt{3}}{48} (r_{12} + r_{13} + r_{14} + r_{23} + r_{24} + r_{34} - 6\,r_0)^2 \right. \right.$$
$$+ \frac{\sqrt{6}}{16} \left[(r_{12} - r_{34})^2 + (r_{13} - r_{24})^2 + (r_{14} - r_{23})^2 \right]$$
$$+ \left. \left. \frac{\sqrt{3}}{16} \left[(r_{12} - r_{13} - r_{24} + r_{34})^2 + \frac{1}{3} (r_{12} + r_{13} - 2\,r_{14} - 2\,r_{23} + r_{24} + r_{34})^2 \right] \right) \right]$$
$$(14)$$

with $\vec{r}_{CM} = \frac{1}{4}(\vec{r}_1 + \vec{r}_2 + \vec{r}_3 + \vec{r}_4)$ and $r_0 = 2^{2/3}\,\omega^{-2/3}$. For all of these states, $\tilde{E}_0 = 9\,2^{-2/3}$ and $\tilde{E}_1 = \frac{1}{4}(6 + \sqrt{48} + \sqrt{54})$ [38].

Involving quadratures over the components of $N-1$ position vectors in an N-particle system, computation of the one-electron densities from these wavefunctions appears quite straightforward at first glance. However, the pertinent integrals are not known in closed forms and their evaluation with the Laplace method [39] cannot be readily generalized to species with more than three electrons. Fortunately, there exists a specialized transformation of coordinates that permits calculation of asymptotically exact one-particle densities from the rovibrational wavefunctions describing arbitrary freely-rotating systems of particles that satisfy two rather unrestrictive conditions, namely the smallness of the classical vibrational amplitudes in comparison to the interparticle distances and the validity of the harmonic approximation for the vibrational motions [40]. Application of this transformation to the Wigner molecules emerging from the three- and four-electron harmonium atoms upon weakening of the confinement produces a universal expression for the one-electron densities that reads

$$\rho_s(R, \theta; \omega) = B\, \omega^{11/6}\, \Xi_s(\theta) \exp\left[-\zeta\omega (R - D\omega^{-2/3})^2 \right], \quad (15)$$

where the constants B, ζ, and D (compiled in Table 12.4) depend only on N, and the angular functions $\Xi_s(\theta)$ (listed in Table 12.5) are state-specific.

TABLE 12.4　The Constants B, ζ, and D for the Three- and Four-Electron Harmonium Atoms [40]

N	B	ζ	D
3	$\frac{3^{25/12}}{8}\,\pi^{-3/2}\left(1+\sqrt{2}+\sqrt{3}\right)^{-1/2}$	$\frac{9}{3+\sqrt{3}+\sqrt{6}}$	$3^{-1/6}$
4	$\frac{4\sqrt{6}}{3}\,\pi^{-3/2}\left(135+54\sqrt{2}+51\sqrt{3}+60\sqrt{6}\right)^{-1/6}$	$\frac{12}{3+\sqrt{3}+2\sqrt{6}}$	$2^{-5/6}\,3^{1/2}$

TABLE 12.5　The Functions $\Xi_s(\theta)$ for Several States of the Harmonium Atoms [40]

N	State	m_L	$\Xi_\alpha(\theta)$	$\Xi_\beta(\theta)$
3	$^2P_-$	0	$\frac{1}{2}\left(1+5\cos^2\theta\right)$	$1-\cos^2\theta$
	$^4P_+$	0	$3\left(1-\cos^2\theta\right)$	n/a
4	$^3P_+$	0	$\frac{1}{4}\left(4-3\cos^2\theta\right)$	$\frac{3}{4}\cos^2\theta$
	$^1S_+$	0	$\frac{1}{2}$	$\frac{1}{2}$
	$^1D_+$	0	$\frac{5}{16}\left(1+6\cos^2\theta-7\cos^4\theta\right)$	$\frac{5}{16}\left(1+6\cos^2\theta-7\cos^4\theta\right)$
	$^1D_+$	2	$\frac{5}{96}\left(9+6\cos^2\theta-7\cos^4\theta\right)$	$\frac{5}{96}\left(9+6\cos^2\theta-7\cos^4\theta\right)$
	$^5S_-$	0	1	n/a

Due to the lower symmetry of the Wigner molecule pertaining to the five-electron harmonium atom, the one-electron density of its $^4S_-$ state is given by a linear combination of two Gaussian functions with [40]

$$\zeta_1 = \left[\frac{1}{5} + \frac{1}{\sqrt{3}\left(2+\tau^2\right)} + \frac{\tau^2\sqrt{54+18\,\tau^2-3\,\tau^3-\tau^5}}{\sqrt{6}\left(2+\tau^2\right)\sqrt{36+\tau^5}} \right.$$
$$\left. + \frac{\sqrt{\frac{3}{2}}\sqrt{54+18\,\tau^2-3\,\tau^3-\tau^5}}{5\sqrt{156-8\,\tau^2-12\,\tau^3+\tau^5}}\right]^{-1} \approx 1.182\,002\,815, \qquad (16)$$

$$D_1 = \frac{2^{1/3}\sqrt{3}}{\sqrt{3+\tau^2}}\left(\frac{18-\tau^3}{12-\tau^3}\right)^{1/3} \approx 1.103\,586\,315, \qquad (17)$$

$$\zeta_2 = \left[\frac{1}{5} + \frac{2\sqrt{6}\sqrt{(3+\tau^2)(18-\tau^3)}}{9(2+\tau^2)\sqrt{36+\tau^5}} \right.$$

$$+ \frac{\sqrt{6}\sqrt{(3+\tau^2)(18-\tau^3)}\,(84+40\,\tau^2-6\,\tau^3-3\,\tau^5+\gamma)^2}{9\gamma\sqrt{-12+80\,\tau^2+6\,\tau^3-5\,\tau^5+\gamma}\,(84-8\,\tau^2-6\,\tau^3+\tau^5+\gamma)}$$

$$+ \frac{\sqrt{6}\sqrt{(3+\tau^2)(18-\tau^3)}\,(-84-40\,\tau^2+6\,\tau^3+3\,\tau^5+\gamma)^2}{9\gamma\sqrt{-12+80\,\tau^2+6\,\tau^3-5\,\tau^5-\gamma}\,(-84+8\,\tau^2+6\,\tau^3-\tau^5+\gamma)}$$

$$\left. + \frac{\sqrt{3}\,\tau^2}{9(2+\tau^2)} \right]^{-1} \approx 1.203\,161\,295, \tag{18}$$

and

$$D_2 = \frac{2^{1/3}\,\tau}{\sqrt{3+\tau^2}} \left(\frac{18-\tau^3}{12-\tau^3} \right)^{1/3} \approx 1.080\,773\,703, \tag{19}$$

where τ is the smaller of the two real-valued solutions of the equation $3888 + 3888\tau^2 - 648\tau^3 + 1296\tau^4 - 648\tau^5 - 261\tau^6 - 216\tau^7 + 27\tau^8 - 24\tau^9 + 9\tau^{10} + \tau^{12} = 0$ and

$$\gamma = \left(7056 - 1344\,\tau^2 - 1008\,\tau^3 + 5824\,\tau^4 + 264\,\tau^5 \right.$$
$$\left. + 36\,\tau^6 - 976\,\tau^7 - 12\,\tau^8 + 41\,\tau^{10} \right)^{1/2}. \tag{20}$$

Inspection of the relative magnitudes of these quantities reveals that the two Gaussian contributions are spatially resolvable only for $\omega \leqslant 2 \cdot 10^{-10}$.

In summary, the one-electron densities of the harmonium atoms within the strong-correlation regime are characterized by the presence of radial maxima at $R \neq 0$. Consequently, these densities exhibit cage points surrounded by critical entities located on spherical surfaces. Depending on the locations and the natures of the angular extrema in $\Xi_s(\theta)$, these entities can be either circles or the entire surfaces (which, occurring only when $\Xi_s(\theta)$ is a constant function, produces the fat attractor [10]).

The striking differences in the topologies of $\rho(R,\theta;\omega)$ pertaining to the weak- and strong-correlation regimes prompt investigation of the one-electron densities arising from the eigenstates of the Hamiltonian (1) at intermediate confinement strengths. Unlike in the two-electron case, the wavefunctions of the many-electron harmonium atoms cannot be determined analytically even for special values of ω. However, their very accurate estimates are available from numerical calculations.

12.2.3 NUMERICAL CALCULATIONS FOR INTERMEDIATE CONFINEMENT STRENGTHS

Variational calculations involving the explicitly correlated ansatz

$$\Psi(1,\ldots,N) = \sum_{I=1}^{M} C_I \, \hat{A} \left[\Theta(\sigma_1,\ldots,\sigma_N) \, \hat{P} \, \chi_I(\vec{r}_1,\ldots,\vec{r}_N) \right], \qquad (21)$$

where $\Theta(\sigma_1,\ldots,\sigma_N)$ is the appropriate spin function, $\hat{A}$ is the antisymmetrizer that ensures proper permutational symmetry, $\hat{P}$ is the spatial symmetry projector, and

$$\chi_I(\vec{r}_1,\ldots,\vec{r}_N) = \exp\left[-\sum_{i=1}^{N} \alpha_{Ii}(\vec{r}_i - \vec{R}_{Ii})^2 - \sum_{i>j=1}^{N} \beta_{Iij}(\vec{r}_i - \vec{r}_j)^2 \right] \qquad (22)$$

is the Ith explicitly correlated Gaussian lobe primitive, are the method of choice for obtaining highly accurate electronic properties of harmonium atoms [22–24, 41]. With moderate numbers of basis functions, they produce energies within single μhartrees from their exact counterparts and energy components of marginally lower accuracy.

The natural spinorbitals and their occupancies $\{v_n(\omega)\}$ are computed from the wavefunctions by diagonalization of the finite-matrix representations $\{\Gamma_{IJ}\}$ of the respective 1-matrices. These representations are produced by accurate Gaussian quadratures of the multiple integrals (note that the wavefunctions are real-valued)

$$\Gamma_{IJ} = \int \Psi(1,\ldots,N)\,\Psi(1',\ldots,N)\,\phi_I(\vec{r}_1)\,\phi_J(\vec{r}_1')\,d\vec{r}_1\,d\vec{r}_1'\,d2\ldots dN, \quad (23)$$

where

$$\phi_I(\vec{r}) = \xi^{3/4} H_{I_x}(\sqrt{\xi}\,x) H_{I_y}(\sqrt{\xi}\,y) H_{I_z}(\sqrt{\xi}\,z)\,\exp(-\xi\,r^2/2). \qquad (24)$$

In Eq. (24), $H_k(x)$ is the kth normalized Hermite polynomial and $I \equiv (I_x, I_y, I_z)$. Optimal quality of the projection (23) is assured by maximization of the trace $\Sigma_I \Gamma_{II}$ with respect to ξ. In practice, employment of the basis functions (24) with $0 \leq I_x, I_y, I_z \leq 10$ (for the total of $11^3 = 1321$ functions) yields sufficiently accurate natural spinorbitals and their occupancies [22–24].

The spin components of the one-electron densities obtain from the natural spinorbitals and their occupancies pertaining to states with well-defined quantum numbers. In the survey that follows, the wavefunctions of the states with $m_L = 0$ are used except for those of the $^1D_+$ singlet of the four-electron harmonium atom, for which $m_L = 2$.

12.3 A SURVEY OF THE ONE-ELECTRON DENSITIES OF THE THREE- AND FOUR-ELECTRON HARMONIUM ATOMS

Although, thanks to their independence of φ [32,33], the spin components $\rho_s(R,\theta;\omega)$ of the one-electron density in terms of the spherical coordinates are functions of two variables, much information of interest is already provided by their one-dimensional representations. One such representation is the angular distribution

$$\Omega_s(\theta; \omega) = \frac{2\pi}{N_s} \int_0^\infty \rho_s(R, \theta; \omega) R^2 \, dR, \qquad (25)$$

where N_s is the number of electrons with the spin s. This distribution, which is normalized according to

$$\int_0^\pi \Omega_s(\theta; \omega) \sin\theta \, d\theta = 1, \qquad (26)$$

simply equals $\frac{1}{2}$ for the S states. On the other hand, for states with nonvanishing angular momenta, it exhibits some properties that seem unexpected at first glance.

The angular dependences of $\Omega_s(\theta;\omega)$ pertaining to the $^2P_-$, $^4P_+$, and $^3P_+$ states of the three- and four-electron harmonium atoms respond quite differently to variations in the confinement strength ω (Figure 12.4) yet in each case they have fixed points of $\Omega_s(\theta_0;\omega) = \frac{1}{2}$ at the "magic angles" of $\theta_0 = \pm\arccos\frac{\sqrt{3}}{3}$. This universality is a direct consequence of the normalization (26) and the limited harmonic content of $\Omega_s(\theta;\omega)$, which in this instance constraints it to a linear function of $\cos^2\theta$ [33]. Consequently,

(a)

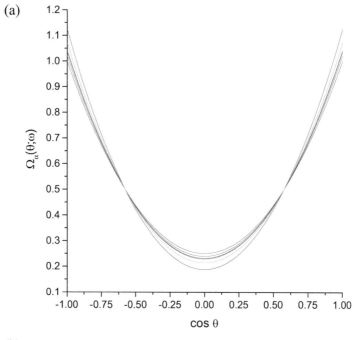

(b)

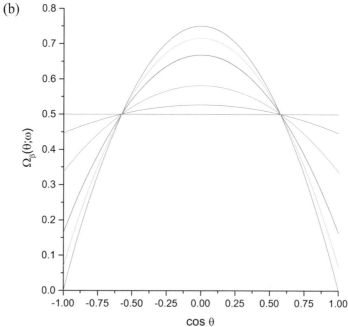

FIGURE 12.4 The angular distributions $\Omega_s(\theta;\omega)$ vs. $\cos\theta$ of the $^2P_-$ (a: α electrons, b: β electrons), $^4P_+$ (c: α electrons), and $^3P_+$ (d: α electrons, e: β electrons) states of the three- and four-electron harmonium atoms for $\omega\to\infty$ (magenta), $\omega = 1$ (blue), $\omega = 0.1$ (orange), $\omega = 0.01$ (purple), $\omega = 0.001$ (green), and $\omega \to 0$ (red) [22, 23].

(c)

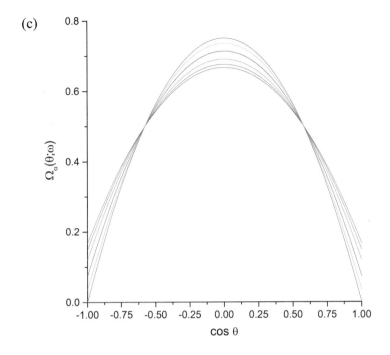

(d)

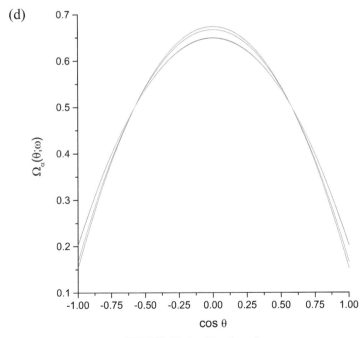

FIGURE 12.4 (Continued).

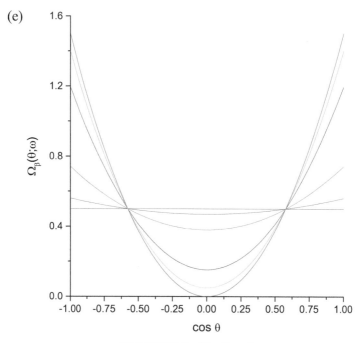

FIGURE 12.4   (Continued).

$$\Omega_s(\theta;\omega) = \frac{1}{2} + \frac{3}{4}\left[2\Omega_s(0;\omega) - 1\right]\left(\cos^2\theta - \frac{1}{3}\right), \qquad (27)$$

which not only demonstrates that $\Omega_s(0;\omega)$ uniquely determines $\Omega_s(\theta;\omega)$ but also explains the origin of the fixed points at the "magic angles". In addition, the variation of the angular dependence of $\Omega_s(\theta;\omega)$ with the confinement strength is elucidated by the dependence of $\Omega_s(0;\omega)$ on ω (Figure 12.5). Thus, the small spread of the parabolic curves in Figures 12.4a, 12.4c, and 12.4d reflects the relative flatness of the $\Omega_s(0;\omega)$ vs. ω plots for the α electrons of the $^2P_-$, $^4P_+$, and $^3P_+$ states of the three- and four-electron harmonium atoms. This flatness, coupled with the presence of a maximum in the respective plot explains the almost complete overlapping of several parabolic curves in Figure 12.4d. The convergence of the angular distributions at $\theta = 0$ to their limiting values compiled in Table 12.6, as well as the equality of these values for the α electrons of the $^3P_+$ state, is noteworthy.

The aforediscussed universality is not preserved for states with higher total angular momenta. In particular, the plots of $\Omega_s(\theta;\omega)$ vs. $\cos\theta$ for the $^1D_+$

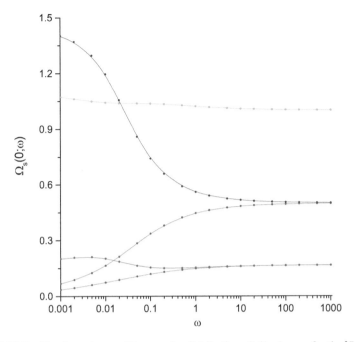

FIGURE 12.5 The dependence of the angular distributions $\Omega_s(0;\omega)$ on ω for the $^2P_-$ (green: α electrons, magenta: β electrons), $^4P_+$ (blue: α electrons), and $^3P_+$ (red: α electrons, purple: β electrons) states of the three- and four-electron harmonium atoms [22, 23].

TABLE 12.6 The $\omega{\to}\infty$ and $\omega{\to}0$ Limits of $\Omega_s(0;\omega)$ for the P States of the Harmonium Atoms

N	State	s	$\Omega_s(0;\infty)$	$\Omega_s(0;0)$
3	$^2P_-$	α	1	$\frac{9}{8}$
		β	$\frac{1}{2}$	0
	$^4P_+$	α	$\frac{1}{6}$	0
4	$^3P_+$	α	$\frac{1}{6}$	$\frac{1}{6}$
		β	$\frac{1}{2}$	$\frac{3}{2}$

state (α or β electrons) of the four-electron harmonium atom (Figure 12.6) exhibit none of the regularities present in their counterparts for the P states. In this case, the angular distributions are quadratic functions of $\cos^2\theta$ [33], which are parameterized by their values at two distinct angles θ, e.g.

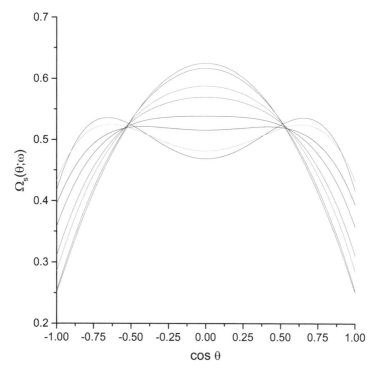

FIGURE 12.6 The angular distributions $\Omega_s(\theta; \omega)$ vs. $\cos \theta$ of the $^1D_+$ state (α or β electrons) of the four-electron harmonium atom for $\omega \to \infty$ (magenta), $\omega = 1$ (blue), $\omega = 0.1$ (orange), $\omega = 0.05$ (dark cyan), $\omega = 0.02$ (pink), $\omega = 0.01$ (purple), $\omega = 0.001$ (green), and $\omega \to 0$ (red) [23].

$$\Omega_s(\theta; w) = \Omega_s\left(\frac{\pi}{2}; w\right) + \frac{3}{4}\left[5 - 2\Omega_s(0; w) - 8\Omega_s\left(\frac{\pi}{2}; w\right)\right] \cos^2 \theta$$
$$+ \frac{5}{4}\left[-3 + 2\Omega_s(0; w) + 4\Omega_s\left(\frac{\pi}{2}; w\right)\right] \cos^4 \theta. \qquad (28)$$

The variations of the quantities $\Omega_s(0;\omega)$, $\Omega_s(\frac{\pi}{2};\omega)$, and the expression $\frac{5}{4}[2\Omega_s(0;\omega) + 4\Omega_s(\frac{\pi}{2};\omega) - 3]$ with ω are displayed in Figure 12.7. The presence of the term quartic in $\cos \theta$ and the sign change (from negative to positive) of the quadratic term upon decreasing ω below a certain value explain the qualitative change (from the single maximum at $\theta = 0$ to two symmetrically located maxima) in the dependences of $\Omega_s(\theta;\omega)$ on θ upon weakening of the confinement.

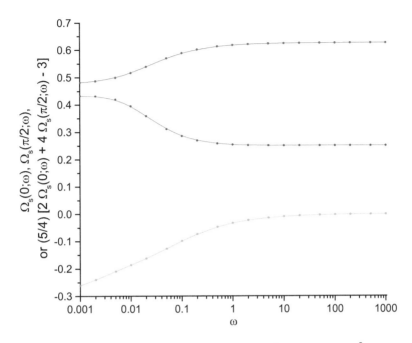

FIGURE 12.7 The dependence of $\Omega_s(0;\omega)$ (blue), $\Omega_s(\frac{\pi}{2};\omega)$ (red), and $\frac{5}{4}[2\Omega_s(0;\omega) + 4\Omega_s(\frac{\pi}{2};\omega) - 3]$ (green) on ω for the $^1D_+$ state (α or β electrons) of the four-electron harmonium atom [23].

The spin components $\rho_s(R,\theta;\omega)$ ($s = \alpha$ or β) of the one-electron densities pertaining to the $^1S_+$ ground state of the two-electron harmonium atom have single attractors at $R = 0$ for all $\omega > \omega_{cr} \approx 0.040116$, whereas for $\omega < \omega_{cr}$ they exhibit "fat attractors" comprising cage points surrounded by spherical surfaces of maximum density [10]. A similar behavior is found for $\rho_s(R,\theta;\omega)$ of the $^1S_+$ singlet ($N = 4$) and $\rho_\beta(R,\theta;\omega)$ of the $^4S_-$ quartet ($N = 5$), for which the values of ω_{cr} are in the vicinities of 0.5 and 0.05, respectively. On the other hand, as predicted by their weak-correlation limits (see Table 12.3), the fat attractors are present in $\rho_\alpha(R,\theta;\omega)$ of both the latter state and its $^5S_-$ counterpart of the four-electron harmonium atom irrespective of the confinement strength. Thus, two distinct patterns in the evolution of the topology of $\rho_s(R,\theta;\omega)$ with ω are encountered in the case of the S states: one characterized by the disappearance of the fat attractor upon strengthening of the confinement and the second devoid of such a catastrophe.

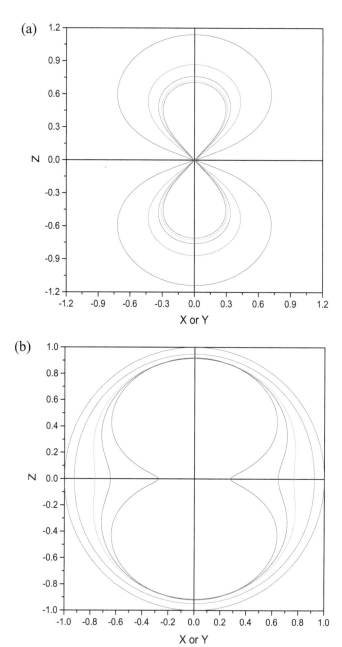

FIGURE 12.8 The xz (or yz) cross sections of the surfaces delineated by the condition $\partial\rho_\alpha(R,\theta;\omega)/\partial R = 0$ for the $^2P_-$ state of the three-electron harmonium atom [22]: (a) (in the length units of $\omega^{-1/2}$) for $\omega \to \infty$ (blue), $\omega = 100$ (green), $\omega = 10$ (magenta), $\omega = 1$ (orange), $\omega = 0.1$ (red) and (b) (in the length units of $3^{-1/6}\,\omega^{-2/3}$) for $\omega = 0.05$ (pink), $\omega = 0.02$ (purple), $\omega = 0.01$ (green), $\omega = 0.001$ (dark cyan), $\omega \to 0$ (blue).

The topologies of the one-electron densities pertaining to states that lack spherical symmetry are more involved. At large values of ω, $\rho_\alpha(R,\theta;\omega)$ of the $^2P_-$ doublet of the three-electron harmonium atom exhibits two attractors located symmetrically on the z axis that are accompanied by a bond point located at $R = 0$ (Figure 12.8a). This topology persists down to ω of the order of 0.1, a further decrease in the confinement strength precipitating a catastrophe that turns the bond point into a cage point at the center of a circle of critical points. The resulting "prolate attractor" becomes progressively more spherical in shape (Figure 12.8b) as the limit of $\omega \rightarrow 0$ is approached. However, even at that limit, the fat attractor is not attained as the one-electron density is not uniform [compare Eq. (15) and Table 12.5] along the gradient paths connecting the axial attractors with the equatorial circle of critical points.

At the limit of $\omega \rightarrow \infty$, $\rho_\beta(R,\theta;\omega)$ of the same state has only one critical point, namely an attractor at $R = 0$. However, lowering ω to ca. 0.05 triggers a catastrophe that creates a "disc attractor" comprising a ring point at the center of a circle of critical points (Figure 12.9). A further weakening of the

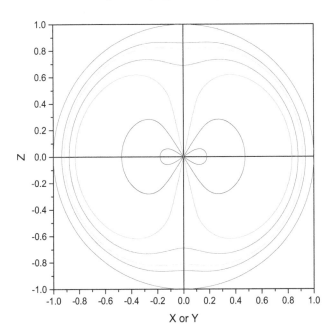

FIGURE 12.9 The xz (or yz) cross sections (in the length units of $3^{-1/6}\,\omega^{-2/3}$) of the surfaces delineated by the condition $\partial\rho_\beta(R,\theta;\omega)/\partial R = 0$ for the $^2P_-$ state of the three-electron harmonium atom [22] for $\omega = 0.05$ (pink), $\omega = 0.02$ (purple), $\omega = 0.01$ (green), $\omega = 0.005$ (olive), $\omega = 0.002$ (cyan), $\omega = 0.001$ (dark cyan), and $\omega \rightarrow 0$ (blue).

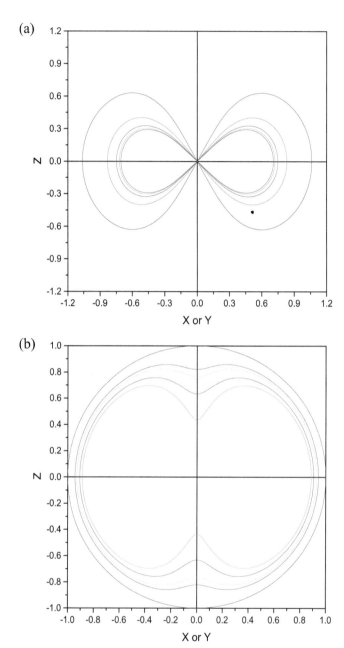

FIGURE 12.10   The xz (or yz) cross sections of the surfaces delineated by the condition $\partial\rho_\alpha(R,\theta;\omega)/\partial R = 0$ for the $^4P_+$ state of the three-electron harmonium atom [22]: (a) (in the length units of $\omega^{-1/2}$) for $\omega \to \infty$ (blue), $\omega = 100$ (green), $\omega = 10$ (magenta), $\omega = 1$ (orange), $\omega = 0.1$ (red) and (b) (in the length units of $3^{-1/6}\,\omega^{-2/3}$) for $\omega = 0.01$ (green), $\omega = 0.005$ (olive), $\omega = 0.002$ (cyan), $\omega = 0.001$ (dark cyan), $\omega \to 0$ (blue).

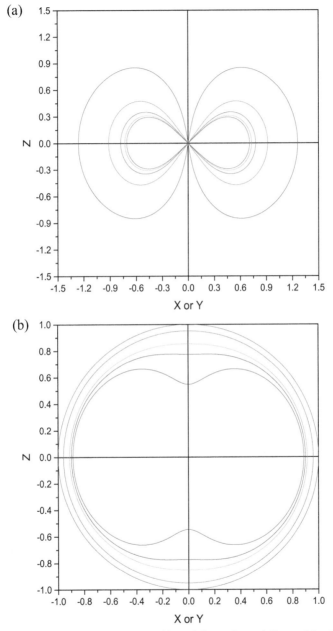

FIGURE 12.11 The xz (or yz) cross sections of the surfaces delineated by the condition $\partial \rho_\alpha(R,\theta;\omega)/\partial R = 0$ for the $^3P_+$ state of the four-electron harmonium atom [23]: (a) (in the length units of $\omega^{-1/2}$) for $\omega \to \infty$ (blue), $\omega = 100$ (green), $\omega = 10$ (magenta), $\omega = 1$ (orange), $\omega = 0.1$ (red) and (b) (in the length units of $2^{-5/6}\, 3^{1/2}\, \omega^{-2/3}$) for $\omega = 0.05$ (pink), $\omega = 0.02$ (purple), $\omega = 0.01$ (green), $\omega = 0.001$ (dark cyan), $\omega \to 0$ (blue).

confinement brings about a second catastrophe at $\omega \approx 0.005$ that involves splitting of the ring point into a set of one centrally located cage point and two axial ring points, which results in formation of an "oblate attractor" (Figure 12.9). Again, as expected from the strong-correlation asymptotics of the one-electron density, the shape of this attractor becomes perfectly spherical at the $\omega \to 0$ limit.

In accordance with the asymptotic predictions (see Table 12.3), the one-electron density of the $^4P_+$ state of the three-electron harmonium atom already has the "disk attractor" topology at strong confinements (Figure 12.10a), which turns into that of the "oblate attractor" at $\omega \approx 0.01$ (Figure 12.10b). The same sequence of topologies is observed for $\rho_\alpha(R,\theta;\omega)$ of the $^3P_+$ triplet of the four-electron harmonium atom (Figure 12.11), the catastrophe occurring at $\omega \approx 0.05$ in this case. On the other hand, the transitions from an ordinary attractor to the set of two axial attractors and a bond point at their midpoint, and then to the "prolate attractor", taking place at the confinement strengths of ca. 0.1 and 0.02, respectively, are observed in the topologies of $\rho_\beta(R,\theta;\omega)$ of the same state (Figure 12.12).

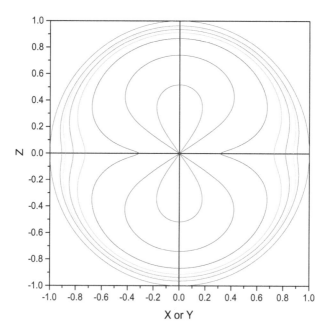

FIGURE 12.12   The xz (or yz) cross sections (in the length units of $2^{-5/6}\,3^{1/2}\,\omega^{-2/3}$) of the surfaces delineated by the condition $\partial\rho_\beta(R,\theta;\omega)/\partial R = 0$ for the $^3P_+$ state of the four-electron harmonium atom [23] for $\omega = 0.1$ (red), $\omega = 0.05$ (pink), $\omega = 0.02$ (purple), $\omega = 0.01$ (green), $\omega = 0.005$ (olive), $\omega = 0.002$ (cyan), $\omega = 0.001$ (dark cyan), and $\omega \to 0$ (blue).

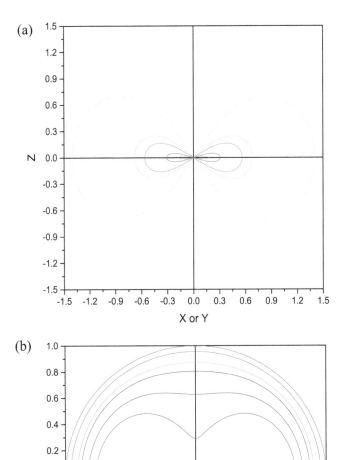

FIGURE 12.13   The xz (or yz) cross sections of the surfaces delineated by the condition $\partial\rho_s(R,\theta;\omega)/\partial R = 0$ ($s = \alpha$ or β) for the $^1D_+$ state of the four-electron harmonium atom [23]: (a) (in the length units of $\omega^{-1/2}$) for $\omega = 100$ (green), $\omega = 10$ (magenta), $\omega = 1$ (orange), $\omega = 0.5$ (light gray), $\omega = 0.2$ (light cyan) and (b) (in the length units of $2^{-5/6}\,3^{1/2}\,\omega^{-2/3}$) for $\omega = 0.1$ (red), $\omega = 0.05$ (pink), $\omega = 0.02$ (purple), $\omega = 0.01$ (green), $\omega = 0.001$ (dark cyan), $\omega \to 0$ (blue).

The sequence of the topologies of $\rho_s(R,\theta;\omega)$ ($s = \alpha$ or β) pertaining to the $^1D_+$ state of the four-electron harmonium atom is a variation on that described above for $\rho_\beta(R,\theta;\omega)$ of the $^2P_-$ doublet. In this case, the ordinary attractor exists only at the limit of $\omega \to \infty$ (note that the presence of a catastrophe at this limit is distinct from its complete absence). At large but finite confinement strengths, it is replaced by the "disk attractor" that steadily dilates (in the length units of $\omega^{-1/2}$) upon weakening of the confinement and finally turns into the "oblate attractor" (possibly with additional rings of critical points) at $\omega \approx 0.1$ (Figure 12.13).

12.4 CONCLUSIONS

Many-electron harmonium atoms are versatile generators of v-representable one-electron densities. These densities, which correspond to ground and excited states of systems with Coulombic interparticle interactions, exhibit a wide range of topologies, reflecting the diversity of the underlying wave-functions. Thanks to the existence of bound states for arbitrary magnitudes of the confinement strength, these wavefunctions involve the entire spectrum of the types and extents of the electron correlation effects encountered in ordinary atoms, molecules, and extended systems. Consequently, electronic properties of the harmonium atoms, including their one-electron densities, are ideal tools for benchmarking of approximate electronic structure methods of quantum chemistry.

The usefulness of these one-electron densities in such applications is further enhanced by the availability of expressions for wavefunctions and the corresponding energies that are asymptotically exact at the weak- and strong-correlation limits. In the former case, the absence of the continuum states allows derivation of closed-form expressions for quantum-mechani-cal observables within the second-order perturbation theory, whereas in the latter one, the suppression of autoionization by the harmonic confinement permits formulation of an asymptotic theory based upon a rovibrational Hamiltonian. When employed in conjunction with the respective energies, the one-electron densities pertaining to those asymptotic regimes impose strict constraints upon approximate entities (such as density functionals) whose performance can be further assessed with the help of electronic properties of harmonium atoms at intermediate confinement strengths. These properties are readily computed with explicitly-correlated numerical approaches that complement the analytic results. Since, due to the absence of singularities in the confining potential, the wavefunctions are devoid of nuclear cusps, the efficiency of such calculations greatly exceeds that observed for fully Coulombic systems.

In summary, the one-electron densities of the many-electron harmonium atoms are expected to appear in many facets of research on approximate electronic structure methods and interpretive tools such as similarity and reactivity indices [42]. More work in this direction, including investigations of the harmonium atoms with greater numbers of electrons, is certainly warranted.

KEYWORDS

- **harmonium atom**
- **unusual attractors**
- **Wigner molecule**

REFERENCES

1. Carbo, R., Arnau, M., & Leyda, L., (1980). How similar is a molecule to another? An electron density measure of similarity between two molecular structures. *Int. J. Quant. Chem., 17,* 1185–1189.
2. Savin, A., Jepsen, O., Flad, J., Andersen, O. K., Preuss, H., & von Schnering, H. G., (1992). Electron localization in solid-state structures of the elements – the diamond structure. *Angew. Chem. Int. Ed. Engl., 31,* 187–188.
3. Bader, R. W. F., (1994). *Atoms in Molecules: A Quantum Theory,* Oxford University Press.
4. Kryashko, E. S., & Ludeña, E. W., (1990). *Energy Density Functional Theory of Many-Electron Systems,* Kluwer, Dordrecht.
5. Macchi, P., Gillet, J.-M., Taulelle, F., Campo, J., Claiser, N., & Lecomte, C., (2015). Modelling the experimental electron density: only the synergy of various approaches can tackle the new challenges. *IUCrJ, 2,* 441–451.
6. Taut, M., (1993). Two electrons in an external oscillator potential: Particular analytic solutions of a Coulomb correlation problem. *Phys. Rev. A, 48,* 3561–3566.
7. Kestner, N. R., & Sinanoğlu, O., (1962). Study of electron correlation in helium-like systems using an exactly soluble model. *Phys. Rev., 128,* 2687–2692.
8. Santos, E., (1968). Calculo aproximado de la energia de correlacion entre dos electrones. *Anal. R. Soc. Esp. Fis. Quim., 64,* 177–193.

9. King, H. F., (1996). The electron correlation cusp. I. Overview and partial wave analysis of the Kais function. *Theor. Chim. Acta., 94,* 345– 381.

10. Cioslowski, J., & Pernal, K., (2000). The Ground State of Harmonium, *J. Chem. Phys., 113,* 8434–8443.

11. Sahni, V., (2010). *Quantal Density Functional Theory II: Approximation Methods and Applications,* Springer-Verlag, Berlin.

12. Gori-Giorgi, P., & Savin, A., (2009). Study of the discontinuity of the exchange-correlation potential in an exactly soluble case. *Int. J. Quant. Chem., 109,* 2410–2415.

13. Zhu, W. M., & Trickey, S. B., (2006). Exact density functionals for two-electron systems in an external magnetic field. *J. Chem. Phys., 125,* 094317/1–12.

14. Hessler, P., Park, J., & Burke, K., (1999). Several theorems in time-dependent density functional theory. *Phys. Rev. Lett., 82,* 378–381.

15. Ivanov, S., Burke, K., & Levy, M., (1999). Exact high-density limit of correlation potential for two-electron density. *J. Chem. Phys., 110,* 10262–10268.

16. Qian, Z., & Sahni, V., (1998). Physics of transformation from Schrödinger theory to Kohn-Sham density-functional theory: Application to an exactly solvable model. *Phys. Rev. A, 57,* 2527–2538.

17. Taut, M., Ernst, A., & Eschrig, H., (1998). Two electrons in an external oscillator potential: exact solution versus one-particle approximations. *J. Phys B, 31,* 2689–2708.

18. Elward, J. M., Hoffman, J., & Chakraborty, A., (2012). Investigation of electron-hole correlation using explicitly correlated configuration interaction method. *Chem. Phys. Lett., 535,* 182–186.

19. Elward, J. M., Thallinger, B., & Chakraborty, A., (2012). Calculation of electron-hole recombination probability using explicitly correlated Hartree-Fock method. *J. Chem. Phys., 136,* 124105/1–10.

20. Glover, W. J., Larsen, R. E., & Schwartz, B. J., (2010). First principles multielectron mixed quantum/classical simulations in the condensed phase. I. An efficient Fourier-grid method for solving the many-electron problem. *J. Chem. Phys., 132,* 144101/1–11.

21. Varga, K., Navratil, P., Usukura, J., & Suzuki, Y., (2001). Stochastic variational approach to few-electron artificial atoms. *Phys. Rev. B, 63,* 205308/1–15.

22. Cioslowski, J., Strasburger, K., & Matito, E., (2012). The three-electron harmonium atom: The lowest-energy doublet and quadruplet states. *J. Chem. Phys., 136,* 194112/1–8.

23. Cioslowski, J., Strasburger, K., & Matito, E., (2014). Benchmark calculations on the lowest-energy singlet, triplet, and quintet states of the four-electron harmonium atom. *J. Chem. Phys., 141,* 044128/1–6.

24. Cioslowski, J., & Strasburger, K. (to be published).

25. Benson, J. M., & Byers Brown, W., (1970). Perturbation energies for the Hooke's law model of the two-electron atom. *J. Chem. Phys., 53,* 3880–3886.

26. White, R. J., & Byers Brown, W., (1970). Perturbation theory of the Hooke's law model for the two-electron atom. *J. Chem. Phys., 53,* 3869–3879.

27. Gill, P. M. W., & O'Neill, D. P., (2005). Electron correlation in Hooke's law atom in the high-density limit. *J. Chem. Phys., 122,* 094110/1–4.

28. Cioslowski, J., & Matito, E., (2011). Note: The weak-correlation limit of the three-electron harmonium atom. *J. Chem. Phys., 134,* 116101/1–2.

29. Petkovsek, M., Wilf, H. S., & Zeilberger, D., (1996). $A = B$, A. K. Peters, Wellesley, Massachusetts.

30. Cioslowski, J., (2013). The weak-correlation limits of few-electron harmonium atoms. *J. Chem. Phys., 139,* 224108/1–5.

31. Cioslowski, J. & Strasburger, K. (2017). Harmonium atoms at weak confinements: The formation of the Wigner molecules. *J. Chem. Phys.*, *146*, 044308/1-8.
32. These spin components of the one-electron density are independent of φ and their dependences on θ are described by finite-degree polynomials in $\cos^2\theta$ [33].
33. Fertig, H. A., & Kohn, W., (2000). Symmetry of the atomic electron density in Hartree, Hartree-Fock, and density-functional theories. *Phys. Rev. A, 62,* 052511/1–10.
34. Bonitz, M., Henning, C., & Block, D., (2010). Complex plasmas: a laboratory for strong correlations. *Rep. Prog. Phys., 73,* 066501/1–29.
35. Cioslowski, J., & Albin, J., (2013). Asymptotic equivalence of the shell-model and local-density descriptions of Coulombic systems confined by radially symmetric potentials in two and three dimensions. *J. Chem. Phys., 139,* 114109/1–6 and the references cited therein.
36. Arp, O., Block, D., Bonitz, M., Fehske, H., Golubnychiy, V., Kosse, S., Ludwig, P., Melzer, A., & Piel, A., (2005). 3D Coulomb balls: experiment and simulation. *J. Phys.: Conf. Ser., 11,* 234–247.
37. Egger, R., Häusler, W., Mak, C.H., & Grabert, H., (1999). Crossover from Fermi liquid to Wigner molecule behavior in quantum dots. *Phys. Rev. Lett., 82,* 3320–3323.
38. Cioslowski, J., (2016). Rovibrational states of Wigner molecules in spherically symmetric confining potentials. *J. Chem. Phys., 145,* 054116/1–12.
39. Cioslowski, J. (2015). One-electron reduced density matrices of strongly correlated harmonium atoms. *J. Chem. Phys., 142,* 114104/1–11.
40. Cioslowski, J. (2017). One-electron densities of freely rotating Wigner molecules. *J. Phys. B: At. Mol. Opt. Phys., 50,* 235102/1-10.
41. Strasburger, K., (2016). The order of three lowest-energy states of the six-electron harmonium at small force constant. *J. Chem. Phys., 144,* 234304/1–8.
42. There is another (rather unscientific) aspect of these densities. When seeing plots quite similar to those presented in Figures 12.8–12.13 of this chapter, an American graduate student attending a lecture by the late R. W. F. Bader exclaimed: "These are vaguely obscene pictures!"

CHAPTER 13

UNDERSTANDING STRUCTURE-PROPERTY RELATIONSHIPS IN OCTAPHYRINS

T. WOLLER,[1] P. GEERLINGS,[1] F. DE PROFT,[1] M. ALONSO,[1] and J. CONTRERAS-GARCÍA[2]

[1]*Eenheid Algemene Chemie (ALGC), Vrije Universiteit Brussel (VUB), Pleinlaan 2, 1050 Brussels, Belgium*

[2]*Sorbonne Universités, UPMC University of Paris 06,CNRS, Laboratoire de Chimie Théorique (LCT), 4 Place Jussieu, 75252 Paris Cedex 05, France*

CONTENTS

13.1 INTRODUCTION

In recent years, expanded porphyrins have emerged as a new class of functional molecules in light of their large conformational flexibility, rich metal coordination behavior, unprecedented chemical reactivities and exceptional nonlinear optical properties [1–3]. Expanded porphyrins are macrocycles consisting of more than four pyrrole rings linked together either directly or through one or more spacer atoms in such a manner that the internal pathway contains a minimum of 17 atoms [4]. Owing to their tunable photophysical and chemical properties with external *stimuli*, expanded porphyrins represent a very promising platform to develop molecular switches for molecular electronic devices [5–7]. Molecular switches are regarded as the most basic component in molecular electronic devices that can reverse from an active/on state to a passive/off state. The switch between two or more states with distinct properties is triggered by external stimuli such as light, pH or voltage [8]. As opposed to normal switches, molecular switches are extremely tiny and their application in nanotechnology, biomedicine and computer chip design opens up brand new horizons. However, despite their potential, the design of molecular switches based on expanded porphyrins has only been scarcely investigated [7, 9, 10].

The most fascinating property of expanded porphyrins lays in their capacity to switch between several distinct π-conjugation topologies encoding different properties [11]. Besides the Hückel topology, expanded porphyrins can adopt a variety of conformations with Möbius and twisted-Hückel topologies that can be interconverted under certain conditions [12, 13]. Recent experimental studies [14, 15] proved that the photophysical and nonlinear optical (NLO) properties of expanded porphyrins strongly depend on their molecular topology and macrocyclic aromaticity. In addition, Alonso and co-workers demonstrated the existence of a triangular relationship between

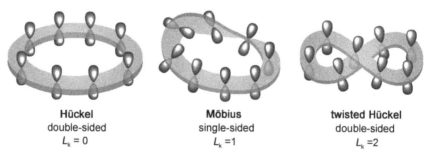

Hückel	Möbius	twisted Hückel
double-sided	single-sided	double-sided
$L_k = 0$	$L_k = 1$	$L_k = 2$

SCHEME 1 Schematic representation of the topologies of p-conjugated electron systems. See text for the definition of L_k. Reproduced from Ref. [8d] with permission from the PCCP Owner Societies.

aromaticity, oxidation state of the macrocycle and topology in selected expanded porphyrins [8–10, 13].

Depending on the topology and the oxidation state of expanded porphyrins, the systems will follow the Hückel's or Möbius rule for aromaticity. According to Hückel's rule, aromatic systems present a full cyclic delocalization, conformational planarity and contain $[4n+2]$ π-electrons in their conjugation pathway [16]. However, Hückel's rule was originally derived for monocyclic systems containing identical atoms, so this rule cannot be strictly applied to porphyrinoids which contain multiple heterocyclic rings. In 1964, theoretical studies on annulenes reported the existence of cyclic π-systems with a half-twist [17]. Those molecules were predicted to follow the reverse Hückel's rule for aromaticity, so singly-twisted compounds were foreseen to be aromatic with $[4n]$ π-electrons. This phenomenon was called Möbius aromaticity in reference to the topology of the system. The range of application of Möbius and Hückel aromaticity was generalized afterwards according to the topology of the conjugated systems. Accordingly, π-conjugated systems with an odd number of half-twists [18] (or more precisely with an even linking number, L_K) follow the Möbius aromaticity while Hückel's rule still holds for predicting the aromaticity of systems with an even number of half-twists (odd values of L_K) [18]. Since expanded porphyrins are the best platform to realize versatile electronic states including Möbius aromatic and antiaromatic species, they provide a good test bed in which the concept of aromaticity can be explored. In fact, very recently, it was found that expanded porphyrins reverse the aromaticity in the lowest-energy triplet excited state compared with its closed-shell singlet ground state, experimentally proving Baird's rule for the first time [19, 20].

Nevertheless, the quantification of aromaticity in expanded porphyrins remains challenging. Aromaticity is currently described as a multidimensional phenomenon, implying that the aromatic character cannot be quantified using a single aromaticity descriptor [21]. In fact, the aromaticity of porphyrinoids was shown to be particularly "multifaceted," resulting in large discrepancies between energetic and magnetic descriptors [22]. Consequently, many authors strongly recommend the use of several descriptors of aromaticity based on different criteria to quantify the aromaticity of such large macrocycles [9, 23].

In previous works, Alonso and co-workers demonstrated that computational chemistry is a powerful tool in aiding the design of viable Möbius expanded porphyrins with high aromatic character [9, 10, 13, 21].

Our initial research focusing on metal-free expanded porphyrins varying in ring sizes and oxidation states revealed that the conformation of the expanded porphyrins is clearly dependent on the oxidation state and the size of the macrocycle. $[4n + 2]$ π-electron expanded porphyrins adopt Hückel conformations, almost planar and highly aromatic, whereas antiaromatic Hückel and Möbius conformers coexist in dynamic equilibrium for $[4n]$ π-electrons expanded porphyrins [9, 10]. The larger [32]heptaphyrin strongly prefers a figure-eight conformation in the neutral state, whereas the Möbius topology becomes the most stable in protonated species [24].

In order to expand our research toward the design of functional molecular switches from expanded porphyrins, we hereby examine the different factors involved in the switching process of [36]octaphyrin(1.1.1.1.1.1.1.1). Octaphyrin macrocycles contain 8 pyrrole rings connected via meso-carbon bridges. This class of expanded porphyrins exhibit interesting properties such as multi-metal coordination, dynamical structures [15], and facile redox reactions. Appealingly, [36]octaphyrin provides Möbius aromatic structures upon protonation leading to large variations in the NLO properties [31].

The aim of this chapter is two-fold. On the one hand, the conformational preferences of [36] octaphyrins in several environments have been thoroughly investigated using DFT calculations, with a particular focus on the influence of meso- and β-substituents (Scheme 2). One the other hand, we have analyzed how the properties of octaphyrins change as a function of the topology, redox reactions and protonation. More specifically, we intend to (i) scrutinize the conformational preferences of octaphyrins in gas-phase and different solvents, (ii) identify the switching mechanism for bistable and tristable octaphyrins, (iii) determine effective stimuli for triggering Hückel-Möbius topological switches and/or aromaticity switches, (iv) establish the structure–property relationship between molecular topology, aromaticity, number of electrons and NLO properties, and (v) qualitatively explore the evolution of the conductance with the topology using the concept of bond metallicity [25].

Besides the fine-tuning of the molecular topology, we also aim at understanding the locality of the electron delocalization changes with respect to topological changes in octaphyrins. It is important to note that such a change of topology is achieved by variation of internal dihedral angles. However, the most common measures of metallicity are non-local. For instance, a metallic bond will be associated with a large local DOS (density of states)

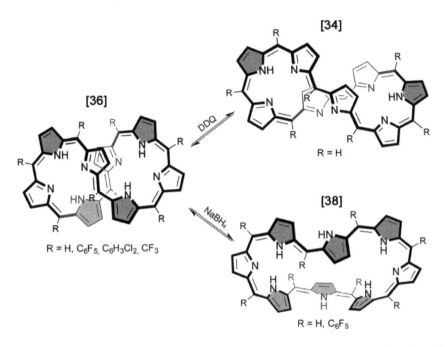

SCHEME 2 Molecular structures of [36]octaphyrins and their redox-triggered interconversions leading to [34] and [38]octaphyrins. Reproduced from Ref. [8d] with permission from the PCCP Owner Societies.

at the Fermi level, with the "nearsightedness" of the off-diagonal term in the first-order density matrix increasing as the band gap does [26–28]. As these forms require non-local information, they cannot provide the connection with real space (molecular structure, bonds and dihedral angles). Therefore, we will apply the concept of bond metallicity, which so far has only been applied to solids, to assess the evolution of the electron delocalization with the topology of octaphyrins [25].

13.2 METHODOLOGY

13.2.1 EVALUATION OF THE STRUCTURAL FACTORS IN OCTAPHYRINS

Due to the poor description of dispersion interactions by most DFT functionals [29], the performance of several exchange-correlation functionals in reproducing the molecular structure of various *meso*-substituted oct-aphyrins was assessed by comparison with X-ray diffraction data [30]. Then, an exhaustive conformational analysis was performed on the unsub-stituted [36]octaphyrin. Since each pyrrole subunit can adopt four different spatial conformations (*cis-cis, cis-trans, trans-cis, trans-trans*), octaphy-rins could theoretically display 256 different conformations. Therefore, we selected a reduced set of conformations based on the reported crystal-lographic structures of neutral, protonated and metalated *meso*-substituted octaphyrins [31].

Since the stability of octaphyrins is influenced by numerous factors [30, 31], we have proposed a set of descriptors to quantify independently the con-tribution of the ring strain, aromaticity and hydrogen bonding. Ring strain is quantified by the torsional descriptor Φ_p, which corresponds to the average dihedral angle between neighboring pyrrole rings. The extent of effective overlap of adjacent p-orbitals is measured by the torsional π-conjugation index (Π), defined as the product of the cosine of the dihedral angles (ϕ_i) along the classical conjugation pathway (CP).

$$\Pi = \prod_i^n \cos(\varphi_i) \tag{1}$$

where Π equals 1 for a completely planar system, it is positive for any Hückel (double-sided) conformation and negative for any Möbius (single-sided) surface. Normally, macrocyclic aromaticity is associated with porphyrinoids having Π values higher than 0.3 [2].

13.2.2 EVALUATION OF NONCOVALENT INTERACTIONS

The hydrogen bonding inside the different macrocycles was assessed with two different methods. On the one hand, the hydrogen-bond index (N_H) qual-itatively provides the number of intramolecular hydrogen bonds stabilizing the conformation and assigns a value of 1 for single N-H$\cdots$N bonds and 1.5 for bifurcated ones. On the other hand, the semi-quantitative noncovalent interaction (NCI) index was used to evaluate the strength and number of hydrogen bonds [33, 34]. The NCI method characterizes noncovalent inter-actions using the electron density and its first derivatives. In this method, the sign of the second eigenvalue (λ_2) of the electron-density Hessian matrix enables to distinguish attractive ($\lambda_2 < 0$) from repulsive interactions ($\lambda_2 > 0$) whereas the magnitude of the density itself provides the strength of the interaction. Weak van der Waals interactions are denoted by values of sign (λ_2)ρ close to 0 whereas stronger hydrogen bonds appear in the region of -0.015 to -0.04 a.u.

13.2.3 QUANTIFICATION OF AROMATICITY

Owing to the multidimensional character of aromaticity, we used a set of energetic, magnetic, reactivity and structural descriptors to quantify the Hückel and Möbius aromaticity of octaphyrins [21, 22].

13.2.3.1 Energetic Criteria

The isomerization method [16] was applied to evaluate the isomerization stabilization energies (ISE), magnetic susceptibility exaltation (Λ) [35] and relative hardness ($\Delta\eta$) of [36]octaphyrins. This method is based on the difference in properties between a methyl derivative of the conjugated system and its non-aromatic exocyclic methylene isomer [23]. The introduction of a methylene substituent disrupts the overall π-conjugation while the methyl compound conserves it. The reactions that were used to compute the aromaticity indices of the twisted-Hückel and Hückel octaphyrins with [36] and [38] π-electrons are shown in Scheme 3.

$$\text{ISE} = E_{CH_2} - E_{CH_3} \qquad (2)$$

The isomerization stabilization energy (ISE) is computed as the energy difference between the methylene isomer (E_{CH_2}) and the methyl derivative (E_{CH_3}) of the systems (Eq. 1). Accordingly, negative ISE are expected for antiaromatic systems while aromatic systems display positive ISE values. However, the isomerization method requires the application of *syn-anti* correction due to the *cis-trans* diene mismatches in the methyl and methylene isomer [36]. The *syn-anti* corrections can be evaluated as the energy difference between the dihydrogen derivative of the *meso*-methyl expanded porphyrin and its respective non-aromatic isomer.

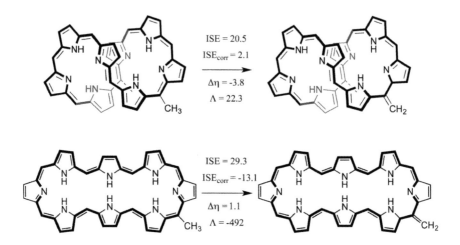

SCHEME 3   Reaction used to evaluate several aromaticity descriptors in octaphyrins. ISE and $\Delta\eta$ are given in kcal mol^{-1} and Λ in ppm cgs. Reproduced from Ref. [8d] with permission from the PCCP Owner Societies.

13.2.3.2   Magnetic Criteria

Magnetic criteria for aromaticity quantify the response of a molecular system to an applied external magnetic field perpendicular to the molecular plane. Aromatic systems present an induced diatropic ring current whereas paratropic ring currents are observed in antiaromatic compounds. The magnetic properties of a molecular system are evaluated either by diamagnetic susceptibility exaltation (Λ) or by proton NMR shift. The magnetic susceptibility exaltation, Λ, is evaluated as the difference between the magnetic susceptibility of a molecular system (χ_M) and that of a reference one without any cyclic electronic delocalization ($\chi_{M'}$) [37]. Here, Λ values were computed via the ISE method.

$$\Lambda = \chi_M - \chi_{M'} \qquad (3)$$

The magnetic susceptibility χ denotes the degree of magnetization of a material in response to an applied external magnetic field. Aromatic systems display negative (diatropic) magnetic susceptibility exaltations while positive exaltations are observed in antiaromatic systems. The second magnetic criterion for aromaticity is based on the ^{1}H chemical shifts [2]. Thanks to the presence of exocyclic and endocyclic hydrogen atoms in expanded porphyrins, the ^{1}H-NMR chemical shifts are frequently used to evaluate the ring current effects in those macrocycles. Aromatic expanded porphyrins are characterized by a large diamagnetic shift on the inner proton signals and large positive shifts for the outer protons. Antiaromatic systems have paratropic ring currents, and the effects on the chemical shifts are reversed. The nucleus-independent chemical shift (NICS) is defined as the negative value of the absolute magnetic shield tensor at the ring center or 1 Å below or above the molecular plane. The NICS(1) values, calculated at 1 Å above the molecular plane, and its corresponding out-of-plane component of the NICS(1) tensor, denoted as $NICS_{zz}(1)$, are considered to better reflect the π-electron effects in organic compounds, such us [n]annulenes [38]. Negative values of NICS are indicative of aromaticity (diatropic ring current) while positive values of NICS are typical of antiaromatic compounds (paratropic ring current) [37].

13.2.3.3   Reactivity Criteria

Recently, Alonso et al. have demonstrated that macrocyclic aromaticity of Hückel and Möbius expanded porphyrins can be quantified using the relative hardness, evaluated via the isomerization method [39]. In conceptual DFT, the chemical hardness, η, is defined as the resistance of the system towards a change in its number of electrons [40]. In the frame of the Koopmans approximation [41], the hardness of a compound corresponds to the energy difference between the frontier orbitals HOMO and LUMO.

$$\eta = \varepsilon_{LUMO} - \varepsilon_{HOMO} \tag{4}$$

Since larger energy gaps are indicative for stability, aromatic systems present larger energy gap than their antiaromatic analogous. This finding is nicely captured in the maximum hardness principle, which states that molecules will arrange themselves to be as hard as possible [42]. However, since the stabilization/destabilization due to the aromaticity is only a small portion of the total energy, the relative hardness was proposed to be a better measure of aromaticity than the absolute hardness [39]. In the frame of the isomerization method, the relative hardness ($\Delta\eta$) corresponds to the difference in hardness between the methyl derivative and the methylene isomer.

$$\Delta\eta = \eta_{CH_3} - \eta_{CH_2} \tag{5}$$

Using the maximum hardness principle, the methyl derivative is expected to be systematically harder than the methylene isomer in aromatic compounds ($\Delta\eta>0$) and *vice versa* in antiaromatic systems ($\Delta\eta<0$).

The harmonic oscillator model of aromaticity (HOMA), defined by Kruszewski and Krygowski (Eq. 6), was used as a structural descriptor of aromaticity [43].

$$\mathrm{HOMA} = 1 - \frac{\alpha}{n}\sum_{i}^{n}\left(R_i - R_{opt}\right)^2 \tag{6}$$

where α is an empirical constant fixed for each type of bond and n corresponds to the number of bonds taken into account in the summation. Consequently, HOMA equals 0 for non-aromatic systems whereas HOMA = 1 for fully aromatic ones with all bonds equal to the optimal value, R_{opt}. R_i denotes the running bond length along the annulene-type conjugation pathway. According to the annulene model, the main conjugation pathway of expanded porphyrins passes through the imine nitrogen group and circumvents the amino NH group [44].

13.2.4 METALLICITY MEASURES

Several approaches to tackle conductivity from local measures have been proposed in the past. The most common approach is to employ density-related values at the bond critical point (*bcp*) as originated from atoms in molecules (AIM) [45–50]. In the frame of AIM, the chemical bonds are characterized by the properties of the electron density at the *bcp*. In an ordinary molecule, the electron density exhibits cusps (associated with maxima) at the nuclei and an exponential decay away from the nuclei. The resulting topology looks like a mountain chain, each of the peaks being commonly identified with an atom. The mountain pass would then represent the bond, and its lowest point is used to identify the presence of the interaction. This point is thus known as bond critical point (*bcp*). Mori-Sanchez and co-workers [51] proposed a method able to distinguish between covalent, ionic and metallic compounds based on the comparison between all critical points in crystals. However, as this method does not provide a truly local measure of metallicity, as related to a given bond, we will concentrate on the measures uniquely related to *bcps*.

In 2009, Jenkins proposed a bond metallicity index based on the electron density and its Laplacian at the *bcp*, $\rho(r_{bcp})$ and $\nabla^2\rho(r_{bcp})$, respectively [52]. The sign of the Laplacian determines the interaction type, distinguishing from closed shell, such as ionic or metallic (positive Laplacian) and shared or covalent interactions (negative Laplacian). Thus, when looking at these quantities for measuring metallicity, we only concentrate on *bcps* with positive Laplacian values [53].

$$\xi_j\left(r_{bcp}\right) = \frac{\rho\left(r_{bcp}\right)}{\nabla^2\rho\left(r_{bcp}\right)} \tag{7}$$

Since ξ_j is neither dimensionless nor defined for singular Laplacian, a revised version, ξ_m, overcoming these flaws was proposed [58]:

$$\xi_m\left(r_{bcp}\right) = \frac{36\left(3\pi^2\right)^{2/3} \rho\left(r_{bcp}\right)^{2/3} \xi_j\left(r_{bcp}\right)}{5} \tag{8}$$

Based on the value of ξ_m or ξ_j at a bcp with positive Laplacian (i.e., at non covalent interactions), metallic bonds can be distinguished from non-metallic bonds. For instance, metallic bonds are denoted by ξ_m greater than 25 whereas ξ_m values between 5 and 25 are indicative of a partial metallic character. Weak metallic bonds are characterized by ξ_m values between 1 and 5. When ξ_m is inferior to 1, the bond exhibits a non-metallic character.

In addition, local metallicity measures have also been introduced in terms of kinetic energy densities and delocalization measures. In our work, we will consider the electron localization function (ELF) and other kinetic energy density based measures.

According to the formulation of Becke and Edgecombe [54], ELF measures the probability of finding an electron in the vicinity of another electron with the same spin.

In Savin's interpretation [55], ELF is constructed from the difference between the monodeterminantal kinetic energy density and Weizsäcker's kinetic energy density (i.e., for a bosonic system), scaled by the Thomas-Fermi kinetic energy. Therefore, the variable χ represents a measure of the effect of the Pauli principle in the kinetic energy density:

$$ELF = \cfrac{1}{\left[1 + \cfrac{\tau(R) - \tau_W(R)}{\tau_{TF}}\right]^2} = \frac{1}{(1 + \chi)^2} \tag{9}$$

$$\tau_{TF}(R) = \frac{3(3\pi^2)^{2/3} \rho(R)^{5/3}}{10} \tag{10}$$

$$\tau_W(R) = \frac{\nabla\rho(R)\nabla\rho(R)}{8\rho(R)} \tag{11}$$

In highly localized regions such as covalent bonds or nuclei, ELF tends to 1 whereas ELF≈0.5 indicates a degree of delocalization similar to the uniform electron gas. ELF can provide insight into the electron delocalization when analyzed in between ELF basins [56] (e.g., around intermolecular *bcps*). In these cases, high ELF values identify delocalization in between localized regions [57]. This explains why ξ_m is positively correlated with the electron localization function at positive Laplacian *bcps* [58].

Moreover, $\tau(R)$ can also be rewritten in terms of the electron density and the local temperature, $\theta(R)$ [59].

$$\chi(r_{bcp}) = \frac{\tau(R) - \tau_W(R)}{\tau_{TF}} = \frac{\frac{3}{2}\rho(R)\theta(R) - \tau_W(R)}{\tau_{TF}} \tag{12}$$

As demonstrated by Jenkins and co-workers, a gradient expansion approximation can be applied in order to explore the link between the different metallicity measures. In the gradient expansion approximation, the kinetic energy density is expanded as the sum of the Thomas-Fermi [60, 61], Weizsäcker's kinetic energy density [61] and the Laplacian [62].

$$\tau = \tau_{TF} + \frac{1}{9}\tau_W - \frac{1}{6}\nabla^2\rho \tag{13}$$

Since the electron delocalization function in ELF can be expressed in terms of the local temperature, it is also possible to establish a link between ELF and the bond metallicity.

$$\chi\left(r_{bcp}\right) \approx 1 + \frac{4}{\xi_m\left(r_{bcp}\right)} \qquad (14)$$

13.3 RESULT AND DISCUSSION

In the first part, the conformational preferences of [36]octaphyrins has been thoroughly investigated, including the substituent effects on the stability of the figure-eight, Möbius and Hückel conformations. Then, we focus on determining effective external *stimuli* for triggering topological/aromaticity switches in octaphyrins. In this sense, the ability of redox reactions and protonation to induce conformational and/or π-electron switching in octaphyrins has been investigated. Once the conformational control in octaphyrins is fully understood, the aromaticity of the twisted-Hückel, Möbius and Hückel conformers with different oxidation and protonation state has been quantified using energetic, magnetic, structural and reactivity criteria. Then, the structure-property relationships between molecular topology, aromaticity and NLO properties have been established. Identifying methods to properly quantify the aromaticity of these unique macrocycles systems might help in the design of octaphyrins with optimal and specific properties.

In the second part, we focus on the structure-property relationship between molecular conformation and molecular conductance in octaphyrins using the concept of bond metallicity. Since this concept so far has only been applied to solids, we first perform a proof of principle on different molecular systems in order to assess their performance and limitations and, of course, their ability to reveal conductance and the subtleties of a topological switch [25]. Finally, these indices will be applied to Hückel and Möbius conformations of octaphyrins in order to understand the locality of the conductivity changes upon dihedral rotation and when this assumption stays valid. This knowledge might help in the design of octaphyrins with optimal properties for molecular switching devices.

13.2.1 CONFORMATIONAL STABILITY OF OCTAPHYRINS UNDER VARIOUS ENVIRONMENT

13.3.1.1 DFT Benchmark Study for Hückel and Möbius

Before the conformational study, we assessed the performance of several exchange-correlation functionals (B3LYP [63], PBE [64], M06 [65], ωB97XD [66], B3LYP-D [67] and BP86 [68]) in reproducing the molecular structure of neutral and diprotonated *meso*-octakis(pentafluorophenyl) [36] octaphyrins. The neutral system exhibits a twisted-Hückel topology whereas a Möbius topology is found for the second system (Figure 13.1) [30, 31]. In this benchmark study, the comparison criteria were based on the root-mean-square deviation (RMS) between DFT-optimized and the X-ray cartesian coordinates as well as the mean unsigned error (MUE) between interatomic distances and dihedral angles.

According to the RMS values depicted in Figure 13.2a, the hybrid meta functional M06 clearly outperforms the functionals without dispersion (B3LYP, PBE, and BP86) and the long-range corrected hybrid functional ωB97XD in the description of the figure-eight conformation of the neutral [36]octaphyrin. Nevertheless, a standard functional such as PBE provides a more accurate description of the degree of bond alternation along the conjugation pathway (Figure 13.2b). In the case of the Möbius diprotonated [36]octaphyrin, M06 ($RMS_{48} = 0.243$) and ωB97XD ($RMS_{48} = 0.332$)

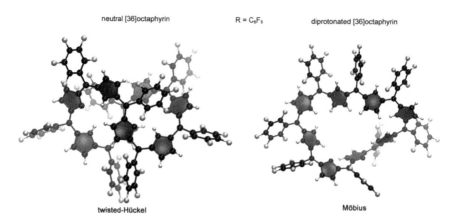

neutral [36]octaphyrin   R = C$_6$F$_5$ diprotonated [36]octaphyrin

twisted-Hückel   Möbius

FIGURE 13.1 X-ray crystal structures of the *meso*-octakis(pentafluorophenyl) [36] octaphyrin(1.1.1.1.1.1.1.1) in the neutral and diprotonated states.

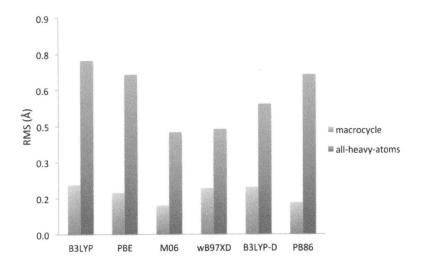

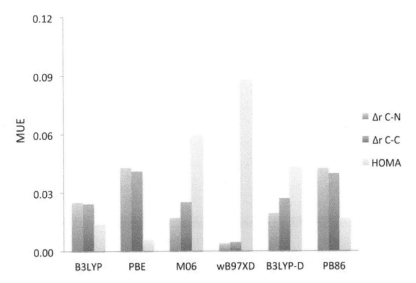

FIGURE 13.2   (a) Root-mean-square deviations (RMS) and (b) mean unsigned errors (MUE) for the bond-length alternation parameters (Δr_{C-X}) of the DFT optimized geometries relative to the X-ray structure of the neutral *meso*-octakis(pentafluorophenyl) [36]octaphyrin. Reproduced from Ref. [8d] with permission from the PCCP Owner Societies.

functionals respectively display the best and the worst overall performance for describing the geometry of the singly-twisted topology. However, the degree of bond-alternation (HOMA = 0.72) is better described by the

M06-optimized geometries (HOMA = 0.70), followed by the B3LYP-optimized ones (HOMA = 0.75).

Recently, Alonso and co-workers observed that the π-π stacking interactions in the figure-eight heptaphyrins were overemphasized in functionals that account for dispersion [69]. Consequently, we have analyzed the performance of the different functionals in describing the different types of π-π stacking interactions present in the twisted-Hückel topology [8]. Interestingly, the M06 functional provides the most accurate geometrical description of the different π-π stacking interactions. Nevertheless, the centroid-centroid distance between the stacked pyrrole rings is shortened from 3.41 Å in the X-ray structure to 3.33 Å with M06. In summary, the M06 hybrid functional presents the best overall performance in describing the geometries of Hückel and Möbius topologies of the [36]octaphyrin. The worst performance is observed for the ωB97XD functional, which underestimates the degree of bond equalization in octaphyrins. Therefore, the conformational analysis of neutral unsubstituted [36]octaphyrin will be performed using the M06 functional.

13.3.2 CONFORMATIONAL ANALYSIS OF NEUTRAL UNSUBSTITUTED [36]OCTAPHYRIN IN GAS PHASE AND DIFFERENT SOLVENTS

Owing to the large amount of possible conformations for octaphyrins, we initially restrict our conformational analysis to eight conformations for which crystallographic structures were available (**a-c, e-i**, Figure 13.3) [2]. Our conformational analysis also encompasses five other conformations (**d, j-m**), which were found in the investigation of the conformational interconversion pathways. Each conformation is described by the topological descriptors Tn^x, which indicates the number of half-twists (n) and the subunits located between two transoid bonds (^{x}) [2]. Moreover, Möbius and Hückel topologies can be distinguished according to the number of formal trans bonds along the smallest macrocyclic pathway. Accordingly, Möbius conformations are characterized by an odd number of *trans* bonds whereas an even number of *trans* bonds denotes a Hückel topology. Since solvent can trigger topological switches in expanded porphyrins, we also investigated the solvation effects on the conformational preferences of unsubstituted neutral [36]octaphyrins **1** using the SMD model [70].

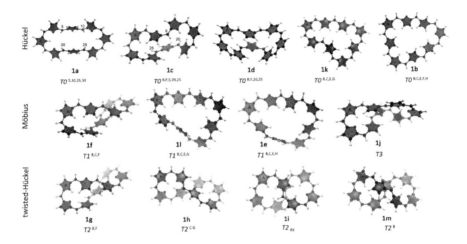

FIGURE 13.3 Hückel, Möbius and twisted-Hückel conformations of neutral unsubstituted [36]octaphyrins **1** (**a-m**). Reproduced from Ref. [8d] with permission from the PCCP Owner Societies.

Table 13.1 reports the relative energies of the different conformations of **1** in gas-phase, dimethylsulfoxide (DMSO), dichloromethane (DCM) and tetrahydrofuran (THF) together with the hydrogen bonding index (N_H) and ring strain descriptor (Φ_p). In spite of its expected antiaromatic behavior according

TABLE 13.1 Relative Energies and Relative Gibbs Free Energies (in kcal mol^{-1}) in Gas-Phase and Different Solvents Together with the Hydrogen Bonding Index (N_H) and Ring Strain (F_p in °) of the Different Conformations of Neutral Unsubstituted [36]octaphyrin **1**[a]

Conformation	Tn^X	E_{rel}	ΔG_{298}	N_H	F_p	ΔG_{THF}	ΔG_{DCM}	ΔG_{DMSO}
1a	$T0^{5,10,25,30}$	23.3	20.0	3	13.0	13.2	12.2	12.1
1b	$T0^{B,C,E,F,H}$	43.7	41.3	0	16.9	47.3	45.7	44.8
1c	$T0^{B,F,5,20,25}$	26.8	25.5	3	23.6	22.4	21.8	22.6
1d	$T0^{B,F,20,25}$	32.0	31.4	3	27.3	28.6	28.0	28.3
1e	$T1^{B,C,E,H}$	47.1	45.0	0	30.3	40.1	38.7	38.4
1f	$T1^{B,C,F}$	26.6	25.4	2.5	31.5	17.9	17.0	16.7
1g	$T2^{B,F}$	17.3	17.3	3	18.4	12.5	12.0	11.3
1h	$T2^{C,G}$	16.3	15.5	3	15.0	11.3	10.8	10.4
1i	$T2_{RX}$	0.0	0.0	4	15.0	0.0	0.0	0.0
1j	$T3$	26.9	26.5	3	31.5	31.5	24.9	24.4
1k	$T0^{B,C,E,G}$	43.9	41.8	1	16.9	16.9	35.2	33.9
1l	$T1^{B,C,E,G}$	46.9	45.2	1	33.9	33.9	38.3	37.0
1m	$T2^B$	11.0	10.6	3.5	16.3	16.3	9.2	8.9

[a] ZPE-corrected relative energies and Gibbs free energies at the M06/6–311+G(d,p)//M06/6–31G(d,p) level of theory. Adapted from Ref. [8d] with permission from the PCCP owner societies.

to Hückel's rule, the figure-eight conformation $T2_{RX}$ is predominant in the neutral unsubstituted [36]octaphyrins **1**, both in gas-phase and solvent. This conformational predominance stems from the intramolecular hydrogen bond network and the lower ring strain of the structure **1i**. Moreover, the distance between the parallel and planar two hemi-porphyrins-like segments of **1i** is about 3.3 Å, which corresponds to the distance of π-π stacking (Figure 13.3).

Compared to the global minimum **1i**, Hückel and Möbius topologies are 20–45 kcal mol^{-1} higher in energy due to their larger ring strain and fewer hydrogen bonds. However, Möbius and Hückel can be further stabilized by polar solvents, especially conformations with outward-pointing pyrrole rings (i.e., **1a** and **1f**). This stabilization results from the formation of inter-molecular hydrogen bonds between outward-pointing NH groups and the solvent molecules. Since all the pyrrolic nitrogen's are pointing inwards, solvent effects are less important for twisted-Hückel topologies. Similarly to other [4n] π-electron expanded porphyrins, the number and the strength of the hydrogen bonds dictates the conformational stability of unsubstituted [36]octaphyrins. Remarkably, each hydrogen bond provides a stabilization around 10 kcal mol^{-1}, which is similar to the stabilization effect previously reported for hexaphyrins and heptaphyrins [9,10]. Although the ring strain does not play a key role in conformational stability of neutral unsubstituted octaphyrins, the instability of Möbius topology is probably related to the higher ring strain and less effective π-conjugation ($-0.35 < \Pi < -0.57$) compared to twisted-Hückel topologies ($0.64 < \Pi < 0.80$).

The role of noncovalent interactions in determining the conformational preferences of neutral [36]octaphyrins was also scrutinized by the noncovalent interaction (NCI) method [33]. The computed $s(\rho)$ diagrams and NCI isosurfaces for the twisted-Huckel **1i** an Möbius **1f** of the neutral [36] octaphyrin are illustrated in Figure 13.4. Aside from the repulsive interaction at the center of each pyrrole ring, both conformations exhibit different number and type of noncovalent interactions. Four single hydrogen bonds are found in the figure-eight topology **1i**, whereas the Möbius topology **1f** is stabilized by one bifurcated and one single hydrogen bond. In **1i**, the four hydrogen bonds are of equivalent strength and thus represented by a unique peak at -0.02 a.u in the $s(\rho)$ plot. In the Möbius topology **1f**, the single hydrogen bond is stronger ($\rho = -0.029$ a.u.) than the hydrogen bonds in **1i**, whereas the bifurcated one is weaker ($\rho = -0.014$ a.u.). Moreover, a delocalized π-π stacking at the central twist is present in the figure-eight **1i** while a weaker and more localized CH-π interaction is found in **1f**.

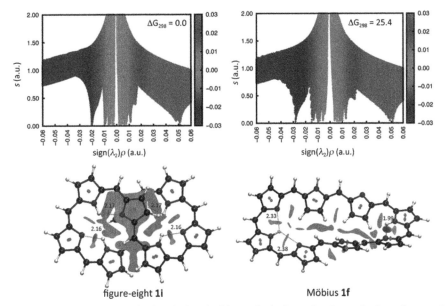

figure-eight **1i**   Möbius **1f**

FIGURE 13.4 NCI analysis of the doubly and singly twisted topologies of neutral unsubstituted [36]octaphyrin **1**. Plots of the reduced density gradient $s(\rho)$ and gradient isosurfaces ($s = 0.5$). The surfaces are colored according to sign(λ_2)ρ over the range -0.03 to 0.03 a.u. The hydrogen bond lengths (in Å) are also shown. Reproduced from Ref. [8d] with permission from the PCCP Owner Societies.

13.3.3 INTERCONVERSION PATHWAYS OF NEUTRAL OCTAPHYRINS

In order to locate the transition states for the different conformational interconversion pathways, we computed a series of 1D and 2D relaxed potential energy surfaces (PES) scans of **1**. Similar to other expanded porphyrins, the interconversion between different π-conjugation topologies is achieved by variation of internal dihedral angles [9, 10]. Figure 13.5 illustrates the surface plot of the energy *versus* the dihedral angles j_1 and j_2 from the global minimum **1i**. The low-energy pathway denotes the inversion of one pyrrole ring ($T2_{RX} \rightarrow T2^B$), leading to the conformation **1m**, which is 10.6 kcal mol^{-1} less stable than **1i**. The activation barrier for the **1i** $\rightarrow$**1m** interconversion is fairly large and a distorted Möbius topology was found as transition state. An alternative pathway corresponds to the interconversion of **1i** into an asymmetric figure-eight conformation **1n** ($T2_{RX} \rightarrow T2$), in which the two hemicycles have different lengths. The

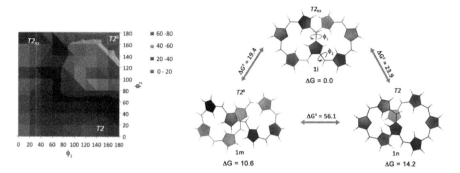

FIGURE 13.5 M06/6–31G(d,p) relaxed potential energy surface for the figure-eight conformation (**1i**) obtained by rotating the dihedral angles f_1 and f_2 (in °). The fully optimized geometries for the different minima and the corresponding Gibbs free energies and activation energies are also shown.

activation barrier is even larger and a Möbius transition state was again located.

Additional calculations showed that the energy barriers for the switching between twisted-Hückel topologies are large ($\Delta G^{\ddagger}$ ranges from 19.4 to 47.7 kcal mol^{-1}). In contrast, the energy barrier is significantly reduced for the interconversion process between figure-eight $T2^{B,F}$ (**1g**) and Möbius conformers $T1^{B,C,F}$(**1f**), as shown in Figure 13.6. Similarly, we observe a low energy barrier for the Hückel-Möbius interconversion $T1^{B,C,E,H}$ (**1e**) → $T0^{B,C,E,G}$(**1k**). These results demonstrate that the figure-eight conformation **1i** is particularly stable in [36]octaphyrin. In fact, the presence of a number of intramolecular hydrogen bonds and π-π stacking hinder the interconversion between several twisted-Hückel topologies. Conversely, the switching between Möbius and Hückel conformers is feasible since the hydrogen bonding interactions are conserved along the interconversion pathways.

13.3.4 MESO-SUBSTITUTION EFFECT ON THE CONFORMATION OF NEUTRAL [36]OCTAPHYRINS

Since peripheral modifications at *meso*- and β-positions are very effective for tuning the conformational properties of expanded porphyrins, [69–72] we have investigated the effect of three substituents (CF_3, C_6F_5,–H,–F) and $C_6H_3Cl_2$) at the *meso*-positions on the conformations of the neutral [36]octaphyrins (Scheme 2). The relative Gibbs free energy of the different neutral

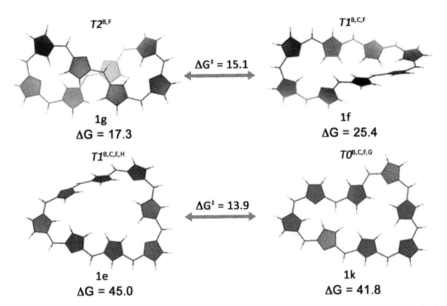

FIGURE 13.6 Activation barriers ($\Delta G^{\ddagger}$ in kcal mol^{-1}) for different topological interconversions. The Gibbs free energies with respect the global minimum **1i** are also shown. Reproduced from Ref. [8d] with permission from the PCCP Owner Societies.

meso- and *β*-substituted [36]octaphyrins are collected in Table 13.2. Due to optimization issues, we could find neither an optimum geometry for $T2^{B,F}$ for aryl-substituted octaphyrins nor a minimum structure for $T0^{B,F}$ [20, 25]. Neutral substituted [36]octaphyrins adopt preferentially the figure-eight structure **i** ($T2_{RX}$) over the other conformations. Our theoretical predictions are in good agreement with the experimental results available for *meso-*octakis(pentafluorophenyl)[36]octaphyrin since the X-ray diffraction analysis revealed a similar figure-eight structure [31]. Nevertheless, in the case of perfluorinated [36]octaphyrin, the crystallographic structure corresponds to the Hückel conformer **a** ($T0^{5,10,25,30}$) [71], in which the central pyrrole rings are largely tilted from the mean plane. These pyrrole rings are almost parallel among them and their nitrogen atoms point inward towards the center of the macrocycle. However, our gas-phase (solvent) Gibbs free energies indicate that the $T2_{RX}$ is 19 (14) kcal mol^{-1} more stable than the crystallographic structure $T0^{5,10,25,30}$ for the perfluorinated [36]octaphyrin.

The incorporation of *meso-*substituents stabilizes the Hückel and Möbius conformers and the magnitude of such stabilization depends on

TABLE 13.2   Relative Gibbs Free Energies (in kcal mol^{-1}) of the Different Conformations of Neutral Meso- and β-Substituted [36]octaphyrins [a]

Conformation	Tn^X	-H	-CF$_3$	-C$_6$H$_3$Cl$_2$	-C$_6$F$_5$, -H	-C$_6$F$_5$,-F	-OMe
a	$T0^{5,10,25,30}$	20.0	6.7	16.6	16.8	19.2	0.0
b	$T0^{B,C,E,F,H}$	41.3	41.2	38.3	41.4	44.2	38.4
c	$T0^{B,F,5,20,25}$	25.5	5.2	8.3	18.3	14.2	15.0
e	$T1^{B,C,E,H}$	45.0	30.1	31.7	40.6	42.4	37.8
f	$T1^{B,C,F}$	25.4	14.8	7.4	10.1	15.7	13.5
g	$T2^{B,F}$	17.3	10.1	–	–	–	3.8
h	$T2^{C,G}$	15.5	3.9	4.0	5.4	11.6	2.5
i	$T2_{RX}$	0.0	0.0	0.0	0.0	0.0	0.0

[a] Gibbs free energies at the M06/6–311+G(d,p)//M06/6–31G(d,p) level of theory.

the nature of the substituents. For instance, the ΔG difference between $T0^{B,F,5,20,25}$ and the $T2_{RX}$ structures is reduced to 5 and 8 kcal mol^{-1} with the bulky CF$_3$ and C$_6$H$_3$Cl$_2$ groups, respectively, whereas it is 18 kcal mol^{-1} for the C$_6$F$_5$ group. For the trifluoromethyl group, the Hückel untwisted conformers **a** and **c** are relatively stable with energy differences with respect to $T2_{RX}$ of 3 and 6 kcal mol^{-1}, respectively. Furthermore, an alternative figure-eight conformation with two inverted pyrrole rings **h** is also viable for the CF$_3$ group. This substituent enhances the ring strain (F$_p$ = 29.3) and reduces importantly the π-conjugation in the doubly-twisted topology **i** (Π = 0.46, Table 13.3).

Thus, the *meso*-trifluoromethyl octaphyrin could exist in solution as equilibrium of several figure-eight and Hückel conformers. Owing to its low macrocyclic strain and the high number of intramolecular hydrogen bonding, the conformation $T2_{RX}$ is predominant in *meso*-C$_6$F$_5$ and C$_6$H$_3$Cl$_2$ neutral [36]octaphyrins (Table 13.3). In comparison to CF$_3$ groups, the variation in ring strain and overall π-conjugation induced by the aryl groups are less pronounced. Interestingly, the Möbius conformer $T1^{B,C,F}$ is further stabilized with the C$_6$H$_3$Cl$_2$ group, which reduces its relative ΔG to 7.4 kcal mol^{-1}. Hence, although neutral substituted [36]octaphyrins prefer a figure-eight conformation, the relative stability of the Hückel and Möbius conformers can be fine-tuned by the *meso*-and β-substituents. In addition, we have investigated how the presence of electron-donating groups (OMe) at *meso* positions modify the conformational preferences of neutral [36]octaphyrin. Due to the presence of additional NH···O hydrogen bonds in the conformer

TABLE 13.3 Relative Energies (E_{rel}), Relative Gibbs Free Energies (ΔG_{298} in kcal mol^{-1}), Hydrogen Bonding Index (N_H), Ring Strain (Φ_p) and p-Conjugation Index (P) of the neutral *meso*-aryl-substituted [36]octaphyrins

Conformation	$-C_6F_5$					$-C_6H_3Cl_2$					$-CF_3$				
	E_{rel}	ΔG_{298}	N_H	Φ_p	Π	E_{rel}	ΔG_{298}	N_H	Φ_p	Π	E_{rel}	ΔG_{298}	N_H	Φ_p	Π
a $T0^{5,10,25,30}$	16.6	16.8	3	30.6	0.38	22.9	16.6	3.0	34.9	0.31	8.8	6.7	3	28.7	0.48
b $T0^{B,C,E,F,H}$	48.7	41.4	0	35.5	0.47	45.4	38.3	0.0	35.1	0.52	49.9	41.2	0	32.9	0.35
c $T0^{5,20,25,B,F}$	20.2	18.3	3	27.1	0.45	12.4	8.3	3.0	35.2	0.38	10.0	5.2	3	31.6	0.30
f $T1^{B,C,F}$	19.3	10.1	2.5	27.1	-0.47	14.0	7.4	2.5	29.8	-0.50	20.0	14.8	2.5	36.8	-0.39
e $T1^{B,C,E,H}$	50.1	40.6	0	33.9	-0.34	42.4	31.7	0.0	35.0	-0.35	39.2	30.1	0	32.2	-0.30
i $T2_{RX}$	0.0	0.0	4	21.6	0.65	0.0	0.0	4.0	21.9	0.61	0.0	0.0	4	29.3	0.46
h $T2^{C,G}$	9.9	5.4	3	22.0	0.60	9.6	4.0	3.0	24.6	0.50	7.3	3.9	3	21.3	0.53

a, the twisted-Hückel conformer **i** ($T2_{RX}$) and the untwisted Hückel conformer **a** ($T0^{5,10,25,30}$) becomes almost isoenergetic.

13.3.5 CONFORMATIONAL CHANGES UPON REDOX REACTIONS AND PROTONATION REACTIONS

As redox reactions and protonation can give rise to conformational changes in several expanded porphyrins [75, 76], we have investigated them as potential chemical triggers to induce conformational and/or aromaticity switches in octaphyrins. According to experimental studies, the change of the oxidation state in [36]octaphyrin is coupled to large changes in the absorption spectra whereas protonation can be used for controlling the molecular topology of [36]octaphyrin [30, 31].

The relative Gibbs free energies of neutral [36] (**1**), [34] (**2**), and [38] (**3**) octaphyrins and diprotonated [36] (**4**) and [38] (**5**) octaphyrins are collected in Table 13.4. Unsubstituted octaphyrins **1**, **3** and **5** adopt preferentially the figure-eight conformation $T2_{RX}$ in gas-phase. For **2** and **4**, an alternative twisted-Hückel structure with two inverted pyrrole rings is preferred. However, the variation of oxidation state and especially protonation reduces the conformational preference of octaphyrins for the figure-eight conformation (Table 13.4).

TABLE 13.4   Relative Gibbs Free Energies (in kcal mol^{-1}) of the Different Conformations of Neutral and Diprotonated Unsubstituted Octaphyrins with Different Oxidation States[a]

Conformation	Tn^X	[36] (1)	[34] (2)	[38] (3)	[36]$^{2+}$ (4)	[38]$^{2+}$ (5)
a	$T0^{5,10,25,30}$	20.0	4.1	14.8	4.3	10.2
b	$T0^{B,C,E,F,H}$	41.3	33.3	48.8	34.7	4.9
c	$T0^{B,F,5,20,25}$	25.5	9.7	16.5	9.7	10.5
d	$T0^{B,F,20,25}$	31.4	14.4	20.4	14.8	13.7
e	$T1^{B,C,E,H}$	45.0	39.6	51.9	29.7	13.0
f	$T1^{B,C,F}$	25.4	16.6	25.7	2.7	9.2
g	$T2^{B,F}$	17.3	0.0	7.9	0.3	2.2
h	$T2^{C,G}$	15.5	0.0	6.8	0.0	4.6
i	$T2_{RX}$	0.0	6.5	0.0	3.4	0.0
	MAD[b]	–	12.2	6.4	14.3	17.0

[a] Gibbs free energies at the M06/6–311+G(d,p)//M06/6–31G(d,p) level of theory. [b] MAD is the mean absolute difference with respect the relative Gibbs free energies of unsubstituted [36]octaphyrin **1**.

With [38] π-electrons, the stability of untwisted Hückel topologies is considerably increased towards its [36]octaphyrin homologous **1**. For instance, the conformers **2a** and **2c** are stabilized by 16 kcal mol^{-1} upon oxidation of the macrocycle. Given the small energy differences between the *T2* and *T0* conformations, we anticipate that both could exist in solution as an equilibrium of Hückel and twisted-Hückel structures. The enhanced stability of the Hückel conformers is connected to their lower ring strain, more effective hydrogen bonding and the presence of aromaticity (Table 13.5). By contrast, the Möbius conformations are very unlikely for **2** because of their high conformational energy.

In **3**, the figure-eight topologies **3g-i** are viable whereas the Gibbs free energies of the Hückel and Möbius topologies exceed 15 kcal mol^{-1}. In general, the reduction of [36]octaphyrin exerts less influence on the conformational stability of the macrocycle than the oxidation. With [38] π-electrons, the Möbius structures are highly destabilized, being 52 and 26 kcal mol^{-1} higher in energy than the figure-eight conformation **3i**. The conformational preference of **3** for the conformation **3i** is related to its low ring strain and effective overlap of the π orbitals (Table 13.5). Therefore, the variation of the oxidation state is not an effective *stimulus* to produce Möbius species in octaphyrins. The conformational stability of both [34] and [38] octaphyrins is still determined by the number of hydrogen bonds, although in a lesser extent than in [36]octaphyrins.

As shown in Table 13.4, protonation seems to be more effective than redox reactions for triggering a conformational switch in unsubstituted [36] octaphyrin. On the one hand, diprotonated [36]octaphyrin **4** could easily adopt either figure-eight or Möbius conformers (**4f**). Besides, the Möbius structures **4e** and **4f** are stabilized by 15 and 23 kcal mol^{-1} in the diprotonated state towards their neutral counterparts. Nevertheless, the figure-eight topology **4h** still remains the global minimum in diprotonated [36]octaphyrin mainly due to its lower ring strain and more effective π network (Table 13.5). Remarkably, a fairly smooth π-conjugation surface ($\Pi = -0.58$) is observed for Möbius topology **4f**. Therefore, we expected a distinct Möbius aromaticity in this conformation. On the other hand, the untwisted Hückel conformers (**a-c**) are more stable than the singly-twisted conformations (**e-f**) in diprotonated [38] octaphyrin **5**. Appealingly, the Hückel conformation with five inverted pyrrole rings (**5b**) is highly stabilized in diprotonated [38] octaphyrins. Because of the lack of intramolecular hydrogen bonds in **5**, the conformational predominance of twisted-Hückel conformers is largely

TABLE 13.5 Hydrogen Bonding Index (N_H), Ring Strain (F_p) and p-Conjugation Index (P) of the Neutral and Diprotonated Unsubstituted Octaphyrins With Different Oxidation Stat

Conformation	[34] (2)			[38] (3)			[36]²⁺ (4)			[38]²⁺ (5)		
	N_H	Φ_p	Π	N_H	Φ_p	Π	N_H	Φ_p	Π	N_H	Φ_p	Π
a $T0^{5,10,25,30}$	3	9.3	0.94	3.0	14.72	0.79	3	16.4	0.74	—	16.4	0.76
b $T0^{B,C,E,F,H}$	0	21.4	0.62	0.0	22.5	0.59	0	26.1	0.52	—	25.1	0.56
c $T0^{5,20,25,B,F}$	3	32.6	0.63	3.0	34.9	0.59	3	28.1	0.60	—	28.6	0.51
e $T1^{B,C,E,H}$	0	33.3	-0.33	0.0	35.1	-0.30	0	38.8	-0.39	—	33.3	-0.29
f $T1^{B,C,F}$	2.5	30.8	-0.56	2.5	28.7	-0.56	2.5	29.5	-0.58	—	28.5	-0.48
g $T2^{B,F}$	3	14.9	0.83	3.0	20.2	0.78	3	15.8	0.81	—	17.5	0.64
h $T2^{C,G}$	3	15.9	0.83	3.0	14.9	0.81	3	17.1	0.80	—	20.2	0.66
i $T2_{RX}$	3	15.9	0.79	3.0	16.0	0.81	3	17.4	0.73	—	14.1	0.79

reduced in this case. The figure-eight topology **5i** is the global minimum for unsubstituted [38]octaphyrins in the diprotonated state due to its low ring strain and the most effective π-conjugation (Table 13.5 and Figure 13.7).

13.3.6 CONFORMATIONAL CHANGES OF MESO-OCTAKIS(PENTAFLUOROPHENYL) OCTAPHYRINS UPON PROTONATION AND REDOX REACTIONS

In the next step, we have investigated the stability of *meso*-octakis(pentafluorophenyl) octaphyrins with different oxidation and protonation states in both gas-phase and solvent. Given the availability of crystallographic structure for **6**, **8** and **9**, studying those structures is an excellent test for our computational methods. The relative Gibbs free energies computed for the *meso*-octakis(pentafluorophenyl) [36] and [38]octaphyrins in neutral and diprotonated states are collected in Table 13.6.

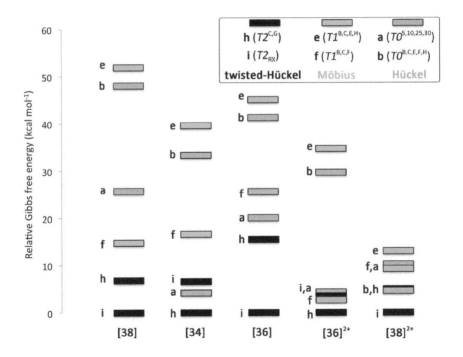

FIGURE 13.7 Relative Gibbs free energies computed for the different structures of unsubstituted octaphyrins with different oxidation and protonation states. Reproduced from Ref. [8d] with permission from the PCCP Owner Societies.

TABLE 13.6 Relative Gibbs Free Energies (in kcal mol^{-1}) of the Different Conformations of Neutral and Diprotonated *Meso*-Octakis(Pentafluorophenyl) [36] and [38]Octaphyrins [a]

Conformation	Tn^X	[36] (6)	[38] (7)	[36]$^{2+}$ (8)	[38]$^{2+}$ (9)
a	$T0^{5,10,25,30}$	16.8 (14.4)	16.7 (17.9)	4.9 (8.6)	6.1 (8.2)
b	$T0^{B,C,E,F,H}$	41.4 (23.4)	39.4 (24.6)	32.6 (23.6)	1.3 (2.4) [c]
c	$T0^{B,F,5,20,25}$	18.3 (16.2)	7.3 (8.7)	8.1 (13.6)	6.5 (12.1)
d	$T0^{B,F,20,25}$	25.4 (23.2)	19.5 (21.4)	21.5 (23.6)	15.6 (20.3)
e	$T1^{B,C,E,H}$	40.6 (24.1)	44.1 (31.2)	29.0 (23.1) [c]	7.0 (5.6)
f	$T1^{B,C,F}$	10.1 (1.8)	12.3 (8.0)	5.0 (2.1)	3.5 (5.7)
h	$T2^{C,G}$	5.4 (1.7)	0.0 (0.0)	0.0 (0.0)	0.0 (0.0)
i	$T2_{RX}$	0.0 (0.0) [c]	7.5 (8.6)	15.8 (11.1)	5.9 (3.0)
	MAD[b]	–	4.7 (4.7)	9.1 (2.9)	12.6 (7.7)

[a] Gibbs free energies at the M06/6-311+G(d,p)//M06/6-31G(d,p) level of theory in gas-phase and dichloromethane/trifluoroacetic acid (in parenthesis). [b] MAD is the mean absolute difference with respect the relative Gibbs free energies of neutral [36]octaphyrin 6. [c] The X-ray structure corresponds to this conformation.

According to Table 13.6, the inclusion of the solvent enhances the stabilization of the conformers, having outward pointing pyrroles (**b** and **e**). In both cases, the stabilization is probably related to the intermolecular hydrogen bonds between the solvent molecules and the inverted pyrrole rings, thus reducing their respective relative ΔG by more than 16 kcal mol^{-1} with respect to the neutral state. According to the relative ΔG computed with M06, the figure-eight conformations are viable for *meso*-octakis(pentafluorophenyl) [36] and [38]octaphyrins in both neutral and diprotonated states. For the neutral [36]octaphyrin **6**, the global minima corresponds to the twisted-Hückel topology **6i** with all the pyrrolic nitrogens pointing inward, in good agreement with the X-ray crystallographic structure [31]. Alternatively, neutral [36]octaphyrins **6** could also display the figure-eight conformation **6h** with two inverted pyrrole rings in solution ($\Delta G = 1.7$ kcal mol^{-1}).

For the *meso*-substituted [38] octaphyrin (**7**), the relative Gibbs free energies indicate the conformational predominance of the figure-eight conformation **7h**. Nevertheless, a Hückel conformation **7c** ($T0^{B,F,5,20,25}$) is highly stabilized upon reduction of the macrocycle, with a relative ΔG of 7.3 and 8.7 kcal mol^{-1} in gas-phase and dichloromethane, respectively. Despite the lack of a crystallographic structure for **7**, a related Hückel conformation $T0^{B,F,20,25}$ (**d**) was proposed by Osuka et al. on the basis of ^{1}H NMR spectroscopy [30]. However, our calculations indicate that the conformer **7d** is not viable neither in gas-phase (ΔG = 19.5 kcal mol^{-1}) nor in solvent (ΔG = 21.4 kcal mol^{-1}). In order to confirm our computational predictions, the ^{1}H-NMR spectra of the three plausible conformations have been computed at the B3LYP/6–311+G(d,p) level of theory in chloroform. The calculated and experimental values of the ^{1}H NMR chemical shifts are shown in Figure 13.8. Importantly, the experimental ^{1}H NMR spectrum indicates the presence of a diatropic ring current by showing β-proton signals in two separated regions: 5.5–7.0 ppm for the outer β-protons and 2.5–3.0 ppm for the inner β-protons attached to the inverted pyrrole rings [30]. Hence, the moderately shielded signals of the ^{1}H NMR spectrum of [38]octaphyrin reveal the presence of a moderate diatropic ring current. The discrepancy between the computed and experimental ^{1}H NMR spectra for the Hückel conformers **7c** and **7d** supports the idea that these Hückel conformers are not the predominant conformations for the [38]octaphyrin. On one side, we observed an important deshielding of the NH protons of the inverted pyrrole rings and the outer pyrrolic β-protons of **7d** compared to the experimental values. On the other side, the computed ^{1}H-NMR spectrum of **7c** displays a

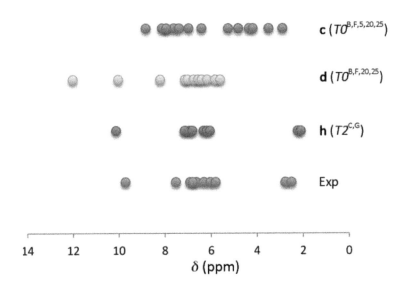

FIGURE 13.8 Experimental and computed ^{1}H NMR shifts of the NH protons and β-protons for the neutral [38] octaphyrin **7** in different conformations. Reproduced from Ref. [8d] with permission from the PCCP Owner Societies.

large number of signals for both the NH protons and the β-protons, as a consequence of the non-symmetric structure. In contrast to the experimental spectrum, several NH signals in **c** appear between 3–5.5 ppm. In addition, the β-protons signals appear in two different regions, around 2 ppm and 6 ppm, in computed ^{1}H NMR spectrum of **7h** and the experimental spectrum. Therefore, on the basis of our calculations, the most plausible structure for the neutral *meso*-octakis(pentafluorophenyl) [38]octaphyrin is the figure-eight conformation **7h** with two inverted pyrrole rings.

Protonation induces a larger change in the ΔG of *meso*-subtituted [36] octaphyrin, as can be inferred from Table 13.6. In the diprotonated state, a figure-eight conformation **h** is the global minimum while the Möbius topology **f** is only 2.1 kcal mol^{-1} higher in energy in trifluoroacetic acid. However, the crystallographic structure of **8** corresponds to a Möbius structure with four inverted pyrrole rings, similar to conformation **e**, which is involved in an extensive network of intermolecular hydrogen bonds with TFA molecules and their counter-anions [30]. According to our calculations, this conformation is 23.1 kcal mol^{-1} higher in energy than the global minimum. The divergence between the experimental and computational results can be associated with the lack of counter-anions in our structures and/or the overestimation of the stability of the diprotonated figure-eight topologies by M06. To discard the second possibility, we have carried out single-point calculations to evaluate the relative ΔG energies at the B3LYP/6–311+G(d,p)//M06/6–31G(d,p). The B3LYP-recomputed ΔG_{solv} (in kcal mol^{-1}) are 0.0 (**8f**), 6.1 (**8e**), 12.2 (**8b**) and 14.1 (**8h**). Importantly, the B3LYP energies point out to the predominance of Möbius structures in diprotonated [36]octaphyrin, which is in better agreement with the experimental observations [30].

According to the relative ΔG of diprotonated *meso*-C$_6$F$_5$ [38] octaphyrin **9**, figure-eight conformers (**9h-i**), Möbius topologies (**9e-f**) and a Hückel conformations (**9b**) are viable considering the small energy differences among these conformations. Despite a conformational preference for the figure-eight conformation **9h**, the Hückel conformer **9b** is greatly stabilized in diprotonated [38]octaphyrin, being just 1.3 kcal mol^{-1} higher in energy than **9h**. Importantly, the X-ray analysis of **9** reveals a Hückel conformation **9b**, where the octaphyrin macrocycle is surrounded by TFA molecules to form an extensive intermolecular hydrogen bonding network [30]. At the B3LYP level, the conformation **9b** corresponds to the most stable structure, followed by the Möbius structure **9e** (ΔG = 6.8 kcal mol^{-1}). Again, the twisted-Hückel topologies raised in energy relative to the ΔGs computed with M06. For instance, the relative ΔG of **9h** increases from 0.0 to 20.2 kcal mol^{-1} as the functional changes from M06 to B3LYP.

In the case of diprotonated [38]octaphyrin, the Hückel **9b** is more stable than the Möbius **9e** due to the expected aromatic stabilization in the Hückel [$4n$ +2] π-electron system. By contrast, the Möbius conformation **8e** displays a higher stability than the Hückel one **8b** in the diprotonated [36]octaphyrin due to the Möbius aromatic stabilization in [$4n$] π-electron system. Thus, aromaticity seems to control the molecular topology of diprotonated octaphyrins, as previously observed experimentally by Osuka *et al.* [30].

Consequently, protonation is an effective method to trigger topological switches in octaphyrins. Whereas diprotonated [36]octaphyrin adopts a Möbius structure, diprotonated [38]octaphyrin prefers a Hückel untwisted conformation. By contrast, the figure-eight topology is conserved after oxidation or reduction of the macrocycle, although two pyrrole rings are inverted upon reduction.

13.4   STRUCTURE-AROMATICITY RELATIONSHIPS IN OCTAPHYRINS

In spite of the multidimensional character of aromaticity [21], most of the experimental studies on expanded porphyrins solely rely on NICS as an aromatic index. Recently, Alonso and co-workers found that the aromaticity of expanded porphyrins is revealed best by the magnetic indices and the relative hardness rather than the structural and energetic descriptors [9, 69, 71]. Their structure-property relationships demonstrated that the aromaticity of expanded porphyrins is highly dependent on the molecular topology and the number of π-electrons. Here, we scrutinize the aromaticity of octaphyrins exhibiting different molecular topologies and different number of π-electrons using energetic, magnetic, structural and reactivity criteria. Additionally, the performance of the different indices to describe Hückel and Möbius aromaticity in octaphyrins is also investigated.

As the B3LYP functional was used previously to evaluate the aromaticity descriptors of penta-, hexa- and heptaphyrins, [9, 69–71] we decided to focus on the B3LYP-computed indices. First, we study the aromaticity of the different conformations of the neutral unsubstituted [36]octaphyrin **1** in order to establish the relationship between the molecular conformation and macrocyclic aromaticity. Table 13.6 reports the energetic (ISE), magnetic

[Λ, NICS(0) and NICS$_{zz}$(1)], structural (HOMA) and reactivity ($\Delta\eta$) indices computed with the B3LYP functional for the Hückel, Möbius and twisted-Hückel conformations of 1.

Table 13.7 points out to the existence of a close relationship between molecular topology (Hückel, Möbius and twisted-Hückel) and aromaticity in [36]octaphyrins. On the one hand, Möbius topologies (1e-f) are clearly aromatic with positive ISE$_{corr}$ and $\Delta\eta$ values and exhibit strong diatropic ring currents. In addition, Möbius topologies are characterized by enhanced degree of bond-equalization, as shown by the HOMA index. On the other hand, Hückel topologies with [36]π-electrons (1a-d) display highly positive values of NICS-based indices and Λ, coupled to negative ISE$_{corr}$ and $\Delta\eta$, which indicates antiaromaticity. Interestingly, the strength of the induced ring current is related to the ring strain and the efficiency of π-conjugation. Hence, the strongest paramagnetic ring current is observed in 1a ($F_p = 13.0$ and $\Pi = 0.83$, Table 13.2). The twisted-Hückel topologies 1g-i display small ISE$_{corr}$ values, negative $\Delta\eta$ and positive Λ, whose magnitude is strongly reduced towards untwisted conformations. The reduced values of the magnetic indices show that the induced paramagnetic ring current diminishes considerably in the figure-eight conformations, especially in the global minimum 1i. In fact, the NICS(0) and Λ supports a non-aromatic character for 1i, in agreement with the ^{1}H NMR spectra of the neutral [36]octaphyrin [31]. Importantly, the NICS$_{zz}$(1) values of the figure-eight conformations are negative indicating aromatic character. In these doubly twisted conformations, the "probe atom" at 1 Å above the mean molecular plane is located near the stacked pyrrole rings, reflecting the local diatropic ring current of the individual pyrrole rings. So, NICS$_{zz}$(1) is not a suitable for the evaluation of the macrocyclic aromaticity in the twisted-Hückel conformations.

From the energetic point of view, the effect of the aromatic stabilization/destabilization in octaphyrins is much weaker than in small monocycles like benzene (ISE$_{corr}$ = 34.3 kcal mol^{-1}). This low aromatic stabilization justifies the viability of antiaromatic [36]octaphyrins. In these large macrocycles, the aromatic destabilization is not strong enough to overcome the most effective hydrogen bonding in the twisted-Hückel topology. Thus, according to the ISE$_{corr}$, the most-stable conformer 1i is destabilized by 2.1 kcal mol^{-1}, but it is overcompensated by the effective intramolecular hydrogen bonds, which provides a stabilization of ca. 10 kcal mol^{-1} per hydrogen bond. This is the reason why aromaticity provides a minor contribution to the relative stability of the conformations of

TABLE 13.7 Energetic, Reactivity, Magnetic, and Structural Indices of Aromaticity of the Different Conformations of the Neutral Unsubstituted [36]octaphyrin 1 [a]

Conformation	Tn^X	ISE	ISEcorr	$\Delta\eta$	Λ	NICS(0)	NICSzz(1)	HOMA
1a	$T0^{5,10,25,30}$	26.6	2.4 [b]	−3.0	422	16.8	51.2	0.73
1b	$T0^{B,C,E,FH}$	18.7	−2.4	−6.2	707	12.3	38.3	0.71
1c	$T0^{B,F,5,20,25}$	12.8	−3.0	−3.0	462	−0.2	37.9	0.73
1d	$T0^{B,F,20,25}$	23.1	−2.4	−3.6	268	5.0	24.4	0.73
1e	$T1^{B,C,E,H}$	22.3	7.9	3.9	−488	−9.3	−22.6	0.79
1f	$T1^{B,C,F}$	24.1	2.7	3.8	−344	−15.6	−32.0	0.85
1g	$T2^{B,F}$	23.3	−0.1	−5.4	193	8.5	−20.2	0.74
1h	$T2^{C,G}$	23.2	1.6 [b]	−6.7	150	10.2	−17.2	0.72
1i	$T2_{RX}$	20.5	2.1 [b]	−3.8	22	−2.6	−19.5	0.75

[a] ISE, ISE_{corr}, and Δh are given in kcal mol^{-1}, L in ppm cgs and NICS indices in ppm. [b] The large flexibility induces topology changes in the dihydrogen derivative of the methylene adducts during the optimization.

neutral [36]octaphyrins, although the photophysical and nonlinear optical properties of octaphyrins are highly dependent on the macrocyclic aromaticity [77, 78].

In the annulene model, the aromaticity of octaphyrins can be predicted according to the number of π-electrons within the annulene-type conjugation pathway [35]. Consequently, redox reactions reverse the aromaticity because it modifies the number of π-electrons within the classical conjugation pathway. In order to demonstrate this assumption, we quantify the degree of aromaticity of neutral unsubstituted [34] and [38]octaphyrins (2 and 3) with structural (HOMA), magnetic (NICS, Λ), reactivity ($\Delta\eta$) and energetic criteria (ISE$_{corr}$) (Table 13.8).

Consistent with the annulene model, the aromaticity of the twisted-Hückel, Hückel and Möbius topologies is totally reversed upon oxidation/reduction of the macrocycle. Therefore, Möbius conformations with [34] and [38] π-electrons are antiaromatic, as shown by the reversed sign of the aromatic descriptors with respect to the Möbius conformation of [36]octaphyrin. The conformers 2f and 3f are characterized by strong paramagnetic ring currents whereas negative NICS and are found for the counterpart 1f. Moreover, the antiaromatic character of 2f and 3f is supported by their negative relative hardness. By contrast, the Hückel and figure-eight conformations become clearly aromatic upon two-electron redox reactions. In [4n+2] π-electron octaphyrins, Hückel conformations are characterized by a strong diatropic current and positive $\Delta\eta$, which denotes aromaticity. Compared to Hückel and twisted-Hückel conformations of [34] and [38]octaphyrins, the HOMA index is reduced in the Möbius topologies. Even though Hückel conformers are aromatic in [34] and [38]octaphyrins, the variation of oxidation state leads to different values of the aromaticity descriptors for equivalent conformations. For instance, the figure-eight conformation 3i is more aromatic than its homologous 2i, according to the Λ, NICS(0), $\Delta\eta$ and HOMA.

Finally, the aromaticity of diprotonated unsubstituted [36] and [38]octaphyrins 4 and 5 has been evaluated using the same descriptors (Table 13.9). In contrast to redox reactions, protonation does not reverse the aromaticity in the diprotonated conformations with respect to the neutral state. This is expected since the number of π-electrons along the conjugation pathway is the same for the neutral and the diprotonated species. Nevertheless, the strength of the induced ring current is enhanced in diprotonated conformers with respect to their neutral homologous. For instance, the aromaticity of the Möbius [36] π-electron conformations e-f significantly raises

TABLE 13.8 Energetic, Reactivity, Magnetic, and Structural Indices of Aromaticity of the Different Conformations of the Neutral Unsubstituted [34]- and [38]Octaphyrins **2** and **3** [a]

Conformation	Tn^X	ISE	ISEcorr	$\Delta\eta$	Λ	NICS(0)	NICSzz(1)	HOMA
2a	$T0^{5,10,25,30}$	27.4	−2.8 [b]	4.1	−445	−15.2	−34.2	0.81
2c	$T0^{B,F,5,20,25}$	23.4	−2.0 [b]	0.5	−192	−21.6	−29.1	0.82
2e	$T1^{B,C,E,H}$	17.4	−0.4	−5.1	349	5.4	19.6	0.71
2f	$T1^{B,C,F}$	15.1	−4.6	−6.2	347	10.8	33.5	0.78
2h	$T2^{C,G}$	25.7	−12.6 [b]	3.6	−127	−12.7	−20.9	0.84
2i	$T2^{RX}$	23.8	11.7	−1.2	−30	−3.2	−9.3	0.85
3a	$T0^{5,10,25,30}$	29.3	−13.5 [b]	1.1	−492	−18.2	−37.3	0.84
3c	$T0^{B,F,5,20,25}$	33.9	−9.8 [b]	9.1	−159	−8.6	−9.0	0.82
3e	$T1^{B,C,E,H}$	28.5	−9.4	4.9	307	5.0	18.3	0.73
3f	$T1^{B,C,F}$	20.8	−6.2	−3.4	325	10.4	34.5	0.83
3h	$T2^{C,G}$	31.9	−12.1 [b]	4.9	−158	−15.8	−41.2	0.87
3i	$T2^{RX}$	27.9	−9.0 [b]	7.9	−107	−14.5	−44.1	0.88

[a] ISE, ISE_{corr} and $\Delta\eta$ are given in kcal mol^{-1}, Λ in ppm cgs and NICS indices in ppm. [b] The large flexibility induces topology changes in the dihydrogen derivative of the methylene adducts of these conformations during the optimization.

TABLE 13.9 Energetic, Reactivity, Magnetic and Structural Indices of Aromaticity of the Different Conformations of the Diprotonated Unsubstituted [36]- and [38]Octaphyrins **4** and **5** [a]

Conformation	Tn^X	ISE	ISEcorr	$\Delta\eta$	Λ	NICS(0)	NICSzz(1)	HOMA
4a	$T0^{5,10,25,30}$	22.1	−13.5	−7.3	881	28.4	81.7	0.69
4c	$T0^{B,F,5,20,25}$	16.1	−15.5	−9.2	479	5.1	37.1	0.80
4e	$T1^{B,C,E,H}$	26.2	0.95	10.8	−559	−10.4	−26.0	0.80
4f	$T1^{B,C,F}$	20.7	3.2	5.1	−361	−15.3	−30.9	0.71
4h	$T2^{C,G}$	20.9	−1.5	−8.0	379	18.6	−17.1	0.78
4i	$T2^{RX}$	25.1	0.3 [b]	−3.4	70	1.2	−25.6	0.77
5a	$T0^{5,10,25,30}$	27.7	−13.0 [b]	7.7	−638	−17.8	−39.2	0.88
5c	$T0^{A,E,5,25,40}$	22.1	−9.6 [b]	6.7	−278	−18.5	−28.0	0.86
5e	$T0^{B,F,5,20,25}$	29.3	−14.2	−2.6	707	12.8	40.5	0.76
5f	$T1^{B,C,E,H}$	15.3	−15.6	−2.7	462	20.8	59.2	0.76
5h	$T1^{B,C,F}$	27.8	−2.4 [b]	8.3	−283	−18.0	−21.1	0.87
5i	$T2^{RX}$	27.6	−13.0 [b]	10.3	−183	−14.4	−33.5	0.88

[a] ISE, ISE$_{corr}$, and $\Delta\eta$ are given in kcal mol^{-1}, Λ in ppm cgs and NICS indices in ppm. [b] The large flexibility induces topology changes in the dihydrogen derivative of the methylene adducts of these conformations during the optimization

upon protonation, as indicated by the larger $\Delta\eta$ and more negative magnetic descriptors for the diprotonated **4e-f**. Likewise, the aromaticity of the Hückel conformations **a-c** with [38] π-electrons increases when the system is diprotonated. Therefore, diprotonated [38]octaphyrins combine the enhanced aromaticity upon protonation with the aromatic character of the Hückel topologies.

Our statistical analysis revealed significant correlations between the magnetic and reactivity descriptors of aromaticity. Interestingly, the isomerization method provides Λ and $\Delta\eta$ values highly correlated with the NICS-based indices (Table 13.10). However, the energetic parameter ISE_{corr} is not correlated with the rest of the indices. For the evaluation of the energetic descriptor, the application of *syn-anti* corrections is mandatory due to the *cis-trans* diene mismatches in the methyl and methylene isomer [36]. Nevertheless, most of the dihydrogen derivatives of the methyl/methylene isomers switch their topology during the optimization step, yielding biased ISE_{corr} values. As the evaluation of Λ and $\Delta\eta$ does not require *syn-anti* corrections, the isomerization method is an effective approach for evaluating Λ and the relative hardness of octaphyrins. In addition, bad correlations are observed for the structural HOMA index. In fact, HOMA is unable to grasp the subtle difference between aromatic and antiaromatic conformations. Therefore, the energetic ISE_{corr} and structural HOMA indices should be applied with caution in octaphyrins, and we strongly recommend to use the $\Delta\eta$, Λ and NICS(0) indices to quantify the aromaticity of octaphyrins.

TABLE 13.10   Correlation Between Several Descriptors of Aromaticity for Unsubstituted Octaphyrins ($n = 25$) [a]

	$\Delta\eta$	Λ	NICS(0)	HOMA	$NICS_{zz}(1)$	ISE_{corr}
$\Delta\eta$	1					
Λ	0.75	1				
NICS(0)	0.83	0.92	1			
HOMA	0.50	0.44	0.54	1		
$NICS(1)_{zz}$	0.89 [b]	0.94 [b]	0.96 [b]	0.49 [b]	1	
ISE_{CORR}	0.01	0.04	0.07	0.04	0.03	1

[a] Conformations with highly distorted methylene and methyl isomers in the isomerization reaction were not taken into account in this statistical analysis.

[b] The figure-eight topologies were left out because their $NICS_{zz}(1)$ values are not associated with the macrocyclic ring current.

Having understood the structure-aromaticity relationships, we finally explored the relationship between NLO properties, aromaticity and the molecular topology of octaphyrins. Although the theoretical study on NLO properties of octaphyrins has not been performed yet, we have used the two-photon absorption (TPA) cross-sections values measured experimentally for neutral and diprotonated[36] and [38]octaphyrins 6 and 7 [30]. As shown in Figure 13.8, the NLO properties of the different octaphyrins are intrinsically coupled to their aromatic and structural properties. The lowest TPA cross-section value (σ_{TPA} = 800 GM) is associated with the neutral [36]octaphyrin, which displays a non-aromatic figure-eight conformation. A much larger two photon absorption cross-section value was found for diprotonated [36]octaphyrin (σ_{TPA} = 5100 GM) [30]. In that case, the preference for a Möbius topology together with the enhanced aromaticity upon protonation explains the much larger NLO properties of diprotonated species. A similar situation is found in [38]octaphyrins, in which the formation of a Hückel aromatic structure upon protonation is accompanied by a large increase of the TPA cross-section (σ_{TPA} = 4600 GM). In contrast, the figure-eight conformation 8h is only weakly aromatic, resulting in a σ_{TPA} = 1800 GM. Again, the presence of a moderate diatropic ring current in the twisted-Hückel conformation with [38] π-electrons justifies the enhanced σ_{TPA} relative to the non-aromatic [36]counterpart. Accordingly, octaphyrins with greater aromatic character give rise to higher TPA cross-section values. Therefore, aromaticity provides a guiding principle for highly efficient two-photon absorption materials from expanded porphyrins.

13.5 EVOLUTION OF THE ELECTRON DELOCALIZATION WITH THE TOPOLOGY IN OCTAPHYRINS

Finally, we explored the evolution of the electron delocalization in two topological interconversions of octaphyrins using the concept of bond metallicity. Since the interconversion between different conformers relies on the variation of one or two dihedral angle(s) [69], the variation of the bond metallicity is local, and thus local metallicity indices are applicable [25]. For several molecular systems, the relationship between the bond metallicity index and conductivity was probed. Our study on p-phenylene and p-xylylene chains revealed that the delocalization indicator is able to provide a qualitative picture of their conductance [25]. Furthermore since Möbius and

Hückel expanded porphyrins exhibit distinct aromatic and nonlinear optical properties, the molecular conductance might also be topology-dependent. Our working hypothesis is that a significant change in the conductance of expanded porphyrins will be observed after the topology switching (Figure 13.9).

Our first study case corresponds to a two-step interconversion between two Möbius structures (**1e, 1l**) involving an intermediate with Hückel topology **1k** (Figure 13.10). As shown above, the switching is feasible with

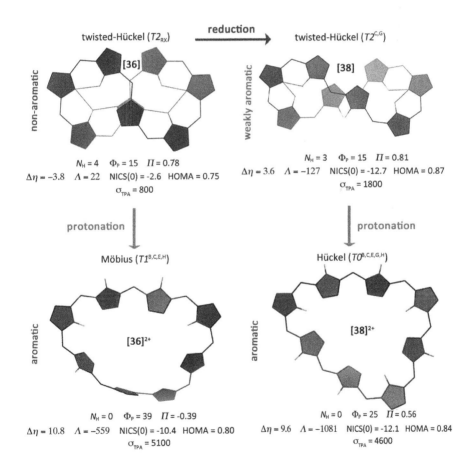

FIGURE 13.9 Conformational and aromaticity changes of the octaphyrin macrocycle upon reduction and protonation. The structural and aromaticity descriptors are shown together with the TPA cross-section values (in GM).

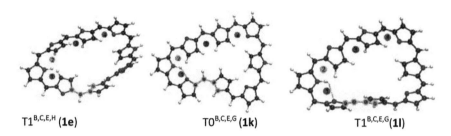

$T1^{B,C,E,H}$ (**1e**) $T0^{B,C,E,G}$ (**1k**) $T1^{B,C,E,G}$(**1l**)

FIGURE 13.10   Investigated bond critical points in the molecular switch, where the rotating dihedral angle is highlighted in light blue. The *bcps* are numbered like in the text and colored like in the graphs. The new hydrogen bond in the 1k structure is highlighted with a red dotted line. Reproduced from Ref. [25] with permission from the PCCP Owner Societies.

energy barriers of 13.9 kcal mol^{-1} for $T1^{B,C,E,H}$ (**1e**) $\rightarrow$ $T0^{B,C,E,G}$(**1k**) and 8.0 kcal mol^{-1} for $T1^{B,C,E,G}$(**1k**)$\rightarrow$ $T0^{B,C,E,G}$(**1l**). Here, the rotation of the *cis* dihedral angle in **1e** induces an additional rotation of the neighboring pyrrole ring, leading to the formation of an additional hydrogen bond in structures **1k** and **1l**. We have only investigated half of the critical points in the octaphyrin macrocycle, focusing on the side of the rotating dihedral angle. In this subsection, the *bcp* next to the perturbation is denoted as 1 and the index keeps increasing in a clock-wise manner. Thus, 2 and 8 are the closest *bcps* to the perturbation, then come 3 and 7, and so on.

Generally, the variation associated with the topological switch is extremely local in nature. As shown in Figure 13.11, the variation of the bond metallicity at *bcps* 2, 3, 4 is negligible with respect to that of *bcp1*. The bond metallicity of *bcp1* rises drastically when the dihedral angle reaches 60° for **1k** and **1l**. This increase in bond metallicity results from the formation of a hydrogen bond denoted in red in Figure 13.10. Despite sharing the same noncovalent interaction framework (**1k** and **1l**), the Hückel topology **1k** leads to a higher bond metallicity and thus better conductance. Nevertheless, we observe the biggest structure-conductance change during the **1e-1k** conformational change. This is specially so if we take into account that *bcp1* was initially the one that was related to the lowest metallicity, and hence, it could have been expected to be the conducting limiting step.

Secondly, we study a one-step Hückel-Möbius interconversion corresponding to the untwisting of the twisted-Hückel conformer **1g** leading to the Möbius conformer **1f** (Figure 13.12).The *bcp* above the changing dihedral angle is characterized as *bcp1* and we follow the same nomenclature we used above. Importantly, the π-π staking interaction that appears at the

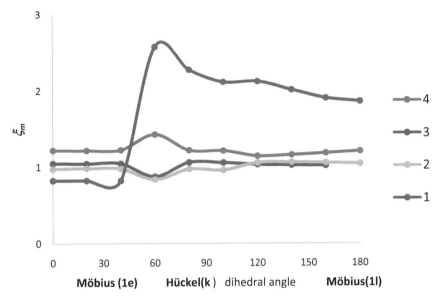

Möbius (1e)   Hückel(k) dihedral angle   Möbius(1l)

FIGURE 13.11 Evolution of the bond metallicity with the amplitude of the dihedral angle for the Hückel-Möbius interconversion. Reproduced from Ref. [25] with permission from the PCCP Owner Societies.

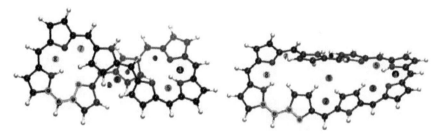

FIGURE 13.12 Twisted-Hückel topology $T2^{B,F}$ (**1g**) and Möbius topology $T1^{B,C,F}$ (**1f**) with the bond critical points colored according to their position the varying dihedral angle is highlighted in light blue. Reproduced from Ref. [25] with permission from the PCCP Owner Societies

half-twist (**1g**) is not conserved in the Mobius topology (**1f**). The *bcps* close to this π-π interaction are named *bcp6* and *bcp9* in our graphs. Although the electron density topological analysis carried out with our modified version of TOPMOD [79] reveals 11 noncovalent *bcps* at the global minimum of the PES (**1g**, $d = 0°$), we only evaluated measures of metallicity at the bcps that can be detected at each step to the interconversion (8). In fact, according to the evolution of the Laplacian during the interconversion, several non-covalent interactions gradually vanish as the amplitude of the dihedral angle increases. Since *bcp1* was soon undetectable by the topological analysis, its measures of metallicity were not included in the analysis of the macrocycle.

The bond metallicity ξ_m changes along the interconversion pathway involving **1g** and **1f** are depicted in Figure 13.13. On one side, the bond critical points lying in the vicinity of the perturbation (*bcp8* and *bcp2*) are more sensitive to the variation of the dihedral angle, highlighting the local nature of the metallicity in this system. On the other side, since the Hückel-Möbius transition involves the creation of a π-π staking interaction, *bcp6* and *bcp9* present a non-negligible variation. In fact, owing to the loss of a π-π stacking interaction, the bond metallicity of *bcp5* and *bcp9* decreases for larger amplitudes of the dihedral angle.

Delocalization at the *bcps* close to the twist (*bcp2*, bcp6 and *bcp8*) increases when going from the twisted-Hückel **1g** to the Möbius topology **1f**. All in all, although the *bcp* closest to the perturbation still presents the highest variation in the measures of metallicity, delocalization is not fully local since important changes also occur associated to the folding. Thus, we have shown that local measures of metallicity can detect the changes in

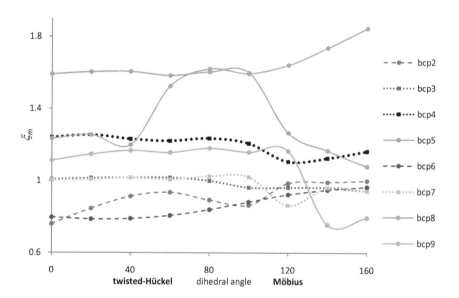

FIGURE 13.13 Evolution of the bond metallicity index at several critical points of the molecule with the amplitude of the dihedral angle for the twisted-Hückel-Möbius interconversion. Reproduced from Ref. [25] with permission from the PCCP Owner Societies

delocalization upon conformational switch in octaphyrins. In those cases where the conformation remains open (so that there is no net change of interactions associated with the twist), the changes remain rather local, which can be used in the qualitative prediction of delocalization and conductivity in porphyrins and expanded porphyrins. In those cases where there is a conformational folding, the change in interactions introduces a non-local behavior, which has to be taken into account.

13.6 CONCLUSIONS

The conformational preferences, aromaticity and the electron delocalization of [36]octaphyrins have been thoroughly investigated using density functional theory calculations and topological analysis. In this work, we have analyzed the influence of solvent, substituents, protonation and oxidation state on the conformation and properties of octaphyrins. Neutral octaphyrins adopt preferentially the non-aromatic figure-eight conformation owing to its more effective intramolecular hydrogen-bonding interactions. Our calculations point to the high stability of the twisted-Hückel conformation in [36]octaphyrins, being predominant in all the solvents and for most substituents. Nevertheless, the balance between twisted-Hückel, Möbius and Hückel conformations in [36]octaphyrins can be fine-tuned by the substituents at *meso-* and *β*-positions.

Different external *stimuli* for triggering a conformational/aromaticity switch in octaphyrins have been scrutinized in our work, including protonation and redox reactions. Whereas protonation induces a topology change from a figure-eight to a Möbius conformation, redox reactions transform non-aromatic $[4n]$ π-electron systems into (weakly) aromatic $[4n + 2]$ π-electron ones. The twisted-Hückel topology remains predominant in the oxidized and reduced species, corresponding to a [34] and a [38] π-electron system, respectively. In addition, the combination of reduction and protonation of the macrocycle yields a highly aromatic Hückel conformation.

The different aromaticity indices reveal a close relationship between the molecular topology, the number of π-electrons and aromaticity. For [36]octaphyrins, Möbius structures are highly aromatic and Hückel conformations are strongly antiaromatic. The neutral twisted-Hückel topology can be classified as non-aromatic due to its reduced paratropicity. Importantly, the aromaticity of all the conformations is totally reversed upon redox reactions, so Hückel structures becomes aromatic in [34] and [38]octaphyrins. By contrast, protonation of the macrocycle enhances the (anti)aromaticity of all the conformations considerably. The aromaticity of octaphyrins is evaluated best by the magnetic indices (Λ and NICS(0)) and the relative hardness rather than the structural and energetic descriptors. Importantly, the NLO properties are greatly influenced by the macrocyclic aromaticity of the π-electron system.

We have reviewed several local indexes of metallicity based on the analysis of electron density, ELF and the kinetic energy density at the bond critical point. Since these indexes provide a qualitative measure of delocalization and conductance in several proof-of-concept molecules, we have carried out a local analysis of the bond metallicity changes along two topological switches based on the [36]octaphyrin. Our results demonstrated that the bond metallicity and delocalization index provide a qualitative description of the electron delocalization in octaphyrins, which is related to conductivity. The measures of bond metallicity indicate a local variation of the electron delocalization when the topological change is restricted to one part of the molecule and semi-local for global topological changes (e.g., folding). In the local cases, our approach enables to identify which conformational switch would be more efficient from an electronic device perspective. In principle, those Hückel-Möbius switches exhibiting the larger changes in bond metallicity will be the most promising conductance switches. Since our indices were able to detect the difference between those two interconversions, they could also be used to perform a qualitative analysis.

In summary, we conclude that [36]octaphyrins are promising platforms for the development of a novel type of molecular switches for nanoelectronic applications as protonation and/or redox reactions induce drastic changes in the aromaticity and NLO properties of the macrocycle. The topology switching might be coupled to changes in the conductance of octaphyrins as revealed by the metallicity indices, which proves the potential use of Hückel-Möbius systems as conductance switches.

KEYWORDS

- **aromaticity**
- **delocalization**
- **expanded porphyrins**

- **metallicity**
- **molecular switches**
- **octaphyrins**

REFERENCES

1. Stępień, M., Szyszko, B., & Latos-Grażyński, L., (2010). *J. Am. Chem. Soc.*, *132*(9), 3140–3152.
2. Stępień, M., Sprutta, N., & Latos-Grażyński, L., (2011). *Angew. Chem. Int. Ed.*, *50*, 4288–4340.
3. Tanaka, T., & Osuka, A., (2017). *Chem. Rev.*, *117* (4), 2584–2640.
4. Chandrashekar, T. K., & Venkatraman, S., (2003). *Acc. Chem. Res*, *36*, 676–691.
5. Pawlicki, M., Collins, H. A. Denning, R. G., & Anderson, H. L., (2009). *Angew. Chem. Int. Ed*, *48*, 3244–3266.
6. Osuka, A., Tanaka, T., & Mori, H., (2013). *J. Mat. Chem. C*, *1*, 2500–2519.
7. Saito, S.,& Osuka, A., (2011). *Angew. Chem. Int.*, *50*, 4342–4373.
8. (a) Feringa, B. L. & Brwone, W. R., (2011). Molecular Switches, Chiroptical Molecular Switches, *1*, 121–122, Wiley-VHC, Weinheim, Germany. (b) Andréasson, J., & Pischel, U., (2010). *Chem. Soc. Rev.*, *39*, 174–188. (c) Gui, B., Meng, X., Chen, Y., Tian, J., Liu, G., Shen, C., et al., (2015). *Chem. Mater*, *27*, 6426–6431. (d) Woller, T., Contreras-García, J., Geerlings, P., De Proft, F., & Alonso, M., (2016). *Phys. Chem. Chem. Phys.*, *18*, 11885–11900.
9. Alonso, M., Geerlings, P., & De Proft, F., (2012). *Chem. Eur. J.*, *18*, 10916–10928.
10. Alonso, M., Geerlings, P., & De Proft, F., (2014). *Phys. Chem. Chem. Phys.*, *16*, 14396–14407.
11. Karthik, G., Min Lim, J., Srinivasan, A., Suresh, C. H., Kim, D., & Chandrashekar, T. K., (2013). *Chem. Eur. J*, *19*, 17011–17020.
12. Yoon, Z. S., Osuka A., & Kim, D., (2009). *Nat. Chem.*,*1*, 113–122.
13. Alonso, M., Pinter, B., & Geerlings, P., & De Proft, F., (2015). *Chem. Eur. J.* 21, 17631–17638.
14. Hrsak, D., Pertejo., Al M. A., Lamshabi, M., Muranaka, A., & Ceulemans, A., (2013). *Chem. Phys. Lett.*, *586*, 148–152.
15. Lim, J. M., Yoon, Z. S., Shin, J-Y., Kim, K. S., Yoon, M-C., & Kim, D., (2009). *Chem. Commun*, 261–273.
16. (a) Garratt, P. J., (1986). *Aromaticity*, John Wiley & Sons,New York. (b) Minkin, V. I., Glukhovtsev M. N., & Simkin, B. Y., (1994). *Aromaticity and Antiaromaticity: Electronic and Structural Aspects,* John Wiley & Sons, New York. (c) Schleyer, P. V. R., (2001). *Chem. Rev.*, *101*, 1115–1117.
17. Heilbronner, E., (1964). *Tetrahedron Lett.*, *5*, 1923–1928.
18. Rappaport S. M., & Rzepa, H. S., (2008). *J. Am. Chem. Soc.*, *130*, 7613–7619.
19. Oh, J., Sung, Y. M., Kim, W., Mori, S., Osuka A., & Kim D., (2016). *Angew. Chem. Int. Ed*, *55*, 6487–6491.
20. (a) Sung, Y. M., Yoon, M-C., Lim, J. M., Rath, H., Naoda, K., Osuka, A., et al., (2015). *Nat. Chem.*, *7*, 418–422. (b) Ottoson, H., & Borbas, K. E., (2015). *Nat. Chem.*,*7*, 373–375.

21. (a) Solà, M., Feixas, F., Jiménez-Halla, J. O. C., Matito E., & Poater, J., (2010). *Symmetry, 2,* 1156–1179. (b) Alonso M., & Herradón, B., (2010). *Phys. Chem. Chem. Phys., 12,* 1305–1317. (c) Cyrański, M. K., Krygowski, T. M., Katritzky A. R., & Schleyer, P. V. R., (2002). *J. Org. Chem., 67,* 1333–1338. (d) Ajami, D., Hess, K., Köhler, F., Näther, C., Oeckler, O., Simon, A., et al., (2006). *Chem. – Eur. J, 12,* 5434–5435. (e) Alonso, M., & Herradón, B., (2009). *J. Comput. Chem., 31,* 917–928.
22. Wu, J. I., Fernandez I., & Schleyer, P. V. R., (2012). *J. Am. Chem. Soc., 315*–321.
23. (a) Wannere, C. S., & Schleyer, P. V. R., (2003). *Org. Lett., 5,* 865–868. (b) Wannere, C. S., Moran, D., Allinger, N. L., Hess, B. A., Schaad, L.-J., & Schleyer, P. V. R., (2003). *Org. Lett., 5,* 2983–2986.
24. Alonso, M., Geerlings P., & De Proft, F., (2013). *Chem. – Eur. J, 19,* 1617–1628.
25. Woller, T., Contreras-García, J., Ramos-Berdullas, N., Mandado, M., De Proft F., & Alonso, M., (2016). *Phys. Chem. Chem. Phys., 18,* 11829–11838.
26. (a) Baer, R., & Head-Gordon, M., (1997). *Phys. Rev. Lett., 79,* 3962–3396. (b) Marino, T., Michelini, M. C., Russo, N., Sicilia, E., & Toscano, M., (2012). *Theor Chem Acc., 131*–141.
27. Kohn, W., (1996). *Phys. Rev. Lett. 76,* 3168–3171.
28. Li, X. P., Nunes, W., & Vanderbilt, D., (1993). *Phys. Rev. B,* 47, 10891–10894.
29. (a) Brauer, B., Kesharwani, M. K., Kozuch S. & Martin, J. M. L., (2016). *Phys. Chem. Chem. Phys., 18,* 20905–20925. (b) Kesharwani, M. K., Karton A., & Martin, J. M. L., (2016). *J. Chem. Theory Comput., 12,* 444–454.
30. Lim, J. M., Shin, J.-Y., Tanaka, Y., Saito, S., Osuka A., & Kim, D., (2010). *J. Am. Chem. Soc., 132,* 3105–3114.
31. (a)Latos-Grażyński, L., (2004). *Angew. Chem., Int. Ed, 43,* 5124–5128. (b) Tanaka, Y., Saito, S., Mori, S., Aratani, N., Shinokubo, H., Shibata, N., et al., (2008). *Angew. Chem. Int. Ed, 47,* 681–684. (c) Tanaka, Y., Shinokubo, H., Yoshimura Y., & Osuka, A., (2009). *Chem. – Eur. J.,15,* 5674–5675.
32. Shin, Y., Furuta, H., Yoza, K., Igarashi, S., & Osuka, A., (2001). *J. Am. Chem. Soc., 123,* 7190–7191.
33. Johnson, E. R., Keinan, S., Mori-Sánchez, P., Contreras-García, J., Cohen A. J., & Yang, W., (2010). *J. Am. Chem. Soc., 132,* 6498–6506.
34. Contreras-García, J., Johnson, E. R., Keinan, S., Chaudret, R., Piquemal, J.-P., Beratan D. N., et al., (2011). *J. Chem. Theory Comput., 7,* 625–632.
35. Dauben, H. J., Wilson J. D., & Layti, J. L., (1968). *J. Am. Chem. Soc., 90,* 811–813.
36. Schleyer, P., & Pühlhofer, F., (2002). *Org. Lett., 4,* 2873–2876.
37. Chen, Z., Wannere, C. S., Corminboeuf, C., Puchta, R., & Schleyer, P. V. R., (2005). *Chem. Rev., 105,* 3842–3888.
38. Corminboeuf, C., Heine, T., Seifert, G., Schleyer, P. V. R., & Weber, J., (2004). *Phys. Chem. Chem. Phys.,* 6, 273–276.
39. De Proft, F., & Geerlings, P., (2004). *Phys. Chem. Chem. Phys., 6,* 242–248.
40. Dauben, H. J., Wilson, J. D., & Layti, J. L., (1968). *J.Am.Chem. Soc., 90,* 811–813.
41. Koopmans, T. A., (1933). *Physica, 1,* 104–113.
42. Pearson, R. G., (1997). *Chemical Hardness,* Wiley, New York.
43. (a) Kruszewski J. & Krygowski, T. M., (1972). *Tetrahedron Lett., 13,* 3839–3842. (b) Krygowski, T. M., (1993). *J. Chem. Inf. Comput. Sci.,33,* 70–78. (c) Krygowski, T. M., Szatylowicz, H., Stasyuk, O. A., Dominikowska, J., & Palusiak, M., (2014). *Chem. Soc. Rev., 114,* 6383–6422.

44. (a) Sondheimer, F., Wolovsky R., & Amiel, Y., (1962). *J. Am. Chem. Soc., 84*, 274–284. (b) Vogel, E., (1993). *Pure Appl. Chem.,65*, 143–152.

45. Bader, R. F. W., (1990). *Atoms in Molecules: A Quantum Theory*, Clarendon, Oxford.

46. Bader, R. F. W., & Nguyendang, T. T., (1981). *Adv. Quantum Chem.,14*, 63–124.

47. Popelier, P. L. A., (2000). *Atoms in Molecules: An Introduction*, Pearson, Harlow.

48. Matta, C. F., Bader, & R. F. W., (2006). *J. Phys. Chem. A. 110*, 6365–6371.

49. Bader, R. F. W., (1998). *J. Phys. Chem. A, 102*, 7314–7323.

50. Bader, R. F. W., Tal, Y., Anderson, S. G., & Nguyen-Dang, T. T., (1980). Isr. J. Chem., 19, 8–29.

51. Mori-Sánchez, P., Martín Pendás,A., & Luaña. V, (2002). *J. Am. Chem. Soc.*, 124, 14721–14723.

52. Jenkins, S., Ayers, P. W., Kirk, S. R., Mori-Sánchez, P., & Martín Pendás, A. (2009). *Chem. Phys. Lett.*, 1(471), 174–177.

53. Jentkins, S., (2012). J. Phys. Condens. Matter., 14, 10251–10263.

54. Becke, A. D., & Edgecombe, K. E., (1990). J. Chem. Phys., 92, 5397–5403.

55. Savin, A., Jepsen, O., Flad, J., Andersen, O. K., Preuss, H., & von Schnering, H. G., (1992). Angew. Chem. Int, 31,187–188.

56. Contreras-García, J., & Recio, J. M., (2011). *Theor. Chem. Acc., 128*, 411–418.

57. Silvi, B., & Gatti, C., (2000). *J. Phys. Chem. A, 104*, 947–953.

58. Jenkins, S., Kirk, S. R., Ayers, P. W., & Kuhs, W. F., (2006). *R. Soc. Chem.*, 265–272.

59. Ghosh, S. K., Berkowitz, M., & Parr, R. G., (1984). *Proc. Natl. Acad. Sci., 81*, 8028–8031.

60. Thomas, L. H., (1927). *Proc. Camb. Philos. Soc., 23*, 542–548.

61. Fermi, E., (1928). *Z. Phys., 48*, 73–79.

62. Tao, J. M., Vignale, G., & Tokatly, I. V., (2008). *Phys. Rev. Lett., 100*, 206405–206410.

63. Becke, A. D., (1993). *J. Chem. Phys., 98*, 5648–5652.

64. Perdew, J. P., Burke K. & Ernzerhof, M., (1997). *Phys. Rev. Lett., 77*(18), 3865–3868.

65. Zhao Y., & Truhlar, D. G., (2008). *Theor. Chem. Acc., 120*, 215–241.

66. Chai J.-D., & Head-Gordon, M., (2008). *Phys. Chem. Chem. Phys., 10*, 6615–6620.

67. (a) Grimme, S., (2004). *J. Comput. Chem.,25*, 1463–1473. (*b*) Ehrlich, J., Moellmann, & Grimme, S., (2012). *Acc. Chem. Res, 46*, 916–926.

68. (a) Becke, A. D., (1988). *Phys. Rev. A, 38*, 3098–3100. (*b*) Perdew, J. P., (1986). *Phys. Rev. B, 33*, 8822–8824.

69. Alonso, M., Geerlings P., & De Proft, F., (2013). *Chem. Eur. J, 19*, 1617–1628.

70. Marenich, A. V., Cramer C. J., & Truhlar, D. G., (2009). *J. Phys. Chem. B, 113*, 6378–6396.

71. Alonso, M., Geerlings P., & De Proft, F., (2014). *Phys. Chem. Chem. Phys., 16*, 14396–14407.

72. (a) Suzuki M., & Osuka, A., (2007). *Chem. – Eur. J, 13*, 196–202. (b) Shimizu, S., Aratani N., & Osuka, A., (2006). *Chem. – Eur. J, 12*, 4909–4918.

73. Marcos, E., Anglada J. M., & Torrent-Sucarrat, M., (2014). *J. Org. Chem., 79*, 5036–5046.

74. Shimizu, S., Shin, J.-Y., Furuta, H., Ismael R., & Osuka, A., (2003). *Angew. Chem. Int. Ed, 42*, 78–82.

75. (a) Saito, S., Shin, J. Y., Lim, J. M., Kim, K. S., Kim D., & Osuka, A., (2008). *Angew. Chem., Int. Ed, 47*, 9657–9660. (b) Shin, J.-Y., Lim, J. M., Yoon, Z. S., Kim, K. S., Yoon, M.-C., Hiroto, S., (2009). *113*, 5794–5802. (c) Karthik, G., Min J., Lim. Srini-

vasan, A., Suresh, C. H., Kim D. & Chandrashekar,T., K., (2013). *Chem. – Eur. J, 19,* 17011–17020.

76. (a) Saito, S., Shin, J. Y., Lim, J. M., Kim, K. S., Kim D., & Osuka, A., (2008). *Angew. Chem. Int. Ed, 47,* 9657–9660. (b) Shin, J.-Y., Lim, J. M., Yoon, Z. S., Kim, K. S., Yoon, M.-C., Hiroto, S., et al., (2009). *J. Phys. Chem. B, 113,* 5794–5802. (c) Karthik, G., Min Lim, J., Srinivasan, A., Suresh, C., H. Kim D., & Chandrashekar, T. K., (2013). *Chem. – Eur. J, 19,* 17011–17020.

77. (a) Osuka, A., & Saito, S., (2011). *Chem. Commun, 47,* 4330–4339. (b) Yoon, Z. S., Cho, D.-G., Kim, K. S., Sessler, J. L., & Kim, D., (2008). *J. Am. Chem. Soc., 130,* 6930–6931. (c) Yoon, M.-C., Cho, S., Suzuki, M., Osuka A., & Kim, D., (2009). *J. Am. Chem. Soc., 131,* 7360–7367. (d) Cho, S., Yoon, Z. S., Kim, K. S., Yoon, M.-C., Cho, D.-G., Sessler, J. L., et al., (2010). *J. Phys. Chem. Lett., 1,* 895–900. (e) Torrent-Sucarrat, M., Anglada J. M., & Luis, J. M., (2012). *J. Chem. Phys., 137,* 184306.

78. Shin, J. Y., Kim, K. S., Yoon, M. C., Lim, J. M., Yoon, Z. S., Osuka A., et al., (2010). *Chem. Soc. Rev., 39,* 2751–2767.

79. Noury, S., Krokidis, X., Fuster, F., & Silvi, B., (1999). *Comput. Chem., 23,* 597–604.

CHAPTER 14

INSIGHTS INTO MOLECULAR ELECTRONIC STRUCTURE FROM DOMAIN-AVERAGED FERMI HOLE (DAFH) AND BOND ORDER ANALYSIS USING CORRELATED DENSITY MATRICES

DAVID L. COOPER[1] and ROBERT PONEC[2]

[1]Department of Chemistry, University of Liverpool, Liverpool L69 7ZD, United Kingdom

[2]Institute of Chemical Process Fundamentals, Czech Academy of Sciences, Prague 6, Suchdol 2, 165 02 Czech Republic

CONTENTS

ABSTRACT

Two families of case studies are used to show how combinations of domain-averaged Fermi hole and bond order analysis provide straightforward links from correlated density matrices (derived from intrinsically multiconfigurational wavefunctions) to intuitive chemical descriptions, including those that focus on the extent of electron sharing. In particular, insights are provided into the electron reorganization that accompanies the making and breaking of chemical bonds, and into the nature of the multicenter bonding in selected molecules.

14.1   INTRODUCTION

The immense impact of the concept of the chemical bond is of course closely tied to the seminal contribution of Gilbert N. Lewis that was presented in a paper submitted for publication in January 1916, more than 100 years ago [1]. Lewis deeply reshaped our understanding of chemical structures, and his association of chemical bonds with shared electron pairs has become one of the cornerstones of chemistry. Although it was initially conceived before the advent of quantum theory, what we now know as Lewis theory has proved to be extremely fruitful and, indeed, resilient. Much of our understanding of molecular structures still relies to a large extent on the classical picture provided by the Lewis electron pair model. Its reconciliation with the quantitative description provided by quantum mechanics still continues to represent challenges for contemporary chemical theory [2–18].

Although the parallel between quantum chemical and classical descriptions of chemical structures could be demonstrated fairly easily using early approaches that were based on the analysis of approximate self-consistent field (SCF) wavefunctions, the extraction of similar chemical interpretations from more sophisticated wavefunctions, such as those resulting from contemporary higher accuracy calculations, has turned out to be more difficult. It is for this reason that the need is increasingly felt to accompany the development of modern sophisticated computational techniques with the design of new auxiliary methods that allow us to transform the structural information 'hidden' in the abstract wavefunctions or electron densities into a language that resembles familiar chemical concepts as bonds, bond orders, valences, and so on. A particularly important example in this respect is Bader's virial partitioning of electron density into contributions associated with individual atoms, i.e., the so-called quantum theory of atoms in molecules (QTAIM) [19], with various very useful auxiliary concepts such as bond critical points and bond paths [20–22]. Other widely used approaches include the analysis of an electron localization function (ELF) [23, 24], of natural bond orbitals (NBO) [25–27], and of the numerical data provided by various bond indices and populations [28–36].

Most of these approaches extract structural information chiefly from the one-electron density matrix. Nonetheless the presence of two-electron terms in the Hamiltonian, the multi-electron nature of electron correlation and, especially, the importance of the electron pairing anticipated by the Lewis model, all suggest the utility of analyzing also the two-electron density matrix (i.e., pair densities). These last can be considered the simplest quantum chemical quantities that describe properly the electron pairs in microscopic systems [37, 38], and we can reasonably anticipate that additional insights into the nature of chemical bonding should emerge from analyzing them. Amongst approaches that pursue such a route, we specifically mention the analysis of the so-called domain-averaged Fermi holes (DAFHs) [39–41]. There are of course also many versions of bond indices that fulfill various roles as theoretical counterparts to the classical concepts of bond order or bond multiplicity [28–36, 42–45].

Appropriate combinations of DAFH analysis and of bond order analysis have proved to be remarkably useful, especially for investigating the electronic structures of molecules with non-trivial bonding patterns. Examples of such applications include studies of multicenter bonding, hypervalence and metal-metal bonding, amongst various others, but the majority of such investigations have been performed only at the SCF and/or Kohn-Sham levels of theory [28, 29, 33, 42, 43, 45–50]. Nonetheless, there have now been a fair number of studies that have successfully used actual correlated pair density matrices that were extracted from more sophisticated, intrinsically multiconfigurational, wavefunctions. The main purpose of the present contribution is to present two families of case studies of this type. First of all, we consider various examples that show how the methodology can be used to provide deep insights into the electron reorganization that accompanies the making and breaking of two-center chemical bonds. We then turn our attention to selected examples of multicenter bonding. It is our intention that the material should be read in the order in which it is presented, because the relevant theoretical details are mostly presented in the sequence in which they are required for the various examples that are discussed.

14.2   CASE STUDIES 1: ELECTRON REORGANIZATION ACCOMPANYING THE MAKING AND BREAKING OF CHEMICAL BONDS

Chemistry is a science of molecular change and so one of the ultimate goals of chemical theory is to provide tools that can link the complex electron reorganization that accompanies chemical reactions to more intuitive chemical descriptions. Lewis's idea of a chemical bond as a shared electron pair is of particular importance, making it entirely natural to attempt to gauge the evolution of the bonding interactions along a reaction path by examining the changes to the extent of electron sharing when interatomic distances are varied. The particular tools of most relevance here are DAFH analysis, using correlated density matrices, and certain definitions of bond order that make use of the same density matrices.

14.2.1   THEORETICAL BACKGROUND

It is useful to open this Section with a brief outline of DAFH analysis [39–41, 51–55], for which a convenient starting point is the following definition of the domain-averaged 'hole' g_Ω for a given domain Ω:

$$g_\Omega\left(r_1,r_1'\right) = \rho^{(1)}\left(r_1,r_1'\right) \int_{\substack{r_2=r_2'\\\Omega}} \rho^{(1)}\left(r_2,r_2'\right)dr_2$$

$$-2\times \int_{\substack{r_2=r_2'\\\Omega}} \rho^{(2)}\left(r_1,r_1'\; ;\; r_2,r_2'\right)dr_2 \tag{1}$$

where $\rho^{(1)}$ and $\rho^{(2)}$ are one- and two-electron densities. Our usual approach for correlated wavefunctions is to expand the spinless 'hole' in the orthonormal basis of (real) natural orbitals f_I, i.e.,

$$g_\Omega\left(r_1,r_1'\right) \equiv \sum_I\sum_J \phi_I\left(r_1\right)G_\Omega\left(I,J\right)\phi_J\left(r_1'\right) \tag{2}$$

It is straightforward to construct this matrix representation, $\boldsymbol{G}_\Omega = \{G_\Omega(I,J)\}$, by taking appropriate combinations of elements of the (spinless) one- and two-electron density matrices, expressed in this natural orbital basis, with so-called domain-condensed overlap integrals:

$$\langle \phi_I \mid \phi_J \rangle_\Omega = \int_\Omega \phi_I(\boldsymbol{r}) \phi_J(\boldsymbol{r}) d\boldsymbol{r} \tag{3}$$

The eigenvectors and eigenvalues of $\boldsymbol{G}_\Omega$ are usually subjected to an isopycnic localization procedure and so it useful at this stage to outline the basic principles and purpose of the isopycnic transformation [56], before commenting on how the domains $\{\Omega\}$ were chosen.

In essence, Cioslowski and Mixon [30] considered the following partitioning of the total number of electrons N for a restricted Hartree-Fock (RHF) wavefunction with singly-occupied orthonormal spin orbitals $|i\rangle$:

$$N = \sum_i 1 = \sum_i \langle i \mid i \rangle \langle i \mid i \rangle = \sum_A \sum_B \sum_i \langle i \mid i \rangle_{\Omega_A} \langle i \mid i \rangle_{\Omega_B} \equiv \sum_A \sum_B C_{AB} \tag{4}$$

in which the non-overlapping domains $\{\Omega_A\}$ partition all space. For more general wavefunctions, in which the occupation of $|i\rangle$ is v_i, the analogous terms take the following form:

$$C_{AB} = \sum_i (v_i)^2 \langle i \mid i \rangle_{\Omega_A} \langle i \mid i \rangle_{\Omega_B} \tag{5}$$

A key aim of the isopycnic transformation [56] is to maximize the sum of all of the one-center terms, C_{AA}, at the expense of the sum of the two-center C_{AB} values, thereby generating a set of more localized orbitals, ψ_p, with associated occupancies λ_p. We note that Cioslowski and Mixon also used the final two-center C_{AB} values, as obtained at the end of the isopycnic localization procedure, to define an index that is now known as the Cioslowski covalent bond order, $C(A,B) = 2\,C_{AB}$ (for $B > A$) [30].

As indicated above, an implementation of the isopycnic localization procedure is used within the DAFH procedures to transform the eigenvalues and eigenvectors of $\boldsymbol{G}_\Omega$ to a set of usually fairly localized DAFH functions $\phi_{p\Omega}(\boldsymbol{r}_1)$ with occupations $n_{p\Omega}$ for each domain Ω. For the most part, insights are obtained from the DAFH analysis for a given molecular system by inspecting the visual depictions of the $\phi_{p\Omega}$ functions for the different domains Ω

in the molecule, alongside an examination of the corresponding occupation numbers, or combinations of them [57, 58]. In practice, DAFH functions (and their associated populations) can usually be associated directly with classical chemical concepts such as bonds, lone pairs and so on, in terms of which chemists continue to interpret and to classify molecular structures. The DAFH analysis provides direct information about how, and to what extent, the electrons in a given domain (usually an atom) are involved in interactions with other domains in the molecule, as well as about the valence state of the atom in the molecule [59–61]. When examining the forms of $\phi_{p\Omega}$ functions, it does of course make sense to focus on those with nontrivial occupations.

It is possible to perform the same sort of analysis also for the more complex domains that can be formed by the union of two or more simpler ones. In such cases, the DAFH analysis yields, alongside information about the interactions with other domains, details about the electron pairs (chemical bonds, lone pairs, …) that are retained within the 'condensed domain'. When carrying out DAFH analysis for such condensed domains, it is still usual to perform the isopycnic transformation using the domain-condensed overlap integrals for the original, uncondensed, atomic domains $\{\Omega\}$. Of course, a very important issue that we have not as yet addressed is the actual choice of these atomic domains.

Although we occasionally make use instead of a somewhat cheaper approach, in which the required domain-condensed overlap integrals are obtained instead using a simple Mulliken-like approximation, our preferred approach when working with correlated density matrices is to employ the very well-known, and very much used, QTAIM approach [19] to partition the given molecule into non-overlapping spatial domains. It has been shown that the pictorial depictions of the DAFH functions $\phi_{p\Omega}$, as well as the sums of complementary occupation numbers $n_{p\Omega}$, are in fact fairly insensitive to the specific choice of AIM scheme of this type [52] (but the same is not usually true for individual values of the occupation numbers). Note that carrying out DAFH analysis for an entire molecule is completely equivalent to applying the isopycnic transformation to the canonical natural orbitals, so as to produce a set of localized natural orbitals; this is because the sum of the G_Ω matrices for all of the individual QTAIM domains is the same as $D^{(1)}$.

In effect, the domain-averaged 'hole' g_Ω for domain Ω is expanded in the following form:

$$g_\Omega\left(r_1,r_1'\right) = \sum_p n_{p\Omega}\varphi_{p\Omega}\left(r_1\right)\varphi_{p\Omega}\left(r_1'\right) \tag{6}$$

A further selective integration of g_Ω over a second (different) domain yields a quantity which is variously known as a delocalization index or as the shared-electron distribution index (SEDI) [35, 44]. Specifically,

$$k_{AB} = \int_{\substack{r_1 = r_1' \\ \Omega_B}} g_{\Omega_A}\left(r_1,r_1'\right)dr_1 \tag{7}$$

and

$$SEDI\left(A,B\right) = k_{AB} + k_{BA} \tag{8}$$

As is implied by the name, the resulting $SEDI(A,B)$ values quantify the extent of the sharing of the electron distribution between domains Ω_A and Ω_B. Monitoring the variations in SEDI values with nuclear separation makes it possible to reveal the key changes to the extent of electron sharing that accompany the dissociation of bonding electron pairs, and thus to obtain intimate insights into the anatomy of bond formation/breaking [44, 57, 58, 62]. Furthermore, values of $SEDI(A,B)$ can of course be resolved into the contributions from individual DAFH functions $\phi_{p\Omega}$ [63].

In spite of the proven utility of $SEDI(A,B)$ values, there are obviously significant advantages to making use instead of quantities that require only the one-electron density matrix. Variants of the Wiberg-Mayer bond order are probably the best known in this category. A Wiberg-Mayer bond index W_{AB} can be defined for correlated wavefunctions in the following manner [64]:

$$W_{AB} = 2\sum_{\mu \in A}\sum_{\nu \in B}\left(P^\alpha S\right)_{\mu\nu}\left(P^\alpha S\right)_{\nu\mu} + \left(P^\beta S\right)_{\mu\nu}\left(P^\beta S\right)_{\nu\mu} \tag{9}$$

in which the various elements of the density (P^α, P^β) and overlap matrices (S) are expressed here in the atomic basis, with the notation $\mu \in A$ signifying that the summation is restricted to those basis functions which are associated with atomic center A. Substitution of the expressions for the spinless density matrix $D^{(1)}$ and for the spin-density matrix P^S, i.e.,

$$D^{(1)} = P^\alpha + P^\beta$$

$$P^S = P^\alpha - P^\beta \tag{10}$$

which leads to:

$$W_{AB} = \sum_{\mu \in A} \sum_{v \in B} \left[\left(D^{(1)} S \right)_{\mu v} \left(D^{(1)} S \right)_{v\mu} + \left(P^S S \right)_{\mu v} \left(P^S S \right)_{v\mu} \right] \qquad (11)$$

Mayer has suggested [64] for singlet-correlated systems that the spin-density matrix P^S should be replaced in this expression by a matrix R, i.e.,

$$W_{AB} = \sum_{\mu \in A} \sum_{v \in B} \left[\left(D^{(1)} S \right)_{\mu v} \left(D^{(1)} S \right)_{v\mu} + \left(RS \right)_{\mu v} \left(RS \right)_{v\mu} \right] \qquad (12)$$

in which R is defined in the following manner [64]:

$$uS = 2D^{(1)}S - \left(D^{(1)}S \right)^2$$
$$u^\lambda = S^{\frac{1}{2}} \left(uS \right) S^{-\frac{1}{2}} \qquad (13)$$
$$R = S^{-\frac{1}{2}} \left(u^\lambda \right)^{\frac{1}{2}} S^{-\frac{1}{2}}$$

When it is expressed in the basis of the (real) orthonormal natural orbitals $\{f_I\}$ with corresponding occupation numbers $\{\omega_I\}$, R turns out to be a diagonal matrix, with elements $R_{II} = [\omega_I (2 - \omega_I)]^{\frac{1}{2}}$. The matrix $D^{(1)}$ is of course also diagonal in this basis, with elements ω_I, and S is a unit matrix. It is then straightforward for a correlated singlet system to obtain the following expression for the QTAIM-generalized [36] Wiberg-Mayer index [65]:

$$W_{AB} = \sum_I \sum_J \left(\omega_I \omega_J + R_{II} R_{JJ} \right) \left\langle \phi_I | \phi_J \right\rangle_{\Omega_A} \left\langle \phi_I | \phi_J \right\rangle_{\Omega_B} \qquad (14)$$

Although this expression can be decomposed in a planar system into the contributions from different symmetries (σ, π, …), it is not usually possible to resolve the Wiberg-Mayer index W_{AB} into contributions from individual orbitals, because of the presence of cross terms. The same is not true of the Cioslowski covalent bond order, mentioned earlier, which involves a simple summation over localized orbitals; it has indeed proved insightful in some studies to examine the relative contributions to values of $C(A,B)$ from the terms involving particular localized natural orbitals (LNOs) [63, 65].

14.2.2 MAKING/BREAKING TWO-CENTER BONDS

As a first example, we show how DAFH analysis can be used to monitor the progress of the dissociation of the simple H_2 molecule. Using a standard cc-pVTZ basis, in spherical form, full configuration interaction (FCI) wavefunctions for H_2 take account of all allowed distributions of the 2 electrons in 28 orbitals. In practice, we chose to generate such FCI/cc-pVTZ descriptions, for a range of nuclear separations, by means of '2 electrons in 28 orbitals' complete active space self-consistent field (CASSCF) calculations [66, 67] using the MOLPRO package [68, 69]. The QTAIM analysis [19] for each of the total electron densities was carried out using the AIMAll program [70], which also provided all of the required domain-condensed overlap integrals. The DAFH analysis, and the computation of bond index values, was carried out using our own codes, with pictorial depictions of the DAFH functions being produced using Virtual Reality Markup Language (VRML) files that were generated with MOLDEN [71].

For each of the geometries that we considered, from 0.4 Å to 2.25 Å, DAFH analysis for the 'hole' averaged over one of the H domains produced in each case only a single DAFH function with a non-trivial occupation number. The shapes of the dominant DAFH functions, which represent the broken or dangling valence of a formally broken H–H bond, are shown in Figure 14.1 (together with the corresponding population) for representative nuclear separations R. We observe that the systematic increase of R has only a marginal effect on the population, but what does change fairly dramatically is the shape of the function. Near the equilibrium geometry of H_2, the dominant DAFH function for this H domain is clearly strongly reminiscent of a slightly asymmetric version of a $1\sigma_g$ molecular orbital so that, when taken together with its symmetry-equivalent counterpart from the other H domain, the resulting description corresponds fairly closely to an almost doubly-occupied $1\sigma_g$ orbital. For comparison, the occupation number of the corresponding natural orbital in the FCI wavefunction is 1.96 at $R = 0.75$ Å. At this nuclear separation, the overlap between the dominant DAFH functions for the two symmetry-equivalent H domains is 0.984.

As can be seen from Figure 14.1, the systematic elongation of the distance between the two atoms increases the degree of asymmetry in the dominant DAFH function for a given H domain, such that the function becomes increasingly localized in that domain. The overlap between the dominant DAFH functions for the two symmetry-equivalent H domains in H_2 is shown in Figure 14.2.

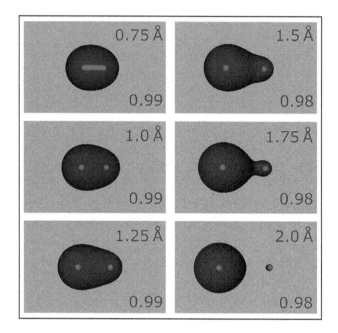

FIGURE 14.1   Geometry dependence of the dominant DAFH function (and its occupation number) for one of the H domains in H_2 (FCI/cc-pVTZ wavefunction).

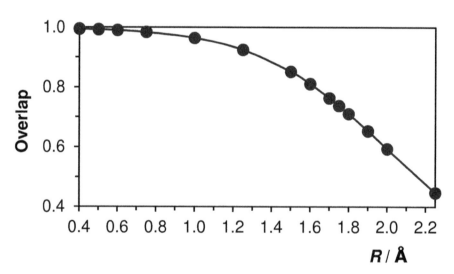

FIGURE 14.2   Geometry dependence of the overlap between the dominant DAFH functions for the two symmetry-equivalent H domains in H_2 (FCI/cc-pVTZ wavefunction).

This overlap initially decreases relatively slowly as the two atoms are moved apart, but the rate of change then becomes somewhat faster at larger values of R, as the bond 'breaks.' According to simple numerical differentiation, the steepest part of the curve shown in Figure 14.2 is located close to $R = 2.0$ Å. It is of course important to supplement this DAFH analysis of the bond breaking process in the FCI/cc-pVTZ description of H_2 with an examination of the various types of bond index that we described above.

The geometry dependence of the values of $SEDI(A,B)$ and of the Wiberg-Mayer index W_{AB} are shown in Figure 14.3(a) for the same FCI/cc-pVTZ wavefunctions. Both of these curves show the expected smooth decrease in the extent of electron sharing as R increases. Simple numerical differentiation places the steepest part of the W_{AB} curve near 1.84 Å and that of the $SEDI(A,B)$ curve near 1.64 Å. Starting at the equilibrium geometry, the values of $SEDI(A,B)$ and of W_{AB} do of course continue to increase as R is reduced, but this is not an indication of greater bonding. It is important in this context to remember that not all electron sharing leads to stabilization [63, 72]: the increased overlap of same-spin orbitals from the two H domains raises the kinetic energy (i.e., leads to Pauli repulsion).

The dependence of the Cioslowski covalent bond order on R is shown in Figure 14.3b. Unlike the smooth curves shown in Figures 14.2 and 14.3a, this quantity shows an abrupt decrease – practically a discontinuity – with the fastest change being close to the location of the steepest part of the W_{AB} curve. As such, it could be tempting to identify the region between 1.8 Å and 1.9 Å as being where the bond really 'breaks', even if the most rapid changes in $SEDI(A,B)$ occur at slightly shorter R.

As a further example, we now report the results of a similar investigation of the dissociation of the N_2 molecule. The DAFH analysis for this archetypal example of a multiple bond was based on correlated density matrices taken from full-valence '10 electrons in 8 orbitals' CASSCF descriptions of N_2 [57], with the 'hole' averaged over the domain of one of the N atoms. As was the case for H_2, the geometry dependence of the DAFH occupation numbers turns out to be fairly marginal, with the progress of the dissociation of the N≡N triple bond being much more clearly reflected in the evolution of the forms of the corresponding functions. Near the equilibrium geometry of N_2, these DAFH functions (dangling valences of formally broken s and p bonds) are again reminiscent of slightly asymmetric versions of the molecular orbitals

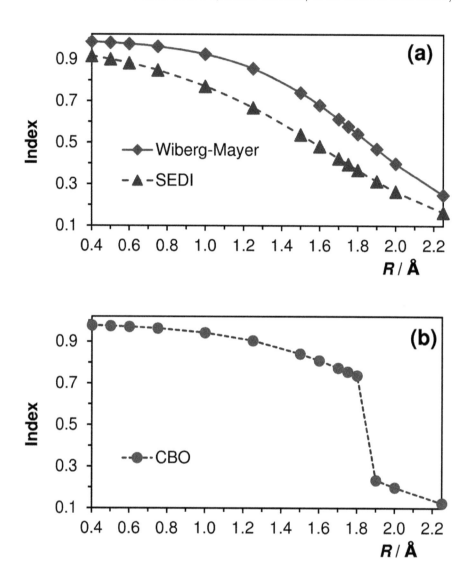

FIGURE 14.3 Geometry dependence of bond index values for H_2 (FCI/cc-pVTZ wavefunction): (a) $SEDI(A,B)$ and Wiberg-Mayer, W_{AB}; (b) the Cioslowski covalent bond order (CBO), $C(A,B)$.

$(3s_g$ and $1p_u)$ but, as R is increased, they transform into the forms that can be anticipated for the eventual dissociation of the molecule into two ground state $N(^4S)$ atoms. As can easily be seen from Figure 14.4, in which we present depictions of the dominant DAFH functions of σ and π symmetry for one of the N domains in this molecule, the two symmetries respond differently to the

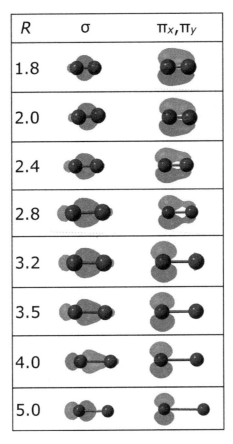

R	σ	πₓ,π_y

Rendered as table header: R | σ | π_x, π_y

FIGURE 14.4 Variation with R (in bohr) of the dominant σ and π DAFH functions for one of the N domains in N_2 (CASSCF(10,8)/cc-pVTZ).

increasing interatomic distance, with the σ bond breaking at somewhat larger R values than does the π bonding. It is interesting in this context to note from Figure 14.5 that the W_{AB} and especially the $SEDI(A,B)$ profiles for this system are rather smooth, reflecting a general reduction of the degree of electron sharing as R is increased; on the other hand, the Cioslowski covalent bond order (CBO) seems to show more subtle effects, exhibiting sharp changes in $C(A,B)$ that appear to reflect the stepwise nature of the splitting of the N≡N triple bond, as was revealed by the DAFH analysis.

Somewhat analogous DAFH and bond order methodology has been used with correlated density matrices to study the anatomy of bond making/breaking in a fair number of mostly diatomic molecules [44, 57, 58, 63], including

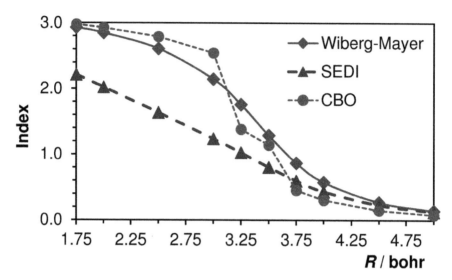

FIGURE 14.5 Geometry dependence of bond index values for N_2 (CASSCF(10,8)/cc-pVTZ): SEDI, Wiberg-Mayer, W_{AB}, and the Cioslowski covalent bond order (CBO), $C(A,B)$.

transition metal systems [73]. In some cases, additional insights were provided by examining also the evolution with geometry of the momentum-space transforms of the usual position-space DAFH functions [62].

14.3   CASE STUDIES 2: THREE-CENTER BONDING

As is well known, there are wide classes of compounds that are usually considered to feature the structural paradigm of multicenter bonding. Typical examples are represented by the broad families of electron deficient boranes and of non-classical carbocations, for which it is usual to invoke the idea of the three-center two-electron bond [74–76]. A somewhat different set of examples is provided by the structures of electron rich molecules such as certain polyhalide anions and neutral hypervalent species (including PF_5, SF_4, etc.) which have often been rationalized using the concept of three-center four-electron bonding [77–80]. Delocalized multicenter bonding also plays an important role in unsaturated (poly)cyclic hydrocarbons and has often being associated with aromaticity [81, 82].

The usefulness of the paradigm of multicenter bonding has stimulated the design of new computational tools and procedures that not only allow critical reevaluations of earlier empirical models of multicenter bonds [74–80], but that also aim to provide the possibility of the direct detection and even quantification of such bonding patterns when analyzing modern sophisticated wave functions. Of special importance in this respect is the use of so-called multicenter bond indices whose introduction was first proposed in the early studies by Sannigrahi and Kar [83–85] and by Giambiagi and coworkers [86–88]. The practicality of these indices for structural elucidation is linked to the nontrivial finding that they attain non-negligible values only for the atoms involved in three-center bonding, whereas the corresponding values practically vanish for other triads of atoms.

14.3.1 THEORETICAL BACKGROUND

A natural starting point for defining multicenter bond indices would be to identify an appropriate quantity $\Delta^{(k)}$ such that

$$\int \Delta^{(k)} \left(r_1 r_2 ... r_N \right) dr_1 dr_2 ... dr_N = N \tag{15}$$

and then to partition this integral into contributions that involve 1, 2, ... k centers. One such approach uses $\Delta^{(k)}$ that is defined in an analogous fashion to cumulants, with the first three of the $\Delta^{(k)}$ quantities being:

$$\Delta^{(1)} \left(r_1 \right) = \rho^{(1)} \left(r_1 \right)$$

$$\Delta^{(2)} \left(r_1, r_2 \right) = \Delta^{(1)} \left(r_1 \right) \Delta^{(1)} \left(r_2 \right) - 2\rho^{(2)} \left(r_1, r_2 \right)$$

$$\Delta^{(3)} \left(r_1, r_2, r_3 \right) = \frac{1}{2} \Big[\Delta^{(1)} \left(r_1 \right) \Delta^{(2)} \left(r_2, r_3 \right) + \Delta^{(1)} \left(r_2 \right) \Delta^{(2)} \left(r_1, r_3 \right) + \Delta^{(1)} \left(r_3 \right) \Delta^{(2)} \left(r_1, r_2 \right)$$

$$-\Delta^{(1)} \left(r_1 \right) \Delta^{(1)} \left(r_2 \right) \Delta^{(1)} \left(r_3 \right) \Big] + 3\rho^{(3)} \left(r_1, r_2, r_3 \right) \tag{16}$$

The direct optimization of spin-coupled (SC) wavefunctions involves density matrices $D^{(n)}$ up to fourth order [89] but for the calculation of three-center indices we require only $D^{(1)}$, $D^{(2)}$ and $D^{(3)}$. This means that it is fairly straightforward to calculate values of $\Delta^{(3)}$ using SC wavefunctions (the SC approach represents a modern development of valence bond theory [90]).

According to the idealized values that can be obtained from very simple three-center three-orbital Hückel-like models [91], our three-center bond indices should be positive for three-center two-electron bonding and they should be negative for three-center four-electron bonding [92]. SC calculations have been carried out for the familiar case of diborane, B_2H_6, treating as active just the four valence electrons of the two BHB bridges [93]. (Results based on a more sophisticated wavefunction will be presented in Section 14.3.2.) For each (B,H,B) triad, the value of the three-center bond index $\Delta^{(3)}$ that we obtained was 0.242, which we can compare with the corresponding RHF value of 0.371 and the 'idealized value' for the corresponding bonding topology [91] of 0.375. Carrying out comparable SC treatments of $B_3H_8^-$ (BHB and BBB units) and of CH_2Li_2 (LiCLi unit), we found that the bond indices computed from $\Delta^{(3)}$ for these various examples of three-center two-electron bonding are always positive, and lie in the range +0.15 to +0.25 [93]. As in the case of B_2H_6, the corresponding RHF values for $B_3H_8^-$ and CH_2Li_2 tended to be a little larger (and somewhat closer to the idealized values). An important message from this study [93] was that, whether by looking directly at the form of the SC wavefunction or by computing bond indices using $\Delta^{(3)}$, it was still entirely straightforward after the inclusion of electron correlation to detect the presence of the non-classical bonding paradigm of three-center two-electron bonding.

The situation did, however, turn out to be somewhat less clear-cut when we looked for the anticipated three-center four-electron bonding units in various electron-rich and hypervalent molecules [94]. In some cases, such as SF_4 and PF_5, we found that switching from a minimal basis set to a more flexible one was associated, even at the RHF level, with the simple three-center four-electron model evolving into a pattern of two more or less normal, albeit rather polar, two-center two-electron bonds. In other systems that we examined, much the same sort of transformation occurred when we included electron correlation via the SC approach. In the case of F_3^-, for example, the three-center bond index $\Delta^{(3)}$ calculated at the SC level using was just −0.077 [94], as compared to −0.341 at the RHF level and an idealized value of −0.375 for this geometrical arrangement [91]. Furthermore, the presence of two fairly localized F−F bonds in F_3^- after the inclusion of correlation was also clearly evident from the pictorial depictions of the corresponding SC orbitals [94]. Similarly, a fairly recent study (based on slightly different methodology) found that the anticipated pattern of multicenter bonding also reduced to a description based on two-center bonds for certain iodanes that feature electron-withdrawing ligands [95].

In spite of their obvious success for detecting three-center two-electron bonding units, there are various obvious objections to the sort of approach that we have described so far for the calculations of three-center indices. Probably the most obvious is that third order density matrices are not widely available for contemporary correlated wavefunctions, although useful results can be obtained with approximations to $D^{(3)}$ or to $\Delta^{(3)}$ [96–101]. Of course, from a more philosophical point of view, it does seem more than a little perverse that the definition of the multicenter bond index for k-center bonding formally requires density matrices up to k^{th} order, even though the Hamiltonian involves no more than two-electron operators. Unsurprisingly, far more popular (and convenient) starting points for defining multicenter bond indices use lower order density matrices. In particular, using only $D^{(1)}$, we may write:

$$\left(\tfrac{1}{2}\right)^{k-1} \text{trace}\left[\left(D^{(1)}S\right)^{k}\right] = \tilde{N} \tag{17}$$

and then partition the left-hand side of Eq. (17) into 1-, 2-, ... k-center quantities [102–104]. (Note that the case with $k = 2$ leads to W_{AB}, described in Section 14.2.1.) Given that $\tilde{N}$ is equal to N at the closed-shell RHF level, but will usually differ from N for intrinsically multiconfigurational correlated wavefunctions, we could choose to use the resulting expressions for bond indices 'as is' or we could include additional terms that may be designed so as to bring $\tilde{N}$ closer to N (cf. Eq. (12) above) and/or we could even consider the scaling of terms by, for example, $N/\tilde{N}$.

In the case of the three-center index, which we denote X_{ABC}, the basic expression (when using $k = 3$) can be written in the following form:

$$X_{ABC} = \tfrac{1}{2}\sum_{\mu \in A}\sum_{\nu \in B}\sum_{\xi \in C}\left[\left(P^{\alpha}S\right)_{\mu\nu}\left(P^{\alpha}S\right)_{\nu\xi}\left(P^{\alpha}S\right)_{\xi\mu} + \left(P^{\beta}S\right)_{\mu\nu}\left(P^{\beta}S\right)_{\nu\xi}\left(P^{\beta}S\right)_{\xi\mu}\right]$$
$$+ \tfrac{1}{2}\sum_{\mu \in A}\sum_{\nu \in C}\sum_{\xi \in B}\left[\left(P^{\alpha}S\right)_{\mu\nu}\left(P^{\alpha}S\right)_{\nu\xi}\left(P^{\alpha}S\right)_{\xi\mu} + \left(P^{\beta}S\right)_{\mu\nu}\left(P^{\beta}S\right)_{\nu\xi}\left(P^{\beta}S\right)_{\xi\mu}\right]$$

$$\tag{18}$$

Substituting the expressions for the spinless density matrix $D^{(1)}$ and for the spin-density matrix P^S (see Eq. (10)) yields the following:

$$
\begin{aligned}
X_{ABC} = \tfrac{1}{8} \sum_{\mu \in A} \sum_{\nu \in B} \sum_{\xi \in C} & \left[\begin{array}{c} \left(D^{(1)}S\right)_{\mu\nu} \left(D^{(1)}S\right)_{\nu\xi} \left(D^{(1)}S\right)_{\xi\mu} + \left(D^{(1)}S\right)_{\mu\nu} \left(P^SS\right)_{\nu\xi} \left(P^SS\right)_{\xi\mu} \\ + \left(P^SS\right)_{\mu\nu} \left(D^{(1)}S\right)_{\nu\xi} \left(P^SS\right)_{\xi\mu} + \left(P^SS\right)_{\mu\nu} \left(P^SS\right)_{\nu\xi} \left(D^{(1)}S\right)_{\xi\mu} \end{array} \right] \\
+ \tfrac{1}{8} \sum_{\mu \in A} \sum_{\nu \in C} \sum_{\xi \in B} & \left[\begin{array}{c} \left(D^{(1)}S\right)_{\mu\nu} \left(D^{(1)}S\right)_{\nu\xi} \left(D^{(1)}S\right)_{\xi\mu} + \left(D^{(1)}S\right)_{\mu\nu} \left(P^SS\right)_{\nu\xi} \left(P^SS\right)_{\xi\mu} \\ + \left(P^SS\right)_{\mu\nu} \left(D^{(1)}S\right)_{\nu\xi} \left(P^SS\right)_{\xi\mu} + \left(P^SS\right)_{\mu\nu} \left(P^SS\right)_{\nu\xi} \left(D^{(1)}S\right)_{\xi\mu} \end{array} \right]
\end{aligned} \quad (19)
$$

In the special case of a correlated singlet system, we may simply replace P^S by R (see Eq. (13)).

As was the case for W_{AB}, described in Section 14.2.1, it is especially convenient to work in the basis of the (real) orthonormal canonical natural orbitals $\{f_I\}$ (with corresponding occupation numbers $\{\omega_I\}$). Switching from Mulliken-like partitioning in the atomic basis (μ,ν,ξ) to the use of QTAIM domains $(\Omega_A, \Omega_B, \Omega_C)$ then leads, after some straightforward algebra, to the following expression:

$$
X_{ABC} = \tfrac{1}{4} \sum_I \sum_J \sum_K \left(\begin{array}{c} \omega_I \omega_J \omega_K + \omega_I R_{JJ} R_{KK} \\ + R_{II} \omega_J R_{KK} + R_{II} R_{JJ} \omega_K \end{array} \right) \langle \phi_I | \phi_K \rangle_{\Omega_A} \langle \phi_I | \phi_J \rangle_{\Omega_B} \langle \phi_J | \phi_K \rangle_{\Omega_C} \quad (20)
$$

Given that the various permutations of the three indices necessarily yield identical values of X_{ABC}, it is convenient when all three indices are different to quote instead a (single) value of $X(A,B,C) = 3! \, X_{ABC}$.

We note in this case that:

$$
\sum_A \sum_B \sum_C X_{ABC} = \tfrac{1}{4} \sum_I \left\{ \left(\omega_I\right)^3 + 3\left(\omega_I\right)^2 \left(2-\omega_I\right) \right\} = \Omega \sum_I \left\{ \left(\omega_I\right)^2 \left(3-\omega_I\right) \right\} \quad (21)
$$

This does of course mean that there is no direct relationship of the sum of all possible X_{ABC} values to the total number of electrons, N. As a possible remedy to this situation, we have also pursued an alternative approach for correlated singlet systems, taking our inspiration from the line of reasoning that led to Eq. (12) and from the general forms of Eq. (13) (see Section 14.2.1). Defining the following equation,

$$tS = 4D^{(1)}S - \left(D^{(1)}S\right)^3$$

$$t^\lambda = S^{1/2}\left(tS\right)S^{-1/2} \tag{22}$$

$$Q = S^{-1/2}\left(t^\lambda\right)^{1/3}S^{-1/2}$$

we may construct the following quantity:

$$Y_{ABC} = \tfrac{1}{8}\sum_{\mu\in Av}\sum_{\in B\xi}\sum_{\in C}\left[\left(D^{(1)}S\right)_{\mu v}\left(D^{(1)}S\right)_{v\xi}\left(D^{(1)}S\right)_{\xi\mu} + (QS)_{\mu v}(QS)_{v\xi}(QS)_{\xi\mu}\right]$$

$$+ \tfrac{1}{8}\sum_{\mu\in Av}\sum_{\in C\xi}\sum_{\in B}\left[\left(D^{(1)}S\right)_{\mu v}\left(D^{(1)}S\right)_{v\xi}\left(D^{(1)}S\right)_{\xi\mu} + (QS)_{\mu v}(QS)_{v\xi}(QS)_{\xi\mu}\right] \tag{23}$$

Notice that Q is simply the diagonal matrix with elements $Q_{II} = [\omega_I(2 + \omega_I)(2 - \omega_I)]^{1/3}$ when it is expressed in the basis of orthonormal canonical natural orbitals. Switching again from Mulliken-like partitioning to the use of QTAIM domains, we obtain the following expression:

$$Y_{ABC} = \tfrac{1}{4}\sum_I\sum_J\sum_K\left(\omega_I\omega_J\omega_K + Q_{II}Q_{JJ}Q_{KK}\right)\langle\phi_I|\phi_K\rangle_{\Omega_A}\langle\phi_I|\phi_J\rangle_{\Omega_B}\langle\phi_J|\phi_K\rangle_{\Omega_C} \tag{24}$$

We note in this case that:

$$\sum_A\sum_B\sum_C Y_{ABC} = \tfrac{1}{4}\sum_I\left\{\left(\omega_I\right)^3 + \omega_I\left(2 + \omega_I\right)\left(2 - \omega_I\right)\right\} = \sum_I\omega_I \tag{25}$$

Although this alternative formulation could be considered slightly *ad hoc*, it does have the advantage that the sum of all possible Y_{ABC} values is equal to N, essentially by construction. Given that the various permutations of the three indices necessarily yield identical values of Y_{ABC}, it is convenient when all three indices are different to quote instead a (single) value of $Y(A,B,C) = 3!\ Y_{ABC}$.

14.3.2   DIBORANE

We now look in somewhat more detail at the bonding in B_2H_6 using a wavefunction in which all of the valence electrons were treated as active, unlike the SC calculations mentioned above. The new electronic structure calculations were performed in D_{2h} symmetry, with the coordinates of the symmetry-unique atoms taking the values listed in Table 14.1. We carried out full-valence CASSCF ('12 electrons in 14 orbitals') calculations using standard cc-pVTZ basis sets in spherical harmonic form. The QTAIM domain-condensed overlap matrices (see Eq. (3)) were generated using the full CASSCF(12,14) wavefunction, but much of our subsequent analysis was then done for the valence space, ignoring the essentially $B(1s^2)$ core orbitals which make negligible contributions to the bonding. It is useful to start our discussion of the bonding in this molecule with an evaluation of the results of the DAFH analysis, based on the correlated density matrices from the CASSCF(12,14) wavefunction, before we look also at the supporting and complementary information that is provided by the analysis of the various bond index values.

We show in Figure 14.6 the dominant DAFH functions for the domains consisting of one of the B atoms (top row) and of one of the adjacent terminal

TABLE 14.1 Coordinates Used for the Symmetry-Unique Atoms in B_2H_6 (D_{2h}), with the Inversion Center Taken as the Origin

Atom	$x/\text{Å}$	$y/\text{Å}$	$z/\text{Å}$
B	0.888375	0	0
H (bridge)	0	0	0.973751
H (terminal)	1.463157	1.035737	0

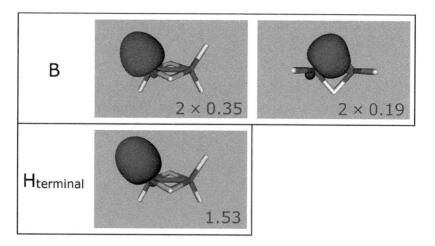

FIGURE 14.6 Symmetry-unique DAFH functions for one of the B atoms (top row) and for one of the adjacent terminal H atoms (bottom row), together with the corresponding occupation numbers.

H atoms (bottom row), together with their occupation numbers. We observe for the boron domain that there are two symmetry-equivalent functions which correspond to dangling valences of formally broken terminal B–H bonds; the associated occupations of 0.35 can be regarded as the contributions from this B atom to the unevenly shared electron pairs of the two terminal B–H bonds. Such an interpretation is straightforwardly corroborated by inspecting the corresponding DAFH function (dangling valence) of the 'hole' averaged over one of the adjacent $H_{terminal}$ atoms; the occupation of 1.53 represents the complementary contribution from this hydrogen atom to the electron pair of this terminal B–H bond. All in all, this picture is entirely consistent with the presence of fairly ordinary but rather polar two-center terminal B–H bonds, as we would expect. We also observe for the B domain that there are two symmetry-equivalent functions with occupation 0.19 which can be characterized as representing the contributions from this B atom to the B–H_{bridge}–B bonding. Although these functions extend over all three centers, they are not symmetric with respect to the mirror plane that is perpendicular to the BB axis. Instead, they show a clear bias in favor of the boron domain to which this function 'belongs'.

We can of course also construct the 'condensed domain' for one of the terminal BH_2 groups, taking the union of the corresponding QTAIM domains or, entirely equivalently, simply adding together the relevant G matrices. The dominant DAFH functions for such a condensed domain are shown in the top row of Figure 14.7. These results now show even more clearly the fairly ordinary character of the two-center B–$H_{terminal}$ bonds. This leaves the symmetry-equivalent functions with occupation 0.25 which can, of course, be characterized as representing the contribution from this terminal BH_2 group to the BH_2–H_{bridge}–BH_2 bonding. The shapes of these functions cannot be distinguished by eye from those for the B domain on its own and, indeed, the overlap integral between the corresponding functions is 0.995.

Turning now to the DAFH analysis for one of the bridging H atoms we observe (bottom row of Figure 14.7) a function with occupation 1.42 that resembles to some extent one of its partner functions from the terminal BH_2 group except that it is now, unsurprisingly, symmetric with respect to reflection in the mirror plane that is perpendicular to the BB axis. Taken together, these various DAFH results are clearly in harmony with traditional notions of three-center two-electron bonding as an electron pair that is shared between all three atoms in each of the B–H–B bridges. Furthermore, much the same sort of description emerges when we look instead at the localized

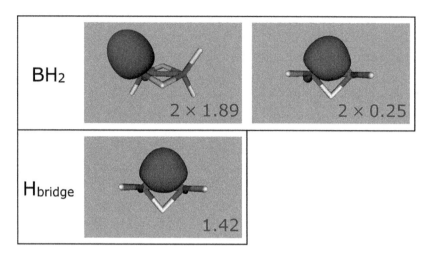

FIGURE 14.7 Symmetry-unique DAFH functions for one of the BH$_2$ condensed domains (top row) and for one of the bridging H atoms (bottom row), together with the corresponding occupation numbers.

natural orbitals (LNOs) that are generated by isopycnic transformation of the canonical natural orbitals for the active space of the same CASSCF(12,14) wavefunction. The forms and occupation numbers of the resulting sym-metry-unique LNOs, which are depicted in Figure 14.8, give a very clear impression of essentially two-center two-electron terminal B–H bonds and three-center two-electron B–H–B bridges.

It is of course useful to complement this DAFH and LNO analysis with an examination of the numerical values of various three-center bond indices. For this purpose, we look here at the values of $X(A,B,C)$ and $Y(A,B,C)$, as defined above in Eqs. (20) and (24), respectively. Nontrivial symmetry-unique

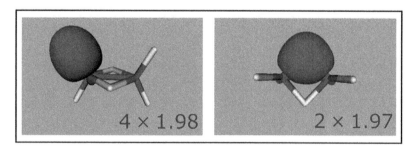

FIGURE 14.8 Symmetry-unique LNOs for the active space of the CASSCF(12,14) wavefunction, together with the corresponding occupation numbers.

'condensed domain' values are reported in Table 14.2, in which we have denoted by $W(A,B,C)$ (=3! W_{ABC}) the corresponding quantities in which any of the terms involving either $\boldsymbol{R}$ or $\boldsymbol{Q}$ have been excluded. We have used the symbol $\tilde{N}$ to signify in Table 14.2 the sum of all possible W_{ABC}, X_{ABC} or Y_{ABC} values. Note that the total number of active electrons is of course $N =$ 12. Somewhat fortuitously, the sum of all X_{ABC} values turns out to be rather close to N in the present case, and so it is not too surprising that the values of $X(A,B,C)$ and $Y(A,B,C)$ match to the number of decimal places that we have quoted in Table 14.2. Indeed, for this particular system, the differences of $X(A,B,C)$ or $Y(A,B,C)$ from the corresponding $W(A,B,C)$ values are in the third decimal place, and could be considered to be of little consequence. Clearly, whichever of these indices we use, the largest values are for the two (BH_2,H_{bridge},BH_2) triads, as we would expect, but we also observe from Table 14.2 that there are nontrivial values for the two $(BH_2,H_{bridge},H_{bridge})$ triads.

In addition to the above convincing picture of the three-center two-electron bonding in the B–H–B bridges, it is also tempting to speculate on the possibilities of generalizing the definition of the intrinsically two-center Cioslowski covalent bond order [30] so as to quantify instead the relative contributions to three-center bonding from the terms involving particular LNOs. For RHF spin orbitals $|i\rangle$ we could consider the following partitioning (*cf.* Eq. (4)) of the total number of electrons, N:

$$N = \sum_A \sum_B \sum_C \sum_i \langle i|i \rangle_{\Omega_A} \langle i|i \rangle_{\Omega_B} \langle i|i \rangle_{\Omega_C} \qquad (26)$$

whereas for more general wavefunctions we might partition instead:

$$\sum_A \sum_B \sum_C \sum_i (v_i)^3 \langle i|i \rangle_{\Omega_A} \langle i|i \rangle_{\Omega_B} \langle i|i \rangle_{\Omega_C} \equiv \sum_A \sum_B \sum_C K_{ABC} \qquad (27)$$

TABLE 14.2   Symmetry-Unique 'Condensed Domain' Values of $W(A,B,C)$, $X(A,B,C)$ and $Y(A,B,C)$ for the Active Space of the CASSCF(12,14) Description of B_2H_6

A	B	C	$W(A,B,C)$	$X(A,B,C)$	$Y(A,B,C)$
BH_2	H_{bridge}	BH_2	0.114	0.115	0.115
BH_2	H_{bridge}	H_{bridge}	0.081	0.085	0.085
		$\tilde{N}$	11.576	11.998	12.000

The $\tilde{N}$ quantities are the sums of all possible W_{ABC}, X_{ABC} or Y_{ABC} values.

By analogy to the usual isopycnic transformation [56], described earlier, we can envisage performing some sort of maximization of the sum of one-center K_{AAA} terms (and possibly also of the sum of the corresponding two-center terms) at the expense of the sum of the three-center K_{ABC} terms. The resulting orbitals and occupation numbers could then be used for a given set of three different centers to evaluate a quantity that resembles $K(A,B,C) = 3! K_{ABC}$.

In the present work, we have considered a much simpler strategy: K_{ABC} was evaluated using the orbitals ψ_p and occupation numbers λ_p from the usual isopycnic transformation [56]. In light of our use of such an approximation, as well as the observation that there is nothing special about the sum of all possible K_{ABC} values, we consider here only the *relative* contributions to $K(A,B,C)$ from the terms involving the different LNOs. In this way, we find that the largest contribution to the bonding in a given (BH_2, H_{bridge}, BH_2) triad comes from the corresponding BH_2–H_{bridge}–BH_2 LNO (92%), with 4% from the other BH_2–H_{bridge}–BH_2 LNO and a further 1% (each) from the four terminal B–H LNOs. Looking instead at the (lower) value that we also find for the $(BH_2, H_{bridge}, H_{bridge})$ triad, we observe contributions of 45% (each) from the two BH_2–H_{bridge}–BH_2 LNOs and a further 5% (each) from both of the relevant terminal B–H LNOs.

14.3.3 BERYLLIUM COMPLEXES

As a somewhat less familiar example we now consider recently isolated neutral complexes [105] that feature a formally zero-valent beryllium atom in a linear C–Be–C unit. The short Be–C distances were said to be indicative of strong multiple bonding and the surprising stability of these complexes was ascribed to strong three-center two-electron π bonding across the C–Be–C moiety [105]. The experimental investigation was accompanied by various calculations for one of the isolated complexes, which they designated as [Be(MeL)$_2$], where MeL is 1-(2,6-diisopropylphenyl)-3,3,5,5-tetramethylpyrrolidine-2-ylidene, and also for a trimmed down model version of it, which they designated as [Be(CAAC^Model)$_2$], in which CAAC signifies a cyclic (alkyl)(amino)carbene. Their various computational results for [Be(MeL)$_2$] and [Be(CAAC^Model)$_2$] were concordant with one another [105].

We show the structure of [Be(CAAC^Model)$_2$] as **3a** in Figure 14.9. We decided to trim down even further the published BP86/def2-TZVPP

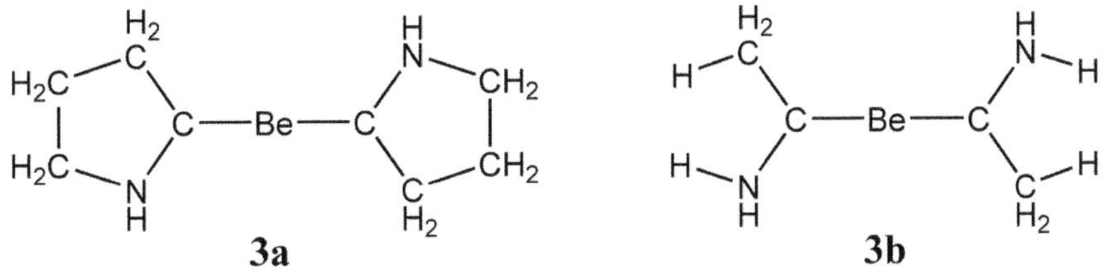

FIGURE 14.9 Structures of [Be(CAAC^{Model})$_2$] (**3a**) [105] and of our more compact model complex (**3b**).

geometry [105] of **3a**, so that we could use our preferred correlated computational approaches to focus directly on the nature of the bonding around the Be atom. In essence, we replaced the ring systems by terminating NH$_2$ and CH$_3$ groups. We did not perform any further geometry optimization because we wanted to retain, as closely as possible, the same coordination around the various heavy atoms kept from **3a**. We replaced the CH$_2$ groups attached to nitrogen by hydrogen atoms, which we moved along the HN−CH$_2$ axes in **3a** until the new N−H bond lengths were identical to the existing ones in **3a**. We used the same 'trick' for the next CH$_2$ group around each ring, this time setting the new C−H bond lengths to be equal to the average of the existing ones at those C atoms in **3a**. The structure of the resulting more compact model complex is shown as **3b** in Figure 14.9. Except for H atoms of the CH$_3$ groups, we notice that **3b** is close to being planar and this made us wonder whether there could be significant contributions from N(2p$_\pi$) orbitals to the claimed π bonding across the central C−Be−C unit.

In order to check the relevance and reliability of our simplified model (**3b**), we carried out preliminary DAFH analysis for the full [Be(^{Me}L)$_2$] complex at the B3LYP/SDD level using the simple Mulliken-like approximation for generating the required domain-condensed overlaps: the results were found to be in harmony with the corresponding ones for our compact model complex **3b**. Furthermore, not only did we find that changing the description of **3b** to the BP86/6−31G** level was of little consequence for the DAFH analysis, but it also turned out that the subsequent use of QTAIM domains in place of the simple Mulliken-like approximation had little effect on the forms of the DAFH functions. Along the way, we also looked at Mulliken-like DAFH analysis of the BP86/def2-tzvpp description of [Be(CAAC^{Model})$_2$] (**3a**) as well as at the Mulliken-like and QTAIM-based DAFH analysis of RHF descriptions of **3b** (with various basis sets). Taken together, all of these various results justify our use of the compact complex **3b** as a sufficient model of the full [Be(^{Me}L)$_2$] system, at least from the point of view of carrying out DAFH analysis. Furthermore, none of these preliminary DAFH results for [Be(^{Me}L)$_2$], [Be(CAAC^{Model})$_2$] (**3a**) or our compact model (**3b**) indicate that there is anything particularly unusual about the various essentially two-center two-electrons bonds that lie outside the central N–C–Be–C–N moiety.

A full-valence CASSCF calculation for the compact model complex **3b** would involve all symmetry-allowed distributions of 38 electrons in 38 orbitals. Such a massive active space is, of course, well beyond our capabilities. Given that our various preliminary results for the various RHF and DFT descriptions are strongly suggestive that it could be valid, and worthwhile, to focus our attention on the bonding amongst the central N–C–Be–C–N atoms of **3b**, we decided to pursue instead a much more tractable CASSCF active space. Fortunately, after switching off the molecular point group symmetry, it proved straightforward at the RHF level to identify Pipek-Mezey localized molecular orbitals [106] that correspond mostly to the various C–H, N–H and C–CH$_3$ bonds. These localized orbitals could then be kept frozen, alongside the various 1s^2-like core orbitals, in a '14 electrons in 14 orbitals' CASSCF construction. In a subsequent step, all of these inactive orbitals were then allowed to relax alongside simultaneous reoptimization of the CASSCF(14,14) active space. We did of course check, by visual inspection of pictorial representations, that there had been no significant swapping of the character of the various active and inactive orbitals. Ultimately, we found that we were able to reproduce exactly the same wavefunction when carrying out appropriate CASSCF(14,14) calculations in C_i symmetry. We used 6–31G** basis sets (albeit now in spherical form) for all of these calculations.

We report in Table 14.3 (in this table, the atoms simply labeled C are those from the central C–Be–C unit) selected QTAIM populations from our RHF/6–31G**, BP86/6–31G** and CASSCF(14,14) descriptions of **3b**, as well as their partitioning into core/valence or active/inactive contributions. Direct comparison of the CASSCF(14,14) N_{active} values for C and for C(CH$_3$), as well as of those for N and for NH$_2$, suggest that our desired partitioning into inactive and active spaces has generally been rather successful. This does of course provide additional confidence that our subsequent DAFH analysis will indeed be focused on the bonding in the central N–C–Be–C–N unit of **3b**, as required. We also notice, reading across the rows, that the changes in N_{total} from RHF to BP86 are not a good predictor of the differences between the corresponding RHF and CASSCF(14,14) values.

Given that the various C–H, N–H and C–CH$_3$ bonds are mostly described by inactive orbitals, and bearing in mind the various N_{active} values that we reported in Table 14.3, it makes most sense in the DAFH analysis of our CASSCF(14,14) description of **3b** to combine the active-space G matrices for certain QTAIM domains. Accordingly, we show in Figure 14.10 the symmetry-unique dominant DAFH functions for the Be domain (top row), for a C(CH$_3$) 'condensed domain' (rows 2 and 3) and for one of the (NH$_2$) condensed domains (bottom two rows). These pictures were produced using the open-source molecular visualization program Molekel [107].

Just as was the case for the other systems that we described above, the DAFH analysis provides direct information about the broken or dangling valences that are created by the formal splitting of the relevant bonds and, of course, about any electron 'pairs' retained with the chosen domains. The first of the DAFH functions for one of the C(CH$_3$) condensed domains (second row of Figure 14.10), with an occupation of 1.77, clearly represents a broken or dangling valence of a formally broken C–Be σ bond. As can be seen from the top row of Figure 14.10, much the same functional forms (but with complementary occupations of 0.14) occur for each of the CBe moieties when performing instead the analysis for the adjacent Be domain. So far, at least, these results are suggestive only of rather polar two-center two-electron C–Be σ bonding.

Turning now to the second of the DAFH functions for one of the C(CH$_3$) condensed domains, we find that it is clearly a π-like function with occupancy 0.95. For the isocontour value that we have used for Figure 14.10, the function appears to be mostly associated with just one of the CBe moieties. However,

TABLE 14.3 Selected QTAIM Populations N_x from RHF, BP86 and CASSCF(14,14) Descriptions of **3b**, As Well As Their Partitioning Into Core/Valence or Active/Inactive Contributions*

Atom(s)	RHF			BP86			CASSCF(14,14)		
	N_{total}	N_{core}	$N_{valence}$	N_{total}	N_{core}	$N_{valence}$	N_{total}	$N_{inactive}$	N_{active}
Be	2.48	1.98	0.50	2.49	1.98	0.50	2.40	1.99	0.41
C	6.24	2.01	4.23	6.37	2.01	4.36	6.32	3.08	3.24
C(CH$_3$)	15.22	4.01	11.21	15.37	4.01	11.36	15.33	11.95	3.38
N	8.40	2.00	6.40	8.17	2.00	6.17	8.31	5.01	3.30
(NH$_2$)	9.54	2.00	7.54	9.38	2.00	7.38	9.46	6.06	3.41

* Atoms simply labeled C are those from the central C–Be–C unit.

it turns out that the overlap between the corresponding DAFH functions from the two symmetry-equivalent $C(CH_3)$ domains is 0.570, suggesting at least some degree of three-center character. We notice that relatively similar functions (with low occupations of 0.07 and 0.09, respectively) arise in the analysis of the Be and NH_2 domains, albeit also with a small out-of-phase $N(2p_\pi)$ component.

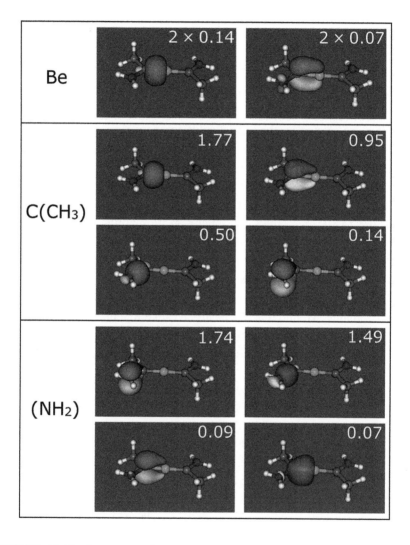

FIGURE 14.10 Symmetry-unique DAFH functions for the Be domain (top row), for a $C(CH_3)$ condensed domain (rows 2 and 3) and for one of the (NH_2) condensed domains (bottom two rows) in **3b**, together with the corresponding occupation numbers.

The third of the DAFH functions for one of the $C(CH_3)$ condensed domains has an occupation of 0.50 and, as can be seen from Figure 14.10, a relatively similar function, with occupancy 1.49, appears in the DAFH analysis of the adjacent NH_2 condensed domain. These two functions can straightforwardly be associated with the dangling valences of a formally broken C–N bond, with the populations representing the contributions from the C and N atoms, respectively, to the unevenly shared electron pair of a more or less ordinary two-center two-electron C–N σ bond. In a similar fashion, the fourth of the DAFH functions (occupancy 0.14) for one of the $C(CH_3)$ condensed domains, when taken together with the first of the DAFH functions (occupancy 1.74) for the adjacent NH_2 condensed domain, describes the slightly less than doubly-occupied $N(2p_\pi)$ lone pair that is deformed in the direction of the adjacent C atom. It is worthwhile to recall at this stage that the NH_2 group is close to being coplanar with the central C–Be–C unit.

Of the various DAFH functions that are shown in Figure 14.10, the only one that we have not yet described is the fourth function for one of the NH_2 condensed domains. It is clearly a σ function with occupancy 0.07 and resembles to some extent a more diffuse version of the C–Be σ functions described above, with a small out-of-phase component in the N–C σ bond. All of the other DAFH functions for the Be, $C(CH_3)$ and NH_2 domains (not shown in Figure 14.10) have occupations less than 0.045.

The description that emerges from this DAFH analysis includes fairly ordinary albeit polar N–C σ bonds but also some interaction of the slightly less than doubly-occupied $N(2p_\pi)$ component with the π system in the C–Be–C unit. Whereas we observe essentially two-center two-electron C–Be σ bonding, there is somewhat clearer evidence for some degree of three-center π bonding across the C–Be–C moiety. In order to gain further insight, we turn now to the values of the $W(A,B,C)$, $X(A,B,C)$, and $Y(A,B,C)$ quantities, which we defined earlier.

Symmetry-unique values of $W(A,B,C)$, $X(A,B,C)$ and $Y(A,B,C)$ for our compact model complex 3b are reported in Table 14.4, where we have focused on the valence space in the case of RHF and BP86 or on the active space in the case of the CASSCF(14,14) description. For the latter, the sum of all possible X_{ABC} values turns out to be fortuitously close to the number of active electrons and the corresponding values of $X(A,B,C)$ and $Y(A,B,C)$ are fairly similar, with differences appearing in the third decimal place. We (mostly) observe a small diminution in the values of $W(A,B,C)$ from RHF to BP86, with a somewhat larger decrease when progressing to CASSCF(14,14). For this system, the CASSCF values of $W(A,B,C)$ typically differ from those of $X(A,B,C)$ or $Y(A,B,C)$ in the second decimal place which, given the magnitude of the various quantities, could be considered significant.

Irrespective of the observed variations of the individual numbers with the level of theory, the most important conclusions emerging from Table 14.4 arise from the significant positive values for all of the indices: not only do such values indicate a non-trivial degree of three-center bonding in the CBeC unit but they are also consistent with the anticipated three-center two-electron nature of such bonding [105]. As can also be seen from Table 14.4, the corresponding values are enhanced when we look instead at the $((CH_3)C,Be,C(CH_3))$ and (LHS,Be,RHS) triads, where LHS and RHS denote the $C(CH_3)(NH_2)$ condensed domains on opposite sides of Be. The various values are, though, also suggestive of some degree of three-center character within each $C(CH_3)(NH_2)$ unit.

The LNOs that are generated by applying the isopycnic transformation [56] to the canonical natural orbitals are shown in Figure 14.11, together with their occupation numbers. As can easily be seen, the forms of these various functions resemble some of those that emerged in the DAFH analysis. With occupations approaching 2, there are LNOs corresponding to C–Be and N–C σ bonding, as well as to a predominantly $N(2p_\pi)$ function that is clearly deformed in the direction of the adjacent C atom. This leaves the two π-like C–Be LNOs with occupations 1.06 that have an overlap of 0.647 with one another. All of the other LNOs, not shown in Figure 14.11, have occupations less than 0.025.

TABLE 14.4 Symmetry-Unique Values of $W(A,B,C)$, $X(A,B,C)$ and $Y(A,B,C)$ for the Valence Space (RHF or BP86) and Active Space (CASSCF) of **3b**. LHS and RHS Denote the $C(CH_3)(NH_2)$ Condensed Domains on Opposite Sides of Be and the $\tilde{N}$ Quantities Are the Sums of All Possible W_{ABC}, X_{ABC} or Y_{ABC} Values

A	B	C	RHF	BP86	CASSCF(14,14)		
			$W(A,B,C)$	$W(A,B,C)$	$W(A,B,C)$	$X(A,B,C)$	$Y(A,B,C)$
C	Be	C	0.199	0.153	0.088	0.107	0.104
(CH_3)C	C	(NH_2)	0.102	0.105	0.059	0.063	0.067
(CH_3)C	Be	$C(CH_3)$	0.223	0.177	0.097	0.118	0.114
LHS	Be	RHS	0.287	0.245	0.124	0.148	0.142
		$\tilde{N}$	38.000	38.000	12.935	14.010	14.000

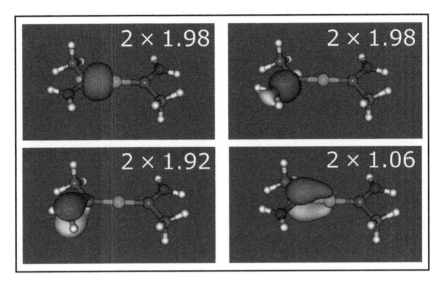

FIGURE 14.11   Symmetry-unique LNOs for **3b**, together with the corresponding occupation numbers.

In much the same way as for B_2H_6, described above, we can aim for some further insight into the nature of the three-center bonding in the central C–Be–C unit by examining the relative contributions to $K(A,B,C)$ from the terms that involve particular LNOs. In the case of the (C,Be,C) triad, for example, terms involving the two LNOs with occupation 1.06 contribute 32% each and a further 18% (each) involves the two σ LNOs with occupation 1.98 that are mostly associated with a CBe moiety. This suggests that 64% of the total three-center bonding can be ascribed to π-like character across the central C–Be–C unit whereas 36% is due instead to σ-like character, with both sets of contributions having the same overall sign. Practically the same percentages arose when we analyzed instead the (CH_3C, Be, CCH_3) and (LHS, Be, RHS) triads, where LHS and RHS denote the $C(CH_3)(NH_2)$ condensed domains on opposite sides of Be.

When we ran DAFH analysis at the RHF or BP86 level, the outcome for the Be domain showed an important difference from the corresponding CASSCF results. Instead of the two equivalent π-like functions with occupation 0.07 (see the top row of Figure 14.10), the corresponding DAFH analysis at the RHF or BP86 levels each produced a single symmetric π-like function, with occupation 0.245 or 0.200, respectively. These functions have predominantly symmetric three-center C–Be–C π-like character, but with small out-of-phase contributions on the two N atoms. Somewhat analogous differences emerged also for the LNOs. Instead of the two equivalent π-like functions with occupation 1.06 (see Figure 14.11), we found a single symmetric π-like function when using instead either RHF or BP86. In each case, this doubly-occupied LNO turned out to be very similar indeed (overlap above 0.96) to the corresponding symmetric DAFH function for the Be domain. Furthermore, it turned out that it was precisely these symmetric π-like LNOs that overwhelmingly dominated (with contributions of 97% and 94%, respectively) the RHF and BP86 values of $K(A,B,C)$ for the (C,Be,C), (CH_3C,Be,CCH_3) or (LHS,Be,RHS) triads.

The authors of the original joint experimental and theoretical study of the $[Be(^{Me}L)_2]$ complex and $[Be(CAAC^{Model})_2]$ (**3a**) model system proposed a relatively simple description of the bonding across the C−Be−C unit [105]. They attributed the existence of a strong three-center two-electron π bond across this central unit to the notion of a significant degree of electron donation from a formally doubly-occupied $Be(2p_\pi)$ orbital into the empty π orbitals on each of the two carbene moieties. Such a simple picture would of course correspond, near the limit of complete electron transfer, to three-center π bonding in which the carbene fragments would now contribute approximately one electron each, with only a relatively small contribution remaining on the Be atom. To a large extent, this is what we have observed in the DAFH, LNO and bond order analysis of our compact model complex **3b**, although there is clearly also some degree of interaction with the occupied $N(2p_\pi)$ orbitals of the (almost coplanar) NH_2 groups

The corresponding situation in the σ system is a little more complicated. The description that was invoked in Ref. [105] for the $[Be(^{Me}L)_2]$ complex and for **3a** involves the donation of electrons from the nonbonding σ pairs of the carbene fragments into suitable empty σ-orbitals on Be. Such a situation could, in principle, be simplified to a three-center three-orbital Hückel-like model [91] which would predict a negative value for $W(A,B,C)$ [92, 94]. Nonetheless, further analysis (not presented here) of actual calculations for **3b** are strongly suggestive instead of a positive three-center four-electron σ contribution that reinforces the corresponding one from the three-center two-electron bonding in the C−Be−C π system. Such a peculiar bonding situation is somewhat unusual but, as is described in detail in a recent paper [108], a positive value for three-center four-electron bonding in the σ system can in fact be rationalized by invoking an appropriate four-orbital Hückel-like model that shares various features with outcomes of our DAFH analysis.

14.4 CONCLUSIONS

This chapter aims to showcase the new insights into molecular electronic structure, including links to intuitive chemical concepts, which can be extracted from correlated electron density matrices by using DAFH analysis alongside an examination of various bond index values. The material is presented as two families of case studies, including appropriate outlines of the theoretical backgrounds to the various methodologies that are used.

The first family of case studies deals with the systematic scrutiny of the electron reorganization that accompanies the splitting and/or formation of two-center chemical bonds. Combinations of the appealing visual insights provided by DAFH analysis with the numerical backup provided by the geometry dependence of various bond index values lend clear support to the intuitive notion of gauging the progress of bond dissociation by monitoring the evolution of the extent of electron sharing.

The second family of case studies focuses on the application of analogous methodology to the non-classical paradigm of three-center bonding, which clearly transcends the familiar Lewis model of well-localized two-center two-electron bonds. The effectiveness of an appropriate combination of DAFH and bond index analysis is demonstrated first using the traditional example of the three-center two-electron BHB bridges in diborane. This is followed by an investigation into the nature of the more complex multicenter bonding across the central CBeC unit in a recently synthesized [105] formally zero-valent beryllium complex.

KEYWORDS

- analysis of correlated density matrices
- bond orders and bond indices
- chemical bonding
- DAFH analysis
- measures of electron sharing
- multicenter bonding

REFERENCES

1. Lewis, G. N., (1916). The atom and the molecule. *J. Am. Chem. Soc., 38*, 762–785.
2. Pauling, L., (1931). The nature of chemical bond. Application of results obtained from the quantum mechanics and from a theory of paramagnetic susceptibility to the structure of molecules. *J. Am. Chem. Soc., 53*, 1367–1400.
3. Pauling, L., (1939). *The Nature of the Chemical Bond*; Cornell University Press: Ithaca.
4. Coulson, C. A., (1961). *Valence,* 2nd ed., Oxford University Press: Oxford.
5. Lennard-Jones, J. E., (1952). The spatial correlation of electrons in molecules. *J. Chem. Phys., 20*, 1024–1029.
6. Berlin, T., (1951). Binding regions in diatomic molecules. *J. Chem. Phys., 19*, 208–213.
7. Ruedenberg, K., (1962). The physical nature of the chemical bond. *Rev. Mod. Phys., 34*, 326–376.
8. Edmiston, C., & Ruedenberg, K., (1963). Localized atomic and molecular orbitals. *Rev. Mod. Phys., 35*, 457–465.
9. Foster, J. M., & Boys, S. F., (1960). Canonical configurational interaction procedure. *Rev. Mod. Phys., 32*, 300–302.
10. Magnasco, V., & Perico, A., (1967). Uniform localization of atomic and molecular orbitals. I. *J. Chem. Phys., 47*, 971–981.
11. Daudel, R., Odiot, S., & Brion, H., (1954). Théorie de la localisabilité des corpuscules. *J. Chim. Phys., 51*, 74–77.
12. Bader, R. F. W., & Stephens, M. E., (1975). Spatial localization of the electronic pair and number distributions in molecules. *J. Am. Chem. Soc., 97*, 7391–7399.
13. Daudel, R., Bader, R. F. W., Stephens, M. E., & Borrett, D. S., (1974). The electron pair in chemistry. *Can. J. Chem., 52*, 1310–1320.
14. Salem, L., (1978). The destiny of electron pairs in chemical reactions. *Nouv. J. Chim., 2*, 559–562.
15. Julg, A., & Julg, P., (1978). Vers une nouvelle interprétation de la liaison chimique? *Int. J. Quantum Chem., 13*, 483–497.
16. Hiberty, P. C., (1981). Analysis of molecular orbital wave function in terms of valence bond functions for molecular fragments I. Theory. *Int. J. Quantum Chem., 19*, 259–269.
17. Luken, W. L., & Beratan, D. N., (1981). Localized orbitals and Fermi hole. *Theor. Chim. Acta, 61*, 265–281.
18. *90 years of chemical bonding.* Frenking, B., Shaik, S., Eds., special issue of *J. Comp. Chem.* (2007). *28.*
19. Bader, R. F. W., (1990). *Atoms in Molecules – A Quantum Theory*; Oxford University Press: Oxford.
20. Bader, R. F. W., Johnson, S., Tang, T. H., & Popelier, P. L. A., (1996). The electron pair. *J. Phys. Chem., 100*, 15398–15415.
21. Bader, R. F. W., (2009). Bond paths are not chemical bonds. *J. Phys. Chem. A, 113*, 10391–10396.
22. Pendás, Á. M., Francisco, E., Blanco, M. A., & Gatti, C., (2007). Bond paths as privileged exchange channels. *Chem. Eur. J., 12*, 9362–9371.
23. Savin, A., Becke, D. A., Flad, J., Nesper J., Preuss, H., & von Schnering, H., (1991). A new look at electron localization. *Angew. Chem. Int. Ed. Eng., 30*, 409–412.
24. Savin, A., & Silvi, B., (1994). Classification of chemical bonds based on topological analysis of electron localization functions. *Nature, 371*, 683–686.

25. Reed, A. E., & Weinhold, F., (1985). Natural population analysis. *J. Chem. Phys., 83*, 735–746.

26. Reed, A. E., & Curtiss, L. A., (1988). Weinhold, F. Intermolecular interactions from a natural bond orbital, donor-acceptor viewpoint. *Chem. Rev., 88*, 899–926.

27. Weinhold, F., & Landis, C., (2005). *Valency and Bonding: A Natural Bond Orbital Donor-Acceptor Perspective*; Cambridge University Press: Cambridge.

28. Wiberg, K. B., (1968). Application of Pople-Santry-Segal CNDO method to cyclopropylcarbinyl and cyclobutylcation and to bicyclobutane. *Tetrahedron, 24*, 1083–1096.

29. Mayer, I., (1983). Charge, bond order and valence in the ab initio theory. *Chem. Phys. Lett., 97*, 270–274.

30. Cioslowski, J., & Mixon, S. T., (1991). Covalent bond orders in the topological theory of atoms in molecules. *J. Am. Chem. Soc., 113*, 4142–4145.

31. Fulton, R., (1993). Sharing of electrons in molecules. *J. Phys. Chem., 97*, 7516–7529.

32. Ponec, R., Uhlík, F., Cooper, D. L., & Jug, K., (1996). On the definition of bond index and valence for correlated wave functions. *Croat. Chem. Acta, 69*, 933–940.

33. Gopinathan, M. S., & Jug, K., (1986). Valency. I. Quantum chemical definition and properties. *Theor. Chim. Acta, 68*, 497–509.

34. Karafiloglou, P., (1990). Chemical structures from multi-electron density operators. *Chem. Phys., 128*, 373–383.

35. Fradera, X., Austen, M. A., & Bader, R. F. W., (1993). The Lewis model and beyond. *J. Phys. Chem. A, 103*, 304–314.

36. Ángyán, J., Loos, M., & Mayer, I., (1994). Covalent bond orders and atomic valence indices in the topological theory of atoms in molecules. *J. Phys. Chem., 98*, 5244–5248.

37. McWeeny, R., (1960). Some recent advances in density matrix theory. *Rev. Mod. Phys, 32*, 335–369.

38. Löwdin, P. O., (1955), Quantum theory of many-particle systems. I. Physical interpretation by means of density matrices, natural spin-orbitals and convergence problems in the method of configuration interaction. *Phys. Rev., 97*, 1474–1489.

39. Ponec, R., (1997). Electron pairing and chemical bonds. Chemical structure, valences and structural similarities from the analysis of the Fermi holes. *J. Math. Chem., 21*, 323–333.

40. Ponec, R., (1998). Electron pairing and chemical bonds. Molecular structure from the analysis of pair densities and related quantities. *J. Math. Chem., 23*, 85–103.

41. Ponec, R., & Duben, A. J., (1999). Electron pairing and chemical bonds: Bonding in hypervalent molecules from analysis of Fermi holes. *J. Comp. Chem., 8*, 760–771.

42. Ponec, R., & Strnad, M., (1994). Population analysis of pair densities: A link between quantum chemical and classical picture of chemical structure. *Int. J. Quantum Chem., 50*, 43–53.

43. Ponec, R., & Bochicchio, R., (1995). Nonlinear population analysis from geminal expansion of pair densities. *Int. J. Quantum Chem., 54*, 99–105.

44. Ponec, R., & Cooper, D. L., (2005). Anatomy of bond formation. Bond length dependence of the extent of electron sharing in chemical bonds. *J. Mol. Struct. Theochem, 727*, 133–138.

45. (a) Mayer, I., (2007). Bond order and valence indices: A personal account. *J. Comp. Chem., 28*, 204–221. (b) Mayer, I., (2017). *Bond Orders and Energy Components: Extracting Chemical Information from Molecular Wave Functions*; CRC Press: Boca Raton.

46. Ponec, R., Yuzhakov, G., & Carbó-Dorca, R., (2003). Chemical structures from the analysis of domain-averaged Fermi holes: multiple metal-metal bonding in transition metal compounds. *J. Comp. Chem., 24*, 1829–1838.

47. Ponec, R., & Yuzhakov, G., (2007). Metal-metal bonding in $Re_2Cl_8^{(2-)}$ from the analysis of domain-averaged Fermi holes. *Theor. Chem. Acc., 118*, 791–797.

48. Ponec, R., & Feixas, F., (2009). Peculiarities of multiple Cr-Cr Bonding. Insights from the analysis of domain-averaged Fermi holes. *J. Phys. Chem. A, 113*, 8394–8400.

49. Ponec, R., Lendvay, G., & Chaves, J., (2008). Structure and bonding in binuclear metal carbonyls from the analysis of domain-averaged Fermi holes. I. $Fe_2(CO)_9$ and $Co_2(CO)_8$. *J. Comp. Chem., 29*, 1387.

50. Ponec, R., (2015). Structure and bonding in binuclear metal carbonyls. Classical paradigms vs. insights from modern theoretical calculations. *Comp. Theor. Chem., 1053*. 195–213.

51. Ponec, R., Cooper, D. L., & Savin, A., (2008). Analytic models of domain-averaged Fermi holes: A new tool for the study of the nature of chemical bonds. *Chem. Eur. J., 14*, 3338–3345.

52. Bultinck, P., Cooper, D. L., & Ponec, R., (2010). Influence of atoms-in-molecules methods on shared-electron distribution indices and domain-averaged Fermi holes. *J. Phys. Chem. A, 114*, 8754–8763.

53. Tiana, D., Francisco, E., Blanco, M. A., Macchi, P., Sironi, A., & Pendás, Á. M., (2011). Restoring orbital thinking from real space descriptions: bonding in classical and non-classical transition metal carbonyls. *Phys. Chem. Chem. Phys., 13*, 5068–5077.

54. Baranov, A. I., Ponec, R., & Kohout, M., (2012). Domain-averaged Fermi-hole analysis for solids. *J. Chem. Phys., 137*, 214109.

55. Francisco, E., Pendás, Á. M., & Costales, A., (2014). On the interpretation of domain averaged Fermi hole analyses of correlated wavefunctions. *Phys. Chem. Chem. Phys., 16*, 4586–4597.

56. Cioslowski, J., (1990). Isopycnic orbital transformation and localization of natural orbitals. *Int. J. Quantum Chem., S24*, 15–28.

57. Ponec, R., & Cooper, D. L., (2007). Anatomy of bond formation. Bond length dependence of the extent of electron sharing in chemical bonds from the analysis of domain-averaged Fermi holes. *Faraday Discuss., 135*, 31–42.

58. Ponec, R., & Cooper, D. L., (2007). Anatomy of bond formation. Domain-averaged Fermi holes as a tool for the study of the nature of the chemical bonding in Li_2, Li_4 and F_2. *J. Phys. Chem. A, 111*, 11294–11301.

59. Parr, R. G., Ayers, P. W., & Nalewajski, R. F., (2005). What is an atom in a molecule? *J. Phys. Chem. A, 109*, 3957–3959.

60. Hybridization, chapter 8, page 205 in Ref. 4.

61. Moffitt, W., (1950). Term values in hybrid states. *Proc. R. Soc. London, A. 202*, 534–547.

62. Cooper, D. L., & Ponec, R., (2009). Anatomy of bond formation: Insights from the analysis of domain-averaged Fermi holes in momentum space. *Int. J. Quantum Chem., 109*, 2383–2392.

63. Cooper, D. L., Ponec, R., & Kohout, M., (2015). Are orbital-resolved shared-electron distribution indices and Cioslowski covalent bond orders useful for molecules? *Mol. Phys., 113*, 1682–1689.

64. Mayer, I., (2012). Improved definition of bond orders for correlated wave functions. *Chem. Phys. Lett., 544*, 83–86.

65. Cooper, D. L., Ponec, R., & Kohout, M., (2016). New insights from domain-averaged Fermi holes and bond order analysis into the bonding conundrum in C_2. *Mol. Phys., 114,* 1270–1284.

66. Werner, H.-J., & Knowles, P. J., (1985). A 2nd order multiconfiguration SCF procedure with optimum convergence. *J. Chem. Phys., 82,* 5053–5063.

67. Knowles, P. J., & Werner, H.-J., (1985). An efficient 2nd-order MC SCF method for long configuration expansions. *Chem. Phys. Lett., 115,* 259–267.

68. Knowles, P. J., Knowles, P. J., Knizia, G., Manby, F. R., & Schütz, M., (2012). Molpro: a general-purpose quantum chemistry program package. *WIREs Comput. Mol. Sci., 2,* 242–253.

69. Werner, H.-J., Knowles, P. J., Knizia, G., Manby, F. R., Schütz, M., & others, (2015). MOLPRO, version 2015.1, a package of ab initio programs: Cardiff, U.K.

70. Keith, T. A., (2012). AIMAll (Version 13.11.04). TK Gristmill Software: Overland Park KS, USA.

71. Schaftenaar, G., & Noordik, J. H., (2000). Molden: a pre- and post-processing program for molecular and electronic structures. *J. Comput.-Aided Mol. Des., 14,* 123–134.

72. García-Revilla, M., Popelier, P. L. A., Francisco, E., & Pendás, Á. M., (2013). Nature of chemical interactions from the profiles of electron delocalization indices. *J. Chem. Theory Comput., 7,* 1704–1711.

73. Cooper, D. L., & Ponec, R., (2013). Bond formation in diatomic transition metal hydrides. Insights from the analysis of domain-averaged Fermi holes. *Int. J. Quantum Chem., 113,* 102–111.

74. Longuet-Higgins, H. C., (1949). Substances hydrogénées avec default d'electrons. *J. Chim. Phys., 46,* 268–275.

75. Lipscomb, W. N., (1977). The boranes and their relatives. *Science, 196,* 1047–1055.

76. Lipscomb, W. N., (1973). Three-center bonds in electron-deficient compounds. The localized molecular orbital approach. *Acc. Chem. Res., 6,* 257–262.

77. Pimentel, G. C., (1951). The bonding of trihalide and bifluoride ions by the molecular orbital method. *J. Chem. Phys., 19,* 446–448.

78. Rundle, R. E., (1963). On the probable structure of XeF_4 and XeF_2. *J. Am. Chem. Soc., 85,* 112–113.

79. Hach, R. J., & Rundle, R. E., (1951). The structure of tetramethylammonium pentaiodide. *J. Am. Chem. Soc., 73,* 4321–4324.

80. Landrum, G. A., Golberg, N., & Hoffmann, R., (1997). Bonding in the trihalides (X_3^-) and mixed trihalides (X_2Y^-) and hydrogen bihalides (X_2H^-). The connection between hypervalent electron-rich three-center, donor-acceptor and strong hydrogen bonding. *J. Chem. Soc. Dalton Trans.,* 3605–3613.

81. Feixas, F., Matito, E., Poater, J., & Solà, M., (2015). Quantifying aromaticity with electron delocalization measures. *Chem. Soc. Rev., 44,* 6434–6451.

82. Mandado, M., & Ponec, R., (2009). Electron reorganization in allowed and forbidden pericyclic reactions: multicenter bond indices as a measure of aromaticity and/or anti-aromaticity in transition states of pericyclic electrocyclizations. *J. Phys. Org. Chem., 22,* 1225–1232.

83. Sanigrahi A. B., & Kar, T., (1990). Three-center bond index. *Chem. Phys. Lett., 173,* 569–572.

84. Sanigrahi, A. B., & Kar, T., (2000). Ab initio theoretical study of three-center bonding on the basis of bond index. *J. Mol. Struct. Theochem. 496,* 1–17.

85. Sanigrahi, A. B., Nandi, P. K., Behera, L., & Kar, T., (1992). Theoretical study of multi-center bonding using a delocalized MO approach. *J. Mol. Struct. Theochem, 276*, 259–278

86. de Giambiagi, M. S., & Giambiagi, M., (1994). The three-center χ-electron chemical bond. *Zeitschr. Für Naturforsch., 49*, 754–758

87. Giambiagi, M., de Giambiagi, M. S., & Mundim, K. C., (1990). Definition of a multi-center bond index. *Struct. Chem., 1*, 423–427

88. Mundim, K. C., Giambiagi, M., & de Giambiagi, M. S., (1994). Multicenter bond index: Grassmann algebra and N-order density functional. *J. Phys. Chem., 98*, 611–6119

89. Cooper, D. L., Gerratt, J., Raimondi, M., Sironi, M., & Thorsteinsson, T., (1993). Expansion of the spin-coupled wavefunction in Slater determinants. *Theor. Chim. Acta, 85*, 261–270.

90. Cooper, D. L., & Karadakov, P. B., (2009). Spin-coupled descriptions of organic reactivity. *Int. Rev. Phys. Chem., 28*, 169–206.

91. Ponec, R., & Mayer I. (1997). Investigation of some properties of multicenter bond indices. *J. Phys. Chem. A, 101*, 1738–1741.

92. Kar, T., & Sanchez-Marcos, E., (1992). Three-center four-electron bonds and their indices. *Chem. Phys. Lett., 192*, 14–20.

93. Ponec, R., & Cooper, D. L., (2004). Generalized population analysis of three-center two-electron bonding. *Int. J. Quantum Chem., 97*, 1002–1011.

94. Ponec, R., Yuzhakov, G., & Cooper, D. L., (2004). Multicenter bonding and the structure of electron-rich molecules. Model of three-center four-electron bonding reconsidered. *Theor. Chem. Acc., 112*, 419–430.

95. de Magalhães, H. P., Lüthi, H. P., & Bultinck, P., (2016). Exploring the role of the 3-center-4-electron bond in hypervalent λ^3-iodanes using the methodology of domain averaged Fermi holes. *Phys. Chem. Chem. Phys., 18*, 846–856.

96. Bochicchio, R., Ponec, R., Torre, A., & Lain, L., (2001). Multicenter bonding within the AIM theory. *Theor. Chem. Acta, 105*, 292–298.

97. Bochicchio, R., Torre, A., Lain, L., & Ponec, R. (2000). Topological population analysis from higher order densities. I. Hartree-Fock level. *J. Math. Chem., 28*, 83–90.

98. Ponec, R., & Uhlík, F., (1996). Multicenter bond indices from the generalized population analysis of higher order densities. *Croat. Chem. Acta, 69*, 941–954.

99. Feixas, F., Rodriguez-Mayorga, M., Matito, E., & Solà, M., (2015). Three-center bonding analyzed from correlated and uncorrelated third-order reduced density matrices. *Comp. Theor. Chem., 1053*, 173–179.

100. Feixas, F., Solà, M., Barroso, J. M., Ugalde, J. M., & Matito, E., (2014). New approximation to the third-order density. Application to the calculation of correlated multicenter indices. *J. Chem. Theor. Comp., 10*, 3055–3065.

101. Francisco, E., Pendás, Á. M., García-Revilla, M., & Alvarez-Boto, R., (2013). A hierarchy of chemical bonding indices in real space from reduced density matrices and cumulants. *Comp. Theor. Chem., 1003*, 71–78.

102. Mandado, M., Gonzáles-Moa, M. J., & Mosquera, R. A., (2007). Chemical graph theory and n-center electron delocalization indices: A study on polycyclic aromatic hydrocarbons. *J. Comp. Chem., 28*, 1625–1633.

103. Bultinck, P., Mandado, M., & Mosquera, R. A., (2008). The pseudo-p method examined for the computation of aromaticity indices. *J. Math. Chem., 43*, 111–118.

104. Lain, L., Torre, A., & Bochicchio, R., (2004). Studies of population analysis at the correlated level: Determination of three-center bond indices. *J. Phys. Chem. A, 108*, 4132–4137.

105. Arrowsmith, M., Braunschweig, H., Celik, M. A., Dellermann, T., Dewhurst, R. D., Ewing, W. C., et al., (2016). Neutral zero-valent s-block complexes with strong multiple bonding. *Nature Chemistry, 8*, 890–894.

106. Pipek, J., & Mezey, P. G., (1989). A fast intrinsic localization procedure applicable for ab initio and semiempirical linear combination of atomic orbital wave functions. *J. Chem. Phys., 90*, 4916–4926.

107. (a) Flükiger, P., Lüthi, H. P., Portmann, S., & Weber, J., (2000-2002). Molekel: Swiss Center for Scientific Computing. (b) Varetto, U., (2009). Molekel version 5.4.0.8: Swiss National Supercomputing Center.

108. Ponec, R., & Cooper, D. L., (2017). Insights from domain-averaged Fermi hole (DAFH) analysis and multicenter bond indices into the nature of Be(0) bonding. *Struct. Chem., 28*, 1033–1043.

CHAPTER 15

BACK TO THE ORIGINS: USING MATRIX FUNCTIONS OF HÜCKEL HAMILTONIAN FOR QUANTUM INTERFERENCE

ERNESTO ESTRADA

Department of Mathematics and Statistics, University of Strathclyde, 26 Richmond Street, Glasgow, G11XH, UK

CONTENTS

15.1   PROLOGUE

When I started my first steps in theoretical chemistry at the beginning of the 1990's in Cuba, I was very excited about the use of graph-theoretic ideas to represent molecules and describe their properties. While attending at a conference I had the opportunity to chat with a well-known theoretical chemist who was visiting Cuba and asking his opinion about "*chemical graph theory.*" His lapidary response was: "*Chemical graph theory cannot go beyond Hückel method, and so it is dead.*" Then, partially because of my ignorance and partially because of the fact that I was listening everyday many things I do not believe in, I decided to give rid of the expert advice and centre my investigation around molecules, the Hückel method and graph theory. I was in my mid 20's '*Non tamen ista meos mutabunt saecula mores: unus quisque sua noverit ire via.*' (But age shall still not change my habits: let each man be allowed to go his own way) [1].

A search on Google Scholar for papers published in the XXI century carried out on the 10th October 2016 has given more than 2,000 papers using the HMO method to study molecules and more than 10,000 ones using the closely related term "tight-binding Hamiltonian." It is not too bad for a method which is dead! Three remarkable examples of how alive the HMO method is in the XXI century are the following. In his 2000 Nobel lecture "The discovery of polyacetylene film: the dawning of an era of conducting polymers" [2] Hideki Shirakawa recognized the value of HMO in determining that "*the difference between the lengths of double and single bonds decreases with increasing the conjugation and that all the bonds tend to be of equal length in an infinitely long polyene.*" In 2007 Kutzelnigg [3] reviewed the HMO method for the *Journal of Computational Chemistry* with the appealing title: "*What I like about Hückel theory*" and in 2012 Ramakrishnan [4] proposed some educational tools for working with the HMO method in a paper published by the *Journal of Chemical Education.*

One of the most attractive aspects of the HMO theory resides in its connection with the mathematical theory of graphs. Graph theory [5], as it is known in mathematics, is today a very well-established mathematical theory with connections to combinatorics, algebra, number theory, topology, theoretical computer sciences, statistical mechanics, quantum gravity, complex systems, quantum information theory, quantum chromodynamics, and of course Chemistry (for a short compilation of such relations see [6]). This connection between HMO and graph theory allows not only the numerical characterization of molecular properties, but more importantly its conceptual understanding in terms of simple molecular terms. In this sense the HMO-graph theory (HMO-GT) marriage responds positively to C. A. Coulson admonition to theoretical chemists: "*Give us insight, not numbers*" [7]. HMO-GT gives numbers and insight at the same time!

A review of the literature of the last part of the XX century shows that most of the works produced in theoretical chemistry were more on the computational side of the problem, giving more numbers and less insight to chemists. There are a few exceptions, of course. And the work of Professor Ramón Carbó-Dorca, our friend Ramón, whose work we celebrate in this book, is one of them.

15.2 HMO AND GRAPH THEORY

In order to describe the electronic structure of conjugated molecular systems the so-called Hückel molecular orbital theory [8, 9], known in physics as the tight-binding approach [10], considers that the interaction between N electrons is determined by a Hamiltonian of the following form:

$$\mathbf{H} = \sum_{n=1}^{N} \left[-\frac{\hbar^2 \tilde{N}_n^2}{2m} + U\left(r_n\right) \right] + \frac{1}{2} \sum_{m \ne n} V\left(r_n - r_m\right) \quad (1)$$

where $U(r_n)$ is an external potential and $V(r_n - r_n)$ is the potential describing the interactions between electrons. Using the second quantization formalism of quantum mechanics this Hamiltonian can be written as:

$$\hat{H} = -\sum_{ij} t_{ij} \hat{c}_i^\dagger \hat{c}_j + \frac{1}{2} \sum_{ijkl} V_{ijkl} \hat{c}_i^\dagger \hat{c}_k^\dagger \hat{c}_l \hat{c}_j \quad (2)$$

where $\hat{c}_i^\dagger$ and $\hat{c}_i$ are 'ladder operators,' t_{ij} and V_{ijkl} are integrals which control the hopping of an electron from one site to another and the interaction between electrons, respectively. They are usually calculated directly from finite basis sets [11].

In the HMO approach for studying solids and conjugated molecules, the interaction between electrons is neglected and $V_{ijkl} = 0$, $\forall i, k, l$. This method can be seen as very drastic in its approximation, but let us think of the physical picture behind it [2, 12]. We concentrate our discussion on alternant-conjugated molecules in which single and double bonds alternate. Consider a molecule like benzene in which every carbon atom has a sp_2 hybridization. The frontal overlapping sp_2–sp_2 of adjacent carbon atoms create very stable α-bonds, while the lateral overlapping p–p between adjacent carbon atoms creates very labile π-bonds. Thus it is clear from the reactivity of this molecule that a σ–π separation is plausible and we can consider that our basis set consists of orbitals centred on the particular carbon atoms in such a way that there is only one orbital per spin state at each site. Then we can write the Hamiltonian of the system as:

$$\hat{H}_{tb} = -\sum_{ij} t_{ij} \hat{c}^\dagger_{ip} \hat{c}_{ip} \tag{3}$$

where $\hat{c}_{ip}^{(\dagger)}$ creates (annihilates) an electron with spin p in a π (or other) orbital centred at the atom i. We can now separate the in-site energy α_i from the transfer energy β_{ij} and write the Hamiltonian as

$$\hat{H}_{tb} = \sum_{ij} \alpha_i \hat{c}^\dagger_{ip} \hat{c}_{ip} + \sum_{\langle ij \rangle p} \beta_{ij} \hat{c}^\dagger_{ip} \hat{c}_{ip} \tag{4}$$

where the second sum is carried out over all pairs of nearest-neighbors. Consequently, in a conjugated molecule or solid with N atoms the Hamiltonian (2.3) is reduced to a $N \times N$ matrix,

$$H_{ij} = \begin{cases} \alpha_i & \text{if } i = j \\ \beta_{ij} & \text{if } i \text{ is connected to } j \\ 0 & \text{otherwise.} \end{cases} \tag{5}$$

Due to the homogeneous geometrical and electronic configuration of many systems analyzed by this method we may take $\alpha_i = \tilde{\alpha}$, $\forall i$ (Fermi energy) and $\beta_{ij} = \tilde{\beta} \approx -2.70 eV$ for all pairs of connected atoms. Thus,

$$\hat{H} = \tilde{\alpha} I + \tilde{\beta} A \tag{6}$$

where I is the identity matrix, and A is a matrix whose entries are defined as

$$A_{ij} = \begin{cases} 1 & \text{if atom } i \text{ is bonded to atom } j \\ 0 & \text{otherwise.} \end{cases} \tag{7}$$

Let us now introduce the following concepts. Let us consider a finite set $V = \{v_1, v_2, ..., v_N\}$ of unspecified elements and let $V \otimes V$ be the set of all ordered pairs $[v_i, v_j]$ of the elements of V. A relation on the set V is any subset $E \subseteq V \otimes V$. The relation E is symmetric if $[v_i, v_j]$ implies $[v_j, v_i] \in E$ and it is reflexive if $\forall v \in V, [v, v] \in E$. The relation E is antireflexive if $[v_i, v_j] \in E$

implies $v_i \neq v_j$. Now let us define a *simple graph* as the pair $G = (V, E)$, where V is a finite set of nodes (also known as vertices), and E is a symmetric and antireflexive relation on V, whose elements are known as the edges or links of the graph [13].

Thus, if in any conjugated molecule we consider every non-hydrogen atom as a vertex, and every covalent bond, excluding those involving hydrogen, as an edge, we have a direct map between a molecule and a graph. These graphs are known as hydrogen-deleted graphs, or simply as molecular graphs. In this context, the matrix A is the adjacency matrix of the graph, which represents the adjacency relation between the nodes of the corresponding graph. Then, the adjacency matrix of the molecular graph and the HMO Hamiltonian has the same eigenfunctions ϕ_j and their eigenvalues are simply related by:

$$\hat{H}\varphi_i = E_i A, \; A\varphi_i = \lambda_i \hat{H}, \; E_i = \tilde{\alpha} + \tilde{\beta}\lambda_i \tag{8}$$

Hence everything we have to do in the analysis of the electronic structure of molecules or solids that can be represented by an HMO Hamiltonian, is to study the spectra of the graphs associated with them. The study of spectral properties of graphs represents an entire area of research in algebraic graph theory. The spectrum of a matrix is the set of eigenvalues of the matrix together with their multiplicities. For the case of the adjacency matrix let $\lambda_1(A) \geq \lambda_2(A) \geq ... \geq \lambda_N(A)$ be the distinct eigenvalues of A and let $m(\lambda_1(A)), m(\lambda_2(A)), ..., m(\lambda_N(A))$ be their algebraic multiplicities, i.e., the number of times each of them appears as an eigenvalue of A. Then the spectrum of A can be written as [13]

$$SpA = \begin{pmatrix} \lambda_1(A) & \lambda_2(A) & \cdots & \lambda_N(A) \\ m(\lambda_1(A)) & m(\lambda_2(A)) & \cdots & m(\lambda_N(A)) \end{pmatrix} \tag{9}$$

The total π (molecular) energy is given by

$$E = \tilde{\alpha} n_e + \tilde{\beta} \sum_{i=1}^{N} g_i \lambda_i \tag{10}$$

where n_e is the number of π-electrons in the molecule and g_j is the occupation number of the j-th molecular orbital. For neutral conjugated systems in their ground state we have [14]

$$E = \begin{cases} 2\sum\limits_{j=1}^{n/2} \lambda_j & n \text{ even,} \\ 2\sum\limits_{j=1}^{(n+1)/2} \lambda_j + \lambda_{(j+1)/2} & n \text{ odd.} \end{cases} \qquad (11)$$

Because an alternant conjugated hydrocarbon has a bipartite molecular graph, i.e., a graph whose set of vertices can be split into two disjoint sets V_1 and V_2, $V = V_1 \bigcup V_2 : \lambda_i = -\lambda_{N-i+1}$ for all $j = 1,2,...,N$. In a few molecular systems the spectrum of the adjacency matrix is known. For instance [2], we have

i)   Polyenes

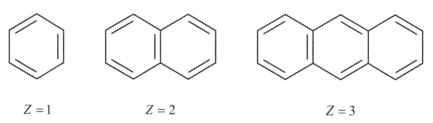

$$\lambda_j(A) = 2\cos\left(\frac{\pi j}{N+1}\right), \; j = 1,...,N \qquad (12)$$

ii)   Cyclic polyenes

$$\lambda_j(A) = 2\cos\left(\frac{2\pi j}{N}\right), \; j = 1,...,N, \; \lambda_i = \lambda_{N-i} \qquad (13)$$

iii)   Linear polyacenes,

$$Z = 1 \qquad\qquad Z = 2 \qquad\qquad Z = 3$$

$$\lambda_r(A) = 1; \lambda_s(A) = -1;$$

$$\lambda_k(A) = \pm\frac{1}{2}\left\{1 \pm \sqrt{9 \pm 8\cos\frac{k\pi}{Z+1}}\right\}, k = 1,\ldots,Z' \tag{14}$$

where all four combinations of signs have to be considered

15.3 MATRIX FUNCTIONS OF THE HMO HAMILTONIAN

There are several alternative definitions of a matrix function. If the matrix $\hat{H}$ is diagonalizable we can express it as $\hat{H} = V\Lambda V^{-1}$, where V is a matrix whose columns are the (orthonormal) eigenvectors of $\hat{H}$ and Λ is a diagonal matrix of eigenvalues. Then, a function of $\hat{H}$ is defined by [15]

$$f(\hat{H}) = V\begin{bmatrix} f(\lambda_1) & \cdots & 0 \\ \vdots & \ddots & \vdots \\ 0 & \cdots & f(\lambda_n) \end{bmatrix}V^{-1} \tag{15}$$

The function $f(\hat{H})$ has the following Taylor expansion:

$$f(\hat{H}) = f(0) + f'(0)\hat{H} + f''(0)\frac{\hat{H}^2}{2!} + \cdots \tag{16}$$

and can be represented by using the Cauchy integral formula [15]:

$$f(\hat{H}) = \frac{1}{2\pi i}\oint_C f(z)(zI - \hat{H})^{-1}dz \tag{17}$$

where f is analytic on and inside the closed contour that encloses the eigenvalues of A.

As we have seen in the previous section there is an identity correspondence between the HMO Hamiltonian and the adjacency matrix of the molecular graph: $\hat{H} = -A$. Then, let us consider that the molecule is submerged into a thermal bath of inverse temperature $\beta = (k_B T)^{-1}$. We are interested in knowing what is the probability that the molecule is found in an energy state $E_j = -\lambda_j$. Assuming, as usual, a Boltzmann distribution we get

$$p_j = \exp(-\beta E_j)/\sum_i \exp(-\beta E_i) \tag{18}$$

The denominator of this expression is known as the electronic partition function of the molecule [16, 17]

$$Z = \sum_i \exp(-\beta E_i) \tag{19}$$

Then, using our equivalence with the adjacency matrix we have

$$p_j = \exp(\beta \lambda_j)/Z \tag{20}$$

It is easy to see that the partition function is

$$Z = Tr \exp(-\beta \hat{H}) = Tr \exp(\beta A) \tag{21}$$

where Tr represents the trace of the matrix, and

$$\exp(\beta A) = I + \beta A + \frac{(\beta A)^2}{2!} + \cdots + \frac{(\beta A)^k}{k!} + \cdots \tag{22}$$

which according to our previous definition is a matrix function of the adjacency matrix of the molecular graph [18, 19].

In chemical graph theory this partition function is typically represented as $Z = E(G,\beta)$ and it is known as the Estrada index of the graph, in particular for $\beta = 1$ (for recent reviews see [20, 21]).

It is well-known that

$$\exp(\beta A) = \sinh(\beta A) + \cosh(\beta A) \tag{23}$$

where

$$\sinh(\beta A) = \sum_{k=0}^{\infty} \frac{(\beta A)^{2k+1}}{(2k+1)!} \tag{24}$$

$$\cosh(\beta A) = \sum_{k=0}^{\infty} \frac{(\beta A)^{2k}}{(2k)!} \tag{25}$$

are two other matrix functions. In bipartite graphs, such as those representing conjugated molecules, $\sinh(\beta A) = 0$ due to the lack of odd cycles in their structures. Thus,

$$Z = EE(G, \beta) = Tr \cosh(\beta A) \qquad (26)$$

There are several other matrix functions, such as trigonometric functions, sign function, ψ-functions, and others. The interested reader can consult the following references for more details [22].

15.4 QUANTUM INTERFERENCE

The phenomenon of quantum interference (QI) consists of a significant reduction of the conductance through the bonds of a molecule to which electrodes have been connected to specific atoms [23, 24]. In conjugated polycyclic compounds, QI controls site-dependent electron transport as corroborated experimentally by different techniques [25–30]. QI has attracted the attention of theorists who have developed different kind of approaches to explain this phenomenon. They include local atom-to-atom transmission [31], the use of simple Hückel molecular orbital (HMO) theory [32], spectral methods [33], graph-theoretic concepts [34] and graphical approaches [35–37]. An example of the last approach is the method developed by Markussen et al. [35], which provides a direct link between QI and the topology of various alternant π systems including *meta*-linked benzene derivatives, anthraquinone derivatives, and cross-conjugated molecules. Such rules can be resumed as follows: (i) Two sites can be connected by a path if they are nearest neighbors. (ii) At all internal sites, i.e., sites other than 1 and N, there is one incoming and one outgoing path. It is now straightforward to show that the condition for complete destructive interference is fulfilled if it is *not* possible to connect the external sites 1 and N by a continuous chain of paths and at the same time fulfill the rules (i) and (ii). On the other hand, if such a continuous path can be drawn, then the condition $\det_{1N}(H_{mol}) = 0$ is not fulfilled and a transmission antiresonance does not occur at the Fermi energy.

Recent results obtained by Xia et al. [38] have challenged some of the theories proposed for the QI effects in conjugated polycyclic compounds. In particular they obtained azulene derivatives containing gold-binding groups at different points of connectivity within the azulene core. By comparing paths through the 5-membered ring, 7-membered ring, and across the long axis of azulene they have found that simple models, such as the one developed by Markussen et al. [35], cannot be used to predict quantum interference characteristics of nonalternant hydrocarbons. In particular, azulene derivatives that are predicted to exhibit destructive interference based on widely accepted atom-counting models show a significant conductance at low biases. In Figure 15.1 we illustrated the predictions made by using Markussen et al. [35] graphical model for 1,3- and 5,7-azulene, which are predicted to display QI but are observed to transmit current through the electrodes. For the sake of comparison the application of the same graphical rules are displayed for naphthalene 1,3- and 2,7-naphthalene.

My interest in this topic has been motivated by a meeting in Glasgow with Professor Roald Hoffmann (Nobel Prize winner in Chemistry, 1982) who introduced me to the topic. Hoffmann et al. [39–42] have used the inverse of the HMO Hamiltonian to account for QI in polycyclic conjugated hydrocarbons. The idea is to consider the Green's function

$$G_{rs}^{(0)}(E_F) = \sum_k \frac{C_{rk} C_{sk}^*}{E_F - \varepsilon_k + i\eta} \tag{27}$$

for which, without any loss of generality, we will consider $\eta = 0$. The Green function can be written in matrix form as $G^0(E_F) = (E_F I - \hat{H})$, where I is the identity matrix. Then, by considering the case $G^0(E_F = 0) = -\hat{H}^{-1}$, QI is predicted for the pair of atoms for which $G_{rs}^{(0)}(E_F = 0) = 0$. This model displays many beautiful and interesting properties, such as mathematical elegance, simplicity and good predictability in alternant conjugated hydrocarbons. However, when applied to azulene derivatives, $G^0(E_F = 0)$ predicts QI for the 5,7-azulene, which has been observed to transmit the electric current.

Here we propose a simple model based on HMO theory that not only predict correctly QI in conjugated (alternant and nonalternant) molecules, but also allows the development of simple graphical rules for predicting QI in these molecules.

15.5   AN HMO MATRIX-FUNCTION MODEL FOR QI

We start by considering the Green function (1) by using the HMO Hamiltonian. Here we consider the following modification of the Hamiltonian

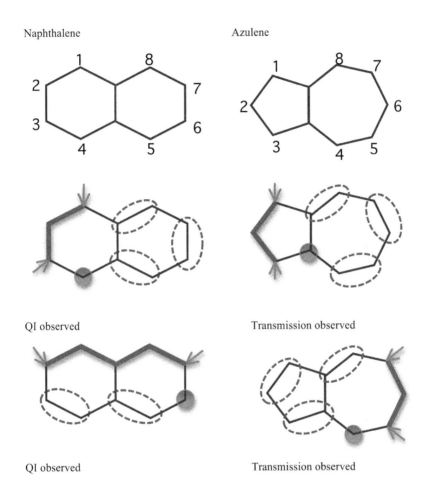

FIGURE 15.1 Illustration of Markussen et al. [35] rules for naphthalene and azulene. The rules predict that there is QI in 1,3- and 5,7-azulene, but transmission has been experimentally observed for these azulene derivatives.

$$\hat{H}^{T} = \hat{H} - \gamma^{2} P \qquad (28)$$

where P represents a small perturbation of the Hückel Hamiltonian based on electron hopping beyond the first nearest-neighbor in the molecule and γ is a parameter that controls the strength of the perturbation (notice that γ^{2} is used to guarantee that this contribution is always negative). It is well-known that the nondiagonal entries of $\hat{H}^{k}$ count the number of walks (hops) of length k between the corresponding pair of atoms. A walk is a sequence of (not necessarily different) atoms and bonds starting at one node and ending at another one. For instance, $(\hat{H}^3)$ gives the number of walks of length 3 between the atoms r and s. Usually such perturbation is accounted for by using a McLaurin series expansion of the Hamiltonian: $P = \kappa^2 \hat{H}^2 + \kappa^3 \hat{H}^3 + \cdots$. Here we introduce the following modification to the perturbation theory of the Hamiltonian. We consider that P has a closed form in order to avoid approximation problems with the power series, such that $\hat{H}^T = \hat{H} - \gamma^2 f(\hat{H})$, where $f(\hat{H})$ is a matrix function of the Hamiltonian. The function $f(\hat{H})$ has to fulfill a few requirements. First, we need to give more weight to the energy levels which are close to the frontier orbitals, which are the ones more involved in the electron conduction (see further). Second, if λ is an eigenvalue of $\hat{H}$ we require that $f(-\lambda) = -f(\lambda)$. This condition is imposed by the observed fact that when the eigenvalues of $\hat{H}$ are symmetric around the Fermi level the transmission spectra is symmetric around zero. These two conditions are fulfilled by the inverse of the Hamiltonian:

$$ P = \frac{(-1)^{N-1}}{\det(\hat{H})} \left(\hat{H}^{N-1} + c_{N-2}\hat{H}^{N-2} + \cdots + c_1 I \right) = \hat{H}^{-1} \tag{29} $$

where $\det(\hat{H})$ is the determinant of $\hat{H}$, which is assumed to be nonzero, N is the number of carbon atoms and I is the identity matrix.

Now we can write the Green's function of $\hat{H}^T = \hat{H} - \gamma^2\hat{H}$ as

$$ G^T\left(E_F\right) = \left(E_F I - \hat{H}^T\right)^{-1} = \left(E_F I - \hat{H} + \gamma^2 \hat{H}^{-1}\right)^{-1} \tag{30} $$

which for $E_F = 0$ reduces to $\mathscr{F} = G^T\left(E_F = 0\right) = \left(-\hat{H} + \gamma\hat{H}^{-1}\right)^{-1}$. Using the property of the inverse of two non-singular matrices: $(AB)^{-1} = B^{-1}A^{-1}$, we can write $\mathscr{F}$ as

$$ \mathscr{F} = \hat{H}\left(\gamma^2 I - \hat{H}^2\right)^{-1} \tag{31} $$

The term in the parenthesis represents a renormalization of the Hamiltonian that folds the original spectrum around the Fermi energy. That is, a Hamiltonian renormalization that places the frontier orbitals (and those close to them) at the bottom of the energy scale, while the orbitals with the highest and lowest energy are placed at the top of this scale. This renormalization is obtained by considering the squared Hamiltonian, such that $\hat{H}\psi = \varepsilon^2\psi$. The 'squared Hamiltonian trick' has been used previously for studying different molecular systems ranging from quasicrystals to doped graphene [43–45]. This renormalization agrees with the idea that the highest contribution to the Green's function is usually made by the frontier orbitals, i.e., those at the middle of the spectrum and thus close to E_F. This, of course, includes the highest occupied molecular orbital (HOMO) and the lowest unoccupied molecular orbital (LUMO). In this case, as analyzed by Yoshizawa et al. [46], the sign of the product $C_{rHOMO}C^*_{sHOMO}$ and of $C_{rLUMO}C^*_{sLUMO}$ mainly determine whether there is transmission between the atoms r and s or there is QI. For instance, if $\text{sgn}(C_{rHOMO}C^*_{sHOMO}) \neq \text{sgn}(C_{rLUMO}C^*_{sLUMO})$ the contributions of the frontier orbitals are enhanced and there is transmission. On the other hand, if $\text{sgn}(C_{rHOMO}C^*_{sHOMO}) = \text{sgn}(C_{rLUMO}C^*_{sLUMO})$ the contributions from these orbitals are cancelled out which may produce QI among this pair (we recall that the contribution of other orbitals is being neglected in this analysis).

For any pair of atoms in a molecule we obtain

$$\mathscr{F}_{rs} = \sum_k \frac{\varepsilon_k C_{rk} C^*_{sk}}{\gamma^2 - \varepsilon_k^2} \tag{32}$$

Obviously $\mathscr{F}$ is just the original Green's function of the Hückel model for the limit $\gamma^2 \to 0$ (no perturbation of the Hamiltonian):

$$\lim_{\gamma \to 0} \mathscr{F} = \lim_{\gamma \to 0} \hat{H}\left(\gamma^2 I - \hat{H}^2\right)^{-1} = -\hat{H}^{-1} = G^0\left(E_F = 0\right) \tag{33}$$

We notice that the function $\mathscr{F} = \hat{H}\left(\gamma^2 I - \hat{H}^2\right)^{-1}$, $0 < \gamma^2 < \min_j \varepsilon_j^2$, always exists. Thus, the condition imposed before that the Hamiltonian needs to be non-singular is now removed. In fact, it exists for any $\gamma^2 \, sp\{\hat{H}\}$, where $sp\{\hat{H}\}$ is the set of eigenvalues of $\hat{H}$. However, for $0 < \gamma^2 < \max_j \varepsilon_j^2$ the function $\mathscr{F}$ is positive, which means that the sign of $\mathscr{F}_{rs}$ is always positive, destroying one of the nice characteristics of $-\hat{H}^{-1}$ as reported by Tsuji et al. [41]. Thus, we will consider here the case when $0 < \gamma^2 < \min_j \varepsilon_j^2$. In this case, the nondiagonal entries of $\mathscr{F}$ indicates whether there is QI or transmission according to the following:

i.   If $\langle r|\mathcal{F}|s\rangle = 0$ there is QI between r and s.

ii.   If $\langle r|\mathcal{F}|s\rangle \neq 0$ there is current transmission between r and s.

As a first example of the potentialities of the current method we study here the transmission through the atoms 2 and 3 of 1,3-butadiene. This molecule was studied by Solomon et al. [47], where it was found that when using Hückel method QI phenomenon is observed between these two atoms. However, when more sophisticated MDE many-body theory is used there is a splitting of the central super-node, a certain level of transmission is observed for this connection (see Figure 15.9 in Solomon et al. [47] paper). The same effect is observed here by considering $(\mathcal{F}_{2,3})^2$ (normalized to one) as the transmission for different values of energy (see Figure 15.2). The structural interpretation of this result is given in the next section.

15.6   RULES FOR QI

We start by expanding the function $\mathcal{F}$ as a power series of the Hamiltonian. We derive here the power series expansion of the function $\mathcal{F}$. Let us use the Cayley-Hamilton for the inverse matrix function $(\gamma^2 I - \hat{H}^2)^{-1}$:

$$\left(\gamma^2 I - \hat{H}^2\right)^{-1}:$$

$$\left(\gamma^2 I - \hat{H}^2\right)^{-1} = \frac{(-1)^{N-1}}{\det\left(\gamma^2 I - \hat{H}^2\right)}\left[\left(\gamma^2 I - \hat{H}^2\right)^{N-1} + c_{N-2}\left(\gamma^2 I - \hat{H}^2\right)^{N-2} + \cdots + c_1 I\right]$$

(34)

First, let us consider the term multiplying the whole parenthesis in (S1). If the number of carbon atoms is even then $(-1)^{N-1} = -1$. For $0 < \gamma^2 < \min \varepsilon_j^2$ all eigenvalues of $\gamma^2 I - \hat{H}^2$ are negative, i.e., it is negative definite. Then, $\det(\gamma^2 I - \hat{H}^2) > 0$. Thus, the term $(-1)^{N-1}/\det(\gamma^2 I - \hat{H}^2) < 0$. Let us replace this term by $-Q$, where $Q \in \mathbb{R}^+$.

Now, we will use the binomial theorem for the power terms of (S1):

$$(x+y)^n = \sum_{k=0}^{n}\binom{n}{k}x^k y^{n-k}$$

(35)

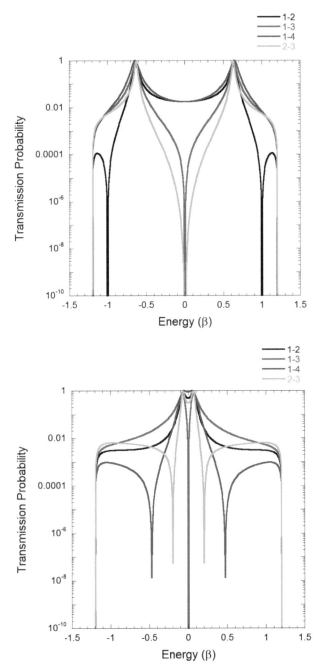

FIGURE 15.2 Illustration of the transmission between the different pairs of atoms in 1,3-butadiene using the Hückel method $-\hat{H}^{-1}$(left) and $\left(\mathscr{F}_{2,3}\right)^2$ (right) using $\gamma^2 = 0.425 \cdot$ The plots are in linear-log scale.

For instance, for benzene these terms become

$$\left(\gamma^2 I - \hat{H}^2\right)^2 = \hat{H}^4 - 2\gamma^2\hat{H}^2 + \gamma^4 \tag{36}$$

$$\left(\gamma^2 I - \hat{H}^2\right)^3 = -\hat{H}^6 + 3\gamma^2\hat{H}^4 - 3\gamma^4\hat{H}^2 + \gamma^6 \tag{37}$$

$$\left(\gamma^2 I - \hat{H}^2\right)^4 = \hat{H}^8 - 4\gamma^2\hat{H}^6 + 6\gamma^4\hat{H}^4 - 4\gamma^6\hat{H}^2 + \gamma^8 \tag{38}$$

$$\left(\gamma^2 I - \hat{H}^2\right)^5 = -\hat{H}^{10} + 5\gamma^2\hat{H}^8 - 10\gamma^4\hat{H}^6 + 10\gamma^6\hat{H}^4 - 5\gamma^8\hat{H}^2 + \gamma^{10} \tag{39}$$

$$\left(\gamma^2 I - \hat{H}^2\right)^6 = \hat{H}^{12} - 6\gamma^2\hat{H}^{10} + 5\gamma^4\hat{H}^8 - 20\gamma^6\hat{H}^6 + 15\gamma^8\hat{H}^4 - 6\gamma^{10}\hat{H}^2 + \gamma^{12} \tag{40}$$

Grouping together all the similar terms in the binomial expansion of the terms in (34) we get

$$\left(\gamma^2 I - \hat{H}^2\right)^{-1} = Q\left[\hat{H}^{2(N-1)} - \vartheta_{N-2}c_{N-2}\hat{H}^{2(N-2)} - \cdots + \vartheta_2 c_2\hat{H}^2 - \alpha_1 c_1 I\right] \tag{41}$$

According to the Cayley-Hamilton theorem, the coefficients c_k are the ones of the characteristic polynomial of $\gamma^2 I - \hat{H}^2$. We use here a result by Brooks [48], which proves that the characteristic polynomial of a given matrix can be expressed as (42)

$$P(x) = x^N - \left[\sum_{\text{sets of 1}}\lambda\right]x^{N-1} + \left[\sum_{\text{sets of 2}}\lambda\lambda\right]x^{N-2} - \cdots + (-1)^k\left[\sum_{\text{sets of k}}\lambda\lambda\cdots\lambda\right]x^{N-k} + \cdots(-1)^N\left[\sum_{\text{sets of n}}\lambda\lambda\cdots\lambda\right]$$

where λ is an eigenvalue of the corresponding matrix. Then, (43)

$$\left(\gamma^2 I - \hat{H}^2\right)^{-1} = Q\left[\hat{H}^{2(N-1)} + \vartheta_{N-2}\left[\sum_{\text{sets of 1}}\lambda\right]\hat{H}^{2(N-2)} - \cdots + \vartheta_2\left[\sum_{\text{sets of } n-2}\lambda\lambda\cdots\lambda\right]\hat{H}^2 - \vartheta_1\left[\sum_{\text{sets of } n-1}\lambda\lambda\cdots\lambda\right]I\right]$$

Because $\gamma^2 I - \hat{H}^2$ is negative definite, i.e., all $\lambda < 0$ we have that

$$\left[\sum_{\text{sets of k}}\lambda\cdots\lambda\right]\begin{cases} < 0 \text{ if } k \text{ is odd} \\ > 0 \text{ if } k \text{ is even} \end{cases} \tag{44}$$

Thus,

$$\left(\gamma^2 I - \hat{H}^2\right)^{-1} = Q\left[\hat{H}^{2(N-1)} - \delta_{N-2}\hat{H}^{2(N-2)} + \cdots - \delta_2\hat{H}^2 + \delta_1 I\right] \quad (45)$$

where δ_k groups all the coefficients. Then, we finally obtain the power series for $\mathcal{F}$:

$$\mathcal{F} = \hat{H}\left(\gamma^2 I - \hat{H}^2\right)^{-1} = Q\left[\hat{H}^{2(N-1)} - \delta_{N-2}\hat{H}^{2(N-2)} + \cdots - \delta_2\hat{H}^2 + \delta_1 I\right] \quad (46)$$

where N is the number of carbon atoms, Q is a positive number and the coefficients δ_k depends on γ^2 and on the coefficients of the Cayley-Hamilton expansion.

Clearly, there are only odd-length walks contributing to $\mathcal{F}$. This means that $\mathcal{F}_{rs} = 0$ if and only if there are no walks of odd length between the atoms r and s. Obviously, the existence of a walk of odd length implies the existence of an odd-length path connecting the two nodes, i.e., a walk in which all atoms and bonds are distinct. Consequently, the rules of QI in conjugated molecules can be formulated as:

i.   There is QI between the atoms r and s if and only if there is no path of odd length connecting them.

ii.   There is transmission between the atoms r and s if and only if there is at least one path of odd length connecting them.

In order to make a clearer interpretation of these results we start by considering an electric circuit formed by the source (electrodes) connected to a pair of atoms and the alternant bond forming resistors connected in series between the two electrodes. As usual in electrical circuit the current flows from the negative pole of the source to the positive one through the wires and resistors. The last ones are polarized in the direction of the current. We consider here that a bond represents a resistor in which the π-electrons at the corresponding atoms have opposite spin. In addition, according to the spin alternation rule, which states that the singlet spin pairing is preferred solely between sites in different subsets, the free valences on the starred and unstarred sites might be identified with "*up*" and "*down*" spin. Let us assume that the polarity of the resistor in the direction of the current implies that the spin of the two electrons in the bond are respectively *down-up* ($\downarrow$–$\uparrow$). Then, there is a current flow between the two electrodes if and only if there is at least one alternant sequence of spins of the form: $\downarrow-\uparrow...\downarrow-\uparrow$. This is only possible if the path connecting the two electrodes is of even length. We assume that if there is at least one of such paths the current will use it to flow from one electrode to the other. Obviously, there is QI when no such path exists between the two electrodes. Notice that all paths should be taken into account (we will illustrate this later). In Figure 15.3 we illustrate the two situations described before.

In an odd cycle there is at least one path of odd length between any pair of atoms. Thus, $\langle r|\mathcal{F}|s\rangle \neq 0$ for every r and s and there is no QI in such molecules. The typical example is azulene (see Figure 15.4). Although it is an alternant conjugated molecule, it is formed by two odd-length cycles and there is always an odd length path between every pair of atoms and consequently a current can be transmitted between them.

In a molecule without odd cycles, i.e., a molecule with bipartite or non-frustrated structure, it is enough to connect the electrodes to a pair of atoms separated by an even number of atoms in order to obtain QI. That is, because there are no odd cycles, all the other paths connecting these two atoms are of even length, such that no alternation of the polarity of the bonds is allowed and the current cannot be transmitted. The prototype of this system is the benzene molecule, which is illustrated in Figure 15.4.

It has been recently remarked [41] that the sign of $G^0(E_F = 0) = -\hat{H}^{-1}$ can play an important role in a yet not reported quantum interference involving

Rules for QI

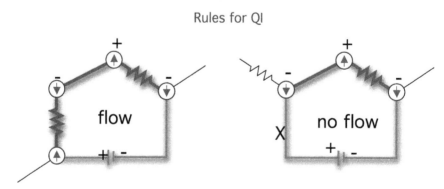

FIGURE 15.3   Schematic representation of the structural rules for the existence of current transmission or QI between a pair of atoms. Flow exists if and only if the polarity of the circuit is not altered as illustrated on the left-hand side of the picture. QI appears when the polarity at the starting and ending points are the same, which impedes the flow of current in the circuit.

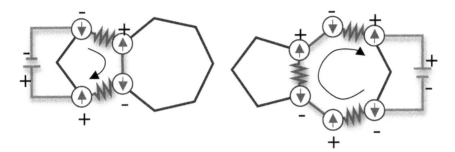

FIGURE 15.4 Illustration of the prediction of current transmission in 1,3- and 5,7-azulene by using the current method. Although there is an even-length path connecting the atoms 1 and 3 (left) and the atoms 5 and 7 (right), the electric current circulates by using any existing odd-length path to avoid the blockage of the current due to lack of polarity alternation (see Figure 15.3).

more than two atoms. Using the function $\mathscr{F}$ it is now easy to formulate some rules about the sign pattern of $\mathscr{F}_{rs}$. That is,

- $\mathscr{F}_{rs} > 0$ if an only if there is at least one path of length $4k+3$, $k = 1,2,...$ between the atoms r and s
- $\mathscr{F}_{rs} > 0$ if an only if there is at least one path of length $4k+1$, $k = 1,2,...$ between the atoms r and s

For instance, in benzene the pair 1,2 is connected by a path of length one and by a path of length 5, which are both of the type $4k+1$. Thus, $\mathscr{F}_{1,2} < 0$. On the other hand, the pair 1,4 is connected by two paths of length 3 and $\mathscr{F}_{1,4} < 0$.

Finally, in Table 15.1 we illustrate the results obtained with $\mathscr{F}_{rs}$ for predicting the conductance in napththalene and azulene and compare them with the experimental values of the conductance and with those obtained by using $G^0_{rs}(E_F = 0)$..

As can be seen $\mathscr{F}_{rs}$ not only predicts correctly the sites where QI appears but also the correct trend of conductance for those sites for which there is a current flow. Notice that although in naphthalene there is a clear trend between conductance and the interatomic distance, which has been previously reported by Taniguchi et al. [27], such trend does not exist for azulene. In fact, for azulene the smallest conductance is observed for the pair 5,7-, which is significantly closer to each other than the pair 2,6-, which displays a significantly higher conductance. The function $\mathscr{F}_{rs}$ reproduces very well this trend, which is not accounted for by the zeroth Green's function $G^0_{rs}(E_F = 0)$.

TABLE 15.1 Results Obtained with $\mathscr{F}_{rs}$ for Predicting the Conductance in Napththalene and Azulene and Compare the Experimental Values Using $G^0_{rs}(E_F = 0)$

Substituent	G_0 (exp.)[a,b]	$[G^0_{rs}(E_F = 0)]^2$	$\left[\mathscr{F}_{rs}\left(\gamma^2 = 0.1\right)\right]^2$
Naphthalene			
1,4	11.0	0.445	0.783
1,5	2.2	0.111	0.274
2,6	1.4	0.111	0.192
2,7	0.1	0	0
Azulene			
2,6	32	0.25	1.430
1,3	32[c]	0.25	0.934
4,7	8	0.25	0.720
5,7	2	0	0.025

[a]Taniguchi et al. [27].

[b]Xia et al. [38].

[c]For the pyridine-linked derivative the value of $G_0 = 9$ is reported in Ref. [38].

15.7 EPILOGUE

At the very dawn of the XXI century, in 2002, Betowski et al. used post-Hartree-Fock calculations with configuration interaction (CI) to study triplet excitation energies of a series of polycyclic aromatic hydrocarbons (PAHs). The study uses a combination of sophisticated *ab initio* techniques with basis sets CIS/6–311G(d,p), CISD/3–21G and UHF-RHF/6–311G(d,p). The paper called my attention because of the calculation time. For perylene, $C_{20}H_{12}$, the calculation took 14 hours when using CIS/6–311G(d,p) on a Cray C94 supercomputer. The time is extended to 24 days if the CISD/3–21G is used. This prompted me to paraphrase the advice that Pliny the Younger gave to his friend Cornelius Rufus [49]: '*I counsel you in that ample and thriving retreat of yours, to hand the repeating and boring calculations over to the computer, and to devote yourself to the study of the mathematical and theoretical aspects of chemistry so as to derive from it something totally your own.*' In Ref. [50], we used the simple HMO-GT approach to study the same problem, which at the end of the day was to understand the phototoxicity of these PAHs. The correlation coefficient obtained for the phototoxicity of PAHs based on the HMO-GT is 0.968, which is very much comparable with that obtained using the CIS/6–311G(d,p) basis (0.978). The HMO-GT method gives much better results than the semiempirical methods such as AM1 (0.899) and PM3

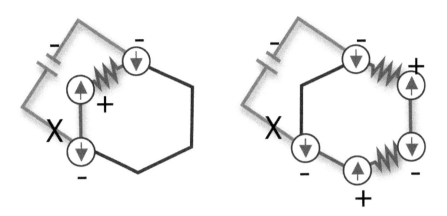

FIGURE 15.5   Illustration of the QI in 1,3-benzene due to the lack of any odd-length path connecting these two atoms. The current flow is impeded in either of the two existing paths that connect the atoms 1 and 3 due to the lack of polarity alternation in the circuit, such that no transmission is possible between these two atoms.

(0.904). It is not needed to say that the calculation time using HMO-GT is just a few seconds using any laptop commercially available today.

The reason for telling this anecdote is to recall the necessity of using an Occam Razor in deciding the complexity of the methods to be used for tackling chemical problems. In the dawn of the XXI century it seems to many that employing brute force by using much faster computers will solve all the problems in modern society. The recently coined and fashionable term "Big Data" is the main paradigm of this trend. But sometimes, simple, mathematically elegant and computationally efficient approaches solve the problem in much easier, efficient and elegant way. Thus, in the dawn of the XXI century we should not get rid of the HMO-GT approach for solving increasingly complex theoretical chemistry problems. The method has passed all the challenging tests during its more than 85 years of existence.

ACKNOWLEDGMENTS

The author thanks Professor Roald Hoffmann for introducing him to this topic as well as for useful discussions. Dr Yuta Tsuji is thanked for the preparation of the Figure 15.2. Professor Michele Benzi is also thanked for useful discussions about the Cayley-Hamilton theorem. The Royal Society of London is thanked for a Wolfson Research Merit Award to the author.

KEYWORDS

- **Matrix functions**
- **HMO Hamiltonian**
- **Hückel molecular orbital theory**
- **quantum interference**

REFERENCES

1. Propertius, S. *Elegiarum Liber Secundus*, Poem XXV, 38.
2. Shirakawa, H., (2001). The discovery of polyacetylene film: the dawning of an era of conducting polymers (Nobel lecture). *Angewandte Chemie International Edition 40*, 2574–2580.
3. Kutzelnigg, W., (2007). What I like about Hückel theory. *Journal of Computational Chemistry, 28*, 25–34.
4. Ramakrishnan, R., (2012). A simple Hückel molecular orbital plotter. *Journal of Chemical Education. 90*, 132–133.
5. Bollobás, B., (1998). *Modern Graph Theory*. Springer-Verlag, New York.
6. Estrada E., (2013). Graph and network theory. In: *Mathematical Tools for Physicists*. M. Grinfeld Ed., Wiley. arXiv preprint arXiv:*1302*.4378. 2013.
7. Coulson, C. A. Proceedings of the Robert A. Welch Foundation Conferences on Chemical Research. XVI. Theoretical Chemistry, Robert A. Welch Foundation, Houston, p. 61.
8. Hückel, E. (1931) Quantentheoretischebeiträgezumbenzolproblem. Zeitschriftfür-Physik A Hadrons and Nuclei 70, 204-286.
9. Yates, K., (2012). *Hückel Molecular Orbital Theory*. Academic Press, New York.
10. Papaconstantopoulos, D. A., (1989). Tight-Binding Hamiltonians. *Alloy Phase Stability*. Springer Netherlands. 351–356.
11. Canadell, E., Doublet, M.-L., & Iung, C., (2012). *Orbital Approach to the Electronic Structure of Solids*, Oxford University Press, Oxford.
12. Powell, B. J., (2009). An introduction to effective low-energy Hamiltonians in condensed matter physics and chemistry. *arXiv preprint arXiv:0906.1640*.
13. Gutman, I., & Polansky, O. E., (2012). *Mathematical Concepts in Organic Chemistry*. Springer Science & Business Media, Berlin.
14. Gutman, I., (2005). Topology and stability of conjugated hydrocarbons. The dependence of total π-electron energy on molecular topology. *J. Serb. Chem. Soc., 70*, 441–456.
15. Higham, N. J., (2008). *Functions of Matrices: Theory and Computation*, SIAM, Philadelphia.
16. Estrada, E., & Hatano, N. (2007). Statistical-mechanical approach to subgraph centrality in complex networks. *Chemical Physics Letters, 439*, 247–251.
17. Estrada, E., (2012). *The Structure of Complex Networks: Theory and Applications*, Oxford University Press, Oxford.

18. Estrada, E., (2000). Characterization of 3D molecular structure. *Chemical Physics Letters, 319*, 713–718.
19. Estrada, E., & Rodriguez-Velazquez, J. A., (2005). Subgraph centrality in complex networks. *Physical Review E 71*, 056103.
20. Deng, H., Radenkovic, S., & Gutman, I., (2009). The Estrada index. *Applications of Graph Spectra, Math. Inst., Belgrade*, 123–140.
21. Gutman, I., Deng, H., & Radenković, S., (2011). The Estrada index: an updated survey. *Selected Topics on Applications of Graph Spectra, Math. Inst., Beograd*, 155–174.
22. Estrada, E., &Higham, D.J., (2010). Network properties revealed through matrix functions. SIAM Review 52, 696–714.
23. Ratner, M., (2013). A brief history of molecular electronics. *Nature Nanotechnology, 8*, 378–381.
24. Aradhya, S. V., & Venkataraman, L., (2013). Single-molecule junctions beyond electronic transport. *Nature Nanotechnology, 8*, 399–410.
25. Fracasso, D., Valkenier, H., Hummelen, J. C., Solomon, G. C., & Chiechi, R. C., (2011). Evidence for quantum interference in SAMs of arylethynylene thiolates in tunneling junctions with eutectic Ga–In (EGaIn) top-contacts. *Journal of the American Chemical Society, 133*, 9556–9563.
26. Guédon, C. M., Valkenier, H., Markussen, T., Thygesen, K. S., Hummelen, J. C., & van der Molen, S. J., (2012). Observation of quantum interference in molecular charge transport. *Nature Nanotechnology, 7*, 305–309.
27. Taniguchi, M., Tsutsui, M., Mogi, R., Sugawara, T., Tsuji, Y., Yoshizawa, K., & Kawai, T., (2011). Dependence of single-molecule conductance on molecule junction symmetry. *Journal of the American Chemical Society, 133*, 11426–11429.
28. Kiguchi, M., Nakamura, H., Takahashi, Y., Takahashi, T., & Ohto, T., (2010). Effect of anchoring group position on formation and conductance of a single disubstituted benzene molecule bridging Au electrodes: change of conductive molecular orbital and electron pathway. *The Journal of Physical Chemistry C, 114*, 22254–22261.
29. Aradhya, S. V., Meisner, J. S., Krikorian, M., Ahn, S., Parameswaran, R., Steigerwald, M. L., et al., (2012). Dissecting contact mechanics from quantum interference in single-molecule junctions of stilbene derivatives. *Nano Letters, 12*, 1643–1647.
30. Arroyo, C. R., Tarkuc, S., Frisenda, R., Seldenthuis, J. S., Woerde, C. H. M., Eelkema, R., et al., (2013). J. Signatures of quantum interference effects on charge transport through a single benzene ring. *Angewandte Chemie International Edition, 52*, 3152–3155.
31. Solomon, G. C., Herrmann, C., Hansen, T., Mujica, V., & Ratner, M. A., (2010). Exploring local currents in molecular junctions. *Nature Chemistry, 2*, 223–228.
32. Yoshizawa, K., (2012). An orbital rule for electron transport in molecules. *Accounts of Chemical Research, 45*, 1612–1621.
33. Ernzerhof, M., Zhuang, M., & Rocheleau, P., (2005). Side-chain effects in molecular electronic devices. *The Journal of Chemical Physics, 123*, 134704.
34. Fowler, P. W., Pickup, B. T., Todorova, T. Z., & Myrvold, W., (2009). Conduction in graphenes. *The Journal of Chemical Physics, 131*, 244110.
35. Markussen, T., Stadler, R., & Thygesen, K. S., (2010). The relation between structure and quantum interference in single molecule junctions. *Nano Letters, 10*, 4260–4265.
36. Stadler, R., & Markussen, T., (2011). Controlling the transmission line shape of molecular t-stubs and potential thermoelectric applications. *The Journal of Chemical Physics, 135*, 154109.

37. Nozaki, D., Sevinçli. H., Avdoshenko, S. M., Gutierrez, R., & Cuniberti, G., (2013). A parabolic model to control quantum interference in T-shaped molecular junctions. *Physical Chemistry Chemical Physics, 15*, 13951–13958.

38. Xia, J., Capozzi, B., Wei, S., Strange, M., Batra, A., Moreno, J. R., et al., (2014). Breakdown of interference rules in azulene, a nonalternant hydrocarbon. *Nano Letters, 14*, 2941–2945.

39. Tsuji, Y., & Hoffmann, R. (2014) Frontier orbital control of molecular conductance and its switching. AngewandteChemie International Edition 53, 4093-4097.

40. Movassagh, Ramis, Yuta Tsuji, & Roald Hoffmann, (2014). The Exact Form of the Green's Function of the Hückel (Tight Binding) Model. *arXiv preprint arXiv:1407.4780.*

41. Tsuji, Y, Hoffmann, R, Movassagh, R, & Datta, S., (2014). Quantum interference in polyenes. The *Journal of Chemical Physics, 141*, 224311.

42. Tsuji, Y., Hoffmann, R., Strange, M., & Solomon, G. C., (2016). Close relation between quantum interference in molecular conductance and diradical existence. *Proceedings of the National Academy of Sciences, 113*, E413–E419.

43. Barrios-Vargas, J. B., & Naumis, G. G., (2011). Doped graphene: the interplay between localization and frustration due to the underlying triangular symmetry. *Journal of Physics: Condensed Matter, 23*, 375501.

44. Naumis, G. G., (2007). Internal mobility edge in doped graphene: frustration in a renormalized lattice. *Physical Review B, 76*, 153403.

45. Naumis, G. G., & Barrio, R. A., (1994). Effects of frustration and localization of states in the Penrose lattice. *Physical Review B, 50*, 9834.

46. Yoshizawa, K., Tada, T., & Staykov, A., (2008). Orbital views of the electron transport in molecular devices. *Journal of the American Chemical Society, 130*, 9406–9413.

47. Solomon, G. C., Bergfield, J. P., Stafford, C. A., & Ratner, M. A., (2011). When "small" terms matter: Coupled interference features in the transport properties of cross-conjugated molecules. *Beilstein Journal of Nanotechnology, 2*, 862–871.

48. Brooks, B. P., (2006). The coefficients of the characteristic polynomial in terms of the eigenvalues and the elements of an $n \times n$ matrix. *Applied Mathematics Letters, 19*, 511–515.

49. The original quote is: '*I counsel you in that ample and thriving retreat of yours, to hand the degrading and abject care of your estates over to those in your employ, and to devote yourself to the study of letters so as to derive from it something totally your own.*' Pliny the Younger, *Epistles*, I, i, no. 3.

50. Estrada, E., & Patlewicz, G., (2004). On the usefulness of graph-theoretic descriptors in predicting theoretical parameters. Phototoxicity of polycyclic aromatic hydrocarbons (PAHs). *Croatica Chemica Acta, 77*, 203–211.

CHAPTER 16

EFFECT OF THE SOLVENT ON THE CONFORMATIONAL BEHAVIOR OF THE ALANINE DIPEPTIDE IN EXPLICIT SOLVENT SIMULATIONS

JAIME RUBIO-MARTINEZ[1] and JUAN JESUS PEREZ[2]

[1] Department of Physical Chemistry, Faculty of Chemistry, Universitat de Barcelona and the Institut de Recerca en Quimica Teorica i Computacional (IQTCUB). Mati iFranques 1–3, 08028 Barcelona, Spain

[2] Department of Chemical Engineering, Universitat Politecnica de Catalunya, Barcelona Tech. Av. Diagonal, 647, 08028, Barcelona, Spain

CONTENTS

16.1 INTRODUCTION

Peptides are important mediators in the regulation of many physiological processes, eliciting actions as hormones, neurotransmitters or immuno-modulators. Accordingly, a deep knowledge of their structural behavior is important to understand their biological activities and is pivotal for designing new peptide surrogates or peptidomimetics that may render new applications and open new frontiers [1].

In contrast to most proteins that exhibit a native structure in solution, the structure of short peptides in solution is normally not well defined and commonly associated to a random coil [2, 3]. However, there is an important difference between "not well-defined" and being "a random coil." The former is the result of an underlying dynamic equilibrium between diverse conformational states, whereas the latter assumes that each residue explores the entire region of the sterically allowed Ramachandran plot irrespective of the behavior and the steric properties of their nearest neighbors. So, despite the associated experimental blurred picture of the conformational profile of a peptide, compelling spectroscopic evidence shows that residues in the peptide chain are more structurally ordered than the random coil model predicts and that unfolded peptides exhibit local order [4]. This experimental evidence supports the theoretical picture that the structure of a peptide in solution is the average of an ensemble of states that can individually be characterized by a combined effort of spectroscopic and computational methods [5].

The application of computational methods to characterize the set of accessible conformations of a peptide at room temperature requires a thorough search of the energy surface. However, this is a difficult task due to its rugged nature with many low energy conformations and barriers that, becomes even more demanding when the peptide is in solution, where the available conformational states are the result of the balance between interactions between the atoms of the peptide (intramolecular) and interactions of the peptide atoms with molecules of the environment (intermolecular). The former dictate the intrinsic conformational features of the peptide that are associated to its amino acid sequence, whereas the latter modify this intrinsic conformational profile and the ensemble distribution.

Due to the number of atoms and degrees of freedom associated, peptides in solution are not amenable to be studied using quantum based methods. Fortunately, classical methods have proven in the last years to be a reliable alternative, despite force fields need to be continuously contrasted when they are used beyond the context where they were developed. The (S)-2-acetylamino-N-methylpropanamide, also known as the alanine dipeptide alanine dipeptide, represents the smallest peptide unit to be used as benchmark for testing force field parameters since the results can be easily being compared with *ab initio* calculations.

The conformational profile of the alanine dipeptide both *in vacuo* as in aqueous solution has been thoroughly investigated in the past using diverse methodologies both, computational and experimental. *In vacuo* both, force field and *ab initio* calculations identify the C_7^{eq} conformation as the lowest energy conformation, followed by the extended C_5 and the C_7^{axial} [6–10]. These results agree well with those provided by microwave spectroscopy that suggests that the alanine dipeptide in the gas phase populates both the C_7^{eq} and the C_5 conformations in a proportion 2:1 [11]. Interestingly, none of these studies identify either the right-handed α-helix or the polyproline II (P_{II}) conformations as low energy minima. This is an intriguing result since the former is often found in helical parts of globular proteins [12] and the latter is known to be the most abundant structure found in solution [13]. Consequently, the environment produced either by the rest of the protein or the solvent modifies the conformational profile of the molecule. On the other hand, studies on the conformational profile of the alanine dipeptide in water identify the P_{II}, C_5 and right-handed α-helix conformations as low energy minima [14]. Unfortunately, different methods and approximations provide a different rank ordering. Experimental studies provided by IR and Raman spectroscopy suggest that the molecule adopts the three conformations mentioned above, being the P_{II} the most abundant [13].

The aim of the present work is to compare the results obtained using classical mechanics with those reported from *ab initio* calculations in the description of the conformational profile of the alanine dipeptide *in vacuo* and in water. For this purpose, we selected two of the most recent versions of the AMBER force field: the ff99SB [15] and ff14SB [16]. Solvent effects on the conformational profile of the molecule were carried out using two different approaches: on the one hand, a generalized Born approach and on the other, a calculation with the peptide soaked in explicit water molecules. Classical calculations were carried out using molecular dynamics. In order to obtain a better sampling of the energy surface of the peptide, multiple molecular dynamics were carried out for each of the systems studied [17]. Thus, in the case of the *in vacuo* and generalized Born calculations four trajectories of 10 μs length were computed, whereas four trajectories of 3 μs length were performed for explicit solvent calculations. The results of these calculations are displayed by means of Ramanchandran plots. Accordingly, in order to classify the diverse conformations identified the partition of the conformational space shown in Figure 16.1 will be used in the present work.

16.2 METHODS

All the calculations reported in the present work were carried out with the AMBER14 software package [18] using the Sander and PMEMD programs in its CPU and GPU versions. The initial 3D structure of the alanine dipeptide was generated in its extended form using the Leap module of Ambertools16 [19].

The potential of mean force of the alanine dipeptide at 300K in water reported in this work was carried out using two methodologies. On the one hand, an explicit representation of solvent molecules and on the other, a generalized Born (GB) procedure that represents solvent effects implicitly. Implicit solvent models have the advantage of dramatically reducing the degrees of freedom that must be sampled by eliminating those associated with the solvent and must be considered an approximation of the more phys- ically rigorous explicit solvent representations.

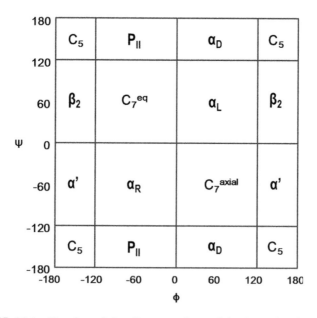

FIGURE 16.1   Notation of the diverse regions of the Ramachandran plot used in the present work.

For the molecular dynamics simulations in explicit solvent, a cubic box of TIP3P waters [20] was created with a minimum distance between any atom of the alanine dipeptide and the edge of the box of 15 Å where water molecules closer than 1.8 Å to any of the atoms were removed. Next, the system was optimized by means of 10,000 steps of the Steepest Descent algorithm to adapt the solvent-solute interface. The minimized structures were heated at 300 K at a constant rate of 30 K/10 ps. Then, 500 ps were performed at constant pressure to increase system density. Finally, four production molecular dynamic simulations of 3 μs length, each of them with different initial velocities, were done under the canonical ensemble, using Langevin [21] thermostat with a collision frequency of 2 ps^{-1} for temperature control. Long-range electrostatic energy was computed using the Particle Mesh Ewald summation method [22] with a cutoff of 10 Å for non-bonded interactions. The SHAKE algorithm [23] was used to constrain bonds involving hydrogen atoms to allow the use of a 2 fs integration step.

For the rest of the calculations, after the minimization step, four production molecular dynamics simulations of 10 μs length, each of them with different initial velocities, were done under the same conditions. However, the cutoff for the non-bonded interactions was increase to 12 Å to assure the inclusion of the maximum atomic distance in the alanine dipeptide in its extended structure. GB calculations were carried out in the present work by means of the neck GB model that adds a geometrically based molecular volume correction term accounting for interstitial high dielectrics to pairwise GB models [24].

Analysis of the results was carried out using one snapshot every other picosecond that yields 500 snapshots/ns, i.e., 6×10^6 snapshots for the solvated MD trajectories (a total of 12 μs length) and 20×10^6 snapshots for the non-solvated MD trajectories (a total of 40 μs length).

16.3   RESULTS AND DISCUSSION

16.3.1   CONFORMATIONAL PROFILE OF THE ALANINE DIPEPTIDE IN VACUO

The potentials of mean force of the alanine dipeptide at 300K *in vacuo*, computed with the AMBER force fields ff99SB [15] and ff14SB [16] are shown in Figure 16.2. Furthermore, low energy minima of the two surfaces are listed in Table 16.1. As can be seen, both force fields are capable to identify in increasing order of energy the C_7^{eq}, the extended C_5 and the C_7^{axial} conforma-tions as minima, with a small difference in the relative energy of the conform-ers between the two calculations. No other minima were identified in these calculations.

Table 16.1 also lists the results obtained from *ab initio* calculations retrieved from the literature [6–10], showing good agreement with classical mechanics results. Specifically, calculations at different levels including the Hartree-Fock method or including electron correlation within the Moller-Plesset methodology at different orders as well as coupled cluster calcula-tions identify these three conformations: C_7^{eq}, the extended C_5 and the C_7^{eq} as the lowest energy conformations with the same rank order as found in classical calculations. Besides, *ab initio* calculations also identify two addi-tional minima: the α_L and the $\beta2$ conformations at much higher energies.

Experimental studies on the conformational profile of the alanine dipep-tide *in vacuo* by means of microwave spectroscopy reveal that in the gas phase the molecule populates both the C_7^{eq} and the C_5 conformations with a population 2:1 approximately, as deduced from the 98 cm^{-1} energy differ-ence between of the two states [11].

Accordingly, experimental results on the conformational profile of the alanine dipeptide *in vacuo* fully agree with *ab initio* calculations as well as those produced with force fields ff99SB and ff14SB. Furthermore, other force fields also reproduce this profile [25].

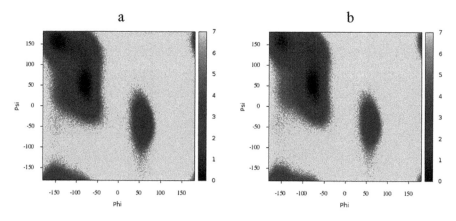

FIGURE 16.2 Potential of mean force of the alanine dipeptide in vacuo. Values of the energy are shown in a grey scale, where the low energy minima are shown in black, whereas high-energy values are shown in white. (a) Map computed with the ff99SB force field; (b) Map computed with the ff14SB force field.

TABLE 16.1 Dihedral Angles and Relative Energies of the Low Energy Minima Identified on the Potential of Mean Force Surface of the Alanine Dipeptide *in vacuo* using the Classical Force Fields ff99SB and ff14SB Compared with *ab initio* Calculations Taken from Refs. [9] and [10] at Different Levels Using a 6-31Gpd Basis Set

	ff99SB			ff14SB			RHF/6-31Gpd (ref.10)			MP2/6-31Gpd (ref.9)			cc-pVTZ//MP2(6-31Gpd) (ref.9)		
	φ	ψ	Energy/ kcal mole^{-1}	φ	ψ	Energy/ kcal mole^{-1}	φ	ψ	Energy/ kcal mole^{-1}	φ	ψ	Energy/ kcal mole^{-1}	φ	ψ	Energy/ kcal mole^{-1}
C_7^{eq}	-80	55	0.00	-80	55	0.00	-86	79	0.00	-82	81	0.00	-82	81	0.00
C_5	-150	155	0.14	-155	155	0.55	-157	159	0.40	-160	159	1.80	-160	159	1.47
C_7^{axial}	60	-45	1.54	60	-45	1.89	76	-55	2.82	76	-63	2.52	76	-63	2.50
$\beta2$							-131	22	4.90	-142	24	3.19	-142	24	3.25
α_L							67	30	4.76	63	35	4.38	53	-133	4.52

16.3.2 *CONFORMATIONAL PROFILE OF THE ALANINE DIPEPTIDE IN WATER*

Figure 16.3 shows the Ramachandran plots that represent the potential of mean force of the alanine dipeptide at 300K computed in the framework of the neck GB procedure using the ff99SB and ff14SB force fields. As anticipated, the most striking difference with the *in vacuo* calculations is the appearance of the polyproline II (P_{II}) and the right-handedα-helix conformations as low energy conformations in both maps.

Despite the qualitative analysis of the two maps suggests that the results of both calculations are similar, the quantitative analysis reveals important differences. Table 16.2 lists the minima identified in both maps. In the ff99SB map the rank order of energies is $C_5 < P_{II} < \alpha_R$. Specifically, the extended C_5 and the P_{II} conformations appear as nearly degenerated, whereas the right-handed α-helix appears around 0.4 kcal/mole above the lowest energy conformation. On the other hand, in the case of the ff14SB map the rank order of energies is $P_{II} < \alpha_R < C_5$. Specifically, the P_{II} appears as the lowest energy minimum followed the right-handed α-helix at around 0.4 kcal/mole above, being the extended C_5 conformation around 0.6 kcal/mole above the lowest energy minimum. At higher energies both maps show the left-handed α-helix conformation around 2.2 kcal/mole above the lowest energy minimum.

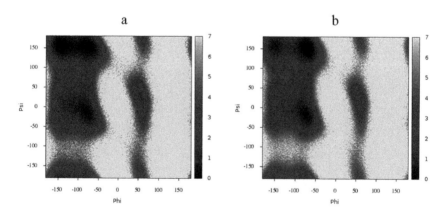

FIGURE 16.3 Potential of mean force of the alanine dipeptide in water computed with a generalized Born method. Values of the energy are shown in a grey scale, where the low energy minima are shown in black, whereas high-energy values are shown in white. (a) Map computed with the ff99SB force field; (b) Map computed with the ff14SB force field.

TABLE 16.2 Dihedral Angles and Relative Energies of the Low Energy Minima Identified on the Potential of Mean Force Surface of the Alanine Dipeptide in Water Computed by Means of a Generalized Born Procedure, Using the Classical Force Fields ff99SB and ff14SB Compared with *ab initio* Calculations Taken from Ref. [9] at Different Levels Using a 6-31Gdp Basis Set

	ff99SB			ff14SB			RHF/6-31Gdp			MP2/6-31Gdp			cc-pVTZ//MP2(6-31Gdp)		
	φ	ψ	Energy/ kcal mole^{-1}	φ	ψ	Energy/ kcal mole^{-1}	φ	ψ	Energy/ kcal mole^{-1}	φ	ψ	Energy/ kcal mole^{-1}	φ	ψ	Energy/ kcal mole^{-1}
C_5	-145	155	0.00	-150	155	0.59	-156	144	0.00	-86	90	0.00	-86	90	0.00
P_{II}	-75	150	0.03	-70	150	0.00	-64	142	0.36	-64	142	0.94	-64	142	0.17
α_R	-75	-20	0.41	-75	-20	0.39	-70	-32	0.36	-70	-32	0.47	-70	-32	0.08
α_L	50	30	1.77	55	25	2.21	59	41	1.90	59	41	1.12	59	41	1.27
α_D	55	175	3.02	60	175	3.48	60	-171	4.34	56	-145	4.05	56	-145	3.59
C_7^{eq}							–	–	–	-86	90	0.85	-86	90	0.92
$\beta2$							-146	27	1.18	-146	27	1.17	-146	27	1.27
C_7^{axial}							75	-54	4.11	75	-54	2.58	75	-54	2.69

Accordingly, the most important discrepancy between the two calculations is the destabilization of the C_5 observed in the ff14SB force field calculation.

In regard to the calculations performed in explicit solvent, Figure 16.4 shows the Ramachandran plots obtained in calculations using the ff99SB and ff14SB force fields and the list of low energy conformation shown in Table 16.3. As can be seen, in the case of the ff99SB the rank order of conformations is $P_{II} < C_5 < \alpha_R$ that represents a reordering of the C_5 and P_{II} conformations compared to the GB calculations, although both conformations can be

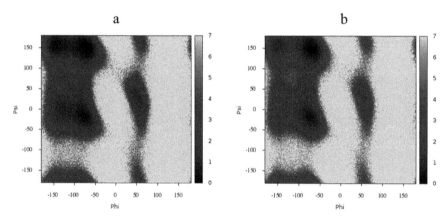

FIGURE 16.4   Potential of mean force of the alanine dipeptide in water computed with explicit solvent molecules. Values of the energy are shown in a grey scale, where the low energy minima are shown in black, whereas high-energy values are shown in white. (a) Map computed with the ff99SB force field; (b) Map computed with the ff14SB force field.

TABLE 16.3   Dihedral Angles and Relative Energies of the Low Energy Minima Identified on the Potential of Mean Force Surface of the Alanine Dipeptide Computed with the Molecule Soaked in a Bath of Explicit TIP3 Waters, Using the Classical Force Fields ff99SB and ff14SB

ff99SB				ff14SB		
	φ	ψ	Energy/kcal mole^{-1}	φ	ψ	Energy/kcal mole^{-1}
P_{II}	−75	150	0.00	−70	150	0.00
C_5	−145	155	0.03	−150	155	0.97
α_R	−80	−15	0.46	−75	−20	0.53
α'	−110	0	1.10	−140	5	1.96
α_L	50	30	1.20	55	25	1.71

considered as degenerated. On the other hand, the ff14SB calculations repro-duce the results of the GB calculations with the rank order of conformations being $P_{II} < \alpha_R < C_5$. Other structures are also sampled but at much higher energies.

Similar results can be obtained with other force fields of diverse lineages [14]. Thus, the same rank order of energies found with the ff99SB force field is observed using Charmm36 [26] and GROMOS 54a7 [27] force fields. Whereas, the same results as the ff14SB force field can be found using the OPLS-AA/L force field [28].

In addition to the traditional ff99SB and ff14SB force fields, we have also tested the performance of the recently developed ff14ipq force field that uses non-polarizable point charges to implicitly represent the energy of polariza-tion for systems in pure water, representing a force field of a new generation [29]. This force field is designed to be used in computations with explicit solvent. The potential of mean force of the alanine dipeptide computed with the ff14ipq force field is shown in Figure 16.5 and the list of low energy conformations is shown in Table 16.4. These calculations identify the C_7^{eq} conformation as low energy minimum in solution. The energy rank order of

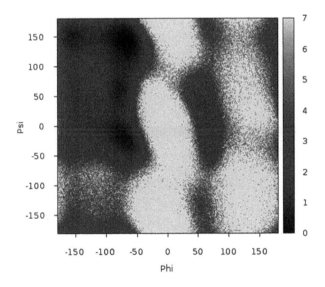

FIGURE 16.5 Potential of mean force of the alanine dipeptide in water computed with the ff14ipq force filed with explicit water molecules. Values of the energy are shown in a grey scale, where the low energy minima are shown in black, whereas high-energy values are shown in white.

TABLE 16.4 Dihedral Angles and Relative Energies of the Low Energy Minima Identified on the Potential of Mean Force Surface of the Alanine Dipeptide Computed with the Molecule Soaked in a Bath of Explicit TIP3 Waters Using the ff14ipq Force Field

	Ff14ipq		
	φ	ψ	Energy/ kcal mole^{-1}
P_{II}	−70	145	0.00
α_R	−70	−25	0.40
C_7^{eq}	−80	60	0.67
C_5	−155	150	0.68
α_L	80	50	0.79
$\beta2$	−150	50	1.64

the different conformations identified in these calculations is: $P_{II} < \alpha_R < C_7^{eq} < C_5$. Specifically, the P_{II} is the lowest energy conformation, followed by the right-handed α-helix 0.4 kcal/mole above. Then, the C_7^{eq} and the extended C_5 conformations appear as nearly degenerated, 0.7 kcal/mole above the lowest energy conformation. The next conformations are the left-handed α-helix found at 0.8 kcal/mole and the $\beta2$ conformation at 1.6 kcal/mole. The most important difference of these results in regard to the previously discussed is the emergence of the C_7^{eq} conformation also found in ab initio calculations, as discussed below.

Ab initio calculations of the alanine dipeptide in water using a polarizable continuum model [30] identify the extended C_5 as the lowest energy conformation, followed by the right-handed α-helix and the P_{II} conformations both exhibiting similar energies (Table 16.2). However, different results are obtained depending of the level of calculation and basis set used [9, 10]. Thus, at the Hartree-Fock level and using a 6-31Gpd basis set the P_{II} and the right-handed α-helix conformations are degenerated at 0.4 kcal/mole. In contrast, when electron correlation is included at the MP2 level, the P_{II} conformation appears 0.3 kcal/mole above the right-handed α-helix and in the coupled-cluster calculations this difference is of only 0.1 kcal/mole. The most striking difference with force field calculations is the appearance of a C_7^{eq} conformation at 0.25 kcal/mole, more stable than the P_{II} conformation at the MP2 and 0.85 kcal/mole above the P_{II} conformation in coupled-cluster calculations. In addition, a low energy $\beta2$ and a left-handed α-helix conformation can also be identified as in the *in vacuo* calculations. At the Hartree-Fock the former is more stable than the latter; at the MP2 level this order is reversed and at the coupled-cluster level they are degenerated.

Experimental studies on the conformational profile of the alanine dipeptide in water using IR and Raman spectroscopy suggest that the P_{II} is the preferred conformation with a 60% population, followed by the extended C_5 conformation with a 29% population and the right-handed α-helix with the remaining 11% population [31–33].

The potential of mean force of the alanine dipeptide in water from diverse calculations reveals important differences in comparison to the *in vacuo* maps. Specifically, these maps show a diminishing of the C_7^{eq} conformation along to the emerging of the P_{II} and the α_R conformations and the shifting of the C_7^{axial} conformation to the α_L. Another interesting feature is that the potential of mean force in water is much flatter than *in vacuo*. This might explain the discrepancies regarding the energy rank order of the low energy minima among the diverse theoretical calculations in the description of the conformational profile of the alanine dipeptide.

As mentioned above, diverse experimental results indicate that the P_{II} conformation is the most favorable in aqueous solution [31–33]. *Ab initio* calculations at different levels fail show this conformation as the lowest energy minimum, showing as the most favorable conformation the extended C_5. Estimation of the entropic contribution to the free energy disfavors the C_7^{eq} conformation, so that the energy rank order provided by these calculations is $C_5 < P_{II} < \alpha_R$ [10]. Accordingly, there is a discrepancy between *ab initio* calculations and available experimental results.

In regard to force field calculations, the results obtained using an implicit model of the solvent show good agreement with those obtained from explicit solvent calculations. Moreover, some of the force fields are able to reproduce the experimental results, while others fail to reproduce them due to an over stabilization of the α_R conformation.

16.4   CONCLUSIONS

The molecular dynamics calculations was carried out to compute the potential of mean force of the alanine dipeptide *in vacuo* and in water using the AMBER force fields ff99SB and ff14SB. Calculations *in vacuo* agree well with the experimental results available from microwave spectroscopy studies. In regard to the calculations in water, two models of the solvent were used: implicit and explicit. The results show good agreement between the two models. Moreover, the ff99SB provides results in agreement with the experimental data available from IR and Raman spectroscopy studies. In contrast, the ff14SB force field fails to give the right energy rank order. We have also studied the recently developed second-generation force field ff14ipq. Unfortunately, the force field fails to show the right rank order and furthermore, identifies the C_7^{eq} conformation, not observed experimentally. Further studies are required to clarify present discrepancies.

ACKNOWLEDGMENTS

The authors are deeply indebted to Prof. Ramon Carbó-Dorca for insightful discussions and friendship in all these years.

KEYWORDS

- Alanine dipeptide
- alanine in water
- peptide conformational profile
- solvent effects

REFERENCES

1. Perez, J. J., Corcho, F. J., & Rubio-Martinez, J., (2010). Design of Peptidomimetics. In *Burger's Medicinal Chemistry, Drug Discovery, and Development.* 7th Edition. Abraham, D. J., & Rotella, D. P., (Eds.), Wiley and Sons, New York, 205–247.
2. Shi, Z., Woody, R. W., & Kallenbach, N. R., (2002). Is polyproline II a major backbone conformation in unfolded proteins? *Adv. Protein Chem. 62*, 163–240.
3. Shi, Z., Olson, C. A., Rose, G. D., Baldwin, R. L., & Kallenbach, N. R., (2002). Polyproline II structure in a sequence of seven alanine residues. *Proc. Natl Acad. Sci. USA. 99*, 9190–9195.
4. Schweitzer-Stenner, R., (2012). Conformational propensities and residual structures in unfolded peptides and proteins. *Mol. BioSyst, 8*, 122–133.
5. Corcho, F., Salvatella, X., Canto, J., Giralt, E., & Perez, J. J., (2007). Structural analysis of substance P using molecular dynamics and NMR spectroscopy. *J. Pept. Sci. 13*, 728–741.
6. Head-Gordon, T., Head-Gordon, M., Frisch, M. J., Brooks, C. L., & Pople, J. A., (1991). Theoretical study of blocked glycine and alanine peptide analogues. *J. Am. Chem. Soc. 113*, 5989–5997.

7. Gould, I. R., Cornell, W. D., & Hillier, I. H., (1994). A quantum Mechanical Investigation of the Conformational Energetics of the Alanine and Glycine Dipeptides in the Gas Phase and in Aqueous Solution. *J. Am. Chem. Soc. 116*, 9250–9256.

8. Beachy, M. D., Chasman, D., Murphy, R. B., Halgren, T. A., & Friesner, R. A., (1997). Accurate ab initio quantum chemical determination of the relative energetics of peptide conformations and assessment of empirical force fields. *J. Am. Chem. Soc. 119*, 5908–5920.

9. Perczel, A., Farkas, D., Jakli, I., Topol, I. A., & Csizmadia. I. G., (2003). Peptide Models. XXXIII. Extrapolation of Low-Level Hartree–Fock Data of Peptide Conformation to Large Basis Set SCF, MP2, DFT, and CCSD(T) Results. The Ramachandran surface of alanine dipeptide computed at various levels of theory. *J. Comput. Chem. 24*, 1026–1042.

10. Wang, Z.-X., & Duan, Y., (2004). Solvation effects on alanine dipeptide: A MP2/cc-pVTZ//MP2/6-31G** Study of (φ, ψ) energy maps and conformers in the gas phase, ether, and water. *J. Comput. Chem. 25*, 1699–1716.

11. Cabezas, C., Varela, M., Cortijo, V., Jimenez, A. I., Pena, I., Daly, A. M., et al., (2013). The alanine model dipeptide Ac-Ala-NH2 exists as a mixture of Ceq7 and C5 conformers. *Phys. Chem. Chem. Phys. 15*, 2580–2585.

12. Novmöller, S., Zhou, T., & Ohlson, T., (2002). Conformations of amino acids in proteins. *Acta Crystallogr, Sect. D. 58*, 768–776.

13. Yung Sam Kim, Jianping Wang, & Robin, M., (2005). Hochstrasser. Two-Dimensional Infrared Spectroscopy of the Alanine Dipeptide in Aqueous Solution. *J. Phys. Chem. B, 109*, 7511–7521.

14. Tzanov, A. T., Cuendet, M. A., & Tuckerman, M. E., (2014). How Accurately Do Current Force Fields Predict Experimental Peptide Conformations? An Adiabatic Free Energy Dynamics Study. *J. Phys. Chem. B. 118*, 6539−6552.

15. Hornak, V., Abel, R., Okur, A., Strockbine, B., Roitberg, A., & Simmerling, C., (2006). Comparison of multiple amber force fields and development of improved protein backbone parameters. *Proteins: Struct. Funct. Bioinf. 65*, 712–725.

16. Maier, J. A., Martinez, C., Kasavajhala, K., Wickstrom, L., & Hauser, J. E., (2015). Simmering, C. f14SB: Improving the accuracy of protein side chain and backbone parameters from ff99SB. *J. Chem. Theory Comput, 11*, 3696–3713.

17. Perez, J. J., Tomas, M. S., & Rubio-Martinez, J., (2016). Assessment of the sampling performance of multiple-copy dynamics versus a unique trajectory. *J. Chem. Inf. Model.* DOI: 10.1021/acs.jcim.6b00347.

18. Case, D. A., Babin, V., Berryman, J., Betz, R., Cai, Q., Cerutti, D., et al., (2014). AMBER14, University of California, San Francisco.

19. Case, D. A., Betz, R. M., Cerutti, D. S., Cheatham, III, T. E., Darden, T. A., Duke, R. E., et al., (2016). AMBER16, University of California, San Francisco.

20. Jorgensen, W. L., Chandrasekhar, J., Madura, J. D., Impey, R. W., & Klein, M. L., (1983). Comparison of simple potential functions for simulating liquid water. *J. Chem. Phys. 79*, 926–935.

21. Uberuaga, B. P., Anghel, M., & Voter, A. F., (2004). Synchronization of trajectories in canonical molecular-dynamics simulations: observation, explanation, and exploitation. *J. Chem. Phys. 120*, 6363–6374.

22. Darden, T., York, D., & Pedersen, L., (1993). Particle Mesh Ewald: An N.Log(N) Method for Ewald Sums in Large Systems. *J. Chem. Phys., 98*, 10089–10092.

23. Ryckaert, J. P., Ciccotti, G., & Berendsen, H. J. C., (1977). Numerical Integration of the Cartesian Equations of Motion of a System with Constraints: Molecular Dynamics of n-alkanes, *J. Comput. Phys. 23*, 327–341.

24. Mongan, J., Simmerling, C., McCammon, J. A., Case, D. A., & Onufriev, A., (2007). Generalized Born Model with a Simple, Robust Molecular Volume Correction. *J. Chem. Theory Comput. 3*, 156–169.

25. Liu, Z., Ensing, B., & Moore, P. B., (2011). Quantitative Assessment of Force Fields on Both Low-Energy Conformational Basins and Transition-State Regions of the (φ-ψ) Space. *J. Chem. Theory Comput. 7*, 402–419.

26. Best, R. B., Zhu, X., Shim, J., Lopes, P. E., Mittal, J., Feig, M. D. Jr. et al., (2012). Optimization of the Additive CHARMM All-Atom Protein Force Field Targeting Improved Sampling of the Backbone φ, ψ and Side-Chain $\chi1$ and $\chi2$ Dihedral Angles. *J. Chem. Theory Comput. 8*, 3257−3273.

27. Schmid, N., Eichenberger, A. P., Choutko, A., Riniker, S., Winger, M., Mark, A. E., et al., (2011). Definition and testing of the GROMOS force-field Versions 54A7 and 54B7. *Eur. Biophys. J. 40*, 843−856.

28. Kaminski, G. A., Friesner, R. A., Tirado-Rives, J., & Jorgensen, W. L., (2001). Evaluation and reparametrization of the OPLS-AA force field for proteins via comparison with accurate quantum chemical calculations on peptides. *J. Phys. Chem. B. 105*, 6474−6487.

29. Cerutti, D. S., Swope, W. C., Rice, J. E., & Case, D. A., (2014). ff14ipq: A self-consistent force field for condensed-phase simulations of proteins. *J. Chem. Theory Comput. 10*, 4515−4534.

30. Cammi, R., Mennucci, B., & Tomasi, J., (2000). Fast evaluation of geometries and properties of excited molecules in solution: a Tamm-Dancoff model with application to 4-dimethylaminobenzonitrile. *J. Phys. Chem. A. 104*, 5631–5637.

31. Kim, Y. S., Wang, J., & Hochstrasser, R. M., (2005). Two-dimensional infrared spectroscopy of the alanine dipeptide in aqueous solution. *J. Phys. Chem. B. 109*, 7511–7521.

32. Grdadolnik, J., Grdadolnik, S. G., & Avbelj, F., (2008). Determination of conformational preferences of dipeptides using vibrational spectroscopy. *J. Phys. Chem. B. 112*, 2712–2718.

33. Grdadolnik, J., Mohacek-Grosev, V., Baldwin, R. L., & Avbelj, F., (2011). Populations of the three major backbone conformations in 19 amino acid dipeptides. *Proc. Natl. Acad. Sci. USA, 108*, 1794–1798.

CHAPTER 17

EXACT ENERGY-DENSITY RELATIONSHIPS FOR SUM OF SCREENED COULOMB POTENTIALS

K. D. SEN and S. MONDAL

School of Chemistry, University of Hyderabad, Hyderabad – 500046, India, E-mail: kds77@uohyd.ac.in

CONTENTS

ABSTRACT

We relate the eigenvalues, $E_{nl}(\lambda)$, corresponding to the sum of the spherical screened Coulomb potentials, $V_m^{Hu}(r) = -Z \sum_{i=1}^{m} \lambda_i \frac{e^{-\lambda_i r}}{1 - e^{-\lambda_i r}}$, $V_m^{Yu}(r) = -\frac{Z}{r} \sum_{i=1}^{m} e^{-\lambda_i r}$ with their one electron density, $\rho_{nl}(r)$ associated with a given $n\ell$-state. The latter is used in terms of the scaled density which is defined as $\eta_{n\ell}(r) = \frac{\rho_{n\ell}(r)}{r^{2\ell}}$. In particular, we derive simple relationships between $E_{nl}(\lambda)$ and the ratio of the second derivative of the scaled density at the origin to the scaled density at the origin, $\frac{\eta''(0)}{\eta(0)}$, for such potentials. Up to three terms in the sum of potentials have been considered with $m = 1$–3. The significance of such relationships are discussed along with the representative numerical tests.

17.1   INTRODUCTION

The screened Coulomb potentials (SCP) constitute a class of spherical model potentials useful in almost all branches of science. With reference to the simple Coulomb potential,

$$V^C(r) = -\frac{Z}{r} \tag{1}$$

two of the common SCP can be defined as the Hulthén [1] potential,

$$V^{Hu}(r) = -Z\lambda \frac{e^{-\lambda r}}{1 - e^{-\lambda r}} \tag{2}$$

and, the Yukawa [2] potential

$$V^{Yu}(r) = -Z \frac{e^{-\lambda r}}{r} \tag{3}$$

The quantities Z and λ, which describe the strength and range of the potentials, assume different physical significance depending upon the application. The behavior of the above potentials for small r (<1 a.u.) resembles that of the Coulomb potential. At large radial distances the screening effect is exhibited more pronouncedly resulting in a more rapid exponential decay. This short-range characteristic of the SCP lead to a finite number of bound states in their eigenspectra which can be obtained from the solutions of Schrödinger equation for the given potential

$$\left[-\frac{1}{2} \frac{d^2}{dr^2} + \frac{\ell(\ell+1)}{2r^2} + V(r) \right] \psi_{n\ell}(r) = E_{n\ell} \psi_{n\ell}(r) \tag{4}$$

While at small values of λ, the SCP has the characteristics of a long ranged potential, for large values of λ the potential becomes short ranged.

Except for the Coulomb potential, Eq. (1), and for the $\ell = 0$ states in case of Hulthén potential, Eq. (2), exact solutions of Eq. (4) are not known. With gradual increase in λ value, a given $n\ell$-state eventually is raised to the zero-energy level at a characteristic critical screening constant $\lambda_{n\ell}^{critical}$, inverse of which defines the scattering length. We should like to make a distinction between the apparently related phenomena of screening and confinement. For example, the spherically confined Coulomb potential defined by an impenetrable cavity [3] with radius R also leads to destabilization of a given $n\ell$-state leading to a critical radius, $R_{n\ell}^{critical}$, at which $E_{n\ell}(R_{n\ell}^{critical})=0$. However, the two phenomena are physically quite different, for example in their influence on the electron-electron interaction.

Most of the reported studies in the literature using the SCP focus on the calculation of energy values for which a wide range of theoretical methods have been developed [4]. There is a need to relate the calculated energy values, $E_{nl}(\lambda)$, with some quantity which is directly related to probability density. In this contribution we have used the scaled density $\eta_{n\ell}(r) = \frac{\rho_{n\ell}(r)}{r^{2\ell}}$ to derive a series of new exact relationships containing $E_{nl}(\lambda)$ corresponding to a set of the sum of SCP represented by

$$V_m^{Hu}(r) = -Z\sum_{i=1}^{m}\lambda_i\frac{e^{-\lambda_i r}}{1-e^{-\lambda_i r}} \tag{5}$$

$$V_m^{Yu}(r) = -\frac{Z}{r}\sum_{i=1}^{m}e^{-\lambda_i r} \tag{6}$$

where in $m = 1, 2$, and 3 are considered in each case. For $m = 1$, the potentials given by Eqs. (2) and (3) are realized. For simplicity, in Eqs. (5)–(6), the screening constants λ_i are taken as $\lambda_1 = \lambda$, $\lambda_2 = 2\lambda$ and $\lambda_3 = 3\lambda$, respectively. All our results in this work are presented in terms of the ratio of second derivative of scaled density at the origin to the scaled density at the origin denoted by $\frac{\eta''(0)}{\eta(0)}$.

The sum of Yukawa potentials finds applications in scattering theory [5] which has motivated us to consider a class of sum of screened Coulomb potentials defined in Eqs. (5)–(6).

17.2   DERIVATION OF EXACT ENERGY-DENSITY RELATIONS

17.2.1   SCALED DENSITY RATIO $\left[\dfrac{\eta''(0)}{\eta(0)}\right]_{n\ell}$ AND ENERGY E_{nl}

FOR A GENERAL SPHERICAL POTENTIAL

The first derivative of the scaled density at origin for any $n\ell$-state is related to the scaled density at the origin according to the Kato cusp [6, 7] condition given by

$$\left[\frac{\eta'(0)}{\eta(0)}\right]_{n\ell} = -\frac{2Z}{(\ell+1)} \tag{7}$$

Now, for a spherical potential of the general form

$$V(r) = -\frac{A}{r} + B + f(r) \tag{8}$$

where $f(r) \to 0$ as $r \to 0$ it has been shown [8] that

$$\left[\frac{\eta''(0)}{\eta(0)}\right]_{n\ell} = \frac{2}{2\ell+3}\left[\frac{A^2}{(\ell+1)^2}(4\ell+5) + 2(B - E_{n\ell})\right] \tag{9}$$

In the following Subsections 17.2.1–17.2.4, the energy-density relationships for the potentials defined in Eqs. (5)–(6) will be considered.

17.2.2   SCREENED COULOMB POTENTIALS OF HULTHÉN SERIES

17.2.2.1   Hulthén Potential

This potential is given by Eq. (1), (or Eq. (5) with $m = 1$),

$$V_{m=1}^{Hul}(r) = -Z\lambda\frac{1}{e^{+\lambda r}-1} = -Z\lambda\frac{1}{[1+\lambda r+\frac{(\lambda r)^2}{2}+.....-1]} = \frac{-Z}{r}\left(1+\frac{\lambda r}{2}+...\right)^{-1} \tag{10}$$

$$V_{m=1}^{Hu1}(r) = \frac{-Z}{r}(1 - \frac{\lambda r}{2} + \frac{(\lambda r)^2}{4}...) \Leftrightarrow \frac{-A}{r} + B + f(r) \qquad (11)$$

where $A = Z$ and $B = +\dfrac{\lambda Z}{2}$ \qquad (12)

$$\therefore \left[\frac{\eta''(0)}{\eta(0)}\right]_{n\ell}^{Hu1} = \frac{2}{2\ell+3}\left[\frac{Z^2}{(\ell+1)^2}(4\ell+5) + (Z\lambda - 2E_{n\ell}^{Hu1})\right] \qquad (13)$$

The Eq. (13) is an exact result which relates $E_{n\ell}^{Hu1}$ with $\left[\dfrac{\eta''(0)}{\eta(0)}\right]_{n\ell}^{Hu1}$ for Hulthén potential in terms of the parameters Z and λ in Eq. (2). In the remaining part of the Section 17.2.2, we shall extend Eq. (13) to the sum of Hulthén potential with $m = 2$–3.

17.2.2.2 Sum of Two Hulthén Potentials

Considering $m = 2$ in Eq. (5) and $\lambda_1 = \lambda$, and $\lambda_2 = 2\lambda$

$$V_{n=2}^{Hu2}(r) = -Z[\frac{\lambda}{e^{+\lambda r} - 1} + \frac{2\lambda}{e^{+2\lambda r} - 1}] \qquad (14)$$

Proceeding as in Eq. (10), we get

$$V_{m=2}^{Hu2}(r) = -Z\left[\frac{\dfrac{\lambda}{\{1 + \lambda r + \dfrac{(\lambda r)^2}{2} + ... - 1\}}} +\dfrac{2\lambda}{\{1 + 2\lambda r + \dfrac{(2\lambda r)^2}{2} + ... - 1\}}\right] \qquad (15)$$

$$V_{m=2}^{Hu2}(r) = \frac{-Z}{r}[(1 + \frac{\lambda r}{2} + ...)^{-1} + (1 + \lambda r + ...)^{-1} \qquad (16)$$

$$V_{m=2}^{Hu2}(r) = \frac{-Z}{r}[(1 - \frac{\lambda r}{2} + \frac{(\lambda r)^2}{4}...) + \{1 - \lambda r + (\lambda r)^2...\}] \qquad (17)$$

$$V_{m=2}^{Hu2}(r) = \frac{-Z}{r}[(2 - \frac{3\lambda r}{2} + \frac{5\lambda^2 r^2}{4}...)] \Leftrightarrow \frac{-A}{r} + B + f(r) \qquad (18)$$

Which gives $A = 2Z$ and $B = +\frac{3\lambda Z}{2}$ (19)

$$\therefore \left[\frac{\eta''(0)}{\eta(0)}\right]_{nl}^{Hu2} = \frac{2}{2\ell + 3}\left[\frac{4Z^2}{(\ell+1)^2}(4\ell + 5) + (3Z\lambda - 2E_{nl}^{Hu2})\right] \qquad (20)$$

17.2.2.3   Sum of Three Hulthén Potentials

Considering $m = 3$ in Eq. (5) and $\lambda_1 = \lambda$, $\lambda_2 = 2\lambda$ and $\lambda_3 = 3\lambda$

$$V_{m=3}^{Hu}(r) = -Z[\frac{\lambda}{e^{+\lambda r}-1} + \frac{2\lambda}{e^{+2\lambda r}-1} + \frac{3\lambda}{e^{+3\lambda r}-1}] \qquad (21)$$

$$V_{m=3}^{Hu}(r) = -Z[\frac{\lambda}{\{1 + \lambda r + \frac{(\lambda r)^2}{2} + ... - 1\}} + \frac{2\lambda}{\{1 + 2\lambda r + \frac{(2\lambda r)^2}{2} + ... - 1\}}$$
$$+ \frac{3\lambda}{\{1 + 3\lambda r + \frac{(3\lambda r)^2}{2} + ... - 1\}}] \qquad (22)$$

$$V_{m=3}^{Hu3}(r) = \frac{-Z}{r}[(1 + \frac{\lambda r}{2} + ...)^{-1} + (1 + \lambda r + ...)^{-1} + (1 + \frac{3\lambda r}{2} + ...)^{-1}] \qquad (23)$$

$$V_{m=3}^{Hu3}(r) = \frac{-Z}{r}[(1 - \frac{\lambda r}{2} + \frac{(\lambda r)^2}{4}...) + \{1 - \lambda r + (\lambda r)^2...\} + \{1 - \frac{3\lambda r}{2} + \frac{(3\lambda r)^2}{4}...\}] \qquad (24)$$

$$V_{m=3}^{Hu3}(r) = \frac{-Z}{r}[(3 - \frac{6\lambda r}{2} + \frac{14\lambda^2 r^2}{4}...)] \Leftrightarrow \frac{-A}{r} + B + f(r) \tag{25}$$

Which gives $A = 3Z$ and $B = +3\lambda Z$ (26)

$$\therefore \left[\frac{\eta''(0)}{\eta(0)}\right]_{n\ell}^{Hu3} = \frac{2}{2\ell+3}\left[\frac{9Z^2}{(\ell+1)^2}(4\ell+5) + (6Z\lambda - 2E_{n\ell}^{Hu3})\right] \tag{27}$$

In order to test the numerical validity of the energy-density relationships derived in Section 7.2.2, we have considered Eq. (13) corresponding to the Hulthén potential given by Eq. (2).

We have used the accurate Lagrange mesh method which has been well described in the literature [9]. Our results are presented in Table 17.1 where the calculated values of E_{nl}^{Hul} and $\left[\frac{\eta''(0)}{\eta(0)}\right]_{n\ell}^{Hul}$ for a set of screening constant have been presented for the 1s, 2p, 3d, and 4f states. The ratio, $\wedge$, of the directly calculated $\left[\frac{\eta''(0)}{\eta(0)}\right]_{n\ell}^{Hul}$ value to the estimated value obtained from Eq. (13) using the calculated E_{nl}^{Hul} is extremely close to unity which establishes the validity of Eq. (13). We have also carried out the numerical tests for the Eqs. (20) and (27), which confirm their validity with similar accuracy as in Table 17.1. However, these results are not reported in order to utilize the available space with more analytic results which follow.

The energy-density relationships derived above can be used in conjunction with the known exact and/or accurate results representing the corresponding eigenvalues. For example, the exact energy, E_{nl}^{Hul} for the $\ell = 0$ states is given by

$$\left.\begin{array}{l} E_{n;\ell=0}^{Hul} = -\frac{\lambda^2}{8n^2}\left[\frac{2Z}{\lambda} - n^2\right]^2 \\ \\ E_{n;\ell=0}^{Hul} = -\left[\frac{Z^2}{2n^2} + \frac{n^2\lambda^2}{8} - \frac{Z\lambda}{2}\right] \end{array}\right\} ; n = 1,2,3... \tag{28}$$

TABLE 17.1 Calculated Values of E_{nl}^{Hul}, and $\left[\dfrac{\eta''(0)}{\eta(0)}\right]_{nl}^{Hul}$, for 1s,2p,3d, and 4f States for $V_{m=1}^{Hu}(r)$, Eq. (2), Over a Set of λ*

nl	λ	E_{nl}^{Hul}	$\left[\dfrac{\eta''(0)}{\eta(0)}\right]_{n\ell}^{Hul}$	Λ	$\dfrac{1}{\Lambda}\left[\dfrac{\eta''(0)}{\eta(0)}\right]_{n\ell}^{Hul}$
1s	0.025	−0.48757813	4.000103965	4.0001042	0.99999995
1s	0.05	−0.47531250	4.000416459	4.0004167	0.99999995
1s	0.075	−0.46320313	4.000937283	4.0009375	0.99999995
1s	0.1	−0.45125000	4.001666436	4.0016667	0.99999994
2p	0.025	−0. 11276046	1.000208368	1.0002084	1
2p	0.05	−0.10104245	1.000833957	1.000834	1
2p	0.075	−0.08984775	1.001878198	1.0018782	1
2p	0.1	−0.07917944	1.003343547	1.0033436	1
3d	0.025	−0.04360305	0.444757293	0.4447573	0.99999999
3d	0.05	−0.03275318	0.445700227	0.4457002	0.99999999
3d	0.075	−0.02303070	0.447287381	0.4472874	0.99999999
3d	0.1	−0.01448423	0.449546537	0.4495465	0.99999999
4f	0.025	−0.01969109	0.250418231	0.2504183	0.99999986
4f	0.05	−0.01006196	0.251694172	0.2516942	0.99999986

*The last column displays the ratio $\left[\dfrac{\eta''(0)}{\eta(0)}\right]_{nl}^{Hul}$ over the estimate $\Lambda = \dfrac{2}{2\ell+3}\left[\dfrac{Z^2}{(\ell+1)^2}(4\ell+5)+2\left(\dfrac{Z\lambda}{2}-E_{nl}^{Hul}\right)\right]$ with $Z=1$ using directly calculated E_{nl}^{Hul}. All values are in a.u.

Which on substitution in Eq. (13) leads to

$$\left[\frac{\eta''(0)}{\eta(0)}\right]_{n,\ell=0}^{Hul} = \frac{2}{3}\left[5Z^2 + \frac{Z}{n^2} + \frac{n^2\lambda^2}{4}\right] \qquad (29)$$

The scaled density Eq. (29), like Eq. (28) for energy is exact for all ℓ of $\left[\dfrac{\eta''(0)}{\eta(0)}\right]_{n,\ell=0}^{Hul}$. However, when compared with E_{nl}^{Hul}, a different dependence on the parameters of the Hulthén [1] potential is observed. Further, for $\ell \neq 0$, it is possible to use, for example, the sixth-order perturbation expansion of E_{nl}^{Hul} due to Lai and Lin [10] given by

$$E_{n\ell}^{Hu1} \approx -\left(\frac{Z^2}{2n^2}\right) + \frac{1}{2}Z\lambda - \frac{1}{24}\lambda^2[3n^2 - \ell(\ell+1)] - \left(\frac{\lambda^4 n^2}{8Z^2}\right)[\frac{1}{24}n^2\ell(\ell+1)$$

$$-\frac{1}{40}\ell^2(\ell^2+1) + \frac{1}{120}\ell(\ell+1)] - \left(\frac{\lambda^6 n^4}{16Z^4}\right)[\frac{1}{192}n^4\ell(\ell+1) \qquad (30)$$

$$+\frac{29}{4320}n^2\ell^2(\ell+1)^2 + \frac{13}{2880}n^2\ell(\ell+1) - \frac{31}{4032}\ell^3(\ell+1)^3$$

$$+\frac{29}{6048}\ell^2(\ell+1)^2 - \frac{1}{504}\ell(\ell+1) + ...]$$

Substituting E_{nl}^{Hu1} from Eq. (30) into Eq. (13) leads to

$$\left[\frac{\eta''(0)}{\eta(0)}\right]_{n,\ell=0}^{Hu1} \approx \frac{2}{(2\ell+3)}\left[\begin{array}{l} \frac{Z^2}{(\ell+1)^2}(4\ell+5) + \left(\frac{Z^2}{n^2}\right) + \frac{1}{12}\lambda^2[3n^2 - \ell(\ell+1)] \\ +\left(\frac{\lambda^4 n^2}{4Z^2}\right)[\frac{1}{24}n^2\ell(\ell+1) - \frac{1}{40}\ell^2(\ell^2+1) + \frac{1}{120}\ell(\ell+1)] \\ +\left(\frac{\lambda^6 n^4}{8Z^4}\right)[\frac{1}{192}n^4\ell(\ell+1) + \frac{29}{4320}n^2\ell^2(\ell+1)^2 + \frac{13}{2880}n^2\ell(\ell+1) \\ -\frac{31}{4032}\ell^3(\ell+1)^3 + \frac{29}{6048}\ell^2(\ell+1)^2 - \frac{1}{504}\ell(\ell+1) + ...] \end{array}\right]$$

$$(31)$$

Eq. (31) provides accurate estimates of the calculated values of the ratio of scaled density to within ~1% for $n < 10$.

17.2.3 SCREENED COULOMB POTENTIALS OF YUKAWA SERIES

17.2.3.1 Yukawa Potential

Starting with the potential,

$$V_{m=1}^{Yu1}(r) = -Z\frac{e^{-\lambda r}}{r} \qquad (32)$$

$$V_{n=1}^{Yu1}(r) = -\frac{Z}{r}[1 - \lambda r + \frac{(\lambda r)^2}{2} + ...] \qquad (33)$$

$$V_{m=1}^{Yu1}(r) = \frac{-Z}{r} + \lambda Z - \frac{Z\lambda^2 r}{2}...) \Leftrightarrow \frac{-A}{r} + B + f(r) \qquad (34)$$

Which gives $A = Z$ and $B = +\lambda Z$ $\qquad\qquad (35)$

$$\therefore \left[\frac{\eta''(0)}{\eta(0)}\right]_{n\ell}^{Yu1} = \frac{2}{2\ell+3}\left[\frac{Z^2}{(\ell+1)^2}(4\ell+5) + 2(Z\lambda - E_{n\ell}^{Yu1})\right] \qquad (36)$$

Eq. (36) is an exact result which relates $E_{n\ell}^{Yu1}$ with $\left[\dfrac{\eta''(0)}{\eta(0)}\right]_{n\ell}^{Yu1}$ for Yukawa

potential defined by Z and λ in Eq. (3). In the remaining part of the Section 17.2.3, we shall develop similar results for sum of Yukawa potential with $m = 2$–3.

17.2.3.2   Sum of Two Yukawa Potentials

We consider a simplified example here with $m = 2$ in Eq. (6) and $\lambda_1 = \lambda$, and $\lambda_2 = 2\lambda$

$$V_{n=2}^{Yu2}(r) = -Z\left[\frac{e^{-\lambda r}}{r} + \frac{e^{-2\lambda r}}{r}\right] \qquad (37)$$

Proceeding as in Eq. (33), we get

$$V_{m=2}^{Yu2}(r) = \frac{-Z}{r}\left[(1 - \lambda r + \frac{(\lambda r)^2}{2} + ...) + (1 - 2\lambda r + \frac{(2\lambda r)^2}{2} + ...)\right] \qquad (38)$$

$$V_{m=2}^{Yu2}(r) = \frac{-Z}{r}(2 - 3\lambda r + \frac{5(\lambda r)^2}{2}...) \qquad (39)$$

$$V_{m=2}^{Yu2}(r) = \frac{-2Z}{r} + 3\lambda Z - \frac{5\lambda^2 r}{2}... \Leftrightarrow \frac{-A}{r} + B + f(r) \qquad (40)$$

Which gives $A = 2Z$ and $B = +3\lambda Z$ $\qquad\qquad (41)$

$$\therefore \left[\frac{\eta''(0)}{\eta(0)}\right]_{n\ell}^{Yu2} = \frac{2}{2\ell+3}\left[\frac{4Z^2}{(\ell+1)^2}(4\ell+5) + (6Z\lambda - 2E_{n\ell}^{Yu2})\right] \qquad (42)$$

17.2.3.3 Sum of Three Yukawa Potentials

Considering $m = 3$ in Eq. (6) and $\lambda_1 = \lambda$, $\lambda_2 = 2\lambda$ and $\lambda_3 = 3\lambda$,

$$V_{m=3}^{Yu3}(r) = -Z[\frac{e^{-\lambda r}}{r} + \frac{e^{-2\lambda r}}{r} + \frac{e^{-3\lambda r}}{r}] \tag{43}$$

$$V_{m=3}^{Yu3}(r) = \frac{-Z}{r}[(1 - \lambda r + \frac{(\lambda r)^2}{2} + ...) + (1 - 2\lambda r + \frac{(2\lambda r)^2}{2} + ...)$$
$$+ (1 - 3\lambda r + \frac{(3\lambda r)^2}{2} + ...)] \tag{44}$$

$$V_{m=3}^{Yu3}(r) = \frac{-Z}{r}(3 - 6\lambda r + \frac{14(\lambda r)^2}{2}...) \tag{45}$$

$$V_{m=3}^{Yu3}(r) = \frac{-3Z}{r} + 6Z\lambda - 7Z\lambda^2 r... \Leftrightarrow \frac{-A}{r} + B + f(r) \tag{46}$$

Which gives $A = 3Z$ and $B = +6\lambda Z$ $\qquad$ (47)

$$\therefore \left[\frac{\eta''(0)}{\eta(0)}\right]_{n\ell}^{Yu3} = \frac{2}{2\ell + 3}\left[\frac{9Z^2}{(\ell+1)^2}(4\ell + 5) + (12\lambda Z - 2E_{n\ell}^{Yu3})\right] \tag{48}$$

In order to test the numerical validity of the energy-density relationships derived in Section 17.2.3, we consider Eq. (36) corresponding to the Yukawa potential given by Eq. (3). Our results are presented in Table 17.2, where the calculated values of E_{nl}^{Yul} and $\left[\frac{\eta''(0)}{\eta(0)}\right]_{n\ell}^{Yul}$ for a set of screening constant have been presented for the 1s,2p,3d, and 4f states. The ratio of the directly calculated $\left[\frac{\eta''(0)}{\eta(0)}\right]_{n\ell}^{Yul}$ value to the estimated value obtained from Eq. (36) using the calculated E_{nl}^{Yul} is extremely close to unity, which establishes the validity of Eq. (36). We have also carried out the numerical tests for the Eq. (42) and Eq. (48), which similarly confirm the validity of the sum of Yukawa

TABLE 17.2 Calculated Values of E_{ni}^{Yul}, and $\left[\dfrac{\eta''(0)}{\eta(0)}\right]_{n\ell}^{Yu1}$ for 1s,2p,3d, and 4f States for $V_{1\ m=}^{Yu}(r)$, Eq. (2), Over a Set of λ*

nl	λ	E_{nl}^{Hul}	$\left[\dfrac{\eta''(0)}{\eta(0)}\right]_{n\ell}^{Hu1}$	Λ	$\dfrac{1}{\Lambda}\left[\dfrac{\eta''(0)}{\eta(0)}\right]_{n\ell}^{Hu1}$
1s	0.025	−0.47546119	4.00061471	4.00061493	0.99999995
1s	0.05	−0.45181643	4.00242165	4.0024219	0.99999994
1s	0.075	−0.42902690	4.00536885	4.0053692	0.99999991
1s	0.1	−0.40705803	4.00941018	4.00941071	0.99999987
2p	0.025	−0.10149246	1.00119397	1.00119397	1
2p	0.05	−0.08074039	1.00459231	1.00459231	1
2p	0.075	−0.06248238	1.0099859	1.0099859	1
2p	0.1	−0.04653439	1.01722751	1.01722751	0.99999999
3d	0.025	−0.03357312	0.44616876	0.44616877	0.99999999
3d	0.05	−0.01691557	0.45093588	0.45093588	0.99999999
3d	0.075	−0.00504507	0.45843845	0.45843846	0.99999999
3d	0.1	0.00127833	0.4691108	0.46911079	1.00000001
4f	0.025	−0.01121821	0.25220806	0.25220809	0.99999986
4f	0.05	5.2394E-05	0.25831012	0.25831005	1.00000029

*The last column displays the ratio $\left[\dfrac{\eta''(0)}{\eta(0)}\right]_{n\ell}^{Yu1}$ over the estimate $\Lambda=\dfrac{2}{2\ell+3}\left[\dfrac{Z^2}{(\ell+1)^2}(4\ell+5)+2(\dfrac{Z\lambda}{2}-E_{n\ell}^{Hul})\right]$ with Z = 1 using directly calculated. EniHul All values are in a.u.

potentials with $m = 2$–3. However, for the sake economy with space, we do not report the numerical data here.

17.3 DISCUSSION

17.3.1 DEPENDENCE OF ENERGY AND DENSITY-RATIO ON λ

We shall now present a few illustrative applications of the energy-density relationships derived above. Assuming $\ell = 0$ and $Z = 1$, Eqs. (13), (20), and (27) for Hulthén potentials reduce to

$$2E_{n\ell=0}^{Hul} + \frac{3}{2}\left[\frac{\eta''(0)}{\eta(0)}\right]_{n\ell=0}^{Hu1} = 5 + \lambda \qquad (49)$$

$$2E_{n\ell=0}^{Hu2} + \frac{3}{2}\left[\frac{\eta''(0)}{\eta(0)}\right]_{n\ell=0}^{Hu2} = 20 + 3\lambda \qquad (50)$$

$$2E_{n\ell=0}^{Hu3} + \frac{3}{2}\left[\frac{\eta''(0)}{\eta(0)}\right]_{n\ell=0}^{Hu3} = 45 + 6\lambda \qquad (51)$$

Eqs. (49)–(51) suggest that a *specific* weighted sum of energy and the density ratio as given by the L.H.S. for *all* the states ($n = 1,2,...$) corresponding to a given ℓ value ($\ell = 0$ in this case) is completely determined by the screening constant λ. For all other values $\ell = 1,2,3...$, identities similar to Eqs. (49)–(51) can be easily obtained from Eqs. (13), (20), and (27).

Similar results for the Yukawa potentials can be derived for $\ell = 0$ and $Z = 1$ from Eqs. (36), (42), and (48) as

$$2E_{n\ell=0}^{Yu1} + \frac{3}{2}\left[\frac{\eta''(0)}{\eta(0)}\right]_{n\ell=0}^{Yu1} = 5 + 2\lambda \qquad (52)$$

$$2E_{n\ell=0}^{Yu2} + \frac{3}{2}\left[\frac{\eta''(0)}{\eta(0)}\right]_{n\ell=0}^{Yu2} = 20 + 6\lambda \qquad (53)$$

$$2E_{n\ell=0}^{Yu3} + \frac{3}{2}\left[\frac{\eta''(0)}{\eta(0)}\right]_{n\ell=0}^{Yu3} = 45 + 12\lambda \qquad (54)$$

A characteristic simple dependence on λ for the specific weighted sum of energy and density ratio as given by the L.H.S. of Eqs. (52)–(54) is found to hold good for the sum of screened Coulomb potentials considered in this work.

17.3.2 DEPENDENCE OF DENSITY-RATIO ON CRITICAL SCREENING CONSTANT

The relationships derived in Section 17.2.3 can be used to obtain the exact value of the ratio of scaled density at the critical screening constant, $\lambda_{n\ell}^{critical}$. Thus, setting $E_{n\ell=0}^{Hu1} = 0, E_{n\ell=0}^{Hu2} = 0$, and $E_{n\ell=0}^{Hu3} = 0$ at their respective values of $\lambda_{n\ell}^{critical}$ in Eqs. (49)–(51),

$$\left[\frac{\eta''(0)}{\eta(0)}\right]_{n\ell=0}^{critical,Hu1}=\frac{2}{3}(5+\lambda_{n\ell=0}^{critical,Hu1})\tag{55}$$

$$\left[\frac{\eta''(0)}{\eta(0)}\right]_{n\ell=0}^{critical,Hu2}=\frac{2}{3}(20+3\lambda_{n\ell=0}^{critical,Hu2})\tag{56}$$

$$\left[\frac{\eta''(0)}{\eta(0)}\right]_{n\ell=0}^{critical,Hu3}=\frac{2}{3}(45+6\lambda_{n\ell=0}^{critical,Hu3})\tag{57}$$

We note here that using a perturbation expansion for the Hulthen potential, Patil [11] has obtained

$$\lambda_{n\ell}^{critical,Hu1}\approx\left[\frac{n}{\sqrt{2}}+0.1654\ell+\frac{0.0983\ell}{n}\right]^{-2}\tag{58}$$

Eq. (13) with $Z=1$ along with Eq. (58) can be used to obtain accurate values of $\left[\frac{\eta''(0)}{\eta(0)}\right]_{n\ell}^{critical,Hu1}$ for any arbitrary $n\ell$-state. For $\ell=0$, Eq. (58) reduces to the exact value of $\lambda_{n\ell}^{critical,Hu1}=\frac{2}{n^2}$

With $Z=1$ and $\ell=0$, Yukawa potentials Eqs. (36), (42), and (48) give

$$\left[\frac{\eta''(0)}{\eta(0)}\right]_{n\ell=0}^{critical,Yu1}=\frac{2}{3}(5+2\lambda_{n\ell=0}^{critical,Yu1})\tag{59}$$

$$\left[\frac{\eta''(0)}{\eta(0)}\right]_{n\ell=0}^{critical,Yu2}=\frac{2}{3}(20+6\lambda_{n\ell=0}^{critical,Yu2})\tag{60}$$

$$\left[\frac{\eta''(0)}{\eta(0)}\right]_{n\ell=0}^{critical,Yu3}=\frac{2}{3}(45+12\lambda_{n\ell=0}^{critical,Yu3})\tag{61}$$

For Yukawa potential with $\ell \neq 0$, Patil [11] has reported

$$\lambda_{n\ell}^{critical,Yu1} \approx \left(n\sqrt{\frac{\pi}{4}} + 0.0154014 + 0.22731\ell \right.$$
$$\left. + \frac{(0.0148326 + 0.16286\ell + 0.053914l^2)}{n} \right)^{-2} \quad (62)$$

The numerical tests suggest that the $\lambda_{n\ell}^{critical}$ values for all states $n \pounds 9$ are predicted by Eq. (62) to within 1% of the exact numerical values.

Similar approximate expression for $\lambda_{n\ell}^{critical}$ due to Green [12] gives

$$\left. \begin{array}{l} Z_l = Z_0(1 + \alpha l + \beta l^2) \\ S_l = S_0(1 + \gamma l + \delta l^2) \\ Z_{nl} = \left\{ \sqrt{Z_l} + [\frac{(n-l-1)}{S_l}] \right\}^2 \end{array} \right| \quad (63)$$

where $Z_0 = 0.839908$, $\alpha = 2.7359$, $\beta = 1.6242$, $\gamma = 0.019102$, $\delta = -0.001684$ & $S_0 = 1.1335$.

$$\lambda_{n\ell}^{critical,Yu1} = \frac{1}{Z_{nl}}$$

which can be substituted in Eq. (36) corresponding to the critical screening parameter $\lambda_{n\ell}^{critical,Hu1}$ given by

$$\left[\frac{\eta''(0)}{\eta(0)} \right]_{n\ell}^{Yu1} = \frac{2}{2\ell+3} \left[\frac{Z^2}{(\ell+1)^2}(4\ell+5) + 2Z\lambda_{n\ell}^{critical,Yu1} \right] \quad (64)$$

Thus, accurate estimates of the ratio of scaled density $\left[\frac{\eta''(0)}{\eta(0)} \right]_{n\ell}^{Yu1}$ can be obtained for any given $n\ell$-state from the knowledge of critical screening constant. Very recently, Roy [4] has reported a compilation of very accurate numerical estimates of the critical screening constants for screened Coulomb potentials for $n \leq 10$. For the Hulthen and Yukawa potentials the estimates of $\left[\dfrac{\eta''(0)}{\eta(0)}\right]_{n\ell}^{Hu1}$ and $\left[\dfrac{\eta''(0)}{\eta(0)}\right]_{n\ell}^{Yu1}$ obtained from Eqs. (55) and (57) using $\lambda_{n\ell}^{critical,Hu1}$ and $\lambda_{n\ell}^{critical,Yu1}$ reported by Roy [4] are in quantitative agreement with our the directly computed values using the wave functions. We note here that the form of the energy-density relationships reported in this paper remain the same in presence of an external confining potential due to an impenetrable spherical wall though of course the individual values of E_{nl} and $\left[\dfrac{\eta''(0)}{\eta(0)}\right]_{n\ell}$ would change.

17.3.3   BOUNDS ON THE DENSITY-RATIO FOR YUKAWA AND HULTHⲚN POTENTIALS

According to the comparison theorem [13], if two potentials follow the behavior $V_1(r) < V_2(r)$ at all radial distances, then their eigenvalues obey $E_1 < E_2$ for all states. For the simple Hulthén and Yukawa potentials we have [14]

$$
\left.
\begin{aligned}
E_{n\ell}^{Hu1} &< E_{n\ell}^{Yu1} \\
E_{n\ell}^{Hu2} &< E_{n\ell}^{Yu2} \\
E_{n\ell}^{Hu3} &< E_{n\ell}^{Yu3}
\end{aligned}
\right]
\tag{65}
$$

Using Eqs. (49)–(54) in conjunction with Eq. (65), it can be shown that

$$
\left.
\begin{aligned}
\left[\frac{\eta''(0)}{\eta(0)}\right]_{n\ell=0}^{Yu1} - \left[\frac{\eta''(0)}{\eta(0)}\right]_{n\ell=0}^{Hu1} &< \frac{2\lambda}{3} \\
\left[\frac{\eta''(0)}{\eta(0)}\right]_{n\ell=0}^{Yu2} - \left[\frac{\eta''(0)}{\eta(0)}\right]_{n\ell=0}^{Hu2} &< 2\lambda \\
\left[\frac{\eta''(0)}{\eta(0)}\right]_{n\ell=0}^{Yu3} - \left[\frac{\eta''(0)}{\eta(0)}\right]_{n\ell=0}^{Hu3} &< 4\lambda
\end{aligned}
\right]
\tag{66}
$$

Similar bounds can be derived for any given $n\ell$-state using the energy-density relationships given by Eqs. (13), (20), (27), (36), (42), and (48).

17.4 SUMMARY

In this chapter, we have derived exact relationships which express the ratio of scaled density at origin, $\left[\dfrac{\eta''(0)}{\eta(0)}\right]$, corresponding to the sum of the spherical Coulomb potentials, $V_m^{Hu}(r)=-Z\displaystyle\sum_{i=1}^{m}\lambda_i\dfrac{e^{-\lambda_i r}}{1-e^{-\lambda_i r}}$, and $V_m^{Yu}(r)=-\dfrac{Z}{r}\displaystyle\sum_{i=1}^{m}e^{-\lambda_i r}$; $m \pounds 3$ in terms of their eigenvalue, $E_{nl}(\lambda)$, Using the energy-density relationships, it has been shown that for a given ℓ, all $n\ell$-states have a fixed weighted sum of energy and the density ratio which is determined similarly by the screening constant λ [Eqs. (49)–(54)]. At the critical screening constant $\lambda_{critical}$, where energy level becomes zero, the density ratio alone shows a common dependence on $\lambda_{critical}$ [Eqs. (55)–(57) and Eqs. (59)–(61)]. Finally, the application of comparison theorem for potentials leads to simple upper bound in terms of λ, for the difference between the density ratios corresponding to the Yukawa and Hulthén potentials [Eq. (66)].

ACKNOWLEDGMENTS

This paper is dedicated to Professor Ramon Carbó-Dorca in celebrating of his scientific contributions which continues to grow at his 75th and beyond. KDS is grateful to Daniel Baye for a copy of the Lagrange Mesh code. This work has been carried out under the scheme of CSIR Emeritus Scientist Fellowship awarded to KDS.

KEYWORDS

- comparison theorem
- energy density relationships
- Hulthén and Yukawa potential
- Kato-Steiner cusp
- Lagrange Mesh
- screened Coulomb potentials
- spherical Coulomb potentials

REFERENCES

1. Hulthén, L. (1942). Arkiv för matematik, astronomi och fysik., *28A*, 1–12.
2. Yukawa, H., (1935). *Proceedings of Mathematical Society of Japan, 17*, 48–57.
3. Michels, A., de Boer, J., & Bijl, A., (1935). *Physica 4*, 981–994.
4. Roy, A. K., (2016). *International Journal of Quantum Chemistry 116*, 953–960 (See Refs. 2–17 for numerical works).
5. Burke, P. J, & Joachain, C. J., (1995). Theory of electron-atom collisions: Part 1 (Eq. (42)), *Potential Scattering*. Plenum Press, NY, p. 37.
6. Kato, T., (1957). *Communications in Pure and Applied Mathematics 10*, 151–177.
7. Nagy, A., & Sen, K. D., (2001). *Journal of Chemical Physics 115*, 6300–6308.
8. Hall, R. H., Saad, N., Sen, K. D, & Hakan, C., (2009). *Physical Review A 80*, 032507.
9. Baye, D., (2015). *Physics Reports. 565*, 1–107.
10. Lai, C. S, & Lin, W. C., (1980). *Physics Letters A 78*, 335–337.
11. Patil, S. H., (1984). *Journal of Physics A 17*, 575–593.
12. Green, A. E. S., (1982). *Physical Review A 26*, 1759–1761.
13. Wang, X. R., (1992). *Physical Review A 46*, 7295–7296.
14. Hall, R. H., (1985). *Physical Review A 32*, 14–18.
15. Sen, K. D., Montgomery, Jr., H. E., & Pupyshev, V. I., (2009). *Advances in Quantum Chemistry, 57*, 25–78.
16. Sen, K. D, & Baye, D. (2008). *Physical Review E 78*, 026701.
17. Steiner, E., (1963). *Journal of Chemical Physics 39*, 2365–2366
18. Stubbins, C., (1993). *Physical Review A 48*, 220–227

CHAPTER 18

THREE-PARTICLE NON-BORN-OPPENHEIMER SYSTEMS

JACEK KARWOWSKI

Institute of Physics, Nicolaus Copernicus University, Toruń, Poland

CONTENTS

ABSTRACT

Bound systems of particles, when described without imposing the Born-Oppenheimer approximation, exhibit many interesting and unexpected features. Accustomed to bound systems composed of electrons and thousands times more massive nuclei, for which the Born-Oppenheimer model is highly accurate, we tend to forget about the non-intuitive and full of surprising phenomena the non-Born-Oppenheimer world. In this chapter, some properties of three-particle non-Born-Oppenheimer systems are described, in particular the dependence of the stability of three-particle Coulomb systems on the masses of their components and near-threshold effects such as the Efimov effect. Finally, an exactly solvable three-particle model is discussed to illustrate the transformation of a shapeless cloud of charge to a linear diatomic molecule while masses of two particles increase.

18.1 INTRODUCTION

The roots of the concepts of wholeness and non-separability may be found in the ancient ideas of Pythagoras and in more recent Leibniz's pre-established harmony. Wholeness is also a central notion of quantum mechanics. The non-separability of spatially separated systems is the essence of the quantum physics. Beautiful and deep poetic reflections of the wholeness have been formulated by Ramon Carbó-Dorca in his *Sis octets d'octets* [1]. Inspired by his poetry, I dedicate this brief review on several effects of non-separability relevant for quantum chemistry to Ramon on the occasion of his 75th birthday.

Theories of many-body systems, in particular quantum chemistry as well as molecular and atomic physics, are based on approximations which impose separability to non-separable multi-dimensional Schrödinger equations. The best known examples are the Hartree-Fock method and the Born-Oppenheimer (BO) approximation. In the resulting models, some features of non-separable effects related to interactions (in the examples mentioned above, the electron correlation and the non-adiabatic effects, respectively) are quenched by the enforced separability.

Atoms, molecules, mesoscopic systems, solids, are composed of electrons and thousands times more massive nuclei. Therefore, many properties of all these systems may be described, with a great accuracy, by the Born-Oppenheimer model, usually referred to as the BO approximation. In this model the external potential, created by a network of the nuclei, confines the motion of electrons. Thus, the only particles described by the rules of quantum mechanics are electrons and the antisymmetric wavefunction describes N electrons moving in a fixed nuclear field. The spatial distribution of the nuclei is, in principle, arbitrary. The energy, as a function of this distribution, defines the potential energy hypersurface. If the nuclei are located in the minima of this hypersurface, then their distribution defines the molecular shape. In this context a specific form of the potential generated by a nucleus is irrelevant. A broad review of this subject has been given some time ago by Andrae [2]. Recently, Besalú and Carbó-Dorca introduced an interesting idea of softened nuclear potential, where the nuclear charge density extends far beyond the limits of the nuclear radius [3].

Let us note that the fundamental chemical notions, like the potential energy surface or the molecular bond length, are hard to define beyond the BO model since a non-BO Hamiltonian of a free molecule is spherically symmetric. The transition between a shapeless structure of the bound system composed of two electrons and two positrons, and the hydrogen molecule with a specific bond length is not only a consequence of changing masses of the particles but also of modifying the way the molecule is described. Electronic, vibrational and rotational degrees of freedom of the BO model, in the non-BO description constitute non-separable components of a complete, non-separable, molecule.

During the last two decades the BO paradigm has been broken on several different motivations. First, due the development of computing facilities with large memories and high speed of operation, non-BO numerical calculations of molecular structure became feasible [4]. Second, the attosecond laser technology opens a way to both tracking and controlling the ultrafast chemical reactions – processes, which cannot be described within the BO paradigm [5, 6]. Third, the development of experimental techniques and mathematical modeling opened a way to studying non-BO near threshold phenomena such as the Efimov effect [7, 8] and the Borromean bounding [9, 10].

In this chapter some properties of three-particle non-BO systems are briefly reviewed. In particular, the dependence of the stability of three-particle Coulomb systems on the masses of their components and near-threshold effects such as the Efimov effect are described. At the end, the transformation between a shapeless cloud of charge to a linear diatomic molecule, while masses of two particles increase, is discussed using an exactly solvable three-particle model.

Hartree atomic units are used in this paper. The reduced Planck's constant $\hbar$, the elementary charge e, the electron mass m_e, and the Coulomb's force constant $k_e = 1/(4\pi\epsilon_0)$ are equal to 1. In particular, the charge of electron is equal to -1. The boldfaced symbols, as $\mathbf{r}$, $\mathbf{u}$ denote vectors and the standard ones, r, u – their lengths.

18.2   STABILITY OF THREE-PARTICLE SYSTEMS

There exists a large set of systems of three particles with charges $e = e_1 = -e_2 = -e_3 = \pm 1$ bound by the Coulomb force. Some of these systems have a stable ground state and some are unstable. For example, $Ps^- = e+ \, e^- \, e^-$ (the negative ion of positronium), $H_2^+ = e^- \, pp$ (the positive ion of the hydrogen molecule), $H^- = pe^- \, e^-$ (the negative ion of the hydrogen atom) are stable while $p\mu^- \, e^-$ (a system composed of proton, muon and electron) is unstable. In particular, every symmetric system ($m_2 = m_3$) is stable.

The dependence of the stability area on the masses of the component particles may be conveniently illustrated by the *stability triangle* [11]. Let us consider an equilateral triangle $A_1A_2A_3$ with unit sides and define a point within this triangle distant from the sides a_i, $i = 1, 2, 3$ by

$$\alpha_i = \frac{1/m_i}{1/m_1 + 1/m_2 + 1/m_3} \tag{1}$$

Obviously, we have $\alpha_1+\alpha_2+\alpha_3 = 1$. The area within the triangle may be divided to three parts: in two of them the system of three particles is unstable and in one, for which $\alpha_2 \sim \alpha_3$, the system is stable. The stability triangle is shown in Figure 18.1.

18.3   THE BORROMEAN SYSTEMS

Three topological circles which are linked in such a way that removing any ring results in two unlinked rings are known as the *Borromean rings*

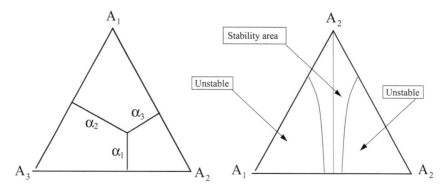

FIGURE 18.1   Stability triangle.

(Figure 18.2). The name "Borromean rings" comes from their use in the coat of arms of an Italian aristocratic Borromeo family. However, similar systems have been known since ancient times. The oldest known images, from about 6 to 7 centuries, were discovered on a stone pillar at Marundeeswarar Temple in Thiruvanmiyur, Chennai, Tamil Nadu (India) and on Norse image stones in Gottland. In atomic and molecular physics, structures where three-particles are bound but any of the two-particle subsystems is not, are known as *quantum Borromean systems*. Their stability depends on masses of the particles and on interactions.

Let us consider a system of three particles with masses $m_1 = m_2 = m$ and m_3 and charges $e = e_1 = e_2 = -e_3 = 1$ interacting by a short-range Yukawa potential

$$V = -\frac{e^{-\lambda r_{13}}}{r_{13}} - \frac{e^{-\lambda r_{23}}}{r_{23}} + \frac{e^{-\lambda r_{12}}}{r_{12}}, \tag{2}$$

where λ is known as the Debye screening parameter.

For a two-particle system interacting by Yukawa potential there is only a finite number of bound states and if λ is larger than a critical value then the two-particle system is unbound. The critical value of λ for binding two particles with reduced mass μ and electric charges ± 1 fulfill a simple relation: $\lambda^\mu_{critical} = \mu\lambda^1_{critical}$. The critical value for $\mu = 1$ is equal to $\lambda^1_{critical} = 1.191$. In particular, for e^+e^-, μ^+e^-, $\pi^+\mu^-$ the critical values of λ are 0.595, 1.185, 140.1, respectively.

FIGURE 18.2 The Borromean rings.

The range of λ for which a three-particle system is Borromean is referred to as the *Borromean window* [9, 10]. The Borromean windows for several systems interacting by the Yukawa potential are collected in Table 18.1. In general, a Borromean system exists if [10]

$$0 < \frac{m_3}{m_1} < 1.668. \tag{3}$$

For $m_3/m_1 = 1.668$ the Borromean window narrows to

$$0.744355797 < \lambda < 0.744355800. \tag{4}$$

18.4 EFIMOV EFFECT

When a slow particle scatters off a sufficiently small object, it cannot resolve its structure since the length of the de Broglie wave is very large and the effects of scattering do not depend on the details of the structure of this object. Scattering of two atoms with the relative position vector $\mathbf{r}$ and with opposite momenta $\pm \hbar k$, at $r \to \infty$, is described by the wavefunction

$$\psi(\mathbf{r}) = e^{ikz} + f_k(\theta)\frac{e^{ikr}}{r} \tag{5}$$

In the low energy limit ($k \to 0$), $e^{ikr} \to 1$ and the scattering becomes isotropic. The low-energy limit of the scattering amplitude $a = -\lim_{k \to 0} f_k(\theta)$ is referred to as the *scattering length*. Particles with short-range interactions and large scattering length have universal low-energy properties that do not depend on the details of their structure or their interactions at short distances.

TABLE 18.1 Borromean Window for Several Yukawa Three-Particle Systems [10]

Range of λ	System
$1.185 < \lambda < 1.333$	$\mu^+ \mu^+ e^-$
$140.112 < \lambda < 140.174$	$\pi^+ \pi^+ \mu^-$
$140.112 < \lambda < 140.115$	$\pi^+ \mu^- \mu^-$
$0.595306 < \lambda < 0.595379$	$e^+ e^- e^-$

If $a > 0$ they form a very weakly bound dimer, while for $a < 0$ their state nearly dissolves in the continuum.

In 1970, Efimov discovered that a three-body system with short range interactions has an infinite series of bound states if the two-body subsystems are exactly at the dissociation threshold [7]. If the two-body subsystems are unbound then three particles may be bound forming the Borromean state. Wavefunctions of all Efimov states have the same angular parts and their energies and sizes form the same geometric progression [7, 8]:

$$E_n = e^{-2\pi sn} E_0, \tag{6}$$

$$\langle r^2 \rangle_n = e^{-2\pi sn} \langle r^2 \rangle_0, \tag{7}$$

where s is a universal parameter. Thus,

$$\frac{E_{n+1}}{E_n} = \frac{\langle r^2 \rangle_{n+1}}{\langle r^2 \rangle_n} = e^{-2\pi s} \approx 22.7. \tag{8}$$

The experimental verification of the existence of the Efimov effect is technically difficult. Therefore the first experiments confirming the existence of this effect were performed nearly 40 years after its theoretical description [12–15]. If the particles have different masses then the geometric scaling factor changes. With large mass ratios it can be much smaller than 22.7 leading to a much denser spectrum of Efimov states. Then a long series of states may be observed [16].

18.5   MUON CATALYZED FUSION – SHAPE OF H_2^+

Let us consider a molecule composed of two heavy, positively-charged particles, with the charge equal to $+e$, and one light, negatively charged with the charge equal to $-e$ and mass equal to m. If the heavy particles are protons and the light one is an electron, then this system is known as the dihydrogen cation or hydrogen molecular ion. If the electron is replaced by a muon, we have the muonic hydrogen molecular ion. Several different equi-charged, isotopic, molecules may be obtained by the replacement of one or two protons by deuterons or tritons. Now, let us select the Hartree atomic units with $m = 1$ and solve the corresponding three-particle Schrödinger equation. We obtain a 'generic' molecule with all properties corresponding to the unit mass of the light particle. The dependence of these properties on the mass m can easily be derived from the scaling properties of the corresponding Schrödinger equation [17]. In the case of Coulombic systems, for example, as the dihydrogen cation, this can also be obtained by an analysis of the way the units of the relevant physical quantities depend on m. In particular, the units of energy, length and mass may be expressed, respectively, as α^2 cm, $\hbar/\alpha$ cm and m, where α is the fine structure constant and c is the velocity of light. By changing the mass of the light particle from 1 to $M > 1$ the energy increases and the length decreases by the factor M. Thus, the bond length, defined as the distance between the maxima of the density distributions of the heavy particles becomes M times shorter. Hence, comparing the usual, electronic, molecule with its muonic analog, we can see that the size of the muonic analog is ~ 207 times smaller and the binding energy ~ 207 times larger than in H_2+. Additionally, the 'effective mass' of the heavy particles (the nuclei) is M times (207 times in the case of the muonic molecule) smaller than in the electronic case. Therefore the nuclear density is more diffuse.

There were some hopes that this phenomenon may open a way to self-supporting muon catalyzed fusion with $d\, t\, \mu$ molecules [18, 19] as a consequence of the following chain of reactions:

$$(DT\mu)^+ \rightarrow {}^4He^{2+} + \mu^- + n + 18 MeV,$$
$$\mu^- + (D_2, DT, T_2) \rightarrow (T\mu) + ...,$$
$$(T\mu) + (D_2, DT, T_2) \rightarrow (DT\mu) + ...,$$

where μ, D, T, n denote, respectively, muon, deuteron, triton, neutron. Both shortening the bond length and changing the nuclear density distribution to more diffuse, strongly increases the overlap of the nuclear densities and, consequently, the probability of fusion. This probability is, indeed, quite large and fusion can easily be observed. Regretfully, the number of muons produced is still too small for the reaction being self-supported. For a chemistry-oriented review of this subject see [20].

18.6 THE EMERGENCE OF MOLECULAR SHAPE

In the BO model the preassumed distribution of the nuclei defines the geometrical shape and, thus, the symmetry group, of the molecule. The Hamiltonian transforms according to the totally symmetric representation of the pertinent symmetry group, e.g., D_{3h} or D_{6h} in the case of the minimum energy configuration of, respectively, ozone, or benzene molecules. The nuclear mass density, in the point nucleus model, are given by properly distributed Dirac delta functions and the electronic mass and electronic charge distributions, except for different units, coincide. The matter is entirely different in the non-BO model. The non-BO Hamiltonian is spherically symmetric. There are several kinds of particles forming the molecule and each of them has its own mass density distribution. The density of mass in a given state is equal to the expectation value of the mass density operator [21–24]. In particular, in the case of three particles with masses m_1, m_2, m_3 the one-particle density of mass operator is equal to

$$\hat{\rho}\left(r; r_1, r_2, r_3\right) = \sum_{i=1}^{3} m_i \, \delta\left(r - r_i\right). \tag{9}$$

If masses of particles 1 and 2 are equal m and they are non-distinguishable but particle 3 is different (e.g., two protons and an electron in the case of H_2^+ or two electrons and a proton in the case of H^-) then, in each state $\Psi\left(r_1, r_2, r_3\right)$, the total mass density is a superposition of two independent mass densities:

$$\rho_{12}(r) = m \left\langle \Psi \,|\, \delta\left(r - r_1\right) + \delta\left(r - r_2\right) \,|\, \Psi \right\rangle \tag{10}$$

and

$$\rho_3(r) = m_3 \left\langle \Psi \,|\, \delta\left(r - r_3\right) \,|\, \Psi \right\rangle. \tag{11}$$

The form of the function which describes the mass distribution depends on the choice of the reference point in the laboratory frame. This problem, appearing also in celestial mechanics, adds another difficulty to the way the transition between free and frozen nuclei models is performed [22, 23]. In this report we assume that the origin of the reference frame coincides with the center of mass of the system.

The transition between a shapeless spherical distribution, as it is for example in the case of the positive ion of positronium, composed of two positrons and an electron, and the hydrogen molecular ion with well defined bond length, may be studied in detail if we replace the real physical system by an exactly-solvable model which retains some important features of the real one, as far as the mass-dependence of the solutions is concerned [21–24]. There exist several separable models of three-particle system. In all these models for two pairs of particles (say $\{1 - 3\}$ and $\{2 - 3\}$) the

Coulomb interaction is replaced by the Hooke law potential. The form of the potential $V(r_{12})$ describing the interaction within the third pair, $\{1-2\}$, does not obstruct separability and may be arbitrary (as long as the degree of its singularity allows for the bound-state solutions). The resulting three-particle Schrödinger equation separates to three one particle equations. The first one describes the free motion of the center of mass. The second one is a spherical harmonic oscillator. The third equation describes a quasi-particle moving in a linear combination of V and a harmonic oscillator potential.

More specifically, the Hamiltonian of a separable system of three inter-acting particles may be expressed as

$$H(\mathbf{r}_1, \mathbf{r}_2, \mathbf{r}_3) = \sum_{i=1}^{3} \frac{\mathbf{p}_i^{\,2}}{2m_i} + W(\mathbf{r}_1, \mathbf{r}_2, \mathbf{r}_3) + V\left(\left|\mathbf{r}_{12}^-\right|\right), \tag{12}$$

where

$$W(\mathbf{r}_1, \mathbf{r}_2, \mathbf{r}_3) = \frac{\omega^2}{2}\left(\mu_{12}\, r_{12}^{-\,2} + \mu_{13}\, r_{13}^{-\,2} + \mu_{23}\, r_{23}^{-\,2}\right), \tag{13}$$

$$\mathbf{r}_{ij}^{\pm} = \mathbf{r}_i \pm \mathbf{r}_j, \tag{14}$$

$$\mu_{ij} = \frac{m_i m_j}{M_{123}}, \tag{15}$$

where $M_{123} = m_1 + m_2 + m_3$, and the remaining symbols have their usual meaning. In the coordinates

$$\mathbf{u}_0 = (m_1\mathbf{r}_1 + m_2\mathbf{r}_2 + m_3\mathbf{r}_3)/M_{123},$$

$$\mathbf{u}_1 = \mathbf{r}_{12}^-, \tag{16}$$

$$\mathbf{u}_2 = \mathbf{r}_{12}^+ - \mathbf{r}_3$$

the three-particle Hamiltonian (12) separates to three one-particle Hamiltonians:

$$H = h_0(\mathbf{u}_0) + h_1(\mathbf{u}_1) + h_2(\mathbf{u}_3). \tag{17}$$

The free motion of the center of mass is described by h_0. The second term, h_1, describes the motion of a particle with mass $\mu_1 = m_1 m_2/(m_1 + m_2)$ in the field of a spherically-symmetric potential

$$v_1(u_1) = \frac{\mu_1 \omega^2}{2} u_1{}^2 + V(u_1). \tag{18}$$

The last term, h_2, is the Hamiltonian of a spherical harmonic oscillator for a particle with mass $\mu_2 = (m_1 + m_2)m_3/M_{123}$ and the potential

$$v_2(u_2) = \frac{\mu_2 \omega^2}{2} u_2{}^2. \tag{19}$$

The theoretical description of this system may be simplified by selecting $m_1 = m_2 \equiv m$ and attaching the origin of the coordinate system to the center of mass, i.e., by setting $\mathbf{u}_0 = 0$. This implies $\mu_1 = m/2$ and

$$m(\mathbf{r}_1 + \mathbf{r}_2) + m_3\,\mathbf{r}_3 = 0. \tag{20}$$

Additionally we assume that

$$V\left(|\mathbf{r}_{\overline{12}}|\right) = \frac{\zeta}{u_1}. \tag{21}$$

Under this assumption the eigenvalue equation of $\mathbf{h}_1$ describes *harmonium*, a quasi-exactly solvable problem [25, 26] where the analytic solutions exist only for some specific sets of the coupling constants ω and ζ. However, if we expand $v_1(u_1)$ around its minimum at

$$u_1 = a_e = \left(\frac{2\zeta}{m\omega^2}\right)^{1/3} \tag{22}$$

to a power series and retain only two first terms of the expansion, then we get the effective potential

$$v_{\text{eff}}(u_1) = \frac{3m\omega^2}{4}\left[a_e{}^2 + (u_1 - a_e)^2\right] \tag{23}$$

for which the low-energy states correctly approximate the ones obtained for the exact potential. The radial Schrödinger equation with potential $v_{\text{eff}}(u_1)$ corresponds to a shifted harmonic oscillator and, for zero angular momentum states, it is exactly solvable [27]. In particular, the radial part of the ground-state wavefunction reads

$$\Phi(u_1) \sim \exp\left[-\frac{\sqrt{3}\,\omega\,m}{4}(u_1 - a_e)^2\right],\tag{24}$$

The eigenvalue equation of $h_2(u_2)$ is exactly solvable for all states [27] and the radial part of the ground state wavefunction is given by

$$\Xi(u_2) \sim \exp\left(-\frac{\omega\mu_2}{2}u_2{}^2\right).\tag{25}$$

For simplicity, the normalization factors have been omitted. Note, that in the case of the ground states, the angular parts of the one-particle wavefunctions are equal to Y_{00}, i.e., are constant. The wavefunction of the ground state of the three-particle system may be expressed as

$$\Psi(\mathbf{r}_1, \mathbf{r}_2, \mathbf{r}_3) = \Phi\left(|\mathbf{r}_{12}^-|\right) \Xi\left(|\mathbf{r}_{12}^+ - \mathbf{r}_3|\right).\tag{26}$$

In order to develop chemical intuitions related to the process of the emergence of the molecular structure from the spherically-symmetric non-BO objects let us trace the behavior of the mass density distribution in the ground state as a function of masses of the component particles. In this way we can see the relation between the degree of the localization of a particle and its mass – an effect crucial for the process of formation of the molecular shapes and for the validity of models based on the BO approximation. By the substitution of the explicit expression for the ground state wavefunction as given by Eqs. (24)–(26) to Eq. (10) we obtain the mass density distribution of the pair of identical particles, {1} and {2}. The density is spherically symmetric. Therefore its structure may be described by exploring the distribution along an axis of the coordinate system. By selecting the axis z, after some algebra (see Ref. [23] for details) we get

$$\rho_{12}(z) \sim \frac{1}{z}\int_0^\infty \sinh(m_z z r)\,\exp(-ar^2 + br - m_z z^2)\,r\,dr,\tag{27}$$

where $a = m\omega(\sqrt{3}/2 + m/\mu_2)$, $b = \sqrt{3}m\omega a_e$ and $m_z = 4m^2\,\omega/\mu_2$.

The integral can be easily evaluated, however the resulting expression is rather complicated and does not add to the understanding of the problem. The explicit expression may be found in Ref. [23]. Also the normalization factor is, for simplicity, omitted. Obviously, according to the definition, the density is normalized to $2m$. A plot of this distribution, for convenience of the presentation normalized to 1 rather than to $2m$, is presented in Figure 18.3. The displayed distributions correspond to $m_3 = 1$, $\omega = 1$ and $\zeta = 1$. Mass m changes from 0.5 to 10 in the upper panel and from 10 to 1900 in the lower panel. The peaks corresponding to particles $\{1\}$ and $\{2\}$ start to be clearly visible for $m > 4m_3$ and transform to high, sharply isolated maxima for large m. Let us note that in the correct normalization the distribution should be multiplied by $2m$. Thus, the maxima of the density for highest displayed peaks are about 10^6 – we can see that the distribution approaches the Dirac delta.

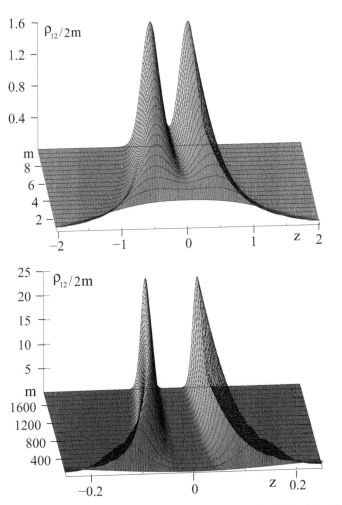

FIGURE 18.3 Distribution of normalized to 1 density of mass of particles $\{1\}$ and $\{2\}$, as a function of m, with $m_3 = 1$, $\omega = 1$ and $\zeta = 1$.

One point which needs some brief analysis is the dependence of the distribution on the mass of the particles. One can see, that the 'bond length' decreases with increasing mass, but much slower that it would apparently result from our analysis performed in the previous section for the hydrogen molecular ion. The bond length is, approximately, determined by the value of a_e [22]. As one can see, a_e is proportional to $m^{-1/3}$. Thus, in this model, also the bond length changes in this way. The correct scaling of a_e, proportional to m^{-1}, would be obtained if ω scaled with mass in the same way as energy, i.e., if ω was set proportional to m.

The distribution of the density of mass of particles {1} and {2} as a function of m_3 is shown in Figure 18.4. As one should expect, two maxima appear only if $m_3 < m$. For larger values of m_3 the system behaves as an atom rather than a molecule and the mass density distribution has only one maximum.

The expression for ρ_3 is much simpler. The substitution of wavefunction (26) to Eq. (11) and some elementary integration yields

$$\rho_3(z) \sim z^2 \exp(-m_z z^2 /4). \tag{28}$$

This distribution has only one maximum at $z_e = 2/\sqrt{m_z}$. As one can see, also in this case the maximum becomes higher and sharper (i.e., the particle becomes more localized) when its mass increases.

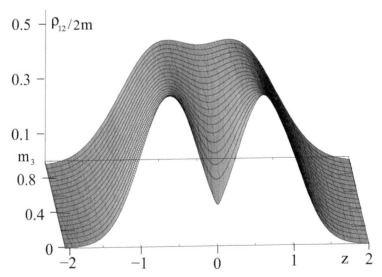

FIGURE 18.4   Distribution of normalized to 1 density of mass of particles {1} and {2}, as a function of m_3, with $m = 1$, $\omega = 1$ and $\zeta = 1$.

18.7 FINAL REMARK

Bound systems of particles, when described without imposing Born-Oppenheimer approximation, exhibit many interesting and unexpected features. Working with relatively strongly bound systems composed of electrons and thousands times more massive nuclei, for which the Born-Oppenheimer model is highly accurate, we frequently forget about the non-intuitive and full of surprising phenomena the non-Born-Oppenheimer world.

KEYWORDS

- **Born-Oppenheimer approximation**
- **Efimov effect**
- **Borromean systems**
- **molecular shape**

REFERENCES

1. Carbó-Dorca, R., (2013). *Sis octets d'octets*, Papers on Demand, Girona.
2. Andrae, D., (2000). Finite nuclear charge density distributions in electronic structure calculations for atoms and molecules, *Phys. Rep., 336*, 413–525.
3. Besalú, E., & Carbó-Dorca, R., (2013). Completely soft molecular electrostatic potentials and total density functions, *J. Math. Chem., 51*, 1772–1783.
4. Stanke, M., & Adamowicz, L., (2013). Molecular relativistic corrections determined in the framework where the Born-Oppenheimer approximation is not assumed, *J. Phys. Chem. A, 117*, 10129–10137.
5. Takatsuka, K., Yonehara, T., Hanasaki, K., & Arasaki, Y., (2015). *Chemical theory beyond the Born-Oppenheimer paradigm*. Non-adiabatic electronic and nuclear dynamics in chemical reactions, World Scientific, Singapore.
6. Yonehara, T., Hanasaki, K., & Takatsuka, K., (2012). Fundamental approaches to non-adiabaticity: toward a chemical theory beyond the Born-Oppenheimer paradigm, *Chem. Rev., 112*, 499–542.
7. Efimov, V., (1970). Energy levels arising from resonant two-body forces in a three-body system, *Phys. Lett. B, 33*, 563–564.
8. Braaten, E., & Hammer, H. W., (2006). Universality in few-body systems with large scattering length, *Phys. Rep., 428*, 259–390.
9. Goy, J., Richard, J. M., & Fleck, S., (1995). Weakly bound three-body systems with no bound subsystems, *Phys. Rev. A, 52*, 3511–3520.

10. Pawlak, M., Bylicki, M., & Mukherjee, P. K., (2014). On the limit of existence of Borromean binding in three-particle systems with screened Coulomb interactions, *J. Phys. B: At. Mol. Opt. Phys., 47*, 095701.

11. Kais, S., & Shi, Q., (2000). Quantum criticality and stability of three-body Coulomb systems, *Phys. Rev. A, 62*, 060502.

12. Kraemer, T., Mark, M., Waldburger, P., Danzl, I. G., Chin, C., Engeser, B., Lange, A. D., Pilch, K., Jaakkola, A., Nägerl, H. C., & Grimm, R., (2006). Evidence for Efimov quantum states in an ultracold gas of caesium atoms, *Nature, 440*, 315–318.

13. Knoop, S., Ferlaino, F., Mark, M., Berninger, M., Schöbel, H., Nägerl, H. C., & Grimm, R., (2009). Observation of an Efimov-like trimer resonance in ultracold atom-dimer scattering, *Nature Physics, 5*, 227–230.

14. Gross, N., Shotan, Z., Kokkelmans, S., & Khaykovich, L., (2009). Observation of universality in ultracold ^{7}Li three-body recombination, *Phys. Rev. Letters, 103*, 163202.

15. Zaccanti, M., Deissler, B., D'Errico, C., Fattori, M., Jona-Lasinio, M., Müller, S., Roati, G., Inguscio, M., & Modugno, G., (2009). Observation of an Efimov spectrum in an atomic system, *Nature Physics, 5*, 586–591.

16. Bo Huang, Sidorenkov, L. A., Grimm, R., Hutson, J. M. (2014). Observation of the second triatomic resonance in Efimov scenario, *Phys. Rev. Letters, 112*, 190401.

17. Karwowski, J., (2005). Influence of confinement on the properties of quantum systems, *J. Mol. Structure (Theochem), 727*, 1–7.

18. Frank, F. C., (1947). Hypothetical alternative energy sources for the 'second meson' events, *Nature, 160*, 525–527.

19. Breunlich, W. H., Kammel, P., Cohen, J. S., & Leon, M., (1989). Muon-catalyzed fusion, *Ann. Rev. Nucl. Part. Sci., 39*, 311–356.

20. Monkhorst, H. J., (2000). On the wanderings of a quantum chemist in the world of fusion, power, and politics, *Intern. J. Quantum Chem., 77*, 468–472.

21. Müller-Herold, U., (2006). On the emergence of molecular structure from atomic shape in the $1/r^2$ harmonium model, *J. Chem. Phys., 124*, 014105(1–5).

22. Mátyus, E., Hutter, J., Müller-Herold, U., & Reiher, M., (2011). On the emergence of molecular structure, *Phys. Rev. A, 83*, 052512(1–5).

23. Karwowski, J., (2013). Some remarks on the mass density distribution, *Croat. Chem. Acta, 86*, 531–539.

24. Karwowski, J., & Szewc, K., (2010). Separable N-particle Hookean models, *J. Phys. Conf. Series, 213*, 0122016(1–13).

25. Taut, M., (1993). Two electrons in an external oscillator potential, particular analytic solutions of a Coulomb correlation problem, *Phys. Rev. A, 48*, 3561–3566.

26. Karwowski, J., & Cyrnek, L., (2004). Harmonium, *Ann. Phys. (Leipzig), 13*, 181–193.

27. Davydov, A. S., (1965). *Quantum Mechanics*, Pergamon Press.

CHAPTER 19

ON THE USE OF QUANTUM MECHANICAL SOLVATION CONTINUUM MODELS IN DRUG DESIGN: IEF/PCM-MST HYDROPHOBIC DESCRIPTORS IN 3D-QSAR ANALYSIS OF AMPA INHIBITORS

TIZIANA GINEX,[1] ENRIC HERRERO,[2] ENRIC GIBERT,[2] and F. JAVIER LUQUE[1]

[1]*Department of Nutrition, Food Sciences and Gastronomy, Campus Torribera, School of Pharmacy and Institute of Biomedicine, University of Barcelona, Av. Prat de la Riba, 171, 08921, Santa Coloma de Gramenet, Spain, E-mail: tiziana.ginex@gmail.com (Tiziana Ginex), fjluque@ub.edu (F. Javier Luque)*

[2]*Pharmacelera S. L., Pl. Pau Vila 1, Edifici Palau de Mar, Sector 1, Nucli C, 08039 Barcelona, Spain*

CONTENTS

19.1 INTRODUCTION

Hydrophobicity is a key physicochemical descriptor to understand the bio-logical profile of (bio)organic compounds, as a broad variety of biochemical, pharmacological, and toxicological processes are ultimately related to the differential solubility of solutes in aqueous and non-aqueous environments [1–3]. The hydrophobicity of a molecule (M) can be quantified by its partition coefficient (P) between water and an organic phase, as this equilibrium thermodynamic property measures the ratio of concentrations of the compound between two immiscible solvents (Eq. 1).

$$P = \frac{[M]_{organic}}{[M]_{water}} \tag{1}$$

The partition coefficient can be related to the transfer free energy ($\Delta G_{tr}^{o/w}$) of the solute between aqueous and organic phases, which in turn can be expressed in terms of the solvation free energy ($\Delta Gsol$) of the compound upon transfer from the gas phase to the condensed phase (Figure 19.1).

$$logP = -\frac{\Delta G_{tr}^{o/w}}{2.303\,R\,T} == \frac{\Delta G_{sol}^{water} - \Delta G_{sol}^{organic}}{2.303\,R\,T} \tag{2}$$

Among the diversity of organic solvents that have been used to measure the partition coefficient of (bio)organic compounds, such as oils, chloroform, or alkanes, octanol has played a fundamental role for applications in chemistry and biology [4–6]. Thus, in the sixties Hansch and Fujita [7, 8] explored a simple additivity scheme to estimate the partition coefficient between octanol and water ($logP_{oct}$) of new compounds. Since then, this parameter has been a cornerstone tool to rationalize the biological behavior of molecules, especially in the context of pharmaceutical research [9–14]. From a computational point of view, a plethora of empirical methods have been developed to estimate the $logP_{oct}$ from contributions determined by molecular fragments or atom types, generally relying on additivity schemes often supplemented by correction rules. A complete description of these

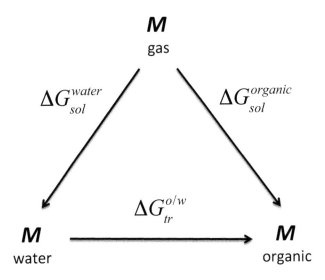

FIGURE 19.1 Thermodynamic cycle used to determine the free energy of transfer of a compound (M) between two immiscible solvents from the solvation free energies.

methods escapes from the purpose of this chapter and we limit ourselves to address the reader to pertinent reviews on this topic [16, 17] and to comparative studies of their performances [18, 19].

Besides conceptual simplicity and low expensiveness, one of the main advantages of fragment-based methods is the fact that the hydrophobicity of a compound, which is a property of the whole molecule, is empirically decomposed into atom/group contributions. This can be utilized to gain insight into the molecular determinants that govern the interactions between bioactive molecules and receptors, if one assumes that the forces that involve the preferential stabilization of a small compound in water and in an organic phase mimics the forces that regulate the affinity for a macromolecular target. Accordingly, the spatial distribution of the empirically determined lipophilicity of molecules, including the differential solvation effects in water and octanol, should provide guidelines about the molecular determinants of ligand binding.

The hydrophobic/hydrophilic complementarity between ligand and receptor is a key requirement for the binding affinity. Indeed, previous studies of target druggability have highlighted the relevance of pocket shape and hydrophobicity in drug binding [19–22]. Among the different contributions that modulate the binding affinity of drugs [23], desolvation has been identified to be largely responsible of the variation in maximal achievable binding energy (ΔG_{MAP}) for a drug-like molecule [24]. Specifically, Chen et al. have modeled the variation in ΔG_{MAP} for an optimized drug-like molecule as largely due to desolvation as noted in Eq. (3) [24],

$$\Delta G_{MAP} \approx \Delta G_{desolvation}^{target} + \Delta G_{desolvation}^{ligand} + \Delta G_{constant} \tag{3}$$

where $\Delta G_{desolvation}$ accounts for the free energy associated to the loss of water molecules from the target binding site and the ligand upon binding, while other binding free energy components are assumed to be constant ($\Delta G_{constant}$). This latter approximation is justified when one considers the binding of drug-like molecules similar in size and number of charges.

If one assumes the validity of the simple desolvation model $\Delta G_{desolvation} = -\gamma A$, where γ stands for the solvent surface tension, and A denotes the relevant solvent-accessible surface area, then the preceding equation can be rewritten as

$$\Delta G_{MAP} \approx -\gamma_{protein}(r)A_{nonpolar}^{target} - \gamma_{ligand}A_{nonpolar}^{ligand} + \Delta G_{constant} \tag{4}$$

Eq. (4) implicitly assumes that the contribution due to polar surfaces makes little contribution to the hydrophobic effect (15), taking advantage of the fact that the electrostatic interaction between polar groups is at large extent compensated by the desolvation term, thus making a small contribution to the maximal affinity. Moreover, a curvature-dependent model for the hydrophobic desolvation is taking into account for the protein.

Finally, normalization for molecular size of the ligands permits to transform the preceding expression into Eq. (5),

$$\Delta G_{MAP} \approx -\gamma_{protein}(r)A_{nonpolar}^{target} \frac{A_{druglike}^{target}}{A_{total}^{target}} + C \tag{5}$$

which includes a term to correct the total SASA of the protein pocket (A_{total}^{target}) by the SASA of the drug ($A_{druglike}^{target}$), and the constant C includes the ligand desolvation cost under the assumption that $A_{nonpolar}^{ligand}$ is a constant.

The druggability model developed by Chen et al. attempts to account for the effect of nonpolar desolvation in drug binding [23]. It intuitively assumes that favorable affinity is largely driven by the hydrophobic effect [26]. This agrees with the view that druggable binding sites appear to be closed and "greasy" cavities, whereas polar interactions are crucial for binding and selectivity [27–29].

19.2   EMPIRICAL APPROACHES TO THE LIPOPHILICITY POTENTIAL

The availability of empirically determined fragment contributions to the $logP_{oct}$ permits the calculation of the 3D hydrophobic distribution of a compound. This has led to the development of the molecular lipophilicity potential (MLP), which takes advantage of the 3D structural information of a molecule (i.e., its bioactive conformation) to determine the hydrophobic pattern implicated in the recognition to the biomolecular target [30–33]. To this end, the MLP combines fragment-based lipophilicity contributions with distance-dependent functions that permit the projection onto the solvent-accessible surface or the surrounding volume of the molecule (Eq. 6).

$$MLP_k = \sum_{i=1}^{N} F_i \cdot f(d_{ik}) \tag{6}$$

where MLP_K stands for the local value of the lipophilicity potential at a given point k, N is the number of atomic or group fragments within the molecule, F_i is the lipophilic fragmental contribution, and $f(d_{ik})$ is a distance function that depends on the separation between a given fragment (i) and any point on the molecular surface or volume (k).

Molecular fields derived from the MLP have been used in the study of a wide range of pharmaceutical applications, including the prediction of skin permeation and distribution of new chemical entities [34], for modeling of peptides and proteins [35, 36], and for structure-activity relationships studies [37–41].

The Hydropathic INTeractions (HINT) method represents another strategy to exploit empirically fragmental contributions to the molecular hydrophobicity for the study of interactions in biomolecules [42, 43]. HINT uses experimental data from solvent partitioning experiments between water and 1-octanol for quantitative scoring of hydropathic interactions. In this scheme, hydropathic attractions between species are related to solvent partitioning phenomena, as it is assumed that dissolution of a ligand in a mixed solvent system (such as water/1-octanol) involves the same fundamental processes and atom–atom interactions as biomolecular interactions within or between proteins and ligands.

The HINT model scores atom–atom interaction between biological molecules as noted in Eq. (7),

$$b_{ij} = a_i S_{iaj} S_j T_{ij} R_{ij} + r_{ij} \qquad (7)$$

where b_{ij} is the interaction score between atoms i and j, a_i and S_i are the hydrophobic constant and SASA of atom i, respectively, T_{ij} is a logical function with values of 1 and -1 depending on the character of the interacting polar atoms, R_{ij} is an exponential function of the distance between the interacting atoms, and finally r_{ij} accounts for a Lennard-Jones potential function. Extending the pairwise function b_{ij} to all pairs of atoms leads to the total interaction score of the interacting partners in the system.

In the HINT method, the hydrophobic atom constants are derived from a functional group primitive set, which is summed and modified by structure-dependent factors related to the connectivity between the group fragments. Since these atomic contributions encode the thermodynamic information from the experimental logP, they reveal the potential type and strength of the interactions formed by such an atom. With this assumption, HINT has been used for a range of diverse applications, including 3D QSAR studies and the analysis of biomolecular complex structures (for instance, see [44–47]).

19.3 QUANTUM MECHANICAL APPROACHES TO HYDROPHOBICITY IN DRUG DESIGN

The relevance of hydrophobicity in molecular recognition justifies the efforts conducted to develop quantum mechanical (QM)-based strategies for the calculation of lipophilic descriptors. Most efforts have relied on the exploitation of the electrostatic potential generated by a molecule on the molecular surface, as illustrated by the heuristic molecular lipophilic potential (HMLP).

The HMLP function is based on the analysis of the electrostatic potential at the molecular surface to provide a unified lipophilicity and hidrophilicity potential [48, 49]. The HMLP function is determined from the QM electrostatic potential, $V(r)$, calculated on the molecular surface, which is in turn used to estimate the molecular lipophilicity potential, $L(r)$, by comparing the local electron density with the electrostatic potential on the surrounding atoms using a screening function.

The $L(r)$ function is defined as

$$L(r) = V(r) \sum_{i \neq a} M_i\left(r, R_i, b_i\right) \tag{8}$$

where r is a point on the molecular surface pertaining to the surface of atom α, and $Mi(r,R_i,b_i)$ is a dimensionless screening function of atom i, which is determined as follows,

$$M_i(r, R_i, b_i) = \zeta \frac{b_i}{\|R_i - r\|^\gamma} \tag{9}$$

where R_i is the nuclear position of atom i, the parameter ζ is defined as the ratio of the γth power of the atomic radius r_o and the electrostatic potential descriptor b_o of the reference atom "o" $\zeta = r_o^\gamma / b_o$, and b_i is defined in Eq. (10).

$$b_i = \sum_{(k \in Si)} V(r_k) \Delta S_k \tag{10}$$

where $V(r_k)$ is the electrostatic potential in the center of the surface element S_k, ΔS_k is the area element on the exposed surface of atom i, and the summation takes place over all the elements that contribute to the exposed surface of atom i.

Finally, the lipophilicity index of atom α can be defined as noted in Eq. (11), which takes into account both the atomic exposed surface area of the atoms, and the electrostatic potential distribution on the atomic surface, as well as the screening effect exerted by the other atoms in the molecule.

$$I_\alpha = \sum_{k \in S_\alpha} L(r) \Delta S_k = \sum_{k \in S_\alpha} V(r) \Delta S_k \sum_{i \neq \alpha} \zeta \frac{b_i}{\|R_i - r\|^\gamma} \tag{11}$$

where recommended values of 1 and 2.5 for ζ and γ have been refined in later studies [50].

If $I_\alpha > 0$, atom α has the same polarity as its environment and is considered to be lipophilic, whereas if $I_\alpha < 0$, the polarity of atom α is opposite to its environment and is considered to be hydrophilic.

Finally, the sum of positive and negative lipophilicity indexes permits to derive molecular lipophilicity (L_M; Eq. 12) and hydrophilicity (H_M; Eq. 13) indexes, respectively, which have been used to determine the lipophilicity and hydrophilicity of the amino acid side chains [51], and in structure-based drug design [52, 53].

$$L_M = \sum_{a,la} {}^{>0^{la}} \tag{12}$$

$$H_M = \sum_{a,la} {}^{<0^{la}} \tag{13}$$

Alternatively, a straightforward strategy for the computation of lipophilicity/hydrophilicity patterns of molecules comes from QM self-consistent reaction field (SCRF) models, which rely on the description of the solvent as a continuum polarizable medium that reacts against the perturbing field created by the charge distribution of the solute [54–56]. The development of refined versions of QM-SCRF methods for a variety of solvents affords a direct procedure to determine the solvation free energy, and hence the partition coefficient (Eq. 2). Furthermore, this quantity can be decomposed into atomic contributions, which then permits to carry out studies about the molecular determinants of bioactivity. In this context, we limit our attention to the works conducted in the framework of the COSMO model, which is briefly outlined here, and the Miertus-Scrocco-Tomasi (MST) method, which will be presented in detail in the next section.

Klamt and coworkers developed the conductor-like screening model (COSMO) to describe molecules in solution (57), which gave rise later to the COSMO model for Real Solvents (COSMO-RS; [58, 59]). COSMO-RS exploits the screening charge densities derived from the COSMO calculation to estimate the chemical potential of the solute in solution via an appropriate mechanical statistical treatment (for a review, see Ref. [60]). For our purposes here, it suffices to remark the concept of the polarization charge density, s, generated on the molecular surface, which represents the polarization charge induced by the solute on a given element of the solute's surface. The one-dimensional histogram distribution of the s values for the whole set of surface elements enclosed in the molecular surface gives rise to the s-profile, which reflects a characteristic signature of the charge distribution of the solute, even though it does not contain information about the spatial distribution of the polarization charge densities.

The information contained in the s-profile may be used to compare the similarity between two molecules. Klamt and coworkers have examined this application in the COSMOsim3D method [61]. Since the standard s-profile lacks spatial 3D information, COSMOsim3D resorts to a grid of local s-profiles, which are generated by projecting the surface charge density, s, of each surface segment onto a regular 3D grid, so that each point of the grid has an associated local s-profile. In other words, this process leads to a four-dimensional histogram, which is defined by the three Cartesian dimensions of the grid point, and the local s-profile as the fourth dimension. By doing this process for two molecules, it is then possible to estimate the overall similarity between their chemical structures. Furthermore, these local s-profiles can be used to generate molecular interactions fields for 3D-QSAR studies [62].

19.4 THE QM CONTINUUM SOLVATION MST MODEL

The MST model [63, 64)] was originally implemented within the formalism of the polarizable continuum model (PCM) [65] and later adapted to the integral equation formalism (IEF-PCM; 66) scheme. This model offers the possibility to partition the overall hydrophobicity of a molecule into atomic contributions, which in turn can be further decomposed into "electrostatic" and "non-electrostatic" contributions [67–69].

In the MST method the solvation free energy (ΔG_{sol}) is calculated by adding three contributions (Eq. 14). The first one is the cavitation term (ΔG_{cav}), which is the work required for creating a cavity shaped to accommodate the solute in the solvent. The second component is the van der Waals term (ΔG_{vW}), which accounts for dispersion-repulsion interactions between solute and solvent. Finally, the third component is the electrostatic term (ΔG_{ele}), which measures the work needed to build up the solute charge distribution in the solvent.

$$\Delta G_{sol} = \Delta G_{ele} + \Delta G_{cav} + \Delta G_{vw} \tag{14}$$

Following the IEF-PCM formalism, the reaction field generated by the solvent consists of a set of imaginary charges located on the solute cavity. This strategy allows to partition ΔG_{ele} into atomic contributions (Eq. 15) following a perturbative description of the solute-solvent electrostatic interaction [70].

$$\Delta G_{ele} = \sum_{i=1}^{N} \Delta G_{ele,i} = \sum_{i=1}^{N} \sum_{\substack{j=1 \\ j \in i}}^{M} \langle \Psi^0 \left| \frac{1}{2} \frac{q_j^{sol}}{|r_j - r_i|} \right| \Psi^0 \rangle \tag{15}$$

where N is the total number of atoms, M is the total number of reaction field charges (q_j^{sol}) spread over the cavity surface, and Ψ^o is the wave function of the solute in the gas phase. The electrostatic contribution of atom i, $\Delta G_{ele,i}$, is calculated from the interaction of the molecular electrostatic potential generated by the whole molecule with the subset of reaction field charges pertaining to the solvent-exposed surface of atom i.

Both ΔG_{cav} and ΔG_{vW} are evaluated using expressions that depend linearly on the solvent-exposed surface of each atom in the molecule, and hence can be directly decomposed into atomic contributions (Eqs. 16 and 17).

$$\Delta G_{cav} = \sum_{i=1}^{N} \Delta G_{cav,i} = \sum_{i=1}^{N} \frac{S_i}{S_T} \Delta G_{P,i} \tag{16}$$

where $\Delta G_{P,i}$ is the cavitation free energy of atom i determined using Pierotti's formalism [71, 72], whose contribution is weighted by the contribution of the solvent-exposed surface (S_i) of atom i to the total surface (S_T).

$$\Delta G_{vW} = \sum_{i=1}^{N} \Delta G_{vW,i} = \sum_{i=1}^{N} \xi_i S_i \tag{17}$$

where ξ_i denotes the atomic surface tension of atom i, which is determined by fitting the experimental free energy of solvation [73].

The total solvation free energy is obtained as a sum of the three contributions for all atoms in the molecule (Eq. 18).

$$\Delta G_{sol} = \sum_{i=1}^{N} \Delta G_{sol,i} = \sum_{i=1}^{N} \left(\Delta G_{ele,i} + \Delta G_{cav,i} + \Delta G_{vW,i} \right) \tag{18}$$

Application of Eq. (5) to the solvation of a given compound in water and octanol leads to octanol/water transfer free energy $(\Delta G_{tr}^{o/w})$, which can be expressed as the sum of atomic contributions (Eq. 19).

$$\Delta G_{tr}^{o/w} = \sum_{i=1}^{N} \Delta G_{tr,i}^{o/w} = \sum_{i=1}^{N} \left(\Delta G_{ele,i}^{o/w} + \Delta G_{cav,i}^{o/w} + \Delta G_{vW,i}^{o/w} \right) \tag{19}$$

Finally, the hydrophobicity of a molecule, expressed as logarithm of the octanol/water partition coefficient (log$P$), can be related to the sum of the atomic contributions (Eqs. 20 and 21).

$$logP = \sum_{i=1}^{N} logP_i = \sum_{i=1}^{N} \left(logP_{ele,i} + logP_{cav,i} + logP_{vW,i} \right) \tag{20}$$

$$logP_X = \sum_{i=1}^{N} logP_{X,i} = \sum_{i=1}^{N} -\frac{\Delta G_{X,i}^{o/w}}{2.303RT} \quad (X = ele, \, cav, \, vW) \qquad (21)$$

Our studies have shown the dependence of the fragmental contributions to the $logP$ on the electronic and steric properties of the substituents attached to the core of the molecule, as well as the variation in the fragmental contributions due to conformational, tautomerism and hydrogen effects [67–69]. From a practical point of view, this opens the way to refine the fragmental contributions in simple additive-based empirical schemes.

Partitioning of the $logP$ into atomic contributions permits to perform the comparison of the hydrophobicity patterns of two molecules and quantify their hydrophobic resemblance. This can be achieved by defining measures of hydrophobic similarity, such as the dot product of the hydrophobicity dipoles [67, 68], or from similarity functions [74], which in turn can be used to derive structure-activity relationships [75].

Recently, the MST-based hydrophobic contributions have been utilized as physicochemical descriptors suitable for 3D-QSAR studies targeting both ligand affinity and target selectivity [76, 77]. By combining the electrostatic and non-electrostatic components of the octanol/water partition coefficient, the 3D-QSAR models were found to have a predictive accuracy that compared well with standard CoMFA [78] and CoMSIA [79] techniques. This can be attributed to the benefits obtained from a more accurate description of the molecular charge distribution for the bioactive conformation of the ligands, although this requires a larger computational cost compared with empirical or force field-based approaches. Finally, the approach also benefits from the usage of descriptors linked to a thermodynamic property that can be measured experimentally and subject to easy to interpretation, thus complementing the analysis performed with electrostatic and steric molecular fields.

19.5 3D-QSAR MST-BASED ANALYSIS OF AMPA INHIBITORS: A TEST CASE

To examine the suitability of the QM MST-based fragmental contributions to the hydrophobicity in structure-activity and drug design studies, we have revisited a set of 57 quinazoline-derived AMPA inhibitors that were examined by standard CoMFA techniques in previous studies [80]. For the sake of the comparison, the set of compounds was partitioned into training and test sets, comprising 52 and 5 compounds, mimicking the protocol adopted in the original study. Inspection of the chemical structures for quinazoline-based AMPA inhibitors permits to identify three main scaffolds (see Chart 19.1).

As shown in Chart 19.1, the first two subsets generally share a 4,5-dihydro-triazolo-[1,5-a]-quinoxaline scaffold (training: mol01–39; test: mol53–55) with loss of planarity related to the replacement of 4-oxogroup with 4-carboxylate for the second one (training: mol40–48; test: mol56). The third subset is characterized by a pyrazolo-[1,5-c]-quinoxaline scaffold(training: mol49–52; test: mol57).

The physicochemical diversity between the compounds is afforded by substitutions at position 2 and 4 (carboxylates and carboxylic esters), and position 8 (nitrogen-containing heterocycles), which confer variability in term of charge and steric volume. Finally, the binding affinity to the AMPA receptor, expressed as *pKi*, covers about 3 log units in the mM range, which makes this system more challenging for 3D-QSAR predictions.

19.5.1 COMPUTATIONAL DETAILS

The software PharmQSAR [81] was used to calculate all the 3D-QSAR models. For comparative purposes, a standard CoMFA model was also generated by combining atomic charges and radii from the TRIPOS force field [82], which were used to compute the electrostatic and steric fields, respectively. The molecules were aligned following the protocol reported in the original paper, and the same alignment was used for the 3D QSAR models derived from MST-based hydrophobic descriptors. Atomic hydrophobic contributions for each molecule were obtained from solvation calculations in water and octanol by using the B3LYP/6–31G(d) version of the IEF/PCM-MST solvation model [63, 64]. For 3D-QSAR models generation, all the aligned set was enclosed in a 1 Å-spaced lattice with boundaries set to 4 Å around the molecules, leading to a total number of 11000 grid points. The classical similarity index function (A^q) implemented in CoMSIA was applied to project the previously calculated atomic hydrophobic contributions (Eq. 22).

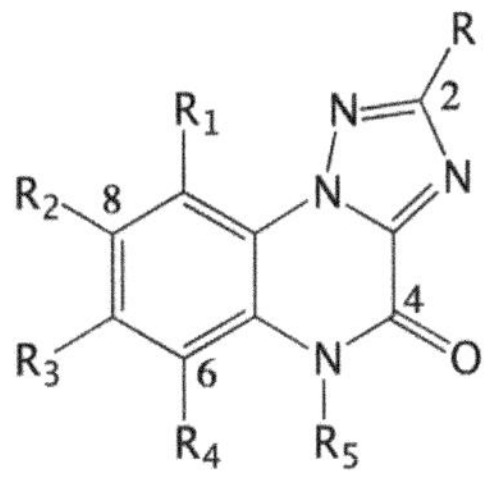

1. Training: mol01-39
   Test: mol53-55

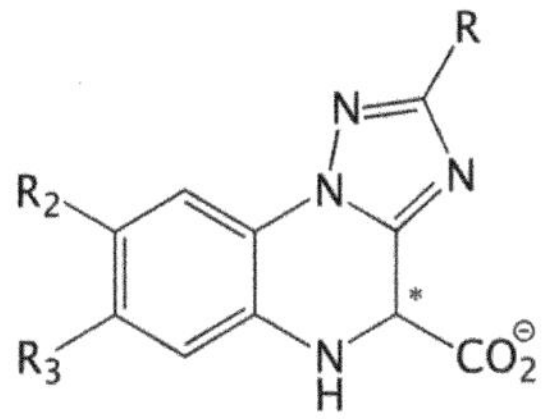

2. Training: mol40-48
   Test: mol56

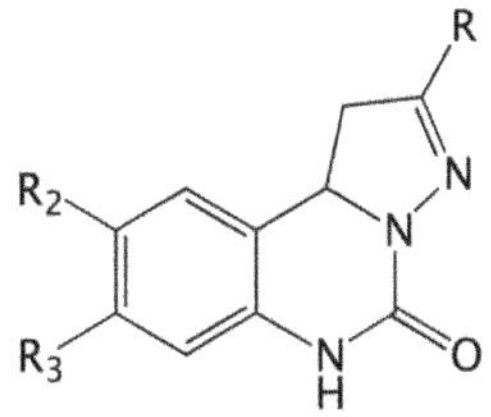

3. Training: mol49-52
   Test: mol57

CHART 19.1 Schematic representation of the three main chemical scaffolds for quinazoline-based AMPA inhibitors (training: mol01–52; test: mol53–57; see Ref. [80] for a complete view of the chemical structures).

$$A^q = \sum_{i=1}^{N} w_{probe}\, e^{-\alpha r_{iq}^2} \qquad (22)$$

where w_i is the actual value of the atomic hydrophobicity for atom i, w_{probe} is the hydrophobicity of the probe atom (+1), α is the attenuation factor, which was set to 0.3 and r_{iq} is the distance between the probe atom at grid point q and atom i of the molecule.

PLS statistical analysis based on the NIPALS algorithm was performed to derive the 3D QSAR models(83). In this regard, each projected field was stored in a MxNg matrix (M is the number of molecules and Ng is the number of grid points). Field values were then centered, scaled to unit variance and columns with a standard deviation lower than a certain threshold (typically 0.1) were excluded. The optimal number of components from PLS analysis was identified following the leave-one-out (LOO) cross validation criterion, in accordance to the lowest standard deviation error in prediction (S_{press}) corrected by the degree of freedom of the model and the predictive ability of the model for the test set (q^2).

19.5.2 OVERALL ANALYSIS OF PREDICTIVE PERFORMANCE

The MST-based partitioning scheme permits to use different combinations of the fractional contributions ($\log P_{ele}$, $\log P_{cav}$ and $\log P_{vW}$) originated from the $\log P_{o/w}$ of the molecule. Nevertheless, to perform a conceptually correct combination, all the involved contributions should be mutually uncorrelated. To verify this last point, the comparison of the values collected for the three previously mentioned contributions was determined, as noted in Figure 19.2.

A reasonable correlation ($r^2 = 0.83$) can be clearly observed and confirmed for $logP_{cav}$ and $logP_{vW}$. This is not unexpected, since these two contributions derive from equations that depend linearly on the solvent-exposed surface of each atom in the molecule (see Eqs. 16 and 17). These two fields are also related to the size and the shape of the molecule, and are thus expected to "similarly" account for its "steric" properties. Conversely, no evident correlation can be seen for $logP_{ele}$ and $logP_{non-ele}$ ($logP_{cav}$ or $logP_{vW}$) contributions ($r^2 = 0.03$). This confirms the lack of redundancy between $logP_{ele}$ and the two previously cited "steric" descriptors, and thus supports the additivity of the information provided by the projection of "electrostatic" and "non-electrostatic" fields.

Statistics for all the generated 3D QSAR models (from standard CoMFA, and MST-based hydrophobic HyPhar models H1-H4) are reported in Table 19.1. The HyPhar models reflect different combinations of electrostatic and non-electrostatic descriptors in order to fully explore the efficacy of the information encoded by each descriptor in modeling of AMPA inhibitors. In this direction, the following combinations have been examined: (i) $logP_{ele}$ and the cube of the atom radii(R^3), which was taken as a simple measure of the atomic size (model H1), (ii) $logP_{ele}$ and $logP_{cav}$ (model H2), (iii) $logP_{ele}$ and $logP_{vW}$ (model H3), and finally (iv) $logP_{ele}$ and the total non-electrostatic component, $logP_{n-ele}$, (model H4), which was obtained as the addition of $logP_{cav}$ and $logP_{vW}$.

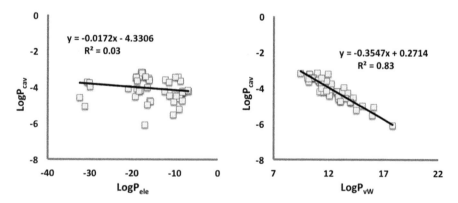

FIGURE 19.2 Orthogonality analysis for $logP_{ele}$, $logP_{cav}$, and $logP_{vW}$. A clear correlation can be found between $logP_{cav}$ and $logP_{vW}$, whereas no evident correlation can be seen for $logP_{ele}$ and $logP_{non-ele}$ ($logP_{cav}$ or $logP_{vW}$) contributions.

Except for the "hybrid" model H1, which gives slightly worse results, all HyPhar models that consider the $logP_{o/w}$-based hydrophobic descriptors to account for both the "electrostatic" and the "non-electrostatic" features perform similarly to classical CoMFA. No significant improvements in predictive power were observed upon inclusion of the two "steric" descriptors, as evidenced from the statistical parameters obtained for models H2, H3 and H4. Looking at the weights determined for each molecular field, there is also large resemblance between the different models.

A satisfactory predictive power is observed for compounds of the test set, with r^2 values ranging from 0.30 to 0.69. Compound 55 is the only exception, as this compound seems to give a leverage effect according to Cook's test (data not shown). Its influence of the regression model is clearly evidenced by the net improvement on all r^2 values re-calculated after its exclusion (see Table 19.1). In this regard, the analysis of the chemical modifications on the quinazoline-derived scaffold reveals that substitutions on position 6 are not explored in the training set. As a consequence, eventual substitutions at this position (as found for compound 55) might not be properly represented in the test set.

TABLE 19.1 Summary of Statistical Performances for Training and Test Sets Obtained for HyPhar (H1-H4) and CoMFA Models*

	CoMFA	H1	H2	H3	H4
Training Set					
r^2	0.87	0.76	0.89	0.83	0.83
S	0.36	0.50	0.33	0.42	0.42
q^2	0.50	0.44	0.51	0.46	0.47
S_{press}	0.54	0.56	0.54	0.56	0.55
Nc^a	5	5	5	5	5
Fields (%) [b]					
	E – 10	$logP_{ele} - 28$	$logP_{ele} - 34$	$logP_{ele} - 39$	$logP_{ele} - 38$
	S – 90	$R^3 - 72$	$logP_{cav} - 66$	$logP_{vW}$ 61	$logP_{n-ele} - 62$
Test Set [c]					
r^2	0.10 (0.29)	0.26 (0.69)	0.39 (0.68)	0.12 (0.38)	0.16 (0.32)
S_{press}	0.38 (0.35)	0.35 (0.23)	0.32 (0.24)	0.38 (0.33)	0.37 (0.35)

*CoMFA analysis was reproduced following the protocol reported in Ref. [80].

[a] Number of principal components.

[b] Fraction of the field (in percentage). E: electrostatic; S: steric; R^3: cube of the atomic radius. $logP_{n-ele}$ correspond to the sum of $logP_{cav}$ and $logP_{vW}$ contributions.

[c] Data in parenthesis refer to statistics calculated without compound 55. Its exclusion was also taken into account according to the Cook's test.

A graphical representation of the statistics of both training (white squares) and test (black triangles) sets of compounds for CoMFA and HyPhar H2-H4 models is shown in Figure 19.3. A satisfactory correlation between experimental (pK_i; mM) and predicted binding affinities for both training and test sets can be generally observed. In particular, H3 and H4 show a very similar trend in correlation.

19.5.3 PHARMACOPHORIC MAPS AND RECEPTOR COMPLEMENTARITY

The graphical representation of the physicochemical properties relevant for biological activity is provided in Figure 19.4. Favorable/unfavorable "polar"

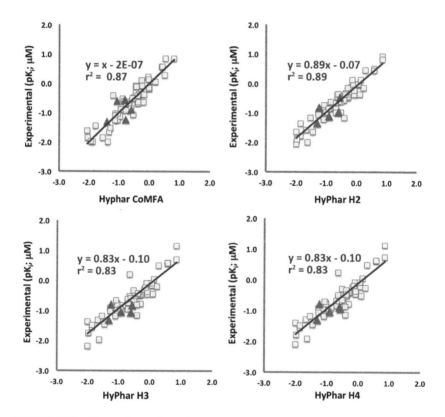

FIGURE 19.3   Comparison of experimental (pKi; mM) and predicted binding affinities for HyPhar CoMFA, H2, H3 and H4 models. The line in black refers to correlation for training set compounds.

(CoMFA: Figure 19.4A; H2: Figure 19.4C) and "non-polar" (CoMFA: Figure 19.4B; H2: Figure 19.4D) isocontours are reported in light grey/ black. The AMPA receptor with PDB ID 1FTL was considered for analysis of the ligand-protein complementarity (see Figure 19.5).

The AMPA-selective compound mol01 was used as model system for the two analyses. Two regions shape the cavity that accommodates the inhibitor (see Figure 19.5). In particular, Glu13, Tyr16, Tyr220, Met196 and Thr174 delimit the first one, whereas Pro89 and Arg96 form the other one. Arg96 may act as hydrogen-bond donor for the 4-oxo group and the nitrogen in position-3 of the triazole moiety. Another polar interaction can be detected

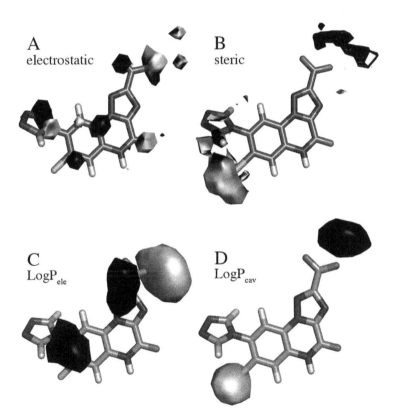

FIGURE 19.4 Representation of isocontour derived from CoMFA (A, B) and HyPhar H2 (C, D) models. Isocontours are expressed as the PLS coefficients corrected by standard deviation and scaled by a factor of 10^5. Favorable/unfavorable electrostatic (CoMFA: electrostatic values: +4 and −25; H2: $\log P_{ele}$ values: +10 and −15) and non-electrostatic (CoMFA: steric values: +900 and −900; H2: $\log P_{cav}$ values: +5 and −4) fields are reported in light grey/black.

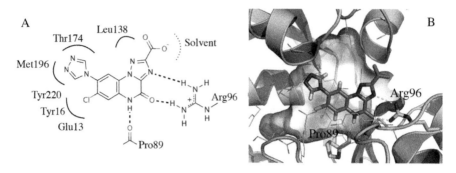

FIGURE 19.5   2D and 3D representation of ligand-protein interactions for the AMPA-selective compound mol01 (AMPA protein, PDB ID 1FTL).

for the backbone oxygen of Pro89, which can act as hydrogen-bond acceptor for the amidic NH in position 5. Taken together, these interactions can be considered to be a common binding motif for the series of compounds included in this study.

The H2 pharmacophoric maps generally are more homogeneous than those derived from CoMFA. Looking at the electrostatic term, both CoMFA and H2 models suggest the favorable presence of polar groups in position 2 of the triazole ring (see Figure 19.4A and 19.4C). This could be related with the possibility for the ligand to form hydrogen-bonding interactions with residues Ser140-Ser142 in the upper part of the ligand-binding cavity of the protein, thus providing an additional anchoring point to the binding pocket. Conversely, black isocontours around positions 7 and 8 of the benzene ring likely reflect the electronic influence due to the presence of electron-withdrawing substituents, such as −Cl or its bioisosteric −CF_3 group in position 7 or nitrogen-containing heterocycles in position 8 (see Figure 19.4A and 19.4C). These groups could be important for inducing an electron redistribution in the benzene ring, which could lead to an electrostatically favorable interaction with the negative electrostatic potential generated by residues Glu13 and Glu193. In this direction, the electro-withdrawing effects exerted by these groups enhance the acidity of the lactam NH, thus reinforcing its interaction with the backbone oxygen of Pro89. This could be related to the better binding observed for the 7-Cl, 8-imidazolyl disubstituted compound mol01 (pK_i = 0.85 mM) or the 7-Cl monosubstituted mol19 (pK_i = 0.1 mM) relative to the 7/8-unsubstituted analogue mol18 (pK_i = −1.13 mM).

The effects of the substituent at position 8 on AMPA inhibitor selectivity were also discussed by Catarzi (84). In this context, the removal of N^3 and N^4 on the imidazole ring at position 8 of mol01 to give the correspondent 8-(pyrrol-1-yl) derivative causes a great reduction of potency (mol02; pK_i = –1.26mM). The 8-nitro (mol08; pK_i = –0.08 mM) as well as the 8-amino (mol10; pK_i = –0.7 mM) groups can act as H-bond acceptor/donor for Thr174, Glu13 or other residues but, for the reasons explained before, a 8-nitro (electron-withdrawing) group seems to be preferred with regard to an 8-amino (electron-donating) one.

For the non-electrostatic term, (see Figure 19.4B and 19.4D), isocontours in light grey suggest a favorable steric contribution around position 7 of the benzene, which reflects the filling of the small pocket formed by Tyr16, Pro89, Met196, and Tyr220 (see Figure 19.5B). On the contrary, black areas around the 2-carboxylate group suggest potential steric clashes induced by size expansion of the ligand, likely reflecting steric clashes with stretch of residues Ser140-Ser142.

19.5.4 STRUCTURE-AIDED ANALYSIS OF AMPA MODELS: THE OUTLIER MOL55

To further clarify the failure in binding affinity predictions for mol55, a rigid docking of mol01, an AMPA-selective inhibitor, and mol55 in AMPA (PDB ID 1FTL) receptor was performed with GLIDE [85]. The predicted binding mode for the two compounds is shown in Figure 19.6.

This analysis showed that mol55 has a higher deviation from the reference quinazoline-derived scaffold compared to mol01 (root-mean square deviation of 1.2 Å for mol55, and only 0.3 Å for mol01). Unlike mol01, which lacks a substituent at position 6, a more pronounced spatial shift is necessary to properly accommodate mol55, and specifically the nitro group present at this position (see Figure 19.6). On the light of these evidences, a standard alignment on the reference scaffold seems to be conceptually incorrect in case of mol55, since ligand-induced conformational modifications in protein environment would be needed to avoid steric clashes with Pro89. This phenomenon, in conjunction with the previously discussed poor description of chemical diversity at position 6 in the training set, justifies the poor prediction of the 3D-QSAR models for compound mol55.

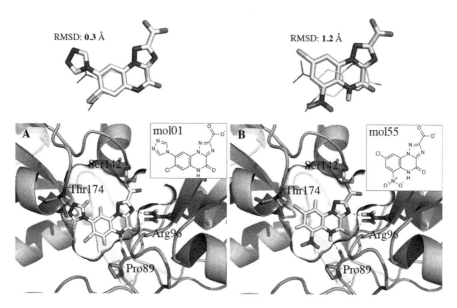

FIGURE 19.6 Docking-predicted binding modes and 2D sketches for mol01 (A) and 55 (B). A direct comparison of binding modes for mol01 and mol55 are also reported. As shown, a spatial shift from the reference crystallographic ligand (DNQ; in white lines) can be observed for mol55 (RMSD: 1.2 Å; in white sticks) respect to mol01 (RMSD: 0.3 Å; in white sticks). This re-orientation is necessary for a suitable placement since there is not enough space around Pro89 to accommodate the additional 6-nitro group of mol55.

19.6 CONCLUSIONS

The main aim of this contribution was to highlight the suitable application of more consistent and accurate molecular descriptors to traditional 3D-QSAR approaches. Given their derivation from quantum mechanical IEF/PCM-MST continuum calculations, these hydrophobic ($logP_{o/w}$-based) descriptors facilitate the projection at the atomic level of the physico-chemical properties that characterize the 3D hydrophobicity pattern of molecules. Furthermore, the partition of the atomic hydrophobicity into "electrostatic" and "non-electrostatic" components makes them suitable for a direct usage within the 3D-QSAR methodology, thus offering a complementary view to the physico-chemical elements mainly involved in ligand recognition and binding affinity.

Contextually to what was reported in this chapter, a set of 57 (52 + 5) quinazoline-derived AMPA inhibitors was examined as a test case (80). The HyPhar models have proven capable to give a good explanation of the relationship between chemical structure and binding affinity. Projection of pharmacophoric maps for electrostatic and non-electrostatic contributions allowed to gain guidelines about the chemical modifications that could be explored to improve the binding affinity. Even if not strictly necessary for a ligand-based approach, this information was discussed in light of the available ligand-surrounding protein environment from available X-ray structures. In this direction, a final docking-assisted analysis gave us the possibility to explain and justify the anomalous behavior observed for mol55.

ACKNOWLEDGMENT

We thank the financial support from the Spanish Ministerio de Economía y Competitividad (SAF2014-57094-R) and the Generalitat de Catalunya (2014SGR1189). FJL acknowledges the support from the Institució Catalana de Recerca i Estudis Avançats (Catalan Institution for Research and Advanced Studies; ICREA Academia).

KEYWORDS

- 3D-QSAR
- fractional contributions
- hydrophobicity
- MST solvation model
- QM calculations
- selectivity

REFERENCES

1. Pliska, V., Testa, B., & van de Waterbeemd, H., (Eds.), (1996). *Lipophilicity in Drug Action and Toxicology*; VCH: Weinheim.
2. Sangster, J., (1997). *Octanol–Water Partition Coefficients: Fundamentals and Physical Chemistry*; Wiley: Chichester.
3. van de Waterbeemd, H., Lennernäs, H., & Artursson, P., (Eds.), (2003). *Drug Bioavailability: Estimation of Solubility, Permeability, Absorption and Bioavailability*; Wiley-VCH: Weinheim.

4. Abraham, M. H., Platts, J. A., Hersey, A., Leo, A. J., & Taft, R. W., (1999). Correlation and estimation of gas-chloroform and water-chloroform partition coefficients by a linear free energy relationship method. *J. Pharm. Sci. 88*, 670–679.

5. Shih, P., Pedersen, L. G., Gibbs, P. R., & Wolfenden, R., (1988). Hydrophobicities of nucleic acid bases: Distribution coefficients from water to cyclohexane. *J. Mol. Biol., 280*, 421–430.

6. Young, R. C., Mitchell, R. C., Brown, T. H., Ganellin, C. R., Griffiths, R., & Jones, M., (1988). Development of a new physicochemical model for brain penetration and its application to the design of centrally acting H2 receptor histamine antagonists. *J. Med. Chem., 31*, 656–671.

7. Hansch, C., & Fujita, T., (1964). π–σ–π Analysis. A method for the correlation of biological activity and chemical structure. *J. Am. Chem. Soc., 86*, 1616–1626.

8. Fujita, T., Iwasa, J., & Hansch, C., (1964). A new substituent constant, π, derived from partition coefficients. *J. Am. Chem. Soc., 86*, 5175–5180.

9. Hansch, C., & Leo, A., (1995). *Exploring QSAR. Fundamentals and Applications in Chemistry and Biology*; American Chemical Society: Washington, DC.

10. Lipinski, C. A., Lombardo, F., Dominy, B. W., & Feeney, P. J., (2001). Experimental and computational approaches to estimate solubility and permeability in drug discovery and development settings. *Adv. Drug Deliv. Rev., 46*, 3–26.

11. Martin, Y. C., (2010). *Quantitative Drug Design: A Critical Introduction.* CRC Press. Boca Raton.

12. Waring, M. J., (2010). Lipophilicity in drug discovery. *Expert Opin. Drug Discov., 5*, 235–248.

13. Hill, P., & Young, R. J., (2010). Getting physical in drug discovery: A contemporary perspective on solubility and hydrophobicity. *Drug Discov. Today, 15*, 648–655.

14. Young, R. J., Green, D. V. S., Luscombe, C. N., & Hill, A. P., (2011). *Drug Discov. Today, 16*, 822–830.

15. Mortenson, P. N., & Murray, C. W., (2011). Assessing the lipophilicity of fragments and early hits. *J. Comput.-Aided Mol. Des., 25*, 663–667.

16. Mannhold, R., & van de Waterbeemd, H., (2001). Substructure and whole molecule approaches for calculating logP. *J. Comput. Aided Mol. Des., 15*, 337–354.

17. Mannhold, R., & Dross, K., (1996). Calculation procedures for molecular lipophilicity: A comparative study. *Quant- Struct.-Act. Relat. 15*, 403–409.

18. Mannhold, R., Poda, G. I., Ostermann, C., & Tetko, I. V., (2009). Calculation of molecular lipophilicity: State-of-the-art and comparison of logP methods on more than 96,000 compounds. *J. Pharm. Sci., 98*, 861–893.

19. Arkin, M. R., & Wells, J. A., (2004). Small-molecule inhibitors of protein-protein interactions: Progressing towards the dream. *Nat. Rev. Drug Disc., 3*, 301–317.

20. Hajduk, P. J., Huth, J. R., & Fesik, S. W., (2005). Druggability indices for protein targets derived from NMR-based screening data. *J. Med. Chem., 48*, 2518–2525.

21. Nayal, M., & Honig, B., (2006). On the nature of cavities on protein surfaces. Application to the identification of drug-binding sites. *Proteins 63*, 892–906.

22. Egner, U., & Hillig, R. C., (2008). A structural biology view of target druggability. *Expert. Opin. Drug Discov., 3*, 391–401.

23. Klebe, G., (2015). Applying thermodynamic profiling in lead finding and optimization. *Nat. Rev. Drug. Discov., 14*, 95–110.

24. Cheng, A. C., Coleman, R. G., Smyth, K. T., Cao, Q., Soulard, P., Caffrey, D. R., et al., (2007). Structure-based maximal affinity model predicts small-molecule druggability. *Nat. Biotechnol., 25*, 71.

25. Karplus, P. A., (1997). Hydrophobicity regained. *Protein Sci., 6,* 1302–1307.
26. Davis, A. M., & Teague, S. J., (1999). Hydrogen bonding, hydrophobic interactions, and failure of the rigid receptor hypothesis. *Angew. Chem. Int. Ed., 38,* 736–749.
27. Schmidtke, P., & Barril, X., (2010). Understanding and predicting druggability. A high-throughput method for detection of drug binding sites. *J. Med. Chem., 53,* 5858–5867.
28. Schmidtke, P., Luque, F. J., Murray, J. B., & Barril, X., (2011). Shielded hydrogen bonds as structural determinants of binding kinetics: Application in drug design. *J. Am. Chem. Soc., 133,* 18903–18910.
29. Alvarez-Garcia, D., & Barril, X., (2014). Molecular simulations with solvent competition quantify water displaceability and provide accurate interaction maps of protein binding sites. *J. Med. Chem., 57,* 8530–8539.
30. Audry, E., Dubost, J.-P-. Colleter, J. C., & Dallet, P., (1986). Une nouvelle approche des relations structure-activité: le "potentiel de lipophilie moléculaire." *Eur J. Med. Chem. Chim. Ther., 21,* 71–72.
31. Furet. P., Sele, A., & Cohen, N. C., (1988). 3D molecular lipophilicity potential profiles: A new tool in molecular modeling. *J. Mol. Graphics, 6,* 182–189.
32. Heiden, W., Moeckel, G., & Brickmann, J., (1993). A new approach to analysis and display of local lipophilicity/hydrophilicity mapped on molecular surfaces. *J. Comput.-Aided Mol. Des., 7,* 503–514.
33. Gaillard, P., Carrupt, P. A., Testa, B., & Boudon, A., (1994). Molecular Lipophilicity Potential, a tool in 3D QSAR: Method and applications. *J. Comput.-Aided Mol. Des., 8,* 83–96.
34. Ottaviani, G., Martel, S., & Carrut, P.-A., (2007). In silico and in vitro filters for the fast estimation of skin permeation and distribution of new chemical entities. *J. Med. Chem., 50,* 742–748.
35. Laguerre, M., Saux, M., Dubost, J., & Carpy, A., (1997). MLPP: A program for the calculation of molecular lipophilicity potential in proteins. *Pharm. Pharmacol. Commun., 3,* 217–222.
36. Efremov, R. G., Chugunov, A. O., Pyrkov, T. V., Priestle, J. P., Arseniev, A. S., & Jacoby, E., (2016). Molecular lipophilicity in protein modeling and drug design. *Curr. Med. Chem., 14,* 393–415.
37. Rozas, I., Du, Q., & Arteca, G. A., (1995). Interrelation between electrostatic and lipophilicity potentials on molecular surfaces. *J. Mol. Graphics, 13,* 98–108.
38. Rozas, I., & Martín, M., (1996). Molecular lipophilic potential on van der Waals surfaces as a toll in the study of 4-alkylpyrazoles. *J. Chem. Inf. Comput. Sci., 36,* 872–878.
39. Nurisso, A., Bravo, J., Carrupt, P.-A., & Daina, A., (2012). Molecular docking using the molecular lipophilicity potential as hydrophobic descriptor: Impact on GOLD docking performance. *J. Chem. Inf. Model., 52,* 1319–1327.
40. Novaroli, L., Daina, A., Favre, E., Bravo, J., Carotti, A., Leonetti, F., Catto, M., et al., (2006). Impact of species-dependent differences on screening, design, and development of MAO B inhibitors. *J. Med. Chem., 49,* 6264–6272.
41. Oberhauser, N., Nurisso, A., & Carrupt, P.-A., (2014). MLP tools: A PyMol plugin for using the molecular lipophilicity potential in computer-aided drug design. *J. Comput.-Aided Mol. Des., 28,* 587–596.
42. Kellogg, G. E., Semus, S. F., & Abraham, D. J., (1991). HINT: A new method of empirical hydrophobic field calculation for CoMFA. *J. Comput.-Aided Mol. Des., 5,* 545–552.
43. Kellogg, G. E., & Abraham, D. J., (2000). Hydrophobicity: Is logP/o/w) more than the sum of its parts? *Eur. J. Med. Chem., 35,* 651–661.

44. Fornabaio, M., Spyrakis, F., Mozzarelli, A., Cozzini, P., Abraham, D. J., & Kellogg, G. E., (2004). Simple, intuitive calculations of free energy of binding for protein-ligand complexes. 3. The free energy contribution of structural waters molecules in HIV-1 protease complexes. *J. Med. Chem.*, *47*, 4507–4516.
45. Amadasi, A., Spyrakis, F., Cozzini, P., Abraham, D. J., Kellogg G. E., & Mozzarelli, A., (2006). Mapping the energetics of water-protein and water-ligands interactions by the HINT "natural" force field: Predictive tools for characterizing the roles of water in biomolecules. *J. Mol. Biol.*, *358*, 289–309.
46. Bayden A. S., Fornabaio, M., Scarsdale, J. N., & Kellog, G. E., (2009). Web application for studying the free energy of binding and protonation states of protein-ligand complexes based on HINT. *J. Comput.-Aided Mol. Des.*, *23*, 621–632.
47. Ahmed, M. H., Koparde, V. N., Safo, M. K., Scarsdale, N., & Kellogg, G. E., (2015). 3D interaction homology: The structurally known rotamers of tyrosine derive from a surprisingly limited set of information-rich hydropathic interaction environments described by maps. *Proteins, 83*, 1118–1136.
48. Du, Q., Liu, P.-J., & Mezey, P. G., (2005). Theoretical derivation of the heuristic lipophilicity potential: A quantum chemical description of molecular solvation. *J. Chem. Inf. Model.*, *45*, 347–353.
49. Du, Q., Arteca, G. A., & Mezey, P. G., (1997). Heuristic lipophilicity potential for computer-aided rational drug design. *J. Comput.-Aided Mol. Des. 11*, 503–515.
50. Du, Q., & Mezey, P. G., (1998). Heuristic lipophilicity potential for computer-aided rational drug design: Optimizations of screening functions and parameters. *J. Comput.-Aded Mol. Des.*, *12*, 451–470.
51. Du, Q.-S., Li, D.-P., He, W.-Z., & Chou, K.-C., (2006). Heuristic molecular lipophilicity potential (HMLP): Lipophilicity and hydrophilicity of amino acid side chains. *J. Comput. Chem.*, *27*, 685–692.
52. Du, Q.-S., Huang, R.-B., Wei, Y.-T., Du, L.-Q., & Chou, K.-C., (2008). Multiple filed three dimensional quantitative structure-activity relationship (MF-3D-QSAR). *J. Comput. Chem.*, *29*, 211–219.
53. Du, Q.-S., Gao, J., Wei, Y.-T., Du, L.-Q., Wang, S.-Q., & Huang, R.-B., (2012). Structure-based and multiple potential three-dimensional quantitative structure-activity relationship (SB-MP-3D-QSAR) for inhibitor design. *J. Chem. Inf. Model.*, *52*, 996–1004.
54. Cramer, C. J., & Truhlar, D. G., (1999). Implicit solvation models: Equilibria, structure, spectra, and dynamics. *Chem. Rev.*, *99*, 2161–2200.
55. Orozco, M., & Luque, F. J., (2000). Theoretical methods for the description of the solvent effect in biomolecular systems. *Chem. Rev.*, *100*, 4187–4226.
56. Tomasi, J., Mennucci, B., & Cammi, R., (2005). Quantum mechanical continuum solvation models. *Chem. Rev.*, *105*, 2999–3094.
57. Klamt, A., (1995). Conductor-like Screening Model for Real Solvents: A New Approach to the Quantitative Calculation of Solvation Phenomena. *J. Phys. Chem.*, *99*, 2224–2235.
58. Klamt, A., & Schüürmann, G., (1993). COSMO: A new approach to dielectric screening in solvents with explicit expressions for the screening energy and its gradient. *J. Chem. Soc. Perkin Trans. II*, 799–805.
59. Klamt, A., Jonas, V., Bürger, T., & Lohrenz, J. C. W., (1998). Refinement and parametrization of COSMO-RS. *J. Phys. Chem. A*, *102*, 5074–5085.
60. Klamt, A., (2011). The COSMO and COSMO-RS solvation models. *WIRES Comput. Mol. Sci.*, *1*, 699–709.

61. Thormann, M., Klamt, A., & Wichmann, K., (2012). COSMOsim3D: 3D-Similarity and alignment based on COSMO polarization charge densities. *J. Chem. Inf. Model.*, *52*, 2149–2156.
62. Klamt, A., Thormann, M., Wichmann, K., & Tosco, P., (2012). COSMOsar3D: Molecular field analysis based on local COSMO σ-profiles. *J. Chem. Inf. Model.*, *52*, 2157–2164.
63. Miertus, S., Scrocco, E., & Tomasi, J., (1981). Electrostatic interaction of a solute with a continuum. A direct utilization of ab initio molecular potentials for the prevision of solvent effects. *Chem. Phys.*, *55*, 117–129.
64. Cancès, E., Mennucci, B., & Tomasi, J., (1997). A new integral equation formalism for the polarizable continuum model: Theoretical background and applications to isotropic and anisotropic dielectrics. *J. Chem. Phys.*, *107*, 3032.
65. Luque, F. J., Curutchet, C., Muñoz-Muriedas, J., Bidon-Chanal, A., Soteras, I., Morreale, A., et al., (2003). Continuum solvation models: Dissecting the free energy of solvation. *Phys. Chem. Chem. Phys.*, *5*, 3827–3836.
66. Soteras, I., Curutchet, C., Bidon-Chanal, A., Orozco, M., & Luque, F. J., (2005). Extension of the MST model to the IEF formalism: HF and B3LYP parametrizations. *J. Mol. Struct. (Theochem)*, *727*, 29–40.
67. Luque, F. J., Barril, X., & Orozco, M., (1999). Fractional description of free energies of solvation. *J. Comput.-Aided Mol. Des.* 13, 139–152.
68. Barril, X., Muñoz, J., Luque, F. J., & Orozco, M., (2000). Simplified descriptions of the topological distribution of hydrophilic/hydrophobic characteristics of molecules. *Phys. Chem. Chem. Phys.*, *2*, 4897–4905.
69. Curutchet, C., Salichs, A., Barril, X., Orozco, M., & Luque, F. J., (2003). Transferability of fragmental contributions to the octanol/water partition coefficient: An NDDO-based MST study. *J. Comput. Chem.*, *24*, 32–45.
70. Luque, F. J., Bofill, J. M. & Orozco, M., (1995). New strategies to incorporate the solvent polarization in self-consistent reaction field and free-energy perturbation simulations. *J. Chem. Phys.*, *103*, 10183.
71. Pierotti, R. A., (1976). A scaled particle theory of aqueous and nonaqueous solutions. *Chem. Rev.*, *76*, 717–726.
72. Claverie, P., (1978). Elaboration of approximate formulas for the interactions between large molecules: Applications in organic chemistry. In: *Intermolecular Interactions: From Diatomics to Biopolymers, Vol. 1*, B. Pullman, (Ed.), Wiley: New York, pp. 69–305.
73. Curutchet, C., Orozco, M., & Luque, F. J., (2001). Solvation in octanol: Parametrization of the continuum MST model. *J. Comput. Chem.*, *22*, 1180–1193.
74. Muñoz, J., Barril, X., Hernández, B., Orozco, M., & Luque, F. J., (2002). Hydrophobic similarity between molecules: A MST-based hydrophobic similarity index. *J. Comput. Chem.*, *23*, 554–563.
75. Muñoz-Muriedas, J., Perspicace, S., Bech, N., Guccione, S., Orozco, M., & Luque, F. J., (2005). Hydrophobic molecular similarity from MST fractional contributions to the octanol/water partition coefficient. *J. Comput.-Aided Mol. Des.*, *19*, 401–419.
76. Ginex, T., Muñoz-Muriedas, J., Herrero, E., Gibert, E., Cozzini, P., & Luque, F. J., (2016). Development and validation of hydrophobic molecular fields derived from the quantum mechanical IEF/^CM-MST solvation model in 3D-QSAR. *J. Comput. Chem.*, *37*, 1147–1162.
77. Ginex, T., Muñoz-Muriedas, J., Herrero, E., Gibert, E., Cozzini, P., & Luque, F. J., (2016). Application of the quantum mechanical IEF/PCM-MST hydrophobic descriptors to selectivity in ligand binding. *J. Mol. Model.*, *22*, 136.

78. Cramer, R. D., III; Patterson, D. E., & Bunce, J. D., (1988). Comparative molecular field analysis (CoMFA). 1. Effect of shape on binding of steroids to carrier proteins. *J. Am. Chem. Soc., 110,* 5959–5967.

79. Klebe, G., Abraham, U., & Mietzner, T., (1994). Molecular similarity indices in a comparative analysis (CoMSIA) of drug molecules to correlate and predict their biological activity. *J. Med. Chem., 37,* 4130.

80. Baskin, I. I., Tikhonova, I. G., Palyulin, V. A., & Zefirov, N. S., (2003). Selectivity fields: comparative molecular field analysis (CoMFA) of the glycine/NMDA and AMPA receptors. *J. Med. Chem., 46,* 4063–4069.

81. PharmQSAR. Pharmacelera; Barcelona, Spain, 2015. (http://www. pharmacelera.com/ pharmqsar/).

82. Clark, M., Cramer, R. D., & Van Opdenbosch, N., (1989). Validation of the general purpose Tripos 5.2 force field. *J. Comput. Chem, 10,* 982–1012.

83. Wold, S., Sjöström, M., & Eriksson, L., (2001). PLS-regression: a basic tool of chemometrics. *Chemom. Intell. Lab. Syst., 58,* 109–130.

84. Catarzi, D., Colotta, V., Varano, F., Filacchioni, G., Galli, A., Costagli, C., et al., (2001). Synthesis, ionotropic glutamate receptor binding affinity, and structure-activity relationships of a new set of 4, 5-dihydro-8-heteroaryl-4-oxo-1, 2, 4-triazolo [1, 5-a.quinoxaline-2-carboxylates analogues of TQX-173. *J. Med. Chem., 44,* 3157–3165.

85. Friesner, R. A., Banks, J. L., Murphy, R. B., Halgren, T. A., Klicic, J. J., Mainz, D. T., et al., (2004). Glide: A new approach for rapid, accurate docking and scoring. 1. Method and assessment of docking accuracy. *J. Med. Chem., 47,* 1739–1749.

86. Klopman, G., & Zhu, H., (2005). Recent methodologies for the estimation of n-cotanol/ water partition coefficients and their use on the prediction of membrane transport properties of drugs. *Mini Rev. Med Chem., 5,* 127–133.

CHAPTER 20

STATISTICALLY INDEPENDENT EFFECTIVE ELECTRONS FOR MULTIDETERMINANT WAVEFUNCTIONS

E. FRANCISCO and A. MARTÍN PENDÁS

Departamento de Química Física y Analtica, Universidad de Oviedo, 33006-Oviedo, Spain

CONTENTS

ABSTRACT

The concept of electron number distribution functions (EDF) of a molecular system has proved to be of great importance in the study of many aspects of chemical bonding theory, and a considerable effort has been made to permit their fast and accurate calculation for multideterminant and single-determinant wavefunctions. In the latter case, we have recently shown that, as long as the molecular wavefunction is expressed in terms of the molecular orbitals (MOs) that diagonalize the domain averaged Fermi hole (DAFH) of a given region of physical space R^3, the resulting EDF can be computed as the direct product of individual electron events. In other words, electrons described with DAFH MOs show counting statistics of independent trial events. Here, we extend this very desirable property (as least, partially) to multideterminant wavefunctions. To achieve this, a linear transformation of the full set of canonical MOs $|\varphi_i\rangle$ ($i = 1, ..., m$) is proposed that transforms it into another set $|\chi_i\rangle$ which is orthonormal in R^3, as well as orthogonal in two arbitrary regions Ω and Ω' that exhaustively partition R^3 ($\Omega \cup \Omega' = R^3$). In addition, the diagonal overlap integral $\mathbf{n}_i = \langle \chi_i | \chi_i \rangle_\Omega$ ($\langle \chi_i | \chi_i \rangle_{\Omega'} = 1 - \mathbf{n}_i$) is the probability that an electron described by the transformed MO χ_i will be within the Ω (Ω') region. Compact expressions are found for the N-electron multideterminant wavefunction $\Psi (1, N)$ and its square in terms of the χ_i's, and for the probability of having exactly v electrons in Ω and $N - v$ electrons in Ω'. Three very simple molecules (H_2, H_2O, LiH), described at the complete active space (CAS) level of calculation have been considered to illustrate our findings.

20.1   INTRODUCTION

Real space techniques in chemical bonding theory provide orbital invariant descriptors based on reduced density matrices. They are slowly, but steadily providing new chemical insight [8]. In many cases, their success is based on examining the distribution of electrons in a coarse version of the physical space. This can be an atomic partition, like that used in the quantum theory of atoms in molecules (QTAIM) of Bader and coworkers [2], or a thinner decomposition, like when using the electron localization function (ELF) [26] or the electron localizability indicator (ELI) [17].

Once such a partition of space is chosen, the closer we can get to analyzing the several reduced density matrices (RDM) that an N-electron system possesses without considering the complex behavior of a multidimensional function lies in condensing these RDM in the different spatial domains in which the system has been divided [19]. With the help of these coarse-grained RDMs we can obtain the probability of finding a given partition of the N electrons in the m domains, $p(n_1, n_2, ..., n_m)$, with $N = n_1 + n_2 + ... + n_m$ [5]. The set of all these values is what we know as the electron number distribution function (EDF) [10–12, 21]. With its help, all chemical bonding indicators built from the chosen spatial partitioning may be written as moments or cumulants of the EDF [14, 15, 21]. However, studying the EDF itself, one can get a wealth of information not contained in the EDF moments or cumulants [20].

EDFs themselves can be used to create a self-consistent space partitioning after the seminal insights of Savin [6]. For instance, if we maximize the probability of finding a given number of electrons in a region, let us say two, we obtain a maximum probability domain (MPD) [22], which is a real space object as close as possible to a Lewis pair. Given the enormous algebraic difficulties to obtain these optimal shapes computationally, only one-domain MPDs have been obtained until now in simple molecules [7, 18, 23], although multi-domain MPDs have been studied in Hubbard models [1].

A very interesting result from the analysis of EDFs for single-determinant functions is that the final electron counting statistics is equivalent to a series of independent Bernoulli trials with different probabilities. In this sense, the final statistics is the same as that provided by a set of "independent" effective electrons, which may be associated to the domain natural one electron functions defined by Ponec [24, 25]. This map is no longer exact in the correlated wave function case, but we show here how an orbital linear transformation allows us to "almost" recover this desirable property even in multi-determinant cases. Although fermions are never statistically independent, we will refer only to their counting statistics in domains when we refer to statistically independent effective electrons in the following section.

20.2   ORTHOGONAL ORBITALS IN R^3, Ω, AND Ω' WITH $\Omega \cup \Omega' = R^3$

Let φ_i ($i = 1, ..., m$) be a set of m molecular orbitals (MO) that are orthonormal in R^3, i.e., $\langle \varphi_i | \varphi_j \rangle = \delta_{ij}$, and $\mathbf{S}$ the overlap matrix defined as

$$\mathbf{S}_{ij} = \langle \varphi_i | \varphi_j \rangle_\Omega, \tag{1}$$

where Ω is an arbitrary region of the physical space, R^3. Although it is not strictly necessary for our purposes, the φ_i's are typically the canonical orbitals resulting from a molecular calculation. Since $\mathbf{S}$ is Hermitian, it can be diagonalized by a unitary transformation $\mathbf{U}$ ($\mathbf{UU}^\dagger = \mathbf{U}^\dagger\mathbf{U} = \mathbf{I}$)

$$\mathbf{SU} = \mathbf{Un} \quad \Leftrightarrow \quad \mathbf{U}^\dagger\mathbf{SU} = \mathbf{n}, \tag{2}$$

where $\mathbf{n}$ is the diagonal matrix of real eigenvalues, $\mathbf{n}_{ij} = n_i \delta_{ij}$. It can be shown that they satisfy $0 \leq n_i \leq 1$. Now, we define the set of transformed orbitals χ by

$$\chi = \varphi\mathbf{U}, \tag{3}$$

that has the following overlap matrix in Ω

$$\tilde{\mathbf{S}} = \langle\chi|\chi\rangle_\Omega = \mathbf{U}^\dagger\langle\varphi|\varphi\rangle_\Omega\mathbf{U} = \mathbf{U}^\dagger\mathbf{SU} = \mathbf{n}. \tag{4}$$

This means that the set χ is simultaneously orthonormal in R^3, since it was obtained from a unitary transformation of the orthonormal φ_i's, and orthogonal in Ω. Moreover, given that $\mathbf{S}' = \langle\varphi|\varphi\rangle_{\Omega'} = \mathbf{I} - \mathbf{S}$, $\mathbf{U}$ is also the eigenvector matrix of $\mathbf{S}'$ and $\mathbf{n}' = \mathbf{I} - \mathbf{n}$ are their eigenvalues, so that the set χ is also orthogonal in Ω' [11, 13].

A particularly relevant case of the above transformation occurs when $m = N/2$, where N is the number of electrons of a closed-shell molecule. It it well known that, in that case, the Slater determinants built with the sets φ_i and χ_i are equal; i.e.,

$$|\chi_1\cdots\chi_m\bar\chi_1\cdots\bar\chi_m| = |\phi_1\cdots\phi_m\bar\phi_1\cdots\bar\phi_m| \tag{5}$$

However, what is less known is that electrons described with the χ_i orbitals are statistically independent. What we mean by this is that the probability of finding $v = 0, 1, 2, ..., N$ electrons in the region Ω (and consequently $N - v = N, N - 2, N - 2, ..., 0$ in Ω') is given by

$$p^\Omega(\nu) = \hat{S}\,[p_1 p_2 \cdots p_N] = \hat{S}\prod_i^N p_i, \tag{6}$$

where $p_i = n_i$ or $p_i = 1 - n_i$ depending on whether the electron i is in Ω or in Ω', and $\hat{S}$ is a symmetrizing operator that takes into account electron indistinguishability. For instance, let us assume that we have 3α electrons occupying the spin-orbitals $\chi_1\alpha$, $\chi_2\alpha$, and $\chi_3\alpha$. Then

$$p^{\Omega}(0) = (1 - n_1)(1 - n_2)(1 - n_3) \tag{7}$$

$$p^{\Omega}(1) = (1 - n_1)(1 - n_2)n_3 + (1 - n_1)n_2(1 - n_3) + n_1(1 - n_2)(1 - n_3) \tag{8}$$

$$p^{\Omega}(2) = (1 - n_1)n_2 n_3 + n_1(1 - n_2)n_3 + n_1 n_2(1 - n_3) \tag{9}$$

$$p^{\Omega}(3) = n_1 n_2 n_3. \tag{10}$$

Since $n_i = \langle \chi_i \mid \chi_i \rangle_{\Omega}$, the probability of finding an electron described by the orbital χ_i in Ω increases as the orbital is more localized in that region. In the extreme cases when the localization of χ_i in Ω is perfect or negligible, we have $n_i = p_i = 1$ and $n_i = p_i = 0$, respectively. The above property of the χ_i's allows for a considerable decrease in the time required to calculate the electron number distribution function (EDF) of the molecule, since those orbitals with $n_i \simeq 1$ or $n_1 \simeq 0$ can be simply ignored.

In the following sections, we show how the orbitals χ_i derived from Eq. (3), which are statistically independent in the sense we have illustrated in the example above, offer a very interesting alternative to the canonical orbitals to obtain the EDF of molecules described with multideterminant wavefunctions.

20.3 Ψ AND $\Psi^*\Psi$ IN TERMS OF STATISTICALLY INDEPENDENT ORBITALS

In this section we want to express the N-electron wave function $\Psi(1, N)$ and its square, $|\Psi(1, N)|^2$, originally given as a linear combination of Slater determinants built in with the φ_i's, in terms of the transformed χ_i's. Before doing this, we re-write Eq. (3) in the form

$$\chi \equiv (\chi_a, \chi_\beta) = (\varphi_a, \varphi_\beta)\,\mathbf{U} \equiv \varphi\mathbf{U}, \tag{11}$$

where χ_a and χ_β are the sets of m α and m β spin-orbitals formed by multiplying χ by the α and β spin functions, respectively (with an analogous definition for φ_a and φ_β), and $\mathbf{U}$ is a $(2m \times 2m)$ block-diagonal matrix formed with two similar $m \times m$ blocks. In the following, the $2m$ eigenvalues of $\mathbf{S}$ will be represented as $\mathbf{n} = \mathrm{diag}(n_1, ..., n_{2m})$ with $n_{i+m} = n_i$ $(1 \leq i \leq m)$.

Let $\Psi(1, N)$ be given as the linear combination

$$\Psi(1, N) = (N!)^{-1/2} \sum_{r=1}^{M} C_r \psi_r(1, N), \tag{12}$$

where $\psi_r(1, N) = \det|\varphi_{r_1}(1)\varphi_{r_2}(2) \dots \varphi_{r_N}(N)|$, and φ_{r_i} $(i = 1, \dots, N)$ is the sub-set of N canonical molecular spin-orbitals (CMSO) that appear in $\psi_r(1, N)$. We will collectively label $(r_1, \dots, r_N)$ as r.

From Eq. (11), we have $\varphi = \chi U^\dagger$, so that φ_{r_i} can be written as

$$\phi_{r_i} = \sum_{j=1}^{2m} \chi_j U_{r_i,j}. \tag{13}$$

Using this equation in the $(N \times N)$ matrix associated to $\psi_r(1, N)$, we have

$$\psi_r = \det \begin{vmatrix} \sum_{j_1} \chi_{j_1}(1) U_{r_1,j_1} & \cdots & \sum_{j_N} \chi_{j_N}(1) U_{r_N,j_N} \\ \vdots & \ddots & \vdots \\ \sum_{j_1} \chi_{j_1}(N) U_{r_1,j_1} & \cdots & \sum_{j_N} \chi_{j_N}(N) U_{r_N,j_N} \end{vmatrix}, \tag{14}$$

where each j_i $(i = 1, \dots, N)$ takes values from 1 to $2m$. Applying a well-known property of determinants, we expand now ψ_r by columns to obtain:

$$\psi_r = \sum_{j_1} \cdots \sum_{j_N} \left[\prod_{k=1}^{N} U_{r_k,j_k} \right] \chi_j, \tag{15}$$

where $j = (j_1, j_2, \dots, j_N)$, and $\chi_j = \det|\chi_{j_1}(1)\chi_{j_2}(2) \dots \chi_{j_N}(N)|$. When the N summations in Eq. (15) are developed, only the terms in which all j_i's are different survive. Otherwise, the determinant has two equal columns and is zero. Furthermore, if $j_1, j_2, \dots, j_N$ is an ordered set $(j_{i+1} > j_i)$ and call D its companion determinant in Eq. (15), all even and odd permutations of $j_1, j_2, \dots, j_N$ are accompanied by a determinant equal to $+D$ and $-D$, respectively. Consequently, Eq. (15) can be recast in the more compact form

$$\psi_r = \sum_j U_{rj} \chi_j, \tag{16}$$

where $U_{rj} = \det[\mathbf{U}_{rj}]$ and $\mathbf{U}_{rj}$ stands for the $(N \times N)$ matrix that results from selecting from the $(2m \times 2m)$ $\mathbf{U}$ matrix (Eq. 11) the rows and columns denoted by vectors r and j, respectively. The summation in Eq. (16) runs, in principle, over all possible sets of N different ordered integers $j_1, j_2, ...,$ j_N taken from the first $2m$ natural numbers. However, the block-diagonal character of $\mathbf{U}$ imposes an additional restriction in the possible j sets. Calling N_α and N_β the number of α and β CMSOs, respectively (obviously, $N_\alpha = N_\beta$ $= N/2$ for a closed-shell wavefuntion), the first N_α elements in j must necessarily be in the range $1 < j_i \le m$ $(0 < i \le N_\alpha)$, and the last N_β elements in the range $m < j_i < 2m$ $(N_\alpha < i \le N)$. This means that $U_{rj} = U^\alpha_{rj} U^\beta_{rj}$, where $U^\sigma_{rj} = \det[\mathbf{U}^\sigma_{rj}]$, with $\mathbf{U}^\sigma_{rj}$ being a $N_\sigma \times N_\sigma$ matrix $(\sigma = (\alpha, \beta))$.

Another interesting point related to Eq. (16) is the following. In general, $\mathbf{U}_{rj}$ is not an unitary matrix and thus $U_{rj} \ne 1$. However, for a single-determinant closed-shell wave function $m = N/2$, j and r can only be $j = r = (1, 2, ...,$ $N)$, and $\mathbf{U}_{rj}$ coincides with $\mathbf{U}$, which is unitary. Consequently, $U_{rj} = 1$ and $\chi_j = \chi_r$ is identical to ψ_r. This shows the well-known invariance of a Slater determinant against an arbitrary and unitary transformation of all of its molecular spin-orbitals.

After substituting Eq. (16) into Eq. (12) one has

$$\Psi(1, N) = (N!)^{-1/2} \sum_j D_j \chi_j, \tag{17}$$

where

$$D_j = \sum_{r=1}^{M} C_r U_{rj}. \tag{18}$$

The number of terms in the expansion 17 is $M' = (m!)^2/[(m - N_\alpha)!(m - N_\beta)! N_\alpha! N_\beta!]$, which should be compared with M, the number of Slater determinants in the original expansion, Eq. (12). In the same way that one can eliminate a ψ_r from Eq. (12) if $C_r < \varepsilon$, where ε is a positive arbitrarily small number, one can decide whether a particular χ_j in Eq. (17) can be neglected or not depending on the value of D_j. Actually, as we will see, even though M' can be greater than M, the interesting orthogonality properties of the χ set commented in Section 20.2 makes Eq. (17) computationally very attractive for obtaining the EDF. Besides this, it will be shown to be particularly illuminating from a physical point of view.

We develop now the expression for $\Psi^*\Psi$ in terms of the χ_i's. Squaring Eq. (17) we have

$$\Psi^*\Psi = (N!)^{-1} \sum_j \sum_k D_j D_k^* \chi_j \chi_k^\dagger, \qquad (19)$$

where, as in the case of j, k represents $(k_1, k_2, ..., k_N)$, with $k_1, ..., k_{N\alpha}$ running over all possible sets of N_α different ordered integers taken from the set (1, ..., m), and $k_{N\alpha} + 1, ..., k_N$ running over all possible sets of N_β different ordered integers taken from the $(m + 1, ..., 2m)$ set. The product $\chi_j \chi_k^\dagger$ is given by

$$\chi_j \chi_k^\dagger = \det[\chi_j] \times \det[\chi_k^\dagger] = \det[\chi_j \chi_k^\dagger] \equiv \det[\chi_j^\dagger \chi_k], \qquad (20)$$

where the (a, b) element of the $\chi_j^\dagger \chi_k$ matrix is

$$[\chi_j^\dagger \chi_k]_{ab} = \sum_{q=1}^N \chi_{ja}(q)\chi_{k_b}(q). \qquad (21)$$

Expanding $\det[\chi_j^\dagger \chi_k]$ by columns and using again the property that a determinant with two equal columns is zero, one obtains

$$\Psi^*\Psi = (N!)^{-1} \sum_j \sum_k D_j D_k^* \sum_q |jkq| \qquad (22)$$

where $q = (q_1, q_2, ..., q_N)$ runs over the $N!$ permutations of $(1, 2, ..., N)$ and

$$|jkq| = \begin{vmatrix} \chi_{j_1}(q_1)\chi_{k_1}(q_1) & \cdots & \chi_{j_1}(q_N)\chi_{k_N}(q_N) \\ \vdots & \ddots & \vdots \\ \chi_{j_N}(q_1)\chi_{k_1}(q_1) & \cdots & \chi_{j_N}(q_N)\chi_{k_N}(q_N) \end{vmatrix}. \qquad (23)$$

We must stress that Eq. (22) is valid for any non-singular transformation of the φ_i orbitals to give the χ_i's. In the following section, we will see how the computation of the probability of having exactly v electrons in a given region $\Omega < R^3$ benefits from the specific transformation defined in Section 20.2, in particular from the fact that the χ_i's are orthogonal both in Ω and in its complementary region $\Omega' = R^3 - \Omega$.

20.4   ELECTRON NUMBER DISTRIBUTION FUNCTIONS (EDF)

Once R^3 has been exhaustively partitioned into two disjoint domains Ω and Ω', the probability of having exactly v electrons in Ω and $N - v$ in Ω' is given by [9, 10]

$$p^{\Omega}(v) = N!\Lambda \int_D \Psi^* \Psi dx_1 \cdots dx_N, \tag{24}$$

where the integration over all spins is implicitly assumed, D is a multidimensional domain in which the first v electrons are integrated in Ω and the last $N - v$ electrons in Ω', and $N!\Lambda = N!/(v!(N-v)!)$ is a combinatorial factor that accounts for electron indistinguishability. Using Eq. (22) in Eq. (24) we have

$$p^{\Omega}(v) = \sum_j \sum_k D_j D_k^* p_{j,k}^{\Omega}(v) \tag{25}$$

where

$$p_{j,k}^{\Omega}(v) = \Lambda \sum_q \int_D |jkq| dx_1 \cdots dx_N. \tag{26}$$

The next step shows the usefulness of the linear transformation described in Section 20.2. When the integration in Eq. (26) is performed, $p_{j,k}^{\Omega}(v)$ becomes a sum of $N!$ determinants built in with overlap integrals within Ω and Ω' between the χ_i's. Since these orbitals are orthogonal in the two regions, only the determinant $j = k$ survives. Besides this, only the overlap integrals in the diagonal of this determinant are non zero. A simple example can illustrate this. Taking $v = 2$, $N = 4$ and $q = (1, 3, 2, 4)$, the integral within the D domain results:

$$\int_D |jkq| dx_1 \cdots dx_N = \begin{vmatrix} \langle j_1|k_1 \rangle_{\Omega} & \langle j_1|k_2 \rangle_{\Omega'} & \langle j_1|k_3 \rangle_{\Omega} & \langle j_1|k_4 \rangle_{\Omega'} \\ \langle j_2|k_1 \rangle_{\Omega} & \langle j_2|k_2 \rangle_{\Omega'} & \langle j_2|k_3 \rangle_{\Omega} & \langle j_2|k_4 \rangle_{\Omega'} \\ \langle j_3|k_1 \rangle_{\Omega} & \langle j_3|k_2 \rangle_{\Omega'} & \langle j_3|k_3 \rangle_{\Omega} & \langle j_3|k_4 \rangle_{\Omega'} \\ \langle j_4|k_1 \rangle_{\Omega} & \langle j_4|k_2 \rangle_{\Omega'} & \langle j_4|k_3 \rangle_{\Omega} & \langle j_4|k_4 \rangle_{\Omega'} \end{vmatrix} \tag{27}$$

where the notation $j_p \equiv \chi_{jp}$ and $k_q \equiv \chi_{kq}$ has been used. Since the elements in j and k are ordered, the above determinant is null unless $j_i = k_i$ for $i = 1, 2,$ ..., N; i.e., unless $j = k$. Moreover, when this happens, only the diagonal elements are non zero since all j_i's are different. Hence, remembering from Eq. (4) that $\langle \chi_i|\chi_i \rangle_{\Omega} = n_i$ and $\langle \chi_i|\chi_i \rangle_{\Omega'} = 1 - n_i$, we have

$$\int_D |\boldsymbol{jkq}| dx_1 \cdots dx_N = n_{j_1}(1 - n_{j_2})n_{j_3}(1 - n_{j_4}) \tag{28}$$

In the general case, using the simpler notation $p_j^{\Omega}(v) \equiv p^{\Omega}_{j,k}(v)$ we obtain

$$p_j^{\Omega}(v) = \Lambda \sum_q \prod_{i=1}^N a_{j_i}(q_i) \left\{ \begin{array}{ll} a_{j_i}(q_i \leq v) & = n_{j_i} \\ a_{j_i}(q_i > v) & = 1 - n_{j_i} \end{array} \right. \tag{29}$$

and Eq. (25) becomes

$$p^{\Omega}(v) = \sum_j |D_j|^2 p_j^{\Omega}(v) = \sum_j d_j p_j^{\Omega}(v), \tag{30}$$

where $d_j = D_j D_j{}^* = |D_j|^2$. Equation (30) shows that each $p^{\Omega}(v)$ is given as a sum of electronic configuration contributions. In the case that $\psi(1, N)$ was given by $\chi_j / \sqrt{N!}$, $p_j^{\Omega}(v)$ would represent the probability of having exactly v electrons in Ω. This means that $\sum_v p_j^{\Omega}(v) = 1$ for every j and, given that the $p^{\Omega}(v)$'s also add to one, the d_j's also fulfill $\sum_j d_j = 1$. The last property may also be derived from $UU^{\dagger} = U^{\dagger}U = I$, and Eqs. (12), (16), (17), and (18).

Eq. (29) for $p_j^{\Omega}(v)$ has a very interesting physical interpretation. Let illustrate it by means of another simple example. Let $N = 3$, $v = 2$, $m = 2$, and $j = (1, 2, 4)$. Then, $\Lambda = 1 = (2!1!) = 1/2$, the eight possible q vectors are (1, 2, 3), (1, 3, 2), (2, 1, 3), (2, 3, 1), (3, 1, 2), and (3, 2, 1), and $p_j^{\Omega}(v)$ is given by:

$$p^{\Omega}_{1,2,4}(2) = n_1 n_2(1 - n_4) + n_1(1 - n_2)n_4 + (1 - n_1)n_2 n_4. \tag{31}$$

The factor $\Lambda = 1/(v!(N-v)!)$ in Eq. (29) always cancels the number of times that each product $\text{II}^N_{i=1} a_{ji}(q_i)$ in this equation is repeated with the same value when the sum over q is carried out. This is that way in all of the cases since any restricted permutation that affects only the ordering of electrons from 1 to v, on the one side, and the ordering of electrons from $v + 1$ to N, on the other side (and there are Λ^{-1} permutations of this type), produces the same $\text{II}^N_{i=1} a_{ji}(q_i)$ value, given that electrons from 1 to v electrons are integrated in Ω and electrons from $v + 1$ to N in Ω'. This means that $p_j^{\Omega}(v)$ may always be written as a sum of $\mathbb{C} = N!/(v!(N-v)!)$ terms of the form $a_{j_1} a_{j_2}, \ldots, a_{j_N}$, being $\mathbb{C}$ the number of possible ways to choose v of the a_{ji}'s equal to n_{ji}, being the other $N - v$ a_{ji}'s equal to $1 - n_{ji}$. Since $n_i = \langle \chi_i | \chi_i \rangle_{\Omega}$ is the probability that an electron described by the spin-orbital χ_i is inside Ω, each $a_{j_1} a_{j_2}, \ldots, a_{j_N}$ product represents the joint probability of a N event process given as the product of the probabilities of N independent events. For instance, $n_1 n_2 (1 - n_4)$ in the above example represents the joint probability that electrons 1 and 2 (described by χ_1 and χ_2, respectively) are in Ω, and electron 3 (described by X_4) is in Ω'. Since electrons are not distinguishable, the above product has to be symmetrized by choosing in $\mathbb{C} = 3!/(2!1!) = 3$ possible ways the distribution of the electrons in Ω and Ω', such that two of them are in Ω and the third one in Ω'. When $\Psi(1, N)$ is a Slater determinant, as in the case of a closed-shell restricted Hartree-Fock (RHF) wavefunction, the linear transformation described in Section 20.2 transforms the N-particle problem of determining the probability of having exactly v electrons in Ω in the N one-particle tasks of computing the probability that a single electron described by χ_i ($i = 1, ...,$ N) is inside Ω. For multi-determinant wavefunctions the above is still true for each configuration j.

All the $p_j^\Omega(n)$'s in Eq. (29) may be computed recursively [6] as follows. Defining $\eta_i = n_{ji}$ and $\lambda_i = 1 - \eta_i$ ($i = 0, 1, ..., N$), one has $p_j^\Omega(n) = f_n^N$ where

$$ f_0^0 = 1 \tag{32} $$

$$ f_k^i (1 \leq i \leq N) = \begin{cases} f_0^i = \lambda_i f_0^{i-1} \\ f_k^i = \eta_i f_{k-1}^{i-1} + \lambda_i f_k^{i-1} \\ \quad\quad 1 \leq k \leq i - 1 \\ f_i^i = \eta_i f_{i-1}^{i-1} \end{cases} \tag{33} $$

20.5 CHEMICAL BONDING INDICATORS

Perhaps the most obvious property that one can calculate for a domain Ω [R^3 is the average number of electrons that it contains, $\langle n^\Omega \rangle$. Within the quantum theory of atoms in molecules (QTAIM), $\langle n^\Omega \rangle$ is usually obtained from the molecular electron density $\rho(r)$ as [2]

$$ \langle n^\Omega \rangle = \int_\Omega \rho(r) dr. \tag{34} $$

However, from a purely probabilistic view point, it is clear that $\langle n^\Omega \rangle$ is also given by

$$\langle n^{\Omega} \rangle = \sum_{\nu=0}^{N} \nu p^{\Omega}(\nu),$$ (35)

and using Eq. (30) in the above equation one obtains

$$\langle n^{\Omega} \rangle = \sum_{j} d_j \sum_{\nu=0}^{N} \nu p_j^{\Omega}(\nu) = \sum_{j} d_j \langle n_j^{\Omega} \rangle.$$ (36)

As expected, $\langle n^{\Omega} \rangle$ is a weighted sum of the average number of electrons in Ω for each electronic configuration. Configurations with great D_j coefficients in the wavefunction (Eq. 17) will be more important in determining $\langle n^{\Omega} \rangle$. The value of $\langle n_j^{\Omega} \rangle$ will be trivially given as the sum of the n_i's of the spin-orbitals χ_i involved in the configuration j. This is so because n_i coincides with the total charge of this spin-orbital within Ω [11], i.e.,

$$n_i = \int_{\Omega} \chi_i^*(r)\chi_i(r)dr.$$ (37)

Thus, we have

$$\langle n_j^{\Omega} \rangle = \sum_{i=1}^{N} n_{j_i} \equiv \sum_{i \in j} n_i,$$ (38)

This equation is also derived in the Appendix. The average number of electrons in Ω, $\langle n^{\Omega} \rangle$, can be expressed in terms of all the n_i's and the γ_i quantities. The latter is defined as

$$\gamma_i = \sum_{j/i \in j} d_j.$$ (39)

Only the d_j coefficients of the configurations where χ_i participates are added. Since $\sum_j d_j = 1$, $\gamma_i < 1$ $\forall_i$ in case of a multi-determinant wavefunction ($N < 2m$), and $\gamma_i = 1$ $\forall_i$ if the wavefunction a Slater determinant. Using this definition and Eq. (38) in Eq. (36) we obtain

$$\langle n^{\Omega} \rangle = \sum_{i}^{2m} \gamma_i n_i.$$ (40)

This expression shows that the average electronic population of a region is given by a sum of all of the n_i's, as given by Eq. (37), each one weighted with a positive coefficient γ_i that measures the relative important of χ_i in $\Psi(1, N)$. We can also define the average electronic population for each χ_i within Ω, $\langle n_i \rangle = \gamma_i n_i = \sum_{j/i \in j} d_j n_i$, so that Eq. (40) becomes

$$\langle n^\Omega \rangle = \sum_i^{2m} \langle n_i \rangle. \tag{41}$$

Within the QTAIM, the bond order between two atoms (or groups of atoms) characterized by the basins (or groups of basins) Ω and Ω' is given by the delocalization index (DI) between these two basins, δ, which is a measure of the number of electron pairs shared by both. δ is usually obtained from the electron density $\rho(\mathbf{r}_1)$ and the two-particle density $\rho_2(\mathbf{r}_1, \mathbf{r}_2)$ as [3, 4]

$$\delta = -2 \int_\Omega \int_{\Omega'} [\rho_2(\mathbf{r}_1, \mathbf{r}_2) - \rho(\mathbf{r}_1)\rho(\mathbf{r}_2)]\, d\mathbf{r}_1 d\mathbf{r}_2. \tag{42}$$

However, it can also be obtained from the covariance of the electron populations of domains Ω and Ω' by using

$$\delta = -2 \left[\langle n^\Omega n^{\Omega'} \rangle - \langle n^\Omega \rangle \langle n^{\Omega'} \rangle \right]. \tag{43}$$

When there are only two regions, $n^\Omega + n^{\Omega'} = N$ and Eq. (43) becomes

$$\delta = 2[\langle (n^\Omega)^2 \rangle - \langle n^\Omega \rangle^2]. \tag{44}$$

The expression for $\langle n^\Omega \rangle$ has been previously derived (Eq. 40), and $\langle (n^\Omega)^2 \rangle$ is given by

$$\langle (n^\Omega)^2 \rangle = \sum_{\nu=0}^N \nu^2 p^\Omega(\nu). \tag{45}$$

Using Eq. (30) for $p^\Omega(\nu)$ we also have

$$\langle (n^\Omega)^2 \rangle = \sum_j d_j \langle (n_j^\Omega)^2 \rangle, \tag{46}$$

where

$$\langle (n_j^\Omega)^2 \rangle = \sum_{\nu=0}^{N} \nu^2 P_j^\Omega(\nu). \qquad (47)$$

It is also shown in the Appendix that the above sum is given by

$$\langle (n_j^\Omega)^2 \rangle = \langle n_j^\Omega \rangle + 2 \sum_{i<k} n_{j_i} n_{j_k}. \qquad (48)$$

In analogy with the definition of γ_i (Eq. 39), we define now the quantity γ_{ik} as

$$\gamma_{ik} = \gamma_{ki} = \sum_{j/i\in j, k\in j} d_j. \qquad (49)$$

Only those j configurations where χ_i and χ_k appear simultaneously contribute to γ_{ik}. From this definition and Eq. (39) we have $\gamma_{ik} \leq \gamma_i$ and $\gamma_{ik} \leq \gamma_k$. Using now Eq. (48) in Eq. (46), and the definitions of $\langle n^\Omega \rangle$ (Eq. 36) and γ_{ik} (Eq. 49), we obtain

$$\langle (n^\Omega)^2 \rangle = \langle n^\Omega \rangle + 2 \sum_{i\neq k} \gamma_{ik} n_i n_k. \qquad (50)$$

Finally, using Eqs. (40) and (50) in Eq. (44), δ becomes

$$\delta = 2 \sum_{i=1}^{2m} n_i \gamma_i (1 - n_i \gamma_i) + 4 \sum_{i>k}^{2m} n_i n_k (\gamma_{ik} - \gamma_i \gamma_k) \qquad (51)$$

$$\delta = \sum_{i=1}^{2m} \delta_i + \sum_{i>k}^{2m} \delta_{ik} = \delta_{\text{intra}} + \delta_{\text{inter}}. \qquad (52)$$

Taking into account that $\gamma_{ii} = \gamma_i$, Eq. (51) can also be recast in the form

$$\delta = 2 \sum_{i=1}^{2m} \gamma_i n_i (1 - n_i) + 2 \sum_{i,k}^{2m} n_i n_k (\gamma_{ik} - \gamma_i \gamma_k) \qquad (53)$$

For a single determinant wavefunction, $\gamma_i = \gamma_k = \gamma_{ik} = 1$ and Eq. (51) takes the simpler form

$$\delta = 2 \sum_{i=1}^{2m} n_i(1 - n_i) = \sum_{i=1}^{2m} \delta_i \quad \text{Slater determinant.} \tag{54}$$

This equation has been derived before and illustrates several important properties of the transformed orbitals χ_i derived in Section 20.2. First of all, it is worth noting that, in this case, these orbitals are those that diagonalize the so-called domain averaged Fermi hole (DAFH) defined by Ponec. Hence, as noted in a previous article, the DAFH orbitals describe statistically independent electrons. The Eq. (54) shows that the statistical independence also holds for the bond order expression, which is a sum of orbital terms. DAFH orbitals very localized in $\Omega (n_i \simeq 1)$ or in $\Omega' (n_i \simeq \Omega')$ give $\delta_i \simeq 0$ while orbitals localized half and half in Ω and $\Omega' (n_i = \frac{1}{2})$ give a maximal contribution $\delta_i = \frac{1}{2}$. For a closed-shell RHF wavefunction $n_i = n_{i+m}$ and δ acquires the form $\delta = 4 \sum_{i=1}^{m} n_i(1 - n_i)$.

Given that $0 \leq n_i \leq 1$ and $0 \leq \gamma_i \leq 1$ the first term in Eq. (51) is always positive. The second term, however, can be positive or negative depending on the relative values of the γ_{ik} and $\gamma_i \gamma_k$ quantities. When χ_i and χ_k appear together in all the j configurations $\gamma_i = \gamma_k = \gamma_{ik}$, so that $\gamma_{ik} - \gamma_i \gamma_k > 0$ and this pair of spin-orbitals increases δ. On the contrary, when the set of determinants in which χ_i appears is disjoint with that involving $\chi_i \gamma_{ik} = 0$, and the pair (i, k) contributes negatively to δ. In summary, pairs of spin-orbitals appearing together or avoiding each other in many configurations tend increase or decrease the bond order, respectively.

20.6   SPIN-RESOLVED EDF

So far we have been involved in computing $p^{\Omega}(v)$; i.e., the probability of having v electrons in Ω and $N - v$ in Ω' regardless the spins of these electrons. Now, we intend to find a richer description of the EDF computing the probability $p^{\Omega}(v_\alpha, v_\beta)$ that exactly $v_\alpha \alpha$ electrons and $v_\beta \beta$ electrons are inside Ω, which implies that $N_\alpha - v_\alpha \alpha$ electrons and $N_\beta - v_\beta \beta$ are in Ω'. The starting point is the equation

$$p^{\Omega}(v_\alpha, v_\beta) = N! \Lambda \int_D |\Psi(1, N)|^2 dx_1 \cdots dx_N, \tag{55}$$

where $\Lambda = \Lambda_\alpha \Lambda_\beta$, $\Lambda_\sigma = [v_\sigma!(N_\sigma - v_\sigma)!]^{-1}$ ($\sigma = \alpha, \beta$), D is a multidimensional domain in which electrons from 1 to v_α are integrated in Ω, from $v_\alpha + 1$ to N_α in Ω', from $N_\alpha + 1$ to $N_\alpha + v_\beta$ in Ω, and from $N_\alpha + v_\beta + 1$ to N in Ω', and the integration over all spins is not implicitly assumed. On the contrary, any overlap integral in Ω or Ω' between two spin-orbitals depending on the coordinates of electrons from $N_\alpha + 1$ to N, or between two β spin-orbitals depending on the coordinates of electrons from 1 to N_α will be 0. As a consequence, only the restricted q permutations in Eq. (26) that contain electrons from 1 to N_α in the first N_α positions and electrons from $N_\alpha +1$ to N in the last N_β positions contribute to $p^\Omega(v_\alpha, v_\beta)$. This fact allows us to write

$$\sum_q \equiv \sum_{q_\alpha} \sum_{q_\beta}, \tag{56}$$

where q_α and q_β run over the $N_\alpha!$ and $N_\beta!$ permutations of $(1, ..., N_\alpha)$ and $(N_\alpha + 1, ..., N)$, respectively. Similarly, remembering that $1 < j_i \leq m$ for $0 < i \leq N_\alpha$ and $m < j_i \leq 2m$ for $N_\alpha < i \leq N$, one can write $j = (j_\alpha, j_\beta)$ and, $\sum_j \equiv \sum_{j_\alpha} \sum_{j_\beta}$ where j_α runs over all possible combinations of N_α elements chosen from the $(1, ..., m)$ set, and j_β over all possible combinations of N_β elements chosen from the $(m + 1, ..., .2m)$ set. After considering all these points, Eq. (29), modified to take into account spin-resolved probabilities can be written as

$$p^\Omega_{j_\alpha, j_\beta}(v_\alpha, v_\beta) = p^\Omega_{j_\alpha}(v_\alpha) \times p^\Omega_{j_\beta}(v_\beta), \tag{57}$$

with

$$p^\Omega_{j_\alpha}(v_\alpha) = \Lambda_\alpha \sum_{q_\alpha} \prod_{i=1}^{N_\alpha} a_{j_i}(q_i) \tag{58}$$

$$p^\Omega_{j_\beta}(v_\beta) = \Lambda_\beta \sum_{q_\beta} \prod_{i=N_\alpha+1}^{N} a_{j_i}(q_i). \tag{59}$$

where $a_{ji}(q_i)$ in Eq. (58) is n_{ji} for $q_i \leq v_\alpha$ and $1 - n_{ji}$ for $q_i > v_\alpha$, and $a_{ji}(q_i)$ in Eq. (59) is n_{ji} for $N_\alpha < q_i \leq N_\alpha + v_\beta$ and $1 - n_{ji}$ for $q_i > N_\alpha + v_\beta$. The final expression for $p^\Omega(v_\alpha, v_\beta)$ is

$$p^\Omega(v_\alpha, v_\beta) = \sum_{j_\alpha} \sum_{j_\beta} d_{j_\alpha, j_\beta} \times p^\Omega_{j_\alpha}(v_\alpha) \times p^\Omega_{j_\beta}(v_\beta). \tag{60}$$

Within a configuration $j = (j_\alpha, j_\beta)$ the spin-unresolved probability of having exactly v electrons in Ω is given in terms of $p_{j\alpha}{}^\Omega(v_\alpha)$ and $p_{j\beta}{}^\Omega(v_\beta)$ by

$$p_j^\Omega(v) = \sum_{\substack{v_\alpha, v_\beta \\ v_\alpha + v_\beta = v}} p_{j\alpha}^\Omega(v_\alpha) \times p_{j\beta}^\Omega(v_\beta). \tag{61}$$

Equation (60) crystal clear shows the statistical independence of the α and β subsets of electrons within each j configuration. For a single-determinant wavefunction, j_α and j_β can only be $(1, 2, ..., N_\alpha)$ and $(N_\alpha, N_\alpha + 1, ..., N)$, respectively, $d_j = 1$, and the statistical independence is complete. Moreover, for a closed shell single-determinant wavefunction $p_{j\alpha}{}^\Omega(v_\alpha)$ and $p_{j\beta}{}^\Omega(v_\beta)$ are the same. As soon as the wavefunction has more than a Slater determinant, the statistical independence breaks down in general, since the d_j coefficient, that can be written as (see Eq. (18)) cannot be factorized into α and β parts.

$$d_j = \left(\sum_{r=1}^M C_r U_{rj}\right)^2 = \left(\sum_{r=1}^M C_r U_{rj}^\alpha U_{rj}^\beta\right)^2, \tag{62}$$

The average number of α electrons is Ω is given by

$$\langle n_\alpha^\Omega \rangle = \sum_{v_\alpha=0}^{N_\alpha} \sum_{v_\beta=0}^{N_\beta} v_\alpha p^\Omega(v_\alpha, v_\beta). \tag{63}$$

Using Eq. (60) in Eq. (63), and remembering that the probabilities of the α and β blocks are individually normalized, i.e., $\sum_{v\beta = 0}^{N\beta} P_{j\beta}{}^\Omega(v_\beta) = 1$, we arrive to

$$\langle n_\alpha^\Omega \rangle = \sum_{j_\alpha} d_j^\alpha \langle n_{j_\alpha}^\Omega \rangle, \tag{64}$$

where

$$d_{j_\alpha} = \sum_{j_\beta} d_j, \tag{65}$$

and $\langle n_{j\alpha}^\Omega \rangle$ is the average number of α electrons in Ω for the j_α configuration, given by

$$\langle n_{j_\alpha}^\Omega \rangle = \sum_{v_\alpha=0}^{N_\alpha} v_\alpha P_{j_\alpha}^\Omega(v_\alpha). \tag{66}$$

This is the analogous to Eq. (38) when spin-resolved probabilities are not considered. In analogy with the definition of γ_i (Eq. 39) we could define $\gamma_i^\alpha = \sum_{ja/ieja}$ However, this definition is unnecessary since γ_i^α is identical to γ_i. It is also easy to see that $\langle n_\alpha^\Omega \rangle$ can be written as

$$\langle n_\alpha^\Omega \rangle = \sum_{i=1}^{m} \gamma_i n_i. \tag{67}$$

Similarly, $\langle n_\beta^\Omega \rangle = \sum_{i=m+1}^{2m} \gamma_i n_i$ with $\langle n_\alpha^\Omega \rangle + \langle n_\beta^\Omega \rangle = \langle n^\Omega \rangle$.

We now analyze the spin-resolved expression for δ. Since $n^\Omega = n_\alpha^\Omega + n_\beta^\Omega$, $\langle n^\Omega \rangle = \langle n_\alpha^\Omega \rangle + \langle n_\beta^\Omega \rangle$, and $n_\sigma^\Omega + n_\sigma^{\Omega'} = N_\sigma (\sigma = \alpha, \beta)$ we have

$$\delta = \sum_{\sigma_1} \sum_{\sigma_2} \delta_{\sigma_1 \sigma_2}, \quad (\sigma_1 \sigma_2) = (\alpha\alpha), (\alpha\beta), (\beta\alpha), (\beta\beta), \tag{68}$$

where

$$\delta_{\sigma_1 \sigma_2} = -2 \left[\langle n_{\sigma_1}^\Omega n_{\sigma_2}^{\Omega'} \rangle - \langle n_{\sigma_1}^\Omega \rangle \langle n_{\sigma_2}^{\Omega'} \rangle \right] = 2 \left[\langle n_{\sigma_1}^\Omega n_{\sigma_2}^\Omega \rangle - \langle n_{\sigma_1}^\Omega \rangle \langle n_{\sigma_2}^\Omega \rangle \right]. \tag{69}$$

The expression for $\langle (n_\alpha^\Omega)^2 \rangle$ is

$$\langle (n_\alpha^\Omega)^2 \rangle = \sum_{\nu_\alpha=0}^{N_\alpha} \sum_{\nu_\beta=0}^{N_\beta} \nu_\alpha^2 p^\Omega(\nu_\alpha, \nu_\beta). \tag{70}$$

After using Eq. (60) in Eq. (70), Eq. (65), $\gamma_i^\alpha = \sum_{ja/ieja} d_{ja} \equiv \gamma_i$ and $\gamma_{ik}^{\alpha\alpha} = \sum_{ja/ieja,keja} d_{ja} \equiv \gamma_{ik}$ the property, $\sum_{n\beta} p_{n\beta}^\Omega(n_\beta) = 1$ and the results of the Appendix we obtain:

$$\delta_{\alpha\alpha} = 2 \sum_{i=1}^{m} n_i \gamma_i (1 - n_i \gamma_i) + 4 \sum_{i>k}^{m} n_i n_k (\gamma_{ik} - \gamma_i \gamma_k). \tag{71}$$

An analogous equation holds for $\delta_{\beta\beta}$ restricting i and k to be in the range $i + 1 \le i, k \le 2m$. Finally, it is trivial to show that $\delta_{\alpha\beta} = \delta_{\beta\alpha}$ is given by

$$\delta_{\alpha\beta} = \delta_{\beta\alpha} = 2 \left[\sum_{i=1}^{m} \sum_{k=m+1}^{2m} n_i n_k (\gamma_{ik} - \gamma_i \gamma_k) \right]. \tag{72}$$

20.7   ILLUSTRATIVE RESULTS AND DISCUSSION

Three very simple examples have been considered to illustrate the type of results that one can obtain related to the expressions defined in the previous sections: H_2, H_2O, and LiH molecules, all of them in the electronic ground state. We have performed complete active space (CAS) calculations with standard 6-311G(d,p) basis sets, placing n electrons into m valence orbitals. The (n, m) pairs are (2, 2), (6, 5), and (2, 2) for H_2, H_2O, and LiH, respectively. In each of the three molecules, the physical space R^3 has been exhaustively partitioned into two domains Ω and $\Omega' = R^3 - \Omega$, according to Bader's QTAIM prescription [2], taking Ω equal to the left hydrogen (say, H_A), the O atom, and the H atom of H_2, H_2O, and LiH molecules, respectively. The S matrix defined in Eq. (1) has been diagonalized to obtain as many n_i eigenvalues as the number of MOs in the corresponding CAS$[n, m]$ wavefunction. From the eigenvectors of S, the transformed orbitals χ_i have been obtained by using Eq. (3), and from these MOs the molecular wavefunction $\Psi(1, N)$ reconstructed as indicated in Eqs. (17) and (18). Finally, the $p(v)$ probabilities and chemical bonding indicators were obtained with the expressions derived in the past three sections.

20.7.1 H_2

At this level, the wavefunction of H_2 is the linear combination

$$\Psi(1, 2) = C_1 |g^+ g^-| + C_2 |u^+ u^-|, \tag{73}$$

where $|g^+ g^-| \equiv |1\sigma_g \alpha 1\sigma_g \beta|$ and $|u^+ u^-| \equiv |1\sigma_u \alpha 1\sigma_u \beta|$ are normalized Slater determinants. At very large interatomic H-H distances (R_{HaHb}), both mixing coefficients are equal, $C_1 = C_2 = 2^{-1/2}$, while $C_1 \gg C_2$ at distances close to equilibrium. Since we are considering a closed system, the overlap matrix S within the left H atom of the pair (say, H_a) is the same for α and β blocks, with $\langle g|g \rangle A = \langle u|u \rangle A = 1/2$ at any R_{HaHb}, and the non-diagonal element, $s = \langle g|u \rangle A$, changes with R_{HaHb} as shown in Figure 20.1a. The two eigenvalues of S ($n_1 = 1/2 - s$ and $n_2 = 1/2 + s$) are also shown in the figure. Since these eigenvalues reach the limits $n_1 \to 0$ and $n_2 \to 1$ at very large internuclear distances, their associated eigenvectors are called $b = 2^{-1/2}(g - u)$ and $a = 2^{-1/2}(g + u)$, because they are highly localized in H_b and

H_a, respectively, this localization becoming perfect in the limit $R_{HaHb} \to \infty$. Despite these names, it is important to remember that $\langle a|b \rangle = 0$. In terms of a and b, Ψ is written as

$$\Psi(1,2) = D_1(|a\bar{b}| + |b\bar{a}|) + D_2(|a\bar{a}| + |b\bar{b}|) \tag{74}$$

$$= D_1\Psi_{cov}(1,2) + D_2\Psi_{ion}(1,2). \tag{75}$$

The coefficients D_1 and D_2 are plotted in Figure 20.1b. It is apparent from Eq. (74) and Figure 20.1b that our results are very similar to a mixing of the covalent $(ab + ba)$ and ionic $(aa + bb)$ structures of the Heitler-London (HL) model for $^1\Sigma_g^+$ H_2 molecule. However, unlike that model, where a and b are atomic orbitals (AO), here they represent orthonormal molecular orbitals, albeit it is true that very localized on both nuclei. Actually, at very long distances, a and b coincide with the HL AOs. This is evident from the limits $n_1 \to 0$ and $n_2 \to 1$ when $R_{HaHb} \to \infty$ observed in Figure 20.1a. Overall, we can say that a and b have almost an atomic character, but each them has a tail that is a small contamination with basis functions of the opposite center to achieve their orthogonality at any distance. This contamination becomes progressively smaller as R_{HaHb} increases and disappears completely in the $R_{HaHb} \to \infty$ limit.

Since there is only a α electron and a β electron and the eigenvalues of **S** are related by $n_1 + n_2 = 1$, all the EDFs can be given simple expressions in terms $n \equiv n_2$. A single α or β electron occupying the orbital a has probabilities n and $1 - n$ of being in H_a and H_b, respectively, with these two

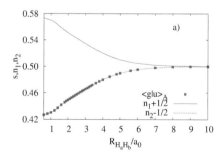
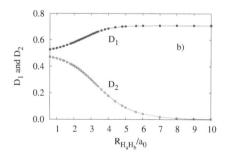

FIGURE 20.1 (a) Overlap integral $s = \langle g|u \rangle_A$ and n_1 and n_2 eigenvalues of H2/CAS[2,2]/6-311G(p) molecule. Note that $s = n_2 - 1/2 = 1/2 - n_1$ and $n_1 + n_2 = 1$. (b) Mixing coefficients D_1 and D_2 of the wavefunction $\Psi(1, 2) = D_1 (|a\bar{b}| + |b\bar{a}|) + D_2(|a\bar{a}| + |b\bar{b}|)$ of H_2/CAS[2,2]/6-311G(p) molecule.

numbers interchanged if the electron is in the orbital b. From this single electron results, the probability of finding both electrons in $\mathbf{H}_a$ or $\mathbf{H}_b$ for the $|a\bar{b}|$ and $|b\bar{a}|$ configurations is $p(2, 0) = p(0, 2) = n(1 - n)$. The probability of having the α electron in $\mathbf{H}_a$ and the β electron in $\mathbf{H}_b$ is $p(1_\alpha, 1_\beta) = n^2$ and $p(1_{\alpha'}, 1_\beta) = (1 - n)^2$ for the $|a\bar{b}|$ and $|b\bar{a}|$ configurations, respectively. When β [$\mathbf{H}_a$ and α [$\mathbf{H}_b$, the probabilities are exactly the opposite, i.e., $(1 - n)^2$ for $|a\bar{b}|$ and n^2 for $|b\bar{a}|$. Finally, ignoring the spin, both $|a\bar{b}|$ and $|b\bar{a}|$, as well as $\Psi_{cov}(1, 2) = (|a\bar{b}| + |b\bar{a}|)$ (the equivalent to the covalent HL structure), give the same EDF: $p(2, 0) = p(0, 2) = n(1 - n)$ and $p(1, 1) = n^2 + (1 - n)^2$. In the ionic HL structure $|a\bar{a}|$, $p(2, 0)$, $p(1, 1)$, and $p(0, 2)$ are n^2, $2n(1 - n)$, and $(1 - n)^2$, with $p(2, 0)$ and $p(0, 2)$ interchanged for $|b\bar{b}|$. As we can see, $p(1, 1)$ for $\Psi_{cov}(1, 2)$ is equal to $p(2, 0) + p(0, 2)$ for $\Psi_{ion}(1, 2)$ and *vice versa*. The evolution of all these probabilities with R_{HaHb} can be seen in Figure 20.2a. $\Psi_{cov}(1, 2)$ gives $p(1, 1) > 0.8$ at any R_{HaHb}, approaching the value 1.0 at very large distances. Consequently, as expected, the ionic structures $(2, 0)$ and $(0, 2)$ become less and less probable for the covalent configuration $\Psi_{cov}(1, 2)$ when R_{HaHb} increases. The results for $\Psi_{ion}(1, 2)$ (not shown in the figure) are the other way around. Since $D_2 \to 0$ at large distances, the $p(1, 1)$ value of $\Psi_{cov}(1, 2)$ approaches the $p(1, 1) = 1.0$ value of $\Psi(1, 2)$ in the limit $R_{HaHb} \to \infty$.

According to the results of the above paragraph, the four n_i's in the order $n_2 = n$, $n_1 = 1 - n$, $n_4 = n$, $n_3 = 1 - n$ correspond to orbitals a, b, $\bar{a}$, and $\bar{b}$, respectively. The four configurations are $|a\bar{b}|$, $|b\bar{a}|$, $|a\bar{a}|$, and $|b\bar{b}|$, with coefficients D_1, D_1, D_2, and D_2. Since, every orbital appears in two configurations, with coefficients D_1 and D_2, from Eq. (39) we have $\gamma_i = D^2_1 + D^2_2 = 1/2 \; \forall_i$. Using

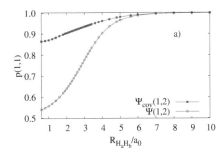 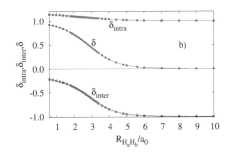

FIGURE 20.2   (a) $p(1, 1)$ probability for the wavefunctions $\Psi_{cov}(1, 2)$ and $\Psi(1, 2) = D_1\Psi_{cov}(1, 2) + D_2\Psi_{ion}(1, 2)$ of H_2//CAS[2,2]//6-311G(p) molecule. The sum $p(2, 0) + p(0, 2)$ of $\Psi_{ion}(1, 2)$ is equal to $p(1, 1)$ of $\Psi_{cov}(1, 2)$, and *viceversa*. (b) $\delta_{intra} = \sum_i \delta_i$ and $\delta_{inter} = \sum_{i>k} \delta_{ik}$ components of the delocalization index δ, Eq. (52).

similar arguments, the γ_{ik} coefficients are $D^2{}_1$ for the pairs $a\bar{b}$ and $b\bar{a}$, and $D^2{}_2$ for the pairs $a\bar{a}$ and $b\bar{b}$. From these n_i's, γ_i's, and γ_{ik}'s the δ_{intra} contribution to the DI δ is given simply by $\delta_{intra} = 1 + 2n - 2n^2$ (see Eq. 52), and the δ_{inter} component results $8n(1 - n)(D_2{}^2 - D^2{}_1) + 4D_2{}^2 - 1$. The total DI δ, as well as, δ_{intra} and δ_{inter}, are plotted in Figure 20.2b. Since $n \rightarrow 1$ and $D_2 \rightarrow 0$ in the limit $R_{HaHb} \rightarrow \infty$, δ_{intra} and δ_{inter} tend to 1.0 and -1.0 at large distances, respectively. However, since n does not change much (for instance, $n = 0.927$ at $R_{HaHb} = 0.60$ bohr), δ_{intra} decreases only slightly in going from $R_{HaHb} = 0.60$ bohr up to ∞, whereas δ_{inter}, always negative, decreases appreciably with the distance, determining the final shape of δ.

20.7.2 H_2O

A summary of the results obtained for water molecule at the equilibrium geometry using a CAS[6,5]/6-311G(d,p) level of calculation appears in Figure 20.3b. The wavefunction expressed in terms of the canonical MOs φ_i is composed of 28 determinants, while it is made of 120 determinants if the transformed MOs χ_i given by Eq. (3) are used instead. However, as it is clear from the absolute values of the mixing coefficients D_j (Eq. 17), ordered by decreasing values in the figure, less than 50% of these determinants have an appreciable influence on the $p(v)$ probabilities. The $p(v)$ values for $v \leq 6$ are completely negligible for both the total wavefunction $\Psi(1, N)$ and each

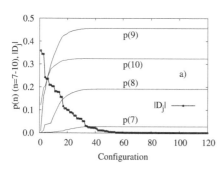
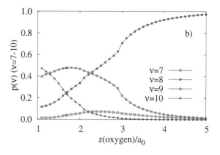

FIGURE 20.3 (a) Probability of the structures O^+ ($p(7)$), O^0 ($p(8)$), O^- ($p(9)$), and O^{2-} ($p(10)$) as more configurations are added to the wavefunction, and absolute value of the mixing coefficient $|D_j|$ of each configuration of H_2 molecule at its equilibrium geometry. (b) Evolution of $p(v)$ ($v = 7 - 10$), probability of the oxygen atom of H_2O molecule having v electrons, as a function of its z coordinate. The positions of the three atoms of the molecule are $(0, -0.715, 0)$, $(0, +0.715, 0)$, and $(0, 0, z)$ in the order H_1, H_2, O.

of the χ_j configurations. This means that resonant structures in which the oxygen atom has a total positive charge greater than +1 do not play any relevant role in the description of the chemical bonding of this molecule. On the other hand, the $p(v)$ probabilities for $v = 7 - 10$ are already saturated with about one third (i.e., 40 configurations) of the total number of χ_j determinants. The most probable oxidation of oxygen atom corresponds to $v = 9$ (O^-), followed by $v = 10$ (O^{2-}) and $v = 8$ (O^0). The associated probabilities (in the same order) are 0.4565, 0.3239, and 0.1914, respectively. We must note that the structure O^+ has a small (0:0269) but non negligible probability.

It is also very illuminating to analyze how the $p(v)$ probabilities change with the molecular geometry. We have performed calculations of water molecule placing both hydrogen atoms at fixed positions (0,–0.715, 0) and (0, +0.715, 0) (in atomic units) and moving the position of oxygen atom along the z axis, with z values changing from 1.0 to 5.0 bohr. The results are shown in Figure 20.3b. It is worth noting several facts. For very large z values, water molecule transforms into a H_2 molecule, with a bond critical point (BCP) between the two **H** atoms, and an (almost) isolated and neutral oxygen atom which, in turn, is connected with the H–HBCP. Due to this, the only probability having an appreciable value in the $z \to \infty$ limit is $p(8)$, corresponding to O^0 (it can be shown that the electronic state of this neutral oxygen is a ^1D), with all the other $p(v)$ with $v \neq 8$ vanishing in this limit. Another interesting point is that O^- and O^{2-} structures are of similar importance at about $z = 1.25$ bohr, with their probabilities being much larger than that of neutral oxygen. For $z < 1.25$, the structure O^{2-} dominates over the structure O^-, while the contrary happens for $z > 1.25$. $p(10)$ decreases continuously with z, crossing with $p(8)$ at a value of z between 1.6 and 1.7. After a maximum at $z = 1.8$, $p(9)$ decreases also monotonically with z, although much more slowly than $p(10)$, and crosses with $p(8)$ at a value of z between 2.2 and 2.3. Finally, it is interesting to observe in Figure 20.3b that the certainly unusual O^+ structure has a probability (with a maximum at $z = 2.5$) greater than that of the more standard one, O^{2-}, for $z > 2.2$.

20.7.3   LiH

As a final example, we analyze the results for the ground state of the LiH molecule at the CAS[2,2]//6-311G(p) level. The wavefunction is composed of four and nine determinants when canonical (φ_i) and transformed (χ_i) MOs are used, respectively. The convergence of $p(v)$ $(v = 0 - 4)$ at the equilibrium Li-H distance as new j configurations are added to the wavefunction is illustrated in Figure 20.4a, and the three eigenvalues of the $\mathbf{S}$ matrix as a function of the Li-H distance are plotted in Figure 20.4b. As we can see, the $|D_j|$ coefficient takes a significant value only for the first four configurations, so that all $p(v)$'s are practically converged at $j = 4$. On the other hand, the lowest eigenvalue of $\mathbf{S}$ (n_1) is very close to zero at any $R_{\text{Li-H}}$. According to the results in Section 4, this means that the transformed MO χ_1 is almost fully localized on the Li atom or, equivalently, that two of the four electrons of the molecule are always in this atom, so that $p(3) \simeq p(4) \simeq 0$ for all Li-H distances. This is corroborated in Figure 20.5, where we observe that only $p(1)$ and $p(2)$, associated to the Li^0H^0 and Li^+H^- structures, respectively, take a significant value; i.e., $p(1) + p(2) \simeq 1.0$. Except at very short Li-H distances the $p(1)$ and $p(2)$ curves are mirror images of each other: For $R_{\text{Li-H}} < 5.78$ bohr the ionic structure Li^+H^- dominates over the neutral one Li^0H^0, and the contrary happens for $R_{\text{Li-H}} > 5.78$. The value $R_{\text{Li-H}} = 5.78$ signals the avoided crossing point between both potential energy surfaces, and closely corresponds to the maximum of the DI δ and to the region where $\delta = 0.5$.

Another example of how well the EDF of this system can be modeled by just two contributions, $p(1)$ and $p(2)$, is the value of the net charge (electronic plus nuclear) of lithium atom, Q_{Li}, that should be equal to $p(2)$ if the equation $p(1) + p(2) = 1$ was exact. Actually, as observed in Figure 20.5, this is almost true for $R_{\text{Li-H}}$ larger than $\simeq 2.5$ bohr.

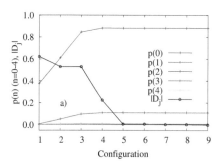
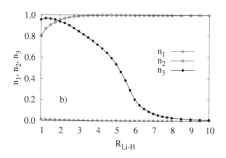

FIGURE 20.4 (a) Evolution of the probability that the hydrogen atom of LiH molecule has a total charge H^{3-} $(p(4))$, H^{2-} $(p(3))$, H^- $(p(2))$, H^0 $(p(1))$, and H^+ $(p(0))$ as more configurations are added to the wavefunction, and absolute value of the mixing coefficient $|D_j|$ of each configuration at the equilibrium $R_{\text{Li-H}}$ distance. (b) Evolution with $R_{\text{Li-H}}$ of the three eigvenvalues of S matrix.

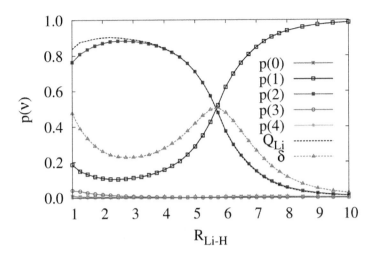

FIGURE 20.5   Evolution of the probability that the hydrogen atom of LiH molecule has a total charge H^+ ($p(0)$), H^0 ($p(1)$), H^{1-} ($p(2)$), H^{2-} ($p(3)$), and H^{3-} ($p(4)$) with R_{Li-H}. Q_{Li} is the net charge of lithium atom.

KEYWORDS

- **domain averaged Fermi hole**
- **electron localizability indicator**
- **electron localization function**
- **electron number distribution functions**
- **maximum probability domain**
- **molecular orbitals**
- **quantum theory of atoms in molecules**

REFERENCES

1. Acke, G., De Baerdemacker, S., Claeys, P. W., Van Raemdonck, M. W. P., & Van Neck, D. P., (2016). *B., Mol. Phys., 114*, 1392.
2. Bader, R. F. W., (1990). *Atoms in Molecules*. Oxford University Press, Oxford.
3. Bader, R. F. W., & Stephens, M. E., (1974). *Chem. Phys. Lett., 26*, 445.
4. Bader, R. F. W., & Stephens, M. E., (1975). *J. Am. Chem. Soc., 97*, 7391.

5. Cancès, E., Keriven, R., Lodier, F., & Savin, A., (2004). *Theor. Chem. Acc., 111*, 373–380.

6. Cancès, E., Keriven, R., Lodier, F., & Savin, A., (2004). *Theor. Chem. Acc. 111*, 373

7. Causà, M., & Savin, A., (2011). *J. Phys. Chem. A, 115*(13), 139.

8. Chauvin, R., Lepetit, C., Silvi, B., & Alikhani, E. (eds.). *Applications of Topological Methods in Molecular Chemistry*. Springer Intl.

9. Daudel, R., (1968). *The Fundamentals of Theoretical Chemistry*. Pergamon Press, Oxford (1968)

10. Francisco, E., Martín Pendás, A., & Blanco, M. A., (2008). *Comp. Phys. Commun., 178*, 621.

11. Francisco, E., Martín Pendás, A., & Blanco, M. A., (2009). *J. Chem. Phys., 131, 124*,125.

12. Francisco, E., Martín Pendás, A., & Blanco, M. A., (2011). *Theor. Chem. Acc., 128*, 433–444.

13. Francisco, E., Martín Pendás, A., & Costales, A., (2014). *Phys. Chem. Chem. Phys., 16*, 4586–4597.

14. Francisco, E., Martín Pendás, A., García-Revilla, M., & Álvarez Boto, R., (2013). *Comput. Theor. Chem., 1003*, 71.

15. Francisco, E., Martín Pendás, A., García-Revilla, M., & Álvarez Boto, R., (2013). *Comput. Theor. Chem., 1003*, 71.

16. Gallegos, A., Carbó-Dorca, R., Lodier, F., Cancès, E., & Savin, A., (2005). *J. Comput. Chem., 26*, 455.

17. Kohout, M., (2004). *Int. J. Quant. Chem., 97*, 651.

18. Mafra Lopes Jr., O., Brada, B., Causà, M., & Savin, A., (2012). *Prog. Theor. Chem. Phys., 22*, 173.

19. Martín Pendás, A., Francisco, E., & Blanco, M. A., (2007). *J. Phys. Chem. A., 111*, 1084.

20. Martín Pendás, A., Francisco, E., & Blanco, M. A., (2007). *Phys. Chem. Chem. Phys., 9*, 1087.

21. Martín Pendás, A., Francisco, E., & Blanco, M. A., (2007). Pauling resonant structures in real space through electron number probability distributions. *J. Phys. Chem. A, 111*, 1084.

22. Menéndez, M., & Martín Pendás, A., (2014). *Theor. Chem. Acc., 133*, 1539.

23. Menéndez, M., Martín Pendás, A., Brada, B., & Savin, A., (2015). *Comput. Theor. Chem., 1053*, 142.

24. Ponec, R., (1997). *J. Math. Chem., 21*, 323.

25. Ponec, R., (1998). *J. Math. Chem., 23*, 85.

26. Silvi, B., & Savin, A., (1994). *Nature, 371*, 683.

APPENDIX

In this appendix we show that $\langle n_j^\Omega \rangle = \sum_{n=0}^N n P_j^\Omega(n)$ and $\langle (n_j^\Omega)^2 \rangle = \sum_{n=0}^N n^2 P_j^\Omega(n)$ are given by $\sum_{i=1}^N n_{ji}$ and $\langle n_j^\Omega \rangle + 2\sum_{i<k} n_{ji} n_{jk}$, respectively. Using Eq. (29) for $P_j^\Omega(n)$, taking into account the arguments that permit to express it in a form analogue to that given in Eq. (31), and considering for simplicity the natural configuration $j = (1, 2, ..., N)$, we can write:

$$
\begin{aligned}
\sum_{n=0}^N n^u P_j^\Omega(n) &= \langle (n_j^\Omega)^u \rangle \\
&= N^u \left[n_1 n_2 n_3 \ldots n_N \right] \\
&+ (N-1)^u \left[(1-n_1) n_2 n_3 \ldots n_N + \ldots \right] \\
&+ (N-2)^u \left[(1-n_1)(1-n_2) n_3 \ldots n_N + \ldots \right] \\
&+ \ldots
\end{aligned}
\tag{76}
$$

After a lengthy but easy manipulation, it turns out that Eq. (76) can be written by means of the following set of coupled equations:

$$
\langle (n_j^\Omega)^u \rangle = \mathcal{P}_N^u \prod_k n_k + \sum_{i_1=1}^N A_{i_1}
\tag{77}
$$

$$
A_{i_1} = \mathcal{P}_{N-1}^u \prod_{k \neq i_1} n_k + \sum_{i_2 > i_1}^N A_{i_2}
\tag{78}
$$

$$
A_{i_2} = \mathcal{P}_{N-2}^u \prod_{k \neq i_1, i_2} n_k + \sum_{i_3 > i_2}^N A_{i_3}
\tag{79}
$$

$$
\vdots
\tag{80}
$$

$$
A_{i_{N-2}} = \mathcal{P}_2^u \prod_{k \neq i_1, \ldots, i_{N-2}} n_k + \sum_{i_{N-1} > i_{N-2}}^N A_{i_{N-1}}
\tag{81}
$$

$$
A_{i_{N-1}} = \mathcal{P}_1^u \prod_{k \neq i_1, \ldots, i_{N-1}} n_k + \mathcal{P}_0^u,
\tag{82}
$$

where

$$\mathcal{P}_K^u = \sum_{i=0}^{K} (-1)^i \binom{K}{i} (K-i)^u \qquad K = 0, \ldots, N. \tag{83}$$

For $u = 1$, $P^u_1 = 1$ and $P^u_{K \neq 1} = 0$, and Eqs. (77)–(82) give

$$\langle n_j^\Omega \rangle = \sum_{i_1} \cdots \sum_{i_{N-1}} \prod n_k, \tag{84}$$

where each $\sum_{i_s}$ runs for $i_s > i_{s-1}$, and the $\prod$ runs for $k \neq i_1, \ldots, i_{N-1}$. After developing the above summations, Eq. (38) is obtained (remember that we are considering the case $j_1 = 1, \ldots, j_N = N$). While developing Eq. (85), the Eq. (48) follows. For $u = 2$, $\mathcal{P}^u_{K \geq 3} = 0$, $\mathcal{P}^u_2 = 2$, $\mathcal{P}^u_1 = 1$, and $\mathcal{P}^u_0 = 0$, and Eqs. (77)–(82) give

$$\langle (n_j^\Omega)^2 \rangle = \sum_{i_1} \cdots \sum_{i_{N-2}} \left[2 \prod_{i_{N-1}} n_k + \sum_{i_{N-1}} \prod n_k \right], \tag{85}$$

where the first and second $\prod$'s run for $k \neq i_1, \ldots, i_{N-2}$ and $k \neq i_1, \ldots, i_{N-1}$, respectively. While developing Eq. (85), Eq. (48) follows.

CHAPTER 21

THE YOTTAFLOP FRONTIER OF ATOMISTIC MOLECULAR DYNAMICS SIMULATIONS

RAMON GOÑI[1,2] and MODESTO OROZCO[2–4]

[1]Life Sciences Department. Barcelona Supercomputing Center, Barcelona, Spain

[2]Joint BSC-IRB Program in Computational Biology, Barcelona, Spain

[3]Institute for Research in Biomedicine (IRB Barcelona), The Barcelona Institute of Science and Technology, Barcelona, Spain

[4]Department of Biochemistry and Biomedicine, Faculty of Biology, University of Barcelona, Barcelona, Spain,
E-mail: modesto.orozco@irbbarcelona.org

CONTENTS

21.1   INTRODUCTION

Classical molecular dynamics (MD) is one of the most widely used biomolecular simulation technique. In the recent years, the increase in the availability of experimental structural data, the massive improvement in the MD codes and the irruption of new computer platforms (massively parallel supercomputers, GPU-based systems and specific purpose computers) produced a revolution in the field, approaching the possibility to simulate with high quality, biologically relevant systems in relevant time scales. Thus, we have recently seen atomistic simulations of molecular systems containing millions of atoms, others extended to the millisecond scale, or dealing with model systems aiming to reproduce the crowded cellular environments. We have seen major rewriting of computer codes originally developed in the seventies, computers created on purpose to perform MD simulations, massive initiatives of distributed computing and the first projects to manage and store the deluge of data provided by MD simulations. Where are all these new innovations going to drive us? Are we going to see in a close future entire-cell simulations? Are we going to analyze second-long trajectories, or perform real-time MD simulations? In this contribution, we will review the technical limits of current state-of-the-art of MD biosimulations and will try to decipher the future in front of us.

21.2   THE MOLECULAR DYNAMICS ALGORITHM

Classical MD is a simulation technique directly derived from the basic formalism of Newtonian dynamics. It can be derived from the basic roots of the time-dependent Schrödinger equation by following a large series of simplifications.

1. Assumption of the Born-Oppenheimer approach and disconnect nuclei and electron movements.
2. Assumption of previous knowledge of the molecular topology (bonding scheme) of the molecule.
3. Assumption that the topology will remain unaltered along the simulation.
4. Substitution of the quantum atoms (nuclei and electrons) by classical beads of spherical symmetry centered at the atom nuclei.
5. Substitution of the quantum Hamiltonian by a classical force-field (the set of equations connecting atom positions with the relative energy of the system) that is supposed to reproduce well the relative conformational energy of the system.
6. Use the classical force-field to obtain forces, from which accelerations are obtained from second Newton's law.

7. Accelerations are integrated to obtain new atomic positions and velocities.

In practice, all existing force-fields contain two parts: the *bonded term*, which represents the energy associated to perturbation of covalent bonds (stretching), angles (*bending*), and torsions, and the *non-bonded* part that accounts for more distant interactions and typically includes a *Coulombic* term, which is used to represent electrostatic interactions between different regions of the molecule, and a *van der Waals* term, which is calibrated to reproduce dispersion interactions as well as short-range repulsion between the classical particles. Additional terms can be added as it will be discussed below.

Technically speaking, the time-consuming part of a MD simulation is the calculation of the non-bonded part, as it requires the evaluation of ($n \cdot (n - 1)/2$) distances (n being the number of particles in the system separated by more than 2 chemical bonds). The slow convergence (r^{-1} dependence) of the *Coulombic* term makes necessary to use very large systems to achieve realistic representations of the solvated systems. Some tricks, such as the use of periodic boundary conditions coupled to Ewald techniques [1] help to alleviate the problem, allowing the simulations of "infinite" (but periodic) systems with a $n \cdot \log(n)$ cost-dependence. However, despite all the efforts to optimize it, the calculation of the non-bonded distances is still the bottleneck of the entire MD simulation. Most effort in MD code developers is focused in trying to accelerate this part of the calculation, taking the most from current computer architectures.

21.3 A SHORT HISTORY OF MOLECULAR DYNAMICS

To our knowledge, the first attempts to reproduce the behavior of a chemical system by integrating Newton's equations of motion should be credited to Hirschfelder, Eyring and Topley who in the middle thirties [2] used it to study the dynamics of the $H+H_2 \rightarrow H_2+H$ reaction. To our knowledge, Alder and Wainwright published in the late fifties the first MD simulation of an ideal gas [3], and in the middle sixties Rahman [4] performed the first MD simulation of a real gas (Argon). The same author published the first simulation of liquid water in 1974, a seminal simulation that was crucial to convince scientists that chemically relevant systems could be subjected to MD simulation [5]. In 1971 Lifson's group developed the first prototype of a force-field able to deal with complex molecules such as proteins [6]. This important development encouraged the work of several groups who realized the possibility to use the analytical derivatives of the potential energy defined by such a force-field to obtain forces, which once transformed into accelerations, could be integrated to derive trajectories of complex biomolecules. To our knowledge, Warshel published the first dynamic study of a biomolecular process [7] and McCammon, Gelin and Karplus [8] published in 1977 the first MD simulation of a protein (bobin prancreatic trypsin inhibitor, BPTI). For current standards, McCammon-Gelin-Karplus simulation would be technically unacceptable: protein was in the gas phase, a nonphysiological ensemble was used and trajectory length was extremely short (10 ps) precluding the sampling of any relevant degree of freedom. However, this simulation was a milestone in the field of molecular simulation and a seminal contribution (collecting around a thousand citations to date) that convinced the structural biology community that MD could be applied to study biological systems. Forty years later, the term "molecular dynamics" or the acronym "MD" appear in 10 [4] publications every year; http://www. ncbi.nlm.nih.gov/pubmed, the numbers being increased every year.

The main ideas behind MD simulations were formulated in the late seventies, but the basic simulation algorithms were defined during the eighties, where MD was properly placed in the field of statistical mechanics [9–12]. Nineties were characterized by the popularization of the technique, the refinement of the force-fields and the systematic application of MD to study a variety of biological processes of increasing complexity. The last decades have been characterized by an explosion of the technique, which has gone parallel to the development of third generation computer programs (Amber, Gromacs, NAMD being the most popular ones) and to the accessibility of better computational platforms. MD is now solidly integrated in the main stream of structural biology, and we are in a course to simulate larger systems for longer periods of time. Software and hardware technology will define the limits of such a race. We will outline in the following our vision of the recent evolution of MD, and which are the technological clues to predict which might be its evolution in the near future.

21.4 EVOLUTION OF MOLECULAR DYNAMICS

We can estimate the impact of MD in the biofield by counting the entries in Pubmed citing "molecular dynamics" and "protein." In 1990 there were only

32 papers (see Figure 21.1), an apparently ridiculous number for today use of the technique. However, analyzing some of the works published that year we could realize the large evolution of the field. For example, in that year Beveridge's group [13] published an interesting simulation of HIV-1 protease dimmer in aqueous solution, which was 10 times longer (in time), and 50 times larger (in size) that MacCammon-Gelin-Karplus 1977 simulation, indicating an increase in the computational demands of MD community. We can roughly estimate such computational demand by multiplying the number of publications by the size (number of atoms) and length of the trajectory (ns) of a representative work. Adopting this scale, the computational demands increased almost 15,000 times from 1977 to 1990, around 30,000,000 times from 1977 to 2002 (using the 2002 gold standard of 50,000 atoms and 1 ns [14]) and they are estimated to be Billion times larger today (using now 100 ns as the standard). According to these numbers we can estimate that in a decade the number of annual publications using MD simulations will increase one order of magnitude, the average size of the simulation systems will at least double and the standard for publication will be 100 times larger

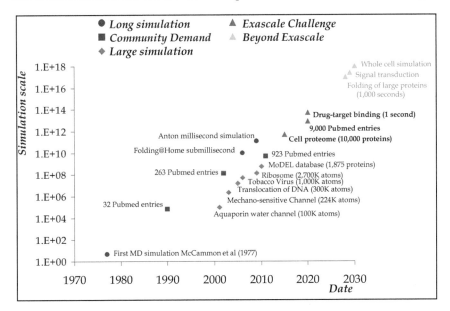

FIGURE 21.1 Simulation scale for long and large simulations is calculated multiplying simulation length (nanoseconds) with size (number of atoms). For community demand we calculate simulation scale multiplying the number of entries in Pubmed related to protein MD by the standards of that year (nanoseconds x number of atoms). Exacale and beyond exascale challenges are based on projections of this data.

than today. The community's computational demand will increase much more than 1,000 times by 2020 horizon, reaching the Exascale computing scale (100–1,000 faster than today's supercomputers capability; see Figure 21.1) [15]. Numbers might be even larger if the basic pair-wise additive Lifson's force-field formalism is abandoned and more complex potential energy functions are used. The simulation frontier twenty years from now will certainly require computer machines out of current pipelines of hardware manufacturers, and newly designed computer programs able to take profit of them.

21.4.1 SYSTEM SIZE AND COMPLEXITY

As described above, new computers and algorithms are challenging the limits of MD simulation in terms of size and complexity of the simulated systems. The increase in the size system to be simulated requires a computer code able to parallelize very well the calculation (NAMD is especially powerful for this type of task) and a very large number of cores. This fits well with the high-performance computer (HPC) paradigm followed by most supercomputer centers in the world. For example, in its current configuration the US Bluewaters supercomputer has 362240 XE Bulldozer cores and 33792 XK Bulldozer cores coupled to 4228 Kepler GPUS, all connected with the Cray Gemini torus network. This makes the computer excellent to deal with systems above one million atoms, as parallelization algorithms break the system in small pieces each of them send to a given core, giving the network the task to interchange energy and force-estimates obtained from the blocks to build the entire calculation.

The improvements in large system simulation can be analyzed by following along long periods of time the work of active research groups such as the one directed by Prof. K. Schulten. They published in 2001 a 1 ns simulation of the Aquaporin water channel, a system of 100,000 atoms [16], a real massive simulation at that time. Three years later, the same group simulated a membrane system nearly three times larger [17], a system with one million atoms in 2006 [18], another containing around 3 million atoms in 2009 [19] and an entire photosynthetic vesicle (100 million atoms) in 2016 [20]. It is obvious that few groups have access to the same computer resources than Klaus, but in 2016 MD simulations in the 10^5–10^6 atoms regime are not rare. How the field is going to evolve, how large are the systems that we will simulate in ten years from now?

We believe than in a decade, simulation of large supramolecular systems, such as complex viruses, polysomes, or long segments of the chromatin fiber coupled to regulatory proteins will be routine. The main limitation that we can envision for this type of simulations is related to the need of a starting configuration for the trajectory, something that may require a deep inter-connection between electron microscopy techniques and MD simulations. We may also study more risky systems, as those mimicking proteins or nucleic acids in real crowded environments [21, 22], approaching to the dream of simulating "whole cells" (approximately 10^{15} atoms for an hepatocyte). This future is still far, since simulation would require computers beyond the Zettaflops level (one million times bigger that current top supercomputers), but it is a nice dream to have in mind.

21.4.2   *EXTENDING THE LENGTH OF THE TRAJECTORIES*

While HPC capabilities are increasing the size of the molecular system of study, increasing the number of processors does not necessarily help in extending the length of the trajectory. For a given system, there is always a limit where increasing the number of processors does not produce any gain in the speed of the simulation. The smaller is the size of the system the sooner this point is reached, and with a few exceptions most of the macromolecular systems that are considered today are not large enough as to take direct profit for more than 1000 core processors working in parallel.

Integration of Newton's equations of motion needs to be done in the femtosecond time scale, which implies that the 10 ps simulation of McCammon-Gelin-Karplus in 1977 required 10^4 force-calculations, an impressive effort at that time which was possible thanks to the CECAM supercomputer. Most of current MD simulations are done in the 10^{12} nanosecond scale, and microsecond trajectories are considered the state-of-the-art for most systems. Certainly, the computational effort behind a microsecond trajectory is impressive (10^9 integrations of Newton's equations of motion), but in practical terms, one microsecond does not cover still the biologically relevant time scale, which in most cases moves from the millisecond to second (or even hours or days). Furthermore, we cannot ignore that most experimental observables are obtained from the statistical average over many individual trajectories, which means that we not only need to simulate beyond millisecond, but that we should obtain many of these trajectories to be able to reproduce the experimental observations.

Extension of simulation times along the years has been continuous, and closely linked to increase in processor and software performance. For example, taking one of the benchmark of one of the most popular MD programs (http://ambermd.org with protein DHFR), we can observe how the simulated time increased every year (see Figure 21.2), roughly doubling every two years (i.e., reducing by the half the computational time required for the same calculation). This rule has been beaten by using computing units less flexible than current multipurpose CPUs, but specialized in performing arithmetic operations. Graphical processing units (GPUs) were originally designed as graphical coprocessors to make more attractive computer games, but in the last years they have evolved by increasing the number of computer cores, by implementing

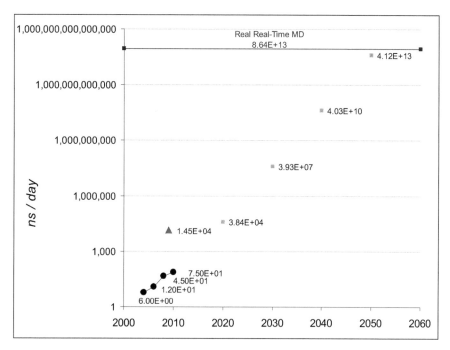

FIGURE 21.2 Improvement of MD ns/day performance. In dark-blue circles, the Amber 8 to 11 benchmark (simulation of DHFR protein, considering only CPU-optimized version of the code). Light-blue dashes represent the projection of performance based on this benchmark (roughly double the ns/day every 2 years). The red triangle is the performance reached by Anton simulating the same protein. The red line represents the Real Real-Time Molecular Dynamics (simulate 1 second in 1 second).

fastest memory and by incorporating the possibility to work with double precision. For example, NVIDIA Kepler GK110 GPU contains more than 1 billion transistors providing each card 3.5 Tflops of peak power, i.e., a single GPU card has the same processing capability than the fastest supercomputer in the world just 15 years ago, and the combined power of all the TOP500 computers 20 years ago. A basic computer node containing 2 GPUs, something accessible even for a very small group would be listed in the TOP500 list just 9 years ago (https://www.top500.org/statistics/perfdevel/), something that nobody could predict in 2007 (and nobody did). GPUs are having a tremendous impact in the MD field, and most of the last-generation of MD codes have a CUDA (the language recognized by the GPUs) versions. Unfortunately, not all CUDA implementations are equally powerful, and often codes ultra-refined for optimum performance in CPU are not so efficient when ported to GPUs (and vice versa). This means that MD user should perform a careful benchmark before selecting the simulation engine to be used. In our hands for a medium sized system (10^4 atoms) the (roughly) same MD throughput can be obtained with a single low-level gaming GPU (GeForce GTX Titan Black) than with 64 Xeon GPU R5–2670 @ 2.60 GHz) working in parallel (using in both cases the most efficient GPU- and CPU- codes that we have available in the group).

The idea to develop computing units, less universal, but more efficient than commodity CPUs to perform MD simulations has led to the development of several projects aiming to create MD-specific processors. Most of them yield only prototypes far from production use, but Shaw's company was successful creating Anthon, a MD-specific computer that provided an increase of three orders of magnitude in the length of the trajectories with respect to the fastest computers at that time [23]. Thanks to these advances, millisecond trajectories have been collected, allowing the study of allosteric transitions, protein folding [24] and the fast binding of model drugs [25]. New generation of these MD-specific computers might make possible the binding/unbinding of real drugs (see Figure 21.1), and hopefully envision folding of large proteins, or even signal transduction events.

21.4.3 TOWARDS REAL-TIME SIMULATION

Current MD protocols follow the "batch mode," where simulations are submitted to a local cluster or a HPC facility, and results are collected after some time (weeks or months). The only intervention of the researcher is limited to set-up de system, maintain the integrity of the trajectory, and check for the lack of artifacts, in other words the human is a mere spectator of the simulation. The ideal paradigm will be to run real-time interactive simulations, matching human and molecular time scales and allowing the user to modify on-the-fly the simulation. Package like YASARA [26] have been pioneer in this approach. Unfortunately, the current simulation-time/real-time ratio is in the order of femtoseconds (10^{-15}) per second (http://www.yasara.org/benchmarks.htm), far away to the ideal unity rate. The challenge is therefore to reach the "real-time" paradigm (simulation subject to real-time constraint) with the capability to match one second of simulation per one molecular-simulation second. The simulation speed would be equivalent to 8.64×10^{13} ns/day. Real real-time MD (RRT-MD) would certainly revolutionize modeling, allowing a direct experiment-simulation interaction, where the research would be able to impact directly on the simulation, fixing possible deviations, and biasing the trajectories based on his/her intuition or available experimental data. As a "back of the envelope" exercise we can try to guess what should be the required computer capability to perform RRT-MD simulations by using:

$$Computing Capability \approx \frac{N \times F}{T} \qquad (1)$$

where T is the integration time step, N is the number of atom-to-atom interactions and F is the number of floating-point operations per interaction. The number of atom-atom interactions (N) in a system is $a \times a$ (being a the number of atoms), but in practice computational time is mainly dependent on the number of pair-wise non-bonded interactions within a cutoff of 5–15 Å [27]. The number of operations per interaction (F) depends on the set up of the simulation, but some authors [28] have suggest a value around 100. Finally, the integration step for simulations is generally set to the femtosecond scale (10^{-15} seconds) for atomistic molecular descriptions. Putting all together, the required computer capability to simulate a typical system such as the Aquaporin water channel ($a = 100,000$ atoms systems) in RRT is in the order of YottaFLOPS.

21.4.4 THE DATA PROBLEM

Current practical use of MD is a heritage from the eighties, where MD simulations were accessible only for a small, well-trained community, with access to powerful-enough computers. Set-up of the simulations took weeks, which was followed by months or even years of calculations. Analysis was done mostly using "in house" software, trajectories were rarely shared, and typically disappeared from the computers after a few months of the publication of the paper. Current scenario is quite different: there is multitude of groups, often with a limited experience, launching MD simulations around the world. GPUs or small CPU cluster makes possible to collect standard (nanosecond scale) trajectories in a reasonable time, even for small research groups with limited funding. The analysis is gaining, however, complexity especially in the context of ensemble simulations (see below). Sharing of trajectories is becoming a common practice [29, 30] and there is an increasing pressure to maintain the trajectories for extended periods of time for further reanalysis [31, 32]. In other words, MD simulations is moving from being a pure FLOPs (floating point operation x second) problem to become a more complex informatics challenge, where data management and data transfer is as important as number crunching.

Within the new MD simulation scenario initiatives have emerged to facilitate the automatic set-up and analysis of the trajectories [31, 33, 34], helping non-expert users to avoid errors in the set-up and early stages of the simulation, as well as to facilitate an efficient and reproducible analysis. Additionally, several initiatives have emerged to collect validated MD simulations, making them accessible to the community and avoiding and endless repetition of the same trajectory by different groups. Initiatives such as MoDEL [31] or dynameomics [35] provide access to curated atomistic trajectories covering a significant part of the proteome, which can be post-processed by a myriad of other techniques [31, 33–37]. As an example, the first release of MoDEL [31] database contained more than 1,800 simulations, covering cluster-90 of cytosolic proteins in the protein data bank (PDB). If a cell has around 10^4 different proteins [38], it is realistic to think that we will be able to simulate the cell proteome soon (see Figure 21.1). Similar data storage and analysis initiatives have emerged for nucleic acids simulation within the Ascona B-DNA consortium (ABC) [29, 30]. Very recently, a database of DNA simulations named BigNAsim [39, 40] have been published collecting already around 100 Terabyte of data, which we believe will rise in a few years to the Petabyte scale. Clearly, as data increases community efforts are required for storage and maintenance. In other words, the field of MD simulations is going to transit the same path previously explored by some areas of particle physics or bioinformatics.

21.5   CLUES FOR THE FUTURE

Making predictions is much more difficult than rationalized the past, especially in a field that is so closely linked with technology developments impossible to anticipate just five or ten years ahead. We will then limit ourselves to discuss what we believe are the clues for the future evolution of the field, letting the reader to guess alternative scenarios a few years from now.

21.5.1   CODE EVOLUTION

Current MD programs are incredible pieces of engineering whose development has involved hundreds of human x year of work. It is difficult for an outsider to understand how further refinement is still possible in these ultra-refined codes. However, major software developers are releasing every year new versions showing always better performance than the previous ones, and this situation is likely to continue in the next years. However, a program cannot be optimal in all the informatics aspects of a calculation, and software developers are forced to make some decisions about the evolution of their codes. Some groups have tailored their programs to be optimal in terms of parallelization, even when this implies some loss of arithmetic efficiency in single cores. These codes are very efficient to study huge systems in massively parallel supercomputers, but might be suboptimal when used to study medium-sized systems in small computer clusters. In the reverse, the optimal code to run simulations in a 500 cores local computer is not going to be efficient to take full profit of the 10^5 cores of a supercomputer. Programs ultra-refined to make the best possible use of a commodity CPU are not easy to adapt to work efficiently in a GPU. Finally, new computer codes developed from scratch to be efficient in GPUs are excellent to obtain very large trajectories of small and medium systems, but parallelization on GPUs is still not mature enough in most codes, which somehow limits the size of the system to be analyzed, or the use of sampling strategies based on the collection of multiple parallel simulations (see below). A tremendous effort is expected from software developers, which should adapt their programs to very fast changes in computer technology (e.g., a simple change in CUDA interface forces a significant refinement of the codes). Infinite gratitude should be expressed to software developers making this effort in codes that are provided free (or nearly free) to the community.

The future is going to continue pressing software developers to maintain their codes updated to the new computing units. The big revolution related to the irruption of massive parallel supercomputers first, and of GPU architectures later mostly removed some well-established codes from the market, and the same situation might happen again if disruptive computer architectures emerge. Co-developing initiatives between software developers and hardware providers would help MD codes to maintain their impressive evolution. Not a surprise then that many of the major hardware developers have already strong links with academic groups responsible for MD software developments.

21.5.2 *IMPROVING THE QUALITY OF THE ENERGY CALCULATION*

All code developers and most of the educated MD users know the intrinsic shortcomings of pair-wise additive Lifson's potentials, but systematically ignore them. Decades of force-field refinement have provided very accurate energy potentials, but room for improvement is now small, and physical limitations of the general force-field formalism are clear. For example, it is physically impossible that a set of point charges could reproduce the charge distribution of a residue when it is placed in the interior of a protein, when exposed to a lipid membrane or when located in an aqueous environment. We can certainly obtain an optimum set of charges to describe the residue charge distribution in the context of a folded protein, but this set will be suboptimal for unfolded states, or for membrane proteins. Similarly, it is impossible with Lifson's force-fields to capture the impact of charge-donating ions, or to capture non-spherically symmetric dispersion interactions, or the generation of quasi-covalent ionic interactions.

The community should make an unpleasant effort to reconsider whether it is justified, with the accessible computer resources, to maintain Lifson's force-fields as the core of MD simulations. Disruptive alternatives would imply moving to quantum descriptions of the molecular interactions [41], but despite recent advances, we are not going to see pure quantum mechanical (QM) or even hybrid quantum mechanical/molecular mechanical (QM/MM) simulations extending to the microsecond regime in the next years. If possible, such calculations will be made using different programs to those ultra-optimized for Lifson's force-fields; a big challenge for software engineers.

A less disruptive approach to go beyond Lifson's potentials is to modify marginally the force-field by including polarization, the most prevalent of the missed interactions [40]. The introduction of these effects can be done following different paradigms, among them the classical Drude oscillator model [42], which is easy to implement in most current MD codes. MacKerell's group has been especially active in pushing polarizable force-fields, developing for the first operational polarizable force-fields for proteins and nucleic acids [43]. In our hands, for a medium sized DNA system, the use of polarized force-field increases by a factor of 10 the cost of the calculations (using the fastest codes for both polarized and pair-wise force-fields), which is a reasonable cost considering current computational power. Unfortunately, introduction of polarization requires a full recalibration of the force-field, or otherwise the unbalance in the interactions can lead to artefactual trajectories [44]. Clearly, more community effort is required to refine polarized force-fields.

21.5.3   NEW PARADIGMS FOR MD SIMULATIONS

Computer architectures are moving towards the massive parallel paradigm. The first computer in the TOP500 list in June 2006 had 131 K cores, and number 5 in the list had 8.7 K cores. Ten years later, Sunway TaihuLight (the 1st computer in the June 2016 TOP500 list) has more than 10 million cores, and the RIKEN supercomputer (the 5th) more than 700 K cores. In other words, a very significant part of the increase in FLOPs in TOP500 supercomputers originates from the accumulation of cores, rather than from a much higher speed of the chips. As shown by Schulten and others (see above), parallelism offers a way to benefit from the large number of cores, allowing the study of bigger systems. However, all current parallelization strategies are based on the partition of the entire system into blocks, for which dedicated cores are assigned. In order to maintain a good load balance between processors the blocks should be similar in size and large enough as to avoid the traffic between cores becoming the rate-limiting step of the calculation. In other words, each block should have a minimum number of atoms, which as discussed above limits the number of processors that can be efficiently used in a typical simulation.

An alternative that is now becoming popular (and we believe it will be dominant in the future in all supercomputer centers) is the "ensemble approach." It is based on "ergodic" condition which most (not all) MD simulations follow. This condition means that time ensemble is a Boltzman's ensemble, and accordingly that in the limit the same information can be obtained from the analysis of N(simulations) each of them M-nanosecond long than from a single trajectory N·M nanoseconds long. Pande and others [45, 46] have shown the power on ensemble simulations to study complex processes, not only in supercomputers, but also in a cloud of personal computers distributed in private houses. We can expect in the future the generation of federations of supercomputer centers, allowing mile-stoning type of simulations. Unfortunately, management of this type of calculations, performed in remote locations and using heterogeneous computer platforms is very difficult, especially without human intervention. Lindahl's group [47] has taken the lead in generating workflows for an intelligent and flexible control of this type of simulations. Their Copernicus software might become a first-generation prototype of applications managing massive ensemble simulations. Copernicus is automatic, but works in a near-human way, using a trained algorithm to decide not only the correct load balancing between processors, but also which of the simulations should be removed, extended or be used as a seed for additional trajectories.

Different authors have shown that ensemble simulations have an additional value to provide the thermodynamics and kinetics of biological processes if post-processed by Markov State models (MSM) [48], opening a new range of possibilities for ensemble simulations. Replica exchange molecular dynamics (REXMD) [49] is a very popular variant of ensemble simulation that guarantee a good sampling of the conformational space with a reduced computational cost. In REXMD calculations many replicas of the same system are lunched, the corresponding trajectories do not run independently, but require a moderate connectivity, making the technique especially suitable for supercomputers. As these machines are becoming larger, the use of REXMD is becoming more popular. We anticipate that this situation is likely to continue in the future for REXMD or related techniques.

21.5.4 THE DELUGE OF DATA

As hardware and software improve the amount of trajectory data generated per day increase, making its analysis, storage and management a major problem that needs to be faced by groups and computer centers. Our original

MODEL database [31, 50] (around 20 Tb), which required more than 2 years of continuum use of the largest supercomputer in Europe at that time, can be now replicated in our local computers in just one month. The analysis, management and storage of data generated in a two-years window is not a dramatic problem, but when the production time reduces to one month, major technical problems appear. Among them the time to analysis, that is still not fully automatic, or the connectivity and storage capabilities, which are evolving not as fast as those of CPUs.

How can MD groups survive from the deluge of data expected in the close future? Certainly, compression strategies can help to reduce the volume of data generated by MD simulations [51], but users should decide in the future between storing and deleting the data. A dramatic alternative follow by some groups is to store just a certain number of snapshots taken after a significant number of integration steps (ex. every 1–10 ns). If a transition between snapshot (i) and snapshot $(i + 1)$ is detected the sub-trajectory $i \rightarrow i + 1$ is redone but collecting now all the intermediate structures for detailed analysis. This approach requires full reproducibility and reversibly in the trajectories and has obvious risks for fast transitions (that might be fully ignored in the first screening), but it is becoming popular for groups using very fast computers. Another approach which is also being adopted is to store locally the data until the cost of repeating the calculation is expected to be less than a given threshold (for example 1 week of the fastest available machine). Finally, other groups, including ours (see above) are trying to store useful trajectories in data warehouse system. In the future those databases can be maintained by international infrastructures such as EUDAT (http://www.eudat.edu) or ELIXIR (http://www.elixir-europe.org). This type of Petabyte-scale databases can provide data extremely useful for post-analysis, for benchmarking of simulation models, and for the development of coarse-grained approaches, but require commitment for maintenance from public institutions. Effort and funding will be also necessary to integrate MD database in wider portals incorporating standard bioinformatics database including sequence information, mutational analysis, pathological information, etc. In this way MD will approach to a huge community not experienced in atomistic simulations.

Forty years after its development MD is facing new and exciting scenarios. It is clear that the technique is here to stay, that has escaped from the hands of highly specialized groups to reach the main stream of research in structural biology. We anticipate that further evolution of the technique will be tightly coupled with technological developments and with new experimental techniques. We speculate that from the myriad of possible evolutions of MD, those adapting better to the new computer architectures, and those providing the most useful type of information for biologist are going to be the dominant ones.

KEYWORDS

- **molecular dynamics**
- **data analysis**
- **high performance computing**
- **computational chemistry and computational biology**
- **molecular simulation**

REFERENCES

1. Darden, T., York, D., & Pedersen, L., (1993). Particle Mesh Ewald: A $N \cdot logN$ method for Ewald sums in large systems. *J. Chem. Phys., 98*, 10089–92.
2. Hirschfelder, J., Eyring, H., & Topley, B., (1936). Reactions involving hydrogen molecules and atoms. *J. Chem. Phys., 4*, 170–177.
3. Alder, B. J., & Wainwright, T. E., (1959). Studies in molecular dynamics. I. General method. *J. Chem. Phys., 31*, 459–466.
4. Rahman, A., (1964). Correlations in the motion of atoms in liquid argon. *Phys. Rev., 136*, A405–A411.
5. Stillinger, F. H., (1974). Rahman: an improved simulation of liquid water by molecular dynamics. *J. Chem. Phys., 60*, 1545–1557.
6. Karplus, S., & Lifson, S., (1971). Consistent force field calculations on 2, 5–diketopiperazine and its 3.6-dimethyl derivatives. *Biopolymers, 10*, 1973–1982.
7. Warshel, A., (1976). Bicycle-pedal model for the first step in the vision process. *Nature, 260*, 679–683.
8. McCammon, J. A., Gelin, B. R., & Karplus, M., (1977). Dynamics of folded proteins. *Nature, 267*, 585–590.
9. Gunsteren, W. F., van, Berendsen, H. J., Hermans, J., Hol, W. G., & Postma, J. P., (1983). Computer simulation of the dynamics of hydrated protein crystals and its comparison with x-ray data. *Proc. Natl. Acad. Sci. USA, 80*, 4315–4319.
10. Berendsen, H. J. C., Postma, J. P. M., van Gunsteren, W. F., DiNola, A., & Haak, J. R. (1984). Molecular dynamics with coupling to an external bath. *J. Chem. Phys, 81*, 3684–3690.
11. Jorgensen, W. L., & Ravimohan, C., (1985). Monte Carlo simulation of differences in free energies of hydration. *J. Chem. Phys. 83*, 3050–3055.

12. Lybrand, T. P., Ghosh, I., & McCammon, J. A., (1985). Hydration of chloride and bromide anions: determination of relative free energy by computer simulation. *J. Am. Chem. Soc. 107*, 7793–7794.

13. Harte, W. E., Swaminathan, S., Mansuri, M. M., Martin, J. C., Rosenberg, I. E., & Beveridge, D. L., (1990). Domain communication in the dynamical structure of human immunodeficiency virus 1 protease. *Proc Natl Acad Sci USA, 87*, 8864–8868.

14. Karplus, M., & McCammon, J. A., (2002). Molecular dynamics simulations of biomolecules. *Nat. Struct. Biol. 9*, 646–652.

15. EESI (2011). European Exascale Software Initiative. Working Group report on life Science and Health activities. http://www.eesi-project.eu/media/download_gallery/EESI_ D3.6_WG3.4-Report_R2.0.pdf.

16. Zhu, F., Tajkhorshid, E., & Schulten, K., (2001). Molecular dynamics study of aquaporin-1 water channel in a lipid bilayer. *FEBS. Lett. 504*, 212–218.

17. Sotomayor, M., & Schulten, K., (2004). Molecular dynamics study of gating in the mechanosensitive channel of small conductance MscS. *Biophys. J. 87*, 3050–3065.

18. Freddolino, P. L., Arkhipov, A. S., Larson, S. B., McPherson, A., & Schulten, K., (2006). Molecular dynamics simulations of the complete satellite tobacco mosaic virus. *Structure 14*, 437–449.

19. Gumbart, J., Trabuco, L. G., Schreiner, E., Villa, E., & Schulten, K., (2009). Regulation of the protein-conducting channel by a bound ribosome. *Structure 17*, 1453–1464.

20. Sener, M., Strumpfer, J., Singharoy, A., Hunter, C. N., & Schulten, K., (2016). Overall energy conversion efficiency of a photosynthetic vesicle. *Elife, 5*, e09541.

21. Harada, R., Sugita, Y., & Feig, M., (2012). Protein crowding affects hydration structure and dynamics. *J. Am. Chem. Soc., 134*, 4842–4849.

22. Candotti, M., & Orozco, M., (2016). The differential response of proteins to macromolecular crowding. *PLOS Comput. Biol., 7*, e1005040.

23. Shaw, D., Dror, R., Salmon, J., Grossman, J., Mackenzie, K., Bank, J., et al. (2009). Millisecond-scale molecular dynamics simulations on anton. In: SC '09: Proceedings of the Conference on High Performance Computing Networking, Storage and Analysis, (Portland, Oregon: ACM), pp. 1–11.

24. Shaw, D. E., Maragakis, P., Lindorff-Larsen, K., Piana, S., Dror, R. O., Eastwood, M., et al., (2010). Atomic-level characterization of the structural dynamics of proteins. *Science, 330*, 341–346.

25. Shan, Y., Kim, E. T., Eastwood, M. P., Dror, R. O., Seeliger, M. A., & Shaw, D. E., (2011). How does a drug molecule find its target binding site? *J. Am. Chem. Soc., 133*, 9181–9183.

26. Krieger, E., Darden, T., Nabuurs, S. B., Finkelstein, A., & Vriend, G., (2004). Making optimal use of empirical energy functions: force-field parameterization in crystal space. *Proteins, 57*, 678–683.

27. Van Der Spoel, D., Lindahl, E., Hess, B., Groenhof, G., Mark, A. E., & Berendsen, H., (2005). J.C. GROMACS: fast, flexible, and free. *J. Comput. Chem., 26*, 1701–1718.

28. Balbuena, P. B., & Seminario, J. M., (1999). *Molecular Dynamics : From Classical to Quantum Methods,* Elsevier: Amsterdam; New York.

29. Beveridge, D. L., Barreiro, G., Byun, K. S., Case, D. A., Cheatham, T. E., Dixit, S. et al. (2004). Molecular dynamics simulations of the 136 unique tetranucleotide sequences of DNA oligonucleotides. I. Research design and results on d (CpG) steps. *Biophys. J., 87*, 3799–3813.

30. Lavery, R., Zakrzewska, K., & Beveridge, D., et al., (2010). A systematic molecular dynamics study of the nearest-neighbor effects on base pair and base step conformations and fluctuations in B-DNA. *Nucleic Acids Res., 38*, 299–313.

31. Meyer, T., D'Abramo, M., Hospital, A., Rueda, M., Ferrer-Costa, C., Pérez, A., et al. (2010). MoDEL (Molecular Dynamics Extended Library): a database of atomistic molecular dynamics trajectories. *Structure, 18*, 1399–1409.

32. Carrillo, O., Laughton, C. A., & Orozco, M., (2012). Fast atomistic molecular dynamics simulations from essential dynamics samplings. *Journal Chemical Theory and Computation, 8*, 792–799.

33. Hospital, A., Andrio, P., Fenollosa, C., Cicin-Sain, D., Orozco, M., & Gelpí, J. L., (2012). MDWeb and MDMoby: an integrated web-based platform for molecular dynamics simulations. *Bioinformatics, 28*, 1278–1279.

34. Hospital, A., Faustino, I., Collepardo-Guevara, R., González, C., Gelpí, J. L., & Orozco, M., (2013). NAFlex: A web server for the study of nucleic acids flexibility. *Nucleic Acids Res. 41*, W47–W55.

35. Kamp, M. W., van der, Schaeffer, R. D., Jonsson, A. L., Scouras, A. D., Simms, A. M., Toofanny, R. D., et al. (2010). Dynameomics: A Comprehensive Database of Protein Dynamics. *Structure, 18*, 423–435.

36. Camps, J., Carrillo, O., Emperador, A., Orellana, L., Hospital, A., Rueda, M., et al., (2009). FlexServ: an integrated tool for the analysis of protein flexibility. *Bioinformatics, 25*, 1709–1710.

37. Chaudhuri, R., Carrillo, O., Laughton, C. A., & Orozco, M., (2012). Application of drug-perturbed essential dynamics/molecular dynamics (ED/MD) to virtual screening and rational drug design. *J. Chem. Theory Comput., 8*, 2204–2214.

38. Lodish, H., & Darnell, J. E., (2003). *Molecular Cell Biology,* W. H.Freeman & Co Ltd New York.

39. Hospital, A., Andrió, P., & Cugnasco, C., (2016). Bignasim: a nosql database structure and analysis portal for nucleic acids simulation data. *Nucleic Acids Res 44*, D272–D278.

40. Ivani, I., Dans, P. D., Noy, A., & Pérez, A., (2016). ParmBSC1: a refined force-field for DNA simulations. *Nature Methods, 13*, 55–58.

41. Dans, P. D., Walther, J., Gómez, H., & Orozco, M., (2016). Multiscale simulation of DNA. *Current Opin. Struct. Biol., 37*, 29–45.

42. Luque, F. J., Dehez, F., Chipot, C., & Orozco, M., (2011) Polarization effects in molecular interactions. *WIREs Comput. Mol. Sci. 1*, 844–854.

43. Lemkul, J. A., Huang, J., Roux, B., & MacKerell, A. D., (2016). An empirical polarizable force field based on the classical Drude oscillator model: development, history and recent applications. *Chem. Rev., 116*, 4963–5013.

44. Dans, P. D., Ivani, I., González, C., & Orozco, M., (2016) How accurate are accurate force-fields for nucleic acids. *Nucleic Acid Res., 45* (7), 4217-4230, 2017

45. Buch, I., Giorgino, T., & De Fabritis, G., (2011). Complete reconstruction of an enzyme-inhibitor binding process by molecular dynamics simulations. *Proc. Natl. Acad. Sci. USA, 108*, 236–241.

46. Kasson, P. M., Kelley, N. W., Singhal, N., Vrljic, M., Brunger, A. T., & Pande, V. S., (2006). Ensemble molecular dynamics yields submillisecond kinetics and intermediates of membrane fusion. *Proc. Natl. Acad. Sci. USA, 103*, 11916–11921.

47. Pronk, S., Pouya, I., Lundborg, M., Rotskoff, Wesén, B., Kasson, P. M., & Lindahl, E., (2015). Molecular simulation workflows as parallel algorithms: The execution engine

of Copernicus, a distributed high-performance computing platform. *J. Chem. Theor. Comput., 11*, 2600–2608.

48. Bowman, G. R., Huang, X., & Pande, V. S., (2009). Using generalized ensemble simulations and Markov state models to identify conformational states. *Methods, 49*, 197–201.

49. Sugita, Y., & Okamoto, Y., (1999). Replica exchange molecular dynamics method for protein folding. *Chem. Phys. Lett., 314*, 141–151.

50. Rueda, M., Ferrer, C., Meyer, T., Pérez, A., Camps, J., Hospital, A., et al., (2007). A consensous view to protein dynamics. *Proc. Natl. Acad. Sci. USA, 104*, 796–801.

51. Meyer, T., Ferrer-Costa, C., Pérez, A., Rueda, M., Bidon-Chanal, A., Luque, F. J., et al., (2006). Essential dynamics: a tool for efficient trajectory compression and management. *J. Chem. Theor. Comput., 2*, 251–258.

CHAPTER 22

THE ELECTRON PAIRING APPROACH IN NATURAL ORBITAL FUNCTIONAL THEORY

MARIO PIRIS

Ikerbasque Research Professor, Donostia International Physics Center (DIPC), 20018 Donostia, Euskadi, Spain; Euskal Herriko Unibertsitatea (UPV/EHU), 20018 Donostia, Euskadi, Spain; Basque Foundation for Science (IKERBASQUE), 48011 Bilbao, Euskadi, Spain, E-mail: mario.piris@ehu.eus

> *Even the formal justification of the electron-pair bond in the simplest cases ... requires a formidable array of symbols and equations.*
> —**Linus Pauling [1]**

CONTENTS

22.1   INTRODUCTION

The electron pairing has been in chemistry for some time. The concept of electron pair was introduced by Lewis as early as 1916 [2] to define the chemical bond in terms of the electronic structure. Pauling [3] provided a rigorous quantum mechanical basis for the electron-pair bonding explaining how the atoms join together to form molecules, which resulted in the valence bond (VB) theory [4].

The wavefunction of an electron pair is often called geminal. The first use of geminals was published by Hurley, Lennard–Jones and Pople [5], an idea suggested by Fock [6]. Geminals can be orthogonal or non-orthogonal, and geminal-based methods can be tailored to be variational as well as size-consistent and size-extensive. There is of course a connection between VB and geminal theories, indeed, the generalized VB (GVB) method [7] can be considered as a special antisymmetrized product of strongly orthogonal geminals (APSG), where each geminal consists only of two spatial orbitals. A comprehensive review of geminal theory can be found in the work of Surján [8].

The electron pairing approach came to the natural orbital functional (NOF) theory [9, 10] with the proposal of PNOF5 [11]. So far this is the only NOF that has been obtained by top–down and bottom–up methods [12]. In the bottom–up method, the functional is generated by progressive inclusion of known necessary N-representability conditions on the two-particle reduced density matrix (2-RDM) [13, 14], whereas the top–down method implies reducing the energy expression generated from an N-particle wavefunction to a functional of the occupation numbers (ONs) and natural orbitals (NOs) [15]. In the case of PNOF5, geminals were obtained [16], and it was realized that the wavefunction is an APSG with the expansion coefficients explicitly expressed by the ONs [17]. The existence of a generating wavefunction confirms that PNOF5 is strictly N-representable, i.e., the functional is derived from a wavefunction that is antisymmetric in N-particles [18].

PNOF5 is a perfect-pairing approximation, therefore, works reasonably well in systems in which the bonding-antibonding coupling is sufficient [19–26]. To improve the functional, it was necessary to extend the PNOF5 including more orbitals in each geminal. This more general ansatz, named extended PNOF5 (PNOF5e) [27], takes into account the important part of the dynamical electron correlation corresponding to the intrapair interactions, and most of the nondynamical effects [27–33]. However, no interpair electron correlation is described by PNOF5 and its extended version [34, 35]. To include the missing electron–electron interactions, two paths have been glimpsed.

One way is to use a multiconfigurational perturbation theory size consistent at the second order (SC2-MCPT) [36] taken as a reference the PNOF5 generating wavefunction which leads to the PNOF5-PT2 method [15, 37]. The other route implies recovering the missing correlation from the outset by introducing interactions between electron pairs in the framework of the NOF theory (NOFT). This has been done [38] within the simple $\mathcal{J}\mathcal{K}\mathcal{L}$-only functional form, and retaining the above mentioned N-representability constraints for the 2-RDM. The resulting functional PNOF6 removes the symmetry-breaking artifacts that are present in independent pairs approaches when treating delocalized systems [38, 39]. This NOF has proved a better treatment of both dynamic and non-dynamic electron correlations than its predecessor PNOF5 [40–43], and is the only known functional that shows a consistent behavior when it is tested with exactly soluble models [44].

This chapter is organized as follows. In Section 22.2, the basic concepts and notations related to the NOFT are discussed. In Section 22.3, the bottom–up method in NOFT is presented. Here, I discuss in details the cumulant of the 2-RDM that leads to the PNOF approximations with pairing restrictions, namely PNOF5 and PNOF6, as well as their extended versions. The Section 22.1.4 is dedicated to the top–down method in NOFT, which so far has been only accomplished for PNOF5e. The chapter is ended with the application of size-consistent perturbation corrections to PNOF5e in order to recover the interpair dynamic correlation. The strengths and weaknesses of the addressed methods are shown by comparing our results with respect to those theoretically obtained using high-level wavefunction methods, and available experimental marks.

22.2 BASIC CONCEPTS AND NOTATIONS

We consider an N-electron molecule described by the Hamiltonian

$$\hat{H} = \sum_{ik} \mathcal{H}_{ki}\hat{\Gamma}_{ki} + \sum_{ijkl} <kl|ij> \hat{D}_{kl,ij} \tag{1}$$

where $\mathbb{H}_{ki}$ denote the one-electron matrix elements of the core-Hamiltonian,

$$\mathcal{H}_{ki} = \int \int d\mathbf{x} \phi_k^* (\mathbf{x}) \left[-\frac{1}{2}\nabla^2 - \sum_I \frac{Z_I}{|\mathbf{r} - \mathbf{r}_I|} \right] \phi_i (\mathbf{x}) \tag{2}$$

and $<kl|ij>$ denote the two-electron matrix elements of the Coulomb interaction

$$< kl|ij >= \int \int d\mathbf{x}_1 d\mathbf{x}_2 \phi_k^* (\mathbf{x}_1) \phi_l^* (\mathbf{x}_2) r_{12}^{-1} \phi_i (\mathbf{x}_1) \phi_j (\mathbf{x}_2) \tag{3}$$

Atomic units are used. Here and in the following $\mathbf{x} \equiv (\mathbf{r}, \mathbf{s})$ stands for the combined spatial and spin coordinates, $\mathbf{r}$ and $\mathbf{s}$, respectively. The spin-orbitals $\{\varphi_i(\mathbf{x})\}$ constitute a complete orthonormal set of single-particle functions,

$$< \phi_k|\phi_i >= \int d\mathbf{x}\phi_k^* (\mathbf{x}) \phi_i (\mathbf{x}) = \delta_{ki} \tag{4}$$

with an obvious meaning of the Kronecker delta δ_{ki}. The spin-orbital set $\{\varphi_i(\mathbf{x})\}$ may be split into two subsets: $\{\phi_p^\alpha (\mathbf{r}) \ \alpha\ (\mathbf{s})\}$ and $\{\phi_p^\beta (\mathbf{r}) \ \beta\ (\mathbf{s})\}$. In order to avoid spin contamination effects, the spin restricted theory is employed, in which a single set of orbitals is used for α and β spins: $\phi_p^\alpha (\mathbf{r})$ $= \phi_p^\beta (\mathbf{r}) = \phi_p(\mathbf{r})$.

The one- and two-particle density matrix operators,

$$\hat{\Gamma}_{ki} = \hat{a}_k^\dagger \hat{a}_i \tag{5}$$

and

$$\hat{D}_{kl,ij} = \left(\frac{1}{2}\right) \hat{a}_k^\dagger \hat{a}_l^\dagger \hat{a}_j \hat{a}_i \tag{6}$$

are constructed from the familiar creation and annihilation operators, $\{\hat{a}_i^\dagger\}$ and, $\{\hat{a}_i\}$ respectively, associated with the set of spin-orbitals $\{\varphi_i(\mathbf{x})\}$. The expectation value of the Hamiltonian (1) for the state Ψ is then

$$E = \sum_{ik} \mathcal{H}_{ki}\Gamma_{ki} + \sum_{ijkl} < kl|ij > D_{kl,ij} \tag{7}$$

where the 1- and 2-RDMs, or briefly the one- and two-matrices, are defined as

$$\Gamma_{ki} =< \Psi|\hat{\Gamma}_{ki}|\Psi > \tag{8}$$

$$D_{kl,ij} = < \Psi | \hat{D}_{kl,ij} | \Psi > \tag{9}$$

We consider $\hat{S}_z$ eigenstates, so only density-matrix blocks that conserve the number of each spin type are non-vanishing. Specifically, the one-matrix has two nonzero blocks $\Gamma^{\alpha\alpha}$ and $\Gamma^{\beta\beta}$, whereas the 2-RDM has three independent nonzero blocks, $\mathbf{D}^{\alpha\alpha}$, $\mathbf{D}^{\alpha\beta}$, and $\mathbf{D}^{\beta\beta}$. The parallel-spin components of the two-matrix must be antisymmetric, but $\mathbf{D}^{\alpha\beta}$ possess no special symmetry [9]. For simplicity, we will address only singlet states in this chapter, so the parallel spin blocks of the RDMs are equal. Moreover, the RDMs satisfy important sum rules [9], for instance, the trace of the 1-RDM equals the number of electrons,

$$\sum_i \Gamma_{ii} = N \tag{10}$$

According to Eq. (7) the energy E of a state Ψ is an exactly and explicitly known functional of Γ and $\mathbf{D}$. Nevertheless, there is an important contraction relation between one- and two-matrices, namely,

$$\Gamma_{ki} = \frac{2}{N-1} \sum_j D_{kj,ij} \tag{11}$$

This implies that the energy functional (7) is just of the two-matrix, because $\mathbf{D}$ determines Γ. Therefore, the functional N-representability which refers to the conditions that guarantee the one-to-one correspondence between $E[\Psi]$ and $E[\mathbf{D}]$, is a related problem to the N-representability of the two-matrix. Attempts to determine the energy by minimizing $E[\mathbf{D}]$ are formidable due to the complexity of the necessary and sufficient conditions for ensuring that the two-matrix corresponds to an N-particle wavefunction [45]. An alternative lies in the development of a functional theory based upon the 1-RDM. Nonetheless, any approximation for the energy functional must comply at least with tractable necessary conditions for the N-representability of the two-matrix. The so called (2,2)-positivity conditions [45] are probably the most familiar, which state that the two-electron density matrix $\mathbf{D}$, the electron-hole density matrix $\mathbf{G}$, and the two-hole density matrix $\mathbf{Q}$, must be positive semi-definite. Precisely, these are the conditions that will require our functionals.

The one-matrix Γ can be diagonalized by a unitary transformation of the spinorbitals $\{\varphi_i(\mathbf{x})\}$ with the eigenvectors being the NOs and the eigenvalues $\{n_i\}$ representing the ONs of the latter,

$$\Gamma_{ki} = n_i \delta_{ki} \tag{12}$$

Restriction of the ONs to the range $0 \leq n_i \leq 1$ represents a necessary and sufficient condition for ensemble N-representability of the 1-RDM [18] under the normalization condition (10).

22.2.1   NATURAL ORBITAL FUNCTIONAL THEORY (NOFT)

The last term in Eq. (7) can be replaced by an unknown functional of the one matrix,

$$E[\Gamma] = \sum_{ik} \mathcal{H}_{ki} \Gamma_{ki} + V_{ee}[\Gamma] \tag{13}$$

The existence of $E[\Gamma]$ is well-established [46, 47]. The major advantage of Eq. (13) is that the kinetic energy is explicitly defined and it does not require the construction of a functional. The unknown functional only needs to incorporate electron correlation by means of $V_{ee}[\Gamma]$. The latter is universal in the sense that it is independent of the external field. However, it is highly difficult to approximate because what we have done is only to change the variational unknown from the complicated many-variable function Ψ to a single one-matrix Γ. Accordingly, the obstacle is the construction of the functional $E[\Gamma]$ capable of describing a quantum-mechanical N-electron system. This functional N-representability is patently related to the N-representability problem of the 2-RDM.

   In the following, all representations used are assumed to refer to the basis in which the one-matrix is diagonal. Accordingly, the energy functional is determined by the NOs and their ONs, i.e., a NOF $E[\{n_i, \varphi_i\}]$. In addition, we assume all NOs to be real.

22.3   THE BOTTOM–UP METHOD IN NOFT

To find approximations for the unknown functional V_{ee}, one can employ the exact functional form (7) with a 2-RDM built using a reconstruction functional. We use a reconstruction functional $\mathbf{D}(\mathbf{n})$, being $\mathbf{n}$ the set of the ONs. We neglect any explicit dependence of $\mathbf{D}$ on the NOs themselves because the energy functional (7) has already a strong dependence on the NOs via the one- and two-electron integrals.

Our reconstruction functional [48] is based on the cumulant expansion [49] of the 2-RDM, namely,

$$D_{kl,ij} = \frac{n_i n_j}{2} \left(\delta_{ki}\delta_{lj} - \delta_{li}\delta_{kj} \right) + \lambda_{kl,ij} \tag{14}$$

Here, the 2-RDM is partitioned into an antisymmetric product of the 1-RDMs, which is simply the Hartree-Fock (HF) approximation, and a correction λ [n] to it. λ is the cumulant matrix of the 2-RDM. It should be noted that matrix elements of λ are non-vanishing only if all its labels refer to partially occupied NOs with ONs different from 0 or 1.

Following the bottom–up recipe, the cumulant matrix must fulfill as many as possible necessary conditions in order to ensure the N-representability of the energy functional. The use of the (2,2)-positivity conditions was proposed in reference [13]. This particular reconstruction is based on the introduction of two auxiliary matrices Δ and Π expressed in terms of the ONs. In a spin restricted formulation, the structure for the two-particle cumulant is

$$\lambda_{pq,rt}^{\sigma\sigma} = -\frac{\Delta_{pq}}{2} \left(\delta_{pr}\delta_{qt} - \delta_{pt}\delta_{qr} \right), \quad \sigma=\alpha,\beta \tag{15}$$

$$\lambda_{pq,rt}^{\alpha\beta} = -\frac{\Delta_{pq}}{2}\delta_{pr}\delta_{qt} + \frac{\Pi_{pr}}{2}\delta_{pq}\delta_{rt} \tag{16}$$

Δ is a real symmetric matrix, whereas Π is a spin-independent Hermitian matrix. The N-representability D and Q conditions of the 2-RDM impose the following inequalities on the spin-independent off-diagonal elements of Δ [13],

$$\Delta_{qp} \leq n_q n_p, \qquad \Delta_{qp} \leq h_q h_p \tag{17}$$

while to fulfill the G condition, the elements of the Π-matrix must satisfy the constraint [51]

$$\Pi_{qp}^2 \leq (n_q h_p + \Delta_{qp})(h_q n_p + \Delta_{qp}) \tag{18}$$

where h_p denotes the hole $1 - h_p$ in the spatial orbital p. Furthermore, the sum rules that must fulfill the blocks of the cumulant yield a sum rule for Δ,

$$\sum_q {}' \Delta_{qp} = n_p h_p \tag{19}$$

The prime indicates that the $q = p$ term is omitted. Within this reconstruction, the energy for singlet states reads as

$$E = \sum_p n_p \left(2\mathcal{H}_{pp} + \mathcal{J}_{pp}\right) + \sum_{pq} {}' \Pi_{qp} \mathcal{L}_{pq}$$
$$+ \sum_{pq} {}' \left(n_q n_p - \Delta_{qp}\right) \left(2\mathcal{J}_{pq} - \mathcal{K}_{pq}\right) \tag{20}$$

where $\mathcal{J}_{pq} = \langle pq|pq \rangle$ and $\mathcal{K}_{pq} = \langle pq|qp \rangle$ are the usual direct and exchange integrals, respectively. $\mathcal{J}_{pp} = \langle pp|pp \rangle$ is the Coulomb interaction between two electrons with opposite spins at the spatial orbital p, whereas $\mathcal{L}_{pq} = \langle pp|qq \rangle$ is the exchange and time-inversion integral [52], so the functional (20) belongs to the $\mathcal{JKL}$-only family of NOFs.

The conservation of the total spin allows to derive the diagonal elements $\Delta_{pp} = n_p^2$ and $\Pi_{pp} = n_p$ [50]. Appropriate forms of matrices Δ and Π lead to different implementations of the NOF known in the literature as PNOFi ($i = 1$–6). The performance of these functionals is comparable to those of best quantum chemistry methods in many cases. The PNOF series has been recently reviewed in Ref. [14].

22.3.1 INDEPENDENT PAIRS

The simplest way to fulfill the sum rule (19) is neglecting all terms Δ_{qp} except one, Δ_{pp}, which will play the leading role in the correlation vector Δ_p. One way to do this is by coupling each orbital g, below the Fermi level ($F = N/2$), with one orbital above it, so all occupancies vanish for $p > 2F$. Accordingly, the ON of the $\tilde{g}$ level coincides with the hole of its coupled state g, namely,

$$n_{\tilde{g}} = h_g \quad (n_{\tilde{g}} + n_g = 1); \quad g = \overline{1, F} \tag{21}$$

It is worth noting that within this ansatz, we look for the pairs of coupled orbitals $(g, \tilde{g})$ which yield the minimum energy for the functional of Eq. (20), however, which are the actual g and $\tilde{g}$ orbitals paired is not constrained to remain fixed along the orbital optimization process. Consequently, the pairing scheme of the orbitals is allowed to vary along the optimization process till the most favorable orbital interactions are found.

In accordance to this assumption, and taking into account the N-representability conditions (17) and (18), the following expressions for matrices Δ and Π were proposed [11],

$$\Delta_{qp} = n_p^2 \delta_{qp} + n_q n_p (1 - \delta_{qp}) \delta_{q\Omega_g} \delta_{p\Omega_g}$$
$$\Pi_{qp} = n_p \delta_{qp} - \sqrt{n_q n_p} (1 - \delta_{qp}) \delta_{q\Omega_g} \delta_{p\Omega_g} \qquad (22)$$

where $\Omega_g \equiv (g,\tilde{g})$ is the subspace containing the orbital g ($g \leq F$) and its coupled orbital g ($\tilde{g} > F$), so $\delta_{q\Omega g} = 1$ if q [Ω_g, or $\delta_{q\Omega g} = 0$, otherwise ($q \ \Omega_g$). This orbital perfect-pairing approach conforms the PNOF5, and the energy (22.20) for the ground singlet-state of any Coulombic system can be cast as

$$E = \sum_{g=1}^{F} E_g + \sum_{f \neq g}^{F} \sum_{p \in \Omega_f} \sum_{q \in \Omega_g} n_q n_p (2\mathcal{J}_{pq} - \mathcal{K}_{pq}) \qquad (23)$$

$$E_g = \sum_{p \in \Omega_g} n_p (2\mathcal{H}_{pp} + \mathcal{J}_{pp}) - 2\sqrt{n_g n_{\tilde{g}}} \mathcal{L}_{\tilde{g}g} \qquad (24)$$

The first term of the energy (23) draws the system as independent $N/2$ electron pairs, whereas the last term contains the contribution to the HF mean-field of the electrons belonging to different pairs. It is clear that the main weaknesses of this ansatz is the absence of the interpair electron correlation.

Several performance tests have shown that PNOF5 yields remarkably accurate descriptions of systems with near-degenerate one-particle states and dissociation processes [11, 19, 23, 24, 22, 26]. In this sense, the results obtained with PNOF5 for the electronic structure of transition metal complexes are probably the most relevant [24]. This functional correctly takes into account the multiconfigurational nature of the ground state of the chromium dimer, known as a benchmark molecule for quantum chemical methods due to the extremely challenging electronic structure of the ground state and potential energy curve.

PNOF5 has also been successfully used to predict vertical ionization potentials and electron affinities of a selected set of organic and inorganic spin-compensated molecules, by means of the extended Koopmans' theorem [28]. The one-electron picture provided by PNOF5 agrees closely with the orbitals provided by the VB method and with those obtained by a standard molecular orbital calculation [20, 21, 22, 25]. The size-consistency property, and the fact that the functional tends to localize spatially the NOs, make PNOF5 an exceptional candidate for fragment calculations. The latter showed a fast convergence, which allowed the treatment of extended system at a fractional cost of the whole calculation [32].

The perfect-pairing approach can be improved by including more orbitals in the description of each pair. This involves coupling each orbital g ($g \leq F$) with N_g orbitals above F, which is reflected in the sum rule for the ONs, namely,

$$\sum_{p \in \Omega_g} n_p = n_g + \sum_{p \in \Omega'_g} n_p = 1; \quad g = \overline{1, F} \tag{25}$$

In Eq. (25), Ω_g is the subspace containing the orbital g and its N_g coupled orbitals above the Fermi level which conform the subspace Ω'_g. Taking into account the spin, notice the total ON for a given subspace is 2. Moreover, we consider that these subspaces are mutually disjoint ($\Omega_{g1} > \Omega_{g2} = \varnothing$), i.e., each orbital belongs only to one subspace Ω_g. Under these conditions, the matrix Δ in the Eq. (22) is unchanged, while the matrix Π is given by the expression

$$\Pi_{qp} = n_p \delta_{qp} + \Pi^g (n_q, n_p) (1 - \delta_{qp}) \delta_{q\Omega_g} \delta_{p\Omega_g} \tag{26}$$

where

$$\Pi^g (n_q, n_p) = \begin{cases} -\sqrt{n_q n_p}, & p = g \text{ or } q = g \\ \sqrt{n_q n_p}, & p, q > F \end{cases} \tag{27}$$

This more general ansatz corresponds to the extended PNOF5 (PNOF5e) [27], with an energy that continues being given by Eq. (23). The difference now is in the energy E_g of the electron pair, which provides a better description of the intrapair electron correlation,

$$E_g = \sum_{p \in \Omega_g} n_p (2\mathcal{H}_{pp} + \mathcal{J}_{pp}) - 2 \sum_{p \in \Omega'_g} \sqrt{n_g n_p} \mathcal{L}_{pg} + \sum_{p,q \in \Omega'_g}' \sqrt{n_q n_p} \mathcal{L}_{pq} \tag{28}$$

In Eq. (28), recall that the prime in the last sum indicates that the $q = p$ term is omitted. In general, each orbital g can be coupled to a different number of orbitals N_g. For simplicity, consider all N_g equal to a fixed number N_c, which gives rise to the functional PNOF5(N_c). If $N_c = 1$, PNOF5(1) corresponds obviously to the simplest formulation PNOF5, whereas $N_c > 1$ leads to different extended formulations of the independent-pair approach PNOF5e.

The performance of the PNOF5(N_c) can be observed in the dissociation of small diatomic molecules: H_2, LiH and Li_2, for which the electron correlation effect is almost entirely intrapair correlation. These dimers comprise different types of bonding character: the prototypical covalent bond of H_2, the highly electrostatic bond of LiH, and the weak covalent bond of the Li_2. In all cases, the dissociation limit corresponds to a two-fold degeneracy with the generation of two doublet atomic states.

It has been observed [11, 27] that different values of N_c produce dissociation curves qualitatively correct, however, a larger value N_c recovers a larger portion of intrapair correlation, so predicts shorter equilibrium bond lengths and larger dissociation energies, getting closer to the experimental data. Similarly, potential energy curves are in better agreement with the experiment according to the resulting harmonic vibrational frequencies and anharmonicities [27].

Table 22.1 lists the errors in total energies as compared against the highly accurate coupled-cluster with singles, doubles, and non-iterative triples (CCSD(T)) calculations, at the equilibrium geometry. The value of N_c corresponds to the maximum value allowed by the employed aug-cc-pVTZ basis set [53]. Inspection of the data reveals that PNOF5(1) stays quite above the CCSD(T) values, whereas PNOF5(N_c) approaches considerably to these values due to the better description of the intrapair correlation. Observe that for H_2, the electron correlation effect is entirely intrapair and PNOF5(N_c) matches the coupled-cluster result. Conversely, Li_2 has greater number of electron pairs than LiH, so its energy is higher as compared against CCSD(T). For these two systems, the error is proportional to the lack of interpair correlation.

TABLE 22.1 Errors in Total Energies, in kcal/mol, As Compared Against CCSD(T) At the Equilibrium Bond Length Using the aug-cc-pVTZ Basis Set

	PNOF5(1)	PNOF5	(N_c)	CCSD(T)
H_2	13.249	0.000	(49)	−1.172756
LiH	19.900	2.302	(39)	−8.048590
Li_2	26.672	4.793	(22)	−14.954066

The CCSD(T) reference values, in Hartrees, are in the last column.

The improvement of PNOF5e over PNOF5 was also observed by visualizing the electron densities by means of the Bader's theory of atoms in molecules in the case of a set of light atomic clusters: Li_2, Li_3^+, Li_4^{2+}, and H_3^+ [27]. In summary, properties that depend directly on the symmetry of the electronic distribution are highly affected by the breakdown of symmetry in PNOF5, whereas PNOF5e is able to give appropriate quantitative results, although the relative energies may be negligible.

22.3.2 INTERACTING PAIRS

To go beyond the independent-pair approximation, let us consider nonzero not only the Δ-elements between orbitals coupled via the condition (22.25), but also those Δ_{qp} related to orbitals belonging to different pairs. After factorizing these elements ($\Delta_{qp} = \Lambda_q \Lambda_p$), meeting the N-representability restrictions (17) and (18), and the corresponding sum rules (22.19), the following ansatz for the off-diagonal elements of Δ and Π matrices may be taken into consideration

Δ_{qp}	Π_{qp}	$Orbitals$
$e^{-2S} h_q h_p$	$-e^{-S}\sqrt{h_q h_p}$	$q \le F, p \le F$
$\frac{\gamma_q \gamma_p}{S_\gamma}$	$-\Pi_{qp}^\gamma$	$q \le F, p > F$
		$q > F, p \le F$
$e^{-2S} n_q n_p$	$e^{-S}\sqrt{n_q n_p}$	$q > F, p > F$

$$(29)$$

where the magnitudes γ and Π^γ are given by

$$\gamma_p = n_p h_p + \alpha_p^2 - \alpha_p S_\alpha, \qquad \alpha_p = \begin{cases} e^{-S} h_p, & p \le F \\ e^{-S} n_p, & p > F \end{cases}$$

$$\Pi_{qp}^\gamma = \left(n_q h_p + \tfrac{\gamma_q \gamma_p}{S_\gamma}\right)^{\frac{1}{2}} \left(h_q n_p + \tfrac{\gamma_q \gamma_p}{S_\gamma}\right)^{\frac{1}{2}}$$

$$(30)$$

$$S = \sum_{q=F+1}^{F+FN_c} n_q, \qquad S_\alpha = \sum_{q=F+1}^{F+FN_c} \alpha_q, \qquad S_\gamma = \sum_{q=F+1}^{F+FN_c} \gamma_q$$

According to Eq. (29), the correlation matrices Δ and Π do not differentiate between orbitals that belong to the same subspace ($q, p \in \Omega_g$), and orbitals belonging to two different subspaces ($q \in \Omega_{g1}, p \in \Omega_{g2}, g1 \ne g2$). Therefore, the interactions between orbitals are independent of the set to which they belong, so the intrapair and interpair correlations are equally balanced. The factor e^{-S} retains the advantages achieved by the forerunner functionals when S is small, while for large values of S, this factor decreases rapidly, bringing about limit values of $\gamma_p = n_p h_p$.

Substituting the new approaches for matrices Δ and Π, we obtain the PNOF6e or PNOF6(N_c) if the number of coupled orbitals (N_g) to each orbital g is a fixed number N_c, namely,

$$E = \sum_{g=1}^{F} E_g + \sum_{f \neq g} \sum_{p \in \Omega_f} \sum_{q \in \Omega_g} E_{pq}^{int} \tag{31}$$

As previously, the first term of the energy (31) draws the system as the sum of $N/2$ electron pairs described now by the following two-electron energy

$$E_g = \sum_{p \in \Omega_g} n_p \left(2\mathcal{H}_{pp} + \mathcal{J}_{pp}\right) + \sum_{p,q \in \Omega_g, p \neq q} E_{pq}^{int} \tag{32}$$

It is worth noting that the last terms of Eqs. (31) and (32), contain the same type of interactions between the electrons in different pairs and inside a pair, respectively. This interaction energy E_{pq}^{int} is given by

$$E_{pq}^{int} = (n_q n_p - \Delta_{qp})(2\mathcal{J}_{pq} - \mathcal{K}_{pq}) + \Pi_{qp}\mathcal{L}_{pq} \tag{33}$$

The simplest pairing, $N_c = 1$, was proposed in [38], while the extended version was soon used to study halogen clusters [40] and hydrogen abstraction reactions [41]. One of the most important results of including the inter-pair electron correlation in PNOF6 is the removal of symmetry-breaking artifacts that are present in independent-pairs approaches such as PNOF5 when treating delocalized systems [37, 39]. The transition of D_{2h} to D_{4h} symmetry in H_4 molecule is among the most simple archetypal examples to illustrate this phenomena.

In Figure 22.1, the relative energies of the H_4 model with respect to the minimum energy for several methods, keeping constant the distance R of each hydrogen atom from the center of mass, and modifying the angle θ subtended at the latter, are shown. PNOF5 shows a spurious cusp at $\theta = 90°$, and the other single-reference post-HF electron correlation methods also fail to qualitatively describe this potential energy surface since two degenerate

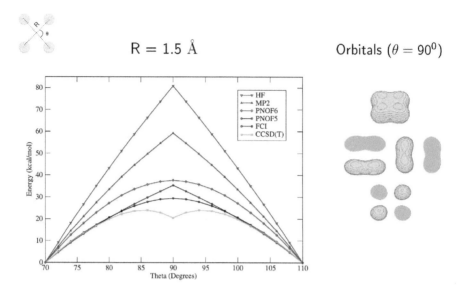

FIGURE 22.1   Relative energies in kcal/mol with respect to the lowest energy along the transition D_{2h} to D_{4h} of H_4.

configurations become equally important. We can observe that the Full Configuration Interaction (FCI) solution is a parabola in black, whereas the PNOF6 solution is the unique other parabola in red. Moreover, the π-orbitals obtained with PNOF6 at 90° are adapted to the D_{4h} symmetry as expected.

A recent example that reveals the success of PNOF6 is the carbon dimer [42]. The inspection of the electronic structure of its ground state, reveals different PNOF5 solutions that expand on increasing values of internuclear distance. The inter-pair correlation effects included by PNOF6 provides a curve with no discontinuities, which resemble closely the solutions obtained with CASSCF(8,8) calculations. Additionally, the obtained NOs carry additional relevant information not contained by the CASSCF, namely, the NOs are uniquely and precisely linked into pairs. The end result is that PNOF6 predicts three overall bonding interactions and one overall antibonding interaction for C_2 which yields an effective bond order of 2.26, intermediate between acetylene and ethylene, reconciling the bond-order linear relation found on calculations based on the bond's force constant, at the minimum energy internuclear distance.

To end this section, let us consider the harmonium atom. The latter is a well-known tool for calibration, testing, and benchmarking of any approximate electronic structure method. In this model system, the electron-nucleus potential is replaced by a harmonic confinement, but the electron-electron Coulomb interaction remains. By varying the strength of the harmonic potential, the correlation regime can be tuned, making possible the transition from the weakly to the strongly correlated regime. Recently, the comparison between the quasi-exact and approximate electron-electron repulsion energy provided by eight known NOFs, in the singlet state of the four-electron harmonium atom with varying confinements, was analyzed in some detail [44]. In Figure 22.2, the electron-electron repulsion energy $V_{ee} - V_{ee}(HF)$ is depicted for the four-electron harmonium atom in its $^1D_+$ ground state. These values are taken from Tables 6 and 10 of Ref. [44]. Among the functionals included, only PNOF6 goes parallel and above the exact solution for all confinements considered.

Furthermore, it is the only one capable of describing accurately the $\omega \to \infty$ limit of this state.

22.4   THE TOP–DOWN METHOD IN NOFT

The most direct method to generate a NOF is by reducing the energy expression obtained from an approximate N-particle wavefunction to a

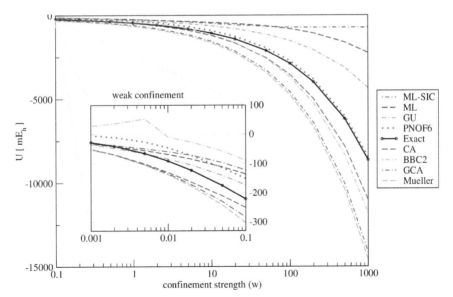

FIGURE 22.2   The electron repulsion energy $V_{ee} - V_{ee}(HF)$ in the $^1D_+$ ground state of the four electron harmonium atom

functional of the ONs and NOs. By doing this, we automatically avoid the
N-representability problem of the 2-RDM, or what is the same, of the functional. However, this is a formidable task that is far from being something
attainable in most cases. So far, and except the simple case of the HF approximation which can be seen as a functional of the 1-RDM, only PNOF5e
has achieved this goal. Indeed, the independent-pair approximation has been
proved equivalent to an APSG [27].

For a spin-compensated system of N particles, the generating wavefunction of the PNOF5e can be defined as

$$|0\rangle = \prod_{g=1}^{F} \hat{\psi}_g^\dagger |vac\rangle \tag{34}$$

where

$$\hat{\psi}_g^\dagger = \sqrt{n_g}\hat{a}_g^\dagger \hat{a}_{\bar{g}}^\dagger - \sum_{p\in\Omega_g'} \sqrt{n_p}\hat{a}_p^\dagger \hat{a}_{\bar{p}}^\dagger \tag{35}$$

is a composite particle creation operator that creates two electrons with oppo-
site spins on a geminal $\Psi_g(x_1, x_2)$. Here $\hat{a}_p^\dagger(\hat{a}_{\bar{p}}^\dagger)$ is a particle creation operator
on the spatial orbital p with spin $\alpha(\beta)$. It is worth noting that the expansion
coefficients of the geminal are expressed through the ONs of orbitals belonging to the subspace Ω_g, which obey the sum rule (25). The latter normalizes
the geminal. It is not difficult to verify that geminals $\{\Psi_g\}$ are orthogonal to
each other due to the orthogonality of the NOs and the spin functions, a condition also known as strong orthogonality since geminals belong to mutually
exclusive subspaces [8].

Significantly, the separable structure of the wavefunction (34) guarantees
the criterion of size-extensivity. The geminal operator (35) can be rewritten as

$$\hat{\psi}_g^\dagger = \sqrt{n_g}\left(1 - \hat{T}_g\right)\hat{a}_g^\dagger \hat{a}_{\bar{g}}^\dagger \tag{36}$$

where

$$\hat{T}_g = \frac{1}{\sqrt{n_g}}\left(\sum_{p\in\Omega_g'} \sqrt{n_p}\hat{a}_p^\dagger \hat{a}_{\bar{p}}^\dagger\right)\hat{a}_{\bar{g}}\hat{a}_g \tag{37}$$

$\hat{T}_g$ annihilates two electrons in the spatial orbital g, and creates them in the
orbital coupled subspace Ω'_g. Accordingly, the state (34) becomes into

$$|0\rangle = \prod_{g=1}^{F} \sqrt{n_g} \left(1 - \hat{T}_g\right) \hat{a}_g^\dagger \hat{a}_{\bar{g}}^\dagger |vac\rangle = \prod_{g=1}^{F} \sqrt{n_g} \left(1 - \hat{T}_g\right) |\Psi_{HF}\rangle \qquad (38)$$

where the $|\Psi_{HF}\rangle$ Fermi vacuum is the HF wavefunction with the one-electron functions representing by the NOs. Eq. (38) shows that our APSG ansatz (34) is a restricted coupled cluster approach. In fact, opening the expression (38), we observe that the state $|0\rangle$ is a sum of Slater determinants which includes the HF, doubly, quadruply, sextuply, up to F-tuply excited determinants, namely,

$$|0\rangle = |\Phi_0\rangle + |\Phi_d\rangle + |\Phi_q\rangle + ... + |\Phi_F\rangle \qquad (39)$$

where

$$
\begin{aligned}
&|\Phi_0\rangle = d_0 |\Psi_{HF}\rangle, \quad d_0 = \sqrt{n_1 n_2 ... n_F} \\
&|\Phi_d\rangle = -\sum_{g=1}^{F} \sum_{p \in \Omega_g'} d_g^p \left|\Psi_{g\bar{g}}^{p\bar{p}}\right\rangle, \quad d_g^p = d_0 \sqrt{\frac{n_p}{n_g}} \\
&|\Phi_q\rangle = \sum_{g<f} \sum_{p \in \Omega_g'} \sum_{q \in \Omega_f'} d_{gf}^{pq} \left|\Psi_{g\bar{g}f\bar{f}}^{p\bar{p}q\bar{q}}\right\rangle, \quad d_{gf}^{pq} = d_0 \sqrt{\frac{n_p n_q}{n_g n_f}} \\
&\qquad\qquad\qquad ... \\
&|\Phi_F\rangle = (-1)^F \sum_{p \in \Omega_1'} \sum_{q \in \Omega_2'} \cdots \sum_{r \in \Omega_F'} d_{12..F}^{pq..r} \left|\Psi_{1\bar{1}2\bar{2}...F\bar{F}}^{p\bar{p}q\bar{q}...r\bar{r}}\right\rangle
\end{aligned}
\qquad (40)
$$

The next step is to determine the 2-RDM of the state $|0\rangle$ to obtain the ground state energy, according to the Eq. (7). The spin-parallel components of the 2-RDM are only intergeminal, and have the same structure as in HF theory, namely,

$$D_{pq,rt} = \tfrac{1}{2} n_p n_q \left(\delta_{pr}\delta_{qt} - \delta_{pt}\delta_{qr}\right) \delta_{p\Omega_f} \delta_{q\Omega_g} \left(1 - \delta_{fg}\right) \qquad (41)$$

while the spin-opposite components read

$$
\begin{aligned}
D_{p\bar{q},r\bar{t}} = &\tfrac{1}{2} n_p n_q \delta_{pr}\delta_{qt}\delta_{p\Omega_f}\delta_{q\Omega_g}\left(1 - \delta_{fg}\right) + \tfrac{1}{2}\big[n_p \delta_{pr} \\
&+ \Pi^g\left(n_p, n_r\right)\left(1 - \delta_{pr}\right)\delta_{p\Omega_g}\delta_{r\Omega_g}\big]\delta_{pq}\delta_{rt}
\end{aligned}
\qquad (42)
$$

In Eq.(42), Π^g is given by Eq. (27), therefore the last term gives the intrageminal contribution to the 2-RDM. Assuming a real set of NOs, the expectation value of the electronic Hamiltonian calculated with the APSG (1.34–1.35) reads then as the PNOF5e energy (23) with E_g given by Eq. (28).

The generating wavefunction (39) has the signs of the expansion coefficients fixed. This allows to express explicitly the energy as a functional of the ONs, in contrast to the general case of an APSG which is a functional of the coefficient amplitudes $\{c_p\}$, but not a NOF. Consequently, the degree of freedom related to the c_p's phases does not exist in the minimization process of the functional, making simpler the optimization process. Evidently, the PNOF5e energy is an upper bound to the APSG energy and the exact energy too.

$$E^{exact} \leq E^{APSG}[\{c_p\},\{\phi_p\}] \leq E^{PNOF5e}[\{n_p\},\{\phi_p\}] \tag{43}$$

This result is remarkable since it demonstrates strictly the N-representability of PNOF5e. This precedent suggests that perhaps the unity of quantum theory on many particle systems can be attained by careful handling of top–down and bottom–up methods, in obtaining functionals of the energy [12].

22.5 INTERPAIR DYNAMIC CORRELATION

Both electron pairing approaches seen so far, PNOF5 and PNOF6, including their extended versions, take into account most of the non-dynamical effects, but also the important part of dynamical electron correlation corresponding to the intrapair interactions. PNOF5 does not describe correlation between electron pairs at all, while PNOF6 seems to describe mostly non-dynamic interpair correlation. Hence, there is a significant lack of interpair dynamic correlation.

The perturbation theory (PT) is probably the most popular strategy followed in quantum chemistry to recover the dynamic correlation. In PT, the Hamiltonian, $\hat{H}$, is divided into two operators: $\hat{H}^0$ and $\hat{V}$. The first is called the zeroth-order Hamiltonian and must afford a complete set of zeroth-order wavefunctions and energies. $\hat{V} = \hat{H} - \hat{H}^0$ corresponds to the perturbation. The partitioning into $\hat{H}^0$ and $\hat{V}$ is the crucial point of all perturbative treatments, especially in multiconfigurational cases.

An active work has been done by several groups on the development of multiconfigurational perturbation theories (MCPT) using the APSG as reference, which is the type of wavefunction behind the PNOF5. While it is true that PNOF6 describes in a more balanced way the static (non-dynamic) and dynamic correlations than PNOF5, we have not yet found the wavefunction that could be behind the interacting-pair approximation in NOFT. The latter seems not to be related to geminal-based theories, although it keeps the orbital-pairing scheme, Eq. (25).

In the current section, the generating PNOF5e wavefunction (34) is used as a reference in the SC2-MCPT developed by the Budapest group [36]. This choice is justified by the simplicity of the SC2-MCPT formulation. There is no need to apply a numerical orthogonalization procedure neither solve a linear system of equations to get the corrections. Furthermore, the structure of the resulting working formulas involve only matrix elements between pure determinants, like occurs in the standard single-reference perturbation theory. Finally, the size-consistency is ensured by omitting projectors from the zero-order Hamiltonian operator and considering energy denominators as differences of one-particle energies. On the other hand, it is important to note that SC2-MCPT is a biorthogonal PT, so the zeroth-order Hamiltonian is not Hermitian. A detailed description of the PNOF5-PT2, i.e., the case when $N_c = 1$, can be found in Ref. [15].

According to the SC2-MCPT, one considers the zero-order ground state $|0\rangle$ as the APSG given by Eq. (39), and seeks perturbative corrections to it. One chooses the vector $|0\rangle$ and the excited determinants $|K\rangle \equiv |\Psi_{K'}\rangle$ with respect to the HF state, as a basis in the full M-dimensional vector space. $|0\rangle$ and $\{|K\rangle, K \neq 0\}$ together span the full space but overlap. To remedy this difficulty, a set of reciprocal (biorthogonal) vectors $\{\langle \tilde{K}|\}$ of the overlapping $\{|K\rangle\}$ is introduced. Due to the simple structure of the metric matrix, the reciprocal vectors read

$$\langle \tilde{K}| = \begin{cases} d_0^{-1} \langle \Psi_{HF}|, & K = 0 \\ \langle K| - d_K \langle \tilde{0}|, & K \neq 0 \end{cases} \qquad (44)$$

It is not difficult to verify that $\langle \tilde{K}| L\rangle = \delta_{KL}$. Having these vectors, one defines a non-Hermitian zero order Hamiltonian through its spectral resolution as

$$\hat{H}^0 = \sum_{K=0}^{M} E_K |K\rangle \langle \tilde{K}| \qquad (45)$$

From the above definition, one can verify by substitution that $\{|K\rangle\}$ are eigenvectors of $\hat{H}^0$ from the right, as $\{\langle \tilde{K}|\}$ are left eigenvectors. In general, the energies E_K-s are free parameters of the theory, so they define the partitioning. In this biorthogonal formulation, E_0 is conveniently taken as the zero-order ground state energy:

$$E_0 = E^{(0)} = \langle \tilde{0} | \hat{H} | 0 \rangle \tag{46}$$

Taking into account Eqs. (39) and (44), and the two-particle nature of the full Hamiltonian $\hat{H}$, we have

$$E_0 = \langle \Psi_{HF} | \hat{H} | \Psi_{HF} \rangle + d_0^{-1} \sum_{g=1}^{F} \sum_{p \in \Omega'_g} d_g^p \langle \Psi_{HF} | \hat{H} | \Psi_{g\bar{g}}^{p\bar{p}} \rangle \tag{47}$$

Considering that $\langle \Psi_{HF} | \hat{H} | \Psi_{g\bar{g}}^{p\bar{p}} \rangle = \langle gg \mid pp \rangle = \mathbb{L}_{gp,}$ the Eq. (47) can be rewritten as

$$E_0 = \tilde{E}_{HF} + \tilde{E}_{corr} \tag{48}$$

$$\tilde{E}_{corr} = - \sum_{g=1}^{F} \sum_{p \in \Omega'_g} \sqrt{\frac{n_p}{n_g}} \mathcal{L}_{gp} \tag{49}$$

where $\tilde{E}_{HF} = \langle \Psi_{HF} | \hat{H} | \Psi_{HF} \rangle$ differs from the true HF energy since we use NOs as the one-electron functions instead of the HF orbitals. By similar arguments, $\tilde{E}_{corr}$ is not the typical definition of the correlation energy. Note also that E_0 is not the ground-state energy (23), which would be given by $\langle 0 | \hat{H} | 0 \rangle$, however its deviation from PNOF5e is small in numerical terms [15].

The zero-order excited energies are chosen in the form $E_K = E_0 + \Delta_K$, where Δ_K are the differences between diagonal elements $\hbar_i$ of the Fock operator, namely,

$$\varepsilon_i = \mathcal{F}_{ii} = \mathcal{H}_{ii} + \sum_{j=1}^{N} [\langle ij| ij \rangle - \langle ij| ji \rangle] \tag{50}$$

$$\Delta_K = \begin{cases} \varepsilon_s - \varepsilon_a & a \leq N \,;\, s > N \\ \varepsilon_s + \varepsilon_u - \varepsilon_a - \varepsilon_b & a, b \leq N \,;\, s, u > N \\ \cdots & \cdots \end{cases} \tag{51}$$

for singles, doubles, etc. This partitioning is different from Møller–Plesset (MP), since the Fockian is non-diagonal. Henceforth, the indexes a, b, ... refer to the spin orbitals occupied in $|\Psi_{HF}\rangle$, while the remaining spin orbitals are labeled by the indexes s, u, ...

Following the standard biorthogonal PT, the lowest order energy corrections can be given as

$$E^{(1)} = \langle \tilde{0}|\hat{V}|0\rangle = \langle \tilde{0}|\hat{H} - \hat{H}^0|0\rangle = 0 \tag{52}$$

$$E^{(2)} = -\sum_{K=1}^{M} \frac{\langle \tilde{0}|\hat{V}|K\rangle \langle \tilde{K}|\hat{V}|0\rangle}{E_K - E_0} = -\sum_{K=1}^{M} \frac{\langle \Psi_{HF}|\hat{H}|K\rangle \langle \tilde{K}|\hat{H}|0\rangle}{d_0 \Delta_K} \tag{53}$$

Due to the two-particle nature of the Hamiltonian, the explicit dependence on $\langle \Psi_{HF}|$ in the numerator of Eq. (53) implies that one needs only to consider single $\{|\Psi_a^s\rangle\}$ and doubly $\{|\Psi_{ab}^{su}\rangle\}$ excited determinants $|K\rangle$ in the summation. Given the ground-state (39), the second-order correction (53) can be expressed in three contributions, namely,

$$E^{(2)} = E_0^{(2)} + E_d^{(2)} + E_q^{(2)} \tag{54}$$

where

$$E_0^{(2)} = -\sum_{K=1}^{M} \Delta_K^{-1} \langle \Psi_{HF}|\hat{H}|K\rangle \langle \tilde{K}|\hat{H}|\Psi_{HF}\rangle \tag{55}$$

$$E_d^{(2)} = \sum_{K=1}^{M} \sum_{g=1}^{F} \sum_{p\in\Omega_g'} \Delta_K^{-1} \sqrt{\frac{n_p}{n_g}} \langle \Psi_{HF}|\hat{H}|K\rangle \langle \tilde{K}|\hat{H}|\Psi_{g\bar{g}}^{p\bar{p}}\rangle \tag{56}$$

$$E_q^{(2)} = -\sum_{K=1}^{M} \sum_{g<f}^{F} \sum_{p\in\Omega_g'} \sum_{q\in\Omega_f'} \Delta_K^{-1} \sqrt{\frac{n_p n_q}{n_g n_f}} \langle \Psi_{HF}|\hat{H}|K\rangle \langle \tilde{K}|\hat{H}|\Psi_{g\bar{g}f\bar{f}}^{p\bar{p}q\bar{q}}\rangle \tag{57}$$

Taking into account that

$$\langle \Psi_{HF}|\hat{H}|\Psi_a^s\rangle = h_{as} + \sum_b \langle ab||sb\rangle = \mathcal{F}_{as} \tag{58}$$

$$\langle \Psi_{HF} | \hat{H} | \Psi_{ab}^{su} \rangle = \langle ab | \, su \rangle - \langle ab | \, us \rangle = \langle ab | \, | su \rangle \tag{59}$$

and, after some algebraic manipulation, the second-order perturbation correction (54) can be cast as

$$E^{(2)} = E_{HF}^{(2)} + E_D^{(2)} + E_Q^{(2)} \tag{60}$$

where

$$E_{HF}^{(2)} = -2 \sum_{g=1}^{F} \sum_{p=F+1}^{nbf} \frac{|\mathcal{F}_{pg}|^2}{\varepsilon_p - \varepsilon_g} - \sum_{g,f=1}^{F} \sum_{p,q=F+1}^{nbf} \frac{\langle gf | \, pq \rangle \, [2 \langle pq | \, gf \rangle - \langle pq | \, fg \rangle]}{\varepsilon_p + \varepsilon_q - \varepsilon_g - \varepsilon_f} \tag{61}$$

$$E_D^{(2)} = 2 \sum_{g=1}^{F} \sum_{p \in \Omega'_g} \sqrt{\frac{n_p}{n_g}} \left\{ \frac{|\mathcal{F}_{pg}|^2}{(\varepsilon_p - \varepsilon_g)} + \sum_{q \in \Omega'_g} \frac{\mathcal{F}_{qg} \langle gq | \, pp \rangle}{(\varepsilon_q - \varepsilon_g)} - \sum_{f=1}^{F} \frac{\langle gg | \, fp \rangle \, \mathcal{F}_{pf}}{(\varepsilon_p - \varepsilon_f)} \right.$$

$$+ \sum_{q \in \Omega'_g} \frac{\langle gg | \, pq \rangle \, \mathcal{F}_{qp}}{(\varepsilon_p + \varepsilon_q - 2\varepsilon_g)} - \sum_{f=1}^{F} \frac{\mathcal{F}_{gf} \langle fg | \, pp \rangle}{(2\varepsilon_p - \varepsilon_g - \varepsilon_f)}$$

$$+ \frac{1}{2} \sum_{q,r \in \Omega'_g} \frac{\langle gg | \, qr \rangle \langle qr | \, pp \rangle}{(\varepsilon_q + \varepsilon_r - 2\varepsilon_g)} + \frac{1}{2} \sum_{f,e=1}^{F} \frac{\langle gg | \, fe \rangle \langle fe | \, pp \rangle}{(2\varepsilon_p - \varepsilon_f - \varepsilon_e)}$$

$$\left. - \sum_{f=1}^{F} \sum_{p \in \Omega'_g} \left[\frac{\langle gf | \, | qp \rangle \langle gq | \, pf \rangle + \langle gf | \, qp \rangle \langle gq | \, fp \rangle + \langle gf | \, pq \rangle \langle gq | \, | fp \rangle}{(\varepsilon_q + \varepsilon_p - \varepsilon_g - \varepsilon_f)} \right] \right\} \tag{62}$$

$$E_Q^{(2)} = \sum_{g,f=1}^{F} {}' \sum_{p \in \Omega'_g} \sum_{q \in \Omega'_f} \sqrt{\frac{n_p n_q}{n_g n_f}} \left\{ \frac{1}{4} \left[\frac{\mathcal{L}_{gp} \mathcal{L}_{fq}}{(\varepsilon_p - \varepsilon_g)} - \frac{\mathcal{L}_{gq} \mathcal{L}_{fp}}{(\varepsilon_q - \varepsilon_g)} \right] + \right.$$

$$\left. \frac{\langle gg | \, pq \rangle \langle ff | \, pq \rangle}{(\varepsilon_q + \varepsilon_p - 2\varepsilon_g)} \right\} + \sum_{g,f=1}^{F} \sum_{p \in \Omega'_g} \sum_{q \in \Omega'_f} {}' \sqrt{\frac{n_p n_q}{n_g n_f}} \frac{\langle gf | \, pp \rangle \langle gf | \, qq \rangle}{(2\varepsilon_p - \varepsilon_f - \varepsilon_g)}$$

$$- \sum_{g,f=1}^{F} \sum_{p \in \Omega'_g} \sum_{q \in \Omega'_f} \sqrt{\frac{n_p n_q}{n_g n_f}} \left[\frac{\langle gf | \, pq \rangle^2 - \langle gf | \, pq \rangle \langle gf | \, qp \rangle + \langle gf | \, qp \rangle^2}{(\varepsilon_q + \varepsilon_p - \varepsilon_f - \varepsilon_g)} \right] \tag{63}$$

The second-order corrections (54) were applied to the simplest formulation PNOF5 ($Nc = 1$) in Ref. [15]. In order to consider only the interpair correlation, the double excitations from different spatial orbitals were exclusively considered. The excited determinants $|\Psi_{g\tilde{g}}^{su}\rangle$ associated to each spatial orbital g were withdraw from $E^{(2)}$. The approach was named as PNOF5-PT2 to differentiate it from the original PNOF5-SC2-MCPT. The performance of the PNOF5-PT2 has been tested in noncovalent interactions, homolytic dissociations and reactivity [15, 37]. The groundstate energy of 36 closed-shell species belonging to the G2/97 test set of molecules was also studied [37]. In summary, PNOF5-PT2 showed a promising performance for a number of illustrative numerical calculations ranging from a vdW system to covalent molecules. At present, a work that takes the PNOF5e as the reference is in progress.

22.6 CLOSING REMARKS

The results achieved in NOFT with electron-pairing approaches show that we can already obtain, in many cases, a degree of accuracy comparable to those provided by the wavefunction-based approximations. So far, only the NOFs that keep these restrictions are able to yield the correct number of electrons in the fragments after a homolytic dissociation. To cite just one example, the dissociation curve of the carbon dimer, at the PNOF5-PT2 and PNOF6 levels of theory, closely resembles that obtained from the optimized CASSCF(8,8) wavefunction [42].

The employed reconstructions for making these functionals, satisfy known conditions for the N-representability of the 2RDM. To some extent, it has been generally assumed that there is no N-representability problem of the functional, as it is believed that only N-representable conditions on the 1-RDM are necessary. The ensemble N-representability constraints for acceptable 1-RDMs are easy to implement, but are insufficient to guarantee that the reconstructed 2-RDM is N-representable, and thereby the functional either. Several proposals have appeared in the literature, but most of them have serious shortcomings related with the violations of sum rules or constraints imposed by the functional N-representability. So far, the only way that exists to guarantee the one-to-one correspondence between $E[\Psi] \equiv E[2RDM]$ and $E[\{n_i, \varphi_i\}]$ is the top–down approach. The results obtained with the PNOF5e support this statement, which is a consequence of treating the electron pairs in a very accurate manner. Nonetheless, the top–down approach is a formidable task that is far from being something attainable for other wavefunctions, therefore, the bottomup method restricted only to some necessary conditions seems to be more plausible approach.

There is also a practical motivation for using pairing approximations. In NOFT, the solution is established optimizing the energy functional with respect to the ONs and to the NOs, separately [54]. The ONs can be conveniently expressed by means of auxiliary variables in order to enforce automatically the N-representability bounds of the 1-RDM. In the case of electron-pairing approaches, the variation can be additionally performed without constraints with respect to these new variables. Accordingly, the constrained nonlinear programming problem for the ONs becomes an unconstrained optimization, for which the efficient conjugated gradient method can be used with the corresponding saving of computational times.

The fact that the functionals lack of the interpair correlation constitutes their major limitation. PNOF5 does not describe correlation between electron pairs at all, while PNOF6 describes mostly non-dynamic interpair correlation. The perturbation theory has been the strategy followed to recover the dynamic correlation in the case of PNOF5, including its extended version. Unfortunately, we have not yet found the wavefunction that could be behind the interacting-pair approximation in NOFT. PNOF6 seems not to be related to geminal-based theories, although it keeps the orbital-pairing scheme.

To end this chapter, a comment about the approximate character of the NOFT is mandatory. The functionals currently in use are only known in the basis where the 1-RDM is diagonal. This implies that they are not functionals explicitly dependent on the 1-RDM and retain some dependence on the 2-RDM. For this reason, it is more appropriate to speak of a NOF rather than a functional of the 1-RDM. In this vein, in the NOFT, the NOs are the orbitals that diagonalize the 1-RDM corresponding to an approximate expression of the energy, like those obtained from an approximate wavefunction. As a consequence, NOFs still depend explicitly on the 2-RDM, so the energy is not invariant with respect to a unitary transformation of the orbitals. The exception is the yet unknown exact ground-state energy functional of the 1-RDM, the only functional for which Gilbert's theorem is fulfilled.

Because of this, the NOFT provides two complementary representations of the one-electron picture, namely, the NO representation and the canonical orbital (CO) representation [25]. The former arises directly from the optimization process solving the corresponding Euler equations, whereas the latter is attained from the diagonalization of the matrix of Lagrangian multipliers obtained in the NO representation. Both set of orbitals represent unique correlated one-electron pictures of the same energy minimization problem, ergo, they complement each other in the analysis of the molecular electronic structure. The orbitals obtained in both representations, using the electron pairing approaches in NOFT, have shown that the electron pairs with opposite spins continue to be the most suitable language for the chemical bond theory.

ACKNOWLEDGMENTS

Financial support comes from Eusko Jaurlaritza (Ref. IT588-13) and Ministerio de Economía y Competitividad (Ref. CTQ2015-67608-P). The SGI/IZO–SGIker UPV/EHU is gratefully acknowledged for generous allocation of computational resources.

KEYWORDS

- first-order reduced density matrix functional theory
- Piris natural orbital functional (PNOF)
- N-representability
- intrapair and interpair electron correlation
- bottom-up and top-down methods

REFERENCES

1. Pauling, L., (1931). "The nature of the chemical bond. Application of results obtained from the quantum mechanics and from a theory of paramagnetic susceptibility to the structure of molecules," *J. Am. Chem. Soc., 53,* 1367.
2. Lewis, G. N., (1916). "The atom and the molecule," *J. Am. Chem. Soc., 38*(4), 762.
3. Pauling, L., (1960). *The Nature of the Chemical Bond and the Structure of Molecules and Crystals: An Introduction to Modern Structural Chemistry.* Cornell University Press.
4. Shaik, S. S., & Hiberty, P. C., (2008). *A Chemist's Guide to Valence Bond Theory.* Hoboken, New Jersey: John Wiley and Sons, Inc.
5. Hurley, A. C., Lennard-Jones, J., & Pople, J. A., (1953). "The molecular orbital theory of chemical valency. XVI. A theory of paired-electrons in polyatomic molecules," *Proc. R. Soc. Lond. A, 220,* 446.
6. Fock, F. A., (1950). *Dokl. Akad. Nauk USSR, 73,* 735.

7. Hunt, W. J., Hay, P. J., & Goddard III, W. A., (1972). "Self-consistent procedures for generalized valence bond wavefunctions applications H_3, BH, H_2O, C_2H_6, and O_2," *J. Chem. Phys., 57*(2), 738.

8. Surjan, P. R., (1999). "An introduction to the theory of geminals," in *Topics in Current Chemistry, 203.* Springer-Verlag Berlin Heidelberg, pp. 63–88.

9. Piris, M., (2007). "Natural orbital functional theory," in *Reduced-Density-Matrix Mechanics: with applications to many-electron atoms and molecules*, Mazziotti, D. A., Ed. Hoboken, New Jersey, USA, John Wiley and Sons, Ch. 14, pp. 387–427.

10. Piris M., & Ugalde, J. M., (2014). "Perspective on natural orbital functional theory," *Int. J. Quantum Chem., 114*(18), 1169.

11. Piris, M., Lopez, X., Ruipérez, F., Matxain, J. M., & Ugalde, J. M., (2011). "A natural orbital functional for multiconfigurational states." *J. Chem. Phys., 134*(16), 164102.

12. Ludena, E. V., Torres, F. J., & Costa, C., (2013). "Functional N-representability in 2-matrix, 1-matrix, and density functional theories," *J. Mod. Phys., 4*(3), 391.

13. Piris, M., (2006). "A New approach for the two-electron cumulant in natural orbital," *Int. J. Quantum Chem., 106*, 1093.

14. Piris, M., (2013). "A natural orbital functional based on an explicit approach of the twoelectron cumulant," *Int. J. Quantum Chem., 113*(5), 620.

15. Piris, M., (2013). "Interpair electron correlation by second-order perturbative corrections to PNOF5," *J. Chem. Phys., 139*(6), 064111.

16. Piris, M., (2013). "Bounds on the PNOF5 natural geminal occupation numbers," *Comp. Theor. Chem., 1003*, 123.

17. Pernal, K., (2013). "The equivalence of the Piris Natural Orbital Functional 5 (PNOF5) and the antisymmetrized product of strongly orthogonal geminal theory," *Comp. Theor. Chem., 1003*, 127.

18. Coleman, A. J., (1963)."Structure of Fermion Density Matrices," *Rev. Mod. Phys., 35*, 668.

19. Matxain, J. M., Piris, M., Ruipérez, F., Lopez, X., & Ugalde, J. M., (2011). "Homolytic molecular dissociation in natural orbital functional theory." *Phys. Chem. Chem. Phys., 13*(45), 20129.

20. Matxain, J. M., Piris, M., Mercero, J. M., Lopez, X., & Ugalde, J. M., (2012). "sp3 Hybrid orbitals and ionization energies of methane from PNOF5," *Chem. Phys. Lett., 531*, 272.

21. Matxain, J. M., Piris, M., Uranga, J., Lopez, X., Merino, G., & Ugalde, J. M., (2012). "The nature of chemical bonds from PNOF5 calculations," *Chem. Phys. Chem., 13*, 2297.

22. Matxain, J. M., Ruipérez, F., Infante, I., Lopez, X., Ugalde, J. M., Merino, G., & Piris, M., (2013). "Communication: chemical bonding in carbon dimer isovalent series from the natural orbital functional theory perspective." *J. Chem. Phys., 138*(15), 151102.

23. Lopez, X., Ruipérez, F., Piris, M., Matxain, J. M., Matito, E., & Ugalde, J. M., (2012). "Performance of PNOF5 natural orbital functional for radical formation reactions: hydrogen atom abstraction and C-C and O-O homolytic bond cleavage in selected molecules." *J. Chem. Theory Comput., 8*, 2646.

24. Ruipérez, F., Piris, M., Ugalde, J. M., & Matxain, J. M., (2013). "The natural orbital functional theory of the bonding in Cr(2), Mo(2) and W(2)." *Phys. Chem. Chem. Phys., 15*(6), 2055.

25. Piris, M., Matxain, J. M., Lopez, X., & Ugalde, J. M., (2013). "The one-electron picture in the Piris natural orbital functional 5 (PNOF5)," *Theor. Chem. Acc., 132*, 1298.

26. Piris M., & March, N. H., (2016). "Potential energy curves for P2 and P2+ constructed from a strictly N-representable natural orbital functional," *Physics and Chemistry of Liquids, 54*(6), 797.

27. Piris, M., Matxain, J. M., & Lopez, X., (2013). "The intrapair electron correlation in natural orbital functional theory," *J. Chem. Phys., 139*(23), 234109.

28. Piris, M., Matxain, J. M., Lopez, X., & Ugalde, J. M., (2012). "The extended Koopmans' theorem: Vertical ionization potentials from natural orbital functional theory," *J. Chem. Phys., 136*(17), 174116.

29. Lopez, X., Piris, M., Nakano, M., & Champagne, B., (2014). "Natural orbital functional calculations of molecular polarizabilities and second hyperpolarizabilities. The hydrogen molecule as a test case," *J. Phys. B: At. Mol. Opt. Phys., 47*, 15101.

30. Piris, M., & March, N., (2014). "Weizsacker inhomogeneity kinetic energy term for the inhomogeneous electron liquid characterising some 30 homonuclear diatomic molecules at equilibrium and insight into Tellers theorem in Thomas-Fermi statistical theory," *Physics and Chemistry of Liquids, 52*(6), 804.

31. Piris, M., & March, N. H., (2015). "Is the Hartree-Fock prediction that the chemical potential of non-relativistic neutral atoms is equal to minus the ionization potential I sensitive to electron correlation?" *Physics and Chemistry of Liquids, 53*(6), 696.

32. Lopez, X., & Piris, M., (2015). "PNOF5 calculations based on the 'thermodynamic fragment energy method': C n H2n+2 (n = 1, 10) and (FH) n (n = 1, 8) as test cases," *Theoretical Chemistry Accounts, 134*(12), 151.

33. Piris, M., & March, N. H., (2016). "Chemical and ionization potentials: Relation via the Pauli potential and NOF theory," *Int. J. Quantum Chem., 116*(11), 805.

34. Matxain, J. M., Ruiperez, F., & Piris, M., (2013). "Computational study of Be2 using Piris natural orbital functionals." *J. Mol. Mod., 19*(5), 1967.

35. Ramos-Cordoba, E., Salvador, P., Piris, M., & Matito, E., (2014). "Two new constraints for the cumulant matrix," *J. Chem. Phys., 141*(23), 234101.

36. Szabados, A., Rolik, Z., Toth, G., & Surjan, P. R., (2005). "Multiconfiguration perturbation theory: size consistency at second order." *J. Chem. Phys., 122*(11), 114104.

37. Piris, M., Ruipérez, F., & Matxain, J., (2014). "Assessment of the second-order perturbative corrections to PNOF5," *Mol. Phys., 112*(5), 711.

38. Piris, M., (2014). "Interacting pairs in natural orbital functional theory," *J. Chem. Phys., 141*(4), 044107.

39. Ramos-Cordoba, E., Lopez, X., Piris, M., & Matito, E., (2015)."H4: A challenging system for natural orbital functional approximations," *J. Chem. Phys., 143*, 164112.

40. Piris, M., & March, N. H., (2015). "Low-lying isomers of free-space halogen clusters with tetrahedral and octahedral symmetry in relation to stable molecules such as SF6," *J. Phys. Chem. A, 119*(40), 10190.

41. Lopez, X., Piris, M., Ruipérez, F., & Ugalde, J. M., (2015). "Performance of PNOF6 for Hydrogen abstraction reactions," *J. Phys. Chem. A, 119*, 6981.

42. Piris, M., Lopez, X., & Ugalde, J. M., (2016). "The bond order of C(2) from an strictly N-representable natural orbital energy functional perspective," *Chemistry - A European Journal, 22*, 4109.

43. Mitxelena, I., & Piris, M., (2016). "Molecular electric moments calculated by using natural orbital functional theory," *J. Chem. Phys., 144*(20), 204108.

44. Cioslowski, J., Piris, M., & Matito, E., (2015). "Robust validation of approximate 1- matrix functionals with few-electron harmonium atoms," *J. Chem. Phys., 143*, 214101.

45. Mazziotti, D. A., (2012). "Structure of fermionic density matrices: complete N-representability conditions," *Phys. Rev. Lett., 108*(26), 263002.

46. Gilbert, T. L., (1975). "Hohenberg-kohn theorem for nonlocal external potentials," *Phys. Rev. B, 12*(6), 2111.

47. Levy, M., (1979). "Universal variational functionals of electron densities, first-order density matrices, and natural spin-orbitals and solution of the v-representability problem," *Proc. Natl. Acad. Sci. USA, 76*(12), 6062.

48. Piris, M., & Otto, P., (2003). "One-particle density matrix functional for correlation in molecular systems," *Int. J. Quantum Chem., 94*(6), 317.

49. Mazziotti, D. A., (1998). "Approximate solution for electron correlation through the use of Schwinger probes," *Chem. Phys. Lett., 289*, 419.

50. Piris, M., Matxain, J. M., Lopez, X., & Ugalde, J. M., (2009). "Spin conserving natural orbital functional theory," *J. Chem. Phys., 131*, 021102.

51. Piris, M., Matxain, J. M., Lopez, X., Ugalde, J. M., (2010). "Communication: The role of the positivity N-representability conditions in natural orbital functional theory," *J. Chem. Phys., 133*, 111101.

52. Piris, M., (1999). "A generalized self-consistent-field procedure in the improved BCS theory," *J. Math. Chem., 25*, 47.

53. Dunning Jr., T. H., (1989). "Gaussian basis sets for use in correlated molecular calculations. I. The atoms boron through neon and hydrogen," *J. Chem. Phys., 90*(2), 1007.

54. Piris, M., & Ugalde, J. M., (2009). "Iterative diagonalization for orbital optimization in natural orbital functional theory," *J. Comput. Chem., 30*, 2078.

CHAPTER 23

MEASURING THE EFFECT OF DENSITY ERRORS WHEN USING DENSITY FUNCTIONAL APPROXIMATIONS

ANDREAS SAVIN

Laboratoire de Chimie Théorique, UMR7616, CNRS, and UPMC, Sorbonne Universités, F-75252 Paris, France

CONTENTS

ABSTRACT

Using two different density functionals (one of them can be the exact one) provide two ground state densities. In order to study whether the difference in energies is due to the form of the functional itself, or the difference between the two densities, one may decompose the changes into that due to a change in the density (for given functional), and a change in the functional (at given density). However, the order in which this changes are made matters, yielding different contributions of the individual terms.

The two ways of computing the changes of the functional at given density (ground state density of each of the functionals) brackets the difference between the energies given by the two functionals.

23.1 INTRODUCTION: DENSITY AND ENERGY ERRORS

A non-relativistic electronic system in its ground state is characterized by the external potential (of the nuclei, v_{ne}), and the number of electrons, N. Once this information is given, the Hamiltonian operator can be constructed, and the Schrödinger equation solved. Density functional theory is based upon the variational principle for obtaining the ground-state energy [1]:

$$E = \min_{n(r)} \left(F[n(r)] + \int v_{ne}(r)n(r)d^3r \right) \tag{1}$$

where $n(r)$ is a positive function integrating to N. As clear from the notation, the computation of the functional F requires only the knowledge of the density, not that of v_{ne}. The minimizing density is the ground state density of the system characterized by v_{ne} and N.

When $F[n]$ is approximated, neither the ground state energy, nor the ground state density is correct. It is commonly believed that both quantities are quite well approximated, although benchmarks, as a rule, concentrate on energetic quantities. It was argued [2] that the success of density functional approximations is due to the sole use of the density, and not of the wave function that has a more difficult structure to catch. An example, the H_2 molecule in a minimal basis set was given. A single Slater determinant (as used in the Hartree-Fock, or in the Kohn-Sham method) is a poor approximation of the wave function. The density it produces is nevertheless exact.

However, it is not at all clear that the errors introduced by using an approximation for $F[n]$ have less importance for the density than for the energy. Could it happen that the opposite is true? In fact, in the early days of density functional theory, such questions have been asked, and it has been noticed that sometimes using density functional approximations give poorer densities than those obtained at Hartree-Fock level (see, e.g., Ref. [3]). A way to understand the source of the problem is to notice that the density is related to the variation of the energy with respect to the potential. Consider, for a trivial analogy, a well know problem from interpolation theory [4]. A polynomial interpolation of exp(x), with $0 \leq x \leq 1$, on m points gives:

$$f(x) \approx \sum_{k=0}^{m-1} a_k x^k$$

Even if the interpolant works quite well for $f(x)$, its quality deteriorates as derivatives are taken; the derivatives of the interpolant do not approximate well the derivative $f'(x)$, cf. Figure 23.1. A derivative of order larger than m vanishes for the polynomial interpolant, while in the example given, all derivatives of $f(x)$ satisfy $f^{(n)}(x) = f(x)$.

Thus, in a recent series of papers (starting with [5]) the following question is raised:

Is the error of a given density functional approximation due to the inability to produce a good energy, or that of generating a correct density?

In order to formulate more precisely this question, let us consider two methods, using the density functionals F_1 and F_2, yielding after the minimization of Eq. (1) the energies E_1, E_2, and ground state densities n_1, n_2, respectively. For example, the index 1 can refer to the exact quantities, and the index 2 to an approximation. Let us consider the expression that is minimized (that is in parentheses) in Eq. (1),

$$\mathcal{E}_i[n(r)] = F_i[n(r)] + \int v_{ne}(r)n(r)d^3r \qquad (2)$$

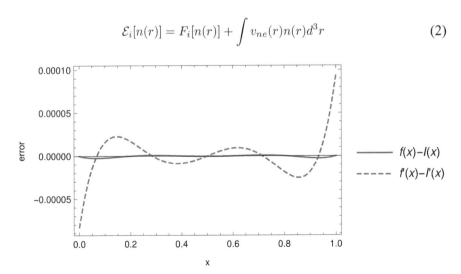

FIGURE 23.1 Errors obtained after interpolating $f(x) = \exp(x)$ (i) on the function (full curve); and (ii) on the first derivative, $f'(x) = f(x)$, dashed curve.

The potential, υ_{ne}, is not explicitly written as argument of ε, as it is considered given in the discussions that follow. ε can be computed in principle for any $n(r)$, in particular, the ground state energy $E_i = \varepsilon_i[n_i]$. In reference [5], the difference between the energies obtained with the two methods is decomposed:

$$E_2 - E_1 = \Delta E_F + \Delta E_D \tag{3}$$

where $\Delta E_F = \varepsilon_2[n_1] - \varepsilon_1[n_1]$, $\Delta E_D = \varepsilon_2[n_2] - \varepsilon_2[n_1]$, and ΔE_F and ΔE_D attributed to errors due to the functional, or due to the density.[1]

In this paper, an alternative decomposition is discussed. It not only provides different numbers for the two contributions, it also provides bounds to the difference $E_2 - E_1$. Furthermore, it will be shown that the origin of effect of the density difference can be traced back to the (exchange-)correlation energy functional.

23.2   TWO PATHS TO THE SAME END, AND ERROR BOUNDS

Let us consider, as above, $E_2 - E_1$. In the discussion below, one can, but is not obliged to associate index 1 to exact quantities, and 2 to approximate quantities. The attribution could be reversed, or both indices could correspond to approximations. Let us present the possibilities $\varepsilon_i[n_j]$, Eq. (2), as a diagram (Figure 23.2). The decomposition of $E_2 - E_1$ described above [5] corresponds to reaching $E_1 = \varepsilon_1[n_1]$ starting from $E_2 = \varepsilon_2[n_2]$ by passing through $\varepsilon_2[n_1]$, i.e., first changing the density from n_2 to n_1, and next changing the method from ε_2 to ε_1 (full arrows in Figure 23.2). Alternatively, one can first change the method, next change the density (dashed arrows in Figure 23.2), i.e., pass through $\varepsilon_1[n_2]$. As in practice it is much easier to compute some approximate density functional with the exact density, i.e., obtain $\varepsilon_2[n_1]$, than to compute the exact energy functional with the approximate density, i.e., to obtain $\varepsilon_1[n_2]$, the choice of reference [5] is understandable. However,

    i)   by virtue of the Hohenberg-Kohn theorem [1], the value of the exact functional for a given density is accessible, see, e.g., [6], and

---

[1] Please notice that ΔE_F can be also written as a difference $F_2[n_1] - F_1[n_1]$, while ΔE_F also contains the effect of the change in density not only in the universal functional, but also a term depending on υ_{ne}

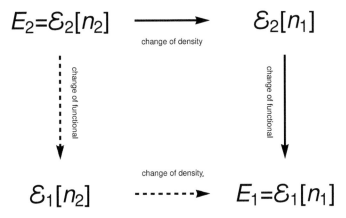

FIGURE 23.2 Different ways to reach E_2 from E_1. (i) First changing the density, next changing the method (as described by the decomposition in ΔE_D, and ΔE_F, Eq. 3): full arrows. (ii) First changing the method, next changing the density: dashed arrows.

ii) ε_i can correspond to different approximations (useful in practice), and in this case the difficulty mentioned does not show up.[2]

We take now into account that n_i is obtained by minimizing $\varepsilon_i[n]$ over n. Thus, by virtue of the variational principle, $\varepsilon_i[n \neq n_i] \geq \varepsilon_i[n_i] = E_i$. This is presented diagrammatically in Figure 23.3. It shows that the effect of changing the density from n_2 to n_1 is different for the ε_i: while the contribution to the density change in Eq. (3), ΔE_D, is nonpositive, that using the other path, is nonnegative:

$$\varepsilon_2[n_2] - \varepsilon_2[n_1] \leq 0 \qquad (4)$$

$$\varepsilon_1[n_2] - \varepsilon_1[n_1] \geq 0$$

We further obtain, by applying the above conditions,

$$\varepsilon_2[n_2] - \varepsilon_1[n_2] \leq E_2 - E_1 \leq \varepsilon_2[n_1] - \varepsilon_1[n_1] \qquad (5)$$

[2] It should be mentioned that the effect of using the Hartree-Fock method constrained to yield the exact density has been also used to provide bonds for the correlation energy [7], and to define and analyze the dynamic/non-dynamic components of the correlation energy [8].

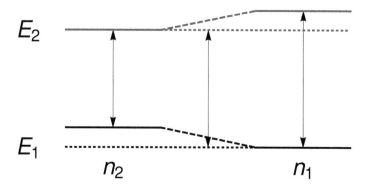

FIGURE 23.3 ε for two different methods, and their ground state densities. The black color is used for ε_1, while the gray color for ε_2. The dashed lines are used to show the raising of ε_i by using a density different from the ground state density. The double arrows shown the quantities showing up in equation 5, $\varepsilon_2[n_2] - \varepsilon_1[n_2], E_2 - E_1 = \varepsilon_2[n_2] - \varepsilon_1[n_1], \varepsilon_2[n_1] - \varepsilon_1[n_1]$.

Thus, bounds that bracket the difference between E_2 and E_1 are provided by the two ways of estimating the changes due to functionals, at fixed densities (the upper bound corresponds to ΔE_F in Eq. 3).

As each of the bounds in Eq. (5) uses the same density, the term containing υ_{ne} in the definition of ε, Eq. (2), disappears from the bounds: only $F_2[n] - F_1[n]$ is used. In practice, most of the density functional calculations are performed using the method of Kohn and Sham [9]. In this method, a kinetic energy functional of the density is defined, $T_s[n]$, subtracted it from $F[n]$, as is the classical Coulomb or Hartree term, $E_h[n]$. These terms are determined by the density alone, and are thus – for a given n – the same no matter what approximation is used for the remaining term, the exchange-correlation energy. We obtain that

$$E_{xc,2}[n_2] - E_{xc,1}[n_2] \leq E_2 - E_1 \leq E_{xc,2}[n_1] - E_{xc,1}[n_1] \qquad (6)$$

In a variant of the Kohn-Sham method, the exchange energy is obtained exactly (like T_s and E_h), and for this category of approximations, only the approximation of the correlation energy produces the inequalities:

$$E_{c,2}[n_2] - E_{c,1}[n_2] \leq E_2 - E_1 \leq E_{c,2}[n_1] - E_{c,1}[n_1] \qquad (7)$$

23.3   ILLUSTRATION: THE HYDROGEN ATOM

In order to illustrate the results obtained, let us now consider some specic example. We choose a very simple one that can be constructed analytically: the hydrogen atom.[3] As further simplifications, the local density approximation (LDA) is used, and a minimal basis set, $\psi = (\zeta^3/\pi)^{1/2} \exp(- \zeta r)$ in which the exponent of the hydrogen wave function, ζ, can be optimized. The optimal value, ζ_{min} depends on the approximation. Please notice that this ansatz corresponds to a scaling of the wave function.

Let us first consider exchange-only approximations (spin-restricted, or unrestricted). The energy expression is given by

$$\frac{1}{2}\zeta^2 - \zeta + \frac{5}{16}\zeta + c_x\zeta \tag{8}$$

The term in ζ^2 comes from the kinetic energy, the linear terms from the interaction with the nucleus, the Hartree, and the exchange term. An exchange term linear in ζ is provided by any approximation that satisfies the exact relationship required by scaling (Eq. (102) of Ref. [10]), in particular by LDA. The value of the coefficient c_x depends on the approximation. In LDA, c_x are known exactly, both for the spin-restricted, or -unrestricted approximation ($c_x \approx - 0.2127$ and $\approx - 0.2680$, respectively).

The exact energy of the hydrogen atom is $E_1 = \varepsilon_1[n_1] = -1/2$ atomic units, and the exact density is obtained by using $\zeta = 1$ in the wave function ψ. The (exchange-only) LDA energy expression using the exact density of the H atom is obtained by choosing $\zeta = 1$ in Eq. (8, $\varepsilon_2[n_1] = -3/16 + c_x$).

As the LDA expression for the energy, Eq. (8), is quadratic, the lowest value is reached when using $\zeta_{min} = 11/16 - c_x$: $E_2 = \varepsilon_2[n_2] = -\zeta^2_{min}/2$.

To obtain the exact energy obtained for the LDA-optimized density, n_2, we consider a one-electron system, and need to find the potential that has the density $n = \psi^2 = (\zeta^3_{min}/\pi)\exp(-2\zeta_{min}r)$. We know it, this solution corresponds to the hydrogen-like atoms, and the potential is $-\zeta_{min}/r$. For this potential, the ground state energy is the same as for the approximation, $\varepsilon_1[n_2] = -\zeta^2_{min}/2$.

The numerical values are given in Table 23.1, where results obtained by adding a correlation functional [11] are also shown. One can notice that due to the incomplete cancellation of the Hartree term by the LDA exchange, the densities become too diffuse ($\zeta_{min} < 1$). The results are also summarized in form of an energy-level diagrams in Figure 23.4. At first sight, the diagrams

TABLE 23.1 Results Obtained by Optimizing the Scaled Density of the H Atom in the Local Density Approximation

Functional approximated	Restricted LDA	Unrestricted LDA
Exchange only	−0.405 (0.900)	−0.456 (0.955)
Exchange and correlation	−0.444 (0.921)	−0.477 (0.965)

The ground state energies are followed by the optimized exponent, ζ_{min}, in parentheses; atomic units used.

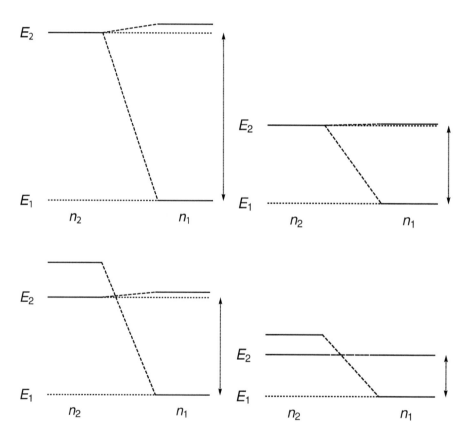

FIGURE 23.4 Energy level diagram, cf. Figure 23.3, comparing the LDA and exact ground state energies (vertically), and the effect of density change, for the H atom, from LDA to the exact (horizontally, in each panel. The results using the exchange-only approximation are on the top panels, the exchange-correlation approximations are in the bottom panels; the spin-restricted form for the left panels, the spin-unrestricted form for the right panels.

in Figures 23.3 and 23.4 look different. However, the bounding properties shown by the inequalities 5 remain valid. The confusion is produced by the degeneracy of the approximate and exact energies for the approximate density (in the exchange only case), and a significant stabilization produced by the correlation functional at LDA level (for the exchange-correlation case) making the lower bound become negative.

We now go a step further, and approach the exact value by modifying the Kohn-Sham model, by treating the contribution of the long-range interaction to exchange and correlation exactly; the long-range part of the interaction is defined by replacing the Coulomb repulsion between electrons by $\mathrm{erf}(\mu r)/r$ [12]. When the parameter μ vanishes, we recover the Kohn-Sham approximation, while when $\mu \to \infty$, the exact Schrödinger equation is obtained. In our hydrogen atom example, this means that the (fictitious self-interaction) Hartree (classical) repulsion between electrons is exactly compensated for the long range. The remaining part uses an LDA-type approximation for exchange and correlation that depends on the parameter μ [12, 13].

The resulting curves of the various energy differences as a function of μ are shown in Figure 23.5. We notice that the attribution to changes due the functional, and due to the density, corresponding to the two paths in Figure 23.2, full vs. dashed curves, are quite different. As show in inequalities 5, a given change of the density, produces opposite effects (different) signs for the contributions to the energy differences corresponding to a change in functionals. The path described by full arrows in the diagram in Figure 23.2, i.e., the path of reference [5], attributes a weak change due to the density, while it is significant for the other path (black curves, full, and dashed, in Figure 23.5). So, on one path the difference between curves is explained mainly by the error in the functional, while on the other, both contributions are significant. The bounds given by the inequalities 5 are shown in gray on Figure 23.5. In accordance with the observation above, the upper bound is tighter than the lower bound.

23.4 CONCLUSION

The question addressed is whether the difference in the energies obtained using two density functional methods (one possibly the exact one) is due to

---

[3]One can imagine this also as a study of the limit of dissociation of the H_2 molecule, and analyze the problems that can occur when (semi-)local approximations are used.

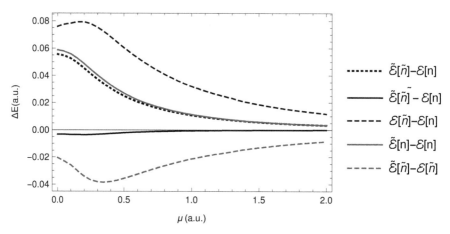

FIGURE 23.5   Evolution of energy differences with the range parameter μ. The difference between the LDA approximation and the exact value is shown as a dotted curve; the effect of changing between the exact and the LDA functional, at given density, is shown in gray, the effect of changing the density, for a given functional in black; the values on the path shown by full arrows in the diagram in Figure 23.2 correspond to full curves, those corresponding to the path shown by dashed arrows in this diagram to dashed curves.

(i) the difference in the (exchange-)correlation functionals used, or (ii) the differences in ground state densities produced by these functionals. It turns out that the order in which this change is made (first change the density, or first change the functional) matters, and a simple example (H atom), shows quite a substantial difference. The signs for the changes due to changing the densities in the two functionals (on the two paths) are different, (cf. inequalities 5). The changes produced by the functionals, when using the corresponding two ground state densities, produce bounds (cf. inequalities 5).

KEYWORDS

- **density functional theory**
- **Hartree-Fock method**
- **Kohn-Sham method**
- **Hohenberg-Kohn theorem**

REFERENCES

1. Hohenberg, P., & Kohn, W., (1964). *Phys. Rev., 136,* B864.
2. Cook, M., & Karplus, M., (1987). *J. Phys. Chem., 91,* 31.
3. Savin, A., Wedig, U., Stoll, H., & Preuss, H., (1982). *Chem. Phys. Letters, 92, 503, Erratum 94,* 536 (1983).
4. Runge, C., (1901). *Zeitschrift füur Mathematik und Physik, 46,* 224.
5. Kim, M. C., Sim, E., & Burke, K., (2013). *Phys. Rev. Lett., 111,* 073003.
6. Colonna, F., & Savin, A., (1999). *J. Chem. Phys., 110,* 2828.
7. Savin, A., Stoll, H., & Preuss, H., (1986). *Theor. Chim. Acta., 70,* 407.
8. Valderrama, E., Ludeña, E. V., & Hinze, J., (1999). *J. Chem. Phys., 110,* 2343.
9. Kohn, W., & Sham, L. J., (1965). *Phys. Rev., 140,* A1133.
10. Levy, M., & Perdew, J. P., (1980). *Phys. Rev. A, 32,* 2010.
11. Vosko, S. H., Wilk, L., & Nusair, M., (1980). *Canadian Journal of Physics, 58,* 1200.
12. Savin, A., (1996). in *Recent Developments and Applications of Modern Density Functional Theory,* edited by Seminario, J. M., Elsevier, Amsterdam.
13. Paziani, S., Moroni, S., Gori-Giorgi, P., & Bachelet, G. B., (2006). *Phys. Rev. B, 73,* 155111.

CHAPTER 24

EXCITED STATE INTRAMOLECULAR PROTON TRANSFER (ESIPT) PROCESS: A BRIEF OVERVIEW OF COMPUTATIONAL ASPECTS, CONFORMATIONAL CHANGES, POLYMORPHISM, AND SOLVENT EFFECTS

ANTON J. STASYUK and MIQUEL SOLÀ

The Institute of Computational Chemistry and Catalysis (IQCC) and Department of Chemistry, University of Girona, C/ Maria Aurèlia Campmany, 69, E-17003-Girona, Catalonia, Spain

CONTENTS

24.1   INTRODUCTION

The proton transfer reaction is considered to be one of the most important and fundamental processes in chemistry and life sciences [1, 2]. In 1950s, Weller observed abnormally huge Stokes shift in salicylic acid, which was not found in very similar *ortho*-methoxybenzoic acid and its other derivatives.

[3, 4]. This finding led him to the detailed study of acid-base properties of molecules in the excited state, which in turn allowed Weller to propose the concept of intramolecular proton transfer in excited state (ESIPT) [5, 6]. It is widely accepted that driving force of ESIPT reaction comes from significant changes in acidity and basicity of involved groups [7, 8]. Proton transfer processes typically take place through the formation of a six-membered ring with strong intramolecular hydrogen bond between proton donor and acceptor groups. Five-membered [9–12] and seven-membered [13, 14] units can also be formed, but much less often. In science and technology, over the past few decades a lot of attention was paid to organic fluorescent molecules undergoing ESIPT due to their unique photophysical properties and numerous practical optoelectronic applications [15–25].

ESIPT is a phototautomerization process that occurs in the electronic excited state of molecule and can be represented by a four-level ($\mathbf{N^0 - N^1 - T^1 - T^0}$) photocycle (Scheme 1).

Usually, ESIPT-capable compounds are most stable in normal ($\mathbf{N^0}$) form in the ground state S_0, while in the first excited state S_1, the transferred form ($\mathbf{T^1}$) is commonly more stable. Excitation of $\mathbf{N^0}$ form leads to Franck-Condon S_1 excited state, which in turn after relaxation to $\mathbf{N^1}$ transforms fast into $\mathbf{T^1}$ by proton transfer. Subsequently, radiative or non-radiative dissipation of energy for $\mathbf{T^1}$ followed by ground-state back proton transfer can be observed. Above-mentioned fast transformation $\mathbf{N}$ form to $\mathbf{T}$ form in the excited state has been confirmed by the fact that measured rate of different ESIPT reactions are usually at pico- to nanosecond time scale.

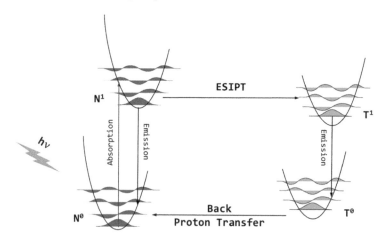

SCHEME 1   General representation of ESIPT.

24.2 EXCITED STATE INTRAMOLECULAR PROTON TRANSFER (ESIPT)

Since the discovery of ESIPT phenomenon, a large number of various compounds undergoing proton transfer process in the excited state have been designed and investigated in detail. These molecules belong such classes of chemical compounds as: benzophenones, flavones, anthraquinones, azoles and their fused analogues, bipyridyls, benzo[*h*]quinolones, and Schiff bases.

For the vast majority of the previously described ESIPT-capable systems, a phenolic OH group acts as a proton donor, whereas a heterocyclic nitrogen atom of pyridine type or a carbonyl group serves as an acceptor of the proton. However, few examples in which acidic C-H bond [26, 27] or primary and secondary amines [28–30] play the role of the proton donor in ESIPT have recently been introduced.

Among various ESIPT-capable systems of OH-type, imidazo[1,2-*a*] pyridines cause the greatest interest in recent years. In 1995, Acuña and co-workers were the first who reported the photoinduced intramolecular proton transfer in phenyl-substituted imidazo[1,2-*a*]pyridine (1,2-HPIP) [31]. Using nuclear Overhauser effect spectroscopy, the authors found that the proton from OH group in 1,2-HPIP is highly deshielded and the NMR spectrum corresponds to a planar ground-state conformation with intramolecular hydrogen bond. At the same time, closest proximity of H-3 proton to H-6' was revealed. The electronic absorption spectrum and fluorescent spectrum of 1,2-HPIP in cyclohexane showed an intensive structured band for absorption and a single structureless band with a large Stokes shift for fluorescence, whereas the emission spectrum in dioxane consisted of two bands: a bathochromically shifted one, similar to that found in cyclohexane, and the "blue" structured band with normal Stokes shift. Redistribution of electronic charges after excitation of 1,2-HPIP causes an increase in acidity of the proton-donating phenolic group and, simultaneously, the increase in basicity of the accepting partner (*i.e.*, nitrogen atom). These changes make ESIPT possible and lead to the formation of zwitterion (Figure 24.1). The authors assumed that efficient radiationless deactivation of the zwitterion probably takes place through intersystem crossing to the triplet manifold(s).

Later, the same authors [32] published a brilliant paper which inherently was a logical continuation of the original 1,2-HPIP study. They put forward the hypothesis of rotational process in the intramolecular proton transfer cycle in 2-(2'-hydroxyphenyl)imidazo[1,2-*a*]pyridine. The authors

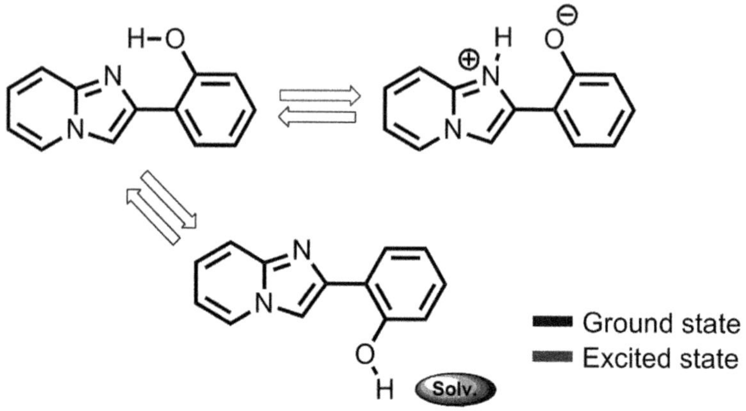

FIGURE 24.1 Proton transfer reaction in 1,2-HPIP in ground and excited states.

also described the light-induced tautomerism of the fluorophore (1,2-HPIP) encapsulated with α–, β– or γ–cyclodextrins (CD). For them, both proton transfer and conformational changes cause spectral shifts, which are directly related to the size of the CD cavity. Based on steady-state experiments, the authors explored and quantified subtle details of the induced fit and cavity diameter.

A considerable amount of accumulated experimental data resulted in that several years later in 1999 Douhal and co-workers [33] published first paper with theoretical studies based on *ab initio* calculations of a simple molecular system – 8-hydroxyimidazo[1,2-*a*]pyridine (8-HIP). The authors analyzed the theoretical results considering two tautomeric structures (normal, **N**, and transferred, **T**, forms) of HIP and the proton-transfer reaction linking them in both ground and first singlet excited state (Figure 24.2). At the initial stage, Douhal and co-workers analyzed the proton transfer reaction in the isolated molecule of 8-HIP. Significant higher energetic barrier for ESIPT process in both ground and first excited state, compare to the number of other studied intramolecular proton transfer processes, was noted.

For explanation of such unusual behavior of energetic barriers, the geometries of three stationary points in S_0 and S_1 was considered (Figure 24.3). The analysis revealed that intramolecular H-bond always has a longer distance in the excited state. At the same time, it was noted that the geometries of the transition states (TS) are quite similar in both states, thus it turned out that the TS in S_1 state involves a more severe distortion

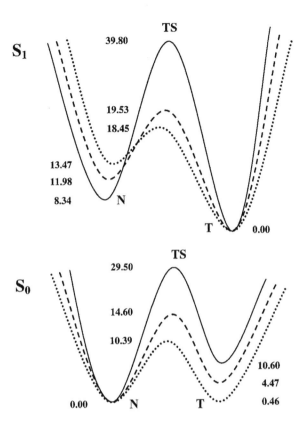

FIGURE 24.2 Schematic energy profiles for the proton transfer process involving N, TS, and T in S_0 and S_1 states. Solid line denotes gas-phase results; dashed line – 8-HIP:H_2O; dotted line – 8-HIP:H_2O with the rest of the solvent estimated through the IPCM. Relative energies are given in kcal/mol. Reprinted with permission from Ref. [33]. Copyright © 2016 American Chemical Society.

from both the reactants and products. Latest leads to higher energy barrier. A nonpolar media effect has been studied by including solvent through the continuum IPCM model. Results suggested a very minor effect in the energy profile and thus the solvent does not affect the picture of the whole process.

To analyze the effect of polar protic solvent on the ESIPT reaction, the authors considered the effect of one discrete water molecule. Based on the previous results for 7-azaindole, hydroxypyridine and other related systems, it is known that the molecule of water can act as bifunctional catalyst connecting proton donor and proton acceptor groups. It has been found that inclusion

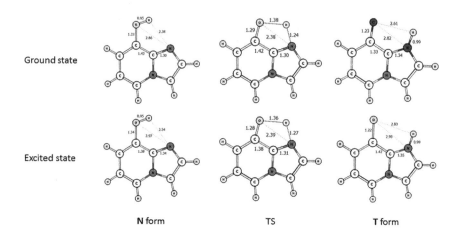

FIGURE 24.3 Geometries of stationary points for N, T, and TS of the proton-transfer process of 8-HIP in the ground S_0 and first excited electronic state S_1. Interatomic distances are given in Å. Adapted with permission from Ref. [33]. Copyright © 2016 American Chemical Society.

of the molecule of water significantly lowers energy barrier for the proton transfer reaction in both ground and excited state. This result is explained by the simultaneous proton donor and acceptor role of water, which promotes the tautomerization of 8-HIP without extensive deformation of the solute.

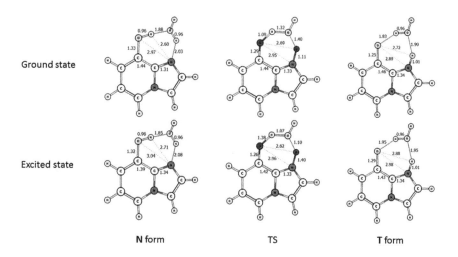

FIGURE 24.4 Geometries of the stationary points for the proton-transfer process in the ground S_0 and first excited state S_1 of the 8-HIP-H_2O complex. Interatomic distances are given in Å. Adapted with permission from Ref. [33]. Copyright © 2016 American Chemical Society.

In this way, water molecule leads to a double proton transfer. However, only one TS has been found for each electronic state. The geometry of the TS indicates that two migrating protons of the complex are not jumping at the same time. In fact, for the **N** tautomer, one of the protons has been almost transferred, whereas the other one is still been in motion. In ground state, the first proton jumps from water molecule to the **N** tautomer adduct, but in excited state, situation is reversed – thus, we can say that 8-HIP acts as a base in S_0 and as an acid in S_1 in respect with water.

Based on theoretical results, Douhal and co-workers were able to explain experimental observations, namely two bands in UV absorption spectrum of 8-HIP in aqueous solution (bands correspond to photoexcitation of **N** and **T** structures) and two bands in fluorescence spectrum. Additionally, they demonstrated that due to the fact that **N** and **T** tautomers absorb at different regions, it is possible to control the fluorescence behavior of the system by tuning excitation wavelength. This demonstrates that theoretical ground could be extremely helpful for understanding such behavior of 8-HIP and related derivatives in various media.

At the same year, Adamowicz and co-workers [34] published a fundamental work that discusses the importance of using the full system to describe ESIPT. The case of two very similar 2-(2'-hydroxyvinyl)benzimidazole (HVBI) and 2-(2'-hydroxyphenyl)imidazole (HPI) ESIPT-capable compounds in the ground state and various excited states ($^1pp^*$, $^1np^*$ and $^3pp^*$) was studied at different levels of theory.

It is well established that an accurate evaluation of the energetic parameters governing the proton transfer requires the use of large basis sets and the inclusion of electron correlation effects. Because of this, theoretical study of experimentally important molecules is usually impractical and requires molecular models of reduced size. Selection of the representative models is extremely important for correct reproducing the properties of interest, i.e., ESIPT properties. For example, 2-(2'-hydroxyphenyl)benzimidazole (HPBI) can be approximated by various systems of different molecular size (Figure 24.5).

1-amino-3-propenal is the simplest approximation model. [35,36] However, the analysis of the intramolecular proton transfer in a larger system as salicylaldimine, which contains an aromatic phenol ring, revealed important differences in the relative stability of two tautomers and in the proton-transfer barriers between 1-amino-3-propenal and salicylaldimine.

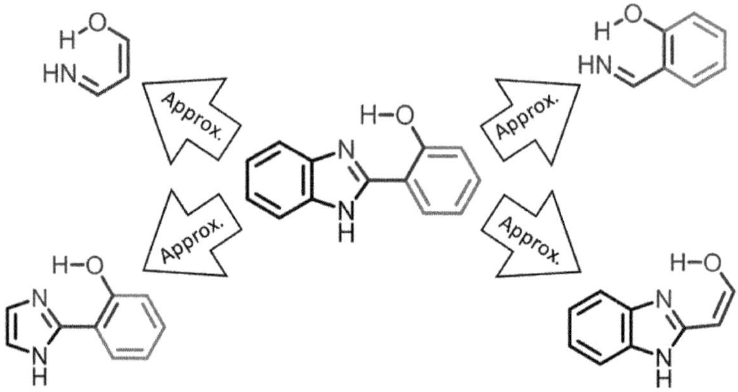

FIGURE 24.5   Molecular models describing intramolecular proton transfer in HPBI.

This indicates that the results obtained using small models are not always comparable to larger systems [35] and can lead to erroneous conclusions.

Comparing the behavior of 2-(2′-hydroxyvinyl)benzimidazole and 2-(2′-hydroxyphenyl)imidazole, the authors found that the phenol ring is essential to describe the intramolecular proton transfer of HPBI in the ground, ^{1}pp*, and ^{3}pp* excited states, while the imidazole is the fragment that has the most influence on the ESIPT of HPBI in the ^{1}np* state.

For the correct description of ESIPT process, the question of the conformational and tautomeric equilibrium of compounds undergoing proton transfer is no less important. Luque, Solà, and co-workers [37] considered in detail the effect of solvent on the ground state Gibbs energies for conformers of HPBI (both enol and keto forms) by means of Monte-Carlo simulations and continuum model calculations (Table 24.1). They tried to find the answer how nature of the solvent would modulate the relative stability of HBPI conformers, which in turn determines the nature of photoinduced ESIPT processes.

It is well known that HPBI may exist in three different enol conformers and one keto tautomer (Figure 24.6). Widely accepted that in S_0 closed enol cis species with intramolecular hydrogen bond is the only form existing in nonpolar and low polar solvents.

At the same time, relative stability of enol cis conformer is noticeably reduced compare to other species with increasing of the hydrogen bonding ability of the solvent. Applying two methods: Monte-Carlo free

TABLE 24.1   Gibbs Energy Differences[a] (kcal/mol) in Water and Chloroform for the Ground-State Enol and Keto Forms of HPBI***

Molecule	Method	MST[b]	MC-FEP
Water			
Closed cis enol	HF	0.0	0.0
	B3LYP	0.0	0.0
		0.0	0.0
Open cis enol	HF	2.8	5.5
	B3LYP	5.8	8.5
		5.5	8.2
Trans enol	HF	1.7	−0.6
	B3LYP	3.9	1.6
		3.4	1.1
Keto	HF	6.3	0.2
	B3LYP	3.8	−2.3
		3.5	−2.6
Chloroform			
Closed cis enol	HF	0.0	0.0
	B3LYP	0.0	0.0
		0.0	0.0
Open cis enol	HF	7.8	8.8
	B3LYP	10.8	11.8
		10.5	11.5
Trans enol	HF	3.6	3.4
	B3LYP	5.8	5.6
		5.3	5.1
Keto	HF	10.7	11.1
	B3LYP	8.2	8.6
		7.9	8.3

[a] Values determined from addition of the relative Gibbs energy differences in gas phase computed at the HF and B3LYP levels to the relative Gibbs energies of solvation determined from MST and MC-FEP simulations. The results obtained from the Gibbs energy differences computed at the B3LYP/D95++** level are given in italics.

[b] The relative Gibbs energy of solvation has been determined as average of the MST-AM1 and MST-HF/6–31G* values.

***Reprinted with permission from Ref. [37]. Copyright © 2016 American Chemical Society.

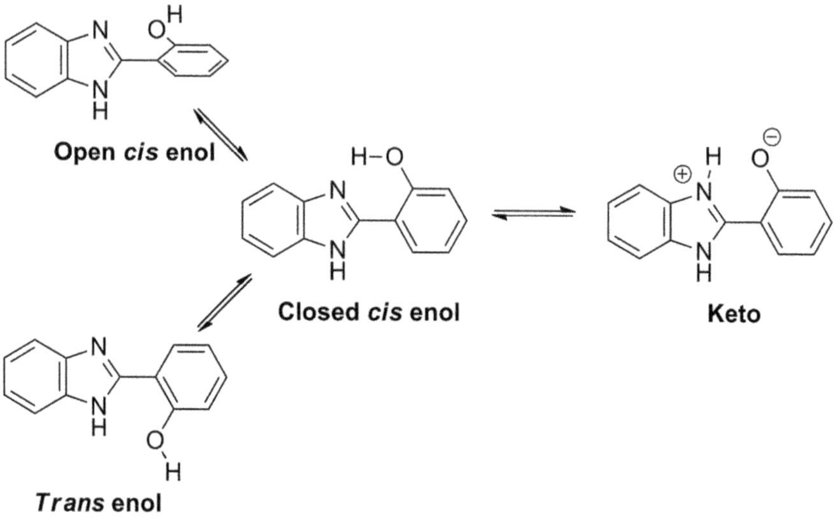

FIGURE 24.6   Conformers of the enol form of HPBI and the keto tautomeric form in ground state.

energy perturbation (MC-FEP) and self-consistent reaction field (SCRF) performed using MST model, the authors demonstrated close agreement between the relative Gibbs energies of solvation of HPBI in chloroform. In water media, qualitative agreement between MC-FEP and MST results is maintained, however some quantitative differences in the relative energies were revealed.

Comparison with the available experimental data allows the authors to speculate that a too simple description of the solute/solvent interface in MST calculations [38, 39] and missing strong specific interactions of water molecules with HPBI, which cannot be properly described in SCRF calculations, might contribute to explanation of such difference.

In the ^1pp first excited state and in the gas phase, the keto form is more stable than each of enol conformers. It was also supported by the phototautomerization observed from the enol to keto structures in the excited state proceeding through ESIPT. Vertical excitation energy for the closed cis form is blue-shifted by 1 kcal/mol in water with respect to the gas phase. At the same time, emission of the keto tautomer is red-shifted compared to the absorption band of the closed cis enol form, and it is largely blue-shifted when the polarity and hydrogen bonding ability of the solvent increase.

Presented results clearly demonstrate that solvent plays a fundamental role in the relative stability of the different rotamers and tautomers for HPBI in the ground state. In gas phase and low-polarity solvents, the closed cis form is the most stable species. In turn, polar protic solvents preferentially stabilize the keto form and trans enol form. This effect is large enough in the magnitude to make the mentioned species to be in equilibrium with the closed cis enol form in aqueous solution. Summarizing results of the work it is necessary to say that solvation adjoined with electronic effects of particular substituents could be an efficient tool to control the nature of the processes that occur in the excited states, especially the photoinduced proton transfer.

However, since 1999 nearly a ten-year lull in the studies of photophysical properties of imidazo[1,2-a]pyridines has been observed. This was apparently caused by a combination of several simultaneously acting factors. For example, low values of the fluorescent quantum yield for the red-shifted band in studied 1,2-HPIPs did not allow to find instantaneous practical applications of such compounds. It is also impossible to overlook the lack of affordable process for the preparation of more complex derivatives of 2-(2'-hydroxyphenyl)imidazo[1,2-a]pyridines. All of the above led expectedly to a decrease of interest also from theoretical researchers.

In 2008, the photochemistry of 2-hydroxyphenyl substituted imidazo[1,2-a]pyridines received a new unexpected stimulus with the paper published by Araki and co-workers [40]. The authors reported a novel and very effective mechanism of switching solid-state luminescence in this group of compounds. Tuning of the luminescence of organic solids, by controlling the mode of molecular packing instead of modification of the chemical structure, is a very promising approach. Araki et al. [40] managed to obtain two crystal polymorphs of 1,2-HPIP exhibiting different colors of photoluminescence with high quantum yields (Figure 24.7).

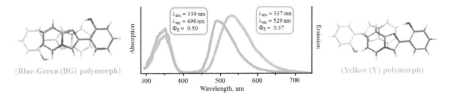

FIGURE 24.7 Absorption and luminescence spectra of blue-green – BG (bluish line) and yellow – Y (yellowish line) polymorphs of 1,2-HPIP as well as top views of the stacked pairs of the denoted polymorphs, respectively.

A few years later, in 2011, Araki and co-workers [41] published another article dealing with excited state intramolecular proton transfer in substituted 1,2-HPIP. Based on the previously obtained results and observations, the authors managed to fabricate a colorless organic white luminescent materials using mixture of proton transfer capable 2-(2'-hydroxyphenyl)-8-phenylimidazo[1,2-a]pyridine (8Ph-1,2-HPIP) and not capable to excited state proton transfer 2-(2'-methoxyphenyl)imidazo[1,2-a]pyridine-6-carbonitrile (6CN-1,2-PIP).

Since the middle of the 1990s, it is well known that 2-(2'-hydroxyphenyl)imidazo[1,2-a]pyridine is only weakly fluorescent compound with large Stokes shift in nonpolar solution, which is attributed to ESIPT emission. Contrariwise, both of its crystal polymorphs exhibit bright ESIPT-based luminescence of different colors (blue-green (BG) and yellow (Y) depending on polymorph). In 2012, Araki and co-workers [42] tried to elucidate the origin of emission enhancement in 1,2-HPIP in the solid state and to describe possible mechanism for the packing-controlled luminescence color tuning based on theoretical studies. The ground state S_0 geometries were optimized using DFT with 6–31G(d,p) basis set in conjunction with B3LYP functional. Additionally, the optimization was performed with state averaged CASSCF(6,6) level of theory over the lowest four states with even weight 0.25, where 6p-electrons were distributed into 3p (HOMO, HOMO−1, HOMO−2) and 3p* orbitals (LUMO, LUMO+1, LUMO+2), using atomic natural orbital small basis set (ANO-S). As the reference, equation-of-motion coupled cluster model with singles and doubles (EOM-CCSD) were employed to estimate the transition energies using cc-pVDZ basis set. Vertical excitation energies were evaluated with time-dependent DFT (TD-DFT) at the TD(B3LYP)/6–31+G(d,p) level. The quantitative vertical excitation energies including dynamical electron correlation were evaluated by multistate CASPT2 (MS-CASPT2).

At the initial stage, the authors analyzed relative enol – keto stability. S_0-enol form was always more stable than S_0-keto. Both S_0-enol and S_0-keto forms of 1,2-HPIP were found existing in the coplanar conformation. The experimentally measured absorption spectra of 1,2-HPIP in different solvents revealed single band maximum at 330 nm. TD-DFT(B3LYP), CASSCF(6,6)/ANO-S, and MS-CASPT2/ANO-L levels of theory were used to estimate $S_0 \rightarrow S_1$ vertical transition energies (Table 24.2). The energy predicted by CASSCF method considerably higher than the experimental value, 4.44 eV $vs.$ 3.68 eV, respectively.

TABLE 24.2 Low-Lying Excited States of the Enol Form of 1,2-HPIP for the S_0 Optimized Geometry (in eV)*

Excited state	TD-B3LYP/6-31+G(d,p)	EOM-CCSD/cc-pVDZ	CASSCF(6,6)/ANO-S	MS-CASPT2(10,9)/ANO-L	Experiment (cyclohexane)
S_1	3.55	4.31	4.44	3.95	3.68
S_2	4.07	4.61	5.71	4.56	
S_3	4.25	4.99	6.38	4.93	
S_4	5.46	5.76	6.63	6.34	
S_5	6.10	5.80	6.94	6.85	

*Reprinted with permission from Ref. [42]. Copyright © 2016 American Chemical Society.

A reason of such overestimation may be a lack of dynamic electron correlation in CASSCF. MS-CASPT2 noticeable lowered overestimated energy to 3.95 eV. EOM-CCSD/cc-pVDZ results were comparable to those obtained by CASSCF for $S_0 \rightarrow S_1$ and to MS-CASPT2 for the higher energies. The authors suggested that more extensive basis such as aug-cc-pVTZ or higher would reduce the different between experimental results and computational estimates.

Excitation to Frank-Condon S_1 state (FC-S_1) can be dominantly described as HOMO – LUMO excitation (Table 24.3). FC-S_1 excitation charge distributions did not show large charge transfer (Figure 24.8a).

TABLE 24.3 Natural Orbital Occupations (Electrons) of the S_0, S_1, and S_2 States of the Coplanar Enol and Keto Forms of 1,2-HPIP[a].*

Excited state	HOMO-1	HOMO	LUMO	LUMO+1
Enol form				
S_0	1.933	1.919	0.072	0.072
S_1	1.768	1.122	0.858	0.221
S_2	1.264	1.828	0.735	0.164
Keto form				
S_0	1.981	1.920	0.074	0.062
S_1	1.904	1.474	0.518	0.103
S_2	1.896	1.059	0.264	0.858

[a] Occupations calculated using 8 state averaged CASSCF(10,9)/ANO-L with even weight 0.125.

*Reprinted with permission from Ref. [42]. Copyright © 2016 American Chemical Society.

a) Enol Form of HPIP

π(HOMO-1)   π(HOMO)   π*(LUMO)   π*(LUMO+1)

b) Keto Form of HPIP

π(HOMO-1)   π(HOMO)   π*(LUMO)   π*(LUMO+1)

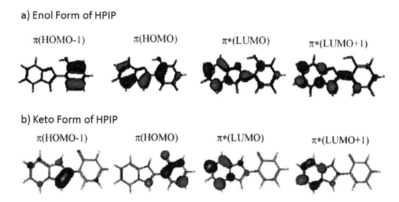

FIGURE 24.8   Natural orbitals of coplanar enol (a) and keto (b) forms of 1,2-HPIP. Adapted with permission from Ref. [42]. Copyright © 2016 American Chemical Society.

The ground state keto form possesses a larger dipole moment than the enol form, owing to the localized electron distribution on the hydroxyphenyl ring. $S_0 \rightarrow S_1$ vertical transition can be described similarly to enol form as HOMO – LUMO transition (Table 24.3). Intramolecular electron transfer from the phenol ring to the imidazopyridine moiety is observed in the transition (Figure 24.8b).

In cyclohexane, the most stable closed-keto form is responsible for the red-shifted emission, while in ethanol, the enol form is stabilized by intra- and intermolecular hydrogen bonds with solvents, and this is responsible for the blue-shifted emission. Geometry optimization under the planar constraint for the S_1 states of the enol and keto forms were performed using CASSCF(6,6)/ANO-S. The coplanar S_1 state of the enol form was found to be almost pure pp*. The coplanar S_1-keto form is more stable (by 7.8 kcal/mol) than the S_1-enol one. This in turn indicates that 1,2-HPIP can easily undergo ESIPT from FC-S_1 state.

Effect of the torsion angle, f, in the S_1-keto state was also studied. Ground state S_0-keto form energy level demonstrates monotonic elevation across a whole range of examined torsion angles. S_1-keto levels exhibit relatively small changes during increase of the torsion angle from 0 to 45°.

At the same time, $S_0 \rightarrow S_1$ energy gap of the keto form became sufficiently small at 60°, and the corresponding crossing point was found at 90°. Conical intersection model (CI) is one of theoretical interpretation of radiationless transitions from the excited state in polyatomic molecules. Conical intersections can be accessed directly or indirectly [43–45] acting as efficient decay funnels from excited states to ground states within a single vibrational period. A critical role of CIs has recently been reported for many compounds [46, 47].

The presence of the S_0/S_1 CI in 1,2-HPIP (Figure 24.9) provided the efficient and fast decay process of the excited state. For deeper understanding a whole 1,2-HPIP S_1 surface including ESIPT, the fully optimized scan of the PES has been performed. Overall results obtained for fully relaxed scan of the PES were similar to those obtained by fixing the torsion angles. Reaction path starts from FC-S_1 enol state and leads to slightly nonplanar conformation (f = 6.6°). The twisting process is associated with the enol to keto tautomerization. The local minimum of the keto-S_1 state was found at the twisting angle of 37.7° and is approximately 0.7 kcal/mol lower in energy than the FC-S_1 level. The intersection point between the S_0 and S_1 states was found at torsion angle of 60°, approximately 4.5 kcal/mol downhill from the keto-S_1

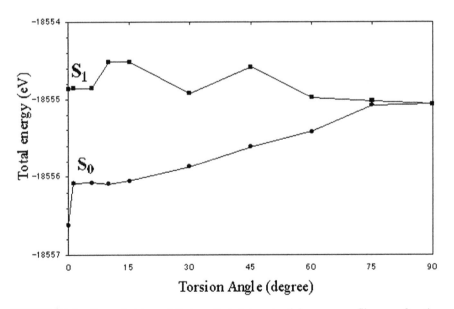

FIGURE 24.9 Ground state and first excited state potential energy profiles as a function of the torsion angle for the keto and enol forms at S_1-optimized geometries (CASSCF(6,6)/ ANO-S). Reprinted with permission from Ref. [42]. Copyright © 2016 American Chemical Society.

minimum at 37.7°. The 1,2-HPIP at the CI switches to the keto-S_0 PES and then goes downhill to the local keto-S_0 minimum at the torsion angle of 10.2°.

As can be clearly seen from Figure 24.9, the energy gap between the S_1-keto and S_0-keto states depends greatly on the torsion angle. The torsion angles f in the crystals are small enough; however, there is a slight difference, i.e., the BG crystal has the torsion angle of 5.8°, and in the Y crystal, the torsion angles of conformers Y1 and Y2 are 1.3° and −1.0°, respectively. However, CASSCF and MS-CASPT2 methods failed to reproduce the observed results. EOM-CCSD/cc-pVDZ computations unexpectedly gave the best absolute agreements with the experiments, despite the compact basis set employed, although the predicted gap between BG and Y1/Y2 is negligibly small (0.01 eV). Summarizing, fluorescence color difference between two crystal polymorphs was not successfully reproduced even at MS-CASPT2 levels of theory. The reproduction of the experiments requires larger scale computations. At the same time, calculations nicely reproduced the experimental photophysical properties of 1,2-HPIP in solution; the coplanar keto form in the S_1 state was indicated to approach smoothly to the S_0/S_1 CI coupled with the twisting motion of the central C–C bond.

Since first discovery of polymorph depending fluorescence for 1,2-HPIP, in 2008 a few dozen imidazo[1,2-a]pyridines demonstrating aggregation induced emission enhancement (AIEE) have been synthesized. Due to the fact that luminescence solids attracted a broad range of interests in science and industrial applications. For example very recently Araki and co-workers [48] demonstrate promising approach to creation of colorless, transparent, dye-doped polymer films exhibiting tunable luminescence color, based on 1,2-HPIP and 6-cyano substituted (2′-hydroxyphenyl)imidazo[1,2-a]pyridine (6CN-1,2-HPIP) prepared by a spin-coating method. By altering of the surrounding matrices, the relative intensity of the two emissions can be tuned (Figure 24.10) as a result of change in the abundance ratio of the two enol forms and subsequently in the luminescent color of the dye-doped polymer films.

Araki et al. [49] have done a decisive effort to elucidate the polymorph-dependent luminescence of 6CN-1,2-HPIP exhibiting three colored luminescence (yellow (Y), orange (O), and red (R)) in the solid state, specific to the corresponding polymorphs (Figure 24.11). Previous attempt to describe different luminescence of two polymorphs of 1,2-HPIP was not successful because of MS-CASPT2 limitations in a direct application to large aggregates. The authors computationally elucidated the

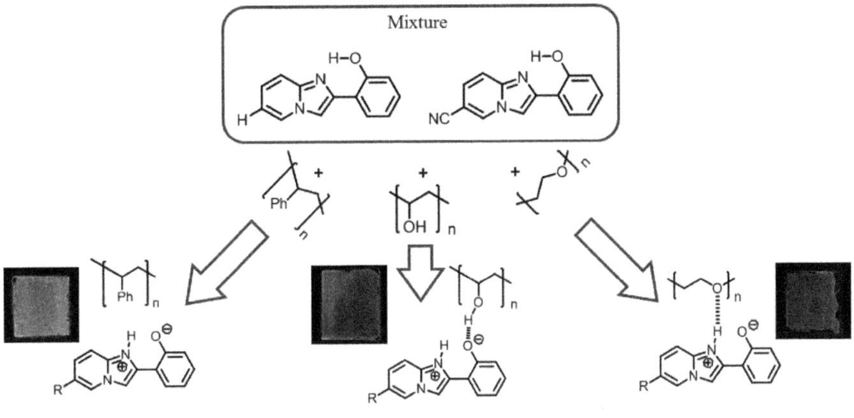

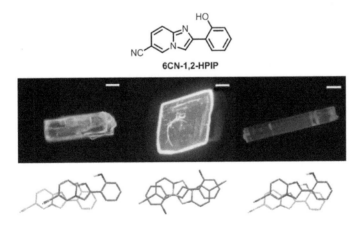

FIGURE 24.10 Proposed hydrogen-bonding mode of intramolecular proton-transferred species and colors of the dye-doped polymer films in PS, PVA, and PEG. Adapted with permission from Ref. [48]. Copyright © 2016 American Chemical Society.

6CN-1,2-HPIP

FIGURE 24.11 Luminescent images under UV lamp and crystal structures (view of the dimer) of polymorphic crystals: Y, O, and R. Reproduced from Ref. [49] with permission from The Royal Society of Chemistry.

emission maxima of the compound in the three different polymorphs using ONIOM [50, 51] and FMO-TDDFT [52–54] approaches applied to the finite cluster models [55].

At the initial stage of the research, the authors prepared a series of computations (TDDFT, CASSCF, and MS-CASPT2) for isolated 6CN-1,2-HPIP molecule. The gap between MS-CASPT2//CASSCF and MS-CASPT2// TDDFT was 31 nm, which means that a qualitatively correct geometry was obtained at TDDFT(B3LYP)/6–31+G(d) level. Using CAM-B3LYP functional instead of B3LYP, which is known to substantially underestimate the $S_1 - S_0$ gap, they achieved quantitative agreement with the previously reported MS-CASPT2 results. As the next step, the ONIOM calculations for Y, O, and R polymorphs were carried out. Calculations were performed separately for monomer (one molecule of keto form of 6CN-1,2-HPIP in excited state (TDDFT): 16 enol 6CN-1,2-HPIPs in ground state geometry (PM3)) and dimer (a pair of keto and enol (TDDFT):15 enols (PM3)). Intermolecular charge transfer excitation and excitonic effects across TDDFT-PM3 regions as well as the crystal field effect from infinite electrostatic interactions were ignored. Values of emission maxima for the studied polymorphs predicted for monomer did not coincidence with the experiment (Table 24.4).

It has to be mentioned that predicted emission wavelength did not coincidence in low energy order with the monomer ONIOM model. Moreover monomer ONIOM model with surrounding 16 enol models taking into account with semi-empirical PM3 exert just 2 nm in comparison with isolated monomer. Latest directly indicate that dimer interaction between keto and enol forms, but not perturbation of surrounding molecules, plays crucial role in the color of the emission of different crystalline. The order of l_{em} was correctly altered for dimer ONIOM model.

Driven by a desire to treat any system with chemical accuracy and by the fact that straightforward TDDFT treatment for whole cluster model is

TABLE 24.4 Computed S_1–S_0 Emission Wavelengths (nm) of the ONIOM Cluster Models for O, Y, and R Polymorphs

Polymorph of 6CN-1,2-HPIP	Isolated		ONIOM		Experimental values
	Monomer[a,b]	Dimer[a,b]	Monomer[b]	Dimer[b]	
Y	542	507	541	506	548
O	593	520	591	518	570
R	557	588	556	586	585

[a] Geometries extracted from the optimized cluster models at the ONIOM(TD(B3LYP)/6–31+G(d):PM3) level. [b] Higher layer TD(CAM-B3LYP)/6–31G(d):lower layer PM3.

*Reproduced from Ref. [49] with permission from The Royal Society of Chemistry.

infeasible, the FMO-TDDFT method (as first FMO1 [53] and second order FMO2 [54]) was applied. The main advantage with respect to TDDFT is that the FMO-TDDFT timing scales approximately linearly in contrast to conventional full TDDFT which suffers from the inherent $O(N^3)$ scaling. FMO-TDDFT calculations were done for ONIOM optimized geometries. Effect of cluster size on $S_1 \rightarrow S_0$ emission was also studied. Analysis showed that intermolecular dimer interactions play a significant role in the emission spectra, while these dimer interactions are strongly depend on the molecular orientation and the mode of packing.

The development of new synthetic methods for the preparation of imidazo[1,2-a]pyridines allows significantly expand a number of possible derivatives of 2-(2'-hydroxyphenyl)imidazo[1,2-a]pyridine. In the first quarter of 2013, Araki and co-workers [56] published an article devoted to the substituent effect on emission properties and tuning of ESIPT fluorescence in 1,2-HPIPs. At the first stage of the study, the authors synthesized a series of 1,2-HPIPs through the reaction of 2-aminopyridine and α-bromo-acetophenones (Scheme 2).

Substitution of HPIP is a simple and useful method for tuning color without damaging the luminescent nature in the solid state. In order to understand

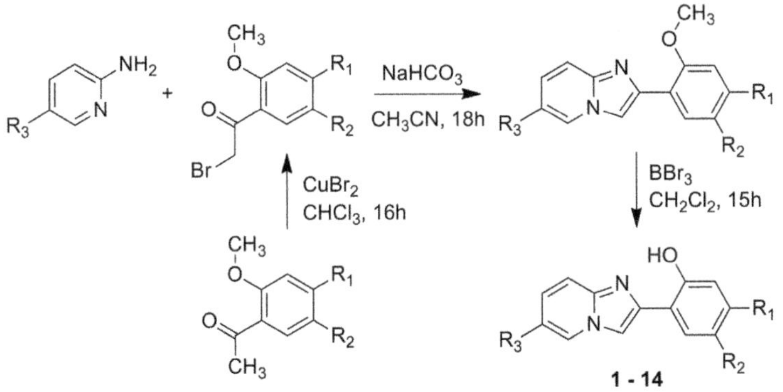

1. R_1 = H, R_2 = H, R_3 = H.
2. R_1 = H, R_2 = OCH_3, R_3 = H.
3. R_1 = H, R_2 = CH_3, R_3 = H.
4. R_1 = H, R_2 = Cl, R_3 = H.
5. R_1 = H, R_2 = Br, R_3 = H.

6. R_1 = H, R_2 = F, R_3 = H.
7. R_1 = H, R_2 = H, R_3 = CH_3.
8. R_1 = H, R_2 = H, R_3 = Cl.
9. R_1 = H, R_2 = H, R_3 = F.
10. R_1 = H, R_2 = H, R_3 = CF_3.

11. R_1 = H, R_2 = H, R_3 = CN.
12. R_1 = OCH_3, R_2 = H, R_3 = H.
13. R_1 = H, R_2 = CH_3, R_3 = Cl.
14. R_1 = H, R_2 = CH_3, R_3 = CF_3.

SCHEME 2   Synthetic route for substituted 1,2-HPIPs.

the substituent effect on emission properties, the electronic states of 1,2-HPIPs were studied using *ab initio* quantum chemical calculations. Geometry optimization was carried out by RHF/6–31G(d) method and results in equilibrium structure with coplanar conformation. The authors demonstrated that the lowest-energy absorption band corresponds to the HOMO–LUMO transition. Both HOMO and LUMO of the enol form were delocalized over a whole molecule. Then the excited electronic state of the PT-form was calculated using the CI-single method (CIS/6–31+G(d)) after geometry optimization at CIS/6–31G(d) level. As a result, the lowest-energy transition band, which is associated with ESIPT fluorescence, was also the HOMO–LUMO transition. However, unlike the enol form, HOMO and LUMO for PT-form were localized mainly on the imidazo[1,2-*a*]pyridine and phenyl rings, respectively (Figure 24.12).

The measured ESIPT emission energies of 1−14 in a PMMA matrix were plotted against the calculated energies of the HOMO(keto)–LUMO(keto) transitions (Figure 24.12). Despite the fact that estimated values of the transition energy were somewhat larger than the observed energies, a linear correlation (R^2 = 0.91) confirms that the quantum chemical simulations are effective for qualitative reproduction of the observed ESIPT fluorescence emissions of 1,2-HPIPs.

Hammett constants [57, 58] (s) could be a useful parameter for explaining the substituent effect on the emission properties of 1,2-HPIPs. The authors plotted calculated energy levels of HOMO(keto) and LUMO(keto)

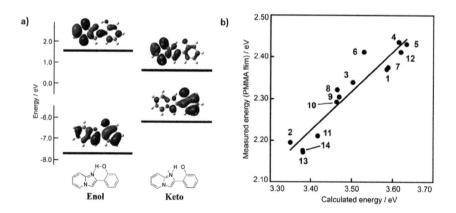

FIGURE 24.12   Molecular orbitals of the enol and keto forms of 1,2-HPIP (a) and plot of the calculated against measured energy of ESIPT fluorescence of 1,2-HPIPs (1−14) (b). Adapted with permission from Ref. [56]. Copyright © 2016 American Chemical Society.

of 5'-substituted and 6-substituted 1,2-HPIPs against the Hammett constants (Figure 24.13).

The Hammett substituent constants showed good correlations with different slopes, that rationally explains the effect of the substituent and its position on the proton transferred state. Thus, the substituent effects on the ESIPT fluorescence properties were successfully reproduced and explained by the electronic configurations and energy levels of HOMO(keto) and LUMO(keto).

Ideologically similar was a paper by Cyrański and co-workers [59] published couple years later and devoted to the study of the effect of HB strength on emission properties in 1,2-HPIP and its 5'-substituted derivatives (Scheme 3). Study has been performed at PBE0/6–311+G(d,p) level of theory.

First, the authors investigated conformational preferences of the studied compounds in the gas phase and in various solutions taking into account solvent effects with the PCM method. In each case, similarly to the previous observations, closed enol form (Conf. 1) was found the most stable. Among keto forms, conformer keto Conf. 1 with torsion angle $f = 0$ in the vacuum and in nonpolar solvent was not observed. The keto Conf. 1 can possibly

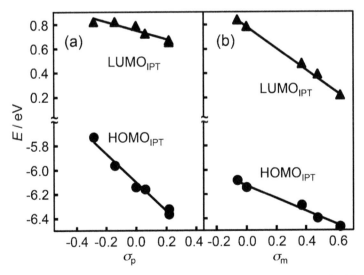

FIGURE 24.13 Plot of the calculated energy levels of HOMO(keto) and LUMO(keto) against the Hammett constants (s_p or s_m) for (a) 5'-substituted 1,2-HPIPs and (b) 6-substituted 1,2-HPIPs. Reprinted with permission from Ref. [56]. Copyright © 2016 American Chemical Society.

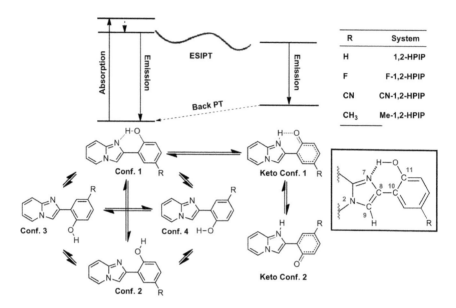

SCHEME 3   Conformations of 2-(2'-hydroxyphenyl)imidazo[1,2-a]pyridines.

exist only in highly polar environments. However, the second of the two possible keto forms, where the formation of the hydrogen bond O–H∥N is not possible (keto Conf. 2), corresponds to the minimum on the PES in the vacuum and in both of the studied solutions. It was found that solvent stabilization effect on the keto form is almost twice bigger than on the enol form.

The authors demonstrated that computationally obtained data for absorption energy for each of the investigated compounds are well correlated with the experimental data. This indicates the existence of effect of substituent in para position to hydroxyl group. In order to study this effect, Cyrański and co-workers examined the strength of the hydrogen bond in 5'-substituted 1,2-HPIPs. As main descriptor of hydrogen bond strength a topology of the electron density ($r(r)$) and potential energy density ($V(r)$) estimated with QTAIM [60,61] was chosen. Relationships between the experimental emission wavelength and both $r(r)$ and $V(r)$ energy showed good correlations ($cc = -0.995$ and 0.986, respectively) in nonpolar solvent (Figure 24.14). Such correlations were noticeable worse in the case of ethanol ($cc = -0.837$ and 0.813 for the electron density $r(r)$ and potential energy density $V(r)$, respectively). The authors rationalized this due to the fact that the experimentally measured values of emission maxima include specific interactions (solute-solvent), while IEFPCM model does not include such interactions.

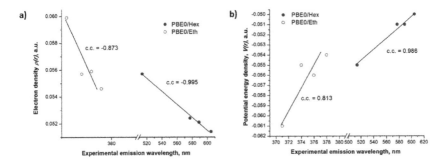

FIGURE 24.14   Correlations between experimental emission wavelength and (a) $\rho(r)$ at BCP of the hydrogen bond and (b) potential energy density $V(r)$ for 5'-substituted 1,2-HPIPs. Reprinted with permission from Ref. [59]. Copyright © 2016 Elsevier B.V.

Additionally, Cyrański and co-workers suggested that aromaticity index HOMA [62, 63] (Harmonic Oscillator Model of Aromaticity) enables to evaluate the extent of p-electron delocalization from molecular geometry. In the studied compounds, four different rings were distinguished. Three of them are pyridine (Ring A), imidazole (Ring B) and phenyl (Ring C) fragments, and the fourth ring is six-membered pseudo ring (Ring D) formed by N7-C8-C10-C11-O-H atoms involved in hydrogen bond (see Scheme 3). The authors found out that greatest substituent effect on the aromaticity is observed for rings directly connected with the substituent. At the same time, changes in aromaticity of the imidazo[1,2-a]pyridine moiety were negligible and do not depend on the substituent or solvent. It is worth to mention that correlations between HOMA index and position of emission maxima were generally worse than those obtained with QTAIM descriptors (cc = 0.900 and 0.834 for hexane and ethanol solutions, respectively).

Very recently new promising aryl-substituted 1,2-HPIPs demonstrating solid-state fluorescence has been published [64]. They exhibit intensified by a p-stacked packing ESIPT luminescence. Series of studied derivatives of 1,2-HPIPs derivatives showing a wide range of luminescence colors from blue (382 nm) to reddish-orange (630 nm). Not less interesting is a new class of ESIPT capable molecules comprise 2-(imidazo[1,2-a]pyridine-2yl) aniline moiety representing new (NH)-type intramolecular hydrogen bonding [28]. In these compounds, differently from other (NH)-type ESIPT systems, ESIPT rate reveals an irregular relationship among ESIPT dynamics, thermodynamics, and H-bond strength.

Although most of fluorescent organic compounds demonstrate noticeable quenching of photoluminescence in solid state, 1,2-HPIPs showed
intense solid-state emission. They demonstrated the reproducible switching
of the polymorph-dependent ESIPT luminescence offering novel concept
for tunable organic luminescence solid. Significant advances in the field of
applied sciences stimulate continued interest of the theoretical scientists,
whose findings allow to reveal characteristics of the behavior of such materials and predict the most effective direction of subsequent modifications.

KEYWORDS

- imidazo[1,2-*a*]pyridine
- proton transfer
- excited state intramolecular proton transfer (ESIPT)
- polymorphism
- solvent effects
- TDDFT

REFERENCES

1. Reece, S. Y., & Nocera, D. G., (2009). Proton-coupled electron transfer in biology results from synergistic studies in natural and model systems. *Annu. Rev. Biochem. 78,* 673.
2. Chaldin, E. F., & Gold, V., Ed., (1975). *Proton-Transfer Reactions,* London: Chapman and Hall.
3. Weller, A., (1961). Fast reactions of excited molecules. *Prog. React. Kinet. Mech. 1,* 187.
4. Weller, A., (1952). Quantitative untersuchungen der fluoreszenzumwandlung bei naph- tholen. *Z. Elektrochem. 56,* 662.
5. Weller, A., (1956). Innermolekularer protonenübergang im angeregten zustand. *Z. Ele- ktrochem Ber. Bunsenges. Phys. Chem. 60,* 1144.
6. Weller, A., (1955). Über die fluoreszenz der salizylsäure und verwandter verbindungen. *Naturwissenschaften. 42,* 175.
7. Barbara, P. F., Walsh, P. K., & Brus, L. E., (1989). Picosecond kinetic and vibrationally resolved spectroscopic studies of intramolecular excited-state hydrogen atom transfer. *J. Phys. Chem. 93,* 29.
8. Douhal, A., Lahmani, F., & Zewail, A. H., (1996). Proton-transfer reaction dynamics. *Chem. Phys., 207,* 477.

9. Bardez, E., Devol, I., Larrey, B., & Valeur, B., (1997). Excited-state processes in 8-hydroxyquinoline: photoinduced tautomerization and solvation effects. *J. Phys. Chem. B 101*, 7786.

10. Zamotaiev, O. M., Postupalenko, V. Y., Shvadchak, V. V., Pivovarenko, V. G., Klymchenko, A. S., & Mely, Y., (2014). Monitoring penetrating interactions with lipid membranes and cell internalization using a new hydration-sensitive fluorescent probe. *Org. Biomol. Chem. 12*, 7036.

11. Woolfe, G. J., & Thistlethwaite, P. J., (1981). Direct observation of excited state intramolecular proton transfer kinetics in 3-hydroxyflavone. *J. Am. Chem. Soc. 103*, 6916.

12. Lin, T.-Y., Tang, K.-C., Yang, S.-H., Shen, J.-Y., Cheng, Y.-M., Pan, H.-A., et al., (2012). The empirical correlation between hydrogen bonding strength and excited-state intramolecular proton transfer in 2-pyridyl pyrazoles. *J. Phys. Chem. A 116*, 4438.

13. Arai, T., Moriyama, M., & Tokumaru, K., (1994). Novel photoinduced hydrogen atom transfer through intramolecular hydrogen bonding coupled with cis-trans isomerization in cis-1-(2-Pyrrolyl)-2-(2-quinolyl)ethene. *J. Am. Chem. Soc. 116*, 3171.

14. Chen, K.-Y., Cheng, Y.-M., Lai, C.-H., Hsu, C.-C., Ho, M.-L., Lee, G.-H., et al., (2007). Ortho green fluorescence protein synthetic chromophore; excited-state intramolecular proton transfer via a seven-membered-ring hydrogen-bonding system. *J. Am. Chem. Soc. 129*, 4534.

15. Feng, L., Liu, Z. M., Hou, J., Lv, X., Ning, J., Ge, G. B., et al., (2015). A highly selective fluorescent ESIPT probe for the detection of Human carboxylesterase 2 and its biological applications. *Biosens. Bioelectron. 65*, 9.

16. Zhang, J. J., Ning, L. L., Liu, J. T., Wang, J. X., Yu, B. F., Liu, X. Y., et al., (2015). Naked-eye and near-infrared fluorescence probe for hydrazine and Its applications in In Vitro and In Vivo bioimaging. *Anal. Chem. 87*, 9101.

17. Mohammed, O. F., Pines, D., Nibbering, E. T. J., & Pines, E., (2007). Base-induced solvent switches in acid–base reactions. *Angew. Chem., Int. Ed. 46*, 1458.

18. Catalan, J., & del Valle, J. C., (1993). Toward the photostability mechanism of intramolecular hydrogen bond systems. The photophysics of 1'-hydroxy-2'-acetonaphthone. *J. Am. Chem. Soc. 115*, 4321.

19. Sobolewski, A. L., & Domcke, W., (2006). Photophysics of intramolecularly hydrogen-bonded aromatic systems: ab initio exploration of the excited-state deactivation mechanisms of salicylic acid. *Phys. Chem. Chem. Phys. 8*, 3410.

20. Zhao, J., Ji, S., Chen, Y., Guo, H., & Yang, P., (2012). Excited state intramolecular proton transfer (ESIPT): from principal photophysics to the development of new chromophores and applications in fluorescent molecular probes and luminescent materials. *Phys. Chem. Chem. Phys. 14*, 8803.

21. Park, S., Kwon, J. E., Kim, S. H., Seo, J., Chung, K., Park, S.-Y., et al., (2009). A white-light-emitting molecule: frustrated energy transfer between constituent emitting centers. *J. Am. Chem. Soc. 131*, 14043.

22. Kim, S., Seo, J., Jung, H. K., Kim, J. J., & Park, S. Y., (2005). White luminescence from polymer thin films containing excited-state intramolecular proton-transfer dyes. *Adv. Mater. 17*, 2077.

23. Park, S., Kwon, O.-H., Lee, Y.-S., Jang, D.-J., & Park, S. Y., (2007). Imidazole-based excited-state intramolecular proton-transfer (ESIPT) materials: observation of thermally activated delayed fluorescence (TDF). *J. Phys. Chem. A 111*, 9649.

24. Chou, P.-T., Martinez, M. L., & Clements, J. H., (1993). The observation of solvent-dependent proton-transfer/charge-transfer lasers from 4'-diethylamino-3-hydroxyflavone. *Chem. Phys. Lett. 204*, 395.

25. Lim, S.-J., Seo, J., & Park, S. Y., (2006). Photochromic switching of excited-state intramolecular proton-transfer (ESIPT) fluorescence: a unique route to high-contrast memory switching and nondestructive readout. *J. Am. Chem. Soc. 128*, 14542.

26. Stasyuk, A. J., Cyrański, M. K., Gryko, D. T., & Solà, M., (2015). Acidic C–H bond as a proton donor in excited state intramolecular proton transfer reactions. *J. Chem. Theory Comput. 11*, 1046.

27. Cuerva, C., Campo, J. A., Cano, M., Sanz, J., Sobrados, I., Diez-Gomez, V., et al., (2016). Water-free proton conduction in discotic pyridylpyrazolate-based Pt(II) and Pd(II) metallomesogens. *Inorg. Chem. 55*, 6995.

28. Stasyuk, A. J., Chen, Y. T., Chen, C. L., Wu, P. J., & Chou, P. T., (2016). A new class of N-H excited-state intramolecular proton transfer (ESIPT) molecules bearing localized zwitterionic tautomers. *Phys. Chem. Chem. Phys. 18*, 24428.

29. Tseng, H. W., Lin, T. C., Chen, C. L., Lin, T. C., Chen, Y. A., Liu, J. Q., et al., (2015). A new class of N-H proton transfer molecules: wide tautomer emission tuning from 590 nm to 770 nm via a facile, single site amino derivatization in 10-aminobenzo[h]quinoline. *Chem. Commun. 51*, 16099.

30. Tseng, H. W., Liu, J. Q., Chen, Y. A., Chao, C. M., Liu, K. M., Chen, C. L., et al., (2015). Harnessing excited-state intramolecular proton-transfer reaction via a series of amino-type hydrogen-bonding molecules. *J. Phys. Chem. Lett. 6*, 1477.

31. Douhal, A., Amat-Guerri, F., & Acuña, A. U., (1995). Photoinduced intramolecular proton transfer and charge redistribution in imidazopyridines. *J. Phys. Chem. 99*, 76.

32. Douhal, A., Amat-Guerri, F., & Acuña, A. U., (1997). Probing nanocavities with proton-transfer fluorescence. *Angew. Chem., Int. Ed. Engl. 36*, 1514.

33. Organero, J. A., Douhal, A., Santos, L., Martinez-Ataz, E., Guallar, V., Moreno, M., et al., (1999). Proton-transfer reaction in isolated and water-complexed 8-hydroxyimidazo[1,2-*a*]pyridine in the S-0 and S-1 electronic states. A theoretical study. *J. Phys. Chem. A 103*, 5301.

34. Forés, M., Duran, M., Solà, M., & Adamowicz, L., (1999). Excited-state intramolecular proton transfer and rotamerism of 2-(2'-hydroxyvinyl)benzimidazole and 2-(2'-hydroxyphenyl)imidazole. *J. Phys. Chem. A 103*, 4413.

35. Forés, M., Duran, M., & Solà, M., (1998). Intramolecular proton transfer in the ground and the two lowest-lying singlet excited states of 1-amino-3-propenal and related species. *Chem. Phys. 234*, 1.

36. Sobolewski, A. L., & Domcke, W., (1993). Evidence for the need of a Non-Born-Oppenheimer description of Eecited-state hydrogen-transfer. *Chem. Phys. Lett. 211*, 82.

37. Forés, M., Duran, M., Solà, M., Orozco, M., & Luque, F. J., (1999). Theoretical evaluation of solvent effects on the conformational and tautomeric equilibria of 2-(2'-Hydroxyphenyl)benzimidazole and on Its absorption and fluorescence spectra. *J. Phys. Chem. A 103*, 4525.

38. Colominas, C., Orozco, M., Luque, F. J., Borrell, J. I., & Teixidó, J., (1998). A priori prediction of substituent and solvent effects in the basicity of nitriles. *J. Org. Chem. 63*, 4947.

39. Grozema, F. C., & van Duijnen, P. T., (1998). Solvent effects on the $\pi^* \leftarrow n$ transition of acetone in various solvents: Direct reaction field calculations. *J. Phys. Chem. A 102*, 7984.

40. Mutai, T., Tomoda, H., Ohkawa, T., Yabe, Y., & Araki, K., (2008). Switching of polymorph-dependent ESIPT luminescence of an imidazo[1,2-*a*]pyridine derivative. *Angew. Chem. Int. Ed., 47*, 9522.

41. Shono, H., Ohkawa, T., Tomoda, H., Mutai, T., & Araki, K., (2011). Fabrication of colorless organic materials exhibiting white luminescence using normal and excited-state intramolecular proton transfer processes. *ACS Appl. Mater. Interfaces, 3*, 654.

42. Shigemitsu, Y., Mutai, T., Houjou, H., & Araki, K., (2012). Excited-state intramolecular proton transfer (ESIPT) emission of hydroxyphenylimidazopyridine: computational study on enhanced and polymorph-dependent luminescence in the solid state. *J. Phys. Chem. A, 116*, 12041.

43. Ismail, N., Blancafort, L., Olivucci, M., Kohler, B., & Robb, M. A., (2002). Ultrafast decay of electronically excited singlet cytosine via π,π^* to n_O,π^* state switch. *J. Am. Chem. Soc., 124*, 6818.

44. Blancafort, L., Cohen, B., Hare, P. M., Kohler, B., & Robb, M. A., (2005). Singlet excited-state dynamics of 5-fluorocytosine and cytosine: An experimental and computational study. *J. Phys. Chem., A 109*, 4431.

45. Sobolewski, A. L., Domcke, W., Dedonder-Lardeux, C., & Jouvet, C., (2002). Excited-state hydrogen detachment and hydrogen transfer driven by repulsive $^1\pi s^*$ states: A new paradigm for nonradiative decay in aromatic biomolecules. *Phys. Chem. Chem. Phys., 4*, 1093.

46. Paterson, M. J., Robb, M. A., Blancafort, L., & DeBellis, A. D., (2004). Theoretical study of benzotriazole UV photostability: Ultrafast deactivation through coupled proton and electron transfer triggered by a charge-transfer state. *J. Am. Chem. Soc., 126*, 2912.

47. Sobolewski, A. L., Domcke, W., & Hattig, C., (2006). Photophysics of organic photostabilizers. Ab initio study of the excited-state deactivation mechanisms of 2-(2'-hydroxyphenyl)benzotriazole. *J. Phys. Chem. A, 110*, 6301.

48. Furukawa, S., Shono, H., Mutai, T., & Araki, K., (2014). Colorless, transparent, dye-doped polymer films exhibiting tunable luminescence color: controlling the dual-color luminescence of 2-(2'-hydroxyphenyl)imidazo[1,2-*a*]pyridine derivatives with the surrounding matrix. *ACS Appl. Mater. Interfaces 6*, 16065.

49. Mutai, T., Shono, H., Shigemitsu, Y., & Araki, K., (2014). Three-color polymorph-dependent luminescence: crystallographic analysis and theoretical study on excited-state intramolecular proton transfer (ESIPT) luminescence of cyano-substituted imidazo[1,2-*a*]pyridine. *Cryst. Eng. Comm., 16*, 3890.

50. Dapprich, S., Komaromi, I., Byun, K. S., Morokuma, K., & Frisch, M. J., (1999). A new ONIOM implementation in Gaussian98. Part I. The calculation of energies, gradients, vibrational frequencies and electric field derivatives. *J. Mol. Struct.: THEOCHEM. 461*, 1.

51. Vreven, T., & Morokuma, K., (2006). Hybrid Methods: ONIOM(QM:MM) and QM/MM. In: *Annual Reports in Computational Chemistry*; David, C. S., Ed., Amsterdam, Elsevier. Vol. 2, p. 35.

52. Chiba M., Fedorov, D. G. and Kitaura, K., (2009). The Fragment molecular orbital–based time-dependent density functional theory for excited states in large systems, p. 91. In *The Fragment Molecular Orbital Method Practical Applications to Large Molecular Systems;* Fedorov, D. G., Kitaura, K., Ed., Boca Raton, FL: CRC Press.

53. Chiba, M., Fedorov, D. G., & Kitaura, K., (2007). Time-dependent density functional theory with the multilayer fragment molecular orbital method. *Chem. Phys. Lett. 444*, 346.

54. Chiba, M., Fedorov, D. G., & Kitaura, K., (2007). Time-dependent density functional theory based upon the fragment molecular orbital method. *J. Chem. Phys. 127*, 104108.

55. Shigemitsu, Y., Mutai, T., Houjou, H., & Araki, K., (2014). Influence of intermolecular interactions on solid state luminescence of imidazopyridines: theoretical interpretations using FMO-TDDFT and ONIOM approaches. *Phys. Chem. Chem. Phys. 16*, 14388.

56. Mutai, T., Sawatani, H., Shida, T., Shono, H., & Araki, K., (2013). Tuning of excited-state intramolecular proton transfer (ESIPT) fluorescence of imidazo[1,2-*a*]pyridine in rigid matrices by substitution effect. *J. Org. Chem. 78*, 2482.

57. Hammett, L. P., (1935). Some Relations between reaction rates and equilibrium constants. *Chem. Rev. 17*, 125.

58. Hansch, C., Leo, A., & Taft, R. W., (1991). A survey of Hammett substituent constants and resonance and field parameters. *Chem. Rev. 91*, 165.

59. Stasyuk, A. J., Bultinck, P., Gryko, D. T., & Cyranski, M. K., (2016). The effect of hydrogen bond strength on emission properties in 2-(2'-hydroxyphenyl)imidazo[1,2-*a*]pyridines. *J. Photochem. Photobiol., A314*, 198.

60. Bader, R. F. W., (1991). A quantum theory of molecular structure and its applications. *Chem. Rev. 91*, 893.

61. Bader, R. F. W., (1990). *Atoms in Molecules: A Quantum Theory*; Oxford University Press.

62. Krygowski, T. M., (1993). Crystallographic studies of inter- and intramolecular interactions reflected in aromatic character of π-electron systems. *J. Chem. Inf. Comput. Sci., 33*, 70.

63. Kruszewski, J., & Krygowski, T. M., (1972). Definition of aromaticity basing on the harmonic oscillator model. *Tetrahedron Lett. 13*, 3839.

64. Mutai, T., Ohkawa, T., Shono, H., & Araki, K., (2016). The development of aryl-substituted 2-phenylimidazo[1,2-*a*]pyridines (PIP) with various colors of excited-state intramolecular proton transfer (ESIPT) luminescence in the solid state. *J. Mater. Chem. C., 4*, 3599.

CHAPTER 25

LOCAL ENVIRONMENTS IN INORGANIC SOLIDS: FROM FAST-ION CONDUCTION TO RADIATION DAMAGE IN OXIDES

NEIL L. ALLAN,[1] ADAM ARCHER,[1] and CHRIS E. MOHN[2]

[1]*School of Chemistry, University of Bristol, Cantock's Close, Bristol BS8 1TS, UK*

[2]*Centre for Earth Evolution and Dynamics, University of Oslo, Postbox 1048, Blindern, N-0315 Oslo, Norway*

CONTENTS

ABSTRACT

In this chapter we highlight the importance of understanding the *local* environment in non-stoichiometric and disordered systems. Using δ-Bi_2O_3 as our example, we demonstrate that to understand the atomic mechanisms underlying fast-ion conduction, a consideration of the *average* structure alone is not sufficient, and potentially misleading. To understand the behavior at the atomic level underlying radiation damage in oxides such as PuO_2 and pyrochlores, we need to consider the healing processes and related mechanisms, rather than, as has traditionally been the case, concentrating on damage creation and the number of displaced atoms. Crucial to healing is ion mobility: in the first stages, oxide ion mobility and then subsequent cation migration in the presence of disorder, reforming the parent compound. In the systems we have examined those which are also fast-ion conductors heal readily. Those that are not fast-ion conductors do not heal but become amorphous. If applicable more generally, these ideas confirmed by further simulation should be valuable in the design of future host ceramics for nuclear waste.

25.1   INTRODUCTION

The underlying theme of this chapter is the *local environment* and its importance in defining solid-state properties. Increasing attention, given advances in methodology and computer hardware, is being paid to the details of the energy landscape, the multi-dimensional dependence of the energy on the variables that define the structure. The lowest point in the deepest valley in the landscape corresponds to the global minimum, the lowest energy stable structure, while other valleys contain local minima corresponding to metastable structures with higher energies. The saddle points separating the valleys are transition states and indicate the lowest energy pathways between different possible structures. The energy barriers between the minima are also important; a structure corresponding to a local minimum will only be metastable if the energy barrier between it and a lower energy structure is sufficiently high and thermally inaccessible. The presence or absence of thermally accessible barriers is crucial for properties apparently as diverse as superionic conductivity and resistance to radiation damage.

We concentrate here on disordered and non-stoichiometric systems. In such cases there are many local minima that are thermally accessible and we shall demonstrate the key features of the energy landscapes which are key to an understanding of their chemical behavior at the atomic level. Our first example is an important and representative fast-ion conductor, δ-Bi_2O_3. The second is a radiation-damaged oxides where the disorder is generated via the radioactive decay of actinides and we discuss how such systems recover and heal from such damage. Most attention has previously based to the prevention of damage, but we shall see that it is the recovery and healing processes that ultimately determine the fate of the system, and there are strong parallels to be drawn between fast ion mobility and the healing process.

25.2 AVERAGE AND LOCAL STRUCTURES IN NON-STOICHIOMETRIC COMPOUNDS

At first sight, compounds containing a large number of vacancies are a close relative of (substitutional) solid solutions. Instead of distributing different atoms on a given sub-lattice, one type of atom and vacancies are distributed together on a sub-lattice. While this can and has served for many years in textbooks as a useful first approach, in practice for oxides we have found these systems are structurally much more closely related to many amorphous systems in that a strong preference for certain coordination polyhedra dictates the local order in the overall disordered state. Structural relaxations accompanying different distributions of these polyhedra may be so large (as in $Ba_2In_2O_5$) [1] that the very concept that vacancies occupy a sublattice is a poor model of the disorder although this is commonly accepted practice. It can be useful to compare this picture with that for an amorphous material such as vitreous silica. In this case the atoms are not distributed on a well-defined sub-lattice but there is considerable short-range order, in the form of SiO_4 tetrahedra, and intermediate-range order arising from the interactions between the polyhedra, revealed by such indices as the Steinhardt order parameter [2]. This type of model is related to the theory of supercooled liquids and the landmark paper of Goldstein [3]. In undercooled liquids the atoms will spend most of their time in an arrangement corresponding to a local minimum in the energy landscape. Occasionally a transition to a neighboring basin will take place. At higher temperatures jumps between basins are more common, but are still rare compared to movements(vibrations) within a particular basin.

For grossly non-stoichiometric compounds, solid solutions or disordered materials more generally, most crystallographic techniques produce an 'average' structure based on partial occupancy of particular sites (see the schematic plot in Figure 25.1). By themselves these yield no information about which local atomic arrangements or configurations give rise to this average!

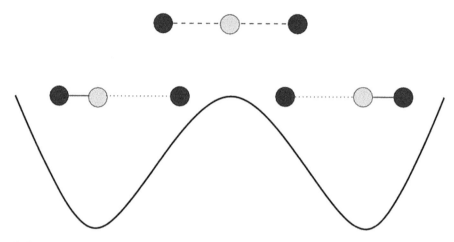

FIGURE 25.1 Schematic picture of two local structures. The "average" structure is a transition state – not a local minimum.

In the absence of more information and in simple statistical mechanical treatments such as mean field theory or regular solution models, it is usual to assume random distributions of cations or vacancies consistent with this average. But such assumptions can fail dramatically.

For example, studies of the energy landscapes in oxygen deficient perovskites such as $Sr_2Fe_2O_5$ and $Ba_2In_2O_5$ reveal a strong energetic preference for just a *few* local structural entities, including $Fe(In)O_4$ tetrahedra, $Fe(In)O_6$ octahedra, and some $Fe(In)O_5$ pyramids. The average structure observed experimentally at elevated temperature can then be interpreted as a spatial average of these local structures involving tetrahedra, octahedral, and pyramids.

This has major implications for understanding both the thermodynamic properties and dynamics. Ion-transport in the fast-ion oxygen-deficient perovskites for example involves transitions between the basins corresponding to different local minima in the energy landscape and thus the interplay of these structural entities.

Thus we are exploring the link between *local* structure and observed physical and materials behavior, including structure, thermodynamics, and transport. This presents many challenges besides the lack of *direct* experimental data for comparison, due to the difficulties associated with the modeling of disorder once one moves away from assuming ideal or regular solutions or defect distributions, and from traditional textbook averaged 'mean-field' treatments which fail to allow for distinct local environments of the different species present. The challenge computationally is that of calculating the energies of a large number of possible atomic arrangements *while taking full account of the changes in local environment of each species from one arrangement to another.* There arises an inevitable trade-off between the computational expense and the number of arrangements considered.

25.3   FAST-ION CONDUCTION: BISMUTH OXIDE

We turn to a specific example of a highly non-stoichiometric compound of considerable current interest. The δ-phase of bismuth oxide, δ-Bi_2O_3, has very high ionic conductivity, but the superionic phase is only stable at high temperatures [4] (approximately 1000–1100 K). The thermodynamically stable bulk phases of bismuth oxide at lower temperatures are ordered and have much lower conductivity. There is a report [5] that the disordered superionic δ-phase can be stabilized even at room temperature by growth on single crystal $SrTiO_3$ and $DyScO_3$.

Just as $Sr_2Fe_2O_5$ and $Ba_2In_2O_5$ can be viewed as oxygen-deficient perovskites, the δ-phase of bismuth oxide can be regarded as a Bi_4O_8 fluorite lattice with one-quarter of the oxygen sites vacant. But once again this simplification can be highly deceptive.

The structure of δ-Bi_2O_3 has received considerable attention recently. Experimental studies have used neutron scattering and analyzed not only the Bragg but the diffuse contributions. This is crucial as such analysis of the total scattering, using reverse Monte Carlo modeling, also provides information about instantaneous correlations between disordered ions and so not only yields a long-range average structure but probes the nature of local deviations from this average [6]. In parallel, computational studies [7, 8] have reported results from configurational averaging and Born-Oppenheimer *ab initio* molecular dynamics. This leads to a radically different picture from the conventional treatment which emphasizes the *average* cubic structure. The new picture highlights the local structural environments of the Bi^{3+}, and closely links the structures of the ordered β- and the disordered δ-phases. Locally, the anion environment of the Bi^{3+} in the δ-phase is very similar to the distorted square pyramidal arrangement present in the ordered β-phase. But in the high-entropy δ-phase these units are found in many different orientations, the *average* of which is the highly symmetric structure. The energy landscape shown schematically in Figure 25.2 is rugged with lots of energy minima separated by barriers, which are thermally accessible. Transitions between the different local, but energetically accessible, asymmetric minima leads to the high oxide mobility seen only in the δ-phase.

Locally the cations and anions are displaced so extensively from their average ideal positions [9] that there is little value in their use. The difference between the local and average structures can be shown simply but dramatically just by considering the bond lengths. In δ-Bi_2O_3 the lattice parameter of the average cubic structure is 5.66 Å with mean Bi-O and Bi-O' distances of ~ 2.45 Å and ~ 2.33 Å respectively [6]. But in the stable low temperature α-and metastable β-phases [6], with the same local environments for Bi as in δ-Bi_2O_3, only one in ten and one in five nearest neighbor bismuth-oxygen distances respectively are in the range 2.40–2.50 Å and there are *none* at all between 2.30 and 2.40 Å! This variation and the distorted local environment reflect the effect of the asymmetric lone pair on Bi. So there are very few bonds in δ-Bi_2O_3 with lengths equal to the average length.

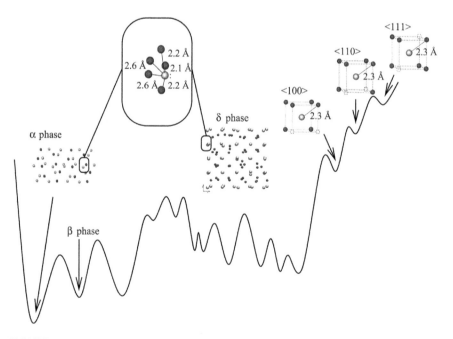

FIGURE 25.2 Key features of the energy landscape of Bi_2O_3. Note the very high energy of the "average" cubic structure.

The average cubic structures are all very high in energy as shown in the energy landscape, and are all thermally inaccessible. In addition these average structures are all metallic or semi-metallic, in complete disagreement with experiment. In contrast the distorted (low energy) structures are all insulators with a calculated GGA band gap of approximately 2.0 eV while experiment indicates a value of 2.6 eV.

One method of stabilization of the δ-phase at room temperature is via replacement of some of the bismuth by lanthanides. However, at high temperatures such systems have lower conductivities than the δ-phase itself. The influence of such substitutions on the conductivity has been studied also by ab initio molecular dynamics studies [10] of δ-Bi_3YO_6. These show that the average occupancy of oxide ions sites, which are next to yttrium increases and the residence times are longer than in the parent compound; oxide ions are effectively trapped in such positions the local environment of the oxide ions is crucial.

Another method of stabilization involves growth of the δ-phase on a suitable substrate. Profit et al. [5] report the stabilization of δ-Bi_2O_3 on the (001) surface of the cubic perovskite $SrTiO_3$ and link this to epitaxial matching between the *average* cubic δ-Bi_2O_3 structure and the surface. Indeed, there appears to be a good match as the Sr-Sr distance in $SrTiO_3$ is 3.90 Å, which is close to the Bi-Bi distance of 4.00 Å in the averaged cubic unit cell of δ-Bi_2O_3. Nevertheless given the comments above regarding the local structures likely to be involved, such arguments must be at least incomplete.

We have examined the energy landscape and the corresponding structures and local Bi environments of Bi_2O_3 adsorbed at the surface to see how these differ from those suggested for the bulk [11]. DFT calculations were carried out with the VASP program [12–15] using the GGA density functional [16] and a plane wave basis set with an energy cutoff of 400 eV, appropriate for the projector-augmented wave (PAW) pseudo-potentials [14, 17]. The SrO-terminated (100) surface of $SrTiO_3$ was represented by a slab of 300 ions, periodically repeated in three dimensions with a gap of 20 Å in the stacking direction to minimize interactions between adjacent slabs. From these studies, it is clear that the low temperature stabilization of the superionic δ-phase of Bi_2O_3 on (100) $SrTiO_3$ cannot be due to epitaxial matching of Bi...Bi with Sr...Sr. It is *not* the matching but the *mismatch* between the Bi-O and Sr-O bond lengths at the surface which is crucial. Small Bi-O islands (domains) form, and these promote the formation of a *disordered* surface of δ-Bi_2O_3. The local environments of the Bi on the surface are irregular and similar to those in the α-phase and metastable β-phase. There is no stabilization of any symmetric structure, but rather a stabilization of *disordered* layers, with associated low oxygen mobility.

25.4 RADIATION DAMAGE IN PuO_2 AND PYROCHLORES

The search for materials not subject to structural failure after prolonged radiation damage is of considerable importance to society and the environment. Despite this, our understanding of the factors which govern the generation of radiation damage at the atomistic level is primitive. When an alpha particle is released by a radioactive atom, conservation of momentum leads to the atom recoiling in the opposite direction to that of the alpha particle, and the kinetic energy of this atom is typically in the range of thousands of electron volts. The recoiling heavy actinide atom is termed a Primary Knock on Atom (PKA), and it is the subsequent movement of this PKA through the material rather than the α-particle which generates substantial disorder and is responsible for most of the damage. The degree of disorder varies markedly from one compound to another and it is this that governs the fate of the sample. A powerful atomic level technique for studying this disorder is molecular dynamics (MD) [18–26]. Due to the large kinetic energy of the recoiling atoms, very large simulation cells are required, often containing as many as several million atoms, and available computational hardware resources are pushed to their limits.

25.4.1 PuO_2

PuO_2 is a radioactive ceramic with the fluorite structure which is used as mixed oxide (MOX) fuel for nuclear reactors. Compounds such as CeO_2, UO_2 and PuO_2 with the fluorite structure are well known to be good fast-ion anion conductors at high temperatures.

We have carried out molecular dynamics simulations of multiple cascade events. We used the DL_POLY 4 code, [27] cubic simulation cells containing 393,216 atoms; this includes ^{360}U ions which are distributed at random. ^{235}U is the alpha decay product from ^{239}Pu, and these atoms serve as the alpha recoil particles. The set of ionic charges and many-body potentials developed by Cooper et al. [28] were used for the non-Columbic interactions between the Pu, U and O, supplemented at short distances (0.2–1.5 Å) by a ZBL repulsive potential to prevent unphysical internuclear separations immediately after each recoil event when just a few atoms have enormous kinetic energies. A U-atom is selected at random and projected in a random direction with a velocity corresponding to a particular kinetic PKA energy (set here to 10 keV). The isothermal-isobaric ensemble is used, allowing the volume to vary throughout the simulation. A Langevin barostat with a friction coefficient of 2 ps^{-1} is used to equalize the pressure while a Nosé-Hoover thermostat with a relaxation time of 31.8 fs is used to regulate the energy density of the simulation cell. A variable time step is employed in order to allow time steps short enough to sample the non-equilibrium events while saving on computational cost when the system is annealing and close to equilibrium.

Our molecular dynamics simulation included 529 decay events with each PKA recoil energy set to 10 keV in order to probe the radiation tolerance of PuO_2. Between each of the decay events, the system was allowed to relax for 3.5 ps. The volume and configurational energy were monitored across the entire simulation in order to compare these properties with any changes in local structure.

Figure 25.3 shows the change in cell volume throughout the simulation with respect to the number of decay events. The volume increases rapidly at first before leveling off somewhat; the final change in lattice parameter (0.7 %) compares favorably with the experimental study [29] in which PuO_2 was exposed to an external (^{238}Pu) alpha source while volume changes are monitored via X-ray diffraction.

To probe possible structural changes further, we have used modified Steinhardt order parameters as implemented by Archer et al. [30] to quantify the amount of damage in the simulation cell at various times throughout the simulation. Order parameters Q_2, Q_4 and Q_{10} for different regions 12 Å in radius in the simulation cell are calculated and compared with the same order parameters from an analogous molecular dynamics simulation on the parent undamaged fluorite structure. In this way regions of the simulation cell can be classified as damaged or undamaged crystal. This technique is more reliable and instructive than simply counting the number of atoms, which have moved a given distance. Such a measure is potentially misleading since it makes no account of regions of the crystal containing atoms which have been displaced by large distances but which have healed back to the parent structure. It also avoids the difficulties of arbitrarily classifying particular ions as interstitials or particular sites as vacant. While such assignments are useful in rather different situations involving point defects (in the

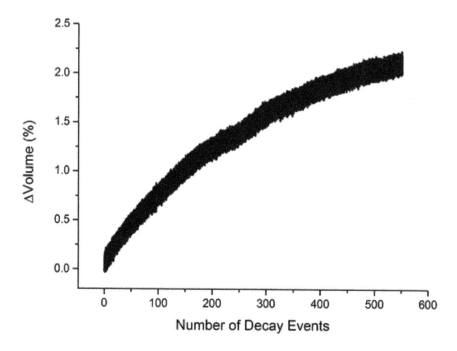

FIGURE 25.3   Variation of percentage volume change with the number of 10 keV alpha decay events.

isolated or dilute limit or close to it) they are less useful when considering large, highly disordered collision cascades.

Figure 25.4 shows the resulting accumulation of damaged regions throughout the 10 keV multiple decay event simulation in PuO_2 as a function of time. There is a general increase in the amount of damage, towards the end of the simulation the rate of the increase of damage increases. Nevertheless the maximum damage is only around 1.7 %. By 176 alpha decay events (Figure 25.5a) there are already several clusters of persistent damage. After 352 decay events, shown in Figure 25.5b, around 0.75% of the cell is made up of damaged regions. Some of the damage clusters visible after 176 decay events are no longer visible at this point, indicating that even large clusters of damage are able to heal over time. At the end of the simulation (Figure 25.5c) the cell is around 1.0% damaged, a marked decrease from the 1.7% maximum seen earlier in the simulation. This shows the operation of powerful healing mechanisms in this material. Underlying this is the rapid oxygen ion mobility responsible also for the fast ion conduction at high temperatures

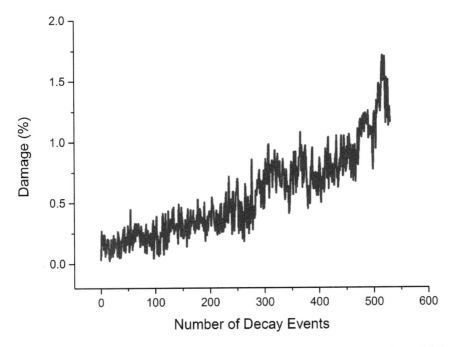

FIGURE 25.4 Variation of the percentage of damaged regions (as identified by the modified Steinhardt order parameters) with the number of decay events.

in the undamaged materials and which allows the lighter anions to move to undamaged lattice site positions as a key step of the healing process. Preliminary studies of cation migration activation energies show that some barriers to cation migration are significantly lowered in the presence of anion disorder. Overall, PuO_2 is very resistant to damage, not despite but due to the ready displacement of the anion sublattice.

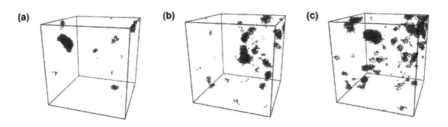

FIGURE 25.5 Damaged regions (red spheres) after (a) 176 (b) 352 and (c) 529 simulated alpha decay events. Regions classified as PuO_2 fluorite are not shown for clarity.

25.4.2 PYROCHLORES

We turn for comparison to crystalline pyrochlores, which are potential hosts for plutonium from the decommissioning of atomic weapons and spent fuel-reprocessing operations. High-level waste (HLW) and Spent Fuel are the greatest liabilities in nuclear waste legacies. The ponds and silos at Sellafield in the UK are major safety and security threats, costing £70M/annum just to maintain their basic condition. Many of these highly active wastes must be recovered and repackaged to facilitate safe storage and reduce risk. The current UK process for this HLW stream is vitrification – the liquid waste is dried to a powder, mixed with glass before heating to 1500 K, solidified and stored. But glass can suffer from low chemical durability, low actinide solubility and low tolerance to high thermal loading. New waste forms, inspired by natural minerals containing radionuclides, use crystalline ceramics as host phases.

The pyrochlore structure (space group $Fd\bar{3}m$) is similar to that of fluorite. The differences are the presence of two different cations (on the 16d and 16c sites, respectively) and the partial occupancy of one anion site (8a) in pyrochlores. Some pyrochlores such as $Gd_2Zr_2O_7$ are fast-ion conductors at high temperature, linked to anion disorder, while others such as the Ti analogue $Gd_2Ti_2O_7$ are not and in which oxygen mobility is lower. There is also a marked difference in resistance to radiation damage in the two compounds. Heavy-ion bombardment of $Gd_2Ti_2O_7$ leads to a transition to an amorphous state [31, 32] and is one of the most readily damaged titanate pyrochlores. But under the same conditions, in marked contrast, $Gd_2Zr_2O_7$ remains crystalline [31], forming a defect fluorite structure, closely related to that of pyrochlore but with a random distribution of the Gd and Ti over the cation sites, and of oxide ions over the 8a and 8b site.

The explanation for such different behavior of two compounds as similar to each other as $Gd_2Ti_2O_7$ and $Gd_2Zr_2O_7$ must ultimately have its origins in the (local) atomic level behavior which we have investigated [33] using multiple cascade simulations similar to those for PuO_2. Once again we used the DL_POLY 4 code, now the interatomic potentials of Gunn et al. [34] and cubic simulation cells containing 360,448 atoms. The initial temperature was set to 300 K, and simulations were in the NPT ensemble with 360 U^{3+} PKAs replacing Gd^{3+} ions. U^{3+} ions were randomly selected in turn to initiate cascades and were projected in a random direction with a recoil energy of 5 keV. The simulations also again used a variable time step, the Langevin barostat (2 ps^{-1}) and the Nosé-Hoover thermostat with a 31.8 fs relaxation time. There were 2,200 alpha decay events in each of the simulations, each separated by 5 ps. The total length of the simulation for each compound was approximately 10 ns. The thermostat relaxation time is chosen so that the system is thermalized between consecutive cascades at an affordable cost in computer time. A smaller relaxation time would produce excessive healing and thus underestimate damage severely; larger times would lead to an unphysical overheating and possible melting. Test runs showed that single cascade damage shape and size were similar to those for the same energy cascade when it took place in a larger system with larger relaxation times. Analysis of the resulting structures was carried out using our modified Steinhardt order parameter technique.

The results for $Gd_2Ti_2O_7$ are very different from those for PuO_2. Anion disorder and cation disorder accumulate and not long after 400 decay events, amorphization of the system starts. Large amorphous clusters form and there is full amorphization after around 1,750 simulated alpha decay events, corresponding to approximately 8 ns. Over the same time period there is a rapid increase in volume of the simulation cell, which accompanies amorphization.

With increasing Zr content, in similar simulations, more atoms are initially displaced from their lattice sites and increasing anion disorder at the start, but this is followed by progressively less amorphization and more healing. The amorphization is delayed and is absent in $Gd_2Zr_2O_7$ itself. All these differences are consistent with enhanced oxygen mobility and the weaker and less directional Zr-O bonds. The overall swelling over the timescale of the simulation is markedly reduced.

Our studies, albeit so far limited, thus not only indicate the importance of the contribution of healing processes to the ultimate extent of radiation resistance, rather than a reduction in initial atomic displacements. They also suggest a strong link between anion mobility and ionic conductivity and resistance to damage. The presence of stronger bonds with larger covalent and directional character such as Ti-O may decrease the number of atomic displacements, but they can also lead to higher activation energies for ion mobility. This hinders the movement of ions back from strained metastable states into lattice positions and thus restricts healing. This is a different picture from those in Refs. [35] and [36] which have concentrated on a correlation between cation antisite energies and radiation damage resistance. Our dynamic rather than static approach emphasizes in contrast the importance of the *anion mobility*. In this context future work on other pyrochlores such as $Gd_2Sn_2O_7$ and on the interplay between anion and cation mobility will both be key. Recent experimental studies [37] indicate the Sn-pyrochlore, like the Zr-, should remain crystalline and not amorphize, consistent with our preliminary simulations on this system which are also revealing marked enhancements in cation mobility accompanying extensive anion disorder.

25.5 CONCLUSIONS

Our case studies in this chapter have highlighted the importance of the *local* environment in disordered systems, whether the system is inherently nonstoichiometric as in fast-ion conductors or the disorder has been generated by radiation damage.

To understand the atomic mechanisms underlying fast-ion conduction, a consideration of the *average* structure alone is not sufficient, and as we have seen can at worst be dramatically highly misleading! There are important consequences for transport and correlated ion motion as well as thermodynamic behavior, electronic properties and solid state reactions such as trace element (dopant) incorporation.

To understand the atomic mechanisms underlying radiation damage in oxides, we need to consider the healing processes and the atomistic mechanisms responsible, rather than, as is traditional, concentrating on damage creation alone. Crucial to the healing processes are ion migration and mobility – in the first stages, oxide ion mobility and then subsequent cation migration in the presence of disorder and in particular cation antisites, reforming the parent compound. Existing correlations between cation anti-site defect energies and susceptibility to damage are useful, but they mask the underlying importance of anion mobility. PuO_2 and $Gd_2Zr_2O_7$ are fast-ion conductors and heal readily in our simulations. In contrast, $Gd_2Ti_2O_7$ is not a fast-ion conductor, does not heal, and becomes amorphous. If established more generally, this correlation should be of considerable help in understanding radiation damage effects in complex materials more broadly and in the development of improved hosts for nuclear waste storage.

With advances in computer power and *ab initio* computational methodologies, as well as experimental techniques for examining local structural variations, the long-term prospects of our work are a step change in our understanding of how the atomic-level behavior governs the complex macroscopic physical and chemical behavior, and of the importance of the local structure and environment in determining materials-related properties of technological interest. Computational and theoretical chemistry is now able to tackle such complex scientific and critical societal problems.

ACKNOWLEDGMENTS

The radiation damage simulations in this work were carried out using the computational facilities of the Advanced Computing Research Centre, University of Bristol - http://www.bris.ac.uk/acrc/. Other computational facilities were made available by NOTUR grants to CEM. AA was supported by EPSRC grant EP/H012230/1 and is grateful for additional funding for the PuO_2 study from AWE, Aldermaston. The Centre for Earth Evolution and Dynamics is funded by CoE grant 223272 from the research council of Norway.

KEYWORDS

- bismuth oxides
- defects
- fast-ion conduction
- inorganic oxides
- local environments
- molecular dynamics
- nuclear waste
- plutonium dioxide
- pyrochlores
- radiation damage
- simulation

REFERENCES

1. Mohn, C. E., Allan, N. L., Freeman, C. L., Ravindran, P., & Stølen, S., (2005). Order in the disordered state: local structural entities in the fast ion conductor $Ba_2In_2O_5$. *J. Solid State Chem. 178*, 346–355.
2. Steinardt; P. J., Nelson, D. R., & Ronchetti, M., (1983). Bond-orientational order in liquids and glasses. *Phys. Rev. B 28*, 784–805.
3. Goldstein, M., (1969). Viscous liquids and glass transition – a potential energy barrier picture. *J. Chem. Phys., 51*, 3728–3739.
4. Harwig A., & Gerards, A. G., (1979). The polymorphism of bismuth sesquioxide. *Thermochim. Acta, 28*, 121–131.
5. Proffit, D. L., Bai, G.-R., Fong, D. D., Fister, T. T., Hruszkewycz, S. O., Highland, M. J., et al., (2010). A. Phase stabilization of δ-Bi_2O_3 nanostructures by epitaxial growth onto single crystal $SrTiO_3$ or $DyScO_3$ substrates. *Appl. Phys. Lett., 96*, 021905.
6. Hull, S., Norberg, S. T., Tucker, M. G., Eriksson, S. G., Mohn, C. E., & Stølen, S. (2009). Neutron total scattering study of the delta and beta phases of Bi_2O_3. *Dalton Trans.*, 8737–8745.
7. Mohn, C. E., Stølen, S., Norberg, S. T., & Hull, S., (2009). Oxide-ion disorder within the high temperature delta phase of Bi_2O_3. *Phys. Rev. Lett., 102*, 155502.
8. Mohn, C. E., Stølen, S., Norberg, S. T., & Hull, S., (2009). Ab initio molecular dynamics simulations of oxide-ion disorder in δ-Bi_2O_3. *Phys. Rev. B., 80*, 024205.
9. Allan, N. L., Stølen, S., & Mohn, C. E., (2008). Think locally-linking structure, thermodynamics and transport in grossly nonstoichiometric compounds and solid solutions, *J. Mat. Chem., 18*, 4124–4132.
10. Krynski, M., Wrobel, W., Mohn, C. E., Dyga, J. R., Malys, M., Krok, F., et al., (2014). Trapping of oxide ions in δ-Bi_3YO_6. *Solid State Ionics, 264*, 49–53.
11. Mohn, C. E., Stein, M. J., & Allan, N. L., (2010). Oxide and halide nanoclusters on ionic substrates: heterofilm formation and lattice mismatch. *J. Mat. Chem. 20*, 10403–10411.
12. Kresse, G., & Hafner J., (1993). Ab initio molecular dynamics for liquid metals. *Phys. Rev. B, 47*, 558–561.
13. Kresse, G., & Hafner J., (1994). Ab initio molecular-dynamics simulation of the liquid-metal-amorphous-semiconductor transition in germanium. *Phys. Rev. B 49*, 14251–14269.
14. Kresse, G., & Furthmüller J., (1996). Efficiency of ab-initio total energy calculations for metals and semiconductors using a plane-wave basis set. *Comput. Mater. Sci., 6*, 15–50.
15. Kresse, G., & Furthmüller J., (1996). Efficient iterative schemes for ab initio total-energy calculations using a plane-wave basis set. *Phys. Rev. B 54*, 11169–11186.
16. Perdew, J. P., Burke K., & Ernzerhof, M., (1996). Generalized gradient approximation made simple. *Phys. Rev. Lett., 77*, 3865–3868.
17. Blöchl, P. E., (1994). Projector augmented-wave method. *Phys. Rev. B 50*, 17953–17979.
18. Purton, J. A., & Allan, N. L., (2002). Displacement cascades in $Gd_2Ti_2O_7$ and $Gd_2Zr_2O_7$: a molecular dynamics study. *J. Mater. Chem., 12*, 2923–2926.
19. Todorov, I. T., Purton, J. A., Allan, N. L., & Dove, M. T. (2006). Simulation of radiation damage in gadolinium pyrochlores. *J. Phys. Condens. Matter, 18*, 2217.

20. Chartier, A., Catillon, G., & Crocombette, J. P., (2009). Key Role of the cation interstitial structure in the radiation resistance of pyrochlores *Phys. Rev. Lett.*, *102*, 155503.

21. Wang, X. J., Xiao, H. Y., Zu, X. T., Zhang, Y., & Weber, W. J., (2013). Ab initio molecular dynamics simulations of ion–solid interactions in $Gd_2Zr_2O_7$ and $Gd_2Ti_2O_7$. *J. Mater. Chem. C., 1*, 1665–1673.

22. Foxhall, H. R., Travis, K. P., & Owens, S. L., (2014). Effect of plutonium doping on radiation damage in zirconolite: A computer simulation study, *J. Nucl. Mat., 444*, 220–228.

23. Trachenko, K., Zarkadoula, E., Todorov, I. T., Dove, M. T., Dunstan, D. J., & Nordlund, K., (2012). Modeling high-energy radiation damage in nuclear and fusion applications. *Nucl. Instrum. Methods Phys. Res. Sect. B-Beam Interact. Mater. At, 277*, 6–13.

24. Aidhy, D. S., Zhang, Y. W., & Weber, W. J., (2015). Radiation damage in cubic ZrO_2 and yttria-stabilized zirconia from molecular dynamics simulations. *Scripta Materialia, 98*, 16–19.

25. Catillon, G., & Chartier, A., (2014). Pressure and temperature phase diagram of Gd_2Ti_2O7 under irradiation. *J. Appl. Phys., 116*, 193502.

26. Zarkadoula, E., Devanathan, R., Weber, W. J., Seaton, M. A., Todorov, I. T., Nordlund, et al., (2014). High-energy radiation damage in zirconia: Modeling results. *J. Appl. Phys., 115*, 083597.

27. Todorov, I. T., Smith, W., Trachenko, K., & Dove, M. T, (2006). DL_POLY_3: new dimensions in molecular dynamics simulations via massive parallelism. *J. Mat. Chem., 16*, 1911–1918.

28. Cooper, M. W. D., Rushton, M. J. D., & Grimes, R. W., (2014). A many-body potential approach to modeling the thermomechanical properties of actinide oxides. *J. Phys. Condens. Mat., 26*. 105401.

29. Weber, W.J., Alpha-irradiation in CeO_2, UO_2 and PuO_2. *Radiat. Eff., 83*, 145-156.

30. Archer, A., Foxhall, H. R., Allan, N. L., Gunn, D. S. D., Harding, J. H., Todorov, I. T., et al., (2014). A. Order parameter and connectivity topology analysis of crystalline ceramics for nuclear waste immobilization. *J. Phys. Condens. Mat., 26*, 485011.

31. Wang, S. X., Begg, B. D., Wang, L. M., Ewing, R. C., Weber, W. J., & Govindan Kutty, K. V., (1999). Radiation stability of gadolinium zirconate: A waste form for plutonium disposition. *J. Mater. Res., 14*, 4470–4473.

32. Lian, J., Wang, L. M., Haire, R. G., Helean, K. B., & Ewing, R. C., (2004). Ion beam irradiation in $La_2Zr_2O_7$–$Ce_2Zr_2O_7$ pyrochlore. *Nucl. Instrum. Methods Phys. Res. Sect. B-Beam Interact. Mater. At., 218*, 236–243.

33. Archer, A., (2016). Radiation damage in ceramic wasteforms. Ph.D. thesis, Bristol University.

34. Gunn, D. S. D., Allan, N. L., Foxhall, H., Harding, J. H., Purton, J. A., Smith, W., et al., (2012). Novel potentials for modelling defect formation and oxygen vacancy migration in $Gd_2Ti_2O_7$ and $Gd_2Zr_2O_7$ pyrochlores. *J. Mater. Chem. 22*, 4675–4680.

35. Sickafus, K. E., Minervini, L., Grimes, R. W., Valdez, J. A., Ishimaru, M., Li F., McClellan, K. J., et al., (2000). Radiation Tolerance of Complex Oxides. *Science, 289*, 748–751.

36. Sickafus, K. E., Grimes, R. W., Valdez, J. A., Cleave, A., Tang, M., Ishimaru, M., et al., (2007). Radiation-induced amorphization resistance and radiation tolerance in structurally related oxides. *Nat. Mat., 6*, 217–223.

37. Yudintsev, S. V., Lizin, A. A., Livshits, T. S., Stefanovksy, S. V., Tomilin, S. V., & Ewing, R. C., (2015). Ion-beam irradiation and 244Cm-doping investigations of the radiation response of actinide-bearing crystalline waste forms. *J. Mater. Res., 30*, 1516–1528.

INDEX